ETHNOZOOLOGY

ETHNOZOOLOGY

ANIMALS IN OUR LIVES

Edited by

RÔMULO ROMEU NÓBREGA ALVES
Departamento de Biologia
Universidade Estadual da Paraíba
Campina Grande, Brazil

ULYSSES PAULINO ALBUQUERQUE
Departamento de Botânica
Universidade Federal de Pernambuco
Recife, Brazil

Academic Press is an imprint of Elsevier
125 London Wall, London EC2Y 5AS, United Kingdom
525 B Street, Suite 1800, San Diego, CA 92101-4495, United States
50 Hampshire Street, 5th Floor, Cambridge, MA 02139, United States
The Boulevard, Langford Lane, Kidlington, Oxford OX5 1GB, United Kingdom

Library of Congress Cataloging-in-Publication Data
A catalog record for this book is available from the Library of Congress

British Library Cataloguing-in-Publication Data
A catalogue record for this book is available from the British Library

ISBN: 978-0-12-809913-1

For information on all Academic Press publications visit our website at
https://www.elsevier.com/books-and-journals

Publisher: Sara Tenney
Acquisition Editor: Kristi Gomez
Editorial Project Manager: Pat Gonazlez
Production Project Manager: Priya Kumaraguruparan
Designer: Matt Limbert

Typeset by TNQ Books and Journals

Contents

26. The Use of Traditional Ecological Knowledge in the Context of Participatory Wildlife Management: Examples From Indigenous Communities in Puerto Nariño, Amazonas-Colombia

NATHALIE VAN VLIET, LAURANE L'HARIDON, JUANITA GOMEZ, LILIANA VANEGAS, FRANÇOIS SANDRIN AND ROBERT NASI

27. Ethnozoology: An Overview and Current Perspectives

RÔMULO ROMEU NÓBREGA ALVES, JOSIVAN SOARES SILVA, LEONARDO DA SILVA CHAVES AND ULYSSES PAULINO ALBUQUERQUE

List of Contributors

Ulysses Paulino Albuquerque Universidade Federal de Pernambuco, Recife, Brazil

Ângelo Giuseppe Chaves Alves Universidade Federal Rural de Pernambuco, Recife, Brazil

Janaina Kelli Gomes Arandas Universidade Federal Rural de Pernambuco, Recife, Brazil

Márcio Luiz Vargas Barbosa Filho Universidade Federal Rural de Pernambuco, Recife, Brazil

Raynner Rilke Duarte Barboza Universidade Estadual da Paraíba, Campina Grande, Brazil

Dandara Monalisa Mariz Bezerra Instituto Federal de Educação, Ciência e Tecnologia da Paraíba, Itabaiana, Brazil

Leonardo da Silva Chaves Universidade Federal Rural de Pernambuco, Recife, Brazil

José da Silva Mourão Universidade Estadual da Paraíba, Campina Grande, Brazil

Thelma Lúcia Pereira Dias Universidade Estadual da Paraíba, Campina Grande, Brazil

Hugo Fernandes-Ferreira Universidade Estadual do Ceará, Quixadá, Brazil

Juanita Gomez Fundación SI, Bogotá, Colombia

Laurane L'haridon Fundación SI, Bogotá, Colombia

Walter Lechner University of Vienna, Vienna, Austria

Sérgio de Faria Lopes Universidade Estadual da Paraíba, Campina Grande, Brazil

Jessica Moreno Fundación SI, Bogotá, Colombia

Ellori Laíse Silva Mota Universidade Estadual da Paraíba, Campina Grande, Brazil; Universidade Federal da Paraíba, João Pessoa, Brazil

Lisa Nagaoka University of North Texas, Denton, TX, United States

Robert Nasi Center for International Forestry Research (CIFOR), Bogor, Indonesia

Lindon Neves de Aquino Universidad Federal do Amazonas, Benjamin Constant, Brazil

Rômulo Romeu Nóbrega Alves Universidade Estadual da Paraíba, Campina Grande, Brazil

Eduardo Silva Oliveira Universidade Federal Rural de Pernambuco, Recife, Brazil

Tacyana Pereira Ribeiro Oliveira Universidade Estadual da Paraíba, João Pessoa, PB, Brazil

Marcia Freire Pinto Universidade Estadual do Ceará, Limoeiro do Norte, CE, Brazil

Iamara da Silva Policarpo Universidade Federal da Paraíba, João Pessoa, Brazil

Pavol Prokop Trnava University, Trnava, Slovakia; Slovak Academy of Sciences, Bratislava, Slovakia

Christoph Randler Didaktik der Biologie, Tübingen, Germany

Maria Norma Ribeiro Universidade Federal Rural de Pernambuco, Recife, Brazil

Luiz Alves Rocha California Academy of Sciences, San Francisco, CA, United States

François Sandrin Fundación SI, Bogotá, Colombia

Josivan Soares Silva Universidade Federal Rural de Pernambuco, Recife, Brazil

Wedson Medeiros Silva Souto Universidade Federal do Piauí (UFPI), Teresina, Brazil

Denise Freitas Torres Universidade Federal Rural de Pernambuco, Recife, Brazil

Maria Franco Trindade Medeiros Universidade Federal de Campina Grande, Cuité, Brazil

Arnold van Huis Wageningen University, Wageningen, The Netherlands

Nathalie van Vliet Center for International Forestry Research (CIFOR), Bogor, Indonesia

Liliana Vanegas Fundación SI, Bogotá, Colombia

Kleber da Silva Vieira Universidade Estadual da Paraíba, Campina Grande, Brazil; Universidade Federal da Paraíba, João Pessoa, Brazil

Washington Luiz Silva Vieira Universidade Federal da Paraíba, João Pessoa, Brazil

Steve Wolverton University of North Texas, Denton, TX, United States

C H A P T E R

1

Introduction: Animals in Our Lives

Rômulo Romeu Nóbrega Alves[1], *Ulysses Paulino Albuquerque*[2]

[1]Universidade Estadual da Paraíba, Campina Grande, Brazil; [2]Universidade Federal de Pernambuco, Recife, Brazil

INTRODUCTION

We are animals, and as such share certain fundamental biological features with a huge diversity of other animal species, which link them to us through an intricate network of ecological and evolutionary relationships. As a natural part of the fauna living on earth, we have interacted with the animals around us throughout our shared coexistence, establishing multiple interrelationships since the origin of our species.

From our long history of coexistence with animals emerges a plethora of complex and multidimensional relationships, which certainly precede any recorded historical evidence. The first testimonies of these relationships are found in the earliest rock paintings and engravings, such as the cave paintings in Lascaux, France, where most of the paintings depict animals such as bisons, horses, mammoths, ibex, deer, lions, aurochs, bears, and wolves. As pointed out by Tedesco (2000), these depictions of animals represent both species that would have been hunted and eaten (such as bison and deer) and those that were feared predators (such as bears, lions, and wolves). This situation most likely reflects a scenario of interactions between humans and fauna that occurred at a time when humans were both predators and prey. Furthermore, it is likely that animals were represented in these paintings because of supernatural/spiritual connections established between these animals and humans (Hurn, 2012).

Independent of the inspiration to represent animal images in prehistoric paintings, and later in other visual works produced by humans, the abundance of representations of fauna clearly indicates their relevance to mankind, and is evidence of the wide range of roles they played in human life since the beginning of their shared history. This relevance is explicit in the earliest written documents, as evidenced by the large number of reports on animals in books, letters, bestiaries, and other various past documents.

Important human activities involving animals, such as hunting, fishing, and domestication, represent practices that have perpetuated themselves over time as forms of subsistence crucial to humanity, and are recognized as having a significant influence on human evolution, both cultural and physical. The dynamic character of human

Ethnozoology
http://dx.doi.org/10.1016/B978-0-12-809913-1.00001-6

activities involving animals, from the oldest to the most contemporary, has resulted in a diversification of human–animal interactions. Certainly what we are today has been profoundly influenced by the relationships established with animals throughout our shared history with them.

From a utilitarian perspective, today, as in the past, humans have long exploited animal products to provide food and materials for making tools, feed, ornaments, medicines, fertilizer, and income, as well as providing agricultural, transport, entertainment, companionship, and religious services (Alves, 2012; Alves et al., 2016; Alves and Rosa, 2013; Bowman, 1977; Figueirêdo et al., 2015; Fitter, 1986). Reinforcing the utilitarian value of fauna, Plous (1993) points out that animal products are also found in the streets of our modern day cities (e.g., in asphalt binders), in the cars we drive (e.g., in brake fluid, upholstery, and car wax), in the walls of our homes (e.g., sheetrock and wallpaper adhesive), on kitchen and bathroom floors (e.g., in ceramic tiles, linoleum, and floor wax), in toiletries (e.g., perfume, deodorants, soap, and cosmetics), and in a variety of paints, plastics, textiles, and machinery oils. Domestic animals provided energy used in a wide range of services for people (Alves, 2016). Wild-caught species have also been tamed for various purposes other than predation, such as alarms for protection, fighting, and as guards, pets, and pack and draught animals (Alves et al., 2013; Fernandes-Ferreira et al., 2012; Gilmore, 1950; Roldán-Clarà et al., 2014; Serpell, 1996; Svanberg and Ståhlberg, 2012). It is clear that human dependence or codependence on faunal resources has intensified over time (Alves, 2012; Alves and Souto, 2015).

In addition to their utilitarian value, animals have had a pronounced participation in the cultural practices and religious beliefs of humankind (Adeola, 1992; Alves et al., 2012a; Kemmerer, 2011; Olupona, 1993). Many myths, proverbs, and stories have been generated from human–animal relationships and passed down from generation-to-generation through oral traditions (Allaby, 2010; Alves et al., 2017, 2012b). Culture guides our attitudes toward animals (Alves et al., 2014; Shanks, 2002), which are also shaped by many other factors, including our experiences with them (perhaps as farmers, hunters, fishermen, pet owners, or experimenters), exposure to biological science, religious beliefs, philosophical presuppositions, and psychological factors, such as our capacity for empathy (Shanks, 2002).

Human interest in animals represents a characteristic of all societies, from those residing in the most isolated locations to those in the most urbanized areas of the world. Due to thousands of years of cohabitation, a strong interdependence exists between humans and other animal species (Alves and Souto, 2015; De Waal, 2009), which generates harmonic and conflictual relationships that can be predatory, competitive, parasitic, mutualistic, and commensal (Baenninger, 1995). In modern societies, our relationship with animals increasingly consolidates human control over fauna, with a view to optimizing the use of faunal resources. Thus, there is an intensification of human interference in the lives of all animals with which humans interact. An example of this is that humans have been responsible for the extinction of several species, while on the other hand they have promoted the geographical spread of several others. In some of these cases, introduced species become pests, creating problems that affect entire ecosystems. It is clear, therefore, that humans are agents that cause impacts that affect not only the species with which they interact directly, but others that are indirectly associated. The pronounced influence of humans and their impacts on the environment characterizes the geological epoch we live in, the Anthropocene, which has increasingly placed the animal world under growing pressure.

This scenario makes the need to understand the complexities of the relationships between humans and other animals increasingly evident, not only from biological and ecological perspectives, but also from historical, social, cultural, economic, psychological, and sociological contexts. Of course, because of the complexity of human–animal interactions, this understanding involves different academic disciplines, especially within the social and biological sciences. Ethnozoology, the theme of this book, arises as a discipline that can provide important contributions and serve as a bridge enabling dialogue between the different academic fields interested in this subject.

INVESTIGATING ANIMALS AND THEIR INTERACTIONS WITH HUMANS

As discussed above, fauna have been of great significance to humans throughout history. Consequently, they represent one of the themes with which most people have the greatest familiarity, interest, or contact (Springer and Holley, 2012). Humans of all societies have always attempted to understand animals by generating a fundamental basis of zoological knowledge by interacting/exploring faunas, which have played a crucial role in the evolution of the human species since its emergence. The earliest humans must have had, and passed on, knowledge about animals to increase their chances of survival, which likely included unsystematic knowledge of animal ecology and anatomy. With domestication of fauna, and the consequent possibility of better observing them, people learned more.

From our ancient and constant interest in fauna emerged the field of zoology, one of the first areas of research of the natural world made by humankind (Springer and Holley, 2012). In addition to information about the animals themselves, zoology and natural science enthusiasts were also interested in the potential uses of animals for humans, in particular the species that were targeted or hunted, as well as the knowledge that native peoples had of the fauna in general. An example of this is evident in the narratives of ancient naturalists who always included information of ethnozoological interest in their writings, among them Linnaeus, Darwin, and Lamarck (Moreira, 2002; Ståhlberg and Svanberg, 2014). This perspective would be the preface for the emergence of academic disciplines that seek to investigate the different aspects of the interactions between animals and humans.

Such interactions are dynamic and multifaceted, which makes it impossible to investigate them through a single area of knowledge. Consequently, several academic disciplines related to the areas of social and biological sciences have sought to investigate the subject, among which are anthropology, sociology, psychology, history, archeology, ethnography, economics, geography, literature, philosophy, zoology, ecology, and conservation biology. As a result, several books have been published, especially in the last decades, which have focused on human–animal interactions in general (Baky, 1980; Bowman, 1977; Gross and Vallely, 2013), as well as more specific treatments such as the utilitarian aspect of fauna, including for medicinal purposes (Alves and Rosa, 2013), as pets (Pręgowski, 2016; Wilson and Turner, 1997), and in service in war (Cooper, 2000; Kistler, 2011; van Vliet, 2007). Some of these publications addressed the social and cultural importance of animals (Allaby, 2010; De Mello, 2016; Kalof, 2007; Kalof and Resl, 2007; Malamud et al., 2007); while others focused on interactions with some specific groups of wild (Dore et al., 2017; Kothari, 2007; Marcum, 2007; Morris and Morris, 1965; Tidemann and Gosler, 2010; Waller, 2016) or domestic (Clutton-Brock, 2007, 2012) animals, to cite a few examples.

These, and many other themes, have also been the focus of a growing number of articles published in numerous scientific journals, including those in the area of ethnobiology (*Ethnobiology and Conservation*, *Journal of Ethnobiology*, *Ethnobiology Letters* and *Journal of Ethnobiology and Ethnomedicine*), those particularly devoted to the examination of different aspects of the relationship between human beings and other animals (e.g., *Anthrozoos; Between the Species; Humanimalia; Journal of Applied Animal Welfare Science, Politics and Animals; Society & Animals*), and several others in areas such as medicine and biodiversity conservation (e.g., *BioScience, Journal of Ethnopharmacology, Biodiversity and Conservation, Conservation Biology, Ecology and Society, Fish and Fisheries*, etc.). This entire arsenal of publications reinforces the fact that animals represent one of the subjects that most attract the attention of human beings. It is not possible to understand our biological and cultural evolution without examining how animals have, and continue to, affect human societies, and vice versa, highlighting the importance of research investigating the connection between animals and human beings (Alves, 2017; Alves and Souto, 2015). Ethnozoology, the theme of this book, joins the different research areas seeking to contribute to the understanding of the complexity and implications of fauna-human interactions. As pointed out by Alves (2017), "Ethnozoology is a hybrid discipline that has been structured with elements from both the natural and social sciences, as it seeks to understand how humans have perceived and interacted with faunal resources throughout history." Together with ethnobotany, ethnozoology forms part of the larger body of science of ethnobiology, which has been increasingly recognized in recent years, especially considering that the world is facing a potentially massive loss of wildlife due to anthropogenic activities (Alves and Albuquerque, 2012). It is ironic that as people impact the ecosystems on which they depend, conserving diversity will increasingly depend on active human involvement (Probst and Crow, 1991). Thus, as humans are the source of problems, as well as the hope for solutions (Saunders, 2003), it is essential to incorporate human dimensions into practices aiming at the conservation of fauna, and in this regard, ethnozoology has a great contribution to offer.

WHY THIS BOOK?

Our history is full of indications of the crucial role that animals have played in human life, resulting in connections that have been debated for centuries by various areas of research, and which have contributed to understanding animal–human relationships from different perspectives. In this scenario, ethnozoology arises as a dynamic and cross-disciplinary area of research that can provide important contributions and serve as a bridge between the different academic fields that seek to investigate human-animal interactions. In this perspective, ethnozoology has attracted researchers from diverse fields of study, which has resulted in a visible increase in the number of publications on the subject in the last decades.

Despite major advances in ethnozoology, there is still a worldwide lack of textbooks that can serve as references for its teaching. Currently, some books have appeared that have filled gaps in ethnobiology for teaching and research (*Ethnobiology, Ethnobotany, Ethnopharmacology*, etc.) (Albuquerque and Alves, 2016; Albuquerque et al., 2014; Anderson et al., 2012; Cunningham, 2001; Heinrich and Jäger, 2015), but there is no single book published in English specifically on ethnozoology. Recognizing this need, *Ethnozoology: Animals in Our Lives* represents the first book about this discipline and provides discussions regarding the key themes of human–animal interactions and their implications.

Following this Introduction, Chapter 2 provides an overview of ethnozoology as a field of study, focusing on its importance, conceptualization, and history, highlighting the role ethnozoology plays as a discipline focusing on the relationships between animals and humans. These interactions are known to be very old according to research in zooarcheology and historical ethnozoology, themes of Chapters 3 and 4, respectively. Chapter 5 discusses the fabulous fauna documented in the reports of ancient travelers and chroniclers and in medieval bestiaries, describing what the authors call "imaginary zoology." Chapter 6 deals with a theme of historical importance in the field of ethnobiology, ethnotaxonomy. The following chapters deal with three of the major human activities involving animals, hunting (Chapter 7), fishing (Chapter 8), and animal domestication (Chapter 9). Subsequent chapters address the use of animals as food, one of the most fundamental uses of fauna by humans. Chapter 10 discusses the importance and diversity of uses of wild animals and the sources of food; and Chapter 11 brings an overview of the practices of harvesting and eating insects in different parts of the world. Reinforcing the importance of fauna for alimentary purposes, Chapter 12 presents an analysis of the trade in wildmeat sales in the Amazon Tri-frontier region between Peru, Brazil, and Colombia. In Chapter 13, a discussion is presented on the intimate relationships between human and animal health, giving examples of the importance of multidisciplinary studies in understanding these connections. An overview of published scientific literature on the use and commercialization of animals for decorative purposes is the subject of Chapter 14, while Chapter 15 discusses the significance of animals in the principal human cultural manifestations of art, literature, symbolism, music, mythology, and religion. Chapter 16 addresses one of the closest human-animal relationships that of animals as pets. The following chapters are devoted to the use of animals in military activities (Chapter 17) and to perform transport and traction services (Chapter 18). Animals kept in zoos and aquariums are the subject of Chapter 19. Chapter 20 discusses the role of animals in entertainment and sport. Chapter 21 brings an interesting, but still little explored, ethnozoological theme, the use of fauna as predictors of climatic and weather events, addressing the role of animals as forecasters. Chapter 22 discusses the complex interactions of conflict between humans and animals and Chapter 23 focuses on the biological influences of human attitudes, as well as on the role of education in forming children's attitudes and perceptions of living creatures. Chapter 24 discusses the importance of local ecological knowledge as a source of scientific insights and complementary academic research. Chapter 25 discusses the implications of human actions on fauna and the role of ethnozoology from a perspective of conservation of animal diversity. In this same vein, Chapter 26 describes two examples of how traditional ecological knowledge may be incorporated in more formal management plans for wildlife. Finally, Chapter 27 presents a global view of scientific production in ethnozoology and its advances, trends, and future perspectives.

We believe that this book, *Ethnozoology: Animals in Our Lives*, significantly expands the knowledge base of ethnozoology by covering a wide range of interactions between humans and animals. Furthermore, we believe it demonstrates the importance of inter and cross-disciplinary approaches for increasing our understanding of the coexistence of humans and other animals. This book will be of value to researchers, students, educators, conservationists, wildlife managers, and policymakers, as well as the general public, with interests in a wide range of areas of science, including zoology, biology, ethnobiology, conservation biology, ecology, anthropology, and sociology, among other related academic disciplines. We also hope that this book will stimulate further research in this fascinating field, thus contributing to the collective understanding of the multidimensional context of interactions between humans and animals, an increasingly pressing need since today, perhaps more than at any other time in human history, we are faced with the challenge of finding forms of exploitation that minimize the impact on animal species; an increasingly clear necessity in the context of animal conservation, as well as human survival.

Acknowledgments

We are grateful to Kristi A.S. Gomez (Senior Acquisitions Editor at Elsevier/Academic Press) and Patricia Gonzalez (Editorial Project Manager/Animal & Plant Sciences/ Elsevier) for entrusting us with the edit of this book and for her constant support.

Special thanks are due to contributors to the book, for their enthusiasm, support, and quality work.

Thanks are also due to CNPq (Conselho Nacional de Desenvolvimento Científico e Tecnológico), for providing us with research productivity scholarships, which helped for the development of our studies on the Ethnobiology.

References

Adeola, M.O., 1992. Importance of wild animals and their parts in the culture, religious festivals, and traditional medicine, of Nigeria. Environmental Conservation 19, 125–134.

Albuquerque, U.P., Alves, R.R.N., 2016. Introduction to ethnobiology. Springer International Publishing, New York/London.

Albuquerque, U.P., Cunha, L.V.F.C., Lucena, R.F.P., Alves, R.R.N., 2014. Methods and Techniques in Ethnobiology and Ethnoecology. Springer, New York.

Allaby, M., 2010. Animals: from Mythology to Zoology. Facts On File, Inc., New York.

Alves, R.R.N., 2012. Relationships between fauna and people and the role of ethnozoology in animal conservation. Ethnobiology and Conservation 1, 1–69.

Alves, R.R.N., 2016. Domestication of animals. In: Introduction to Ethnobiology. Springer, Albuquerque, UP, Switzerland, pp. 221–225.

Alves, R.R.N., 2017. Ethnozoology. In: Fuentes, A. (Ed.), The International Encyclopedia of Primatology. Wiley.

Alves, R.R.N., Albuquerque, U.P., 2012. Ethnobiology and conservation: why do we need a new journal? Ethnobiology and Conservation 1, 1–3.

Alves, R.R.N., Feijó, A., Barboza, R.R.D., Souto, W.M.S., Fernandes-Ferreira, H., Cordeiro-Estrela, P., Langguth, A., 2016. Game mammals of the Caatinga biome. Ethnobiology and Conservation 5, 1–51.

Alves, R.R.N., Lima, J.R.F., Araújo, H.F., 2013. The live bird trade in Brazil and its conservation implications: an overview. Bird Conservation International 23, 53–65.

Alves, R.R.N., Rosa, I.L., 2013. Animals in Traditional Folk Medicine: Implications for Conservation. Springer-Verlag, Berlin, Heidelberg.

Alves, R.R.N., Rosa, I.L., Léo Neto, N.A., Voeks, R., 2012a. Animals for the gods: magical and religious faunal use and trade in Brazil. Human Ecology 40, 751–780.

Alves, R.R.N., Vieira, K.S., Santana, G.G., Vieira, W.L.S., Almeida, W.O., Souto, W.M.S., Montenegro, P.F.G.P., Pezzuti, J.C.B., 2012b. A review on human attitudes towards reptiles in Brazil. Environmental Monitoring and Assessment 184, 6877–6901.

Alves, R.R.N., Silva, V.N., Trovão, D.M.B.M., Oliveira, J.V., Mourão, J.S., Dias, T.L.P., Alves, A.G.C., Lucena, R.F.P., Barboza, R.R.D., Montenegro, P.F.G.P., 2014. Students' attitudes toward and knowledge about snakes in the semiarid region of Northeastern Brazil. Journal of Ethnobiology and Ethnomedicine 10, 1–8.

Alves, R.R.N., Souto, W.M.S., 2015. Ethnozoology: a brief introduction. Ethnobiology and Conservation 4, 1–13.

Alves, R.R.N., Souto, W.M.S., Barboza, R.R.D., 2017. Primates in mythology. In: Fuentes, A. (Ed.), The International Encyclopedia of Primatology. John Wiley & Sons, Inc., pp. 1–6.

Anderson, E.N., Pearsall, D., Hunn, E., Turner, N., 2012. Ethnobiology. Wiley-Blackwell.

Baenninger, R., 1995. Some consequences of animal domestication for humans. Anthrozoos 8, 69–77.

Baky, J.S., 1980. Humans and Animals. HW Wilson Co.

Bowman, J.C., 1977. Animals for Man. Edward Arnold, London.

Clutton-Brock, J., 2007. How Domestic Animals Have Shaped the Development of Human Societies.

Clutton-Brock, J., 2012. Animals as Domesticates: a World View Through History. MSU Press.

Cooper, J., 2000. Animals in War. Corgi Books, London.

Cunningham, A.B., 2001. Applied Ethnobotany: People, Wild Plant Use and Conservation. Earthscan/James & James.

De Mello, M., 2016. Mourning Animals: Rituals and Practices Surrounding Animal Death. MSU Press.

De Waal, F., 2009. The Age of Empathy. Harmony, New York.

Dore, K.M., Riley, E.P., Fuentes, A., 2017. Ethnoprimatology: A Practical Guide to Research at the Human-nonhuman Primate Interface. Cambridge University Press.

Fernandes-Ferreira, H., Mendonça, S.V., Albano, C., Ferreira, F.S., Alves, R.R.N., 2012. Hunting, use and conservation of birds in Northeast Brazil. Biodiversity and Conservation 221–244.

Figueirêdo, R.E.C.R., Vasconcellos, A., Policarpo, I.S., Alves, R.R.N., 2015. Edible and medicinal termites: a global overview. Journal of Ethnobiology and Ethnomedicine 11, 1–7.

Fitter, R.S.R., 1986. Wildlife for Man: How and Why we Should Conserve our Species. Collins, London.

Gilmore, R., 1950. Fauna and ethnozoology of South America. In: Steward, J. (Ed.), Handbook of South American Indians, pp. 345–463 Washington, DC.

Gross, A., Vallely, A., 2013. Animals and the Human Imagination: a Companion to Animal Studies. Columbia University Press, New York.

Heinrich, M., Jäger, A.K., 2015. Ethnopharmacology. John Wiley & Sons.

Hurn, S., 2012. Humans and Other Animals: Cross-cultural Perspectives on Human-animal Interactions. PlutoPress, London.

Kalof, L., 2007. Looking at Animals in Human History. Reaktion Books.

Kalof, L., Resl, B., 2007. A Cultural History of Animals: in Antiquity. Berg.

Kemmerer, L., 2011. Animals and World Religions. Oxford University Press.

Kistler, J.M., 2011. Animals in the Military: from Hannibal's Elephants to the Dolphins of the US Navy. ABC-CLIO.

Kothari, A., 2007. Birds in our Lives. Universities Press.

Malamud, R., Kalof, L., Pohl-Resl, B., 2007. A Cultural History of Animals in the Modern Age. Berg.

Marcum, 2007. Living with Animals: Snakes and Humans, Encyclopedia of Human-animal Relationships: a Global Exploration of our Connections with Animals. Greenwood Press, Westport, pp. 1181–1184.

Moreira, I.C., 2002. O escravo do naturalista – O papel do conhecimento nativo nas viagens científicas do século 19. Ciência Hoje 31, 40–48.

Morris, R., Morris, D., 1965. Men and Snakes. McGraw-Hill.

Olupona, J.K., 1993. Some notes on animal symbolism in African religion and culture. Anthropology and Humanism 18, 3–12.

Plous, S., 1993. The role of animals in human society. Journal of Social Issues 49, 1–9.

Pręgowski, M.P., 2016. Companion Animals in Everyday Life: Situating Human-animal Engagement Within Cultures. Springer.

Probst, J.R., Crow, T.R., 1991. Integrating biological diversity and resource management. Journal of Forestry 89, 12–17.

Roldán-Clarà, B., Lopez-Medellín, X., Espejel, I., Arellano, E., 2014. Literature review of the use of birds as pets in Latin-America, with a detailed perspective on Mexico. Ethnobiology and Conservation 3, 1–18.

Saunders, C.D., 2003. The emerging field of conservation psychology. Human Ecology LReview 10, 137–149.

Serpell, J., 1996. In the Company of Animals: a Study of Human-animal Relationships. Cambridge University Press, Cambridge.

Shanks, N., 2002. Animals and Science: a Guide to the Debates. ABC-CLIO.

Springer, J.T., Holley, D., 2012. An Introduction to Zoology: Investigating the Animal World. Jones & Bartlett Learnign.

Ståhlberg, S., Svanberg, I., 2014. Among fishermen and horse Nomads. In: Svanberg, I., Łuczaj, Ł. (Eds.), Pioneers in European Ethnobiology. Uppsala University Library, Uppsala, Sweden, pp. 73–98.

Svanberg, I., Ståhlberg, S., 2012. Wild animals in Russian, Siberian and Central Asian households according to eighteenth-century travel reports. Journal de la Société Finno-Ougrienne 93, 455–486.

Tedesco, L.A., 2000. Lascaux (ca. 15,000 B.C.). In: Heilbrunn Timeline of Art History. The Metropolitan Museum of Art, New York.

Tidemann, S., Gosler, A., 2010. Ethno-ornithology: birds, indigenous people. In: Culture and Society. first ed. Earthscan/James & James.

van Vliet, D., 2007. Animals and war. In: Heathcote-James, E. (Ed.), Psychic Pets – How Animal Intuition and Perception Has Changed Human Lives. John Blake, London, England.

Waller, M.T., 2016. Ethnoprimatology: Primate Conservation in the 21st Century. Springer, Switzerland.

Wilson, C.C., Turner, D.C., 1997. Companion Animals in Human Health. Sage Publications.

CHAPTER

2

Ethnozoology: Conceptual and Historical Aspects*

Rômulo Romeu Nóbrega Alves[1], *Wedson Medeiros Silva Souto*[2], *Ulysses Paulino Albuquerque*[3]

[1]Universidade Estadual da Paraíba, Campina Grande, Brazil; [2]Universidade Federal do Piauí (UFPI), Teresina, Brazil; [3]Universidade Federal de Pernambuco, Recife, Brazil

INTRODUCTION

Extremely close connections have existed between humans and animals throughout history (Alves, 2012; Kalof, 2007; Kalof and Resl, 2007). Humans have always attempted to understand animals, enslave them, and capture their strength and power (Holley, 2009). Archeological researchers have determined that humans have consumed a wide variety of fish, mollusks, birds, mammals, reptiles, and amphibians for at least 1500 years (Emery, 2007; Foster and James, 2002; Hamblin, 1985; Kyselý, 2008; Masson, 1999; Masson and Peraza Lope, 2008; McKillop, 1984, 1985; Pohl, 1976, 1981) and perhaps as many as 4000 years (Jorgenson, 1998). Other evidence of ancient human–animal relationships can be seen in rock paintings that depict wild animals such as bison, horses, and deer being hunted by human figures.

Hunting is one of oldest known human activities, and animals have been hunted for utilitarian reasons, as well as for defense against large predators (Alves, 2012). Similarly, the importance of fishing for mankind has a long history. Faunal-derived products are used in many ways, especially as food, but also as clothing and tools, and for medicinal and magic and religious purposes (Alvard et al., 1997; Alves et al., 2007, 2009, 2012a,; Inskip and Zimmermann, 2009; Léo Neto et al., 2009; Prins et al., 2000). This enduring relationship of dependence has also contributed to the formation of affective links with certain animals, and many species are kept as pets, especially birds and mammals, and now, reptiles and amphibians too (Alves et al., 2010d, 2012b; Franke and Telecky, 2001; Hoover, 1998). These relationships with animals go beyond simple utilitarian considerations; for there have been strong supernatural relationships between the worlds of humans and animals since remote

*This chapter represents a revised and updated version of the article "Alves RRN, Souto WMS (2015) Ethnozoology: A Brief Introduction. Ethnobiology and Conservation 4:1–13."

Ethnozoology
http://dx.doi.org/10.1016/B978-0-12-809913-1.00002-8

times (Alves, 2012). All human cultures have mythologies, and all of them show close integration and connections with animals, and totemic, ancestral, or mythological (imaginary) animals or animal-gods have been present throughout human history (Allaby, 2010; Alves et al., 2012a).

Interactions between humans and animals have given rise to activities such as fishing and hunting, which are among the most ancient human practices, and they continue to play important roles in the survival and evolution of humanity (Alves, 2012). The domestication of animals is another excellent example of the relevance of the animals in human history. This process allowed early human societies to enrich their diets with regular sources of meat and milk. Later, certain domesticated animals provided new sources of muscular energy as pack and mounted animals or for the traction of plows and wagons—thus, multiplying the productive capacity of men, as well as their spatial mobility (Ribeiro, 1998).

From a historical perspective, the domestication of biological resources resulted in extremely important modifications of human lifestyles, allowing populations to abandon nomadic practices (carried out since the very beginning of human evolution), become sedentary, and occupy certain territories where plants could be cultivated and animals domesticated. This situation progressively decreased the previous complete human dependence on gathering activities, hunting, and fishing (Alves, 2016), although these activities have perpetuated and still contribute significantly to human food supplies.

The connection between animals and humans dates back to thousands of years, and cultures all over the world have developed characteristic ways of interacting with their regional fauna over time. Animals have played a wide range of roles in human life from the earliest days of recorded history, resulting in many kinds of interactions with other animals, including interactions that are predatory, competitive, parasitic, mutualistic, and commensal (Baenninger, 1995). The variety of interactions (both past and present) that human cultures maintain with animals is the subject matter of ethnozoology, a discipline that has its roots as deep within the past as the first relationship between humans and other animals. Sax (2002) noted that human attitudes about animals evolved long before their first attempts to represent them in the arts and history, and only much later did people begin to study them scientifically. As such, the origin of traditional zoological knowledge, one of the focal points of the study of ethnozoology, can be thought of as coinciding with the origin of humans and with the first contacts between our species and other animals.

The existence of interrelationship between humans and animals has a wide variety of implications, depending on the perspective adopted. On one hand, there are many human societies that promote a deep respect for animals, as these creatures are important actors in their spiritual traditions due to their utilitarian or spiritual value. Societies in Asia, Africa, and Latin America frequently established sacred localities with inherent spiritual or religious significance, and they were frequently also natural sanctuaries of biodiversity. Many traditional cultures still consider certain animal species sacred and foster their conservation (even though that is not their primary motivation) (McNeely, 2001). On the other hand, animals and animal organs are universally utilized in many different ways by human groups, and anthropogenic activities can exert great direct or indirect influence on the local fauna (especially target species), and these interactions must be taken into account when conservation actions are being considered (Alves et al., 2008, 2010b, 2010f).

The conservation of natural resources and biodiversity is indispensable, not only to preserve genetic diversity but also to guarantee the subsistence of large numbers of humans throughout the world (Alves and Souto, 2010), and it will not be possible to create meaningful animal conservation strategies without considering the interaction of humans with animals—the focus of ethnozoological studies (Alves, 2012). As such, the present chapter discusses numerous conceptual and historical aspects of ethnozoology–a discipline that seeks to investigate the complex but important interrelationship between humans and the animals with which we share this earth.

HISTORICAL CONSIDERATIONS ABOUT HUMAN AND FAUNA INTERACTIONS

Rock paintings and archeological inscriptions provide clear evidence of the antiquity of interactions between humans and other animals (Baker, 1941; Martínez, 2008; Russell, 2012), representing important records of how the first humans related to their regional faunas. Ancient cave paintings often contain animal and/or human figures and show interactions between them—indicating the utilitarian and symbolic importance of animals dating back to prehistorical times (Alves et al., 2010g; Martínez, 2008; Russell, 2012). Although rock paintings and archeological inscriptions can be considered the first human–fauna interaction records (see Baker, 1941), written documents have more precisely recorded information about the interaction of ancient humans groups with their regional fauna and their uses of those animals (Alves and Souto, 2015). In every ancient culture with a written language, people have recorded useful knowledge about animals, plants, and environments (Svanberg et al., 2011).

Animals were linked to people in many ways in the cultural conceptions of the time, and contributed to defining royal institutions, as well as solidifying emergent cosmologies that linked humans to celestial orbs, the earth, and the gods. These views were preserved in hieroglyphs, papyrus documents, and other records left behind by ancient civilizations (Alves and Souto, 2015). In ancient Egypt, for example, royal hunts of wild bulls were well documented during the reign of Amenophis III during the later part of the Eighteenth Dynasty (ca. 3300 BC) when these animals apparently became locally extinct (Dodd, 1993). These Old World civilizations had (often exaggerated) beliefs that certain species of animals shared important characteristics with humans, and cattle, horses, and snakes, for example, became symbols that were closely associated with power/domination or libido/fertility (Dodd, 1993).

The antiquity in the use of medicinal animals is another important evidence recorded in historical texts (Alves et al., 2013b). Some of these texts have been preserved (Svanberg et al., 2011), such as Assyrian, Egyptian, and Greek historical documents, which bear witness to extensive knowledge about how animal and plant products could be utilized (MacKinney, 1946; Raven, 2000). Our understanding of Egyptian medicine and pharmacology, for example, is based on inscriptions on monuments and graves and papyrus rolls, the most important one being the Papyrus Ebers. It is thought to have been written around 1550 BC (Lidgard, 2005) and includes 800 or more prescriptions comprising various herbs, animals, and minerals; a considerable number contain matter derived from, both wild and domestic, insects, reptiles, and fish (Alves et al., 2013b; MacKinney, 1946).

While animals and humans have shared a very long history, and humans have been accumulating knowledge for untold generations about the fauna with which they interact, the origin of ethnozoology (like many other academic disciplines) is more closely linked to the naturalists and explorers who spread out about the world starting in the 16th century. Until the beginning of the 19th century, most of our knowledge concerning the world's biota, including its fauna, was derived from the reports of naturalists and explorers who recorded information about activities such as hunting and fishing (Alves, 2017; D'Ambrosio, 2015). These documents include the works of naturalists who demonstrated interest in the fauna, as well as the zoological knowledge of native residents. These naturalists generally compiled lists of native animals together with their regional and scientific names and descriptions of their uses (Sillitoe, 2006). Information concerning the use of animals by indigenous populations in the New World have been accumulating ever since the first voyage of Columbus (Castetter, 1944). This tradition continued through the 19th and 20th centuries, as exemplified by the voyage of Darwin on the *HMS Beagle* during which he recorded biological information about regional ecosystems, and the work of Wallace during his stay in the Malaysian Archipelago (now Indonesia). The zoological information contained in these pioneering works was likewise codependent on the work of Linnaeus—one of the most notable naturalists of that time (Ellen, 2004). We can thus interpret these works as the roots of what was to become the science of ethnozoology, as these European naturalists and explorers not only sought to learn about new regions of the world but also to take advantage of those new natural resources by identifying the animal species found there and documenting their uses. Some pioneer ethnographers, such as Alfred Cort Haddon and Franz Uri Boas, however, were more interested in studying the local communities that they encountered rather than their surrounding environments (Sillitoe, 2006).

The interests of naturalists went well beyond simply recording the uses of the fauna by the native populations, and the direct or indirect help of these local populations was indispensable to discovering thousands of additional animal species. As was exemplified by Moreira (2002), 19th century naturalists spread out over the planet and enormously amplified the scientific knowledge of the time—and the success of their scientific expeditions were often greatly dependent on the collaboration of native or resident communities and their traditional knowledge. This traditional knowledge was systematized by the naturalists, filtered by the scientific outlook predominant at that time, and subsequently incorporated into the growing universal scientific pool. In the specific case of zoology, the aid of the local populations was critical in many ways, especially in terms of locating, collecting, and naming animals; preparing and preserving the specimens; discovering "new" species; analyzing their habits and utilitarian features; domesticating certain "wild" animals; and in developing techniques and tools for capturing and preserving them.

Moreira (2002) illustrates in a very interesting article the importance that native populations had for the natural sciences by citing the examples of three notable naturalists, the Englishmen Alfred R. Wallace (1823–1913) and H.W. Bates (1825–1892) and the Swiss explorer Louis Agassiz (1807–1873), who all undertook expeditions to Brazil during the 19th century. These scientists were very successful during their expeditions and made enormous contributions to zoology through their descriptions of thousands of species. Henry Walter Bates, for example, collected 14,712 different species (mostly insects)—of which 8000 were new to science—during 11 years in the Amazon region. His records of his trips throughout the Amazon region cite about 135 different people (most of them by name) from all walks of life, who helped during the fieldwork and in the localization and capture of specimens: businessmen, farmers, workers, slaves, military personnel, Amerindians, and hunters. Similarly, many parts of the travel logs of Wallace in Brazil and many of the scientific articles based on these expeditions record the participation of local inhabitants in collecting specimens and mapping the Negro River. Wallace often noted in his records the importance of the native knowledge of the flora and fauna and their geographic distributions. Likewise, Louis Agassiz (who led the Thayer expedition from 1865 to 1866) repeatedly pointed out that the contributions of the local habitants were essential to the success of the fieldwork program, like locating and capturing Amazonian fish and describing their behavior.

This traditional or local zoological knowledge[1] (one of the focal points of ethnozoological studies) has been extremely useful to zoological studies of numerous species, and many descriptions have likewise been based on specimens collected by native hunters/fishers accompanying naturalists during their expeditions (Alves and Souto, 2015). It can thus be seen that the histories of zoology and ethnozoology overlap—although the roles of native populations were not always fully recognized. Moreira (2002) pointed out that although there were many diverse references in the travel logs and letters of naturalists to the essential aid provided by local habitants, this information was rarely widely disseminated due to the usually concise nature of scientific publications (books, reports, articles). This situation contributed, among other factors, to the emergence of the image of scientists as "hero-explorers" who survived enormous dangers almost alone through herculean efforts, "discovering" large numbers of new species of animals and plants. It was often emphasized that these scientists had encountered hostile relationships with indigenous groups (which probably only rarely occurred), but little note was otherwise made of the existence of these people, or that their support and knowledge had been extremely important to the success of their scientific quests.

It is important to remember that local populations continue to provide much more than basic logistic support to zoologists and ecologists (research areas that commonly count on the cooperation and assistance of local inhabitants)—including indicating sites that are best for mounting collecting equipment (and many times even directly collecting the specimens themselves)—thus perpetuating the roles and practices that were available to early naturalists. Additionally, the information retained in traditional zoological knowledge has inspired many hypotheses and has been used to complement ecological and zoological studies (see Chapter 23, in this book–The Role of Ethnozoology in Animal Studies).

THE ORIGIN AND HISTORY OF ETHNOZOOLOGY

The first written documents that recorded connections between mankind and nature (e.g., those produced in Greece, Egypt, and Asia) (Alves et al., 2013b; Bala, 1985) represent historic records of the even more ancient relationships between our species and other living creatures. Although these writings can be interpreted as the first ethnozoological records, ethnozoology is a new formal science. These ancient records were produced within different academic disciplines such as anthropology, ethnology, archeology, and medicine. Clément (1998) noted that "*the origin of ethnobiology lato sensu lies deep in the mists of time, when the first hominids took an interest in plants and animals; it can rightly be argued that the foundations of ethnobiology were laid long before the 19th century, and are to be sought in the sacred texts, oral or written, that form the substructures of many civilizations.*" Nonetheless, from an academic perspective, the beginnings of what would become ethnobiological studies occurred only in 1874 (Hidayati et al., 2015), when Stephen Powers first used the term "aboriginal botany" to refer to the uses of plants by Amerindians in California (Powers, 1874). At about the same time, other researchers were likewise becoming interested in the uses of biodiversity in the American west by its native inhabitants (Brown, 1868; Palmer, 1871, 1878; Stearns, 1889).

The history of ethnobiology and its principal subdivisions (ethnobotany and ethnozoology) overlap, as has been addressed by

[1] For a discussion about the use of the terms "local knowledge" or "traditional Knowledge" see Alves and Albuquerque (2010).

a number of different authors who generally rely on the historiography proposed by Clément (1998) that divides their development into three phases. The first phase (1860–99), known as the preclassical phase, corresponds to the period during which terms such as ethnobotany and ethnozoology were first coined; the second phase (1950–80s), known as the classical period, was characterized by a number of linguistic studies and ethnobiological classifications; the third phase, known as the postclassical period (1990s), was characterized by true partnerships between Western scientists and local populations, with greater recognition and respect for traditional ecological knowledge.

Hunn (2007) added one more phase to those proposed by Clement. For Hunn (2007), Phase 1 (1895–1950) was characterized by the documentation of "useful" plants and animals; Phase 2 (1954–70s) corresponds to the phase of "cognitive ethnobiology" or "ethnoscience"; Phase 3 (1970–80s) saw the rise of ethnobiological research with an ecological focus; in Phase 4 (1990s), collaborative research arose, emphasizing the rights of local populations (Hunn, 2002).

Wolverton (2013) suggested a fifth phase, although not indicating the exact year it was initiated (although it is presumed to have begun in 2001 and continues to the present day). This phase is characterized by interdisciplinary studies and the recognition of the importance of ethnobiological research in the context of complex environmental and cultural changes. We believe that another aspect has become important in this last phase—the use of folk ecological knowledge as a complement to traditional ecological/zoological research (see Chapter 23, in this book)—and the appreciation of this local knowledge has noticeably grown among researchers who are not necessarily ethnobiologists (Alves and Nishida, 2002; Anadón et al., 2010; Beaudreau and Levin, 2014; Kotschwar Logan et al., 2015; Rasalato et al., 2010; Ziembicki et al., 2013; Zuercher et al., 2003).

Curiously, although ethnobotany and ethnozoology are considered subdivisions of ethnobiology, these terms appeared before the recognition of the latter discipline. The term "ethnobiology" was first used in the United States in 1935 (Castetter, 1935) as a fusion of two elements, "ethnos" and "biology"—after the fashion of many similar terms formed since the words "ethnography" and "ethnology" were first coined in the late 18th century (Clément, 1998). Ethnobotany received its widely accepted name from Hershberger in 1895 (Hershberger, 1885). The first academic publication with an ethnozoological orientation (using the prefix "ethno") was prepared by Stearns (1889), who discussed "ethno-conchology" (which would now be placed within the sub-area of ethnomalacology) in his study of the use of shell money—even before the term "ethnobotany" was coined. The term ethnozoology first appeared in 1899, as a branch of zootechnology (Mason, 1899) and, somewhat later, Henderson and Harrington (1914) referred to ethnozoology as the study of existing cultures and their relationships with the animals in their surrounding environments. In 1944, Castetter coined the term "ethnobiology" to signify the use of plants and animals by "primitive" peoples (Castetter, 1944). In addition to the abovementioned disciplines, ethnobiology encompasses a wide range of subdisciplines such as ethnoecology, ethnopharmacology, ethnomedicine, ethnomycology, and ethnoveterinary (Hidayati et al., 2015). A timeline of important events in ethnozoology history can be seen in Fig. 2.1.

Although animals have played important roles in all human cultures since remote times, specific studies concerning the uses of animals have consistently lagged behind similar studies devoted to plants. Although the fields of ethnobotany and ethnozoology appeared at approximately the same time, the former grew more rapidly in terms of its bibliographic production, which resulted in its greater consolidation. The first publications in ethnozoology did not

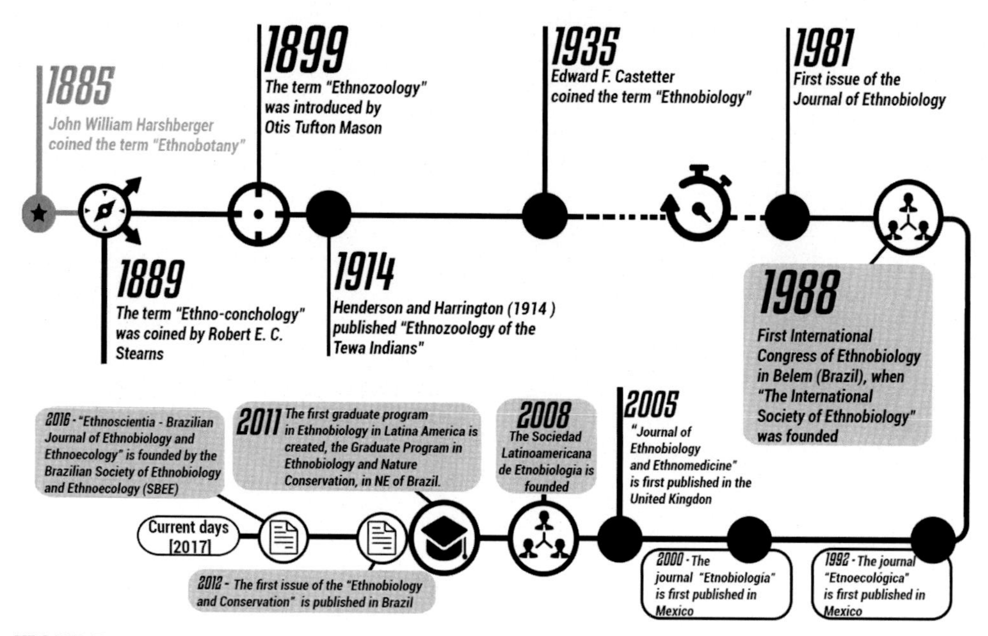

FIGURE 2.1 A timeline of important events in ethnozoology history, with emphasis on Latin America.

stimulate much additional research, although a considerable body of knowledge concerning the utilization of animals by traditional societies can be found in a variety of earlier publications not produced with a specific focus on ethnozoology (Birket-Smith, 1976; Hornaday, 1889; Merriam, 1905; Steensby, 1917).

After the publication of the first articles clearly focusing on ethnozoology (Mason, 1899; Stearns, 1889), later publications became concentrated on utilitarian aspects and folk classifications of animals (Sturtevant, 1964). Following the preclassical tendencies of ethnobiology in 1908, Chamberlin listed the common names of animals used by the Goshute Amerindians of North America (Chamberlin, 1908). In a later phase, ethnozoological researchers focused on the perceptions and classifications of animals. Within this context, the works of Malkin concerning the ethnozoology of the Seri, Sumu, and Cora peoples in Mexico (Malkin, 1956a,b; 1958) called attention to the high numbers of taxa in folk taxonomies and to native knowledge about themes such as sexual differentiation, development, and feeding habits among animals. Various other ethnozoological studies were undertaken during this period that revealed the grasp of traditional societies of the principals of classification, nomenclature, and species identifications (Berlin et al., 1973; Bright and Bright, 1965; Diamond, 1966).

Ethnozoologists and other researchers are currently concentrating their efforts on research areas that include (1) cultural perception and zoological classification systems (Fleck et al., 1999; Holman, 2005; Mourão et al., 2006; Posey, 1982); (2) importance and presence of animals in stories, myths, and beliefs (Descola, 1998; Léo Neto et al., 2009; Lewis, 1991); (3) biological and cultural aspects of animal use by human societies (Dias et al., 2011; Gunnthorsdottir, 2001; Posey, 1978); (4) methods of obtaining and preparing organic substances extracted from animals (for cosmetic, ritualistic, medicinal, or food uses, etc.) (Alves, 2006, 2009; Alves et al., 2007; Barboza et al., 2007; Costa-Neto and Oliveira,

2000; Lev, 2003, 2006; Rocha, 2007; Rocha et al., 2008; Rosa et al., 2011; Vázquez et al., 2006); (5) domestication, examining the cultural bases and the biological consequences of long-term faunal resource management (Digard, 1992; Haudricourt, 1977); (6) biological heterogeneity and the cognitive processes involved in the management and conservation of natural resources (Alves and Nishida, 2002; Fleck and Harder, 2000); and (7) collection techniques and their impacts on animal populations (Alves et al., 2009; Balée, 1985; Bezerra et al., 2012; Nishida et al., 2006; Nordi et al., 2009; Quijano-Hernández and Calmé, 2002; Souto, 2007).

The number of publications concerning ethnozoology has steadily increased, reflecting the interests of researchers in diverse areas. It is important to note that interactions between humans and animals are investigated in many different disciplines that have obvious links with ethnozoology. One example is zooarchaeology (which focuses on studies of the remains of animals found in archaeological sites), which is important for understanding ancient relationships between humans and their environments (especially between humans and animal populations; see Reitz and Wing, 2008). Some authors have adopted the term ethnozooarchaeology (Albarella and Trentacoste, 2011; Broderick, 2016) to describe the study of human–animal relationships from the examination of animal remains found at archeological sites. Another discipline having clear intersections with ethnozoology is anthrozoology, which, according to Bradshaw (2010), consists of the study of animal–human interactions. This discipline generally focuses more directly on human interactions with domestic animals, investigating themes such as keeping pets (especially dogs and cats) and animal welfare. Another discipline that demonstrates historical links with ethnozoology is ethnozootechny, which investigates local knowledge, practices, and beliefs related to animal husbandry (Alves et al., 2010a); also closely related is ethnoveterinary, which studies knowledge, skills, methods, practices, and beliefs concerning animal care (Alves et al., 2010c; Mathias-Mundy and McCorkle, 1989; Souto et al., 2013). When the use of animals for medicinal purposes is considered, ethnozoology has close links with two other areas: ethnopharmacology and ethnomedicine.

Ethnozoology can now be further subdivided into taxon-specific fields of interest (Alves, 2017) (Fig. 2.2). A good example is ethnoprimatology, a field that focuses on ecological and cultural interconnections between human and nonhuman primates (Riley, 2006; Sponsel, 1997; Waller, 2016). Different subdivisions of ethnozoology have similarly arisen to examine human interactions with other important animal taxa (Alves, 2017) such as insects (ethnoentomology), mollusks (ethnomalacology), fish (ethnoichthyology), birds (ethnoornithology), mammals (ethnomastozoology), and reptiles/amphibians (ethnoherpetology) (Alves, 2017). It should be noted that these subdivisions are related to animal groups of established importance to humans and, in some cases, they have been further detailed to consider specific animal groups such as elephants and primates, designating areas such as ethnoelephantology (Locke, 2013) and ethnoprimatology (Sponsel, 1997), as subdivisions of ethnomastozoology; snakes (ethno-ophiology) (Joshi and Joshi, 2010) as a subdivision of ethnoherpetology; and butterflies (ethnolepidopterology) (Parsons, 1991) as a subdivision of ethnoentomology. There are even groups with lesser utilitarian importance that have nonetheless inspired the creation of further subsections of ethnozoology, such as sponges (ethnospongiology) (Docio et al., 2013).

In spite of this increase in the numbers of subdivisions within ethnozoology, theoretical and/or methodological advances have not necessarily followed—only an inflationary tendency of ethnobiological terminology. This tendency can be viewed on one hand as a simple effort to organize the "taxonomy" of the various fields within ethnozoology, or simply as an excessive growth of terminologies that are not fully justified epistemologically.

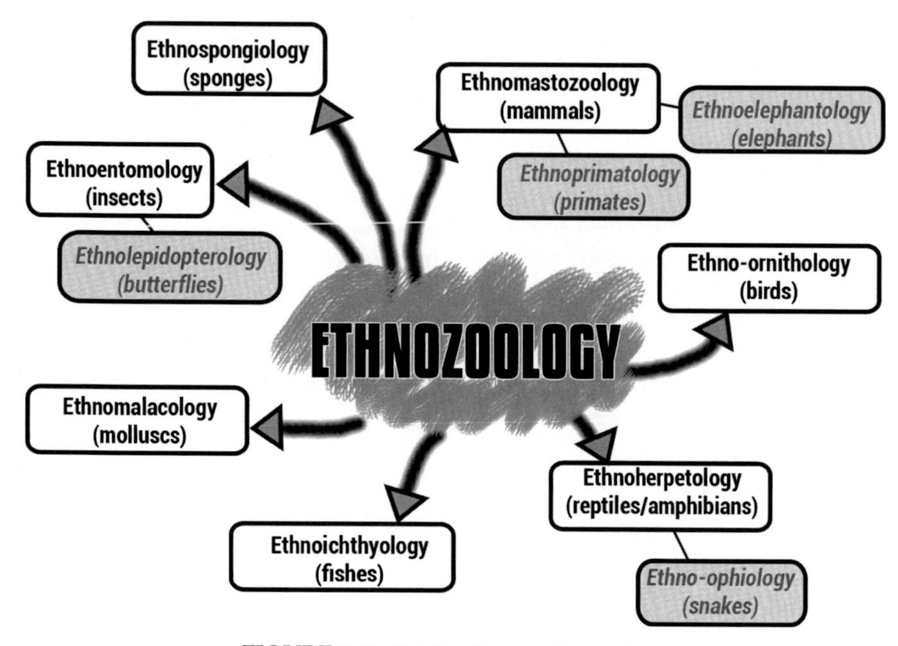

FIGURE 2.2 Subdivisions of ethnozoology.

ETHNOZOOLOGY: WHAT ARE THE MARCHING ORDERS?

Ethnozoology has drawn the attention of researchers, in the biological sciences, who have found that such studies can aid in evaluating the impact of human populations on other animal species, and in the development of sustainable management plans—and are therefore fundamental to conservation efforts. Ethnozoological publications have greatly increased now as judged by reviews published in Latin America (Albuquerque et al., 2013), although the component studies have largely been concentrated in Brazil (Alves and Souto, 2011) and Mexico (Santos-Fita et al., 2012). Importantly, these publications generally cite high proportions of national research efforts, as opposed to international studies, indicating a tendency toward insularity (Campos et al., 2016). This same phenomenon is usually associated with low levels of international cooperation (as can be seen in Latin America; Albuquerque et al. (2013)), which can lead to theoretical stagnation and reduced originality and innovation.

Many works have considered themes such as hunting, fishing, and the use and commerce of the native fauna. On a global scale, some disciplines have demonstrated strong growth, such as ethnoprimatology. The term ethnoprimatology, first coined by Leslie Sponsel (1997), has been a field of growing interest, as demonstrated by the publications of books (Fuentes and Wolfe, 2002; Waller, 2016; Dore et al., 2017) and special issues on ethnoprimatology in important primatology journals (Fuentes and Hockings, 2010). In spite of these advances, however, the study of ethnozoology now requires a process of self-criticism to evaluate its advances. While large volumes of information have been produced concerning interaction between humans and animals, there have not been corresponding advances in theoretical terms, and relatively few works have been guided by theories or hypotheses—which has led to its criticism as a weak and immature science. It should be noted that ethnobotany has suffered the same criticism, even though this sister discipline is older and has been considerably more productive (see Albuquerque and Hanazaki (2009)).

Unsurprisingly, many ethnozoological studies have been associated with two central activities undertaken by local communities: hunting and fishing (Alves, 2012)—although ethnozoological studies can also be undertaken in urban environments, especially in localities such as zoos, schools, and traditional markets (Alves et al., 2013c, 2007, 2014; Marques and Guerreiro, 2007). In pointing out that the study of ethnozoology "begins at home," Overal (1990) has called attention to the study of ethnozoological phenomena within our own culture (as opposed to examining culturally distant societies)—mentioning groups and/or phenomena that could be studied from an ethnozoological perspective in both Western and traditional cultures, such as: animal trainers; farmers who "call in" cattle, pigs, and other animals; breeders of dogs and other pets; urban hunters; and breeders of fighting cocks and other animals kept for "sport" and betting purposes.

Public markets likewise present excellent opportunities for developing ethnozoological studies in urban areas (Alves, 2006; Alves et al., 2010d, 2010e, 2013a, 2013c; Alves and Rosa, 2008; Apaza et al., 2003; Fernandes-Ferreira et al., 2012; Ferreira et al., 2013; Noss, 1998; Oliveira et al., 2010; Williams et al., 2013) as many live animals (both wild and domesticated) and various products of animal origin can be found there; and these are traditional sites for exchanging and acquiring cultural information. Depending on their size, these public markets commonly have specific areas that sell animals and animal parts, and the vendors can provide important information about the different origins of those resources (Alves and Rosa, 2007). Information about the exotic and native fauna of a region obtained in public markets should be very useful when evaluating conservation plans for those same natural resources (Almeida and Albuquerque, 2002; Alves and Pereira Filho, 2007; Alves and Rosa, 2008; Alves and Santana, 2008; Broad, 2001; Yi-Ming et al., 2000).

In spite of their cultural and economic importance, however, very few ethnozoological investigations have examined these public markets in any depth (Alves et al., 2013c). In many countries (such as Brazil) legal implications related to the commercialization of wild animals (especially those listed as threatened with extinction) contribute greatly to the difficulty of freely obtaining ethnozoological information in public places.

In many countries, especially those located in tropical regions that have great faunal diversity, the illegal commerce in wild animals removes many species from their natural environments. This is certainly one of the gravest threats to many populations of native species, and ethnozoological studies constitute an invaluable tool for understanding the socioeconomic and cultural context into which the commercialization of the wild fauna is embedded—an essential aspect to the elaboration of conservation proposals.

FINAL CONSIDERATIONS

The connections between humans and other species of animals involve predatory and symbiotic relations established in remote times, but academic scholars have begun to examine this theme with intensity. Human communities have developed a huge store of knowledge about animals down through the centuries (passed from generation to generation, largely through oral traditions) that is closely integrated with many other cultural aspects. This zoological knowledge is an important part of the human cultural heritage that has been accumulating for millennia.

Interactions between people and animals are extremely varied and can be explored through disciplines associated with the social and natural sciences, fostering the emergence of ethnozoological approaches—fundamental areas of scientific research that examine historical, economic, sociological, anthropological, and environmental aspects of the relationships between humans and animals. These studies can aid in the evaluation of the impacts of human populations on other animal species and in the development of sustainable management plans—and are thus fundamental to conservation efforts. Additionally, local or traditional knowledge about the fauna can be important to academic research projects, and it offers the possibility of significant savings in comparison to the costs involved with conventional methodologies. Ethnozoological studies also facilitate communication between researchers and the agents responsible for the elaboration of management plans and local human populations—which is of fundamental importance to the development of efficient conservation strategies.

In spite of its historical roots and many contributions, Clément (1998) observed that a large proportion of the contemporary scientific production in ethnozoology still mirrors its classic phase. This implies that the field of ethnozoology has not accompanied the advances adopted by its sister disciplines of ethnobotany and ethnoecology, although those areas are likewise passing through continuous processes of reflection and self-criticism.

References

Albarella, U., Trentacoste, A., 2011. Ethnozooarchaeology: The Present and Past of Human–animal Relationships, first ed. Oxbow Books, Oxford, UK.

Albuquerque, U.P., Hanazaki, N., 2009. Five problems in current ethnobotanical research—and some suggestions for strengthening them. Human Ecology 37, 653–661.

Albuquerque, U.P., Silva, J.S., Campos, J.L.A., Sousa, R.S., Silva, T.C., Alves, R.R.N., 2013. The current status of ethnobiological research in Latin America: gaps and perspectives. Journal of Ethnobiology and Ethnomedicine 9, 1–9.

Allaby, M., 2010. Animals: From Mythology to Zoology. Facts On File, Inc., New York.

Almeida, C.F.C.B.R., Albuquerque, U.P., 2002. Uso e conservação de plantas e animais medicinais no Estado de Pernambuco (Nordeste do Brasil): Um estudo de caso. Interciencia 27, 276–285.

Alvard, M.S., Robinson, J.G., Redford, K.H., Kaplan, H., 1997. The sustainability of subsistence hunting in the neotropics. Conservation Biology 11, 977–982.

Alves, A.G.C., Albuquerque, U.P., 2010. "Ethno what?" Terminological problems in ethnoscience with a special emphasis on the Brazilian context. In: Albuquerque, U.P., Hanazaki, N. (Eds.), Recent Developments and Case Studies in Ethnobotany. Nupeea, Recife, Brazil.

Alves, A.G.C., Pires, D.A.F., Ribeiro, M.N., 2010a. Conhecimento local e produção animal: uma perspectiva baseada na Etnozootecnia. Archivos de Zootecnia 59, 45–56.

Alves, R.R.N., 2006. Uso e comércio de animais para fins medicinais e mágico-religiosos no Norte e Nordeste do Brasil, Programa de Pós-Graduação em Ciências Biológicas. Universidade Federal da Paraíba, João Pessoa, Paraíba. 252 pp.

Alves, R.R.N., 2009. Fauna used in popular medicine in Northeast Brazil. Journal of Ethnobiology and Ethnomedicine 5, 1–30.

Alves, R.R.N., 2012. Relationships between fauna and people and the role of ethnozoology in animal conservation. Ethnobiology and Conservation 1, 1–69.

Alves, R.R.N., 2016. Domestication of animals. In: Albuquerque, U.P., Alves, R.R.N. (Eds.), Introduction to Ethnobiology. Springer, Cham, Switzerland, pp. 221–225.

Alves, R.R.N., 2017. Ethnozoology. In: Fuentes, A. (Ed.), The International Encyclopedia of Primatology. Wiley, New Jersey.

Alves, R.R.N., Barboza, R.R.D., Souto, W.M.S., 2010b. A global overview of canids used in traditional medicines. Biodiversity and Conservation 19, 1513–1522.

Alves, R.R.N., Barboza, R.R.D., Souto, W.M.S., 2010c. Plants used in animal health care in south and latin America: an overview. In: Katerere, R.D., Luseba, D. (Eds.), Ethnoveterinary Botanical Medicine: Herbal Medicines for Animal Health. CRC Press, New York, pp. 231–256.

Alves, R.R.N., Lima, J.R.F., Araújo, H.F., 2013a. The live bird trade in Brazil and its conservation implications: an overview. Bird Conservation International 23, 53–65.

Alves, R.R.N., Medeiros, M.F.T., Albuquerque, U.P., Rosa, I.L., 2013b. From past to present: medicinal animals in a historical perspective. In: Alves, R.R.N., Rosa, I.L. (Eds.), Animals in Traditional Folk Medicine: Implications for Conservation. Springer, Heidelberg, Germany, pp. 11–23.

Alves, R.R.N., Mendonça, L.E.T., Confessor, M.V.A., Vieira, W.L.S., Lopez, L.C.S., 2009. Hunting strategies used in the semi-arid region of Northeastern Brazil. Journal of Ethnobiology and Ethnomedicine 5, 1–50.

Alves, R.R.N., Nishida, A.K., 2002. A ecdise do caranguejo-uçá, Ucides cordatus L. (Decapoda, Brachyura) na visão dos caranguejeiros. Interciencia 27, 110–117.

Alves, R.R.N., Nogueira, E., Araujo, H., Brooks, S., 2010d. Birdkeeping in the Caatinga, NE Brazil. Human Ecology 38, 147–156.

Alves, R.R.N., Oliveira, M.G.G., Barboza, R.R.D., Lopez, L.C.S., 2010e. An ethnozoological survey of medicinal animals commercialized in the markets of Campina Grande, NE Brazil. Human Ecology Review 17, 11–17.

Alves, R.R.N., Pereira Filho, G.A., 2007. Commercialization and use of snakes in North and Northeastern Brazil: implications for conservation and management. Biodivers Conserv 16, 969–985.

Alves, R.R.N., Rosa, I.L., 2007. Zootherapy goes to town: the use of animal-based remedies in urban areas of NE and N Brazil. Journal of Ethnopharmacology 113, 541–555.

Alves, R.R.N., Rosa, I.L., 2008. Use of tucuxi dolphin sotalia fluviatilis for medicinal and magic/religious purposes in North of Brazil. Human Ecology 36, 443–447.

Alves, R.R.N., Rosa, I.L., Albuquerque, U.P., Cunningham, A.B., 2013c. Medicine from the wild: an overview of the use and trade of animal products in traditional medicines. In: Alves, R.R.N., Rosa, I.L. (Eds.), Animals in Traditional Folk Medicine. Springer, Heidelberg, Germany, pp. 25–42.

Alves, R.R.N., Rosa, I.L., Léo Neto, N.A., Voeks, R., 2012a. Animals for the gods: magical and religious faunal use and trade in Brazil. Human Ecology 40, 751–780.

Alves, R.R.N., Rosa, I.L., Santana, G.G., 2007. The role of animal-derived remedies as complementary medicine in Brazil. BioScience 57, 949–955.

Alves, R.R.N., Santana, G.G., 2008. Use and commercialization of Podocnemis expansa (Schweiger 1812) (Testudines: Podocnemididae) for medicinal purposes in two communities in North of Brazil. Journal of Ethnobiology and Ethnomedicine 4, 1–6.

Alves, R.R.N., Silva, V.N., Trovão, D.M.B.M., Oliveira, J.V., Mourão, J.S., Dias, T.L.P., Alves, A.G.C., Lucena, R.F.P., Barboza, R.R.D., Montenegro, P.F.G.P., 2014. Students' attitudes toward and knowledge about snakes in the semiarid region of Northeastern Brazil. Journal of Ethnobiology and Ethnomedicine 10, 1–8.

Alves, R.R.N., Souto, W.M.S., 2010. Etnozoologia: conceitos, considerações históricas e importância. In: Alves, R.R.N., Souto, W.M.S., Mourão, J.S. (Eds.), A Etnozoologia no Brasil: Importância, Status atual e Perspectivas. NUPEEA, Recife, PE, Brazil, pp. 19–40.

Alves, R.R.N., Souto, W.M.S., 2011. Ethnozoology in Brazil: current status and perspectives. Journal of Ethnobiology and Ethnomedicine 7, 1–18.

Alves, R.R.N., Souto, W.M.S., 2015. Ethnozoology: a brief introduction. Ethnobiology and Conservation 4, 1–13.

Alves, R.R.N., Souto, W.M.S., Barboza, R.R.D., 2010f. Primates in traditional folk medicine: a world overview. Mammal Review 40, 155–180.

Alves, R.R.N., Souto, W.M.S., Mourão, J.S., 2010g. A Etnozoologia no Brasil: Importância, Status atual e Perspectivas, first ed. NUPEEA, Recife, Brazil.

Alves, R.R.N., Vieira, K.S., Santana, G.G., Vieira, W.L.S., Almeida, W.O., Souto, W.M.S., Montenegro, P.F.G.P., Pezzuti, J.C.B., 2012b. A review on human attitudes towards reptiles in Brazil. Environmental Monitoring and Assessment 184, 6877–6901.

Alves, R.R.N., Vieira, W.L.S., Santana, G.G., 2008. Reptiles used in traditional folk medicine: conservation implications. Biodiversity and Conservation 17, 2037–2049.

Anadón, J.D., Giménez, A., Ballestar, R., 2010. Linking local ecological knowledge and habitat modelling to predict absolute species abundance on large scales. Biodiversity and Conservation 19, 1443–1454.

Apaza, L., Godoy, R., Wilkie, D., Byron, E., Huanca, T., Leonard, W.R., Peréz, E., Reyes-García, V., Vadez, V., 2003. Markets and the use of wild animals for traditional medicine: a case study among the Tsimané ameridians of the Bolivian rain forest. Journal of Ethnobiology 23, 47–64.

Baenninger, R., 1995. Some consequences of animal domestication for humans. Anthrozoos 8, 69–77.

Baker, F.C., 1941. A study of ethnozoology of the prehistoric Indians of Illinois. Transactions of the American Philosophical Society 32, 51–77.

Bala, P., 1985. Indigenous medicine and the state in ancient India. Ancient Science of Life 5, 1–4.

Balée, W., 1985. Ka'apor ritual hunting. Human Ecology 13, 485–510.

Barboza, R.R.D., Souto, W.M.S., Mourão, J.S., 2007. The use of zootherapeutics in folk veterinary medicine in the district of Cubati, Paraíba State, Brazil. Journal of Ethnobiology and Ethnomedicine 3, 1–14.

Beaudreau, A.H., Levin, P.S., 2014. Advancing the use of local ecological knowledge for assessing data-poor species in coastal ecosystems. Ecological Applications 24, 244–256.

Berlin, B., Breedlove, D.E., Raven, P.H., 1973. General principles of classification and nomenclature in folk biology. American Anthropologist 75, 214–242.

Bezerra, D.M.M., Araujo, H.F.P., Alves, R.R.N., 2012. Captura de aves silvestres no semiárido brasileiro: técnicas cinegéticas e implicações para conservação. Tropical Conservation Science 5, 50–66.

Birket-Smith, K., 1976. Ethnographical Collections from the Northwest Passage. AMS Press, New York.

Bradshaw, J.W.S., 2010. Anthrozoology. In: Mills, D.S., Marchant-Forde, J.N. (Eds.), The Encyclopedia of Applied Animal Behaviour and Welfare. CABI, Wallingford, UK, pp. 28–30.

Bright, J.O., Bright, W., 1965. Semantic structures in Northwestern California and the sapir-whorf hypothesis. American Anthropologist 67, 249–258.

Broad, S., 2001. The Nature and Extent of Legal and Illegal Trade in Wildlife. Hughes Hall, International and Africa Resources Trust, Seminar on Wildlife Trade Regulation and Enforcement. TRAFFIC, Cambridge, pp. 17–18.

Broderick, L., 2016. People with Animals: Perspectives and Studies in Ethnozooarchaeology. Oxbow Books, Oxford, UK.

Brown, R., 1868. On the vegetable products used by the North-west American Indians as food and medicine, in the arts, and in superstitious rites. Transactions and Proceedings of the Botanical Society of Edinburgh 9, 378–396.

Campos, J.L.A., Sobral, A., Silva, J.S., Araújo, T.A.S., Ferreira-Júnior, W.S., Santoro, F.R., Santos, G.C., Albuquerque, U.P., 2016. Insularity and citation behavior of scientific articles in young fields: the case of ethnobiology. Scientometrics 109, 1037–1055.

Castetter, E.F., 1935. Uncultivated native plants used as sources of food. University of New Mexico Bulletin 4, 1–63.

Castetter, E.F., 1944. The domain of ethnobiology. American Naturalist 78, 158–170.

Chamberlin, R.V., 1908. Animal names and anatomical terms of the Goshute Indians. Proceedings of the Academy of Natural Sciences of Philadelphia 60, 74–103.

Clément, D., 1998. The historical foundations of ethnobiology (1860–1899). Journal of Ethnobiology 18, 161.

Costa-Neto, E.M., Oliveira, M.V.M., 2000. Cockroach is good for Asthma: Zootherapeutic practices in northeastern Brazil. Human Ecology Review 7, 41–51.

D'Ambrosio, U., 2015. Theoretical reflections on ethnobiology in the third millennium. Contributions to Science 10, 49–64.

Descola, P., 1998. Estrutura ou sentimento: a relação com o animal na Amazônia. Mana 4, 23–45.

Diamond, J.M., 1966. Zoological classification system of a primitive people. Science 151, 1102–1104.

Dias, T.L.P., Leo Neto, N.A., Alves, R.R.N., 2011. Molluscs in the marine curio and souvenir trade in NE Brazil: species composition and implications for their conservation and management. Biodiversity and Conservation 20, 2393–2405.

Digard, J.P., 1992. Un aspect méconnu de l'histoire de l'Amérique: la domestication des animaux. L'Homme 32, 253–270.

Docio, L., Costa-Neto, E.M., Pinheiro, U.S., Schiavetti, A., 2013. Folk classification of sea sponges (Animalia, Porifera) by artisanal fishermen of a traditional fishing community at Camamu Bay, Bahia, Brazil. Interciencia 38, 60–66.

Dodd Jr., C.K., 1993. Strategies for Snake Conservation, Ecology and Behavior. McGraw-Hill, Inc., New York.

Dore, K.M., Riley, E.P., Fuentes, A., 2017. Ethnoprimatology: A Practical Guide to Research at the Human-nonhuman Primate Interface. Cambridge University Press.

Ellen, R., 2004. From ethno-science to science, or 'What the indigenous knowledge debate tells us about how scientists define their project'. Journal of Cognition and Culture 43, 409–450.

Emery, K.F., 2007. Assessing the impact of ancient Maya animal use. Journal for Nature Conservation 15, 184–195.

Fernandes-Ferreira, H., Mendonça, S.V., Albano, C., Ferreira, F.S., Alves, R.R.N., 2012. Hunting, use and conservation of birds in Northeast Brazil. Biodiversity and Conservation 21, 221–244.

Ferreira, F.S., Fernandes-Ferreira, H., Leo Neto, N., Brito, S.V., Alves, R.R.N., 2013. The trade of medicinal animals in Brazil: current status and perspectives. Biodiversity and Conservation 22, 839–870.

Fleck, D.W., Harder, J.D., 2000. Matses Indians rainforest habitat classifi cation and mammalian diversity in Amazonian Peru. Journal of Ethnobiology 20, 1–36.

Fleck, D.W., Voss, R.S., Patton, J.L., 1999. Biological basis of saki (Pithecia) folk species recognized by the Matses Indians of Amazonian Peru. International Journal of Primatology 20, 1005–1027.

Foster, M.S., James, S.R., 2002. Dogs, deer, or guanacos: zoomorphic figurines from pueblo grande, central Arizona. Journal of Field Archaeology 29, 165–176.

Franke, J., Telecky, T.M., 2001. Reptiles as Pets: An Examination of the Trade in Live Reptiles in the United States. Humane Society of the United States, Washington.

Fuentes, A., Hockings, K.J., 2010. The ethnoprimatological approach in primatology. American Journal of Primatology 72, 841–847.

Fuentes, A., Wolfe, L.D., 2002. Primates Face to Face: Conservation Implications of Human-nonhuman Primate Interconnections. Cambridge Univerity Press, Cambridge, UK.

Gunnthorsdottir, A., 2001. Physical attractiveness of an animal species as a decision factor for its preservation. Anthrozoös 14, 204–215.

Hamblin, N.L., 1985. The role of marine resources in the Maya economy: a case study from Cozumel, Mexico. In: Pohl, M. (Ed.), Prehistoric Lowland Maya Environment and Subsistence Economy. Harvard University Press, Cambridge, MA, USA, pp. 159–173.

Haudricourt, A.G., 1977. Note d'ethnozoologie. Le rôle des excrétats dans la domestication. L'Homme 17, 125–126.

Henderson, J., Harrington, J.P., 1914. Ethnozoology of the Tewa Indians. Bureau of American Ethnology Bulletin 56, 1–76.

Hershberger, J.W., 1885. The purpose of ethnobotany. Botanical Gazette 21, 146–164.
Hidayati, S., Franco, F.M., Bussmann, R.W., 2015. Ready for phase 5-current status of ethnobiology in Southeast Asia. Journal of Ethnobiology and Ethnomedicine 11, 2–8.
Holley, D., 2009. The History of Modern Zoology. Available at: http://suite101.com/article/the-history-of-modern-zoology-a135787.
Holman, E.W., 2005. Domain-specific and general properties of folk classifications. Journal of Ethnobiology 25, 71–91.
Hoover, C., 1998. The US Role in the International Live Reptile Trade: Amazon Tree Boas to Zululand Dwarf Chameleons. TRAFFIC North America, Washington, USA.
Hornaday, W.T., 1889. The Extermination of the American Bison, with a Sketch of its Discovery and Life History. Report of National Museum, Washington, USA.
Hunn, E., 2007. Ethnobiology in four phases. Journal of Ethnobiology 27, 1–10.
Hunn, E.S., 2002. Traditional environmental knowledge: alienable or inalienable intellectual property. In: Stepp, J.R., Wyndham, R.S., Zarger, R.K. (Eds.), Ethnobiology and Biocultural Diversity. ISE/University of Georgia Press, Athens, Georgia, pp. 3–10.
Inskip, C., Zimmermann, A., 2009. Human-felid conflict: a review of patterns and priorities worldwide. Oryx 43, 18–34.
Jorgenson, J.P., 1998. The Impact of Hunting on Wildlife in the Maya Forest of Mexico, Timber, Tourists and Temples: Conservation and Development in the Maya Forests of Belize, Guatemala and Mexico. Island Press, Washington D.C, pp. 179–194.
Joshi, T., Joshi, M., 2010. Ethno-ophiology-A traditional knowledge among tribes and non-tribes of Bastar, Chhattisgarh. Indian Journal of Traditional Knowledge 9, 137–139.
Kalof, L., 2007. Looking at Animals in Human History, first ed. Reaktion Books, London.
Kalof, L., Resl, B., 2007. A Cultural History of Animals in Antiquity, first ed. Berg publishers, Oxford, UK/New York.
Kotschwar Logan, M., Gerber, B.D., Karpanty, S.M., Justin, S., Rabenahy, F.N., 2015. Assessing carnivore distribution from local knowledge across a human-dominated landscape in Central-Southeastern Madagascar. Animal Conservation 18, 82–91.
Kyselý, R., 2008. Frogs as a part of the Eneolithic diet. Archaeozoological records from the Czech Republic (Kutná Hora-Denemark site, Rivnác Culture). Journal of Archaeological Science 35, 143–157.
Léo Neto, N.A., Brooks, S.E., Alves, R.R.N., 2009. From Eshu to Obatala: animals used in sacrificial rituals at Candomble "terreiros" in Brazil. Journal of Ethnobiology and Ethnomedicine 5, 1–23.
Lev, E., 2003. Traditional healing with animals (zootherapy): medieval to present-day Levantine practice. Journal of Ethnopharmacology 85, 107–118.
Lev, E., 2006. Ethno-diversity within current ethno-pharmacology as part of Israeli traditional medicine A review. Journal of Ethnobiology and Ethnomedicine 2, 1–12.
Lewis, I.M., 1991. The spider and the pangolin. Man 26, 513–525.
Lidgard, H., 2005. Quackery or Complementary Medicine - a Historical Approach to the Present Situation. University of Lund, Lund, Sweden.
Locke, P., 2013. Explorations in ethnoelephantology: social, historical, and ecological intersections between Asian elephants and humans. Environment and Society: Advances in Research 4, 79–97.
MacKinney, L.C., 1946. Animal substances in materia medica. Journal of the History of Medicine and Allied Sciences 1, 149–170.
Malkin, B., 1956a. Seri ethnozoology: a preliminary report. Davidson Journal of Anthropology 2, 73–83.
Malkin, B., 1956b. Sumu ethnozoology: herpetological knowledge. Davidson Journal of Anthropology 2, 165–180.
Malkin, B., 1958. Cora ethnozoology, herpetological knowledge; a bioecological and cross cultural approach. Anthropological Quarterly 31, 73–90.
Marques, J.G.W., Guerreiro, W., 2007. Répteis em uma Feira Nordestina (Feira de Santana, Bahia). Contextualização Progressiva e Análise Conexivo-Tipológica. Sitientibus Série Ciências Biológicas 7, 289–295.
Martínez, C., 2008. Las aves como recurso curativo en el México antiguo y sus posibles evidencias en la arqueozoología. Arqueobios 11–18.
Mason, O.T., 1899. Aboriginal american zoötechny. American Anthropolosgist 1, 45–81.
Masson, M.A., 1999. Animal resource manipulation in ritual and domestic contexts at postclassic maya communities. World Archaeology 31, 93–120.
Masson, M.A., Peraza Lope, C., 2008. Animal use at the Postclassic Maya center of Mayapán. Quaternary International 191, 170–183.
Mathias-Mundy, E., McCorkle, M.C., 1989. Ethnoveterinary Medicine: An Annotated Bibliography. Bibliographies in Technology and Social Change. Technology and Social Change Programme. IOWA State University, Ames, IOWA, EUA.
McKillop, H., 1984. Prehistoric Maya reliance on marine resources: analysis of a midden from Moho Cay, Belize. Journal of Field Archaeology 11, 25–35.
McKillop, H.I., 1985. Prehistoric exploitation of the manatee in the Maya and circum-Caribbean areas. World Archaeology 16, 337–353.
McNeely, J.A., 2001. Religions, traditions and biodiversity. COMPAS Magazine 20–22.

Merriam, C.H., 1905. The indian population of California. American Anthropologist 7, 594–606.
Moreira, I.C., 2002. O escravo do naturalista - O papel do conhecimento nativo nas viagens científicas do século 19. Ciência Hoje 31, 40–48.
Mourão, J.S., Araujo, H.F.P., Almeida, F.S., 2006. Ethnotaxonomy of mastofauna as practised by hunters of the municipality of Paulista, state of Paraíba-Brazil. Journal of Ethnobiology and Ethnomedicine 2, 1–7.
Nishida, A.K., Nordi, N., Alves, R.R.N., 2006. Mollusc gathering in Northeast Brazil: an ethnoecological approach. Human Ecology 34, 133–145.
Nordi, N., Nishida, A.K., Alves, R.R.N., 2009. Effectiveness of two gathering techniques for ucides cordatus in Northeast Brazil: implications for the sustainability of mangrove ecosystems. Human Ecology 37, 121–127.
Noss, A.J., 1998. Cable snares and bushmeat markets in a central African forest. Environmental Conservation 25, 228–233.
Oliveira, E.S., Torres, D.F., Brooks, S.E., Alves, R.R.N., 2010. The medicinal animal markets in the metropolitan region of Natal City, Northeastern Brazil. Journal of Ethnopharmacology 130, 54–60.
Overal, W.L., 1990. Introduction to ethnozoology: what it is or could be. In: Posey, D.A., Overal, W.L. (Eds.), Ethnobiology: Implications and Applications. MPEG, Belém, Brasil, pp. 127–129.
Palmer, E., 1871. Food Products of the North American Indians. Report to the Commissioner of Agriculture for 1870. (Washington, DC).
Palmer, E., 1878. Plants used by the Indians of the United States. American Naturalist 12, 593–606.
Parsons, M., 1991. Butterflies of the Bulolo-wau Valley. Bishop Museum Press, Honolulu, Hawaii.
Pohl, M., 1976. Ethnozoology of the Maya: An Analysis of Fauna from Five Sites in Petén, Guatemala. Harvard University, Boston.
Pohl, M., 1981. Ritual continuity and transformation in Mesoamerica: reconstructing the ancient Maya cuch ritual. American Antiquity 46, 513–529.
Posey, D.A., 1978. Ethnoentomological survey of Amerind Groups in lowland Latin America. The Florida Entomologist 61, 225–229.
Posey, D.A., 1982. The importance of bees to Kayapo Indians of the Brazilian Amazon. The Florida Entomologist 65, 452–458.
Powers, S., 1874. Aboriginal botany. Proceedings of the California Academy of Sciences 5, 373–379.
Prins, H.H.T., Grootenhuis, J.G., Dolan, T.T., 2000. Wildlife Conservation by Sustainable Use. Kluwer Academic Publishers, Dordrecht, the Netherlands.
Quijano-Hernández, E., Calmé, S., 2002. Patrones de cacería y conservación de la fauna silvestre en una comunidad Maya de Quintana Roo, México. Etnobiología 2, 1–18.
Rasalato, E., Maginnity, V., Brunnschweiler, J.M., 2010. Using local ecological knowledge to identify shark river habitats in Fiji (South Pacific). Environmental Conservation 37, 90–97.
Raven, J.E., 2000. Plants and Plant Lore in Ancient Greece. Leopard's Head Press, Oxford.
Reitz, E.J., Wing, E.S., 2008. Zooarchaeology. Cambridge University Press, Cambride, UK.
Ribeiro, D., 1998. O processo civilizatório: etapas da evolução sociocultural. Editora Companhia das Letras.
Riley, E.P., 2006. Ethnoprimatology: toward reconciliation of biological and cultural anthropology. Ecological and Environmental Anthropology 2, 75–86.
Rocha, M.S.P., 2007. O Uso dos recursos biológicos pelas comunidades de Barra de Mamanguape e Lagoa da Praia, no Estuário do Rio Mamanguape, litoral Norte do Estado da Paraíba: Um enfoque Etnoecológcio. Departamento de Biologia. Universidade Estadual de Paraíba, Campina Grande, Brazil.
Rocha, M.S.P., Mourão, J.S., Souto, W.M.S., Barboza, R.R.D., Alves, R.R.N., 2008. Uso dos recursos pesqueiros no Estuário do Rio Mamanguape, Estado da Paraíba. Brasil. Interciencia 33, 903–909.
Rosa, I.L., Oliveira, T.P.R., Osório, F.M., Moraes, L.E., Castro, A.L.C., Barros, G.M.L., Alves, R.R.N., 2011. Fisheries and trade of seahorses in Brazil: historical perspective, current trends, and future directions. Biodiversity and Conservation 20, 1951–1971.
Russell, N., 2012. Social Zooarchaeology: Humans and Animals in Prehistory. Cambridge University Press, Cambrige, UK.
Santos-Fita, D., Villamar, A.A., Domínguez, M.A., Martínez, M.Q., 2012. La etnozoología en México: la producción bibliográfica del siglo XXI (2000–2011). Etnobiología 10, 41–51.
Sax, B., 2002. The Mythological Zoo: An Encyclopedia of Animals in World Myth, Legend and Literature. ABC-CLIO, Inc., Santa Barbara, USA.
Sillitoe, P., 2006. Ethnobiology and applied anthropology: rapprochement of the academic with the practical. Journal of the Royal Anthropological Institute 12, S119–S142.
Souto, F.J.B., 2007. Uma abordagem etnoecológica da pesca do caranguejo, *Ucides cordatus*, Linnaeus, 1763 (Decapoda: Brachyura), no manguezal do Distrito de Acupe (Santo Amaro-BA). Biotemas 20, 69–80.
Souto, W.M.S., Pinto, L.C., Mendonça, L.E.T., Mourão, J.S., Vieira, W.L.S., Montenegro, P.F.G.P., Alves, R.R.N., 2013. Medicinal animals in ethnoveterinary practices: a world overview. In: Alves, R.R.N., Rosa, I.L. (Eds.), Animals in Traditional Folk Medicine. Springer, Heidelberg, Germany, pp. 43–66.
Sponsel, L.E., 1997. The human niche in Amazonia: explorations in ethnoprimatology. In: Kinzey, W.G. (Ed.), New World Primates: Ecology, Evolution and Behaviour. Aldine e Gruyter, New York, pp. 143–165.

Stearns, R.E.C., 1889. Ethno-conchology: A Study of Primitive Money. Report of the United States National Museum for 1887. United States Government official press, Washington, USA.

Steensby, H.P., 1917. An anthropological study of the origin of the Eskimo Culture. Meddelelser Om Grønland 53, 39–228.

Sturtevant, W.C., 1964. Studies in ethnoscience. American Anthropologist 66, 99–131.

Svanberg, I., Łuczaj, Ł., Pardo-de-Santayana, M., Pieroni, A., 2011. History and current trends of ethnobiological research in Europe. In: Anderson, E.N., Pearsall, D.M., Hunn, E.S., Turner, N.J. (Eds.), Ethnobiology. Wiley-Blackwell, New Jersey, pp. 189–212.

Vázquez, P.E., Méndez, R.M., Guiascón, Ó.G.R., Piñera, E.J.N., 2006. Uso medicinal de la fauna silvestre en los Altos de Chiapas, México. Interciencia 31, 491–499.

Waller, M.T., 2016. Ethnoprimatology: Primate Conservation in the 21st Century, first ed. Springer, Cham, Switzerland.

Williams, V.L., Cunningham, A.B., Bruyns, R.K., Kemp, A.C., 2013. Birds of a feather: quantitative assessments of the diversity and levels of threat to birds used in African traditional medicine. In: Alves, R.R.N., Rosa, I.L. (Eds.), Animals in Traditional Folk Medicine: Implications for Conservation. Springer, Heidelberg, Germany, pp. 383–420.

Wolverton, S., 2013. Ethnobiology 5: interdisciplinarity in an era of rapid environmental change. Ethnobiology Letters 4, 21–25.

Yi-Ming, L., Zenxiang, G., Xinhai, L., Sung, W., Niemelä, J., 2000. Illegal wildlife trade in the Himalayan region of China. Biodiversity and Conservation 9, 901–918.

Ziembicki, M.R., Woinarski, J.C.Z., Mackey, B., 2013. Evaluating the status of species using Indigenous knowledge: novel evidence for major native mammal declines in northern Australia. Biological Conservation 157, 78–92.

Zuercher, G.L., Gipson, P.S., Stewart, G.C., 2003. Identification of carnivore feces by local peoples and molecular analyses. Wildlife Society Bulletin 31, 961–970.

CHAPTER

3

Zooarcheology: Investigating Past Interactions Between Humans and Other Animals

Steve Wolverton, Lisa Nagaoka

University of North Texas, Denton, TX, United States

INTRODUCTION

Zooarcheology is the study of animal remains, such as whole and fragmented bone, teeth, antler, horn, shell, and other animal tissues, from archeological sites. It is a field that fits well under the umbrella of ethnobiology, which is the study of human–biota and human–environmental interactions. Two research topics commonly addressed in zooarcheology that are ethnobiological include the study of past human diets (or subsistence strategies) and the study of past environments or landscapes that people lived within. Zooarcheology also addresses other topics, such as ritual and medicinal uses of animals, within social archeology (Miller and Sykes, 2016; Russell, 2011). Here we concentrate on the topics of subsistence and human–environmental interactions over time.

In addition to the breadth of topics that are traditionally addressed in zooarcheology, there is increasing methodological diversity within archeological chemistry, such as the study of biomolecular remains and stable isotopes from tissues and artifacts. In this chapter, we focus on macrozoological remains, such as bones, antlers, horns, teeth, shells, and fragments thereof. In so doing, we hold four ontological positions about how zooarcheology ought to be done. First, laboratory practices in zooarcheology must formally relate to decision rules and verification procedures that fall within quality assurance. In addition, taphonomic analysis is an important component of data quality. Second, quantitative, analytical approaches in zooarcheology should be conservative due to the nature of sampling in archeology. Third, zooarcheologists should adopt a theoretical framework for addressing research questions about subsistence and past human–environmental interactions. Researchers may adopt such a framework from a number of perspectives in anthropology and ecology; a theoretical framework from human behavioral ecology, particularly from optimal foraging theory, is adopted in the case study presented here. Fourth, zooarcheological data have

Ethnozoology
http://dx.doi.org/10.1016/B978-0-12-809913-1.00003-X

important environmental management implications in contemporary conservation biology and restoration ecology.

The structure of this chapter follows the four positions outlined in the preceding paragraph. First, data quality and taphonomy are addressed; second, quantitative analysis of faunal remains from archeological contexts is addressed; third, a case study using optimal foraging theory is summarized; and the chapter concludes with a section on applied zooarcheology in conservation science. Many of the points that are made in this chapter have been summarized in other chapters, articles, and books by zooarcheologists, but we do not cite them extensively here. Primary sources that we draw extensively upon include but are not limited to Broughton and Miller (2016), Driver (2011), Lyman (1994, 2008), Nagaoka (2001, 2002a,b), and Stiner (1994) (see Tables 3.1 through 3.3).

DATA QUALITY

Zooarcheologists may or may not be part of a larger team that works with research design to recover archeological materials. In either case, zooarcheologists receive faunal remains to analyze after the excavation process, and even though participation in fieldwork can provide important information on recovery context, identification of remains takes place in a laboratory setting with the use of a comparative skeletal collection. Thus, the quality of data produced by the zooarcheologist and the validity of analyses depend heavily on procedures that take place after recovery (Dibble, 2015). Assuming detailed knowledge of the archeological context from which faunal remains were recovered, the zooarcheologist can directly influence data quality in two ways: (1) employing a quality assurance process; and (2) understanding the taphonomic history of the remains that are analyzed. Selected sources on data quality and taphonomy are presented in Table 3.1.

Quality Assurance

The process that ensures production of high quality data is termed quality assurance, which can be divided into two components—quality control (QC) and quality assessment (QA). QC concerns the laboratory processes and procedures that comprise basic principles and a series of decision rules for identifying faunal remains. Previous papers on QC in zooarcheology by Driver (2011) and Wolverton (2013a) have delineated five decision rules for identifying and reporting the results of faunal analysis that aid in ensuring the validity of data. Here, we add a sixth decision rule (see Rule 6 below). Use of these rules refers to two fundamental principles about the identification process. First, fragments of bone (or other tissues) rarely retain morphological characteristics, such as diagnostic features, that enable the analyst to identify them to fine taxonomic categories, such as to the species level. A second guiding principle is that all instances of identification are hypothetical, thus the criteria for making the identification need to be explicated, recorded, and justified. The six decision rules are guided by these two framing principles.

Rule 1: Each Specimen Should be Identified on its Own Merits

This rule refers to an unfortunate tendency to practise "identification by association," a process by which if one analyzes remains from the same context and if it appears that all of the specimens, those with highly diagnostic features and those without, appear to be from the same species (or other taxonomic category, such as genus) then because they are from the same context they can all receive the same taxonomic designation. Identification by association can also take place at the element scale, such that one might assume that midshaft fragments of a long bone with few or no diagnostic features can be identified to a particular skeletal element (e.g., the femur) if they were recovered in association

TABLE 3.1 Selected Sources on Data Quality, Taphonomy, and Quantification in Zooarcheology[a]

DATA QUALITY
Driver, J.C., 2011. Identification, classification, and zooarchaeology (with comments). Ethnobiology Letters 2, 19–39.
Gobalet, K.W., 2001. A critique of faunal analysis; inconsistency among experts in blind tests. Journal of Archaeological Science 28, 377–386.
Lyman, R.L., 2002. Taxonomic identification of zooarchaeological remains. The Review of Archaeology 23, 13–20.
Wolverton, S., 2013. Data quality in zooarchaeological faunal identification. Journal of Archaeological Method and Theory 20, 381–396.
TAPHONOMY
Brain, C.K., 1983. The Hunters or the Hunted? An Introduction to African Cave Taphonomy. University of Chicago Press, Chicago.
Haynes, G., 1983. A guide for differentiating mammalian carnivore taxa responsible for gnaw damage to herbivore limb bones. Paleobiology 9, 164–172.
Lyman, R.L., 1994. Vertebrate Taphonomy. Cambridge University Press, New York.
Lyman, R.L., 2010. What taphonomy is, what it isn't, and why taphonomists should care about the difference. Journal of Taphonomy 8, 1–16.
Marean, C.W., Spencer, L.M., 1991. Impact of carnivore ravaging on zooarchaeological measures of element abundance. American Antiquity 56, 645–658.
Nagaoka, L., Wolverton, S., Fullerton, B., 2008. Taphonomic analysis of the Twilight Beach seals. In: Clark, G., Leach, F., O'Connor, S., (Eds.), Islands of inquiry: colonisation, seafaring, and the archaeology of Maritime Landscapes, Terra Australis vol. 29, pp. 475–498.
Stiner, M.C., 1994. Honor Among Thieves: A Zooarchaeological Study of Neanderthal Ecology. Princeton University Press, Princeton, NJ.
Wolverton, S., Randklev, C.R., Kennedy, J.H., 2010. A conceptual model for freshwater mussel (family: *Unionidae*) remain preservation in zooarchaeological assemblages. Journal of Archaeological Science 37, 164–173.
QUANTIFICATION
Grayson, D.K., 1979. On the quantification of vertebrate archaeofaunas. Advances in Archaeological Method and Theory 2, 199–237.
Grayson, D.K., 1984. Quantitative Zooarchaeology: Topics in the Analysis of Archaeological Faunas. Academic Press, New York.
Grayson, D.K., Frey, C.J., 2004. Measuring skeletal part representation in archeological faunas. Journal of Taphonomy 2, 27–42.
Lyman, R.L., 1994. Quantitative units and terminology in zooarchaeology. American Antiquity 59, 36–71.
Lyman, R.L., 2008. Quantitative Paleozoology. Cambridge University Press, New York.
Marean, C.W., Kim, S.Y., 1998. Mousterian large-mammal remains from Kobeh Cave behavioral implications for Neanderthals and early modern humans. Current Anthropology 39, S79–S114.
Wolverton, S., Dombrosky, J., Lyman, R. L., 2016. Practical significance: Ordinal scale data and effect size in zooarchaeology. International Journal of Osteoarchaeology 26, 255–265.

[a] *This table is not intended to extensively cover the literature in these areas but represents readings we assign on these topics in our zooarcheology classes.*

with fragments with diagnostic characteristics. The problem is that one makes an assumption that because remains were recovered in close proximity, they are from the same individual animal or portion of that animal's carcass. This assumption may be founded, but it is an archaeological conclusion to be made with data external to the zooarcheological laboratory process. The faunal analyst should refer the long-bone midshaft remains only to the taxonomic category that each specimen can defensibly be identified to based on the morphological characteristics of that specimen alone—for example, identification to long-bone midshaft of "medium artiodactyl" or "large mammal" rather than to femur of *Odocoileus* (the genus of white-tailed deer or mule deer). An interpretation related to identification by association can be noted in the identification comments for the specimen. This simple step ensures the validity of the identification. Alternatively, and permitting available time, resources, and personnel, the analyst may seek to refit fragmented specimens (e.g., Marean and Kim, 1998), such that those without diagnostics features are matched to those with such features.

Rule 2: Set the Universe

Faunal analysis focuses on identification of remains from a particular geographic location and period of time. Setting the universe requires the zooarcheologist to determine what species of animals are to be expected in the region surrounding and the time period represented by the archeological site from which remains were recovered. The zooarcheologist must have contemporary information on the wildlife ecology and biogeography of animal species likely to be represented in the fauna, and comparative reference skeletons (for vertebrates, exoskeletons for invertebrates) will be required to make element and taxonomic identifications. The researcher must also have knowledge of the paleontological record and whether or not biogeography of animal communities has changed since the period of occupation corresponding to a zooarcheological sample.

Rule 3: Set Diagnostic Criteria

The faunal analyst must study osteological reference materials, as well as published keys and guides for criteria that enable separation of remains by element and taxon. A specimen's identification is a hypothesis; the zooarcheologist is assessing if its osteological morphology relates more closely to that of animals in one or another taxonomic category (e.g., species, genus, family, or higher-level units). Through comparison, potential species (or higher-level units) are ruled out. There are multiple ways that zooarcheologists engage this process: (1) the most common means is to use reference skeletal specimens of known species and element; (2) published guides and keys are available for some species in some regions; and (3) the analyst may have to develop and carefully record morphological or biometric criteria. Development of criteria, use of published standards, and use of reference materials constitute QC because they represent protocols for the laboratory. However, whether or not such standards produce valid identification of remains from archeological sites is a different question, one that requires research focusing on QA (see below).

Recording diagnostic criteria (as well as assessment of those criteria) is particularly important when the faunal analyst encounters remains from what appears to be a rare species or one for which the biogeographic range has shifted. In addition, reference collections rarely include individuals that capture intraspecific variability in osteology (e.g., age and sex variation). Thus, analysts may need to employ more than one collection from multiple institutions (e.g., universities or museums) to verify their identifications. Increasingly, identifications may be verified using biomolecular approaches, such as ancient DNA and ancient protein analysis.

Rule 4: Anticipate Difficult to Separate Taxa

Recording diagnostic criteria is most important when studying closely related species with similar osteological morphology. For example, in the

American Southwest, members of the genus *Lepus* (jackrabbits and hares) are commonly not identified to species because of overlapping morphology. Indeed, it may be difficult to separate remains between the genus *Sylvilagus* (cottontail rabbits) and the genus *Lepus* within the family *Leporidae*. Large cottontails and small jackrabbits are difficult to separate because there is overlap in size. In the same region two members of the genus *Odocoileus* (mule deer and white-tailed deer) are thought to have occurred in the past. For this genus, cranial remains are easier to distinguish than postcranial remains (Jacobson, 2003). Regardless of which region one works in, the zooarcheologist must become familiar with not just the universe of species as candidates for identification but also which species are closely related in terms of evolutionary biology and whether or not shared biology leads to similar osteological morphology.

Rule 5: Write a Systematic Paleontology

A systematic paleontology is a report that presents a detailed description of how the identification process unfolds for a particular fauna (Lyman, 2002). This report provides information on how QC was accomplished during faunal analysis in such a manner that other archeologists seeking to use the data can assess its quality. Ideally, the report provides information on the reference collections that were used, the details of morphological and biometric criteria used to make element and taxonomic identifications, as well as justifications for why identification to one or another taxon is warranted. Such reports are the place where QC procedures are made clear, but they are typically difficult to publish in scholarly journals. The systematic paleontology may be published as a chapter or appendix in a thesis or dissertation, could be included as online supplemental material, might be self-archived through scholarly networking web services (e.g., researchgate.net or academia.edu), and/or could be curated digitally with data produced from the analysis using digital library and archiving resources (e.g., Open Context).

Rule 6: Make Data Accessible

Resources are increasingly available for making data generated through faunal analysis available to the scholarly community (Kansa, 2015). For example, Open Context archives data from archeological projects in a format that makes data generally accessible to other researchers. Data archiving services are increasingly provided by universities; what is required to use Open Context or university services is a data management plan crafted during research design, which includes budgeting funds to pay for this type of service. Alternatively, researchers can make datasets available as supplemental material when scholarly research is published in peer-reviewed journals. If such resources are not available, either because a journal does not support publication of datasets or due to lack of financial support, researchers can post datasets through self-archiving web services, such as researchgate.net and academia.edu. One advantage of more comprehensive services, such as Open Context or digital archives at universities, is that a systematic paleontology report can easily be archived with datasets.

Moving From Quality Control to Quality Assessment

QC is a critically important process that relates to how confidently one can assess the validity of published studies. To demonstrate that QC is important in scholarly articles and conference presentations, authors do not have to report their full QC protocol; rather, they can simply cite important papers that document the general rules for insuring QC to illustrate that their laboratory process addresses concerns of data quality. The adoption of QC rules in zooarcheology facilitates confidence in the validity of identifications that support raw data underlying scholarly research; however, a separate step is required to address whether or not QC protocols are meaningful—QA.

QA includes practices—indeed full studies—that verify the integrity of the protocols of QC. Each rule discussed above can be studied in terms of QA. In general, laboratory procedures might be reviewed independently by colleagues to ensure that components of QC are included in the process. Moreover, there are a number of specific practices that can be adopted to achieve QA.

An increasing number of zooarcheologists reanalyze a random subsample of each assemblage they study to determine an identification error rate. This is particularly useful for early career zooarcheologists as it is well-known that ability to identify faunal remains improves with experience. Zooarcheologists have remarked that with experience identification becomes less taxonomically specific as analysts become progressively more aware of which species are too difficult to distinguish based on morphological and/or biometric criteria (see Rule 4 above).

QA includes probabilistic assessment of morphological and/or biometric identification criteria to verify their accuracy (e.g., Hager and Cosentino, 2006; Jacobson, 2003; Munro et al., 2011). Ideally, zooarcheologists survey osteological comparative collections and assess such criteria using skeletal specimens of known species, age, and sex. Such studies consider a variety of analytical problems and scales; for example, for a single species, morphological or biometric criteria on paired elements can be used to develop approaches for determining the minimum number of individuals (MNI) represented in a faunal assemblage. Alternatively, subtle morphological criteria can be studied to determine whether or not closely related species can be reliably separated. Biometric analysis of size and shape of particular elements to support taxonomic identification can also be assessed in a similar manner. Zooarcheologists can more confidently incorporate criteria into QC once they have been assessed using osteological specimens of known identity. Care must be taken, however, as intraspecific osteological variability is difficult to represent in reference collections from zooarcheology laboratories and natural history museums.

It is common in fields such as environmental chemistry to have QC procedures audited by an independent researcher. Similarly, entomologists engage third-party analysts to verify insect identifications, a practice that is increasingly common among zooarcheologists, particularly for remains from closely related species that are difficult to separate. In addition, biomolecular approaches, such as paleoproteomics and analysis of ancient DNA, can be employed either to assess or verify identification of remains based on morphological or biometric criteria. Such approaches are now being used to produce primary data in faunal analysis, that is, to identify fragments of bone that cannot be identified using conventional approaches (e.g., Welker et al., 2015). QC and QA practices in zooarcheology will undoubtedly grow and change as the field continues to evolve; an important component of zooarcheology that relates to quality assurance is to analyze the taphonomic history of animal remains from faunal assemblages.

TAPHONOMIC ANALYSIS

Taphonomy is the study of the transition of organic matter from the biosphere to the lithosphere, and the word literally means "burial studies" (see Lyman, 1994 for thorough discussion; see Broughton and Miller, 2016 for a basic summary). Zooarcheologsts and paleontologists recognize that this transition can involve a substantial variety of processes and effects that relate to an array of taphonomic agents. Such processes can be additive—meaning that signatures of processes, such as weathering, butchery, or animal gnawing (e.g., rodent or carnivore), might be left on bone—telling the zooarcheologist about the history of taphonomic damage after the animal died and prior to excavation. Alternatively, taphonomic agents can be subtractive; for example, gnawing and digestion of bone by carnivores can destroy portions of skeletal elements or complete bones. The magnitude of such damage varies among species of carnivores and may even differ within the same species depending on environmental and behavioral settings [e.g., how productive the environment is or how hungry the carnivore is; see Nagaoka (2015)].

The taphonomist—in this case the zooarcheologist who is studying the taphonomy of remains from an archeological site—is writing a detailed narrative (a taphonomic history) of the processes that influenced the skeletal and taxonomic composition of the remains she/he is studying. In some cases, taphonomists are able to apply general rules of thumb. For example, zooarcheologists can easily determine whether or not a faunal assemblage was influenced by "density mediated destruction" of remains. In such cases, low-density (and thus less robust) bones and parts of bone are of lower abundance or are absent compared to portions and elements that are higher density (more robust). A number of processes can lead to differential destruction of low-density portions and elements, such as animal gnaw damage, weathering, and even hydrological transport, which produces separate lag and transported deposits. Similarly, the shape of skeletal (or exoskeletal) elements can influence whether or not destruction or preservation occurs (Darwent and Lyman, 2002).

In other cases, the taphonomic history of an assemblage may be quite unique, related either to exceptional preservation or destruction conditions. For example, natural trap caves may attract and accumulate high numbers of carnivores (and their remains), compared to midden deposits near prehistoric villages and camps. If a carnivore is trapped in a cave but survives for a period of time, it is likely that any previously deposited animal remains will be extensively gnawed. In contrast, animal remains deposited in trash middens near archeological sites, may or may not have attracted carnivores, but may also have been differentially exposed to weathering from wind, precipitation, and soil conditions. If people cooked meat without removing bones, or if bones were smashed to increase surface area for processing within-bone nutrients, such as grease, then yet a different taphonomic narrative will emerge during analysis.

Correspondingly, taphonomy is an iterative process that balances the study of what are known to be general, identifiable effects with the unique historical contingencies of animal death, butchery, transport, consumption, discard, burial, and preservation. Identifiable taphonomic effects might include marks from hammering and fracturing bone to remove marrow or carnivore tooth marks from gnawing. However, the degree to which each (or any other taphonomic agent) influences the character of a zooarcheological assemblage of remains depends on the cultural and environmental setting—or the configuration—of the taphonomic history. Striking this iterative balance is intimately related to assessing data quality because studying the taphonomic history of one or another assemblage of animal remains may lead to the conclusion that it is useful for answering one type of archeological research question but not others. Because taphonomic histories of faunal assemblages influence what a particular dataset represents about prehistory, it is important to make conservative statistical assumptions when quantifying zooarcheological remains.

QUANTIFICATION

There are two aspects of quantification covered here: (1) two fundamental quantitative units; and (2) descriptive and inferential statistics. Selected sources on quantification in zooarcheology are presented in Table 3.1. The basic quantitative unit employed in zooarcheological research is the tally of animal remains, such as whole bones and fragments of bones, which have been identified to taxon and to skeletal element. Each fragment or whole element represents a specimen, and thus the number of identified specimens (NISP) for an assemblage is the complete tally of identified remains.

Zooarcheologists use NISP of faunal remains to contrast the taxonomic abundance of various animals (e.g., species) from an assemblage. Comparatively, one might explore whether or not the taxonomic abundance of a species (or higher-level taxonomic category) differs between assemblages or changes over time, which could indicate a change in subsistence strategy or local environment. In addition, the zooarcheologist might also tally the NISP of particular skeletal parts from a single type of animal in an assemblage to address research questions about butchery patterns, carcass processing, and/or transport of animal resources from hunting locations to camps or villages. An inherent weakness of NISP, because it is a tally of fragments and whole elements, is that differential fragmentation can lead to inflated counts in one or a few categories and deflated counts in others, particularly at ratio scale. This is known as the problem of specimen interdependence, which amounts to double- (or triple- or greater) counting. Because of the potential deflation and inflation effects of interdependence, some quantification experts in zooarcheology have concluded that NISP is at best an ordinal scale as opposed to ratio-scale measure of taxonomic or element abundance (e.g., Grayson, 1984; Lyman, 2008). Ordinal-scale measures convey more-than and less-than relationships but do not provide the precision to determine the magnitude of difference.

The MNI is calculated to overcome the problem of interdependence. In a faunal assemblage, MNI is the number of individual animals represented by the tallied fragments and whole bones. Specimens are visually inspected in terms of morphology to determine if more than a single element from a single animal of the species (or higher-level taxon) is represented. For paired elements (e.g., humeri, femora, tibiae, and the like), this inspection is done with left- and right-side elements and portions of elements. The element that is most commonly occurring in the assemblage represents the MNI. Thus, MNI solves the problem of multicounting or interdependence inherent with NISP. Its use, however, introduces a separate problem termed the effect of aggregation. Depending on how the researcher separates or aggregates assemblages into larger or smaller subsamples, the most abundant element is likely to change. One might study a faunal assemblage from a stratum with three excavation levels; the stratum as a whole might have humeri of a species as the most abundant element, with perhaps 9 lefts and 7 rights represented, equaling MNI of 9. Broken down by level, however, MNI could change. If Level I had 4 lefts and 1 right, Level II had 2 lefts and 6 rights, and Level III had 3 lefts and 0 rights, the MNI for Level I would equal 4, that for Level II would equal 6, and that for Level III would equal 3. MNI for the three levels added together would then be 13 instead of 9 for the stratum depending on whether or not one aggregated the stratum as a whole or summed the MNI for the level assemblages. Thus, NISP and MNI have contrasting strengths and weaknesses as quantitative units; MNI overcomes interdependence but introduces effects of aggregation, whereas NISP overcomes problems of aggregation but introduces a problem of interdependence. Fortunately, multiple studies have shown that at ordinal scale, MNI and NISP tend to correlate to one another, and in such cases, aggregation is a bigger problem than interdependence. Because MNI is derived from NISP, we advocate that zooarcheologists always report NISP and that if MNI is used researchers clearly present how it is calculated.

Regardless of which quantitative unit one uses, MNI or NISP, to compare taxonomic and/or element abundances, quantitative zooarcheological data are at best ordinal scale. Zooarcheological data are produced from observations of faunal remains from assemblages that were not randomly sampled from prehistoric animal populations, passing through the so-called "cultural filter" (Daly, 1969). Faunal remains also pass through a taphonomic history, which may or may not reduce confidence in the representativeness of past animal populations. The zooarcheologist may be interested in target variables that relate to the life assemblage (the past living population) of prehistoric animals, such as the population's age structure. Alternatively, target variables might relate to the death assemblage, or what was killed and exploited by hunters in the past. Because the life assemblage and death assemblages are not sampled directly, it is imperative to understand the taphonomic history of the faunal assemblage, but it also critical that the zooarcheologist adopt conservative, ordinal-scale statistical assumptions.

A conservative approach can be accomplished by adopting nonparametric (or ordinal-scale) descriptive statistics, such as using the median (nonparametric) as a measure of central tendency as opposed to the mean (parametric) or the interquartile range (nonparametric) as opposed to the standard deviation or variance (parametric). In addition, there are several conservative nonparametric inferential approaches that are cousins to well-known parametric tests. For example, the independent two-sample *t* test, which assesses a difference in means, can be replaced with the Mann–Whitney *U* test, which assesses medians. Spearman's rho correlation (rank order nonparametric correlation) is a more conservative approach than Pearson's *r* correlation. The reason these descriptive and inferential tests are conservative is that they avoid the assumption of normality, which directly concerns the precise, ratio-scale variability of individuals within samples around means through reliance on the central limit theorem. Normality requires assumptions about the way error behaves in random, representative samples at ratio scale. Zooarcheologists, indeed archeologists in general, study samples that are of unknown representativeness and that are certainly not random, making our data ordinal scale at best, necessitating this shift away from assuming normality. For more detailed information on descriptive and inferential statistics in zooarcheology, see Wolverton et al. (2016). The statistical assumptions we do or do not make influences the way that we logically frame research questions using theoretical frameworks.

A THEORETICAL FRAMEWORK

Anthropology and ecology provide a number of theoretical perspectives with which to frame zooarcheological research. The case study we present in this chapter is framed within optimal foraging theory (OFT) from evolutionary ecology. As with any theory that concerns exploitation of animal populations and human subsistence in the past, OFT requires fundamental assumptions about human behavioral ecology. We first discuss the basic premises of OFT and then summarize a zooarcheological case study on prehistoric hunting at the Shag Mouth site in New Zealand. Comprehensive treatment of OFT in ecology can be found in Stephens and Krebs (1986), and many details of its application within anthropology can be found in papers in Smith and Winterhalder (1992). In addition, Broughton and Cannon (2010) edited a volume of important archaeological papers. We rely on a large body of ecological and zooarcheological literature to frame OFT; we present resources that we consider to be essential reading in Table 3.2, rather than cite multiple studies in this section.

OFT developed within evolutionary ecology in the 1960s and 1970s. A commonly used model framed within OFT is the *prey choice model*, which states that foraging decisions about which animals to hunt during foraging bouts are adaptive in the short term (see The Optimality Assumption section). Foraging efficiency, or the net return of hunting and gathering, is assumed to be maximized. In terms of diet breadth, new species are only added to the dietary array once foraging efficiency declines to the point that it is optimal to include a less desirable (lower ranked) species. Thus, hunters are optimizers who maximize foraging efficiency during hunting trips.

TABLE 3.2 Selected Zooarcheological and Ecological Sources on Optimal Foraging Theory[a]

ZOOARCHEOLOGY
Broughton, J.M., 1994. Late Holocene resource intensification in the Sacramento Valley, California: the vertebrate evidence. Journal of archaeological Science 21, 501–514.
Broughton, J.M., 1997. Widening diet breadth, declining foraging efficiency, and prehistoric harvest pressure: ichthyofaunal evidence from the Emeryville Shellmound, California. Antiquity 71, 845–862.
Broughton, J.M., Cannon, M.D. (Eds.), 2010. Evolutionary Ecology and Archaeology: Applications to Problems in Human Evolution and Prehistory. University of Utah Press, Salt Lake City.
Broughton, J.M., Cannon, M.D., Bartelink, E.J., 2010. Evolutionary ecology, resource depression, and niche construction theory: applications to central California hunter-gatherers and Mimbres-Mogollon agriculturalists. Journal of Archaeological Method and Theory 17, 371–421.
Butler, V.L., 2000. Resource depression on the northwest coast of North America. Antiquity 74, 649–661.
Cannon, M.D., 2003. A model of central place forager prey choice and an application to faunal remains from the Mimbres Valley, New Mexico. Journal of Anthropological Archaeology 22, 1–25.
Jones, E.L., 2006. Prey choice, mass collecting, and the wild European rabbit (*Oryctolagus cuniculus*). Journal of Anthropological Archaeology 25, 275–289.
Munro, N.D., 2004. Zooarchaeological measures of hunting pressure and occupation intensity in the Natufian. Current Anthropology 45, S5–S34.
Nagaoka, L., 2001. Using diversity indices to measure changes in prey choice at the Shag River Mouth site, southern New Zealand. International Journal of Osteoarchaeology 11, 101–111.
Nagaoka, L., 2002. The effects of resource depression on foraging efficiency, diet breadth, and patch use in southern New Zealand. Journal of Anthropological Archaeology 21, 419–442.
Stiner, M.C., Munro, N.D., Surovell, T.A., Tchernov, E., Bar-Yosef, O., 1999. Paleolithic population growth pulses evidenced by small animal exploitation. Science 283, 190–194.
Ugan, A., 2005. Does size matter? Body size, mass collecting, and their implications for understanding prehistoric foraging behavior. American Antiquity 70, 75–89.
ECOLOGY
Charnov, E.L., 1976. Optimal foraging, the marginal value theorem. Theoretical Population Biology 9, 129–136.
Charnov, E.L., Orians, G.H., Hyatt, K., 1976. Ecological implications of resource depression. The American Naturalist 110, 247–259.
MacArthur, R.H., Pianka, E.R., 1966. On optimal use of a patchy environment. The American Naturalist 100, 603–609.
Orians, G.H., Pearson, N.E., 1979. On the theory of central place foraging. In: Horn, D.J., Stairs, G.R., Mitchell, R.D. (Eds.), Analysis of Ecological Systems, Ohio State University Press, Columbus, pp. 155–177.
Schoener, T.W., 1971. Theory of feeding strategies. Annual Review of Ecology and Systematics 2, 369–404.
Schoener, T.W., 1979. Generality of the size–distance relation in models of optimal feeding. The American Naturalist 114, 902–914.
Southwood, T.R.E., 1977. Habitat, the templet for ecological strategies? The Journal of Animal Ecology 46, 337–365.
Stephens, D.W., Krebs, J.R., 1986. Foraging Theory. Princeton University Press, Princeton, NJ.

[a] *This table is not intended to extensively cover the literature in these areas but represents readings we assign on these topics in our zooarcheology classes.*

Many zooarcheological studies document resource depression, which is a decline in the availability of a prey population caused by actions of the predator. Charnov et al. (1976) envisioned three types of resource depression: exploitation depression related to mortality caused by predation (harvest); behavioral depression due to changes in prey behavior in proximity to predators (e.g., hiding); and microhabitat depression in which prey animals leave an area to avoid predators. Resource depression is at times misunderstood to mean "any decline in prey populations"; however, animal resources (or other types of resources) can decline in availability because of environmental change, which is an important alternative hypothesis to resource depression.

Other zooarcheological studies employ the *patch choice model*, which shifts the geographic scale of foraging efficiency to the types of habitats that are exploited. Resource patches become the spatial unit of foraging rather than the location of individual prey animals, and decisions about which ones to exploit are based on the net returns of the patch instead of the net returns of the animal. Charnov (1976) framed the marginal value theorem (MVT) to conceptualize how foragers make decisions when resource depression takes place within patches as occurred in the New Zealand example near the Shag Mouth site (see below). New patches may be added if the average net return rate for other available patches decreases to the point that the new patches are then profitable to exploit.

Application of OFT in zooarcheology is often more nuanced than simple use of these models, though they have been applied widely. Theoretical expectations have been developed related to foraging from central places, such as base camps. The MVT has also been applied at a finer scale in butchery and transport studies. Ranking of prey species in terms of foraging returns often uses the proxy of body size, but other currencies of net return have been conceptualized such as prey speed. Clearly, humans are innovators who improve technological means for hunting, fishing, gathering, and producing food, and these processes have also received attention in archeological studies. In literature, this is playing out in terms of implementing niche construction theory (NCT) in archeology, which holds that humans (and other animals) are capable of modifying the selective environment including the possibility of increasing net return rates of resources. In sum, multivariate approaches have been developed for application of OFT in zooarcheology. A common characteristic of zooarcheological data is that samples are time-averaged palimpsests representing multiple foraging bouts, which we believe has led some skeptics to conclude that researchers who employ OFT in zooarcheology may ignore the subtle ecologies (sensu Wyndham, 2009) of human foraging behaviors. We believe this skepticism is based on a misunderstanding of the optimality assumption as invoked in archeology.

The Optimality Assumption

A critical assumption of OFT is that decisions about prey selection are based on economic optimization. Species that are included in the diet are taken upon encounter because it is adaptive or optimal to do so. *What is meant by optimality*? According to famed British ecologist Southwood (1977), an optimal decision occurs when the energetic returns of, say *hunting here/now* outweigh those of *waiting here*, or *waiting to go there/later* reflecting the constraints of time and space contingencies on decision-making (Fig. 3.1). Alternatively, it may be favorable to *wait here* or to *forage there later* if local resources are depleted. In terms of NCT, some organisms (humans in particular) can influence resource return rates through modification of environments, transforming a previously undesirable resource into a more optimal one. For example, agricultural intensification using the logic of Southwood's simple model (Fig. 3.1), would transform investment through technological innovation and

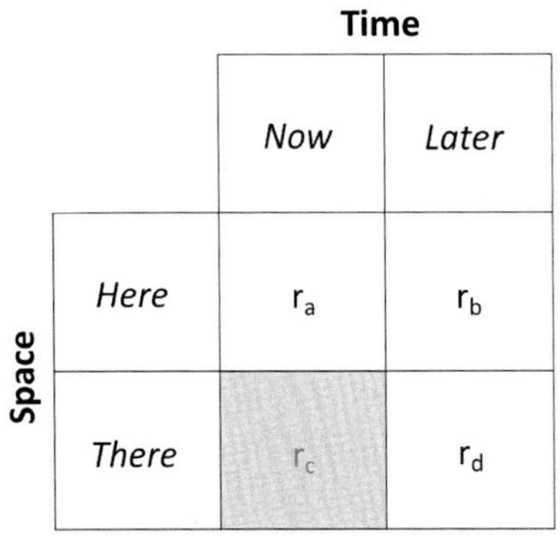

		Time	
		Now	*Later*
Space	*Here*	r_a	r_b
	There	r_c	r_d

FIGURE 3.1 A conceptual model of patch use from Southwood (1977:339, after figure 1), which he entitled the "favorableness matrix," where r refers to the "rate of increase" related to each strategy. The square with r_c is in gray, because it is impossible for a forager to be there now.

plant tending in the *here* or *there patch* with the intention of providing increased returns *later*. Trapping, as another example, might be investment in the *there patch* converting what might previously have been hunted resources into collected ones—*later*. Other examples include construction of clam gardens (Groesbeck et al., 2014), feeding of animals, and technology employed in mass harvest, such as game drives (see also Madsen and Schmitt, 1998).

If the optimality assumption holds, then decisions about what prey to take (and when) will change based on the availability of preferable prey animals (and other desired resources). In the short term, people clearly make decisions that may not be optimal for the long term, thus "sustainability" continues to be an issue in many human societies. If the short term and long term are not distinguished by zooarcheologists, it is a problem of scale. What evolutionary ecologists consider are the short-term contingencies and decisions that took place on foraging bouts; long-term effects are not what are being assumed to be optimized.

The optimality assumption continues to come into question because it is easy to imagine unrealistic scenarios; clearly foragers did not do econometric cost–benefit analyses during foraging bouts. Nor was decision-making an acultural black-box-response to environmental stimuli. Rather, foraging decisions likely reflected deep traditional ecological knowledge (TEK) comprising "subtle ecologies" of what resources constituted reliable and efficient foods within a suite of environmental constraints. Subtle ecologies, which are difficult to observe and quantify even in extant human cultures, comprise "slow relations that rely on diffuse causalities and micro-effects related to invisible or fleeting action" (Wyndham, 2009, p. 272). In times of change, use of alternative resources came from similar TEK. OFT models in archeology simply seek to record the results of these short-term decisions and to study if and how foraging behaviors changed over time. What we often observe in faunal data are palimpsests of remains accumulated from many foraging bouts. This may, in fact, be advantageous because for a foraging strategy to persist we are arguing that it must have been on-average adaptive. Less optimal behaviors likely occurred, and are part of the variability in this long-term average. Whether or not zooarcheological palimpsests are representative of this foraging average is a question of data quality and taphonomy (see earlier sections).

Suspicion that OFT is teleological concerns how to explain change. How could have people made optimal decisions that lead to resource depression? This is where long-term and short-term scales become conflated. Optimality only refers to the short term, but foraging efficiency can change over time. What makes OFT robust instead of teleological is that it aids in spelling out alternative hypotheses for explaining changes in foraging efficiency over longer periods represented in the archeological record. If humans caused a reduction in prey abundance, then resource depression occurred as an average influence of foraging bouts over time. If habitat productivity for high-ranked prey declined, resource depression did not occur, but the environmental constraints on foraging would have changed over time. Alternatively, if foragers began to manage resources through niche construction, then previously lower-ranked foods could become more accessible, which is also a shift in environmental constraints. In addition, improvements in harvest technology could increase the return rates of particular types of resources. OFT does not determine which of these alternatives best explains changes (or lack thereof) in human foraging behaviors; the archeologist must test these hypotheses using multiple lines of evidence to draw those types of conclusions.

Prey populations were in ecological relationships with humans. How might those populations change under consistent harvest pressure due to decisions made in the short term, which eventually translated into an amalgamation of effects over time? It is helpful here to introduce a concept from contemporary environmental impact assessment where nuanced causes are often attributed to long-term effects, termed *cumulative effects* (Atkinson and Canter, 2011). An example from environmental science is extirpations in a fish community attributed to decades-long influences of channelization, toxic contamination, overharvest, and/or nutrient release from farm fields. No single cause can be reliably attributed, but the cumulative effect is still apparent. In time-averaged assemblages we are assuming that various forms of TEK were part of foraging bouts, and that these influenced and were incorporated into what was optimal. It takes knowledge to develop a foraging strategy. Thus, we are examining in one period or another, the cumulative effects of such knowledge-based decisions.

Foraging at the Shag River Mouth Site

The Shag River Mouth site is located on the east coast of the South Island of New Zealand (Fig. 3.2); it was a village site occupied by hunter-gatherers for roughly 200 years from AD 1250–1450. As a stratified multicomponent site, faunal assemblages from Shag Mouth are ideal for addressing zooarcheological research questions framed within OFT (Nagaoka, 2001, 2002a,b, 2005, 2006). Nagaoka conceived of three resource patches that would have been exploited by hunter-gatherers who occupied the site—the inland patch, the coastal patch, and the offshore patch. The inland patch comprised birds; the coastal patch included dogs living near the site, pinnipeds along the coast, and nearshore fishes; and the offshore patch incorporated fishes, such as barracouta. The offshore patch, importantly, was the only one that did not incorporate a large-bodied, high-net-return animal, such as seals or moa. Exploitation of the offshore patch would have provided relatively low net returns compared to the other patches. Early in the sequence at Shag Mouth, it is clear that animal exploitation focused on the inland and coastal patches, for which high-ranked resources were moa and pinnipeds. Analysis of prey choice shows that over time these high-ranked resources became rare; that is, foraging efficiency declined. For the inland patch, Nagaoka was able to demonstrate that the shift in moa exploitation was not likely due to habitat change. She was also able to show that diet breadth in the inland and coastal patches was significantly higher in the upper layers at Shag Mouth when lower-ranked resources were incorporated as foraging efficiency declined. These changes were not driven by taphonomic effects, such as differences in fragmentation rates between lower and upper layers. Fishing in the offshore patch increased in importance over time as well. Travel and fishing technology costs would have been higher for the offshore patch, and this patch did not become a focus of foraging until the average returns of the inland and coastal patches declined—as expected from the MVT.

Using the central place foraging model, Nagaoka considered the carcasses of large animals (moa and seals) as resource patches. Once procured, the hunter had to make choices about how to process and transport meat and other products from the carcass back to the central

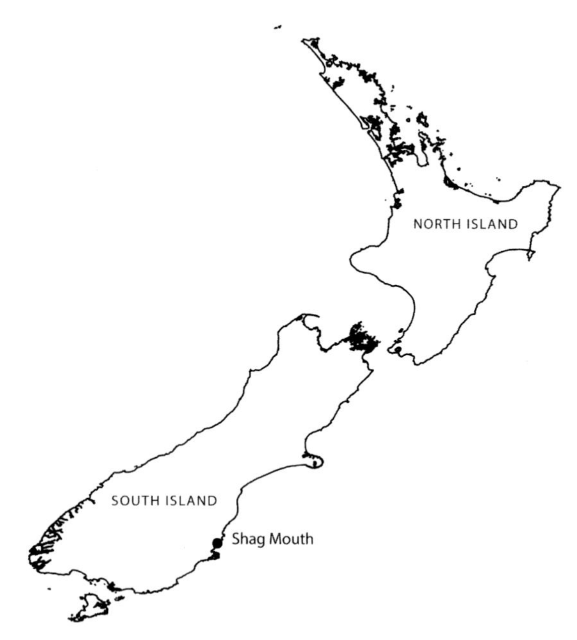

FIGURE 3.2 A map of New Zealand showing the location of the Shag River Mouth site on the east coast of the South Island.

place (the village at Shag Mouth). The economic value (or utility) in terms of meat and within-bone nutrients (e.g., bone grease, oil, and marrow) varies within each carcass but in similar ways for members of the same species (or closely related ones). Thus, the forager is balancing the cost of transport, which increases with distance from the central place (here the Shag River Mouth site), against the benefits of incorporating carcass parts of diminishing economic value. If the forager is hunting near the central place, it is economically viable to exploit more of the carcass because the cost of transport is low. As distance increases, however, the cost of transport rises. As large prey animals become rare on the landscape, the foraging radius grows as foraging efficiency decreases, and intensification of carcass exploitation may be rewarded. The carcass exploitation strategy of seals and moa at Shag Mouth was measured in two ways. First the average utility (or food value) of carcass parts was analyzed to determine whether or not lower food value parts were transported back to the site later in time. Second, bone fragmentation rates were analyzed to determine if within-bone nutrient exploitation shifted.

For moa hunting, as foraging efficiency declined, transport distance increased, and butchery strategy shifted to maximize net returns of each foraging bout. That is, there was an increasing emphasis on butchery and processing of moa carcasses at the kill site to maximize the efficiency of transport across farther distances back to Shag Mouth. This is visible as a substantial increase in the average utility of moa carcass parts over time. In addition, foragers intensified their use of within-bone nutrients for those carcass parts that were brought back to the site; for example, there was increasing exploitation of moa bones with small (low caloric value) marrow cavities over time. In addition, there was intensified fragmentation of bone in general, perhaps to increase the surface to volume ratio of bone fragments for grease extraction.

Unlike for moa, the transport distance for exploitation of pinniped carcasses did not increase, which relates to the site's location on the coast and proximity to seal colonies and haul outs. It is clear, however, that prey choice shifted over time as foraging efficiency decreased and pinnipeds became rare. Without the constraint of an increase in the foraging radius, the contingency of reduced foraging efficiency on seal carcass exploitation resulted in butchery intensification; thus, there was a substantial reduction in the average utility of seal carcass parts that were exploited later in time. As seals and moa became rare, it was important to exploit the seal carcasses intensively when encountered. Unlike for moa, the intensity of seal within-bone nutrient exploitation did not increase, which relates to two factors. First, seal bones do not contain marrow, but instead contain oil, which is a different resource. Second, oil and grease from pinniped bones may not have been a high-value resource because their carcasses also contain blubber, which is a much more accessible form of fat and oil. Given that transport distance did not increase for seals, foraging efficiency at Shag Mouth does not appear to have declined to an extent that it was worth the effort to exploit oil and grease from seal bones.

The Shag Mouth site was a nexus on the landscape among three animal resource patches. OFT provides a logically consistent set of expectations for studying what drove shifts in hunting over time. The prey choice model predicts that if exploitation of a resource patch led to resource depression—a decline in foraging efficiency caused by the hunter—then there should have been shifts to lower-ranked animal resources. This happened at Shag Mouth; as the abundance of moa declined, the abundance of low-ranked prey remains increased. Further, additional low-ranked species were incorporated into the diet breadth in the inland and coastal patches as resource depression of moa and pinnipeds occurred. Late in the sequence the average returns of those patches declined to the point that patch choice was affected. Hunter-gatherers shifted toward fishing in the offshore patch. Within the inland and coastal patches, different patterns emerge. Resource depression of moa led to an increase in the foraging radius and an increase in travel costs; thus, butchery was done in such a manner that maximized the net returns of each foraging bout, what Cannon (2003) terms the *delivery rate*. In addition, those moa bones that were transported to Shag Mouth were more intensively exploited for marrow and bone grease over time. Resource depression of pinnipeds (e.g., seals) also occurred, but Shag Mouth is located within the coastal resource patch, thus travel costs did not increase. As pinnipeds became less abundant, however, carcass exploitation intensified, as even low-utility carcass parts were used. Foraging efficiency never declined to the point that within-bone nutrients from pinnipeds were intensively exploited, which likely relates to the availability of blubber from these species.

Together these multiple lines of evidence are well integrated using OFT, which illustrates the cumulative effects of foraging at Shag Mouth. Conceptualizing animal resource exploitation using the prey choice and patch choice models provides clear expectations about the conditions under which foraging efficiency changes. If foraging efficiency declines, lower-ranked prey taxa should be incorporated in to diet breadth, and the delivery rate of carcass exploitation should shift. If travel costs increase, average utility should increase as hunters butcher carcasses to maximize the net returns from each bout. If travel costs are not an issue, then during periods of low foraging efficiency, average utility should decrease as hunters would have been seeking to retrieve all possible returns from a declining resource. If the average returns of a patch decrease substantially, not only will lower-return species be added to the diet, hunters will eventually turn toward lower-ranked patches, perhaps shifting from hunting to offshore fishing as at Shag Mouth. Whether or not resource depression occurred may or may not be concluded by the archeologist depending on the weight of evidence. At Shag Mouth, several lines of evidence highlight that there was a decrease in foraging efficiency over time; this change does not appear to have been driven by environmental change and was not a product of taphonomic effects. Therefore, it is likely that the decrease in foraging efficiency is an example of resource depression.

APPLIED ZOOARCHEOLOGY

From the perspective of historical ecology, humans as hunters have always been embedded within the ecosystem processes of the landscapes they occupy. Adoption of OFT as a theoretical framework emphasizes that the human component of ecosystem processes was held as TEK and that foraging efficiency could shift either due to new knowledge, a changing environment, or resource depression of important prey species. Anecdotally, zooarcheological research provides an important record of human–environment interactions through time across diverse landscapes. This is an important lesson for ecology, one that emphasizes that humans are not separate from ecosystems processes. In addition to this general benefit of zooarcheological research in ecology, an increasing number of case studies highlight the importance of zooarcheology in conservation biology. This applied movement within zooarcheology has been referred to as applied paleozoology, conservation paleozoology, conservation paleobiology, conservation archeobiology, as well as applied zooarcheology. We use the moniker applied zooarcheology to characterize products of zooarcheological research that provide information for conservation biology and environmental management.

Many case studies have now been published that emphasize the benefits of zooarcheological data and perspectives in conservation. Such data provide time depth that can be used to provide conservation baselines, which is a set of expectations about what should be conserved or restored (e.g., species composition of a location or region). In addition, historical ecological data, including those from zooarcheology, provide the only long-term record of human environmental interactions, and—as zooarcheologist R. Lee Lyman points out—represent a set of ecological experiments that have already been done. The issue of temporal scale is an important one in that a conservation target too far back in time may represent change in biota at an evolutionary scale and thus be inaccurate, but targets that are two shallow in time may not represent sufficient variability in ecological processes.

We hold that the strongest examples of applied zooarcheology concern the past biogeography of species whose ranges have been expanded or curtailed due to contemporary modification of environments by people. Our own research and that of our graduate students focus on pre-EuroAmerican presence of aquatic species in river basins of western North America during the late Holocene, such as Charles Randklev's and Traci Popejoy's research on mussels in Texas and Jonathan Dombrosky's research on fishes of the northern Rio Grande in New Mexico (Dombrosky et al., 2016; Popejoy et al., 2017; Randklev et al., 2010). There are many other published examples of such research focusing on the conservation implications of zooarcheological biogeographic data from many parts of the world. We list several important applied zooarcheology studies in Table 3.3.

The issue of data quality in zooarcheology is particularly important when engaging conservation research. To publish in conservation science academic journals, such as *Environmental Management*, *Conservation Biology*, or *Hydrobiologia*, biologists must be convinced that our data are high quality—meaning that we must present evidence for the validity of our taxonomic identifications and the geographic and temporal representativeness of our sampling. Thus, the topics covered early in this chapter are of critical importance in applied zooarcheology. We refer the reader to fundamental sources presented in Table 3.3 for additional information.

CONCLUSION

In this chapter we presented components of data quality, taphonomy, and quantification in zooarcheology, which are important irrespective of the theoretical framework or the research question one adopts. That said, we presented a case study on the use of OFT in zooarcheology that reflects our personal research interests. We acknowledge that contemporary zooarcheological research spans from postprocessual approaches through anthropological archeological inquiry to ecological research. One's focus within zooarcheology depends on one's

TABLE 3.3 Selected Sources for Applied Zoorchaeology[a]

BOOKS
Dietl, G.P., Flessa, K.W. (Eds.), 2009. Conservation paleobiology: The use of the past to manage for the future. Paleontological Society Papers, vol. 15. Boulder, CO.
Louys, J. (Ed.), 2012. Paleontology in Ecology and Conservation. Springer-Verlag, Berlin.
Lyman, R.L., 1998. White Goats, White Lies: The Misuse of Science in Olympic National Park. University of Utah Press, Salt Lake City.
Lyman, R.L., Cannon, K.P. (Eds.), 2004. Zooarchaeology and Conservation Biology. University of Utah Press, Salt Lake City.
Wolverton, S., Lyman, R.L. (Eds.), 2012. Conservation Biology and Applied Zooarchaeology. University of Arizona Press, Tucson.
Wolverton, S., Nagaoka, L., Rick, T.C., 2016. Applied Zooarchaeology: Five Case Studies. Eliot Werner Publications, Clinton Corners, NY.
REVIEW PAPERS
Dietl, G.P., Kidwell, S.M., Brenner, M., Burney, D.A., Flessa, K.W., Jackson, S.T., Koch, P.L., 2015. Conservation paleobiology: leveraging knowledge of the past to inform conservation and restoration. Annual Review of Earth and Planetary Sciences 43, 79.
Lyman, R.L., 1996. Applied zooarchaeology: The relevance of faunal analysis to wildlife management. World Archaeology 28, 110–125.
Lyman, R.L., 2012. A warrant for applied palaeozoology. Biological Reviews 87, 513–525.
Lyman, R.L., 2016. A conservation paleobiology perspective on reintroduction. In:Jachowski, D.S., Millspaugh, J.J., Angermeier, P.L., Slotow, R. (Eds.), Reintroduction of Fish and Wildlife Populations. University of California Press, Oakland, pp. 29–51.
Rick, T.C., Lockwood, R., 2013. Integrating paleobiology, archeology, and history to inform biological conservation. Conservation Biology 27, 45–54.
Wolverton, S., Randklev, C.R., Barker, A., 2011. Ethnobiology as a bridge between science and ethics: an applied paleozoological perspective. In: Anderson, E., Pearsall, D.M., Hunn, E.S., Turner, N.J. Ethnobiology, Wiley Blackwell, Hoboken, NJ, pp. 115–132.
CASE STUDIES
Dombrosky, J., Wolverton, S., Nagaoka, L., 2016. Archaeological data suggest broader early historic distribution for blue sucker (*Cycleptus elongatus*, Actinopterygii, Catostomidae) in New Mexico. Hydrobiologia 771, 255–263.
Grayson, D.K., 2005. A brief history of great basin pikas. Journal of Biogeography 32, 2103–2111.
Lyman, R.L., 1988. Zoogeography of Oregon coast marine mammals: The last 3000 years. Marine Mammal Science 4, 247–264.
Peacock, E., Randklev, C.R., Wolverton, S., Palmer, R.A., Zaleski, S., 2012. The "cultural filter," human transport of mussel shell, and the applied potential of zooarchaeological data. Ecological Applications 22, 1446–1459.
Wolverton, S., Kennedy, J.H., Cornelius, J.D., 2007. A paleozoological perspective on white-tailed deer (*Odocoileus virginianus texana*) population density and body size in central Texas. Environmental Management 39, 545–552.

[a] *This table is not intended to extensively cover the literature in these areas but represents readings we assign on these topics in our zooarcheology classes.*

perspective about how archeology ought to be done. It also depends upon the taphonomic history and data quality of the archeological fauna that is being researched. One advantage of embedding zooarcheology within ethnobiology is that research tends to center on findings rather than disciplinary epistemology (Nagaoka and Wolverton, 2016); there is a greater emphasis within anthropological archeology upon disciplinary epistemology. What we have presented here is not simply a rundown of how to use OFT in zooarcheology, but more importantly, a means to cast research within a theoretical framework. We find that ecological research in zooarcheology fits well within ethnobiology, which focuses on human–biota interactions in different cultural contexts, be those spatially or temporally distinctive.

Applied zooarcheology has direct ecological implications in conservation science. Holistically, applied zooarcheology (as with other types of environmental archeology) is an example of historical ecology. The direct implications of applied zooarcheological research are often quite simple in that our data are useful for setting biogeographic conservation baselines. The zooarcheologist must acknowledge, however, that the validity of any particular baseline is embedded within the larger context of political ecology—the social, economic, and political contexts in which environmental management occurs. Applied ethnobiological research in these political ecological contexts—of which applied zooarcheological is an excellent example—is indicative of what scholars have termed the fifth phase of ethnobiology (aka ethnobiology 5) (Wolverton, 2013b). This contemporary phase seeks to leverage the merits of ethnobiological data in the phase of current and growing global and local environmental and cultural crises. Zooarcheology has much to offer ethnobiology 5.

References[1]

Atkinson, S.F., Canter, L.W., 2011. Assessing the cumulative effects of projects using geographic information systems. Environmental Impact Assessment Review 31, 457–464.

Broughton, J.M., Cannon, M.D. (Eds.), 2010. Evolutionary Ecology and Archaeology: Applications to Problems in Human Evolution and Prehistory. University of Utah Press, Salt Lake City.

Broughton, J.M., Miller, S.D., 2016. Zooarchaeology and Field Ecology: A Photographic Atlas. University of Utah Press, Salt Lake City.

Cannon, M.D., 2003. A model of central place forager prey choice and an application to faunal remains from the Mimbres Valley, New Mexico. Journal of Anthropological Archaeology 22, 1–25.

Charnov, E.L., 1976. Optimal foraging, the marginal value theorem. Theoretical Population Biology 9, 129–136.

Charnov, E.L., Orians, G.H., Hyatt, K., 1976. Ecological implications of resource depression. The American Naturalist 110, 247–259.

Daly, P., 1969. Approaches to faunal analysis in archaeology. American Antiquity 34, 146–153.

Darwent, C.M., Lyman, R.L., 2002. Detecting the postburial fragmentation of carpals, tarsals, and phalanges. In: Haglund, W.D., Sorg, M.H. (Eds.), Advances in Forensic Taphonomy: Method, Theory and Archaeological Perspectives. CRC Press, Boca Raton, pp. 356–377.

Dibble, W.F., 2015. Data collection in zooarchaeology: incorporating touch-screen, speech-recognition, barcodes, and GIS. Ethnobiology Letters 6, 249–257.

Dombrosky, J., Wolverton, S., Nagaoka, L., 2016. Archaeological data suggest broader early historic distribution for blue sucker (Cycleptus elongatus, Actinopterygii, Catostomidae) in New Mexico. Hydrobiologia 771, 255–263.

Driver, J.C., 2011. Identification, classification and zooarchaeology. Ethnobiology Letters 2, 19–39.

Grayson, D.K., 1984. Quantitative Zooarchaeology: Topics in the Analysis of Archaeological Faunas. Academic Press, New York.

Groesbeck, A.S., Rowell, K., Lepofsky, D., Salomon, A.K., 2014. Ancient clam gardens increased shellfish production: adaptive strategies from the past can inform food security today. PLoS One 9, e91235.

Hager, S.B., Cosentino, B.J., 2006. An identification key to rodent prey in owl pellets from the northwestern and southeastern United States: incisor size to distinguish among genera. The American Biology Teacher 68, e135–e144.

[1] The sources presented as references are cited directly in the text of the paper and may overlap with suggested readings from Tables 3.1–3.3.

Jacobson, J.A., 2003. Identification of mule deer (*Odocoileus hemionus*) and white-tailed deer (*Odocoileus virginianus*) postcranial remains as a means of determining human subsistence strategies. Plains Anthropologist 48, 287.

Kansa, S.W., 2015. Using linked open data to improve data reuse in zooarchaeology. Ethnobiology Letters 6, 224–231.

Lyman, R.L., 1994. Vertebrate Taphonomy. Cambridge University Press, New York.

Lyman, R.L., 2002. Taxonomic identification of zooarchaeological remains. The Review of Archaeology 23, 13–20.

Lyman, R.L., 2008. Quantitative Paleozoology. Cambridge University Press, New York.

Madsen, D.B., Schmitt, D.N., 1998. Mass collecting and the diet breadth model: a Great Basin example. Journal of Archaeological Science 25, 445–455.

Marean, C.W., Kim, S.Y., 1998. Mousterian large-mammal remains from Kobeh Cave behavioral implications for Neanderthals and early modern humans. Current Anthropology 39 (S1), S79–S114.

Miller, H., Sykes, N., 2016. Zootherapy in archaeology: the case of the fallow deer (Dama dama dama). Journal of Ethnobiology 36, 257–276.

Munro, N.D., Bar-Oz, G., Hill, A.C., 2011. An exploration of character traits and linear measurements for sexing mountain gazelle (Gazella gazella) skeletons. Journal of Archaeological Science 38, 1253–1265.

Nagaoka, L., 2001. Using diversity indices to measure changes in prey choice at the Shag River Mouth site, southern New Zealand. International Journal of Osteoarchaeology 11, 101–111.

Nagaoka, L., 2002a. The effects of resource depression on foraging efficiency, diet breadth, and patch use in southern New Zealand. Journal of Anthropological Archaeology 21, 419–442.

Nagaoka, L., 2002b. Explaining subsistence change in southern New Zealand using foraging theory models. World Archaeology 34, 84–102.

Nagaoka, L., 2005. Declining foraging efficiency and moa carcass exploitation in southern New Zealand. Journal of Archaeological Science 32, 1328–1338.

Nagaoka, L., 2006. Prehistoric seal carcass exploitation at the Shag Mouth site, New Zealand. Journal of Archaeological Science 33, 1474–1481.

Nagaoka, L., 2015. Differential carnivore damage as a potential indicator of resource availability and foraging efficiency. Journal of Archaeological Method and Theory 22, 828–856.

Nagaoka, L., Wolverton, S., 2016. Archaeology as ethnobiology. Journal of Ethnobiology 36, 473–475..

Popejoy, T., Randklev, C.R., Wolverton, S., Nagaoka, L., 2017. Conservation implications of late Holocene freshwater mussel remains of the Leon River in central Texas. Hydrobiologia (in press).

Randklev, C.R., Wolverton, S., Lundeen, B., Kennedy, J.H., 2010. A paleozoological perspective on unionid (Mollusca: Unionidae) zoogeography in the upper Trinity River basin, Texas. Ecological Applications 20, 2359–2368.

Russell, N., 2011. Social Zooarchaeology: Humans and Animals in Prehistory. Cambridge University Press, New York.

Smith, E.A., Winterhalder, B. (Eds.), 1992. Evolutionary Ecology and Human Behavior. Aldine De Gruyter, New York.

Southwood, T.R.E., 1977. Habitat, the templet for ecological strategies? The Journal of Animal Ecology 46, 337–365.

Stephens, D.W., Krebs, J.R., 1986. Foraging Theory. Princeton University Press, Princeton, NJ.

Stiner, M.C., 1994. Honor Among Thieves: A Zooarchaeological Study of Neanderthal Ecology. Princeton University Press, Princeton, NJ.

Welker, F., Soressi, M., Rendu, W., Hublin, J.J., Collins, M., 2015. Using ZooMS to identify fragmentary bone from the late middle/early upper palaeolithic sequence of Les Cottes, France. Journal of Archaeological Science 54, 279–286.

Wolverton, S., 2013a. Data quality in zooarchaeological faunal identification. Journal of Archaeological Method and Theory 20, 381–396.

Wolverton, S., 2013b. Ethnobiology 5: interdisciplinarity in an era of rapid environmental change. Ethnobiology Letters 4, 21–25.

Wolverton, S., Dombrosky, J., Lyman, R.L., 2016. Practical significance: ordinal scale data and effect size in zooarchaeology. International Journal of Osteoarchaeology 26, 255–265.

Wyndham, F.S., 2009. Spheres of relations, lines of interaction: subtle ecologies of the Rarámuri landscape in northern Mexico. Journal of Ethnobiology 29, 271–295.

CHAPTER

4

Studying Ethnozoology in Historical Documents

Maria Franco Trindade Medeiros[1], *Rômulo Romeu Nóbrega Alves*[2]

[1]Universidade Federal de Campina Grande, Cuité, Brazil; [2]Universidade Estadual da Paraíba, Campina Grande, Brazil

INTRODUCTION

The socioeconomic and technical–scientific history of humans and our relations with animals, biodiversity, and natural resources, in general, have always have been intertwined. In the particular case of the world's fauna, strong connections have been established with humans who have outlived the historical moments experienced by human society, and these connections have survived even with changes in modes of production.

The use of natural biological resources has resulted in human interventions in natural environments that have left permanent marks on essentially all global ecosystems. The gradual but tenacious actions of human interventions in natural environments have left cumulative records, altering and disturbing the sustainability of complex ecosystems, a process that has been accelerated by modern scientific and technological developments (Alves and Albuquerque, 2012; Dias, 2011). Human interventions in natural ecosystems have increased substantially in recent years, resulting in expressive changes in the dynamics of ecological equilibriums throughout the world, which has resulted in reflection by the scientific community and efforts from the perspective of multidisciplinary studies to examine those changes—as is the case of ethnozoology field.

Faunal elements historically have always had central roles in human societies, having importance in hunting and fishing as well as in their domestication (Alves, 2012; Alves and Souto, 2015). The relevance of these activities has been noted in many historical records that can furnish valuable information and contribute to our understanding of the complex history of human–animal relations, allowing us, in many cases, to understand the importance of faunal elements in human survival and development that have persisted to modern times. This approach is the focus of historical ethnozoology, a branch of ethnozoology that investigates historical relations between humans and animals as registered in ancient documents and which will be briefly discussed in this chapter.

Ethnozoology
http://dx.doi.org/10.1016/B978-0-12-809913-1.00004-1

HISTORICAL RECORDS OF THE RELATIONSHIPS BETWEEN HUMAN SOCIETY AND ANIMALS

Records of the importance of animals to humans exist from before the time of written languages, as humans used drawings as a form of communication well before writing systems were developed. As was pointed out by Justamand (2004), primitive art has had an important social function in cultural transmission since ancient times, positively contributing to relationships between humans and between them and natural resources—especially animals. Cave drawings therefore furnish in numerous elements that can be used to infer historical panoramas characterizing ancient relationships between people and the regional fauna.

The vast majority of prehistoric cave drawings depict animals (Kalof, 2007), showing the importance of the fauna to humans since the beginning of history. Cave drawings frequently record large animals (either predators or animals hunted for food), zoomorphic representations, as well as hunting scenes—historical ethnozoological scenes that demonstrate the social and cultural importance of animals to humans. In addition to cave paintings, animal remains encountered in archeological sites are also valuable sources of information that can aid in our understandings of historical interactions between humans and other animals. These remains and clues are studied within the realm of zooarcheology and are discussed in Chapter 3 of this book.

Although rock paintings and zooarcheological records allow us to reconstruct historical interactions between humans and animals and in some cases even analyze the impact of humans on those animal species (Russell, 2012), written documents often record those relations in much more detail. The Greek philosopher Aristotle (384–322 BC), considered the father of zoology, was interested in the knowledge accumulated by people who directly exploited animal resources, recognizing that they had accumulated important information concerning the ecology of those species. According to Chansigaud (2009), Aristotle was interested in the experiences of hunters, fishermen, and farmers who could report details about the distributions, diets, songs, or wintering habits of birds, and he sorted birds into several groups based on the morphologies of their extremities (clawed feet, webbed feet, or separate digits) and their feeding habits (insectivorous, granivorous, or carnivorous). Martins (2015) noted that through conversation with fishermen, Aristotle discovered that when dolphins remain trapped in fishing nets below the water surface, they could drown because they need to breathe air to survive. Early explorers and naturalists, for example, Charles Darwin, Alfred R. Wallace, Louis Agassiz, and Henry Walter Bates, likewise took advantage of local native knowledge during their travels throughout the world.

From cave paintings to historical documents, the central role of animals in the lives of humans is quite evident, reflecting a wide diversity of established interactions over time (Alves et al., 2013; Martínez, 2008; Sutton, 1995). Written documents are the principal sources of historical ethnozoology, a branch of ethnozoology that investigates historical relations between animals and humans. These types of investigations require certain basic methods and care that will be discussed in the following pages.

HOW CAN ETHNOZOOLOGICAL INFORMATION BE FOUND IN HISTORICAL DOCUMENTS?

The starting point for encountering ethnobiological information, whether ethnozoological or ethnobotanical, consists of locating potential material to be consulted. Historical documents that could form the basis of these types of research are found in recorded memories, which can be symbolic or physical. These records include both symbolic and informational functions, combining the goal of a social memory with a recall of facts, places, and people through evidence that can be found in these locations (Medeiros, 2014). In this context, memories are recorded in physical locations such as archives, libraries, and museums, constantly transforming locations that can provide individual and/or collective memories.

Recorded memories can be loosely classified as texts; visual, audio, or three-dimensional records; or mixtures of these categories (Medeiros, 2009). In establishing a line of historical ethnozoological research, researchers are often anxious to focus on a specific theme and to rapidly find answers to questions such as the following: Where can documents pertinent to a given subject be found? Who can provide information? How can the information collected be analyzed and developed?

A justified anxiety exists when determining if elements relevant to a research theme, once located, can, in fact, be accessed. Determining precisely what information should be collected and how much of it should be interpreted involves complex questions that can only be measured indirectly by the quantity and quality of the new scholarship to be produced.

In fact, documentary memory can be used traditionally, such as in coordinate indexing with "uniterms," or electronically, in files located in searchable bibliographic databases. In this sense, we also make use of guides and catalogs, usually organized by topics of interest or fields. This is, therefore, a technique for locating sources that may prove useful in our research.

By establishing a generative theme, a key species, or a document, we begin establishing the relations among representations that will be relevant to the focus of our research. This implies that what is being represented in the documentary memories will be gradually uncovered by the researcher, who will determine which pieces of information are relevant during the search. Juxtaposed with the capacity for an item to expand the field of reference, the dissemination of information, paradoxically, makes use of reductive strategies to screen out undesired data. These strategies, considered "filters," consist of choosing informative nuclei that are representatives of the aspects of the universe being researched.

The procedure constantly used during the research process consists of prereading, followed by in-depth reading that includes analysis and synthesis of the data. During these processes of document analysis, the establishment of logical causal relations or signifier-signified relations among the documents under consideration (Medeiros, 2016) is crucial. Analysis of the informational attributes of documents is based on identification of essential attributes, identification of the relations among informational attributes specific to the documents, and the informational conditions of the subject and space–time.

Thus, connecting the attributes requires situating them in a larger order of events, which implies reading the document itself, considering the need for peripheral information to form the unique identity of that memory unit being accessed. Furthermore, the implicit or explicit intentionality in the creation and maintenance of the document and the influences of thought in the temporal dimension both at the moment of its production and over time should be included.

With this basic understanding of the methodological aspects specific to historical ethnozoology, we turn to considerations about the context of the interactions between human societies and fauna.

INVESTIGATING THE HISTORICAL IMPLICATIONS OF HUMAN EXPLOITATION OF FAUNA

Most of the scholarship on the relation of human societies with fauna and, more broadly, with nature, which has grown in both quantitative and quantitative terms, especially since the 1960s, reveals that intensive human interventions have had an increasingly serious impact on the dynamics of animal populations, turning the natural environment into a degraded environment due to constant and intense economic exploitation. The direct exploitation of faunal resources through unregulated hunting and fishing has placed pressure on the populations of various exploited animals and, indirectly, on the species that make up their network of interaction. In addition to faunal resources being exploited for human consumption, other human activities such as illegal animal trafficking and the introduction of exotic species, for example, have negatively affected animal biodiversity.

In the 19th century, developmentalist ideas about continuous progress, fueled by natural resources, dominated. Starting in the second half of the 20th century, studies in different fields showed that the human presence in both terrestrial and oceanic realms was characterized primarily by predatory intervention, aimed at controlling natural resources and informed by an economic and rational worldview.

Over time, human intervention has been degrading ecosystems, hindering their balance and putting various animal and plant species at risk of extinction. Over the past 10,000 years, human activities have caused profound changes in global ecosystems. Expansion during the settlement process, creation of pastures and agricultural fields, ceaseless deforestation of wild areas, etc., have gradually reduced the natural habitats of animals and plant species. Deliberate hunting of animals for food, skins, and other parts prized by humans, along with recreational hunting, has caused a significant reduction in the populations of many species (Ponting, 1995).

Therefore, the motives and intentions of humans came to determine the needs and activities of people in the environment and in the perception of wildlife as a resource. As a dynamic process, the use of these resources was also shaped by time, historic moments, and economic interests. Especially after the 16th century, human interventions became omnipresent, resulting in changes in animal populations and the physical–natural world as a whole, which became increasingly integrally related. One of the milestones for the negative impact on wildlife and the environment, on a global scale, was the new paradigm brought about by the Industrial Revolution in the 18th century, based on appropriation of natural resources and large-scale consumption, which changed the mode and relations of production and also brought changes in the relations between human populations of modern society and natural resources, including animals.

Thus, although the extinction of animal (and plant) species may result, in some situations, from centuries-old natural and evolutionary factors (Russell, 2012), human action has had a remarkable impact in the process (Alves, 2012). The case of the extinction of megafauna illustrates this situation (Malhi et al., 2016). The extinction of some megafauna species has been associated with the arrival of *Homo sapiens* (Barnosky et al., 2004; Sandom et al., 2014), considered an effective, generalist superpredator capable of hunting even large mammals that, until then, had faced little pressure from predators (Malhi et al., 2016), which may have caused a cascade effect of extinctions, considering these animals' central role in the trophic chains (Koch and Barnosky, 2006).

Recent human interference in ecosystems around the world and the disappearance and reduction in the number of animal species is a cumulative process that has been accelerated since the 16th century with the expansion of the European economy and culture. Over the past few centuries, various examples indicate that the population decline of several animal species has been accelerated by the process of industrialization. An example of this situation is the catching and industrialization of whales, which caused a marked decline in the population density of these animals (McCauley et al., 2015). In situations like this, an historical ethnozoological study of documents from the past can provide important information that allows us to infer the scenario leading to the exploitation, decline, and even extinction of animal species. The next section of this chapter presents a brief history of whaling in Brazil as a case study that illustrates how historical documents can enable the reconstruction of past scenarios of animal resource exploitation.

WHALING IN BRAZIL

Economic activities have contributed to changes in Brazilian natural environments. Predatory hunting on land and fishing of the seas are part of the economic exploitations undertaken and the environmental degradation caused by Portuguese settlers beginning in the 16th century. Whaling along the Brazilian coastline is a representative example of an economic activity that was important to the maintenance and domination of the colony's coastal areas. This example illustrates the main point of this chapter: the establishment is possible of a research goal and subject, starting with the cultural traits related to the animals and based on historical facts, which will be revealed through multiple sources. If we consider the case of whaling, we see documentary evidence of the attack undertaken by humans against this animal species and unearth details about the history of the exploitation of this resource, including the people; the whale species involved; the locations where the whales were hunted and their products processed, sold, and consumed; the motivations and business logic behind whaling; etc. This research perspective means that even for an extinct animal species, the location of data is possible to show the animal's existence and traces of its exploitation, revealing, in the present, the practices by which the species was exterminated and the speed, scale, and extent of the activity, so a comparison with current practices can be drawn. Thus, an understanding is developed of the history of the relation of people with animals and nature.

In Brazil, except perhaps for indigenous coastal inhabitants taking advantage of accidentally beached whales (Singarajah, 1985), the hunting of these cetaceans began during the colonial period (Tavares, 1916). Documents left by chroniclers are very useful for understanding the historical process of whaling in Brazil. Among these is the "History of Brazil," written in 1627 by the Bahian Friar Vicente do Salvador (1564–1635) (Salvador, 1627). Among other impressions, Friar Vincente left a written record of facts related to whaling along the coast of the state of Bahia. His information focuses on how whaling began in the early 17th century. In 1817, another chronicler, the traveler Louse-François of Tollenare (1780–1853), recorded how whaling was viewed as a spectacle in the Baía de Todos os Santos (State of Bahia) (Tollenare, 1905). These two examples highlight how the contact these men had with whaling on the coast of colonial Brazil yielded records of their impressions of the purpose and techniques used in this economic activity on the coast of Brazil.

Similarly, others from the colonial period and other historical eras have recounted how all the parts of each animal were used, including the fat, bones, meat, etc. They also described new uses found for whale by-products, what hunting practices were used, to which markets the whale products were sent, whether the demand produced a change in the market, what labor force was employed in the business, and how the social and productive nucleus associated with whaling was structured.

Another report on the subject of whaling that deserves comment is José Bonifácio de Andrade e Silva's "Memória sobre a pesca da baleia e a extração do seu azeite" (Memory of whaling and extraction of whale oil), published in 1790 by the Lisbon Academy of Sciences (Silva, 1790). An intellectual associated with the University of Coimbra, José Bonifácio included harsh criticism in his work of the way in which natural resources were being exploited, both in Portugal and in its overseas territories. Pádua's (2004) analysis of José Bonifácio's report underscores the elements that run through this intellectual's work: a worldview based on the economy of nature; a defense of economic progress as a civilizing force; and a defense of rationalizing techniques of production through the application of scientific knowledge. These elements of intellectual criticism in an era heavily influenced by the Enlightenment are fundamental to understanding the activities related to the exploitation of fauna in Brazil because these elements bring us face-to-face with the logic underlying the practices supported by the Portuguese Metropole. In handling documents from this period, the researcher has before his eyes the learned discourse that could be found in Brazil in the late 18th and early 19th centuries, which took a critical stance with respect to hunting activities, portraying them as the predatory and irrational exploitation of natural resources, especially regarding the use of forests and soil and the extermination of animals and plants. On the other hand, while reading official documents such as the Royal Edicts, the scholar of historical ethnozoology encounters discourses of exploitation concerned with mercantilist interests and the colonial economic activities associated with slavery.

Another body of documents that may serve as the genesis for a detailed research project would be the texts of authors such as Manuel Arruda da Câmara (1752–1810), Alexandre Rodrigues Ferreira (1756–1815), Joaquim Nabuco (1849–1910), and André Pinto Rebouças (1838–98); these texts reveal an element of concern for the environment that was present in the work of Luso-Brazilian intellectuals. In these sources, the first stirrings can be seen of a movement in opposition to a practice that is still sharply criticized today: the indiscriminate hunting of some animal species, the control of species of greater economic viability by the countries considered richer and the extinction of species.

More recent information on 20th-century whaling can be obtained from documents of whaling companies, old reports from local newspapers, and scientific articles. These documents reveal that whale catches in Brazil increased in the early 20th century with the installation of whaling stations in Paraíba and later in Cabo Frio, as well as the use of modern whaling techniques (Fig. 4.1 and 4.2), and that whaling exploitation in Brazil regained prominence on the international stage after having been in decline since 1773 (Toledo, 2009).

Therefore, this survey of sources makes it possible to identify elements in hunting and fishing activities that are the precursors to social movements that defend the rational use of faunal resources, as well as to identify pioneering ideas concerning patterns of faunal exploitation in various ecosystems.

Taking Brazil as an example, the untrammeled process of reduction and extinction of faunal resources has been occurring since the moment the Portuguese colonial project began in 1500. The way by which the colonizer viewed the colonized and the faunal (and floral) resources was historically characterized by transformative actions. These actions wrought decisive changes in natural environments, including

FIGURE 4.1 Whaling vessel used by the COPESBRA—Northern Fishing Company of Brazil. *Source: NDIHR-Núcleo de Documentação Histórica e Regional.*

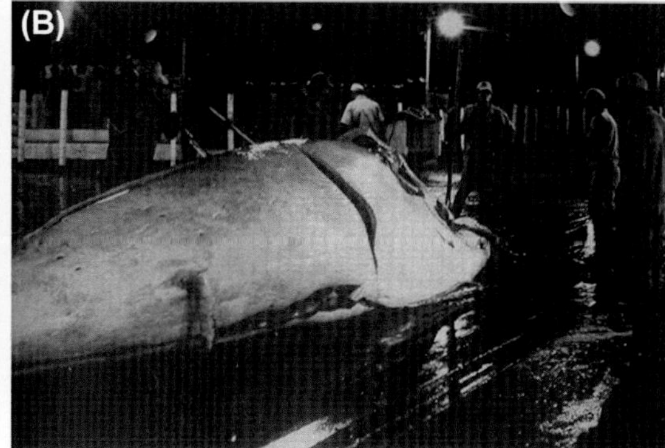

FIGURE 4.2 Cutting procedure (using saw machine) (A) and slash up with machetes (B) the whales captured in the State of Paraíba, Brazil, in the early 20th century. *Source: NDIHR- Núcleo de Documentação Histórica e Regional.*

the ecological balance of these environments, throughout the process of the development of each region of Brazil. In this way, by destroying the chains and processes specific to each ecosystem, human actions and interventions have exacerbated the vulnerability and extinction of several species of Brazilian fauna (and flora).

REFLECTION ON HISTORICAL ETHNOZOOLOGY

The legacy we receive from the past is understood as being the patrimony we experience in the present and bequeath to future generations. Through this patrimony, we are able to reflect on the past with memories and see the effects of the various actions of multiple cultures to the situation that is our patrimony. Historical sources are thus a collective good of extreme value to society. These sources contain positions, views, and feelings, as well as the symbols and meanings of cultural references (Luca, 2008).

Studying ethnozoology through historical documents means dealing with this patrimony, using the document as source for questioning the intentions of the author, and employing a critical eye to put the document in context (Bacellar, 2008). The documentary source allows the researcher to perceive different relations of society, power, identity, and formation through animal resources. Historical ethnozoology is thus tied to use the document as the basic source for research and for the narrative.

When developing a narrative in ethnozoology using historical documents, one must, importantly, construct a forum for an interdisciplinary discussion that highlights the importance of various sources and languages, whether these sources are images, sounds, or written and oral sources, into a narrative fabric of equally varied contours and contexts.

In this chapter, our goal was to provoke reflection about the role of documentary sources in constructing the narrative in historical ethnozoology, for which the accurate use of documents is an essential condition for the researcher. Research in this field is vast and rich in details to be uncovered, pondered, and understood, and those interested in this subject must make an effort to contribute to the enrichment of human history. Ethnozoology, therefore, is a promising field because we must address a documentary patrimony related to ethnozoology as a legacy received from the past, which the postmodern man experiences in the present moment and will bequeath to future generations.

References

Alves, R.R.N., 2012. Relationships between fauna and people and the role of ethnozoology in animal conservation. Ethnobiology and Conservation 1, 1–69.

Alves, R.R.N., Albuquerque, U.P., 2012. Ethnobiology and conservation: why do we need a new journal? Ethnobiology and Conservation 1, 1–3.

Alves, R.R.N., Medeiros, M.F.T., Albuquerque, U.P., Rosa, I.L., 2013. From past to present: medicinal animals in a historical perspective. In: Alves, R.R.N., Rosa, I.L. (Eds.), Animals in Traditional Folk Medicine: Implications for Conservation. Springer, pp. 11–23.

Alves, R.R.N., Souto, W.M.S., 2015. Ethnozoology: a brief introduction. Ethnobiology and Conservation 4, 1–13.

Bacellar, C., 2008. Fontes documentais: uso e mau uso dos arquivos. In: Pinsky, C.B. (Ed.), Fontes Históricas. Contexto, São Paulo, pp. 23–79.

Barnosky, A.D., Koch, P.L., Feranec, R.S., Wing, S.L., Shabel, A.B., 2004. Assessing the causes of Late Pleistocene extinctions on the continents. Science 306, 70–75.

Chansigaud, V., 2009. The History of Ornithology. New Holland, London.

Dias, R., 2011. Gestão ambiental: responsabilidade social e sustentabilidade. Atlas.

Justamand, M., 2004. As pinturas rupestres do Brasil: educação para a vida até hoje. 41 Revista Espaço Acadêmico.

Kalof, L., 2007. Looking at Animals in Human History. Reaktion Books.

Koch, P.L., Barnosky, A.D., 2006. Late Quaternary extinctions: state of the debate. Annual Review of Ecology, Evolution, and Systematics 37, 215–250.

Luca, T.R., 2008. De Fontes impressas: história dos, nos e por meio dos periódicos. In: Pinsky, C.B. (Ed.), Fontes Históricas. Contexto, São Paulo, pp. 111–154.

Malhi, Y., Doughty, C.E., Galetti, M., Smith, F.A., Svenning, J.-C., Terborgh, J.W., 2016. Megafauna and ecosystem function from the Pleistocene to the Anthropocene. Proceedings of the National Academy of Sciences 113, 838–846.

Martínez, C., 2008. Las aves como recurso curativo en el México antiguo y sus posibles evidencias en la arqueozoología. Arqueobios 11–18.

Martins, R.A., 2015. Aristóteles e o estudo dos seres vivos. Editora livraria da fisica.

McCauley, D.J., Pinsky, M.L., Palumbi, S.R., Estes, J.A., Joyce, F.H., Warner, R.R., 2015. Marine defaunation: animal loss in the global ocean. Science 347, 1255641.

Medeiros, M.F.T., 2009. Etnobotânica histórica: princípios e procedimentos. NUPEEA, Recife.

Medeiros, M.F.T., 2014. Procedures for documentary analysis in the establishment of ethnobiological information. In: Albuquerque, U.P., Cruz da Cunha, L.V.F., Lucena, R.F.P., Alves, R.R.N. (Eds.), Methods and Techniques in Ethnobiology and Ethnoecology. Springer, New York, pp. 75–86.

Medeiros, M.F.T., 2016. Historical ethnobiology. In: Albuquerque, U.P., Alves, R.R.N. (Eds.), Introduction to Ethnobiology. Springer, New York, pp. 19–24.

Pádua, J.A., 2004. Um sopro de destruição – pensamento político e crítica ambiental no Brasil escravista, 1786–1888. Zahar, Rio de Janeiro.

Ponting, C., 1995. Uma história verde do mundo. Civilização Brasileira, Rio de Janeiro.

Russell, N., 2012. Social Zooarchaeology: Humans and Animals in Prehistory. Cambridge University Press.

Salvador, V.d.F., 1627. História do Brasil: 1500–1627.

Sandom, C., Faurby, S., Sandel, B., Svenning, J.-C., 2014. Global late Quaternary megafauna extinctions linked to humans, not climate change. Proceedings of the Royal Society 20133254.

Silva, J.B.A., 1790. Memória sobre a pesca das baleias. In: Obras científicas, políticas e sociais de José Bonifácio de Andrada e Silva. Santos: Imprensa Oficial, 1963 [1790].

Singarajah, K.V., 1985. A review of Brazilian whaling: aspects of biology, exploitation and utilization. In: Proceedings of the Symposium of Endangered Marine Animals and Marine Parks, Cochin, India, pp. 131–148.

Sutton, M.Q., 1995. Archaeological aspects of insect use. Journal of Archaeological Method and Theory 2, 253–298.

Tavares, J.S., 1916. A pesca da baleia no Brazil. Brotéria, Revista de Sciencias Naturaes: Série de Vulgarização Scientifica 14, 69–81.

Toledo, G.A.C., 2009. O homem e a baleia: aspectos históricos, biológicos, sociais e econômicos da caça na Paraíba. Universidade Federal da Paraíba, João Pessoa.

Tollenare, L.F., 1905. de Notas dominicaes tomadas durante uma residencia em Portugal e no Brasil nos annos 1816, 1817 e 1818. Parte relativa a Pernambuco traduzida do manuscripto francez inedito por Alfredo de Carvalho com um prefacio de M. de Oliveira Lima. Impreza do Jornal do Recife, Recife.

CHAPTER

5

Imaginary Zoology: Mysterious Fauna in the Reports of Ancient Travelers and Chroniclers

Kleber da Silva Vieira[1,2], *Washington Luiz Silva Vieira*[2], *Rômulo Romeu Nóbrega Alves*[1]

[1]**Universidade Estadual da Paraíba, Campina Grande, Brazil;** [2]**Universidade Federal da Paraíba, João Pessoa, Brazil**

INTRODUCTION

A time existed when the world seemed larger and vaster, filled with mysteries, unknown places, and monstrous beings. At that time, many writers and travelers, including both merchants and mere adventurers, reported exotic places, strange people, and fabulous creatures.

In those days, the world was yet to be discovered. Promises existed of gold and untold riches in distant lands, located beyond fierce seas, endless deserts, dark woods, or mystical mountains. The dream of changing one's life and becoming wealthy was one of the main reasons for so many men being led to their death or to face the unknown in such places, only to tell us what they had found there after surviving many dangers.

These adventurers' chronicles reveal a bestiary composed of veritable imaginary fauna in both real and idealized places. Strange beings were born of men's madness and fantasy over time and space. These creatures have always attracted our attention, which is why areas of human knowledge such as zoology, paleontology, and anthropology have acquired a certain grace and charm through their history. The study of these monsters and beasts even came to inspire the creation of a parallel and entirely new branch of study, popularly known as cryptozoology.

Cryptozoology is often defined as "the study of animal species whose existence is not supported by empirical evidence, but rather hypothesized via indirect and uncertain information, including oral traditions, eyewitness accounts, and inconclusive physical evidence" (Rossi, 2016). Cryptozoologists consider any figure from folklore to be a "cryptid" (from the Greek κρύπτω, *krypto*, meaning "hide") or "hidden animal." A cryptid is therefore a creature that arises from human observation or imagination and thus represents an ethnozoological entity.

Ethnozoology
http://dx.doi.org/10.1016/B978-0-12-809913-1.00005-3

Naturally, these mysterious beings are of interest to ethnozoology too, although in these cases, we take a more ethnozoological than mythological perspective.

Based on this historical legacy of accounts and chronicles, two types of geographic narratives can be identified: one legendary, the other modern. A clear line between them is difficult to distinguish, but some authors agree that the reports left by Venetian merchants Nicola Mafeo and Marco Polo of their travels in the 13th century clearly differ from those of older times (Hyde, 1982), as is quite clear to anyone who has ventured to read the *Book of Wonders*.

What is curious about these narratives, whether they be ancient, medieval, or modern, is the length to which the copyists went to assemble reports about certain places, their geography, people, local fauna, and flora in order to make them known, either out of strictly academic interest or merely to establish trade relations through the opening of new trade routes.

Reading some of these fabulous descriptions takes us back to an almost immemorial age, both extraordinary and fascinating, but also makes it clear how misleading and unbelievable lists of fauna based solely on second-hand or even third-hand accounts can be, especially due to the lack of precision in the accounts or to the desire of the storytellers and calligraphers to embellish the creatures of a given place to render them more fantastical or remarkable. Out of the intentional or accidental distortions of these ancient "naturalist geographers," so to speak, we seek to produce a small sample of a fabulous fauna, always respecting the original text's message, to compose what we term an imaginary zoology, both to honor the memory of these intrepid and creative adventurers or copyists and to recall the infancy of what in the future would become geography and zoology.

A GARDEN IN THE ORIENT

Although the records are not at all specific, we can induce that the fauna were varied and perhaps even picturesque in what is known in the Judeo-Christian tradition as Eden (*edinu in* Akkadian, *edin* in Sumerian, or *Kden* of Hebrew), a word meaning pleasure or delight (Wikicristiano, 2016).

Eden is the mythological name of the volcanic plateau bordered by the rivers Tigris,[1] Euphrates, Pisom, and Gihon, where Yahweh planted a garden to the east (Autores et al., 1990). According to the tradition, the *Book of Genesis* tells the story of "Moses," about the origins of the world, of man, and of living beings. The place was, in theory, located in the valleys of modern-day Iraq and environs, which used to be called Mesopotamia[2] (Hamblin, 1987). According to *Genesis* 1:20–24, animals were abundant in the garden of Eden:

> And God said: Let the waters bring forth abundantly the moving creature that hath life, and fowl that may fly above the earth in the open firmament of heaven. And God created great whales, and every living creature that moveth, which the waters brought forth abundantly, after their kind, and every winged fowl after his kind: and God saw that it was good. And God blessed them, saying, Be fruitful and multiply, and let fowl multiply in the earth. And the evening and the morning were the fifth day. And God said: Let the Earth bring forth the living creature after his kind, cattle, and creeping thing and beast of the earth after his kind; and it was so.

According to the biblical account, the garden located east of Eden was filled with aquatic reptiles; various flying birds; cattle; terrestrial beasts; and also whales–apparently swimming in the Persian Gulf–as well as, of course, the human species itself. Medieval representations of this garden (Figs. 5.1 and 5.2), such as those painted by Hieronymus Bosch (c. 1450–1516)

[1] Heb. jiddeqel, or Tigra in ancient Persian and gr. Tigris: fast Tigris; heb. Perâth; pers. Ufr ~ tu; gr. Eufráts: large; heb. Pîshôn: flowing and heb. Gîjôn: surging.

[2] From ancient Greek Μεσοποταμία; composed of μέσος, "middle," and ποταμός, "river," i.e., "[land] between rivers."

FIGURE 5.1 *Garden of Earthly Delights* by Hieronymus Bosch (c. 1450–1516). The country scene portrayed by the Dutch artist shows various strange and even deformed or hybrid animals. Prado Museum, Madrid.

FIGURE 5.2 Eden in the eyes of German artist Lucas Cranach (1472–1553). Although many familiar animals are depicted, the scene also includes other very odd ones. Gemäldegalerie Alte Meister (1908 A).

and Lucas Cranach (1472–1553), depicted highly unusual creatures like the unicorn and other very strange or even deformed creatures constituting the fauna of this mythical place.

Beyond Eden, mysterious wilderness creatures, such as the Djinn, lemurs, and giants, appear in Semitic mythology, as well as in Islamic,

FIGURE 5.3 Heraldry of Simurgh, Simorgh, Senmurv, a winged hybrid beast of Persian mythology, described as large enough to carry an elephant or a whale in its lion's claws.

Arabic, and Persian mythology (Adam, 1881; El Hayek, 2006). According to legend, these beings could have inhabited the Land of Nod, to the east of Eden (Gn. 4:16), as well as the Kāf Mountains, places where strange beings had lived for millennia, long before the birth of Adam (Sayce, 1888).

According to ancient Muslim tradition, the Kāf Mountains (from the Persian کوه قاف *qaafkuh* or the Arabic قاف جبل *jabal qaf*), were the most distant points on Earth and home to the Simurgh (Figs. 5.3 and 5.4) and Roc (de Vasconcellos, 2008; Muhawi and Kanaana, 1989), huge winged creatures often related to the eagle, oddly similar to the Phoenix (Ethiopia), Garuda (India), Peng (China), and thunderbirds (North America).

> The Roc birds (…) when there appeared in the air, at a considerable distance from us, two great clouds. (…) each of them carried between their talons, stones, or rather rocks, of a monstrous size. When they came directly over my ship, they hovered, and one of them let fail a stone (…) threw the stone so exactly upon the middle of the ship, that it split in a thousand pieces. The mariners and passengers were all killed by the stone, or sunk. ***Fifth Voyage of Sinbad, the Sailor.***

FIGURE 5.4 A Simurgh in the form of a phoenix in a decorative motif on the exterior of the madrassa Nadir Divan-Beghi Uzbekztão, Bukhara.

The Kāf Mountains seem to correspond to the mountain range located in the far north of the Caucasus, northeast of Turkey and northwest of Iran, including the mythical Mt. Alborz (Hara Berezaiti?), which, coincidentally, may be the same place where Prometheus was chained and subjected to an eagle (Ethon) feeding on his liver every day until finally being freed by Heracles (Abrão et al., 2000).

SOMETHING MORE IN SONG OF THE MUSES

Myths suggest the existence of many other places beyond paradisiacal Eden, including far more dangerous places filled with equally terrible beasts. In ancient times, to oppose such creatures was not merely a heroic deed but a necessity, as they represented true public disasters. *The Labours of Heracles* (Fig. 5.5), or Hercle in Etrusca (Gual, 2003), for example, suggests the presence of fabulous animals that inhabited the Peloponnese, in particular, coming into serious conflict with the local population. A certain dose of exaggeration aside, the existence of this creature seems quite believable as an ancient Hellenic fauna, which probably existed between approximately 1200 and 700 BC, as once noted by the Austrian writer Norman Douglas (1928).

FIGURE 5.5 Hercules facing the Nemean lion, statue in marble. Rome, 2nd century. The State Hermitage Museum, Saint Petersburg. Photo: S. Sosnovskiy.

The mythological accounts of the deeds of Heracles depict, for example, the presence of lions in Argolis (Nemea), boars and deer in Arcadia (Erymanthus and Ceryneia) and strange birds of prey in Lake Stymphalia (Arcadia). Other tales tell of bulls and wild goats in Crete (Minus, Aix, and Amalteia); great mountain pigs in Aetolia (Meleager); bears in Arcadia (Callisto); and strange giant snakes dwelling in caves and mountains in Cilicia, the Peloponnese or Syria (Echidna), Mount Parnassus (Python), or even in Delphi and Libya (Lamia).

One creature that was rather notorious in Greek culture, as well as in some other mythologies, with an apparently wide territorial distribution, is the griffin (Fig. 5.6) or the *uccello grifone* (griffin bird) according to Marco Polo

FIGURE 5.6 Fresco depicting a griffin. Royal Throne Room, Palace of Knossos, Bronze Age, Crete.

(1985), the Venetian merchant who ended up confusing it with the Roc bird while in the area of Madeigastar (Madagascar).

> (…) in those other Islands (…), which the ships are unable to visit because this strong current prevents their return, is found the bird Gryphon, which appears there at certain seasons. The description given of it is however entirely different from what our stories and pictures make it. For persons who had been there and had seen it told Messer Marco Polo that it was for all the world like an eagle, but one indeed of enormous size; so big in fact that its wings covered an extent of 30 paces, and its quills were 12 paces long, and thick in proportion. And it is so strong that it will seize an elephant in its talons and carry him high into the air, and drop him so that he is smashed to pieces; having so killed him the bird gryphon swoops down on him and eats him at leisure. ***Marco Polo, The Book of India, Part XXIII, 192.***

Obviously, the griffin described by Marco Polo bears a greater resemblance to the birds of the Kāf Mountains, perhaps even to Haast's eagle (*Harpagornis moorei*), than to the actual griffin of Greco-Roman mythology (Jonston, 1808). Although it has been associated with fossils of *Protoceratops andrewsi* (Mayor, 2000), the griffin was formerly seen as a type of bird sacred to the god Apollo, originally inhabiting areas slightly south of Hyperborea (whose people had a close relationship with the gods), a region located somewhere in the northern extremities of the Earth (Heródoto, 2006), rather than in the far south as reported by Marco Polo.

> Caüstrobius (…) visited the Issedones; beyond these (he said) live the one-eyed Arimaspians, beyond whom are the griffins that guard gold, and beyond these again the Hyperboreans, whose territory reaches to the sea. ***Herodotus, Book IV, XIII, XXVII.***

According to Mandeville (1900), the animal was abundant in regions of Bacharia (Bactria), today in Afghanistan, where it cohabited with weird hippotaynes (hippopotami).

> In that country (Bacharia) be many hippotaynes that dwell sometime in the water and sometime on the land. And they be half man and half horse (…). And they eat men when they may take them. And there be rivers of waters that be full bitter, three sithes more than is the water of the sea. In that country be many griffins, more plenty than in any other country. Some men say that they have the body upward as an eagle and beneath as a lion; and truly they say sooth, that they be of that shape. But one griffin hath the body more great and is more strong than eight lions, of such lions as be on this half, and more great and stronger than an hundred eagles such as we have amongst us. For one griffin there will bear, flying to his nest, a great horse, if he may find him at the point, or two oxen yoked together as they go at the plough. For he hath his talons so long and so large and great upon his feet, as though they were horns of great oxen or of bugles or of kine, so that men make cups of them to drink of. And of their ribs and of the pens of their wings, men make bows, full strong, to shoot with arrows and quarrels. ***Sr. John Mandeville, Cap. XXIX, 177.***

Descriptions of hybrid animals, such as Mandeville's account of hippotaynes, are always very curious and also raise similarly interesting questions about creatures of this nature: If the hippotaynes were seen as half man and half horse, then to what do accounts of griffins, striges, lemures, and sphynxes actually refer?

Human faces, especially female faces, seen in certain animals, may in fact suggest the image of faces with almond-shaped, disproportionately

FIGURE 5.7 Winged creature with a female face carrying off a small child. Bas-relief of Xanthus, Lycia.

FIGURE 5.8 Harpy according to Jonston's Theathrum Universale (1808), Tab. 62.

large eyes with elongated outer corners and eyebrows (similar to Egyptian eye makeup) on a fairly flat forehead. This bears many similarities to owls, for example, or even many monkeys and medium-sized or large wild cats, whose manes may have been mistaken for long locks hanging from their delicate heads in the eyes of a very creative or very near-sighted observer.

In light of this, in many regions of ancient Rome, reports exist of a flying beast that preyed on children; the beast was similar to an owl but had the head of a woman (Figs. 5.7 and 5.8) and was known as a strige (Latin *Strix*; gr. *Strígks, ggós*). Similar to Greek harpies and sometimes associated with the Jewish mythological figure Lilith or even the Sumerian Inanna (Hurwitz, 2009), the striges were said to have long beaks and sharp talons that were used to suck the blood and tear the flesh of their victims, children in cradles (Abrão et al., 2000). Many accounts depicted them as having bat wings or four legs and inhabiting ruins and abandoned places where they would rest for 3 days after feeding.

The unicorn is another creature that seems to enjoy wide popularity (Fig. 5.9). Although represented in the paradisiacal gardens of Mesopotamia, the unicorn is known through mythical accounts since 400 BC. The Greek author Ctesias reported such an animal inhabiting the kingdoms of Hindustan (India). His account is considered the first Western description of this fabulous creature (Llewellyn-Jones and Robson, 2010).

> The Indians have wild asses … the body is white in colour, the head purple, the eyes dark blue. On the forehead they have a horn, one cubit long, the lower part of which … is pure white, while the uppermost part, which is pointed, is dark purple, and the middle is black. ***Indica 48b, p. 33.***

His typical depiction, popularized in the Middle Ages, is of a small horse with the hind legs of an antelope, the beard of a goat and a long spiral horn on his forehead (O'Connell

FIGURE 5.9 Maiden with unicorn. Tapestry, 15th century. Cluny Museum, Paris.

and Airey, 2009). In his *Natural History*, Pliny describes it in a somewhat nontraditional way:

> There are in India oxen also with solid hoofs and a single horn and a wild beast called the axis, which has a skin like that of a fawn, but with numerous spots on it, and whiter; this animal is looked upon as sacred to Bacchus. The Orsasan Indians hunt down a kind of ape, which has the body white all over; as well as a very fierce animal called the monoceros, which has the head of the stag, the feet of the elephant, and the tail of the boar, while the rest of the body is like that of the horse; it makes a deep lowing noise, and has a single black horn, which projects from the middle of its forehead, two cubits in length. This animal, it is said, cannot be taken alive. ***Pliny the Elder, Livro III, Cap. 31.***

In popular belief, healing power was attributed to the powder of the unicorn's horn and heart, which is why its figure is shown in some insignia related to pharmacies (de Rosa, 2009; Jackson, 2004).

In many ancient and medieval texts, the unicorn is described as a specific, though unique, animal, but in his *Anthologie raisonnée de la litterature chinoise*, Margoulès (1948) provides a vague description of this creature according to a 19th-century writer:

FIGURE 5.10 The k'i-lin visits Yan Zheng Zai, Confucius's mother and first teacher, shortly before the future philosopher's conception.

> (...) even children and village women know that the unicorn constitutes a favorable presage. But this animal does not figure among the domestic beasts, it is not always easy to find, it does not lend itself to classification. It is not like the horse or the bull, the wolf or the deer. In such conditions, we could be face to face with a unicorn and not know for certain what it was. We know that such and such an animal with horns is a bull. But we do not know what the unicorn is like.

As with other fabulous creatures, descriptions of the unicorn almost always differ from each other, which is curious. If rather than attributing these features to the daydreams of their narrators, we take differences in descriptions as morphological variations, we could assume the existence of at least three different varieties (species) of unicorns in antiquity, with distribution from continental Europe to the eastern regions of Asia. This distribution and these descriptions lead some to assert that the unicorn may have been a real animal, the result of confusing local reports of the *Rhinoceros unicornis or* even the *Oryx leucoryx* (Davies, 2015; Jackson, 2004).

In northernmost lands, a variant existed of the unicorn that was called a k'i-lin, which according to legend, may have inhabited the regions of Tibet, China, and Mongolia (Fig. 5.10). Apparently, the k'i-lin is not much different from a common unicorn except for a few peculiar characteristics, in particular, the nature of its horn and the scales that cover a good part of its body (Borges, 2007):

> The k'i-lin has the body of a deer, the tail of an ox, and the hooves of a horse. Its short horn, which grows out of its forehead, is made of flesh; its coat, on its back, is of five mixed colors, while its belly is brown or yellow. It is so gentle that when it walks it is careful not to tread on the tiniest living creature and will not even eat live grass but only what is dead.

> (...) the spirits of the five planets brought her an animal with the shape of a cow, scales of a dragon, and a horn on its forehead. ***The Birth of Confucius.***

Travelers to the Orient have always described it as a place remarkable for its striking beauty, the oddness of its peoples, and the existence of equally exotic creatures. According to the Hellenistic ethnographer and explorer Megasthenes (McCrindle, 1877), in his work *Indika*, in addition to describing unicorns, tapirs, monkeys, tigers, and elephants stronger than those of Libya, also described a type of strange winged scorpion, a flying snake, and the satyr that could be found there (Ashton, 1890).

> There are winged scorpions in India of enormous size, which sting Europeans and natives alike. There are also serpents which are likewise winged. These do not go abroad during the day, but by night, when they let fall urine, which if it lights upon any one's skin at once raises putrid sores thereon. ***Fragm. XIV. Ælian, Hist. Anim. XVI.41. Aelianus (1832)***

> Among the mountainous districts of the eastern parts of India, in what is called the country of the Catharcludi, we find the Satyr, an animal of extraordinary swiftness. They go sometimes on four feet, and sometimes walk erect; they have, also, the features of a human being. On account of their swiftness, these creatures are never to be caught, except when they are aged, or sickly (...) the Satyr stow away food in the pouches of their cheeks, after which they will take out piece by piece in their hands, and eat it. ***John Ashton (1890, p. 53)***

FIGURE 5.11 Edward Topsell's satyr, *The History of Four-footed Beasts* (1658, p. 8).

Indian satyrs, like their African, Mediterranean, or Polynesians counterparts, were a sort of savage men whose descriptions varied according to the region where they were sighted, the witnesses or the chroniclers. Although descriptions vary, males and females of all "species" were frequently described as being covered with hair (Figs. 5.11 and 5.12). Topsell (1658) draws a direct relation between Indian satyrs and those of classic Greco-Roman mythology, giving the impression of some type of hostile and aggressive wild creature:

> The Satyrcs are in the Islands Satiridae, which are three in number, right over against India on the farther side of the Ganges; of which Euphemus Car rehearseth this history: that when he sailed unto Italy, by the rage of winde and evill weather, they were

FIGURE 5.12 The classic Greco-Roman satyr as shown by John Ashton (1890, p. 55).

driven to a coast unnavigable, where were many desart Islandes, inhabited of wild men, and the marriners refused to land upon some Islands, having heretofore had triall of the inhumaine and uncivill behaviour of the inhabitants, so that they brought us to the Satyrian Islands, where we saw the inhabitants red, and had tayles joyned to their backs, not much lesse than horsses. These, being perceived by the marriners to run to the shippes, and lay hold on the women that were in them, the shipmen, for fearc, took one of the Barbarian women, and set her on the land among them, whom in most odious and filthy manner, they abused, whereby they found them to be very bruit beasts.

In later texts, it becomes evident that these wild men, the satyrs, actually have a strange relation with certain types of monkeys:

Under the Equinoctiall toward the East and South, there is a kind of ape called Aegopithecus, an Ape like a goate. For there are apes like beares, called Arctopitheci, and some like lyons, called Leontopitheci, and some like dogs, called Cynocephali.

Amongst the rest there is a beast called Pan; who in his head, face, horns, legs, and from the loynes downward resembleth a goat, but in his belly, breast, and armes, an ape: such a one was sent by the King of Indians to Constantine, which, being shut up in a cave or close place, by reason of the wildnesse thereof, lived there but a season, and when it was dead and bowelled, they pouldred it with spices, and carried it to be scene at Constantinople: the which beast having beene scene of the ancient Grecians, were so amazed at the strangenesse thereof, that they received it for a god, as they did a Satyre, and other strange beasts.

The curious relation drawn by Topsell between wild men, such as satyrs, and certain varieties of monkeys is also noted by John Ashton, who graciously concedes the detail of the former's notes and descriptions.

THE GIFTS OF THE NILE

When one advances up the Nile to the east, traveling through the Empire of Axum, inhabited by the heirs of Solomon and the Queen of Sheba, and passing through the realm of Cuche, inhabited by the sons of Khan or the men with burnt faces, strange people can be seen who have animal heads, in this case, the head of a dog.

They state also that the island of Gagaudes lies at an equal distance from Syene and Meroe, and that it is at this place that the bird called the parrot was first seen; while at another island called Articula, the animal known as the sphingium was first discovered by them, and after passing Tergedus, the cynocephalus.
Pliny the Elder, Book VI, Cap. 35.

Some notes in Pliny's *Natural History* refer to the Cynocephali as being in truth baboons (*Papio cynocephalus*), while the sphyngium (sphynga) were small monkeys favored as pets by Roman

FIGURE 5.13 The sphyngium was a simian that could appear at times to have a rounded face and a bust similar to a woman. Illustration presented by John Ashton (1890, p. 62).

FIGURE 5.14 Depiction of Cynocephali in a psalter, Kiev, 1397.

ladies (Fig. 5.13), who paid high prices for them in the market (Ashton, 1890; Pliny, 1855). These animals were thus very similar to the Asian cynomolgus monkey (*Macaca fascicularis*) and were described at different times as having human or canine faces and were sometimes confused with the Saudi bird of the same name, which builds its nest from branches of the cinnamon tree (Pliny, 1855; Solinus, 1895).

It is clear that the Cynocephali (Fig. 5.14) and Pliny's sphynx were creatures with "human or canine heads" and were, despite their similarities, distinct from the figure of the Kalystrii described by Ctesias and Megastenes in India (Migne, 1859):

> (…) although they have canine features and speak no language, they communicate very well among themselves, fully understand the languages of other Indians and interact with them through gestures. They have laws, survive by hunting and have never worked at anything else. They bake the specimens they catch under the sun, drink goat milk and include fruit in their diet, especially some sweet fruit that they dry and store in baskets together with some purple flowers that they then sell every year to Indians or (…) offer to the king in exchange for flour, bread, cotton, swords, bows and arrows that they use to hunt. They live in rugged mountains (…) and in caves. Women bathe only once a month, after their period. The men never bathe and wash only their hands. However, three times a year they anoint their bodies with an oil extracted from milk. They wear garments made of skin, identical for both sexes and the richer ones (who are few in number) wear linen garments. All of them sleep on leaves. The one with the most camels is considered the richest, but the majority of their property is divided communally. Men and women have a tail, like dogs, but longer and furrier. They have sexual relations from behind like dogs, and it is considered dishonourable to do it differently. They are pleasant, love justice and live long lives, sometimes as long as two hundred years. ***Migne, Patrologiae Graeca 103, col. 222s (text adapted).***

These dog-headed creatures are in no way similar to the Hundigar of Germanic medieval legend, nor even to Hesiod's Hemican, and thus should not be confused with them, although they are referred to as Cynocephali in various texts and reports (Anônimo, séc. XIII, 2009; Baring-Gould, 2008; Tolkien, 2009):

> (...) Then Sigmund rushed at him so hard that he staggered, and Sigmund bit him in the throat. Now that day they might not come out of their wolf-skins: but Sigmund lays the other on his back, and bears him home to the house, and cursed the wolf-gears and gave them to the trolls. ***Völsunga Saga*, Ch. 8.**

> (...) but the land west of this (Triton River), where the farmers live, is exceedingly mountainous and wooded and full of wild beasts. In that country are the huge snakes and the lions, and the elephants and bears and asps, the horned asses, the dog-headed and the headless men that have their eyes in their chests, as the Libyans say, and the wild men and women, besides many other creatures not fabulous. ***Herodotus*, *Book IV, CXCI.***

Ancient citations exist for many creatures with canine heads in classical texts found well beyond the area of Egypt, for example, in Serbia, the mythical island of Nacumera (in the Atlantic Ocean), Burgundy, Arcadia, Iceland, and Norway. In most cases, the descriptions can be attributed to birth defects (Eberly, 1988), folklore, and misidentification of certain animals, rather than to true primate species, as in the example of Pliny.

Deep within Ethiopia, in the lands from which the Nile originates another, more terrible animal exists, whose gaze is able to kill any person who looks directly at it (Fig. 5.15):

FIGURE 5.15 Catoblepas, Der Naturen Bloeme manuscript, 1350. National Library of the Netherlands.

> Among the Hesperian Aethiopians is the fountain of Nigris, by many, supposed to be the head of the Nile. I have already mentioned the arguments by which this opinion is supported. Near this fountain, there is found a wild beast, which is called the catoblepas[3]; an animal of moderate size, and in other respects sluggish in the movement of the rest of its limbs; its head is remarkably heavy, and it only carries it with the greatest difficulty, being always bent down towards the earth. Were it not for this circumstance, it would prove the destruction of the human race; for all who behold its eyes, fall dead upon the spot. ***Pliny the Elder, Book VIII, Cap. 32.***

Pliny's description of the Catoblepas can be augmented by additional features, such as those described by Flaubert (1904):

> The Catoblepas appears, a black buffalo, with a pig's head hanging to the earth, and connected with his shoulders by a slender neck, long and flabby as an empty gut. He is wallowing on the ground; and his feet disappear under the enormous mane of hard hairs that descend over his face. ***Gustave Flaubert, The Temptation of Saint Antony, p. 151.***

Because of its peculiar appearance, except for the fantastical powers attributed to it by classic texts, Cuvier believed the Catoblepas was actually a species of wild African bovine–the gnu (*Connochaetes gnou*)—mixed with the image of the Basilisk (Borges, 2007). More detailed information about the Catoblepas can be found in the writings of Aelianus (1832) in his book *Animals*.

In the mysterious and distant lands beyond the sources of the Nile, where the mountains faded into an almost infinite indigo horizon, peoples ceased to be human and became bestial and savage, and animals even more dangerous

[3]From the ancient Greek καταβλεπω, "to look downwards."

FIGURE 5.16 Manticore according to Edward Topsell's *The History of Four-footed Beasts* (1658).

than the Catoblepas were seen and described by adventurous chroniclers who risked crossing the plains in the heart of the black continent. In these regions of the fantastic Ethiopia, an animal known to the locals as Crocota or Leucrocota (Fig. 5.16) lived; this animal was much better known to the Persians as the Manticore (Topsell, 1658).

> By the union of the hyaena with the Aethiopian lioness, the corocotta is produced, which has the same faculty of imitating the voices of men and cattle. Its gaze is always fixed and immoveable; it has no gums in either of its jaws, and the teeth are one continuous piece of bone; they are enclosed in a sort of box as it were, that they may not be blunted by rubbing against each other. Juba informs us, that the mantichora of Aethiopia can also imitate the human speech. ***Pliny the Elder, Book VIII, Cap. 45.***

> The corocotta, an animal which looks as though it had been produced by the union of the wolf and the dog, for it can break any thing with its teeth, and instantly on swallowing it digest it with the stomach; (…) The leucrocotta, a wild beast of extraordinary swiftness, the size of the wild ass, with the legs of a stag, the neck, tail, and breast of a lion, the head of a badger, a cloven hoof, the mouth slit up as far as the ears, and one continuous bone instead of teeth; it is said, too, that this animal can imitate the human voice. ***Pliny the Elder, Book VIII, Cap. 30.***

The region known as Ethiopia, while beautifully fantastic and fabulous in mythical descriptions, lacks historical precision (Dworacki, 2009). For the Mediterranean peoples of the

FIGURE 5.17 *Oedipus and the Sphinx* by Gustave Moreau, Metropolitan Museum of Art.

4th century BC, Ethiopia (Αιθιοψ) could be any country inhabited by black people along the east bank of the Nile (Snowden, 1948) or even a specific place associated with a hero of a similar name with ties to the peculiar island of Lesbos in the Aegean Sea (Safo, 2011).

In this exotic region of Ethiopia, strange hybrid animals similar to the Manticore were first seen but were never found. Among these animals were the strange Sphinx and the deadly Basilisk, which resembled the Catoblepas.

Several varieties of Sphinxes were said to be distributed throughout the Mediterranean (Thebes), Mesopotamia (Assyria), and the "north–central and northeast" of Africa (Ethiopia). The Greek Sphinx (Fig. 5.17), best known for having been killed by Oedipus (Sófocles, 1990), had the head and breasts of a woman, the wings of a bird, and the feet and body of a lion. Some even claim it had the tail of a snake and the body of a dog. The African or Egyptian Sphinx, however, also called the Androsphynx by Herodotus, was a lion with the head of a man, very similar to the Middle Eastern version.

> Aethiopia produces the lynx in abundance, and the sphinx which has brown hair and two mammae on the breast, as well as many monstrous kinds of a similar nature; horses with wings, and armed with horns, which are called pegasi (...) There are oxen, too, like those of India, some with one horn, and others with three. ***Pliny the Elder, Book VIII, Cap. 30.***

Like other hybrid beasts of antiquity, Sphinxes could be the result of confusing real animals, especially some medium-sized and large monkeys or felines found throughout the ancient territories of Africa, Europe, and Asia. This possibility can be perceived in some relief images in which the Sphinx has no wings and more closely resembles a stately lion, although with a human head.

While Perseus was carrying the head of the deadly Gorgon in his hand, the blood of the monster generated other monsters equally terrible as the blood fell to the ground and, thus, were born all the snakes of Libya: the asp, the amphisbaena, and the ammodytes were among them (Lucan, 1820). Luckily, this fact was noticed by the young hero before he was dazzled by the beauty of Andromeda, who was chained to a rock in the land of Ethiopia with the waves bathing her beautiful feet (Vernant, 2008). This mythological explanation for the origin of the snakes of the Old World may also be the origin of many dragons and other fierce beasts that managed to survive into postdiluvian times. This beautiful and lyrical origin did not go unnoticed by the ancient scholars, which is worth noting.

FIGURE 5.18 Medieval conception of the basilisk in its form as a rooster published by Johann Heiden for a German version of Pliny's *Historia Naturalis* and illustrated by Jost Amman (1584).

In addition to common snakes and to the Chrysaor and Pegasus, yet another creature generated by the blood of Medusa was the Basilisk (Fig. 5.18), whose power to kill with a glance may have been inherited from its mother. The Basilisk is so named because it was considered the king of snakes ("little king") and it was described as a snake, albeit a fabulous one, by Pliny:

> There is the same power (as the catoblepas) also in the serpent called the basilisk. It is produced in the province of Gyrene,[4] being not more than twelve fingers in length. It has a white spot on the head, strongly resembling a sort of a diadem. When it hisses, all the other serpents fly from it: and it does not advance its body, like the others, by a succession of folds, but moves along upright and erect upon the middle. It destroys all shrubs, not only by its contact, but those even that it has breathed upon; it burns up all the grass too, and breaks the stones, so tremendous is its noxious influence. It was formerly a general belief that if a man on horseback killed one of these animals with a spear, the poison would run up the weapon and kill, not only the rider, but the horse as well. To this dreadful monster the effluvium of the weasel is fatal, a thing that has been tried with

[4] An ancient Greek colony in modern Libya. This colony is said to have been founded by Apollo in tribute to Cyrene, daughter of Hypseus, kidnapped by the god because of the young girl's magnificent beauty.

success, for kings have often desired to see its body when killed; so true is it that it has pleased Nature that there should be nothing without its antidote. The animal is thrown into the hole of the basilisk, which is easily known from the soil around it being infected. The weasel destroys the basilisk by its odour, but dies itself in this struggle of nature against its own self.
Pliny the Elder, Book VIII, Cap. 33.

The image described by Pliny underwent profound changes over time, and by the Middle Ages, the Basilisk had gradually lost its traditional form and more closely resembled a four-legged rooster or even an eight-legged scaly beast (Aldrovandi, 1640).

Many Christian encyclopedists sought rational explanations for the origin of the Basilisk and completely rejected the ideas contained in the *Pharsalia*. These scholars were unable to discard their belief in this creature, as the Latin Vulgate tied it closely to the Tsepha (venomous reptile), so they suggested the basilisk must have been born from a deformed or counterfeit egg laid by a rooster and hatched by a frog or a snake (Borges, 2007).

The Basilisk was considered a very dangerous animal, mainly because everything around it was laid waste, died or rotted, and the rivers from which it drank were poisoned for centuries. Therefore, travelers were advised to carry a rooster with them, or at least a mirror, when passing through unknown lands because the basilisk was known to detest its own reflected image and the sound of a cock's crow; these beliefs later melded to yield the terms (or new animals) basilicock, cockatrice, and cocodrille (Sánchez, 1978).

Leaving aside exaggerations and eccentricities, Pliny may, in fact, have been referring to some type of African snake of the genus *Naja* when he described the Basilisk. Some species of mongoose are known to fight some varieties of snakes. Furthermore, references in the Roman naturalist's text indicate the use and application of musk from certain mammal species as an effective way to repel or even kill snakes, and especially Basilisks, a word that seems to be used as a collective term referring to any venomous serpent rather than to a specific animal.

A WORLD OF IMAGINATION AND FANTASY

When we think of pioneering conquerors, we generally imagine them as intrepid men, brave and thirsting for adventure, and this is certainly the romanticized portrayal produced by the popular Western imagination for the purpose of transforming their past ghosts into bold heroes. This portrayal is as mythical and fabulous as the very monsters and adventures depicted in the oldest tales, behaviour that can perhaps only be explained by the analytical psychology of Jungian or Freudian psychoanalysis. Seen thus, this portrayal could be one more partially frustrated attempt to confront the terrors of the unconscious (Salles, 1998).

This portrayal of fearless characters does not fit, for example, with the terror that was cultivated with regard to the equatorial regions or the ends of the earth, where it was believed that damnation lay, where one might be consumed by flames or tumble into a vast and deep abyss. This same fear could at times only be overcome by the feeling of hope or greed that sought enrichment and a drastic change in life's oppressive and miserable conditions (Loyn, 1997). Therefore, many explorers threw themselves into danger, motivated not by a taste for adventure or courage but by dreams of enrichment or the simple human desire for a brighter future for themselves or their families, whether through commerce or less noble means.

In many cases, then, the chronicles—whether classical, medieval, or contemporary—were narrated and written by men guided by the fantastical, feeding their hyperbolic distortions with improbable facts, creating beings and things so out of proportion that only in the wildest of dreams could such absurdities make sense

(Waugh, 2010). This is how apes turn into hairy wild men, serpents become famished and are seen as terribly poisonous dragons, humans take on canine visages, real animals become bloodthirsty creatures, and deformities are interpreted as divine punishments or sorcerers' and witches' spells; these tales were not limited to ignorant people, for learned men were the principal consumers and transmitters of fabulous tales, and they gave these compendia of beasts their "rational" explanations (Del Priore, 2000; Sánchez, 1978).

The bestiaries—for they were not yet called zoological treati—sesillustrate the outcome of many of these expeditions and contain descriptions of creatures and people who lived far beyond the pale of the "civilized" world. These encyclopedic compendia attempts to make known what was inaccessible even to the doctors themselves, much less the copyists, because much of what we observe in these classical texts are copies, notes, and interpretations of things seen and witnessed by others in far-off places.

Even in the later voyages of the great 15th-century and 16th-century expeditions, more than two millennia after Herodotus, we can still find monsters inhabiting what came to be called the New World. This can be observed, for example, in the writings of Cardim (1583–1601), Gabriel Soares de Sousa (1587), and Gândavo (1576) with regard to the sea men and strange marine creatures of Brazil (Fig. 5.19):

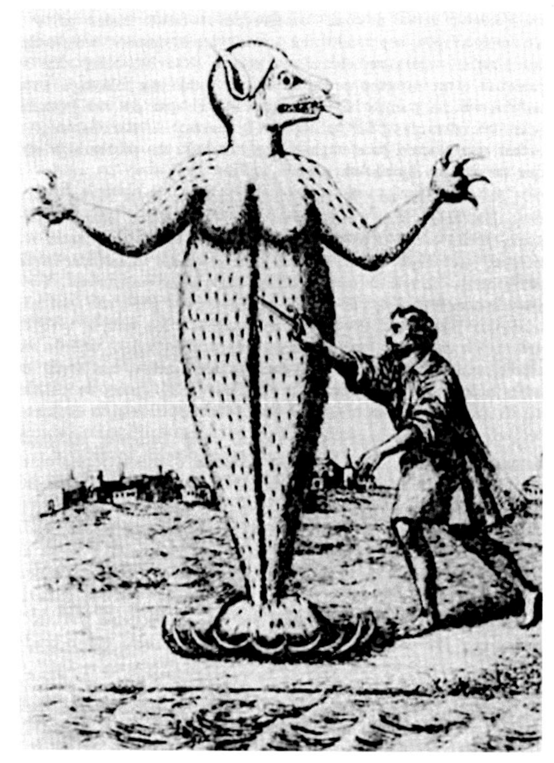

FIGURE 5.19 Sea monster killed by Baltasar Ferreira in the province of Vera Cruz in 1564. Gândavo—History of the Province of Vera Cruz.

> These sea men are called Igpupiara in the local language; the natives are in such fear of them that many die just upon seeing them, and none that see them escape; some die immediately, and when asked about the cause, they say they have seen this monster; they have the appearance of reasonably tall men, but with very deep-set eyes. The females have the appearance of women with long hair and are beautiful; these monsters are found on the sandbars at the mouths of freshwater rivers. Many of them have been found in Jagoarigpe, seven or eight leagues from Bahia; and in the year '82, an Indian who had gone fishing was chased by one, and taking refuge on his raft, he told his master about it; in order to encourage the Indian, his master wanted to go and see the monster, and while carelessly hanging a hand outside the canoe, he was seized, taken and never again seen; in the same year Francisco Lourenço Caeiro's other Indian died. Several have been seen in Porto Seguro, and they have killed several Indians. The way it kills is by embracing the person very strongly, kissing them and squeezing them so tightly that it breaks them without tearing them apart; when it senses they are dead, it emits several groans, as if from emotion, and flees, leaving them behind. If it takes any, it eats only the eyes, nose, fingertips and toes, and genitalia, and so they are found on the beach looking ordinary except for the missing parts. ***Fernão Cardim, pp. 151–152.***

> (...) it also happened to some negroes from Guinea; at times the ghosts or sea men killed five of my Indians; and once a monster took two Indian fishermen off a raft and one was taken away and the other escaped but was frightened to death, and some

of them do die of it. And a sugar master at my mill stated that he was looking out the window of the mill which is above the river and that some negro women who were cleaning sugar moulds screamed, and he saw a shadow bigger than a man on the bank of the river but it soon dove in; and the negro women told the sugar master that the ghost had come to get them and that it was the sea man, of which they were frightened for days; and that these happenings occur a lot in the summer; in the winter, no negroes are ever missing. ***Gabriel Soares de Sousa, p. 268.***

In the Captaincy of São Vicente, it being already late at night when everyone would surrender themselves to sleep, an Indian woman slave of the captain decided to leave the house; casting her eyes on the floodplain that lies close to the sea and close to the village of the captaincy, she saw a monster moving about from one side to the other, with unusual gait and movements and from time to time letting out such ugly screams that she was terrified and, beside herself, she went to the son of this same captain, whose name was Baltasar Ferreira, and told him what she had seen, which seemed to her some sort of diabolic vision. But as he was no less stern than he was industrious, and since these local people are not very credible, he did not pay much heed to her words but remained in bed and ordered her to go back out and make certain what it was. And, obeying his order, the Indian woman went out, and she came back even more frightened, repeatedly affirming that something so ugly was walking around that it could only be a demon. ***Pedro de Magalhães de Gândavo, pp. 117–118.***

According to a note by Ricardo Martins Valle (p. 119, the 2008 edition), the terrible demonic sea monster described in Gândavo's work was probably a sea lion, a carnivorous pinniped that occasionally came ashore following a cold current from the South Atlantic.

Even today, in the age of technology and information, we find reports very similar to the ones transcribed above, so that they would be almost indistinguishable from those of a millennium ago:

Looking at us was this thing (Sasquatch) that had the appearance of man, although it was three times the size of the average man (...). It turned to face us, staring into the headlights (...). It was covered with hair and there was a flat-profiled face (...). The most striking feature was the creature' s eyes. They were really sunk in. ***John Bindernagel, p. 58.***

(...) a Mapinguari, that huge monkey, hairy as a spider monkey, with feet of a donkey, turned backwards, carried under its arm it poor companion, dead, strangled, dripping blood. With claws like a jaguar, the monster began to rip pieces from the unfortunate fellow, stuffing them into its big maw, torn up to the stomach. ***Câmara Cascudo, p. 427, about the version of Mapinguari of the Rio Purus, Amazonas.***

The identification of these fabulous creatures in this New World is not strange. After all, far beyond the Pillars of Hercules, past the mystical Indika or the marvelous Aethiopia, far to the west, lay a magical land inhabited by fantastic beings, called centuries earlier in the Gaelic-Irish tradition by the name Hy-Breasil, well before being changed by European conquistadors to simply Brasis, the place of the red wood (Tolkien, 1964).

A LOOK AT PARADISE

Thinking in a fabulous way is without doubt a human trait. To say that this virtue or virtuosity is among the main characteristics that separate us from other animals is not an exaggeration. Obviously, other animals do not speak, at least not like us. However, if they were able to tell stories, what would they say? We suppose that they would embellish their own deeds and that they would distort, intentionally or not, what they saw and did in their own stories, as do we.

Every event that caused someone astonishment would undergo various distortions, to a greater or lesser extent, as it was narrated or described, such that the story's construction would be able to transmit a message capable of stirring in the listener equally startling sensations and perceptions; otherwise, they would not be fabulous in and of themselves, unless nature proved more fantastic than human fantasy, which is not entirely untrue.

Human perception and ingenious ways of seeing the world gave birth to imaginary zoology in all cultures, thereby giving rise to frightening creatures of the most varied and incredible sort. From Amerindian totemic animals, to the beasts of classical Western mythology and the mystical Orient, to the would-be scientific cryptozoology of the 21st century, we see fantastic but credible beings inhabiting no less fabulous landscapes, principally when we propose to understand them through an open and honest gaze, glimpsing them through the paradise of human fantasy.

From the times before Pliny until the present day, humanity has been fascinated by the unknown, and even with the entire technological revolution of the modern age, imaginary creatures, unlike real animals, run no risk of becoming an endangered species. Some distant unexplored place will always exist that is inhabited by savage people, monsters, and fierce beasts, all of them manifestations of the human psyche, the externalization of our own perplexity at what cannot be internalized or adequately explained by science or common sense.

References

Abrão, B.S., Coscodai, M.U., Sargadoy, W., 2000. Dicionário de Mitologia. Editora Nova Cultural, São Paulo.

Adam, C., 1881. The Holy Bible, Old and New Testament Including the Marginal Readings and Parallel Texts with a Commentary and Critical Notes. Ward, Lock & Co., Warwick House, London.

Aelianus, C., 1832. Aeliani de Natura Animalium Libri xvii. Verba ad fidem librorum MSS. F. Jacobs Jenae Imprensis Friderici Frommanni.

Aldrovandi, U., 1640. Serpentum, et draconum historiæ libri duo. Bononiæ, apud C. Ferronium, Bolonha.

Anônimo, séc. XIII, 2009. Sagas Islandesas: Saga dos Volsungos. Hedra, São Paulo.

Ashton, J., 1890. Curious Creatures in Zoology. John C. Nimmo, London.

Autores, V., et al., 1990. Bíblia Sagrada —Edição Pastoral. Paulus, São Paulo.

Baring-Gould, S., 2008. O Livro Dos Lobisomens. Editora Aleph, São Paulo.

Borges, J.L., 2007. El Libro De Los Seres Imaginarios. Destino, Barcelona.

Cardim, F., 1583–1601. Tratados da Terra e Gente do Brasil. Hedra, São Paulo.

Davies, E., 2015. Meet Ten Animals that Look Like Real-Life Unicorns. BBC Earth, BBC, London.

de Rosa, M.C.A., 2009. Dicionário de Símbolos: o alfabeto da linguagem interior. Editora Escala, São Paulo.

de Sousa, G.S., 1587. Tratado Descritivo do Brasil em 1587. Hedra, São Paulo.

de Vasconcellos, P.S., 2008. As Mil e Uma Noites: contos selecionados. Editora Sol-Objetivo, São Paulo.

Del Priore, M., 2000. Esquecidos por Deus: monstros no mundo europeu e ibero-americano (séculos XVI–XVIII). Companhia das Letras, São Paulo.

Douglas, N., 1928. Birds and Beasts of the Greek Anthology. Chapman and Hall Ltd., London.

Dworacki, S., 2009. Peculiarities of the non-Greek world in Heliodorus' Aethiopica. Symbolae Philologorum Posnaniensium Graecae et Latina XIX, 135–141.

Eberly, S.S., 1988. Fairies and the folklore of disability: changelings, hybrids and the solitary fairy. Folklore 99, 58–77.

El Hayek, S., 2006. O Sagrado Alcorão. LCC Publicações Eletrônicas, Foz do Iguaçu, Paraná.

Flaubert, G., 1904. The Temptation of St. Antony or a Revelation of Soul. Dodo Press, Great Britain.

Gândavo, P.D.M.D., 1576. História da Província de Santa Cruz. Hedra, São Paulo.

Gual, C.G., 2003. Diccionario de Mitos. Siglo Veintiuno de España Editores, s.a., Madrid.

Hamblin, D.J., 1987. Has the Garden of Eden Been Located at Last?, Smithsonian Magazine. Smithsonian Institute, Washington, DC, p. 7.

Heródoto, 2006. História, vols. I e II. Ed. eBooksBrasil.

Hurwitz, S., 2009. Lilith – The First Eve: Historical and Psychological Aspects of the Dark Feminine, third ed. Daimon Verlag, Einsiedeln, Switzerland.

Hyde, J.K., 1982. Real and imaginary journeys in the later middle ages. Bulletin of the John Rylands Library 65, 125–147.

Jackson, W., 2004. The use of unicorn horn in medicine. The Pharmaceutical Journal 273, 925–927.

Jonston, J., 1808. Theathrum Universale Omnium Animalium Piscim, Avium, Qoadrupèdum, Exanguium, Aquaticorum, Insectorum, et Angium. Proflat apud R. & G. Wetstenios, Amstelodami.

Llewellyn-Jones, L., Robson, J., 2010. Ctesias' History of Persia – Tales of the Orient. Routledge Taylor and Francis Group, London and New York.

Loyn, H.R., 1997. Dicionário da Idade Média. Jorge Zahar Ed., Rio de Janeiro.

Lucan, F.v.O., 1820. Pharsalia. Cumtibus Rodwell et Martin [etc.] excudit T. Davison.

Mandeville, J., 1900. The Travels of Sir John Mandeville: The Version of the Cotton Manuscript in Modern Spelling, with Three Narratives, in Illustration of It, from Hakluyts "Navigations, Voyages & Discoveries". Macmillan and Co. Limited/The Macmillan Company, New York.

Margoulès, G., 1948. Anthologie Raisonnée de la Litterature Chinoise. Payot, Paris.
Mayor, A., 2000. The First Fossil Hunters: Paleontology in Greek and Roman Times, first ed. Princeton University Press, Princeton.
McCrindle, J.W., 1877. Ancient India as Described by Megasthenes and Arrian. Trübner& CK, London.
Migne, J.-P., 1859. Patrologiæ Græca. Bibliotecæ Cleri Universæ, Paris.
Muhawi, I., Kanaana, S., 1989. Speak, Bird, Speak Again: Palestinian Arab Folk Tales. University of California Press, Berkeley; Los Angeles.
O'Connell, M., Airey, R., 2009. The Complete Encyclopedia of Signs & Symbols. Anness Publishing Ltd., United Kingdom.
Pliny, T.E., 1855. The Natural History of Pliny Bostock. John Riley, H T, London.
Polo, M., 1985. O Livro das Maravilhas, first ed. L&PM Editores, Porto Alegre.
Rossi, L., 2016. A review of cryptozoology: towards a scientific approach to the study of "hidden animals". In: Angelici, F.M, (Eds.), Problematic Wildlife. Springer International Publishing, Switzerland, pp. 573–588.
Safo, 2011. Hino a Afrodite e Outros Poemas. Hedra, São Paulo.
Salles, C.A.C., 1998. Somos Feitos da Matéria dos Sonhos: Uma nova visão da masculinidade. Editora Rosa dos Tempos, Rio de Janeiro.
Sánchez, G.B., 1978. Ontogenia y Filogenia del Basilisco. El Basilisco N°1 64–79.
Sayce, A.H., 1888. Records of the Past of the Ancient Monuments of Egypt and Western Asia. Samuel Bagster and Sons, Limited, London.
Snowden Jr., F.M., 1948. The Negro in ancient Greek. American anthropologist. New Series 50, 31–44.
Sófocles, 1990. A Trilogia Tebana: Édipo Rei; Édipo em Colono; Antígona. Zahar, Rio de Janeiro.
Solinus, C.J., 1895. Collectanea Rerum Memorabilium. Weidmann, Berolinum.
Tolkien, J.R.R., 1964. Tree and Leaf. George Allen & Unwin, United Kingdom.
Tolkien, J.R.R., 2009. The Legend of Sigurd and Gudrún. HarperCollins, New York.
Topsell, E., 1658. The History of Four-Footed Beasts and Serpents. E. Cotes, London.
Vernant, J.-P., 2008. O Universo os Deuses e os Homens. Companhia das Letras/Editora Scwarcz Ltda., São Paulo.
Waugh, D.C., 2010. The silk roads in history. Expedition 52, 9–22.
Wikicristiano, 2016. Diccionario Biblico Cristiano Gratis—Concordancia Biblica Online. Wikicristiano.

Further Reading

Bindernagel, J.A., 2004. The sasquatch: an unwelcome and premature zoological discovery? Journal of Scientific Exploration 18, 53–64.
Cascudo, L.D.C., 2012. Dicionário do Folcore Brasileiro, twelfth ed. Global, São Paulo.
Gândavo, P.D.M.D., 2004. A Primeira História do Brasil: história da província Santa Cruz a que vulgarmente chamamos Brasil, second ed. Jorge Zahar Editor, Rio de Janeiro.

CHAPTER

6

Ethnotaxomy as a Methodological Tool for Studies of the Ichthyofauna and Its Conservation Implications: A Review

José da Silva Mourão[1], *Márcio Luiz Vargas Barbosa Filho*[2]

[1]Universidade Estadual da Paraíba, Campina Grande, Brazil; [2]Universidade Federal Rural de Pernambuco, Recife, Brazil

INTRODUCTION

Archeological studies have revealed that the first hominids (*Homo habilis* and *Homo erectus*) fished in freshwater lakes (Stewart, 1994). Their activities probably furnished information that allowed them to group the fish and create categories—which could be interpreted as prototaxonomy. This categorizing appears to represent an innate human characteristic, with our brains having evolved to create groups and categories based on our sensorial perceptions of the world around us (Mishler and Donoghue, 1982). Based on this presumption, Raven et al. (1971) noted that discontinuities in nature can be perceived by the human mind based on similarities and differences between objects and events in the physical environment. Sharing these thoughts, Frazão (2001) argues that classification, as a form of understanding and ordering external material elements of the world, exposes the notion of the existence of a universal mental structure linking various cognitive domains, which in this specific case refers to the existence of universal logical constructs that generate classifying behavior.

On the other hand, the need to classify and order nature does not appear to be an innate characteristic of the human mind but, rather to the contrary, nature itself shows clear differences between life-forms that allow such a comprehension and interpretation. This idea is shared by other authors such as Berlin et al. (1973, 1974) and Berlin (1992) who defend the idea that the discontinuities found in nature appear so evident to humans that it is difficult to imagine that they are not likewise perceived by different cultures. This gives rise to the necessity of humans to impose a certain order on the universe, as only through some form of order can they understand, refer to, and adapt themselves to the world around them (Lévi-Strauss, 1997). Berlin (1974) considered local classification

Ethnozoology
http://dx.doi.org/10.1016/B978-0-12-809913-1.00006-5

systems as prototaxonomic systems, developed principally from simple morphological distinctions between species. This implies that local classification systems should be understood as products of the processes linked to human survival. From this perspective, taxonomies can be viewed as reflections of human thought, with each taxonomic system presenting theories concerning the creatures being classified (Gould, 2001).

Daly (1998) emphasized that the names given to animals or plants indicate a concept, category, or taxon, which represents a natural history archive full of information that can reveal (or sometimes obscure) processes of perception, identification, and naming. Lévi-Strauss (1997) pointed out that naming is a process that gives contextual significance to the continuity and discontinuity of objective characteristics. From that perspective, patterns of linguistic expression demonstrate ethnosemantic regularities that incorporate holistic and mutual links between humans, plants, animals, landscapes, and the supernatural (Greene, 2007). The naming of living organisms in folk taxonomies is a process that confers contextual significance to the objective continuities and discontinuities in nature, which are essentially similar in all languages and can be described based on a small number of nomenclatural principals (Berlin et al., 1973).

To Atran (1990), the creation of a vocabulary (lexicon) by a given local population would be the first step toward gathering information about the diverse cognitive domains that compose the human mind, and also a way of indirectly approaching the formation and diffusion of concepts related to the taxonomy folk. Authors such as Lévi-Strauss (1997), Hunn (1982), and Hays (1983) stated that the processes of categorizing are culturally influenced (cognitive categories) and organized in logical patterns (taxonomic structures) distinct to each society. Conklin (1962) noted that the botanical nomenclature used by the Hanunóo (Philippines) presented a hierarchical taxonomy with various interrelated and inclusive levels. Frazão (2001), in turn, noted that studies of lexicons, as performed by Conklin, opened the way for research concerning the existence of universal biological taxonomic structures.

From a conceptual point of view, folk taxonomy is a metaphorical hierarchy involving notions linked to categories, levels, and contrasts and represents a way of understanding natural systems that are encountered in all cultures and serve as guides to interpret the natural world, involving mental processes similar to those used in modern science (Hunn, 1975; Maffi, 1999; Medin and Atran, 2004). Atran (1998) investigated the nature of human knowledge and the relationships between human understanding of the environment and adaptation to it. The totality of this knowledge results in a true warehouse of information concerning biology, ecology, and the etiology of diverse groups of plants and animals. Additionally, folk taxonomies reflect not only the manners in which humans observe environmental components but also how they perceive natural systems as a whole, reflecting the availability of animals and plants in the surrounding environment.

Berlin (1973) was one of the pioneers of studies of folk taxonomies and viewed that field as being concerned with the elucidation of the underlying principles of prescientific human classification, and the naming and identification of living creatures. Based on those studies, Berlin et al. (1973) established three principal areas of ethnobiological systematics: classification, which refers to the set of principles by which classes of organisms are naturally organized in the prescientific mind; nomenclature, which refers to the descriptions of linguistic principles used for naming the organized classes of living organisms in a given idiom; and identification, which refers to the physical characteristics of the distinctive traits (physical, or not) that denote an "ethnospecies" in a given group.

The proposal of Berlin (1992) for ethnobiological classifications is based on the hypothesis of universality among different cultures, expecting that "there are regularities in the classification and naming of plants and animals among members of traditional nonliterate societies, regularities that persist beyond the local environment, culture, society, and language." The basis for this argumentation is that the patterns observed can be explained by wide (but unconscious) similarities between humans that perceive and appreciate natural affinities between groups of plants and animals: people recognize and name groups of organisms independent of their actual or potential utility or symbolic significance. The Berlinian model of ethnobiological classification contains categorizing and nomenclatural principles that allow the recognition of the linguistic, semantic, and taxonomic criteria used in naming. One particular advantage of the Berlinian model is its usefulness for understanding the internal structures of the hierarchical categories, which avoids ethnobiological classification becoming a simple cataloging of "ethnospecies" associated with their scientific equivalents—making it possible to interpret them within a logical structural arrangement.

This chapter aimed to carry out an ethnotaxonomic review, based on the Berlinian classification model.

HISTORICAL ASPECTS OF ETHNOTAXONOMY

Biodiversity is threatened globally, and there is no current perspective at all that a complete survey of life-forms can be completed before great numbers of them disappear (Wilson, 1997). This same author understands that taxonomists and systematists are becoming increasingly rare, and the problems of tropical conservation are exacerbated by the lack of taxonomic knowledge and the reduced numbers of researchers in that area—making the gathering of local information concerning the fauna and flora increasingly important—not only to better understand biodiversity but also to gain information about local environments that scientists often ignore (Leff, 2001).

Ethnoscience represents, according to Atran (1990), the search for the cognitive fundamentals of human knowledge representing all of the cultural and linguistic variations found throughout the world. That author views the study of the vocabularies (lexicons) adopted by local populations as the first step toward accessing information concerning the diverse cognitive domains that compose the human mind, as well as a form of indirect approximation to the formation and diffusion of concepts related to the world around us. One of the objectives of ethnobiology is to understand how living creatures are perceived, recognized, and classified by diverse human cultures. Ethnobiological classifications contain important descriptions of living organisms, contributing important information about local and/or regional diversity that can often allow various types of analyses of nomenclatural designations—whether hierarchical or inserted within different cultural contexts, such as taboos, ceremonial languages, and myths.

The first attempts at ethnobiological classifications were simply composed of lists of the names of plants and animals, with their descriptions and the manners in which they were utilized. Zeisberger (1887) recorded this type of information in your dictionary. Later, these inventories helped stimulate the first works exclusively considering ethnobiological classifications (Dennler, 1939; Wyman and Harris, 1941). A review of the historical foundations of ethnobiology was published by Clément (1998), who identified three major periods. The first period is considered the "preclassical" phase, which extended until 1950 and was marked by the appearance of numerous branches of ethnobiology, such as ethnobotany and ethnozoology. The second period is considered the "classical" phase extended until the 1980s, with the attentions of ethnobiologists being directed toward studies based on the unique local perceptions of folk classifications of nature. The third, or "postclassical" period, has been characterized as the emergence of joint efforts between western researchers and indigenous populations.

The recognition of the value of ethnobiological knowledge, once it has been contextualized and put up for discussion in universities, should allow the introduction of a new form of integrative thought in terms of both research and teaching, with the inclusion of cultural diversity being considered an aid to the maintenance of biodiversity and nature conservation. Diegues (2000) noted that this conservationist vision should integrate not only the knowledge of natural scientists but also local specialists who have accumulated knowledge for generations concerning ecosystems and their variations—in light of the overwhelming necessity of their contributions to the planning and execution of conservation efforts. Similar arguments were put forward by Gragson and Blount (1999), who stressed the importance of comparisons between ethnoecological knowledge and scientific knowledge to gain a better overall comprehension of natural ecosystems and the interrelationship between human and nature, to be able to establish policies for the control, development, and/or conservation of natural resources.

Studies of ethnobiological classifications turned a corner in the 1950s with the publication of the classical work in ethnobotany with the Hanunóo in the Philippines, by Harold Conklin (1954). The 1960s and 1970s were of great importance to the development of the science of ethnobiology, energized by the works of Lévi-Strauss (1997); Conklin (1962); Berlin and Kay (1969); Berlin (1972, 1973, 1976); and Berlin et al. (1973, 1974). In the last decades, large numbers of researchers have dedicated themselves to studies of folk systematics, establishing that humans in varied diverse regions of the world utilize similar strategies to classify living organisms and organize biological concepts (Bulmer, 1974; Berlin et al., 1973; Hunn, 1977, 1982; Hays, 1983; Brown, 1984; Atran, 1990, 1998; Marques, 1991; Berlin, 1992; Ellen, 1993; Clément, 1995; Paz and Begossi, 1996; Holman, 2005).

Ethnotaxonomic studies not only have recorded how plants, animals, and fungi are classified but have also related those processes to indigenous forms of resource management and conservation (Nazarea, 1999). Numerous experiences have likewise demonstrated the viability of training local populations as parataxonomists who can collaborate directly with efforts to evaluate and document regional diversity (Basset et al., 2000; Sheil and Lawrence, 2004), principally in tropical countries where research demands far exceed the numbers of professionals in that area of research (Agnarsson and Kuntner, 2007).

Studies of ethnobiological classifications follow certain lines of thought. Durkheim and Mauss (1905), for example, in their studies of rudimentary classifications developed by humans, observed that those systems reflect groupings of natural objects that were related to social groupings. As such, those authors concluded that those classifications represented contributions of collective representations and were not solely the product of individual activities. There is a line of thought that humans, in all parts of the world, recognize the structure and order inherent in the biological world, independent of any practical value that those living organisms may offer (Lévi-Strauss, 1997; Berlin, 1992). Another line of thought, considered utilitarian, affirms that the principal objective of ethnobiological classification has been to help human populations adjust to their respective habitats, giving distinct names only those organisms that have practical consequences to human survival (Hunn, 1982; Ellen, 1993). Nazarea (1999) argued that this debate is essentially fruitless, because humans can quite easily operate at both levels, suggesting that the discussion should be redirected to focus on the connections between plant classification, for example, and the conservation of plant genetic resources, or even to the cultural concepts of landscapes and the management of shared resources. Nazarea (1999) concluded by pointing out that ethnoecology has a wide potential for contributing to this debate, at both theoretical and practical levels.

In Brazil, according to Carrara (1997), pioneering works in ethnobiological classification were undertaken with various indigenous Amerindian groups, including the Tupi-Guaranis, the Kaingang of Paraná State, and the Canela do Maranhão as published by Ihering (1904), Baldus (1947), and Vanzolini (1956–1958), respectively. At the end of the 1960s, Hartmann (1967) studied the botanical nomenclature of the Bororó Amerindians in Mato Grosso State, describing and linguistically analyzing their names for plants and focusing on the principles that oriented their classification and categorization of the plant species in the *cerrado* (neotropical savanna) environment. This study decisively influenced subsequent ethnobiological work in Brazil.

Jensen (1985) studied indigenous bird classification based on a description and analysis of the Wayãpi system, focusing on the fundamental principles that structured their nomenclature and taxonomy of the regional avifauna. The studies of Posey (1987) concerning indigenous taxonomies indicated that their patterns of classification and nomenclature "helped reveal the underlying principles of indigenous logic." In a study of the bird classification and cosmology of the Kayapó-Xikrin tribe, Gianinni (1991) described how those Amerindians named and classified birds and other animals as well as the principles that oriented those names and classifications.

The coastal populations also use a wide variety of common names for the fish species they catch (Freire and Pauly, 2005)—which represents another complication for the adequate monitoring of artisanal fishing and, consequently, for managing national fisheries (Freire and Pauly, 2003, 2005). Within this context, numerous workers have focused on the classification systems used by Brazilian fishermen since the 1990s (Marques, 1991; Begossi and Figueiredo, 1995; Paz and Begossi, 1996; Costa-Neto and Marques, 2000; Mourão and Nordi, 2002a, b; Souza and Begossi, 2007; Begossi et al., 2008; Beaudreau et al., 2011; Pinto et al., 2013, 2016; Previero et al., 2013) in an attempt to integrate that knowledge into workable management schemes.

THE HIERARCHIZATION OF ETHNOTAXONOMY

Studies of ethnotaxonomy were kick-started by the pioneering work of Conklin (1954), which resulted in a list of 1625 types of plants recognized by the Hanunóo people in the Philippines. That author classified plants as those called exclusively by a basic name (type 1) and those known by a basic name with one or more added attributes (type 2)—suggesting the existence of a type of hierarchy in that folk taxonomy. Later, this same author presented, for the first time, the idea that hierarchical arrangements in folk taxonomies are probably universal. A decade later, Berlin et al. (1973) published the general principles of folk biological classification and nomenclature, describing important aspects of how humans organize and classify the natural world around them. After more than 20 years revising those classification principles (which were developed exclusively based on ethnobotanical studies), Berlin (1992) began incorporating ethnozoological knowledge within the general principles of ethnobiological categorizing and nomenclature.

Among the categorizing principles, especially those relative to hierarchical structures, is that conceptual recognition is attributed to a subset of the existing fauna and flora. This subset comprises species that are biologically more distinct (salience) in the local habitat and are organized conceptually in a shallow hierarchical structure, comparable to Linnean taxonomy, forming decreasing inclusive taxonomic classes. According to Berlin (1992), this hierarchy structural is formed by the following levels or ranks:

1. **Kingdom:** The rank of kingdom is unique in that it includes but a single member. Taxonomically, the kingdom incorporates all taxa of lesser rank. For ethnobotanical classification, the kingdom corresponds approximately to the biological taxon Plantae; in ethnozoology the corresponding biological taxon is Animalia.
2. **Life-form:** Taxa of the life-form rank mark a small number of highly distinctive morphotypes based on the recognition of the strong correlation of gross morphological structure and ecological adaptation. Life-form taxa are broadly polytypic and incorporate the majority of taxa of lesser rank.
3. **Intermediate:** Taxa of intermediate rank are found most commonly as members of life-form taxa and are comprised of small member of folk generics that show marked perceptual similarities with one another. Data are inadequate to indicate the relative numbers of such taxa in actual systems of ethnobiological classification.
4. **Generic:** The most numerous taxa in folk biological taxonomies will be taxa of generic rank. In both ethnobotanical and ethnozoological systems of classification, the number of folk generics reaches an upper limit at about 500–600 taxa in systems typical of tropical horticulturalists. Roughly 80% of folk generic taxa folk systems are monotypic and include no taxa of lesser rank. While most folk generics are taxonomically included in taxa of life-form rank, a small number is conceptually unaffiliated due to morphological uniqueness or, in some cases, economic importance. Generic taxa are among the first taxa learned by children because they acquire their society's system of biological classification.
5. **Specific:** Taxa of the rank of folk species partition folk generic taxa into two or more members. In those systems where they occur, folk varietals further subdivide folk species. Subgeneric taxa are less numerous than folk generics in all systems examined to date. There is some evidence to suggest that the recognition of subgeneric taxa is loosely associated with a society's form of subsistence. The conceptual recognition of subgeneric taxa appears to be motivated in part by cultural considerations, in that a major proportion refer to domesticated species of plants and animals. There is some evidence that foraging societies have poorly developed or lack entirely taxa of specific rank. No foraging society will exhibit taxa of varietal rank.

In terms of the principles to which the **nomenclature** system refers, Berlin et al. (1973) and Berlin (1992) proposed that the taxa of the **life-form** rank could also not be named, but when those taxa are named they frequently demonstrated polysemic relationships with the taxa of subordinated ranks. The plant and animal names demonstrate a lexical structure of one of two universal lexicon types that can be called primary and secondary name. These types can be recognized by linguistic, semantic, and taxonomic criteria. Primary names are of three subtypes: simple (fish), productive (catfish), and unproductive (horse-eye jack), with generally specifiable exceptions, occur only in contrast sets whose members share a constituent that refers to the taxon that immediately includes them (e.g., maple). The taxa **life-form** and **generic** are designated by primary names; the subgeneric taxa are generally designated by secondary names.

There are two well-defined conditions under which a subgeneric taxon can be assigned a primary name, although these two conditions do not apply to all empirically observed data. The first condition occurs when the name of the prototypical subgeneric is polysemous with its superordinate generic, in which the ambiguity of the polysemy is resolved by the optional occurrence of a modification interpreted as a "genuine" or an "ideal" type. The second condition occurs when nonprototypical subgenerics refer to subgeneric taxa of great cultural importance.

APPLICATION OF THE BERLINIAN MODEL

The specialized literature indicates that fishermen employ various criteria when classifying, identifying, and naming fish, such as morphology, ecological niches, and their behavior (social, trophic, reproductive) (Maranhão, 1975; Begossi and Garavello, 1990; Marques, 1991; Paz and Begossi, 1996; Costa Neto and Marques, 2000; Mourão and Nordi, 2002a,b; Souza and Begossi, 2007; Ramires et al., 2007; Begossi et al., 2008; Beaudreau et al., 2011; Pinto et al., 2013, 2016; Previero et al., 2013).

Morphological characteristics can include color, body forms, or characteristic body elements; fish size or the size of a certain part of their body; and types of scales, fins, and eyes (Begossi and Silvano, 2008; Costa Neto, 2001; Costa-Neto and Marques, 2000; Marques, 1991; Paz and Begossi, 1996; Beaudreau et al., 2011; Pinto et al., 2013), and Table 6.1 cites some examples of the morphological characteristics used by the fisherman (Mourão and Nordi, 2002a,b). In terms of the ecological criteria employed, these characteristics principally refer to the types of habitat or substrate where the fish are encountered, for example, longnose stingray (Dasyatidae), southern eagle ray (Myliobatidae), guri sea catfish (Ariidae) (Mourão and Nordi, 2002a, b). Other authors likewise have cited examples of the use of morphological characteristics (Morril, 1967; Diamond, 1989; Marques, 1991; Paz and Begossi, 1996; Costa-Neto and Marques, 2000; Costa Neto, 2001; Begossi and Silvano, 2008; Pinto et al., 2013; Pinto et al., 2016). Some fish are named by the fishermen in relation to their behaviors, which present useful information for systemization, for example, the live sharksucker (*Echeneis naucrates*), which carries that name because it "lives holding onto another species of fish" (the external bodies of fish such as shark, rays, barracudas, and some species of Carangidae, Serranidae, and Lutjanidae) (Menezes and Figueiredo, 1980). These facultative biotic interactions are usually considered forms of commensalism and are related to the feeding habits of the live sharksucker (*E. naucrates*), which scavenges the remains of food eaten by the above cited fish. Another example is the *Brazilian* electric ray or trembling fish (*Narcine brasiliensis*), which carries that name because of its common behavior of trembling its body to produce electrical discharges between 14 and 37V. Contact with the skin can produce a severe electric shock. In addition to the main electric organ, this species possesses a bilateral accessory electric organ speculated to have a possible role in social communication.

Fish names can also reflect analogies in relation to domestic animals or objects: rock hind (*Epinephelus adscensionis*), lookdown (*Selene vomer*), molly miller (*Scartella cristata*), and ballyhoo halfbeak (*Hemiramphus brasiliensis*), being some examples (Mourão and Nordi, 2002a).

TABLE 6.1 Morphological Characteristics Used by Fishermen at the Mamanguape River Estuary to Identify and Name Fish

Morphological Characteristics	Folk Species
Color	Acoupa weakfish
Body shape	Common snook
Mouth size	Zabaleta anchovy
Types of scales	River Plate sprat
Types of fins	Gafftopsail sea catfish
Head size	Drums or croakers

Source: Mourão, J.S., Nordi, N., 2002a. Comparação entre as taxonomias folk e científica para Peixes do Estuário do Rio Mamanguape, Paraíba-Brasil. Interciencia, 27, 607–612.

Other examples were cited by Begossi and Garavello (1990), who noted that fishermen in the midcourse of the Tocantins River also made allusions to characteristics common to fruits and other animal species in naming fish, as did Pinto et al. (2016) in their studies of fishermen at Tamandaré (Pernambuco) and Batoque (Ceara), Brazil. Marques (1991) studied fishermen in the estuary–lake complex at Mundaú-Manguaba and concluded that the process of identification used there included an information set of general morphological characters that considered the shapes of the fish or their body parts and/or specific morphological characteristics made through analogies with other animals; the same authors also noted the use of ecological characteristics, with emphasis on aspects of the spatial distributions (habitats) of the fish. The ethnozoological classification of the Montagnais people (Clément, 1995) also use morphological (four-legged, feathered, small, and large), ecological (land and water), miscellaneous (wild, domesticated, harmful), and behavioral criteria in naming. They also classify animals as edible (red meat and white meat) and inedible.

Studies undertaken with the Wayampi Amazonian Amerindians (Jensen, 1985) showed that they retain a vast knowledge of their living world, naming at least 3500 biological species there. These denominations considered details of their morphology, relationships, and behavior. Jansen (1985) studying the avifauna classification used by that indigenous group in the Brazilian Amazon likewise observed a preponderance of those same criteria used in naming, with additional concepts that indicated relationships with the spiritual world.

According to Berlin (1992), the ethnobiological lexicon related to naming plants and animals can be considered in terms of two types of names: primary and secondary. Primary names are generally simple denominations, only sometimes compound. Generally, the life-form and generic ranks receive simple primary names, such as snook, sardines, and ray, whereas dog snapper, horse-eye jack, hairy blenny, and yellow jack represent compound or complex primary names. Productive and unproductive lexemes can be found among the latter group. In the case of productive lexemes, one of the constituents of the compound name indicates a taxonomic category superordinate to that of the form in question. In the examples above (plain soapfish and full moonfish), one of the constituents of the name ("fish") indicates a superordinate category, being, therefore, a complex productive primary lexeme (Mourão and Nordi, 2002a). Among the examples above (horse-eye jack), the constituents of the complex primary name do not indicate a superior category and are clearly unproductive (Mourão and Nordi, 2002a,b).

Tables 6.2 and 6.3 present various examples of primary and secondary names used by the fishermen at Mamanguape River Estuary (MRE) to denominate fish. Secondary names are utilized to identify different fish that belong to the same generic group. Secondary names can also occur as contrasting sets, sharing the same superordinate constituents (Mourão and Nordi, 2002a,b). Hays (1976, 1979) defined two taxa as being members of the same contrasting set if they were immediately included in the same superordinate taxon. White sardine and blue sardine, for example, are specific denominations, and both are included in the superordinate generic taxon sardine (sardine). The secondary names will be necessarily productive, because one of their constituents will always indicate a superior taxonomic category (Mourão and Nordi, 2002a,b).

According to Mourão and Nordi (2002a,b), in addition to recognizing the fish category, the fishermen at the MRE also recognize other categories such as shellfish (Veneridae), oyster (Ostreidae), and crabs (Ocypodidae). All of this biological diversity is included in a larger category, not verbally defined, which corresponds to the animal kingdom within the Berlinian classification. Frequently, other aquatic vertebrates (manatees, dolphins, and whales) and some of

TABLE 6.2 Examples of Primary Lexemes of Fish at the Mamanguape River Estuary

Simple Primary Names	Primary Complex Productive Names	Primary Complex Unproductive Names
Cobia	Rock hind fish	Dog snapper
Barbu	Full moonfish	Horse-eye jack
Mojarra	Flying gurnard fish	Hairy blenny
Snapper		Yellow jack
Acoupa		
Angelfish		
Tarpon		
Lookdown		

Source: Mourão, J.S., Nordi, N., 2002a. Comparação entre as taxonomias folk e científica para Peixes do Estuário do Rio Mamanguape, Paraíba-Brasil. Interciencia, 27, 607–612.

TABLE 6.3 Examples of Secondary Lexemes of Fish at the Mamanguape River Estuary

Superordinate Category	Secondary Name
Ray (simple primary name)	Longnose stingray
	Smooth butterfly ray
	Southern eagle ray
Catfish (simple primary name)	Gafftopsail sea catfish
	Guri sea catfish
	Madamango sea catfish
	Pemecou sea catfish

Source: Mourão, J.S., Nordi, N., 2002a. Comparação entre as taxonomias folk e científica para Peixes do Estuário do Rio Mamanguape, Paraíba-Brasil. Interciencia, 27, 607–612.

the invertebrates mentioned above (shrimp, and sometimes crabs) are classified as fish. The classification of some aquatic mammals and invertebrates in the fish category is due to the fact that the fishermen grouped these organisms not only due to possible physical resemblances but also because they share the same habitat. Different from the observations reported by Marques (1991) and Paz and Begossi (1996), the fish called ocellated moray (*Gymnothorax ocellatus*) and moray eel (Gymnothorax) are not included in the categories of insect and snake. The fish called toadfish (*Thalassophyrne punctata*) is often referred to as a "pest" due to its sharp spine (Mourão and Nordi, 2002a).

Marques (1991) also considered the "fish" category to be elastic, with the fishermen from Alagoas state including in it groups such as dolphins, whales, and caimans. According to that author, some invertebrates are circumstantially considered fish, as in the cases of mollusks and crustaceans, which are cyclically consumed according to a religious calendar ("fish that are consumed during Holy Week"). Berlin and Berlin (1983), in their study of the Aguaruna and Huambisa people (Peruvian Jívaros), noted that dolphins are not considered edible and are not included in the "fish" category. Clément (1995) studied the folk zoology of the Montagnais and found that the "fish" category was quite elastic and included shrimp, lobsters, crabs, and mollusks. Costa-Neto and Marques (2000) reported that sea turtles, dolphins, whales, and anacondas were considered to be fish according to the fishermen along the northern coast of Bahia State, Brazil. Paz and Begossi (1996) noted that the fishermen of Gamboa and Baía de Septiba considered the "fish" category to include a variety of aquatic organisms including turtles—but excluded the moray eel, which was not considered a fish by most fishermen as it "bites just like a snake."

According to Mourão and Nordi (2002a,b), the fishermen in the study grouped the fish according to their similarities or differences, using a hierarchical system compatible with that described above and forming, in descending order of taxonomic pertinence, the following ranks: kingdom, life-form, generic, and specific (Fig. 6.1); an intermediate rank was only observed in one category. Jansen (1985) reported a classification with four hierarchical ranks among the Wayãpi Amerindians, which he designated as ethnoclass, ethnofamily, ethnogenus, and ethnospecies. Anderson (1967) studied the ethnoichthyology of the Cantonese people of Hong Kong and encountered a hierarchical classification composed of only three levels, corresponding to life-form, generic, and specific. The hierarchization in that study could be clearly seen in expressions such as the "sardines (Clupeidae) come in numerous types"; "Barbu (*Polydactilus virginicus*) come in various qualities"; "Florida pompano (*Trachinotus carolinus*) come in two types"; and "Eyed flounder (*Bothus ocellatus*) has three appearances." Marques (1991) observed that a similar hierarchization in the classification used by fishermen in Alagoas State, who employed expressions such as "there are lots of kinds" and "there's a lot of diversity." Costa-Neto and Marques (2000) noted that Siribinha fishermen created subcategories of fish in their use of expressions such as "it's the same as"; "it's in the same family"; and "it's the same thing."

Not all of the names attributed to fish by the fishermen at MRE characterized hierarchization (Mourão and Nordi, 2002a). The various generic as well as family names (catfish family, acoupa family, sardines family, ray family, common snook family, puffers family) do not clearly indicate a subcategorization, being more closely related to the cultural and economic importance of those generic categories. The term family was also used to group fish into larger sets based on

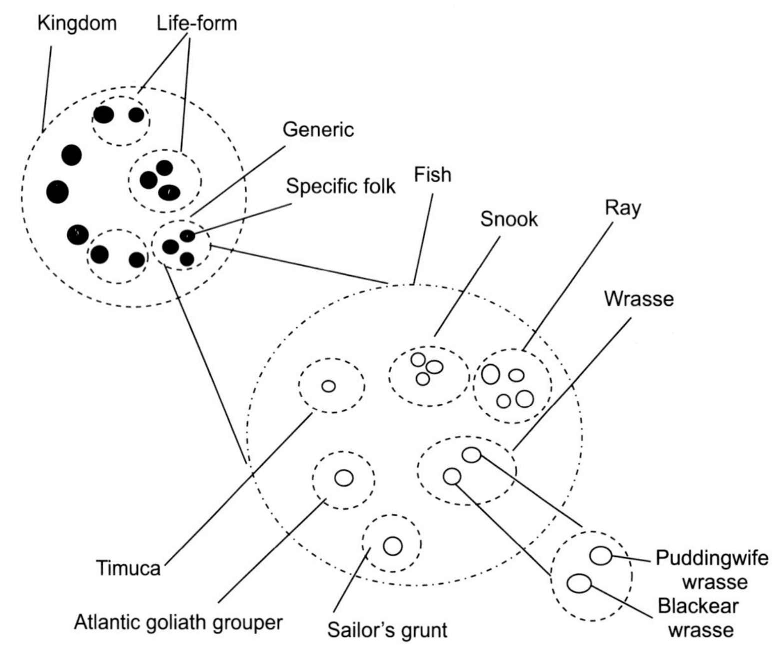

FIGURE 6.1 Schematic representation of the four ethnobiological classification categories and their respective taxa in comparison with the hierarchical classification of the MER [BERLIN model (1992)]. *Source: Pinto, M.F., Mourão, J.S., Alves, R.R., 2016. How do artisanal fishermen name Fish? An ethnotaxonomic study in northeastern Brazil. Journal of Ethnobiology 36, 348–381.*

habitat usage, for example, "rockfish family" and "mudfish family." Different from the fish cited earlier, which are grouped principally in relation to morphological and behavioral similarities, the latter fish are grouped in relation to their similarities in terms of habitat use (ecological criteria). The Wayampi Amerindians studied by Jansen (1985) distinguished two large ethnofamilies of birds among the 27 classified based on ecological criteria (trophic relationships)—uniting those that fed on fruits of the wasei palm (*Euterpe oleracea*) as distinct from those that did not. Paz and Begossi (1996) reported that the fishermen at Gamboa distinguished various families of fish based on morphological, behavioral, and ecological criteria, the quality of their meat, and their economic value; examples include the families of *sharks*, *rays*, and others. Berlin (1992) observed that the terms "related to" or "comrade" are used by the Tzetal Maya of Mexico to identify very similar species. In an analogous situation, according to that same author, species showing similarities were called "brothers" or "members" of the same family by the Aguaruna and Jívaro Amerindians of Peru. The fishermen studied there called fish belonging to the same family "relatives."

Denominations such as caícos fish and sarandaio fish refer only to small fish of lesser economic value. White mullet/sauna (*Mugil curema*), lebranche mullet (*Mugil liza*), horse-eye jack (*Caranx latus*), tarpon (*Megalops atlanticus*), and goliath grouper (*Epinephelus itajara*) correspond to immature fish of the generic folk white mullet (*M. curema*), lebranche mullet (*M. liza*), horse-eye jack (*C. latus*), tarpon (*M. atlanticus*), and goliath grouper (*E. itajara*) (Mourão and Nordi, 2002a,b). Maranhão (1975) reported use of the denomination "curimaí" for immature fish of the generic folk "lebranche mullet." Marques (1991) and Paz and Begossi (1996) likewise observed strictly semantic names in relation to fish sizes, with a continuous distribution. Taxonomic and nontaxonomic relationships were reported by Forth (1995) in a study of the ethnozoological classification system of the Nage in western Indonesia. In terms of the Berlinian model, purely semantic names, or those with purely cultural or economic connotation, should not be used to include groupings in hierarchical levels.

Members of the generic ethnobiological level usually are included in the **life-form** level, referring to small natural discontinuities that are easily recognized as distinct morphological characteristics. The taxa in the folk generic category "are the most numerous and are considered the core of any ethnobiological classification system" (Berlin, 1992) and frequently correspond to scientific species in the Linnaean system, composing the folk hierarchy most easily recognized and learned by children in traditional societies (Stross, 1973); the generic grouping was the most notable among the hierarchical categories recognized in the community studied.

Berlin (1992) reported that most of the generic taxa in folk taxonomy are monotypic, without any taxon from inferior levels. Polytypic generic taxa that are subdivided into specifics invariably refer to classes of organisms that are culturally important. Additionally, the recognition of generic polytypic levels reflects the biological diversity of some regions. Table 6.4 lists generic monotypic and polytypic animals named by the MRE fishermen. In the ethnobiological classification furnished by those fishermen, 77% of the generics are monotypic and 23% polytypic, confirming the view of Berlin. Table 6.5 presents comparisons between the relative proportions of monotypic and polytypic genera identified in the MRE and other ethnozoological classification systems (Mourão and Nordi, 2002a,b). Monotypic generics represent the most numerous taxa in Tamandaré and Batoque, reinforcing the patterns encountered in other fishing communities in Brazil (Begossi and Silvano, 2008; Costa-Neto and Marques, 2000; Marques, 1991; Paz and Begossi, 1996).

Mourão and Nordi (2002a) noted that polytypics represented 13 generic taxa that were subdivided into 56 specific taxa (ethnospecies),

TABLE 6.4 List of the Monotypic and Polytypic Folk Generic Named by the Fishermen at the Mamanguape River Estuary

Monotypic Generics	Polytypic Generics
Mutton snapper	Goby
Atlantic goliath grouper	Moray
Cobia	Ray
Sailor's grunt	Catfish
Barbu	Puffer
Irish mojarra	Wrasse
Sergeant major	Timucu
Largehead hairtail	Snook
Hairy blenny	Pompano
Gulf kingcroaker	Acoupa
Spotted snake eel	Sardine
Twinspot bass	Flounder
Brazilian flamefish	Mullet
Plain soapfish	
Ground croaker	
Spotted scorpionfish	
Tarpon	
Molly miller	
Cocoa damselfish	
Barbfish	
Rock hind	
Pluma porgy	
Live sharksucker	
Atlantic bumper	
Flying gurnard	
Black margate	
Bonefish	
Dog snapper	
Blue runner	
Roughneck grunt	

with polytypic generics being either more or less diverse. Sardines (Clupeidae), for example, include 11 specific taxa; rays (Dasyatidae) comprise 2, and the darter goby (Gobiidae) comprise 2. According to Berlin (1992), specific taxa are very similar, differing only in small distinctive morphological details, many of which can be rapidly perceived and often verbalized. The ethnospecies of each of the polytypic generics encountered in the MRE were very similar, differing only in a few characteristics such as color, types of scales, and mouth size. The pemecou sea catfish and the guri sea catfish, for example, differed principally by their colors. The difference between the Atlantic thread herring and the false herring resided in the texture of their scales. The importance of social factors in recognizing specific taxa leads to the conclusion that many of the distinctions are culturally constructed as a result of their direct manipulation by humans due to their social and economic importance. To verify the proportions of monotypic/polytypic generics in Brazilian ethnoichthyological research texts, we analyzed the names with primary and secondary lexemes encountered in the works of Marques (1991), Begossi and Figueiredo (1995), and Costa-Neto and Marques (2000). The high percentage of binomial generic names cited by Marques only allowed inferences concerning the presence of various polytypic generics associated with specific folk categories. In spite of providing a complete list of the generic names, it was not possible to determine if the various primary lexemes formed groupings, or not, in the work of Begossi and Figueiredo (1995), due to the fact that there were no detailed descriptions of the relationships of those names. Costa-Neto and Marques (2000) and Pinto et al. (2016), in addition to providing lists, provided various descriptions that allowed a preliminary analysis of the proportions of monotypic and polytypic generics. The numbers of binomial generics encountered in the MRE were high, although representing only 23% of the total ichthyofaunal diversity. The same polytypic generic taxon can comprise one or more species around which similar species

TABLE 6.5 Relative Proportions of Monotypic and Polytypic Generic Taxa in Some Ethnozoological Classification Systems

Group	Monotypic	Polytypic	Total	% Polytypic
Mamanguape River Estuary	44	13	57	23
Tzeltal	280	55	335	16
Ndumba	156	30	186	16
Wayãpi	489	100	589	17
Tobelo	342	78	420	19
Huambisa peixe	52	18	70	25
Siribinha-BA	47	7	54	13

Sources: Berlin, B., 1992, Ethnobiological Classification: Principles of Categorization of Plants and Animals in Traditional Societies, first ed. Princeton University Press, Princeton; Costa Neto, E.M., 1998. Folk taxonomy and cultural significance of "abeia" (Insecta, Hymenoptera) to the Pankararé, north-eastern Bahia State, Brazil. Journal of Ethnobiology 18 (1), 1–13; Mourão, J.S., Nordi, N., 2002a. Comparação entre as taxonomias folk e científica para Peixes do Estuário do Rio Mamanguape, Paraíba-Brasil. Interciencia, 27, 607–612.

can be grouped. Species that nucleate groups of other species represent the generic folk categories to which they belong or are more outstanding in cultural or economic terms—being denominated prototype species. When there is more than one prototype species, the folk genera can be divided into subclasses. Among the fishermen of the MER, prototypicality, with more than one typical species, was clearly observed in the generic folk "sardinha" (sardine). Fig. 6.2 schematically presents the internal structure of the folk generic category. The prototypic members can be observed, with or without subdivisions into subclasses, in Fig. 6.2B and C, respectively.

Prototypicality was only slightly evident in the generic polytypic taxa catfish, acoupa, and common snook, and the data available were not sufficient to affirm its existence. There was no evidence of the existence of typical species in the other polytypic folk generic taxa encountered. Future research with more adequate methodologies should, however, be undertaken to confirm these observations.

This model characterized by focal species was reported by Forth (1995) in the folk classification of snakes used by the Nage, and by Marques (1991) in descriptions of the concentric system for the generic taxa "catfish" and "lebranche mullet" within the folk systematics of the fishermen. There are only binomial designations within the model described for "catfish," whereas simple or primary names were used together with binomial designations for the generic "lebranche mullet" (*Mugil liza*). This is in agreement with the fourth principle of Berlinian nomenclature, which indicates two possible conditions under which subgeneric taxa could be assigned two primary names: when it is considered a prototype of the generic taxa, or when the nonprototype refers to a taxon of great cultural importance.

It is precisely within the polytypic generic denominations that the groups of major economic, cultural, and psychological importance are encountered. In the case of the MER fishermen, the "sardine" (Clupeidae), "acoupa" (Scianidae), "ray" (Dasyatidae), "common" (Centropomidae), and "catfish" (Ariidae) represent some of the polytypic generic taxa. Table 6.6 demonstrates the diversity of the specific folk taxa of the 13 polytypic generic taxa observed in the naming of the ichthyofauna encountered by the MER fishermen. "Catfish," "sardine," and "acoupa" were the most

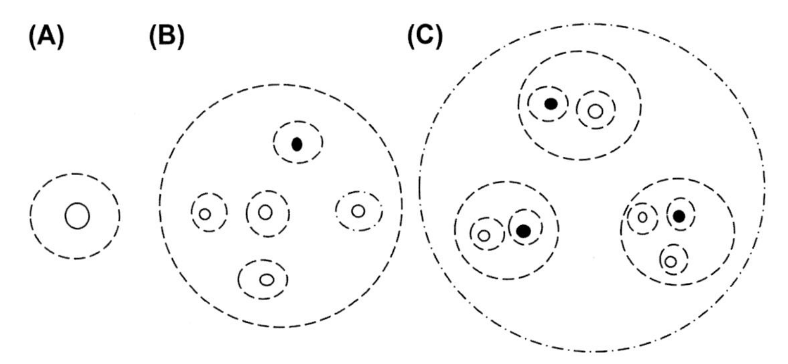

FIGURE 6.2 Schematic representation of the internal structure of the category "folk generic": (A) monotypic folk generic, (B) polytypic folk generic (without subclass), and (C) polytypic folk generic (with subclasses).

diverse polytypic generic taxa, with seven and more than seven specific folk taxa, respectively.

Color appears to be a very frequent character in binomial naming of the taxa of different groupings. When a group includes two taxa, color is frequently predominant in distinguishing between them, for example, "drums or croakers," "darter goby," and "emerald goby." As the number of specific folk taxa in the grouping increases, other characteristics appear in their designations, such as body shape—"tonguefishes" (*Symphurus tessellatus*), "bigtooth flat" (*Isopisthus parvipinnis*), common snook (*Centropomus undecimalis*), and "puddingwife wrasse" (*Halichoeres radiatus*); mouth size—zabaleta anchovy; and types of scales—"River Plate sprat," "Guiana longfin herring" (*Odontognathus mucronatus*). In addition to these characteristics, fishermen can use other information, principally related to habitat "guri sea catfish," "longnose stingray," southern eagle ray (*Myliobatis australis*) or behavior "seamstress sardine" (*Anchovia clupeoides*), depending on the size of the grouping and the contrast chosen.

ANALOGIES BETWEEN FOLK AND SCIENTIFIC TAXONOMIC CLASSIFICATION SYSTEMS

According to Hennig (1966), biological classifications form catalogs of global biological diversity and are true repositories of information about the natural world. Ethnobiological classifications are extremely informative because they contain an enormous richness of information concerning the biology, ecology, the etiology of diverse groups of plants and animals—but they have few hierarchical levels as compared to scientific systems, possibly due to the artificial nature of the latter and the precision requirements of taxonomists. Amorim (1997) argued that the lack of knowledge of nonacademic individuals concerning biological diversity tends to generate only limited numbers of taxonomic levels as compared to those produced by professional researchers—but the reduced numbers of levels in folk classifications do not necessarily indicate a lack of knowledge by traditional cultures, but rather reflect their holistic view of the natural world. Comparisons between the patterns detected in folk taxonomy with formal (scientific) biological taxa would be, according to Berlin (1992), a step forward in understanding the analogies between the two classification systems.

Because of their practicality in displaying taxonomic relationships between folk and scientific classifications, generic polytypic (Mourão and Nordi, 2002a, b) will be presented here using Venn diagrams as introduced by Hunn (1975) and Gardner (1976).

Venn diagrams have been used since the beginning of the 1960s by Simpson (1961) in the "*Principles of Animal Taxonomy*" to demonstrate

TABLE 6.6 Comparisons of the Generic Folk Polytypic Richness Observed in the Present Study With Other Ethnozoological Classification Systems

	Number of Specific Folk Taxa						
Generics	**2**	**3**	**4**	**5**	**6**	**7**	**>7**
Moray		X					
Ray				X			
Catfish							X
Puffer	X						
Wrasse	X						
Drums	X						
Snook				X			
Pompano	X						
Acoupa							X
Sardines							X
Flounder		X					
Mullet	X						
Goby	X						

biological relatedness and the relationships of inclusion in groups. Berlin (1992) adopted the use of Venn diagrams as the best way to present Berlinian models of folk classification because of at least four advantages: they allow the explicit indication of the biological and ethnobiological taxa by means of distinct circles, allowing a discussion of the biological series of the folk taxa; they explicitly indicate the prototype members; they allow the possibility of demonstrating true affinities between members of the specific folk taxa; and they display the folk and scientific names together.

Figs. 6.3 through 6.8 are Venn diagrams showing the specific folk taxa of the principal generic polytypic identified by the fishermen at MRE and compare them with their equivalent scientific taxa (Mourão and Nordi, 2002a). Fig. 6.3 presents the specific folk taxa of the generic Florida pompano and their corresponding taxa according to scientific classification. According to the fishermen, the generic Florida pompano can be distinguished by body shape and color; the Florida pompano—soft-headed pampo, for example, has a long body with part of the ventral region being yellow and the dorsal region bluish. The Great pompano has a wide, yellow body. According to Menezes and Figueiredo (1980), color is an important characteristic in the identification of *Trachinotus* species: *Trachinotus goodei* has splotches or dark vertical stripes on the sides of its body, whereas *Trachinotus carolinus* has no splotches or vertical stripes. The members of the generic *wrasse*, and their corresponding Linnean taxa, are presented in Fig. 6.4. The specific folk blackear wrasse and *puddingwife wrasse* differ from each other in terms of their body shapes and coloring: *puddingwife wrasse* is wider and purplish colored, whereas the *blackear wrasse* is green. Similarly, the species of *Halichoeres* are distinguished principally by their colors in their scientific descriptions:

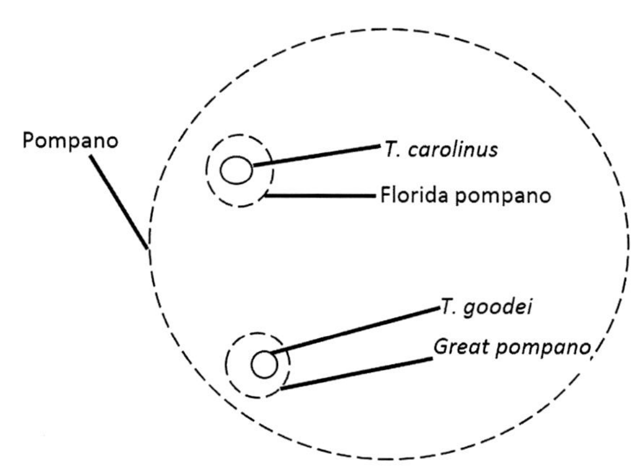

FIGURE 6.3 Specific folk taxa of the generic Florida pompano/pampo and their corresponding taxa according to scientific classification.

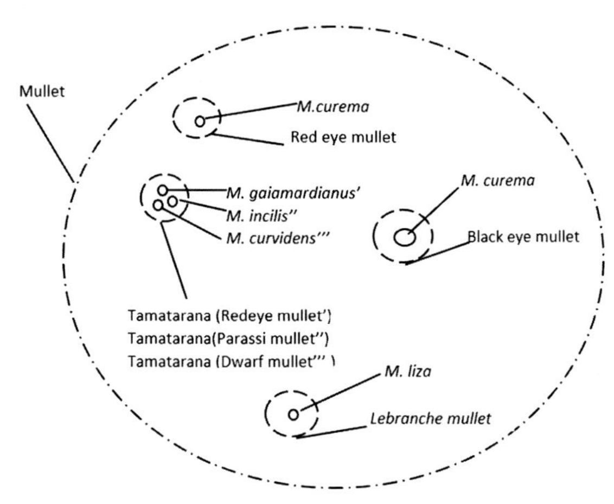

FIGURE 6.5 Specific folk of the mullets family and the scientific equivalents.

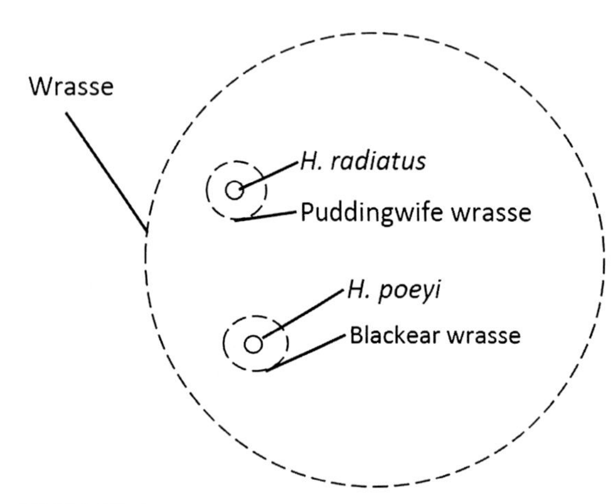

FIGURE 6.4 Folk specific folk generic wrasse and their equivalents in scientific classification.

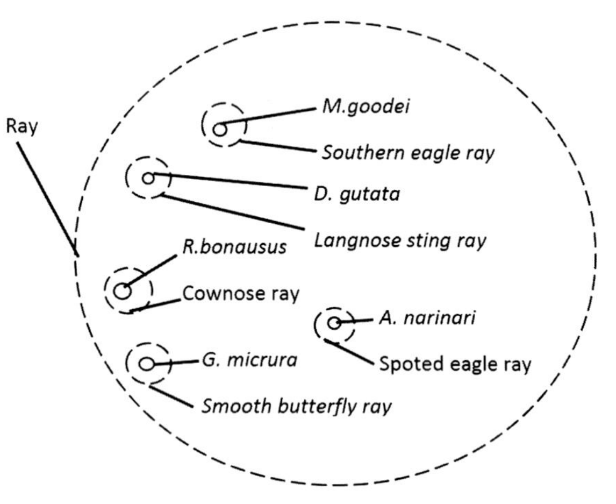

FIGURE 6.6 Specific folk taxa of the generic ray and their corresponding taxa according to scientific classification.

Halichoeres poeyi is greenish yellow, with an oval-shaped black spot just behind each eye that is of the same size or slightly larger than the pupils; *H. radiatus* has an overall greenish yellow body color and no black spot behind the eyes. While the specific folk taxa of the generic Rake stardrum are differentiated by fishermen based on the shape of the head.

Comparisons between the classification systems in the case of the white mullet/fish are presented in Fig. 6.5. In describing "the mullets/family," the fishermen of the MRE reported that this grouping is based around the generic folk white mullet that serves as a standard reference, to which the other generic taxa are compared and defined. Costa-Neto and Marques (2000) likewise encountered a generic folk *white mullet* considering a prototype among the fishermen at Siribinha. Fishermen in estuaries in Alagoas State studied by Marques (1991) considered the *lebranche mullet* as the "head of the family". Morphological characteristics are

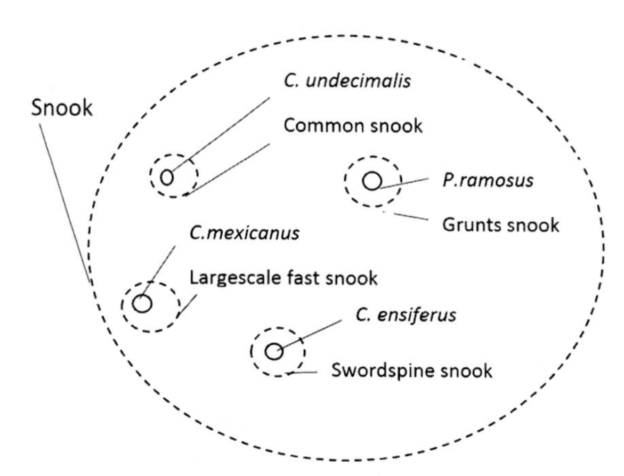

FIGURE 6.7 Specific folk taxa of the generic snook and their corresponding taxa according to scientific classification.

most frequently used by the MRE fishermen to classify and identify fish belonged to the "mullets/family." The principal criteria employed are head and body shape, the thicknesses of the scales, and the size and colors of the eyes (the latter reflecting the denominations "mullet red" and "mullet black"). Analogous names based on eye color were also reported by Marques (1991) and Costa-Neto and Marques (2000).

The MRE fishermen recognize the generic folk *dwarf mullet* by a behavioral characteristic ("a strong smell"), this being the only criteria that distinguishes it from the *tainha*. In São Luis (Maranhão), this same species is called tainha peitiu *mullet smelly* (Martins-Jurus et al.

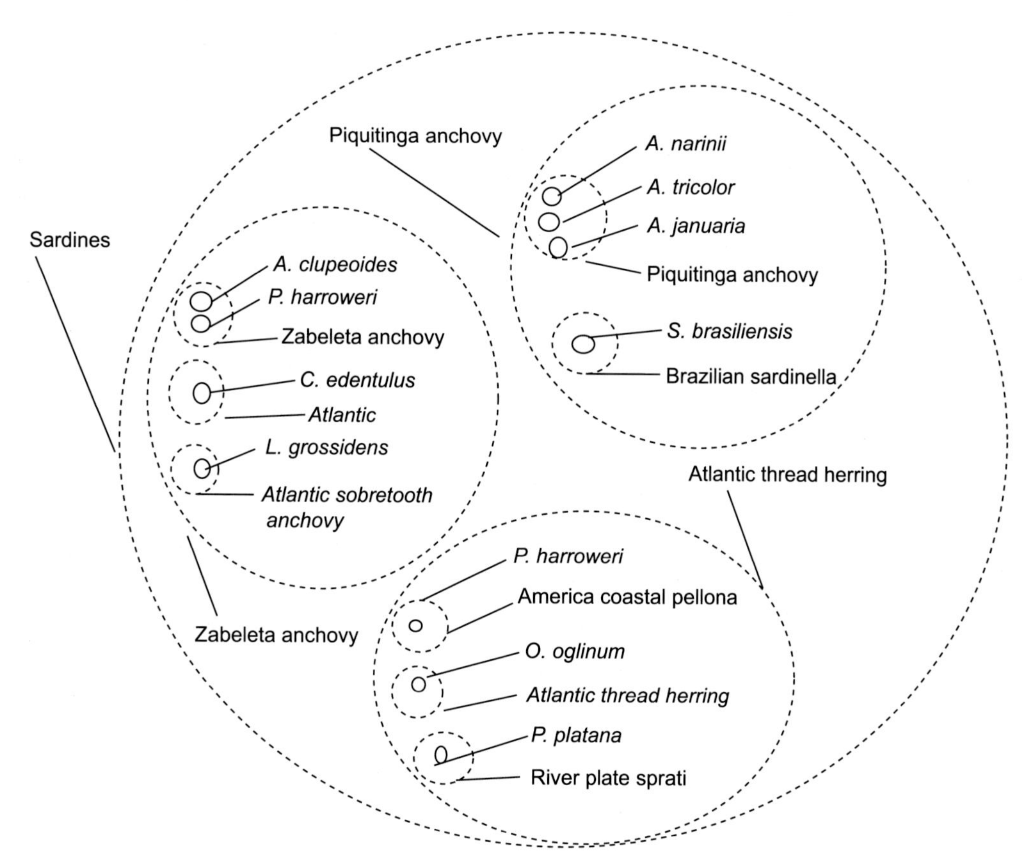

FIGURE 6.8 Specific folk taxa of the generic sardines and their corresponding taxa according to scientific classification.

apud Marques, 1991); according to the Câmara Cascudo dictionary, *peitiú* comes from the Tupi-Guarani Amerindian name "peitiú," which signifies "smelly." Menezes and Figueiredo (1985) reported that the species *M. curema* and *Mugil gaiamardianus* have very similar general characteristics, differing only by the fact that *M. gaiamardianus* has reddish eyes. The descriptions provided by Menezes and Figueiredo, as cited above, for the species *Mugil curvidens* (*dwarf mullet*) and *Mugil incilis* (parassi mullet) (which are considered as one by the fishermen in the generic folk *tamatarana*), differ only in terms of the numbers of spines in the anal fin and their scales being arranged in lateral series. The morphological criteria used by the fishermen to differentiate between white mullet and lebranche mullet are shapes of the head and body and the thicknesses of the scales. The head of the *mullet* has a "rounded" shape; its scales are not very thick; and its body is "slightly flattened," "not very rounded." The lebranche mullet (*M. liza*), on the other hand, has a "flattened" head; its scales are quite thick; and its body is longer and more cylindrical.

According to the MRE fishermen, all of the members of the "*mullet* family" have a "navel," which is described in the following manner: "it's kind of round, kind of hard, when you cut into it mud comes out—because the mud has to go through the navel first, to first go through the fish's anal channel. The fish eats mud and it stays in its intestines, and from the anal channel it goes through the navel and then goes out." The fishermen studied by Marques likewise grouped the "white mullet" and "lebranche mullet" into a folk family denominated "fish with navels."

The similarities between the folk and scientific classification systems for the generic ray are shown in Fig. 6.6. The generic folk ray is recognized using the following morphological characteristics: head shape ("rounded, with a beak," "flat with two lips"); body shape ("rounded"); tail size ("tail short or long"); color ("white, yellow, gray, green, black with white spots, or bluish"). The trophic levels, habitats, and behavioral aspects of those fish are also used to distinguish between them. The specific folk *smooth butterfly ray* has a "short tail, doesn't have a spine and its body is round, but it is different from the others, it looks like some kind of bird, its fins look like wings." Figueiredo (1977) reported that *Gymnura micrura* is a Brazilian species of the Gymnuridae without spines and with a very short tail. Ziswiller (1976) reported that this species has a body that is wider than long and extends its pectoral fins like wings. The specific folk *longnose stingray* draws its name as an allusion to the "croa" microhabitat where it is said to feed. These rays are also identified by the fishermen through their dimorphic sexual characteristics, denominated "*espigão*" and "*catocos*": "the male has two claspers near its tail," and the "female has the sign of its sex, which is wide." According to Figueiredo (1977), the males of the Chondrichthyes have an appendage sustained by cartilage that is called a clasper and develops on the internal margin of each pelvic fin. The fishermen recognize the "spotted eagle ray" (*Aetobatus narinari*) principally by its color ("black, with a white spot"), its snout ("in the form of a beak"), the presence of "rocks in its mouth" (dental plates), and the spines ("it has five stingers").

The generic folk *common snook* is described by the fishermen as a fish with a "long face or thin beak"; "the chin between the fin looks like a saw blade"; "a spine together with the tail, under its belly"; "a dark stripe on either side of the body," "white body" and "wide body." Menezes and Figueiredo (1980) described the Centropomidae as fish with elongated bodies, generally with accentuated convex dorsal profiles; preoperculum with serrated posterior margin, dorsal fins separate, anal fin with three spines, the second being stronger and more developed; lateral line extending to the extremities of the median region of the tail fin. Comparisons between the members of the generic common snook with their scientific equivalents are presented in Fig. 6.7. The generic folk *common snook* corresponds to two Linnean families: Centropomidae and Haemulidae. The species *C. undecimalis* and *Pomadasys ramosus* have the greatest numbers of ethnonames or synonyms among the fish species known to that group of fishermen.

As was observed earlier, the sardines (Fig. 6.8) are subdivided into three subclasses or subgeneric taxa denominated "Atlantic thread herring," "zabaleta anchovy," and "Marini's anchovy" around which are grouped a set of morphologically similar specific folk taxa. According to the scientific literature, the species most closely related to the zabaleta anchovy (*Opisthonema oglinum*) are very similar morphologically, differing basically in terms of the numbers of rays in the anal fin (Figueiredo and Menezes, 1978). The taxa of the subgeneric *Marini's anchovy*, which corresponds to the Linnean genera *Cetengraulis* and *Anchovia*, are also very similar among themselves in terms of their descriptions in the scientific literature, differing only in smaller details, for example, the position of the anal fin (Figueiredo and Menezes, 1978).

The Marini's anchovy and Brazilian sardinella (specific folk taxa of the folk subgeneric *Piquitinga anchovy*) differ in a single morphological detail on their ventral region. According to the descriptions provided by the fishermen, the *Brazilian sardinella* differs from the *P. anchovy* due to the "saw under its belly." Figueiredo and Menezes (1978) described one of the characteristics of the family Clupeidae (*Sardinella brasiliensis—Brazilian sardinella*) as the presence of a median ventral keel formed by modified scales. The *P. anchovy* (Engraulidae) corresponds to the Linnean species *Anchoa narinii, Anchoa januaria,* and *Anchoa tricolor*, which are very similar terms of their scientific descriptions, differing only in terms of the diameters of their eyes.

The fishermen interviewed in the MRE affirmed that all of the fish that have three spines ("espetos") on their dorsal and lateral regions are called catfish—those spines therefore represent the unifying morphological feature of the generic folk catfish grouping. The fishermen interviewed by Marques (1991) described the act of "brooding in their mouths" as a unifying character for identifying fish of the family catfish. The fishermen also identified various phases of the reproductive process of the pemecou sea catfish *Sciades herzbergii*, naming them and generating a very detailed classification of their reproductive events, which Marques (1991) described as cyclic. The catfish seek out the mouths of rivers or lake regions during spawning; the males (and more rarely the females) incubate the eggs in their buccal cavities until they have completed their developmental cycle (Breeder and Rosen, 1966; Mishima and Tanji, 1985; Barbieri et al., 1992; Chaves, 1994).

One very important ecological attribute described by the fishermen for the pemecou sea catfish (the ethnoname of the species *S. herzbergii*) was its trophic function. This species is recognized by all of the fishermen by this particular characteristic—being "the most disgusting and slimy fish"; for this reason it is considered as a prototype species. In the work of Marques (1991) this same fish is called "bagre marruá" (the "principal" or "true" catfish), which indicates it as a representative of the generic folk catfish group.

The members of the generic folk moray, according to the fishermen, are similar to "snakes," except for the fact that moray eels are "flat" while snakes are "round." The specific folk taxa differ by simple color patterns. In the same way, the species of the genus *Gymnothorax* described by Figueiredo and Menezes (1978) have colors that vary from uniform dark green in *Gymnothorax funebris* to yellowish brown in *Gymnothorax ocellatus*, and yellowish but densely covered with dark spots in *Gymnothorax vicinus*. The moray eel (Muraenidae) is related to two other generic folk taxa: ocellated moray and snake eels (Ophichitidae), due to their very similar body shapes. In the studies undertaken by Marques (1991), these fish were included in a single family, "mororó."

CORRESPONDENCE BETWEEN GENERIC FOLK TAXA AND SCIENTIFIC SPECIES

Atran et al. (1997) noted that generic folk taxa frequently correspond to a genus or scientific species, although, according to Berlin (1973), any comparisons between folk and Linnean systematics must look for correspondence between "generic folk taxa and scientific species, because the latter is the basic scientific taxonomic unit, while the generic is the primary unit of the folk system." Comparisons show at least three types of correspondence: 1:1 correspondence; overdifferentiation; and subdifferentiation of types 1 and 2.

A one-to-one (1:1) correspondence can be observed when a single generic folk taxon refers to only a single scientific species. Overdifferentiation occurs when two or more generic folk taxa refer to a single scientific species. Subdifferentiation occurs when a single generic folk taxon corresponds to two or more scientific species. Type I subdifferentiation occurs when a single generic folk taxon refers to two or more species of the same scientific genus; type II subdifferentiation occurs when a single generic folk taxon refers to two or more species of two or more scientific genera.

The comparisons between the folk and scientific systematic systems showed in terms of percentages can be seen that there is 71% of 1:1 correspondence, 13.45% subdifferentiated correspondence of types 1 and 2, and only 1.93% of overdifferentiation (see Mourão and Nordi, 2002a). Pinto et al. (2016) studied two fishing communities along the northeastern coast of Brazil and encountered the following correspondence between specific folk taxa and scientific species: 49% of 1:1 type correspondence in Tamandaré and 45% in Batoque; 34% subdifferentiation in Tamandaré and 30% in Batoque; 12% type 1 subdifferentiation in Tamandaré and 4.6% type I subdifferentiation in Batoque. Berlin (1973) reported 61% of 1:1 correspondence, 36% subdifferentiation, and 3% correspondence of types 1 and 2 overdifferentiation in an ethnobotanical survey among the Tzeltal in Mexico. Studies undertaken with fishermen along the southeastern coast of Brazil by Begossi et al. (2008) showed a 1:1 correspondence at the generic level instead of the specific level, although among the Ketengban in Papua New Guinea, the correspondence was 1:1 for the entire spectrum of 115 local names of birds (Diamond and Bishop, 1999); to Begossi et al. (2008) this apparently demonstrated that the Ketengban are more integrated into nature than the artisanal fishermen described here.

According to Marques (1991), a crucial question in terms of a folk taxonomic system is its ability to be compared with Linnean taxonomy and legitimized by confirmed scientific validity. Jensen (1985) encountered 65% correspondence of the scientific families/ethnofamilies among the Wayampi Amerindians. In terms of 1:1 correspondence, Hays (1983) reported a lack of universal correspondence between folk and Linnean categories among the Ndumba (New Guinea). Berlin (1973), however, observed that only precision in defining comparative units guarantees adequate comparisons, suggesting that some discrepant results may reflect imprecisions in choosing the appropriate units. Morril (1967), while studying the ethnoichthyology of the Cha-Cha, observed that in only one case did the Cha-Cha employee refined taxonomic distinctions recognized by professional ichthyologists—which led that author to conclude that the taxonomy of the Cha-Cha was not very elaborate or precise when compared to scientific classifications. Marques (1991) disagreed with the conclusions of Morril (1967), however, observing that the pioneering nature of his work, allied to the methodological limitations of that time, could have tainted his conclusions. The latter author observed that the "scientific" nature of the ichthyological systematics of the fishermen studied was taxonomically refined, resulting in an intellectual ordering of nature that was elaborate, accurate, and precise. These same characteristics were observed with the ethnobiological systematics of the MRE fishermen.

FINAL CONSIDERATIONS

Analogies between folk and Linnean classification systems demonstrated that the criteria used to differentiate between species were, in large part, concordant, because morphological criteria were largely used in the identification, naming, and classification in all of the cultures studied. The patterns of categorization (hierarchies) and naming, as well as the fact that generic folk taxa are largely monotypes, are in conformity with the proposed Berlinian principles. The subdivisions of generics (polytypic) of specific folk taxa were encountered, notably, among those fish having economic or cultural importance. The high equivalence observed between the generic folk taxa and scientific species (71%) indicated that fish taxonomy based on the local knowledge gave a very strong approximation of the biological richness of the fish communities in the environments examined here. This local knowledge can provide important information about the perception of risks of extinction and the stocks of economically and culturally important species, and it also expresses fundamental differences of how individuals perceive biological diversity. In addition to the information provided here, the local artisanal fishermen retain additional detailed knowledge of regional biology and their environment that could have important implications for the management of fishing resources. Other important aspects related to ethnotaxonomy can aid university and institutional researchers by furnishing in-the-field perceptions about the patterns of nature and provide different perceptions about how biological diversity is organized.

References

Agnarsson, I., Kuntner, M., 2007. Taxonomy in a changing world: seeking solutions for a science in crisis. Systematic Biology 56 (3), 531–539.

Amorim, D.S., 1997. Elementos Básicos de Sistemática Filogenética, first ed. Holos, Ribeirão Preto.

Anderson, E., 1967. The Etnoichthyology of the Hong-kong Boat People (Master thesis). University of California, Berkeley.

Atran, S., 1990. Cognitive Fundations of Natural History, second ed. Cambridge University Press, London.

Atran, S., 1998. Folk biology and the anthropology of science: cognitive universals and cultural particulars. Behavioral and Brain Sciences 21, 547–609.

Atran, S., Estin, P., Coley, J.D., Medin, D.L., 1997. Generic species and basic levels: essence and appearance in folk biology. Journal of Ethnobiology 17, 22–45.

Baldus, H., 1947. Vocabulário zoológico Kaingang. Arquivos do Museu Paranaense 6, 149–160.

Barbieri, L.R., dos Santos, R.P.J., Andreata, J.V., 1992. Reproductive biology of the marine catfish, *Genidens genidens* (Siluriformes, Ariidae), in the Jacarepaguá lagoon system, Rio de Janeiro, Brazil. Environmental Biology of Fishes 35, 23–35.

Basset, Y., Novotny, V., Miller, S.E., Pyle, R., 2000. Quantifying biodiversity: experience with parataxonomists and digital photography in Papua New Guinea and Guyana. BioScience 50, 899–908.

Beaudreau, A.H., Levin, P.S., Norman, K.C., 2011. Using folk taxonomies to understand stakeholder perceptions for species conservation for species conservation. Conservation Letters 4, 451–463.

Begossi, A., Garavello, J.C., 1990. Notes on the ethnoichthyology of fishermen from the Tocantins river (Brazil). Acta Amazônia 20, 341–351.

Begossi, A., Figueiredo, J.L., 1995. Ethnoichthyology of southern coastal fishermen: cases from Búzios Island and Sepetiba Bay (Brazil). Bulletin of Marine Science 56, 682–689.

Begossi, A., Silvano, R.A.M., 2008. Ecology and ethnoecology of dusky grouper garoupa, *Epinephelus marginatus* (Lowe, 1834) along the coast of Brazil. Journal of Ethnobiology and Ethnomedicine 4, 20.

Begossi, A., Clauzet, M., Figueiredo, J.L., Garuana, L.R.V., Lima, P.F., Maccord, P.F., Ramires, M., Silva, A.L., Silvano, R.A.M., 2008. Are biological species and higher-ranking categories real? Fish folk taxonomy on Brazil's Atlantic Forest and in the Amazon. Current Anthropology 49, 1–16.

Berlin, B., 1972. Speculations on the growth of ethnobotanical and nomenclature. Language in Society 1, 51–86.

Berlin, B., 1973. The relation of folk systematic to biological classification and nomenclature. Annual Review of Ecology and Systematics 4, 259–271.

Berlin, B., 1974. Further notes on covert categories and folk taxonomies. American Anthropologist 76, 327–331.

Berlin, B., 1976. The concept of rank in ethnobiological classification: some evidence from Aguaruna folk biology. American Ethnologist 3, 381–399.

Berlin, B., 1992. Ethnobiological Classification: Principles of Categorization of Plants and Animals in Traditional Societies, first ed. Princeton University Press, Princeton.

Berlin, B., Berlin, E.A., 1983. Adaptation and ethnozoological classification: theoretical implications of animal resources and diet of the Aguaruna and Huambisa. In: Hames, R.B., Vickers, W.T. (Eds.), Adaptative Responses of Native Amazonians. Academica Press, New York, pp. 301–327.

Berlin, B., Kay, P., 1969. Basic Color Terms: Their Universality and Evolution, first ed. University of California Press, Berkeley.

Berlin, B., Breedlove, D.E., Haven, P.H., 1973. General principles of classification and nomenclature in folk biology. American Anthropologist 75, 214–242.

Berlin, B., Breedlove, D.E., Haven, P.H., 1974. Principles of Tzeltal Plant Classification, second ed. Academic Press, New York.

Breeder, J.R.C.M., Rosen, D.E., 1966. Modes of Reproduction in Fishes, first ed. American Museum of Natural History by the Natural History Press, New York.

Brown, C.H., 1984. Language and Living Things: Uniformities in Folk Classification and Naming. Rutgers University Press, New Brunswick.

Bulmer, R.N.H., 1974. Folk biology in the New Guinea Highlands. International Social Sciences Council 13, 9–28.

Carrara, E., 1997. Tsi tewara – um vôo sobre o Cerrado Xavante (Master thesis). Universidade de São Paulo, São Paulo.

Chaves, P.T., 1994. A incubação de ovos e larvas em *Genidens genidens* (Valenciennes) (Siluriformes, Ariidae) da Baía de Guaratuba, Paraná, Brasil. Revista Brasileira de Zoologia 11, 641–648.

Clément, D., 1995. Why is taxonomy utilitarian? Journal of Ethnobiology 15, 1–44.

Clément, D., 1998. The historical foundations of ethnobiology (1860–1899). Journal of Ethnobiology 18, 161–187.

Conklin, H.C., 1954. In ethnoecological approach to shifting agriculture. Transactions of the New York Academy of Sciences 17, 133–142.

Conklin, H.C., 1962. Lexicographical treatment of folk taxonomies. International Journal of American Linguistics 28, 119–141.

Costa Neto, E.M., 2001. A cultura pesqueira do litoral norte da Bahia: etnoictiologia, desenvolvimento e sustentabilidade. EDUFBA, Salvador; EDUFAL, Maceió, Brasil, pp. 159.

Costa-Neto, E.M., Marques, J.G.W., 2000. A Etnotaxonomia de recursos ictiofaunísticos pelos pescadores da comunidade de Siribinha, Norte do Estado da Bahia, Brasil. Biociências 8, 61–76.

Daly, D.C., 1998. Systematics and ethnobotany: what's in a name? In: Fonseca, V.S., Silva, I.M., Sá (Orgs.), C.F.C. (Eds.), Etnobotânica: bases para conservação. EDUR, Seropédica, pp. 50–68.

Dennler, J.G., 1939. Los nombres indígenas en guaraní de los mamíferos de la Argentina y países limítrofes y su importancia para la sistemática. Physis 16, 225–244.

Diamond, J., 1989. The ethnobiogist's dilemma. Natural History 6, 26–30.

Diamond, J., Bishop, K.D., 1999. Ethno-ornithology of the Ketengban people, Indonesian New Guinea. In: Medin, D., Atran, S. (Eds.), Folkbiology. MIT Press, Cambridge, pp. 17–46.

Diegues, A.C., 2000. Etnoconservação da natureza: enfoques alternativos. In: Diegues (Org.), A.C. (Ed.), Etnoconservação: novos rumos para a proteção da natureza nos trópicos. Hucitec, São Paulo, pp. 1–46.

Durkheim, É., Mauss, M., 1905. De Quelques Formes Primitives de classification: contribution a l'étude des représentations. L'Année Sociologique 6, 1–72.

Ellen, R., 1993. In: The Cultural Relations of Classifications. Cambridge University Press, Cambridge.

Figueiredo, J.L., 1977. Manual de Peixes Marinhos do Sudeste do Brasil. I. Introdução. Cações, raias e quimeras, first ed. Museu de Zoologia da USP, São Paulo.

Figueiredo, J.L., Menezes, A., 1978. Manual de Peixes Marinhos do Sudeste do Brasil. III- Teleostei, first ed. Museu de Zoologia de USP, São Paulo.

Forth, G., 1995. Ethnozoological classification and classificatory language among the Nage of Eastern Indonesia. Journal of Ethnobiology 15, 45–69.

Frazão, A., 2001. As classificações botânicas Nalu (Guiné-Bissau): consensos e variabilidades. Etnográfica 1, 131–155.

Freire, K.M.F., Pauly, D., 2003. What is in there: common names of Brazilian marine fishes. Fisheries Centre Research Report 11, 1439–1444.

Freire, K.M.F., Pauly, D., 2005. Richness of common names of Brazilian marine fishes and its effect on catch statistics. Journal of Ethnobiology 25, 279–296.

Gardner, P.M., 1976. Birds, words, and a requiem for the omniscient informant. American Ethnologist 3, 446–468.

Gianinni, I.V., 1991. A ave resgatada: a impossibilidade de leveza do Ser (Master thesis). Universidade de São Paulo, São Paulo.

Gould, S.J., 2001. Lance de dados: a ideia de evolução de Platão a Darwin, first ed. Record, Rio de Janeiro.

Gragson, T.L., Blount, B.G., 1999. Ethnoecology: Knowledge, Resources, and Rights. University of Georgia Press, Athens.

Greene, J., 2007. Why are VMPFC patients more utilitarian? A dual-process theory of moral judgment explains. Trends in Cognitive Science 11, 322–323.

Hartmann, T., 1967. A nomenclatura botânica Borôro (materiais para um ensaio etnobotânico). Instituto de Estudos Brasileiros/USP, São Paulo.

Hays, T.E., 1976. An empirical method for the identification of covert categories in ethnobiology. American Ethnologist 3, 489–507.

Hays, T.E., 1979. Plants classification and nomenclature in Ndumba, Papua New Guinea Highlands. Ethnology 18, 253–270.

Hays, T.E., 1983. Ndumba folk biology and general principles of ethnobotanical classification and nomenclature. American Anthropologist 85, 592–611.

Hennig, W., 1966. Phylogenetic Systematic. University of Ilinois Press, URBANA, III.

Holman, E.W., 2005. Domain-specific and general properties of folk classifications. Journal of Ethnobiology 25, 71–91.

Hunn, E.S., 1975. A measure of the degree of correspondence of folk to scientific biological classification. American Ethnobiology 2, 309–327.

Hunn, E.S., 1977. Tzeltal Folk Zoology: The Classification of Discontinuities in Nature. Academic Press, New York.

Hunn, E.S., 1982. The utilitarian factor in folk biological classification. American Anthropologist 84, 830–847.

Ihering, H.V., 1904. As abelhas sociais do Brasil e suas denominações tupis. Revista do Instituto Histórico e Geographico de São Paulo 1503 376–388.

Jensen, A.A., 1985. Sistemas Indígenas de Classificação de Aves: Aspectos Comparativos, Ecológicos e Evolutivos (Doctoral thesis). UNICAMP, Campinas.

Leff, E., 2001. Epistemologia Ambiental, first ed. Cortez, São Paulo.

Lévi-Strauss, C., 1997. O Pensamento Selvagem, second ed. Papiros, Campinas.

Maffi, L., 1999. Language and the environment. In: Posey, D.A. (Ed.), Cultural and Spiritual Values of Biodiversity. ITP, London, pp. 22–29.

Maranhão, T., 1975. Náutica e classificação ictiológica em Icaraí, Ceará: um estudo em antropologia cognitiva (Master thesis). Universidade de Brasília, Brasília, Distrito federal.

Marques, J.G.W., 1991. Aspectos Ecológicos na Etnoictiologia dos Pescadores do ComplexoEstuarino-Lagunar Mundaú-Manguaba, Alagoas (Doctoral thesis). UNICAMP, Campinas.

Medin, D.L., Atran, S., 2004. The native mind: biological categorization and reasoning in development and across cultures. Psychological Review 111, 960–983.

Menezes, N.A., Figueiredo, J.L., 1980. Manual de Peixes Marinhos do Sudeste do Brasil V. Teleostei (3), first ed. Museu de Zoologia da USP, São Paulo.

Menezes, N.A., Figueiredo, J.L., 1985. Manual de Peixes Marinhos do Sudeste do Brasil V. Teleostei (4). Museu de Zoologia da USP, São Paulo.

Mishima, M., Tanji, S., 1985. Fecundidade e Incubação dos bagres marinhos (Osteichthyes, Ariidae) do Complexo Estuarino Lagunar de Cananéia. Boletim do Instituto da Pesca de São Paulo 12, 77–85.

Mishler, B.D., Donoghue, J.M., 1982. Species concepts: a case for pluralism. Systematic Zoology 3, 491–503.

Morril, W.T., 1967. Ethnoichthyology of the Cha-Cha. Ethnology 6, 405–417.

Mourão, J.S., Nordi, N., 2002a. Comparação entre as taxonomias folk e científica para Peixes do Estuário do Rio Mamanguape, Paraíba-Brasil. Interciencia 27, 607–612.

Mourão, J.S., Nordi, N., 2002b. Principais critérios utilizados por pescadores artesanais na taxonomia folk dos peixes do estuário do Rio Mamanguape, Paraíba - Brasil. Interciencia 27, 1–7.

Nazarea, V.D., 1999. Ethnoecology. Situated Knowledge/ Located Lives. University of Arizona Press, Tucson.

Paz, V.A., Begossi, A., 1996. Ethnoichthyology of Gamboa fishermen of Sepetiba Bay, Brazil. Jounal of Ethnobiology 16, 157–168.

Pinto, M.F., Mourão, J.S., Alves, R.R., 2013. Ethnotaxonomical considerations and usage of ichthyofauna in a fishing community in Ceará state, Northeast Brazil. Journal of Ethnobiology and Ethnomedicine 9, 17–25.

Pinto, M.F., Mourão, J.S., Alves, R.R., 2016. How do artisanal fishermen name Fish? An ethnotaxonomic study in Northeastern Brazil. Journal of Ethnobiology 36, 348–381.

Posey, D.A., 1987. Introdução - Etnobiologia: Teoria e Prática. In: Ribeiro (Coord.), B.G. (Ed.). Ribeiro (Coord.), B.G. (Ed.), SUMA Etnológica Brasileira, vol. 1. Vozes, Petrópolis, pp. 15–25.

Previero, M., Minte-Vera, C.V., Moura, R.L., 2013. Fisheries monitoring in Babel: fish ethnotaxonomy in a hotspot of common names. Neotropical Ichthyology 11, 467–476.

Ramires, M., Molina, S.M.G., Hanazaki, N., 2007. Etnoecologia caiçara: o conhecimento dos pescadores artesanais sobre aspectos ecológicos da pesca. Revista Biotemas 20, 101–103.

Raven, P., Berlin, B., Breedlove, D., 1971. The origins of taxonomy. Science 174, 1210–1213.

Sheil, D., Lawrence, A., 2004. Tropical biologists, local people and conservation: new opportunities for collaboration. Trends in Ecology & Evolution 19, 634–638.

Simpson, G.G., 1961. Principles of Animal Taxonomy. Columbia University Press, New York.

Souza, S., Begossi, A., 2007. Whales, dolphins or fishes? The ethnotaxonomy of cetaceans in São Sebastião, Brazil. Journal of Ethnobiology and Ethnomedicine 3, 1–15.

Stewart, K.M., 1994. Early hominid utilisation of fish resources and implications for seasonality and behavior. Journal of Human Evolution 27, 229–245.

Stross, B., 1973. Acquision of botanical terminology by Tzeltal Children. In: Edmonson, B.S. (Ed.), Meaning in Mayan Languages. Mouton, The Hauge, pp. 107–141.

Vanzolini, P.E., 1956–1958. Notas sobre a zoologia dos índios Canela. Revista do Museu Paulista 10, 155–171.

Wilson, E.O., 1997. A situação atual da diversidade biológica. In: Wilson, E.O. (Ed.), Biodiversidade. Nova Fronteira, Rio de Janeiro, pp. 3–24.

Wyman, L.C., Harris, S.K., 1941. Navajo Indian Medical Ethnobotany. The University of New Mexico Bulletin, Albuquerque.

Zeisberger, D., 1887. Zeisberger's Indian Dictionary. John Wilson and Son, Cambridge.

Ziswiller, V., 1976. Wirbeltiere: Spezielle Zoologie. Band I: Anamnia. Georg Thieme Verlag, Stuttgart.

CHAPTER

7

The Importance of Hunting in Human Societies

Rômulo Romeu Nóbrega Alves[1], *Wedson Medeiros Silva Souto*[2], *Hugo Fernandes-Ferreira*[3], *Dandara Monalisa Mariz Bezerra*[4], *Raynner Rilke Duarte Barboza*[1], *Washington Luiz Silva Vieira*[5]

[1]Universidade Estadual da Paraíba, Campina Grande, Brazil; [2]Universidade Federal do Piauí (UFPI), Teresina, Brazil; [3]Universidade Estadual do Ceará, Quixadá, Brazil; [4]Instituto Federal de Educação, Ciência e Tecnologia da Paraíba, Itabaiana, Brazil; [5]Universidade Federal da Paraíba, João Pessoa, Brazil

INTRODUCTION

Meat is an animal product and an essential component of the diet of various carnivore and omnivore species. Among primates, hunting is not exclusively performed by humans, although this was considered to be the case until a few decades ago. Primatology and animal behavior studies have deconstructed this myth by reporting nonhuman primate hunting activities (Stanford, 1999). Although hunting is an activity performed by various animals, most anthropologists and biologists have suggested different characteristics of the hunting performed by humans and some of their ancestors (verHill, 1982). Many factors, such as frequent tool usage, group coordination for hunting journeys, and daily presence of meat in diets, indicate that hunting is possibly accountable for the evolution of various human characters (Dart, 1953). However, some characteristics of hunting activities, such as the use of tools, seem to be common among humans and nonhuman primate species, indicating that hunting is remarkable among primates, in that it differs from other animals' predatory activities (Stanford, 1999).

The killing of other animal species for food and other uses is the primeval way of human life (Hughes, 2007) and characterizes hunting as one of the oldest known human activities (Alves, 2012), which has been perpetuated until the present time. Hunting has been disseminated worldwide with different motivations depending on geographic area and several biological, socioeconomic, political, and institutional factors. Hunting presents several implications for humans who exploit game animals and for the exploited species and their environment. This

Ethnozoology
http://dx.doi.org/10.1016/B978-0-12-809913-1.00007-7

chapter discusses the role of hunting in human society, highlighting its sociocultural implications and effects on the conservation of faunal biodiversity.

BRIEF HISTORY OF HUNTING

The advent of omnivorism through hunting was the determinant of the evolutionary success of hominids. Animal protein is rapidly absorbed by humans and saves energy spent on digestion, the surplus of which is mainly directed for brain development. Due to this fact the brain of *Homo sapiens* is 3.5 times bigger than that of the first hominids (Hill, 1982; Speth, 2010; Washburn and Lancaster, 1968). Furthermore, the exploitation of different food resources led to the selection of more intelligent individuals, who employed different hunting strategies depending on the target animal. Hunting also led to the evolution of social relationships by demanding a complex communication structure and better mutual collaboration among the individuals in a group (Henshilwood and Marean, 2003). Alongside hominid evolution to the present day, wild fauna became increasingly important for humans (Crosby, 1993; Kalof, 2007a) and human dependence or codependence on faunal resources became increasingly evident (Alves, 2012; Alves and Souto, 2015).

Meat has higher energy and protein content than almost all known vegetables (Souto, 2014). The methods of scavenging and hunting are used to obtain meat from terrestrial vertebrates (Hawkes et al., 2001; Hill and Kintigh, 2009). However, no information is available on the methods used by ancient hominids for obtaining meat (Speth, 1989). Analysis of archeological sites supports the argument that scavenging (of opportunistically found animals or those abandoned by predators) was the main meat source for australopithecines, *Homo erectus* (perhaps), and even ancient premodern humans, indicating that hunting initially played a limited role as a method for obtaining animal protein (Blumenschine, 1986; Bunn et al., 1980; Isaac and Crader, 1981; Trinkaus, 1987).

Shipman (1986) reported that an advantage of obtaining meat through scavenging is the relatively lower energy expense associated with it compared to hunting activities. Counterarguments state that the excessive dependence on scavenging of wild animal carcasses would provide a limited meat amount and would expose ancient hominids to attack by opportunistic predators (Mundy et al., 1993; Russell, 2012). Despite the contribution of hunting in providing meat at a relatively higher frequency than the other animal protein obtainment strategies, various lines of evidence indicate that hunting was present from the beginning of human evolution. Even nonhuman primates actively pursue and capture wild animals (ver Boesch, 1994; Mitani and Watts, 2001); thus it is almost certain that hunting has been performed since times older than the fossil reports were able to detect (Souto, 2014).

From the first hominids to the *Homo* species, the increase in meat consumption has been directly related to hunting activities. Therefore, hunting for meat obtainment might be the most important interaction between our evolutionary lineage and the exploited wild fauna. For forager groups, an increasing dependency on game meat over a few thousand years favored the expansion of the *Homo* genus beyond African endemism centers. One of the most accepted hypotheses about human geographic expansion states that the colonization of Northern Africa and some European and Asian areas 1 million years ago is attributed to the hunting abilities of *Homo* species (Glambe, 2003).

Although the extinct hunting and scavenging societies presented a relatively feasible system for maintaining low demographic populations, they needed to move frequently for obtaining resources and withstanding limitations imposed by adverse climatic and ecological conditions (Stearns et al., 2011). Hunting, one of the most basic subsistence strategies, permitted human groups to survive with scarcity of some food types. In African savannas with limited vegetable food resources (e.g., fruits), *H. erectus* has been known to actively hunt large mammals 1.5 million years ago. Some studies have reported that the evolution and dispersion of this species resulted due to its capability of hunting large mammals (Brand-Miller et al., 2009; O'Connell et al., 1999). Neanderthals mainly depended on game meat and spread along thousands of kilometers of icy areas, from Europe to Asia (Sørensen, 2009).

Taphonomic analysis at various stratigraphic levels indicated that hunting was an increasingly important activity for subsistence during the Pleistocene and Holocene epochs (Blasco and Fernández Peris, 2012; Hockett and Haws, 2002). Cutting and burning marks and bone fragmentation patterns have been essential for reconstructing human hunting history, providing data beyond faunal composition and permitting understanding of how hunting occurred, how meat was extracted, which other animal parts might have been used, and which tools were used for hunting (Cochard et al., 2012; Konner and Eaton, 2010). Within this context, it is known that the horse (*Equus* sp.), red deer (*Cervus* sp.), woolly rhinoceros (*Coelodonta antiquitatis*), woolly mammoth (*Mammuthus primigenius*), and bison (*Bison* sp.), in addition to small species, provided sufficient energy and protein content for Neanderthal needs. Some of these large animals, especially *M. primigenius*, were hunting targets for ancient human species (Sørensen, 2009). In Europe, sites from the early Middle Pleistocene have presented remains of various game species, including the straight-tusked elephant (*Palaeoloxodon antiquus*), deer (*Cervus elaphus*, *Dama* sp.), auroch (*Bos primigenius*), horse (*Equus caballus torralbae*), and leporids (Cochard et al., 2012; Costamagno and Laroulandie, 2004; Villa and Lenoir, 2009). This indicates that humans obtained game meat from a wide variety of prey species, ranging from large animals considered extremely dangerous to medium-sized and small animals, since prehistoric ages.

With an increase in hunting success through different periods and ecological conditions, game meat has been an advantageous resource for human subsistence. The Pleistocene epoch presented a rich large mammal fauna; a single mammoth could feed a group of 50 individuals for at least 3 months (Brand-Miller et al., 2009), indicating the importance of wild fauna for ancient human groups. Currently, various examples indicate that game meat is extremely relevant as a protein source. For Guayaki-Aches, wild meat is a substantial energy source. Six out of the nine main species hunted by this group provided more than 5000 cal/h, whereas oranges were the main vegetable resource, providing 5071 cal/h (Hill, 1982). At present, foragers from high-altitude regions, who almost exclusively depend on game meat for subsistence (Dunbar, 1977; Mittermeier et al., 1994; Richerson et al., 1996), provide a proof of the important role of wild animal meat during the most extreme periods of human evolution, including the glacial ages.

Hunting for meat obtainment was not only one of the most important activities for survival of the evolutionary lineage groups of *H. sapiens*, but also the reason for several behavior traits, including individual interactions permitting strong social relations. Since its origin, the hunting activity has been performed by groups and thus required well-planned coordination of individuals (Frison, 1987) to compensate for the absence of weapons for hunting large mammals. In this sense, hunting possibly enabled improvements in cognition and communication among human groups. Hawkes (2001) emphasized that hominids would never have become skilled in hunting large animals without language. As a unique human characteristic, language coevolved with a series of morphological modifications, e.g., increase in brain size (MacWhinney, 2002). A strong argument states that the selective pressure for language resulted because of the requirements for survival, e.g., obtaining meat, under brutal conditions, rather than human–social interactions (Calvin and Bickerton, 2000).

MOTIVATION TO HUNT: BEYOND MEAT

Hunting is one of the most ancient human activities and has perpetuated through history to a greater or lesser extent in almost all countries worldwide for various reasons. There is a consensus that hunting has been performed for defense against wild animals and food procurement (Fig. 7.1). However, it has been emphasized that in addition to food procurement, hunting provides animal products for various uses; this, in turn, provided an additional stimulus for this activity.

Faunal resources have presented various uses. Although meat was the main exploited product, diverse faunal resources have been used for various hominid and human requirements since prehistoric ages. An important use of faunal resources was protection against cold weather. Although fire was one of the most

FIGURE 7.1 Wild meat on the menu. (A) A man holds up bushmeat, Papua, Indonesia, (B) wild rats roasted at fire for consumption in Lindi, Tanzania, (C) market bushmeat in Laos; (D) turtle killed for meal in Limbe, Cameroon, West Africa. *Photo credits: (A) Agus Andrianto for Center for International Forestry Research (CIFOR); (B), (C) and (D) Anthony B. Cunningham.*

common resources for protection against cold weather, various clothing types were manufactured using the leather and skin of various animal species (Bordes and Thibault, 1977; Coon, 1971). In addition to protection, these clothes provided warmth outside shelters, thus permitting movement through cold regions for performing daily activities. Use of leather varied among different populations, periods, and locations; however, the use of the skin of animals, e.g., bears, foxes, wolves, lagomorphs, and rodents, has been reported for large portions of the Old World, indicating that humans used these resources for thermal insulation as well as esthetic items (David, 2011; Fairnell, 2003; Klein, 1971). In general, leather and skin typically characterized Paleolithic clothing; hunting for obtaining meat as well as complex clothing (composed by more than one layer) might be directly related to the human capability of living in cold locations (Adams, 2011; Hoffecker, 2005). In addition, migrating human groups combined leather with plant material, stone, or bone to construct temporary shelters (McCartney, 1984; Stearns et al., 2011; Tyldesley and Bahn, 1983).

Wild fauna also became a source for boat construction materials, which permitted *H. sapiens* to effectively explore aquatic environments (Cordain, 1999). Evidence indicates that sea adaptation (including oceanic traveling) favored demographic and geographic expansion 150,000 years ago (Erlandson, 2001). In the Upper Paleolithic period, boats made of skin were used for colonizing North American regions through migration routes from Asia. Such boats permitted the hunting of aquatic mammals in regions where land was occupied by icy layers, which hindered the use of terrestrial floristic and faunal resources (Bradley and Stanford, 2004). Based on similarities between Solutrean (European) and Pre-Clovis/Clovis (American) tools (Bradley and Stanford, 2004), it was emphasized that seal hunting by marine mammal hunters might have stimulated human groups to follow the migration patterns of these animals, eventually allowing some humans to reach islands and locations on the North American west coast.

Use of animals or their parts for medical applications is another ancient practice that became widespread among various known cultures and is still used by traditional medical systems to the present day (see Chapter 16). Bone marrow and fat from vertebrates were found to be the most ancient products for therapeutic use. Furthermore, the topical use of these products has been reported in the majority of contemporary Eastern and Western pharmacopoeias, which indicates the perpetuation of their use for such purposes over time (e.g., Alves and Rosa, 2013; Ashwell and Walston, 2008; El-Kamali, 2000; Lev, 2003; Mahawar and Jaroli, 2008; Quave et al., 2010; Souto et al., 2011). Although plausible, this assumption was not easily confirmed using paleontological–archeological approaches, as bone marrow and fat do not undergo fossilization; however, their exploitation might be inferred by analyzing other animal parts, e.g., broken bones (Outram, 2001; Rustioni et al., 2007; Sørensen, 2009). Furthermore, fat and bone marrow were more appreciated by hunters and scavengers as energetic foods (deFrance, 2009; Reitz and Wing, 2008; Russell, 2012); thus it was difficult to identify other uses of these products by paleo-zooarcheological investigations. It is worth noting that at the dawn of the recorded history, early man was known to often eat or wear some portion of an animal that was thought to have a healing or protective effect (MacKinney, 1946). This aspect highlights the fact that the medicinal use of faunal elements was intertwined with their use as food (Alves et al., 2013a,b,c).

Later in human history, several wild fauna species were domesticated. Currently, pets might be domesticated or wild animals (Russell, 2012); however, in both cases they involved a part of the history of human hunting activities. Dogs, for example, were the first animals to be domesticated 14,000 years ago (Sablin and Khlopachev, 2002). Current assumptions state that this process was favored by similar social behaviors presented by humans and gray wolves, which frequently fed on carcasses remaining from human hunting or scavenging activities (Kalof, 2007b). Although dogs are commonly used as pets, their domestication occurred mainly due to their hunting abilities (Clutton-Brock, 1999; Gupta, 2004; Morey, 1994); thus, dogs have been supporting hunting activities since ancient ages.

Various animals have been kept as pets by humans, especially birds, to which individuals have been attracted by a combination of factors, such as their singing and flashy feathers (Alves et al., 2013a,b,c; Fernandes-Ferreira et al., 2012; Nash, 1993). Capturing and keeping animals as pets was a widely disseminated practice among cultures worldwide, probably without a single origin. Ancient Greeks skillfully bred birds using cages and offered them as gifts (Kitchell, 2011). There are reports from India about parakeets, mynahs, and Columbidae birds being kept as pets since ancient times, mainly by noble families (Ahmed, 1997). Juan and Ulloa (1758) were astonished by the demonstrations of bonds and affection between South American natives and wild animals kept as pets. Long before, Europeans observed, while occupying America, that wild animal breeding was a common practice among natives. The memoirs of Bernal Diaz del Castillo' about Mexican occupation and colonization highlight that the emperor Montezuma II maintained an aviary composed of various bird species, including eagles, psittacidae, and passeriformes, some of which were sourced from long distances by Aztec merchants (del Castillo, [1570] 1996). Among Incans, a traditional practice consisted of nobles gifting the emperor with feline, nonhuman primates, and wild birds for breeding purposes (de la Vega, [1612] 1961). In practical terms, although the objective was to keep these specimens alive, this type of faunal exploitation was not different from their capture and slaughter for other uses, as specimens were mostly isolated from the wild population, from a reproductive perspective (Sick, 2001). Currently, the use of live animals as pets is one of the most disseminated breeding practices worldwide and involves increasing taxonomic diversity (Alves, 2012; Mitchell and Tully, 2009). Millions of people worldwide interact with a wide variety of wild and domestic taxa as pets (Alves, 2012; Alves et al., 2010; Beck and Katcher, 1996; Mitchell and Tully, 2009). Although mammals (mainly dogs and cats) are predominantly used as pets, other animals, such as reptiles, amphibians, and some invertebrate groups, are becoming increasingly common among breeders (Alves, 2012).

Another hunting stimulus is the sale of animal products for different purposes. Historic reports suggest that the commerce of wild fauna is an ancient practice (Khorenatsi, 1978; Vivian, 1991). Currently, various studies indicate that legal and illegal trade of wild fauna products is disseminated worldwide, involving different taxa and their products, and even live animals, thus providing a strong hunting stimulus at various locations (Alves et al., 2013a,b,c; Roldán-Clarà et al., 2014; Stuart, 2004; Vanstreels et al., 2010; Williamson, 2004; Wu, 2007).

Although various animal products were already being used, ancient humans could have become prey for wild animals. Naturally, defense against wild predators has been, and still is, a source of motivation for hunting. The advent of domestication intensified the motivation for hunting, because humans already concerned with defense against predators needed to protect their herds from the attack of natural predators. Furthermore, animals attacking crops also became hunting targets. Thus, conflicts between humans and wild animals have been considered one of the main reasons for slaughtering various taxa and are a serious problem for wildlife conservation (Conover, 2002; Fernandes-Ferreira et al., 2013; Garcia-Alaniz et al., 2010; Graham et al., 2005; Marshall et al., 2007). Pitman et al. (2002), for example, reported that carnivore mammals forced to coexist with domestic animals might experience conflict with farmers, thereby increasing wild fauna mortality rates. For this reason, the carnivore group has been harassed worldwide and has extremely low natural populations (Cavalcanti, 2006). Furthermore, this motivation also stimulates hunting of the other taxa, e.g., snakes, falcons, and hawks (Alves et al., 2012, 2014; Fernandes-Ferreira et al., 2012).

Another hunting stimulus is the status of this activity in some societies, which acts as a marker to distinguish social strata, ranging from indigenous to royal societies (Allsen, 2006). Among ancient emperors, hunting was a recreational activity and a status symbol. These aspects, in association with other hunting activities, have been a strong stimulus for the slaughter of wild species worldwide (Allsen, 2006; Alves et al., 2009, 2016; Barboza et al., 2016; El Bizri et al., 2015; Fernandes-Ferreira et al., 2012). This type of hunting activity encourages competition among hunters for larger prey animals or larger amounts of slaughtered species (Fig. 7.2). This activity is referred to as "trophy hunting" and is observed worldwide, mainly in locations where hunting is legal and results in various controversial discussions among researchers regarding its sustainability. Hell and Find'o (1999) analyzed the population density of brown bears (*Ursus arctos*) in Czech Republic and found that trophy hunting of large males affected their population structure. As a result, hunting of bears larger than 150 kg was strongly restricted and lately prohibited. Madhusudan and Karanth (2005) reported that, in precolonial India, hunting for subsistence and as a feudal sport was common and based on traditional techniques and abilities. After the establishment of colonial potencies and introduction of firearms, trophy hunting exceeded the yearly permitted extent of slaughter in the country.

FIGURE 7.2 Trophy hunting of elephants in Africa. *Photo credits: Alan Gardiner.*

In addition to the examples of trophy hunting as an environmental impact factor, this activity is also condemned by animal rights groups. Although there is no consensus among environmentalists (Lindsey et al., 2007), it is important to point out that well-conducted trophy hunting is considered an important conservation tool because of its economic value and might be used for conserving wild areas. Lindsey et al. (2006) reported that trophy hunting is an important economic activity in Africa and generates significant revenue for wildlife at various locations. Dickman (2008) highlighted that Tanzania had its abundant wildlife explored by foreign hunters for generating revenue. As per this researcher's data, biodiversity accounted for approximately 16% (US$ 52 million) of the country's gross domestic product (GDP) in 2006. Especially, trophy hunting was considered an important revenue source, which generated US$13 million during that year. A frequent issue is that only a small part of this revenue is allocated for environmental organizations and local communities (Balmford et al., 2015; Humavindu and Barnes, 2003; Leader-Williams et al., 2009). However, some positive examples have also been observed, such as in Namibia, where an expressive proportion of the legal and regulated hunting revenue is transferred to the traditional populations, thus generating income and increasing abundance of various species (Booth, 2010).

Di Minin et al. (2016) evaluated negative and positive examples of trophy hunting and recognized that several African countries do not establish slaughter quotas based on the scientific studies, and animal ages are usually not established, among several other economic and ecological management problems. However, the authors state that banning trophy hunting might result in important biodiversity losses, firstly due to economic factors, because resources for conservation are usually limited in highly biodiverse countries, even when considering the biotourism revenues. Furthermore, when compared with ecotourism, hunting presents a reduced ecological footprint, i.e., reduced carbon emission, personnel requirements, and infrastructure. Thus, the authors concluded that proper and regulated management through appropriate ecologic studies is essential for preserving the hunting activity with conservation-promoting results.

HUNTING TECHNIQUES

Despite lacking typical carnivore morphology, humans became one of the most efficient predators worldwide (Ojasti, 2000). Gathering faunal resources requires hunters to develop strategies and tools for capturing different-sized animals living in various environments (Alves et al., 2009; Bezerra et al., 2012; Fernandes-Ferreira et al., 2012). Development of sophisticated hunting strategies turned humans into the most efficient hunters among all animals. A recent study conducted by Darimont et al. (2015) suggested that humans function as unsustainable "superpredators," which kill wild herbivores at about the same rate as other predators such as lions, wolves, and grizzly bears; however, humans kill large carnivores at nine times the rate of the other predators. The impact caused by the success of humans as "superpredators" led humans to slaughter adult animals, unlike other animals, which prefer younger (and easier-to-slaughter) prey animals. This, in turn, led to the increased extinction rates due to elimination of animals at peak reproductive age (Darimont et al., 2015).

In addition to requiring mastery of different hunting strategies, hunters must develop detailed knowledge about the habits of game species, an important factor for hunting success. After choosing the prey, hunters must know where they are located—generally, at the same location where they eat, drink, or sleep. In this sense, the knowledge about the game species results in adaptive human behavior toward the environment to improve their hunting efficiency. Thus, the knowledge about species is essential for developing hunting activities.

As pointed out by Ross et al. (1978), an important part of the adaptive hunting is the coevolution of hunting technology and general procurement strategies. A variety of hunting methods are available to exploit the faunal resources, and individual hunters commonly use more than one technique. The possibility of using various hunting techniques permits adaptation to the varying availability/accessibility of the game animals (Alves et al., 2009).

The choice of hunting technique depends on the prey type, its habits, and the environment where it lives, in addition to the intended use of the animal. In general, humans perform ambushes, speed chases, and/or teamwork, among active techniques that require the actual presence of a hunter. Alternatively, they may adopt passive techniques, which do not require the actual presence of a hunter.

One of the most common active techniques is known as the "waiting game," in which a hunter camouflages and ambushes the prey (Fig. 7.3). Camouflaging involves hiding behind vegetation or building a structure made of branches, which is installed in an area close to where the prey animals eat or drink; these structures might be built on the top of trees in areas with tall plants. Once the ambush is set up, the camouflaged hunter stands by and slaughters the approaching animals, using firearms or other hunting tools such as arrows and spears.

Another important strategy to improve hunting success involves approaching prey animals

FIGURE 7.3 Representations of waiting/ambush hunting strategy, in which a hunter camouflages himself and ambushes the prey. *Illustration: Washington Vieira.*

FIGURE 7.4 Hunting by calling, a strategy in which hunters make sounds by voice or with call mechanisms (whistles, for example) as similar as possible to prey's vocalization to attract them to be killed. *Illustration: Washington Vieira.*

FIGURE 7.5 Whistles used by hunters to attract preys (game vertebrates) in the semiarid region of Brazil. *Photo credits: Lívia E. T. Mendonça.*

very closely, or attracting them nearby, and then slaughtering or capturing them immediately. In the Brazilian semiarid region, hunters imitate the voices of animals (mainly of birds) to attract them. Hunters simulate singing using their own lips or manufactured whistles (Figs. 7.4 and 7.5). Prey animals are slaughtered as they approach the hunter's range (Alves et al., 2009). Species of the family Anatidae and Tinamidade are hunted using these strategies in several locations worldwide. Another example is of the Venezuelan Achagua people, who imitate tapir sounds to attract the animals, which are then slaughtered by the arrows coated with plant poison (Gumilla, 1686–1750). Jaguars (*Panthera onca*) have been captured by some South American populations using an instrument simulating a cuíca (a Brazilian friction drum, with a wooden rod attached to a leather membrane from the inside) to imitate the animal's roar and lure it to a position within their range.

Another method for approaching the prey involves hunter camouflaging to avoid detection. Normally, clothes are used for camouflaging because of their fabric color or the use of natural elements, e.g., branches and leaves. In duck hunting performed by the natives from the Maracaibo Lake (Venezuela) and the Brazilian states of Mato Grosso, Pará, and Amazonas, hunters go into the water wearing a gourd (popular designation for *Lagenaria* and *Cucurbita* fruits) with holes (for enabling vision) on their heads (Fig. 7.6). They approach the ducks and then rapidly pull them underwater until drowning (Pereira, 1967; Walker, 1822).

FIGURE 7.6 Duck hunting in which hunters approach the prey by camouflaging himself to avoid frighten them. *Illustration: Washington Vieira.*

The same technique was adopted by Apaches from Arizona and New Mexico (Cremony, 2016). Apaches also hunted antelopes by approaching them, while wearing an antelope's skin with head and horns (Cremony, 2016).

In addition, the hunters might make use of the social behavior of prey species to attract them. FitzGibbon (1998) reported that if Piro hunters in Peru catch a social primate that is still alive, they try to force it to make distress calls for attracting other members of the troop. Once these individuals are lured into their range, they can then be killed. A similar technique is used to catch capybaras (*Hydrochoerus hydrochaeris*); one hunter carries a young capybara as a lure whereas another walks ahead with a torch and a club to intercept any capybaras that approach. Another example is of bison (commonly called a buffalo) hunting by American Indian people. To attract these animals, a young man would dress up like a buffalo calf. He would then approach the herd, mimicking calf behavior. He would have to make the calls of a buffalo calf—not just any calls—but also those made by a calf in distress. Other young men dressed as wolves and mimicking their behavior would appear in the prairie behind the herd and create an illusion of danger (Fig. 7.7). Slowly and patiently, these buffalo runners would lead the herd toward the killing site (Ojibwa, 2012).

Persistence hunting (also called endurance hunting or cursorial hunting) is believed to be one of the most ancient hunting strategies. In this type of hunting, a hunter uses a combination

FIGURE 7.7 Bison hunting by American Indians, in which animals are strategically attracted by hunters on a buffalo calf skin apparel and others hunters dressed on wolves skins. *Illustration: Washington Vieira.*

of running, walking, and tracking techniques to pursue prey until it is exhausted (Liebenberg, 2006). Persistence hunting was likely one of the tactics used by early hominins (Carrier et al., 1984; Liebenberg, 2008) and is still performed in various locations worldwide, as reported by the indigenous hunter-gatherer people of Southern Africa (Attenborough, 2003). Native American tribes also had various traditions of chasing down animals on foot (Heinrich, 2001; Nobokov, 1981). Tarahumara chased deer through the mountains of northern Mexico until the animals collapsed from exhaustion, and then throttled them by hand (Bennett and Zingg, 1935). Aborigines of North Western Australia are also known to have hunted kangaroos by chasing them (McCarthy, 1957; Sollas, 1915).

Throughout history, persistence hunting might have been supported by animals, e.g., dogs, falcons, and horses (Fig. 7.8). Dogs are the most ancient and widespread hunting support for humans. Dog domestication first occurred approximately 14,000 years ago for protection and hunting optimization (Morey, 1994). Since then, various cross-breeding procedures to promote the morphological specialization of breeds completely adapted to hunting have been performed independently and collaboratively across the world. During hunting, the hunter follows predefined trails, whereas the dog tracks adjacent areas. Although prey slaughter by dogs is common and might characterize this technique as a direct capture method, this technique is essentially used for locating and approaching game species. It is important to highlight that this technique significantly improves the capture rates. Redford and Robinson (1987) reported that the use of dogs for hunting usually results in an increase in the number of prey animals captured. Koster (2008) conducted research on the use of dogs for hunting in Nicaragua and found that hunters who used this technique obtained nine times the success rate of hunters without dogs. Furthermore, prey animals captured by dogs were more diverse, and some were captured exclusively when this technique was used. In Panama, Ventocilla et al. (1995) reported that hunting supported by dogs was the main reason for the local extinction of several species.

Prey tracking is also a common technique used for hunting at various locations (Fig. 7.9). It consists of identifying the prey's trail, which leads

FIGURE 7.8 Bison and kangaroo hunting supported by horses and dogs. *Illustration: Washington Vieira.*

FIGURE 7.9 Prey tracking, a widespread hunting technique. *Illustration: Washington Vieira.*

the hunters to their location; prey animals are then slaughtered as soon as they are found, usually using firearms. Tracking hunters must know the life habits of prey animals and should be able to identify them by their trails (Alves et al., 2009). Species that make trails through vegetation or use burrows, particularly simple and single-exit ones, are more vulnerable to snaring than those species that have less-defined movement patterns. For example, one of the reasons that the four-toed elephant shrews (*Petrodomus tetradactylus*) are caught much more frequently than the golden-rumped species (*Rhynchocyon chrysopygus*) in Arabuko-Sokoke Forest, Kenya, is that the former make trails through the leaf litter along which the snares can be set (Fitzgibbon et al., 1995). Pigs are also vulnerable because they travel along trails (M. Alvard, cited in FitzGibbon (1998)).

Several animals are nocturnal; thus, flashlights might be used for hunting. In the Brazilian semiarid region, hunters use flashlights to obfuscate the prey animals, usually birds on branches (Fig. 7.10). Birds are unable to take off due to impaired vision and are captured by hunters (Alves et al., 2009; Bezerra et al., 2012). Similarly, hunters in Arabuko-Sokoke Forest, Kenya, use strong flashlights to temporarily blind duikers (FitzGibbon, 1998).

Some techniques promote the reckless slaughter of prey species by using fire and poison.

FIGURE 7.10 Hunting using flashlights to obfuscate the game animals. *Illustration: Washington Vieira.*

Fire is used for various purposes, e.g., management of landscapes and native plants for food or manufacturing, as observed in the Brazilian Midwest region (Miranda et al., 2009). This technique involves first scaring the animals and then directing them to a location where traps are set and firearms are ready for use. It is also used to capture ungulate herbivores, especially tapir and deer, which visit burned areas during seedling growth as fire is a great stimulus for the germination of herbaceous and subshrub plants as well as for the synchronization of flower production and cross-pollination (Melo and Saito, 2011).

Use of poison consists of adopting natural or artificial chemical compounds for capturing or eliminating wild animals. A wide variety of natural compounds have been used as poisons by traditional populations for centuries; however, artificial poisons, e.g., strychnine, might be used. For example, the South American communities traditionally use native plants for producing toxic compounds for hunting and fishing. In Brazil, the most commonly used plants for this purpose include *Paullinia* spp., *Serjania* spp. (Sapindaceae), *Tephrosia toxicaria*, *Enterolobium timbouva*, *Derris elliptica* (Fabaceae), *Palicourea marcgravii* (Rubiaceae), *Mascagnia rigida* (Malpighiaceae), and *Manihot* spp. (Euphorbiaceae) (Heizer and Ribeiro, 1987; Pezzuti and Chaves, 2009; Tokarnia et al., 2000). In some cases, poison is applied to the arrowheads used for animal slaughtering. For example, Indians of northern South America use the skin secretion of frogs of the genus *Phyllobates* to poison their arrows for hunting (Myers et al., 1978). Similarly, the Abor people of northeastern India use poison arrows that can kill tigers, buffaloes, and elephants in hunting and warfare (Jones, 2009).

Use of firearms is considered the most important animal slaughter method, because it permits the slaughtering of all main species instantaneously, effectively, and from long distances. Although some human communities still do not have access to this technology, the preference for firearms is considered a global standard

(Anderson, 1985). Firearms accurately hit animals and prevent issues associated with melee combats, e.g., the wounds that a tiger receives while hunting boars (Darimont et al., 2015). Revolvers are sporadically used by some hunters; however, rifles are the weapons of choice among firearms users. The most commonly used model is a shotgun, breech-loaded or centerfire types, invented by a Belgian military engineer called Berminmolin, who adapted a model by Lefauchex in 1852. The advent of the two World Wars, in 1914 and 1945, disseminated these firearms worldwide. Shooting techniques strongly depend on the hunting target and its behavior. Body and firearm positions, point of aim adjustment, and shot momentum might change depending on the prey's direction of flight or sprint as well as the environmental characteristics (thicket, open field, or water bodies), presence of dogs or other hunters, and various other particularities. The main body regions that guarantee an ideal prey slaughter are auricular, scapular, and frontal cranial regions (Buys, 1934; Santos, 1950).

TRAPS

The use of traps is ancient and perpetuates to the present. They are used for capturing a wide variety of prey of different sizes and taxa. An advantage of using traps is the time and energy saved by the hunter, because it can be set up and then checked after some time. Traps of varied types, sizes, complexity, and functionality exist globally (Fig. 7.11 and 7.12). According to Fernandes-Ferreira (2014), hunting

FIGURE 7.11 Examples of hunting traps. *Illustration: Washington Vieira.*

FIGURE 7.12 Trap from bamboo stakes used for large ungulates, DR Congo (above) and trap for small rodents, NW Zambia (below). *Photos: Anthony B. Cunningham.*

techniques can be categorized in various ways. Based on their lethality, they are categorized as lethal and nonlethal; thus, the hunter is able to choose between slaughtering the animal or not. Selective traps permit the hunters to choose which animal will be captured, e.g., "armadillo cages," which are funnel-shaped traps used for capturing armadillos in South America (Alves et al., 2009; Santos, 1950). Contrarily, randomized capture traps do not permit the hunters to choose their prey. Additionally, the traps might be categorized by their quantitative potential of capture: individual, if only a single specimen is to be captured, e.g., snare traps, or collective, if it is possible to capture several individuals, e.g., deadfalls.

Using deadfall traps is one of the most rudimentary hunting techniques established by the ancient Hominidae members. To the present, this method is widely used by various ethnic groups because of its low cost. The depth of the deadfall trap depends on the size of the species to be captured, hunting area, local culture, and other factors that might affect the use of other elements, e.g., sharp sticks in the ground, coverage with plant parts for camouflage, and even a lid for preventing escape. A similar, usually nonlethal, technique involving the capture of entire animals is the use of cages made of wood, metal, or other materials. The cages are usually associated with the baits for attracting animals, which are rapidly captured by the trap's funneling or shutting systems. A similar strategy is the use of large labyrinth-shaped fences or fences with shutting systems for capturing medium and large terrestrial social mammals.

Other techniques involve capturing animals by hitting a body part, usually using invasive and potentially lethal methods. This category includes snare traps, mouse traps, hooks, and body grip traps. Crushing traps outscore these lethal techniques and consist of hanging stones, spears, or wooden logs that collapse over the animal as it reaches inside the trap. Branches and other flexible structures might be twisted and tensioned, and combined with a trigger system for the same purpose. These methods are observed in practically all traditional hunting cultures worldwide. The second group of lethal methods comprises more complex projectile shooters. Spears, arrows, and bullets are shot using primitive equipment, e.g., crossbows and handcrafted shotguns, or industrialized firearms.

IMPORTANCE AND IMPLICATIONS OF HUNTING

Hunting has been undoubtedly crucial for the evolution of humans, who have directly or indirectly depended on its products throughout history. Human species, through their cultural and technologic progress, became the most efficient predators worldwide. They pursued hunting to meet the large demand for animal products (Darimont et al., 2015) and consequently intensified faunal exploitation led to the unsustainable rates of wild animal hunting in some cases (Bennett and Robinson, 2000). Darimont et al. (2015) compared human predatory patterns with those of the other predators that compete for prey animals (terrestrial mammals, and river and sea fishes) and concluded that humans slaughter adult prey animals, which are the reproductive assets of a population, at a rate 14 times higher than the other predators. This competitive domain and other exclusively human predatory behaviors led specialists to define humans as unsustainable "superpredators," who will continue to globally modify ecological and evolutionary processes, unless extreme and improbable behavior modifications occur.

Mammals and birds have been the animals most hunted in the world (Figs. 7.13–7.15), and hunting by humans represents an important extinction threat to several species. Reptiles also are an important source of protein for human populations in many locations (Fig. 7.16). Increase in human population, introduction of modern hunting techniques, and increase in the trading of faunal species have intensified the pressure on wild populations (Milner-Gulland and Bennett, 2003; Robinson and Bennett,

FIGURE 7.13 On the right, an antelope captured by a local villager in Cameroon, Africa *(Photo by Edmond Dounias for Center for International Forestry Research (CIFOR))*, and, on the left, a wild boar meat (*Sus scrofa*), an important game mammal for Boepe village residents, Merauke, Papua, Indonesia *(Photo by Michael Padmanaba for CIFOR)*.

FIGURE 7.14 Game animals slaughtered and trade market of bushmeat. (A) Kids shouldering bushmeat in Brazil, (B) skinned antelope for sale in Guinea, Africa, (C) wild boar caught in a hunter's trap, Pengerak village, West Kalimantan, Indonesia, and (D) bushmeat sold in the local market at Ebolowa, Cameroon, Africa. *Photo credits: (A) Luke Parry for Center for International Forestry Research (CIFOR), (B) Terry Sunderland for CIFOR, (C) Ramadian Bachtiar for CIFOR, and (D) Colince Menel for CIFOR.*

2000). Hunting is currently recognized as the second greatest threat, after habitat loss, to the wild fauna (Vié et al., 2009), and its effects have been recorded for thousands of years. Some zooarcheological and paleontological studies indicate that during the late Pleistocene and early Holocene (between 12,000 and 7000 years ago), hunting was one of the main reasons for the extinction of the Neotropical realm's megafauna, e.g., ground sloths (*Megatherium* and *Glossotherium*) and notoungulates (*Toxodon*) (Barnosky et al., 2004; Borrero, 1999; Gutiérrez and Martínez, 2008). However, wild fauna extinction rates were extremely high after the 17th century (Dirzo et al., 2014; Redford and Robinson, 1991; Vié et al., 2009). Dirzo et al. (2014) indicated that from the 1500s to the present, 322 terrestrial vertebrate species became extinct, and the remaining presented at least an average 25% population decline during the last decades. These authors also reported serious knowledge gaps about the consequences of defaunation; such gaps hinder an accurate evaluation of defaunation.

FIGURE 7.15 An armadillo and a parrot captured by hunter in Colombia (above) and an exotic bird killed by poachers in Cameroon (below). *Photo credits: Barbara Fraser/CIFOR and Terry Sunderland/CIFOR, respectively.*

Various studies indicate that, in general, hunting activities are unsustainably performed worldwide (Dirzo et al., 2014; Fitzgerald, 1994; Hill and Padwe, 2000; Peres, 2000). Peres (2000) observed that the rural populations from various Amazon regions annually consume a total of 9.6–23.5 million reptiles, birds, and mammals, which represents a total estimated biomass of 67,173–164,692 ton, with a 36,392–89,224 ton consumption of wild meat. The author also reports that the small (<1 kg) and medium (1–5 kg) species did not present variations per hunting rates, whereas the large (5–15 kg) and very large species (>15 kg) presented a decrease in biomass and density due to the intensification of hunting activities.

Furthermore, some authors have stated that sustainability might be achieved in areas with appropriate management of the game species (Alvard et al., 1997; Bodmer et al., 1994; Bodmer and Puertas, 2000). For example, Bodmer et al. (1994) reported that establishing management strategies for the Tamshiyacu Tahuayo Regional Conservation Area, Peru, resulted in a 35% decrease in the total biomass of captured mammals.

Excessive capturing of the nature's specimens is considered one of the most important threats against global fauna (Bennett and Robinson, 2000; García-Moreno et al., 2007; Redford and Robinson, 1991), in addition to other factors, e.g., habitat destruction and fragmentation, climate change, introduction of exotic species, pollution, natural disasters, and roadkill (Alves and Albuquerque, 2012; Barlow et al., 2002; Dirzo et al., 2014; Hengemühle and Cademartori, 2008; Peres and Nascimento, 2006). Therefore, the causes of defaunation must be addressed, considering a wide variety of factors. However, it is necessary to understand that hunting (even in an individual manner) comprises the power for potential large-scale extinction and might be pointed out, in various situations, as the primordial motivation for the population depletion of game species (Canale et al., 2012; Cullen et al., 2000; Peres, 2001). Consequences of the hunting activity are widely reported, for example, the elimination of seed spreaders and pollinators, which are important for maintaining floristic diversity, a factor that is also threatened by the predation of large herbivores participating in forest regeneration processes. Furthermore, the depletion of top predators promotes an uncontrolled increase in the population of mesopredators, which in turn causes a severe food chain imbalance and increases threats against nongame species, in addition to promoting prejudice for human health and agriculture (Culot et al., 2013; Dirzo et al., 2014; Giacomini and Galetti, 2013).

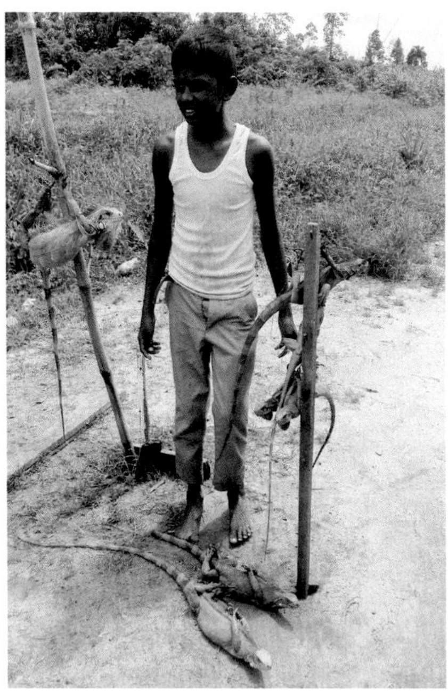

FIGURE 7.16 Reptiles as food. Common tortoise (*Kinixys erosa*) sold for bushmeat in Cameroon (right) and boy selling iguanas (left). *Photo credits: Verina Ingram/CIFOR and François Sandrin/CIFOR, respectively.*

FINAL CONSIDERATION

Hunting is performed by various animal groups; however, humans have become the most efficient hunters on earth. This success was essential for human survival, and hunting has perpetuated and diversified throughout history, with extremely relevant social and cultural roles in various human communities. Besides being used for obtaining animal protein for subsistence, hunting activities have been used for obtaining diverse products for a wide variety of uses and have expressive cultural relevance in various human communities.

The development of hunting activities by human societies up to the present time provides evidence of an evolutionary process that incorporated several biological, environmental, cultural, social, economic, political, and technological factors. Technological advances of tools, techniques, traps, and strategies used for hunting potentiated the capture of a large number of various prey species of different sizes within different environments. This led to the overexploitation of various species, causing extinction to some and population depletion of others.

This reality raises the challenge of searching for other methods that minimize the overexploitation of game species and guarantee the subsistence of human communities that exploit these resources. Thus, it is important to understand the multidimensional context in which hunting activities take place—a scenario in which the contribution of ethnozoological research is essential.

References

Adams, L.S., 2011. A History of Western Art, fifth ed. McGraw-Hill, New York.

Ahmed, A., 1997. Live Bird Trade in Northern India, first ed. TRAFFIC India, New Delhi, India.

Allsen, T.T., 2006. The Royal Hunt in Eurasian History. Cambridge Univ Press.

Alvard, M.S., Robinson, J.G., Redford, K.H., Kaplan, H., 1997. The sustainability of subsistence hunting in the neotropics. Conservation Biology 11, 977–982.

Alves, R.R.N., 2012. Relationships between fauna and people and the role of ethnozoology in animal conservation. Ethnobiology and Conservation 1, 1–69.

Alves, R.R.N., Albuquerque, U.P., 2012. Ethnobiology and conservation: why do we need a new journal? Ethnobiology and Conservation 1, 1–3.

Alves, R.R.N., Feijó, A., Barboza, R.R.D., Souto, W.M.S., Fernandes-Ferreira, H., Cordeiro-Estrela, P., Langguth, A., 2016. Game mammals of the Caatinga biome. Ethnobiology and Conservation 5, 1–51.

Alves, R.R.N., Leite, R.C., Souto, W.M.S., Bezerra, D.M.M., Loures-Ribeiro, A., 2013a. Ethno-ornithology and conservation of wild birds in the semi-arid Caatinga of northeastern Brazil. Journal of Ethnobiology and Ethnomedicine 9, 1–12.

Alves, R.R.N., Lima, J.R.F., Araújo, H.F., 2013b. The live bird trade in Brazil and its conservation implications: an overview. Bird Conservation International 23, 53–65.

Alves, R.R.N., Medeiros, M.F.T., Albuquerque, U.P., Rosa, I.L., 2013c. From past to present: medicinal animals in a historical perspective. In: Alves, R.R.N., Rosa, I.L. (Eds.), Animals in Traditional Folk Medicine: Implications for Conservation. Springer, pp. 11–23.

Alves, R.R.N., Mendonça, L.E.T., Confessor, M.V.A., Vieira, W.L.S., Lopez, L.C.S., 2009. Hunting strategies used in the semi-arid region of northeastern Brazil. Journal of Ethnobiology and Ethnomedicine 5, 1–50.

Alves, R.R.N., Nogueira, E., Araujo, H., Brooks, S., 2010. Bird-keeping in the Caatinga, NE Brazil. Human Ecology 38, 147–156.

Alves, R.R.N., Pereira Filho, G.A., Silva Vieira, K., Souto, W.M.S., Mendonças, L.E.T., Montenegro, P.F.G.P., Almeida, W.O., Vieira, W.L.S., 2012. A zoological catalogue of hunted reptiles in the semiarid region of Brazil. Journal of Ethnobiology and Ethnomedicine 8, 1–29.

Alves, R.R.N., Rosa, I.L., 2013. Animals in Traditional Folk Medicine: Implications for Conservation, first ed. Springer-Verlag, Heidelberg/New York/Dordrecht/London.

Alves, R.R.N., Silva, V.N., Trovão, D.M.B.M., Oliveira, J.V., Mourão, J.S., Dias, T.L.P., Alves, A.G.C., Lucena, R.F.P., Barboza, R.R.D., Montenegro, P.F.G.P., 2014. Students' attitudes toward and knowledge about snakes in the semiarid region of Northeastern Brazil. Journal of Ethnobiology and Ethnomedicine 10, 1–8.

Alves, R.R.N., Souto, W.M.S., 2015. Ethnozoology: a brief introduction. Ethnobiology and Conservation 4, 1–13.

Anderson, J.K., 1985. Hunting in the Ancient World. University of California Press, Berkeley.

Ashwell, D., Walston, N., 2008. An Overview of the Use and Trade of Plants and Animals in Traditional Medicine Systems in Cambodia, first ed. TRAFFIC Southeast Asia, Greater Mekong Programme, Hanoi, Vietnam.

Attenborough, D., 2003. Human Mammal, Human Hunter, the Life of Mammals.

Balmford, A., Green, J.M.H., Anderson, M., Beresford, J., Huang, C., Naidoo, R., Walpole, M.J., Manica, A., 2015. Walk on the wild side: estimating the global magnitude of visits to protected areas. PLoS Biology 13, e1002074.

Barboza, R.R.D., Lopes, S.F., Souto, W.M.S., Fernandes-Ferreira, H., Alves, R.R.N., 2016. The role of game mammals as bushmeat in the Caatinga, northeast Brazil. Ecology and Society 21, 1–11.

Barlow, J., Haugaasen, T., Peres, C.A., 2002. Effects of ground fires on understorey bird assemblages in Amazonian forests. Biological Conservation 105, 157–169.

Barnosky, A.D., Koch, P.L., Feranec, R.S., Wing, S.L., Shabel, A.B., 2004. Assessing the causes of Late Pleistocene extinctions on the continents. Science 306, 70–75.

Beck, A.M., Katcher, A.H., 1996. Between Pets and People: The Importance of Animal Companionship. Purdue Univ Pr.

Bennett, E.L., Robinson, J.G., 2000. Hunting of Wildlife in Tropical Forests. The World Bank Environment Department Papers. , pp. 1–42.

Bennett, W.C., Zingg, R.M., 1935. The Tarahumara, an Indian Tribe of Northern Mexico. University of Chicago Press, Chicago.

Bezerra, D.M.M., Araujo, H.F.P., Alves, R.R.N., 2012. Captura de aves silvestres no semiárido brasileiro: técnicas cinegéticas e implicações para conservação. Tropical Conservation Science 5, 50–66.

Blasco, R., Fernández Peris, J., 2012. A uniquely broad spectrum diet during the Middle Pleistocene at Bolomor Cave (Valencia, Spain). Quaternary International 252, 16–31.

Blumenschine, R.J., 1986. Carcass consumption sequences and the archaeological distinction of scavenging and hunting. Journal of Human Evolution 15, 639–659.

Bodmer, R.E., Fang, T.G., Moya, L., Gill, R., 1994. Managing wildlife to conserve Amazonian forests: population biology and economic considerations of game hunting. Biological Conservation 67, 29–35.

Bodmer, R.E., Puertas, P., 2000. Community-based co-management of wildlife in the Peruvian Amazon. In: Robinson, J.G., Bennett, E.L. (Eds.), Hunting for Sustainability in Tropical Forests. Columbia University Press, New York, pp. 395–409.

Boesch, C., 1994. Cooperative hunting in wild chimpanzees. Animal Behaviour 48, 653–667.

Booth, V.R., 2010. The Contribution of Hunting Tourism: How Significant Is This to National Economies?, Contribution of Wildlife to National Economies. Joint Publication of Food and Agricultural Organization and the International Council for Game and Wildlife Conservation, Budapest, Hungary, pp. 7–36.

Bordes, F., Thibault, C., 1977. Thoughts on the initial adaptation of Hominids to European glacial climates. Quaternary Research 8, 115–127.

Borrero, L.A., 1999. The prehistoric exploration and colonization of Fuego-Patagonia. Journal of World Prehistory 13, 321–355.

Bradley, B., Stanford, D., 2004. The North Atlantic ice-edge corridor: a possible Palaeolithic route to the New World. World Archaeology 36, 459–478.

Brand-Miller, J., Mann, N., Cordain, L., 2009. Paleolithic nutrition: what did our ancestors eat? In: Selinger, A., Green, A. (Eds.), ISS 2009 Genes to Galaxies. The Science Foundation for Physics. University of Sydney. University Publishing Service, University of Sydney, Sydney.

Bunn, H., Harris, J.W.K., Isaac, G., Kaufulu, Z., Kroll, E., Schick, K., Toth, N., Behrensmeyer, A.K., 1980. FxJj50: an early Pleistocene site in northern Kenya. World Archaeology 12, 109–136.

Buys, C.F., 1934. Armas e munições de caça. Livraria Globo, Porto Alegre.

Calvin, W.H., Bickerton, D., 2000. Lingua Ex Machina: Reconciling Darwin and Chomsky with the Human Brain, first ed. MIT Press, Cambridge, MA, USA.

Canale, G.R., Peres, C.A., Guidorizzi, C.E., Gatto, C.A.F., Kierulff, M.C.M., 2012. Pervasive defaunation of forest remnants in a tropical biodiversity hotspot. PLoS One 7, e41671.

Carrier, D.R., Kapoor, A.K., Kimura, T., Nickels, M.K., Satwanti, E.C.S., So, J.K., Trinkaus, E., 1984. The energetic paradox of human running and hominid evolution. Current Anthropology 483–495.

Cavalcanti, S.M.C., 2006. Manejo e controle de danos causados por predadores. In: Cullen Jr., L., Rudran, R., Valladares-Padua, C. (Eds.), Métodos de estudos em Biologia da Conservação e Manejo da Biodiversidade. UFPR, Curitiba.

Clutton-Brock, J., 1999. A Natural History of Domesticated Mammals, second ed. Cambridge University Press, Cambridge, UK.

Cochard, D., Brugal, J.-P., Morin, E., Meignen, L., 2012. Evidence of small fast game exploitation in the Middle Paleolithic of Les Canalettes Aveyron, France. Quaternary International 264, 32–51.

Conover, M.R., 2002. Resolving Human-wildlife Conflicts: The Science of Wildlife Damage Management. CRC.

Coon, C.S., 1971. The Hunting Peoples. Nick Lyons Books, Boston.

Cordain, L., 1999. Cereal grains: humanity's double-edged sword. In: Simopoulos, A.P. (Ed.), Evolutionary Aspects of Nutrition and Health: Diet, Exercise, Genetics and Chronic Disease, first ed. Karger Publishers, Berlin, pp. 19–73.

Costamagno, S., Laroulandie, V., 2004. L'exploitation des petits vertébrés dans les Pyrénées françaises du Paléolithique au Mésolithique: un inventaire taphonomique et archéozoologique. In: Brugal, J.-P., Desse, J. (Eds.), Petits Animaux et Sociétés Humaines. Du Complément Alimentaire aux Ressources Utilitaires, first ed. APCDA, Antibes, France, pp. 369–382.

Cremony, J., 2016. Apache method of Hunting Geese and Ducks.

Crosby, A.W., 1993. Imperialismo Ecológico: A expansão biológica da Europa, 900–1900, first ed. Companhia das Letras, São Paulo.

Cullen Jr., L., Bodmer, R.E., Padua, C.V., 2000. Effects of hunting in habitat fragments of the Atlantic forests, Brazil. Biological Conservation 95, 49–56.

Culot, L., Bovy, E., Vaz-de-Mello, F.Z., Guevara, R., Galetti, M., 2013. Selective defaunation affects dung beetle communities in continuous Atlantic rainforest. Biological Conservation 163, 79–89.

Darimont, C.T., Fox, C.H., Bryan, H.M., Reimchen, T.E., 2015. The unique ecology of human predators. Science 349, 858–860.

Dart, R.A., 1953. The predatory transition from ape to man. International Anthropological and Linguistic Review 1, 201–218.

David, A., May 2011. Introductions to Heritage Assets: Caves, Fissures and Rockshelters. English Heritage, pp. 1–6.

de la Vega, G., [1612] 1961. The Incas, the Royal Commentaries of the Inca Garcilaso de la Vega 1539–1616. Orion Press, New York.

deFrance, S.D., 2009. Zooarchaeology in complex societies: political economy, status, and Ideology. Journal of Archaeological Research 17, 105–168.

del Castillo, B.D., [1570] 1996. The Discovery and Conquest of Mexico, first ed. Da Capo Press, Cambridge, Massachusetts.

Di Minin, E., Leader-Williams, N., Bradshaw, C.J.A., 2016. Banning trophy hunting will exacerbate biodiversity loss. Trends in Ecology & Evolution 31, 99–102.

Dickman, A.J., 2008. Key Determinants of Conflict between People and Wildlife, Particularly Large Carnivores, Around Ruaha National Park. University College London, Tanzania.

Dirzo, R., Young, H.S., Galetti, M., Ceballos, G., Isaac, N.J.B., Collen, B., 2014. Defaunation in the Anthropocene. Science 401, 345.

Dunbar, R.I.M., 1977. The Gelada baboon: status and conservation. In: Prince Rainier III, H.S.H., Bourne, G.H. (Eds.), Primate Conservation. Academic Press, New York, pp. 363–383.

El-Kamali, H.H., 2000. Folk medicinal use of some animal products in Central Sudan. Journal of Ethnopharmacology 72, 279–282.

El Bizri, H.R., Morcatty, T.Q., Lima, J.J.S., Valsecchi, J., 2015. The thrill of the chase: uncovering illegal sport hunting in Brazil through YouTube™ posts. Ecology and Society 20, 30.

Erlandson, J.M., 2001. The archaeology of aquatic adaptations: paradigms for a new millennium. Journal of Archaeological Research 9, 287–350.

Fairnell, E.H., 2003. The Utilisation of Fur-bearing Animals in the British Isles: A Zooarchaeological Hunt for Data. University of York, York, Canada, p. 194.

Fernandes-Ferreira, H., 2014. A caça no Brasil: Panorama histórico e atual. Universidade Federal da Paraíba.

Fernandes-Ferreira, H., Mendonça, S.V., Albano, C., Ferreira, F.S., Alves, R.R.N., 2012. Hunting, use and conservation of birds in Northeast Brazil. Biodiversity and Conservation 221–244.

Fernandes-Ferreira, H., Mendonca, S.V., Cruz, R.L., Borges-Nojosa, D.M., Alves, R.R.N., 2013. Hunting of herpetofauna in montane, coastal, and dryland areas of Northeastern Brazil. Herpetological Conservation and Biology 8, 652–666.

Fitzgerald, L.A., 1994. Tupinambis lizards and people: a sustainable use approach to conservation and development. Conservation Biology 8, 12–15.

FitzGibbon, C., 1998. The management of subsistence harvesting: behavioral ecology of hunters and their mammalian prey. In: Caro, T. (Ed.), Behavioural Ecology and Conservation Biology. Oxford University Press, Oxford, pp. 449–474.

Fitzgibbon, C.D., Hezron, M., Fanshawe, J.H., 1995. Subsistence hunting in Arabuko-Sokoke forest, Kenya, and its effects on mammal populations. Conservation Biology 9, 1116–1126.

Frison, G.C., 1987. Prehistoric hunting strategies. In: Nitecki, M.H., Nitecki, D.V. (Eds.), The Evolution of Human Hunting, first ed. Plenum Press, New York, pp. 177–223.

Garcia-Alaniz, N., Naranjo, E.J., Mallory, F.F., 2010. Human-felid interactions in three Mestizo communities of the Selva Lacandona, Chiapas, Mexico: Benefits, conflicts and traditional uses of species. Human Ecology 38, 451–457.

García-Moreno, J., Clay, R.P., Ríos-Muñoz, C.A., 2007. The importance of birds for conservation in the Neotropical region. Journal of Ornithology 148, 321–326.

Giacomini, H.C., Galetti, M., 2013. An index for defaunation. Biological Conservation 163, 33–41.

Glambe, C., 2003. Human evolution: the last one million years. In: Ingold, T. (Ed.), Companion Encyclopedia of Anthropology: Humanity, Culture and Social Life. Taylor & Francis e-Library, London/New York, pp. 79–107.

Graham, K., Beckerman, A.P., Thirgood, S., 2005. Human-predator-prey conflicts: ecological correlates, prey losses and patterns of management. Biological Conservation 122, 159–171.

Gumilla, J., 1686–1750. El Orinoco ilustrado, y defendido, historia natural, civil y geographica de este gran rio, y sus caudalosas vertientes, govierno, usos y costumes de los índios sus habitadores, Tomo Segundo, Segunda Impression. Madrid, Manuel Fernandez. 1745, p. 428.

Gupta, A.K., 2004. Origin of agriculture and domestication of plants and animals linked to early Holocene climate amelioration. Current Science 87, 54–59.

Gutiérrez, M.A., Martínez, G.A., 2008. Trends in the faunal human exploitation during the late pleistocene and early Holocene in the Pampean region (Argentina). Quaternary International 191, 53–68.

Hawkes, K., 2001. Is meat the Hunter's property? big game, ownership, and explanations of hunting and sharing. In: Stanford, C.B., Bunn, H.T. (Eds.), Meat-eating & Human Evolution, first ed. Oxford University Press, USA, Oxford, UK/New York, pp. 215–236.

Hawkes, K., O'Connell, J.F., Blurton Jones, N.G., 2001. Hunting and nuclear families. Current Anthropology 42, 681–709.

Heinrich, B., 2001. Why We Run. Harper Collins, New York.

Heizer, R.F., Ribeiro, D., 1987. Venenos de pesca. In: Ribeiro, D. (Ed.), Suma Etnológica Brasileira, Edição Atualizada Do Handbook of South American Indians. Vozes, FINEP, Rio de Janeiro, pp. 189–233.

Hell, P., Find'o, S., 1999. Status and management of the brown bear in Slovakia. In: Servheen, C., Herrero, S., Peyton, B. (Eds.), Bears. Status Survey and Conservation Action Plan. IUCN/SSC Bear and Polar Bear Specialist Groups, IUCN, Gland, Switzerland and Cambridge, UK, pp. 96–100.

Hengemühle, A., Cademartori, C.V., 2008. Levantamento de mortes de vertebrados silvestres devido a atropelamento em um trecho da estrada do mar (RS-389). Biodiversidade Pampeanda 6, 4–10.

Henshilwood, C.S., Marean, C.W., 2003. The origin of modern human behavior. Current Anthropology 44, 627–651.

Hill, K., 1982. Hunting and human evolution. Journal of Human Evolution 11, 521–544.

Hill, K., Kintigh, K., 2009. Can anthropologists distinguish good and poor hunters? Implications for hunting hypotheses, sharing conventions, and cultural transmission. Current Anthropology 50, 369–377.

Hill, K., Padwe, J., 2000. Sustainability of Aché hunting in the Mbaracayu reserve, Paraguay. In: Robinson, J.G., Bennett, E.L. (Eds.), Hunting for Sustainability in Tropical Forests. Columbia University Press, New York, pp. 79–105.

Hockett, B., Haws, J.A., 2002. Taphonomic and methodological perspectives of leporid hunting during the Upper Paleolithic of the western Mediterranean Basin. Journal of Archaeological Method and Theory 9, 269–302.

Hoffecker, J.F., 2005. Innovation and technological knowledge in the Upper Paleolithic of northern Eurasia. Evolutionary Anthropology: Issues, News, and Reviews 14, 186–198.

Hughes, J.D., 2007. Hunting in the ancient Mediterranean world. In: Kalof, L. (Ed.), A Cultural History of Animals in Antiquity. Berg, Nwe York, pp. 47–70.

Humavindu, M.N., Barnes, J.I., 2003. Trophy hunting in the Namibian economy: an assessment. South African Journal of Wildlife Research 33, 65–70.

Isaac, G.L., Crader, D.C., 1981. To what extent were early hominids carnivorous? In: Harding, R.S.O., Teleki, G. (Eds.), Omnivorous Primates, first ed. Columbia University Press, New York, pp. 37–103.

Jones, D.E., 2009. Poison Arrows: North American Indian Hunting and Warfare. University of Texas Press.

Juan, G., Ulloa, A., 1758. A Voyage to South America: Describing at Large the Spanish Cities, Towns, Provinces, &c. on that Extensive Continent. Interspersed throughout with Reflections on the Genius, Customs, Manners, and Trade of the Inhabitants: Together with the Natural History of the Country. And an Account of Their Gold and Silver Mines. Undertaken by Command of His Majesty the King of Spain, first ed. L. Davis & C. Reymers, London, UK.

Kalof, L., 2007a. A Cultural History of Animals in Antiquity, first ed. Berg publishers, Oxford, UK/New York.

Kalof, L., 2007b. Looking at Animals in Human History, first ed. Reaktion Books, London, UK.

Khorenatsi, M., 1978. History of the Armenians (Robert W. Thomson, Trans.). Harvard University Press, Cambridge.

Kitchell, K., 2011. Penelope's geese: pets of the ancient greeks. Expedition 53, 14–23.

Klein, R.G., 1971. The pleistocene prehistory of Siberia. Quaternary Research 1, 133–161.

Konner, M., Eaton, S.B., 2010. Paleolithic nutrition: twenty-five years later. Nutrition in Clinical Practice 25, 594–602.

Koster, J., 2008. The impact of hunting with dogs on wildlife harvests in the Bosawas Reserve, Nicaragua. Environmental Conservation 35, 211–220.

Leader-Williams, N., Baldus, R.D., Smith, R.J., 2009. The influence of corruption on the conduct of recreational hunting. In: Dickson, B., Hutton, J., Adams, W.A. (Eds.), Recreational Hunting, Conservation and Rural Livelihoods: Science and Practice. Wiley-Blackwell, Oxford, pp. 296–316.

Lev, E., 2003. Traditional healing with animals (zootherapy): medieval to present-day Levantine practice. Journal of Ethnopharmacology 85, 107–118.

Liebenberg, L., 2006. Persistence hunting by modern hunter-gatherers. Current Anthropology 47, 1017–1026.

Liebenberg, L., 2008. The relevance of persistence hunting to human evolution. Journal of Human Evolution 55, 1156–1159.

Lindsey, P.A., Alexander, R., Frank, L.G., Mathieson, A., Romanach, S.S., 2006. Potential of trophy hunting to create incentives for wildlife conservation in Africa where alternative wildlife-based land uses may not be viable. Animal Conservation 9, 283–291.

Lindsey, P.A., Roulet, P.A., Romanach, S.S., 2007. Economic and conservation significance of the trophy hunting industry in sub-Saharan Africa. Biological Conservation 134, 455–469.

MacKinney, L.C., 1946. Animal substances in materia medica. Journal of the History of Medicine and Allied Sciences 1, 149–170.

MacWhinney, B., 2002. The gradual emergence of language. In: Givón, T., Malle, B. (Eds.), The Evolution of Language Out of Pre-language. John Benjamins Publishing, Amsterdam, pp. 231–263.

Madhusudan, M.D., Karanth, K.U., 2005. Hunting for an Answer: Is Local Hunting Compatible with Large Mammal Conservation in India? Columbia University Press, New York.

Mahawar, M.M., Jaroli, D.P., 2008. Traditional zootherapeutic studies in India: a review. Journal of Ethnobiology and Ethnomedicine 4, 17.

Marshall, K., White, R., Fischer, A., 2007. Conflicts between humans over wildlife management: on the diversity of stakeholder attitudes and implications for conflict management. Biodiversity and Conservation 16, 3129–3146.

McCarthy, F.D., 1957. Australia's Aborigines: Their Life and Culture. Colorgravure Publications, Melbourne.

McCartney, A.P., 1984. Prehistory of the Aleutian region. Handbook of North American Indians 5, 119–135.

Melo, M.M., Saito, C.H., 2011. Regime de queima das caçadas com uso do fogo realizadas pelos Xavante no cerrado. Biodiversidade Brasileira 97–109.

Milner-Gulland, E.J., Bennett, E.L., 2003. Wild meat: the bigger picture. Trends in Ecology & Evolution 18, 351–357.

Miranda, H.S., Sato, M.N., Neto, W.N., Aires, F.S., 2009. Fires in the Cerrado, the Brazilian Savanna, Tropical Fire Ecology. Springer, Berlin Heidelberg, pp. 427–450.

Mitani, J.C., Watts, D.P., 2001. Why do chimpanzees hunt and share meat? Animal Behaviour 61, 915–924.

Mitchell, M., Tully Jr., T.N., 2009. Manual of Exotic Pet Practice. Elsevier Health Sciences, St. Louis, Missouri.

Mittermeier, R.A., Tattersall, I., Konstant, W.R., Meyers, D.M., Mast, R.B., 1994. Lemurs of Madagascar. Conservation International, Washington DC.

Morey, D.F., 1994. The early evolution of the domestic dog. American Scientist 82, 336–347.

Mundy, P., Bunchart, D., Ledger, J., Piper, S., 1993. The Vultures of Africa, first ed. Acorn Books and Russel Friedman Books, Randburg, South Africa.

Myers, C.W., Daly, J.W., Malkin, B., 1978. A dangerously toxic new frog (Phyllobates) used by EmberÃ¡ Indians of western Colombia, with discussion of blowgun fabrication and dart poisoning. Bulletin of the American Museum of Natural History 161, 309–365.

Nash, S.V., 1993. Sold for a Song: The Trade in Southeast Asian Non-CITES Birds, first ed. TRAFFIC Intenational, Cambridge, UK.

Nobokov, P., 1981. Indian Running: Native American History and Tradition. Aneburt City Press, Santa Fe.

O'Connell, J.F., Hawkes, K., Blurton Jones, N.G., 1999. Grandmothering and the evolution of Homo erectus. Journal of Human Evolution 36, 461–485.

Ojasti, J., 2000. Manejo de Fauna Silvestre Neotropical. In: Dallmeier, F. (Ed.), SI/MAB Series. Smithsonian Institution/MAB Biodiversity Program, Washington D.C, p. 290.

Ojibwa, 2012. Ancient America: The Buffalo Hunt.

Outram, A.K., 2001. A new approach to identifying bone marrow and grease exploitation: why the "Indeterminate" fragments should not be ignored. Journal of Archaeological Science 28, 401–410.

Pereira, M.N., 1967. Moronguetá: um Decameron indígena. Civilização Brasileira, Rio de Janeiro.

Peres, C., Nascimento, H., 2006. Impact of game hunting by the Kayapó of south-eastern Amazonia: implications for wildlife conservation in tropical forest indigenous reserves. Biodiversity and Conservation 15, 2627–2653.

Peres, C.A., 2000. Effects of subsistence hunting on vertebrate community structure in Amazonian forests. Conservation Biology 14, 240–253.

Peres, C.A., 2001. Synergistic effects of subsistence hunting and habitat fragmentation on Amazonian forest vertebrates. Conservation Biology 15, 1490–1505.

Pezzuti, J., Chaves, R.P., 2009. Ethnography and natural resources management by the Deni Indians, Amazonas, Brazil. Acta Amazonica 39, 121–138.

Pitman, M.R.P.L., Oliveira, T.G., Paula, R.C., Indrusiak, C., 2002. Manual de identificação, prevenção e controle de predação por carnívoros. Edições IBAMA, Brasília.

Quave, C.L., Lohani, U., Verde, A., Fajardo, J., Rivera, D., Obón, C., Valdes, A., Pieroni, A., 2010. A comparative assessment of zootherapeutic remedies from selected areas in Albania, Italy, Spain and Nepal. Journal of Ethnobiology 30, 92–125.

Redford, K.H., Robinson, J.G., 1987. The game of choice: patterns of indian and colonist hunting in the Neotropics. American Anthropologist 89, 650–667.

Redford, K.H., Robinson, J.G., 1991. Subsistence and commercial uses of wildlife in Latin America. In: Robinson, J.G., Redford, K.H. (Eds.), Neotropical Wildlife Use and Conservation, pp. 6–23.

Reitz, E.J., Wing, E.S., 2008. Zooarchaeology, second ed. Cambridge University Press, Cambridge, UK/New York.

Richerson, P.J., Mulder, M.B., Vila, B.J., 1996. Principles of Human Ecology, second ed. Pearson Custom Publishing, New York.

Robinson, J.G., Bennett, E.L., 2000. Hunting for Sustainability in Tropical Forests, first ed. Columbia University Press, New York :.

Roldán-Clarà, B., Lopez-Medellín, X., Espejel, I., Arellano, E., 2014. Literature review of the use of birds as pets in Latin-America, with a detailed perspective on Mexico. Ethnobiology And Conservation 3, 1–18.

Ross, E.B., Arnott, M.L., Basso, E.B., Beckerman, S., Carneiro, R.L., Forbis, R.G., Good, K.R., Jensen, K.-E., Johnson, A., Kaplinski, J., Khare, R.S., Linares, O.F., Martin, P.S., Nietschmann, B., Nurse, G.T., Pollock, N.J., Sahai, I., Kenneth Clarkson, T., Turton, D., Vickers, W.T., Wetterstrom, W.E., 1978. Food Taboos, diet, and hunting strategy: the adaptation to animals in Amazon cultural ecology [and comments and reply]. Current Anthropology 19, 1–36.

Russell, N., 2012. Social Zooarchaeology: Humans and Animals in Prehistory, first ed. Cambridge University Press, Cambridge, UK.

Rustioni, M., Mazza, P., Magnatti, M., 2007. Multivariable analysis of an Italian late Neolithic archaeofauna. Journal of Archaeological Science 34, 723–738.

Sablin, M.V., Khlopachev, G.A., 2002. The earliest ice age dogs: evidence from Eliseevichi 1. Current Anthropology 43, 795–799.

Santos, E., 1950. Caças e Caçadas. F. Briguiet & Cia, Rio de Janeiro.

Shipman, P., 1986. Scavenging or hunting in early hominids: theoretical framework and tests. American Anthropologist 88, 27–43.

Sick, H., 2001. Ornitologia Brasileira (Edição revista e ampliada por José Fernando Pacheco), third ed. Editora Nova Fronteira, Rio de Janeiro.

Sollas, W.J., 1915. Ancient hunters and their modern representatives. Macmillan and Company.

Sørensen, B., 2009. Energy use by Eem Neanderthals. Journal of Archaeological Science 36, 2201–2205.

Souto, W.M.S., 2014. Atividades cinegéticas, usos locais e tradicionais da fauna por povos do semiárido paraibano. Universidade Federal da Paraíba.

Souto, W.M.S., Mourão, J.S., Barboza, R.R.D., Alves, R.R.N., 2011. Parallels between zootherapeutic practices in ethnoveterinary and human complementary medicine in NE Brazil. Journal of Ethnopharmacology 134, 753–767.

Speth, J.D., 1989. Early hominid hunting and scavenging: the role of meat as an energy source. Journal of Human Evolution 18, 329–343.

Speth, J.D., 2010. The Paleoanthropology and Archaeology of Big-game Hunting, ProteIn: Fat, or Politics? Springer, New York, Heidelberg, Dordrecht, London.

Stanford, C.B., 1999. The Hunting Apes: Meat Eating and the Origins of Human Behavior. Princeton University Press.

Stearns, P.N., Adas, M.B., Schwartz, S.B., Gilbert, M.J., 2011. World Civilizations: The Global Experience (Combined Volume), first ed. Pearson, Upper Saddle River, USA.

Stuart, B.L., 2004. The harvest and trade of reptiles at U Minh Thuong National Park, southern Viet Nam. Traffic Bulletin 20, 25–34.

Tokarnia, C.H., Dobereiner, J., Peixoto, P.V., 2000. In: Plantas Tóxicas Do Brasil. Helianthus, Rio de Janeiro.

Trinkaus, E., 1987. Bodies, brawn, brains and noses: human ancestors and human predation. In: Nitecki, M.H., Nitecki, D.V. (Eds.), The Evolution of Human Hunting, first ed. Plenum Press, New York, pp. 107–145.

Tyldesley, J.A., Bahn, P.G., 1983. Use of plants in the European Palaeolithic: a review of the evidence. Quaternary Science Reviews 2, 53–81.

Vanstreels, R.E.T., Teixeira, R.H.F., Camargo, L.C., Nunes, A.L.V., Matushima, E.R., 2010. Impacts of animal traffic on the Brazilian Amazon parrots (Amazona species) collection of the Quinzinho de Barros Municipal Zoological Park, Brazil, 1986–2007. Zoo Biology 28, 1–15.

Ventocilla, J., Herrera, H., Nuñez, V., 1995. Plants and Animals in the Life of the Kuna Austin, TX, USA. University of Texas Press.

Vié, J.-C., Hilton-Taylor, C., Stuart, S.N., 2009. Wildlife in a Changing World – an Analysis of the 2008 IUCN Red List of Threatened Species, first ed. Lynx Edicions, Barcelona, Spain, p. 180.

Villa, P., Lenoir, M., 2009. Hunting and hunting weapons of the Lower and Middle Paleolithic of Europe. In: Hublin, J.-J., Richards, M.P. (Eds.), The Evolution of Hominin Diets. Springer, Netherlands, Amsterdan, pp. 59–85.

Vivian, K., 1991. The Georgian Chronicle: The Period of Giorgi Lasha. Adolf M. Hakkert, Amsterdam.

Walker, A., 1822. Colombia: Being a Geographical, Statistical, Agricultural, Commercial, and Political Account of that Country, Adapted for the General Reader, the Merchant, and the Colonist. Baldwin, Crodock, and Joy, London.

Washburn, S., Lancaster, C., 1968. The evolution of hunting. In: Lee, R.B., Devore, I. (Eds.), Man the Hunter. Aldine, Chicago, IL, USA, pp. 293–303.

Williamson, D.F., 2004. Tackling the Ivories: The Status of the US Trade in Elephant and Hippo Ivory. TRAFFIC North America. World Wildlife Fund, Washington, DC.

Wu, J., 2007. World without borders: wildlife trade on the Chinese-language internet. Traffic Bulletin 21, 75–84.

CHAPTER

8

People and Fishery Resources: A Multidimensional Approach

Marcia Freire Pinto[1], *Tacyana Pereira Ribeiro Oliveira*[2], *Luiz Alves Rocha*[3], *Rômulo Romeu Nóbrega Alves*[4]

[1]Universidade Estadual do Ceará, Limoeiro do Norte, CE, Brazil; [2]Universidade Estadual da Paraíba, João Pessoa, PB, Brazil; [3]California Academy of Sciences, San Francisco, CA, United States; [4]Universidade Estadual da Paraíba, Campina Grande, Brazil

INTRODUCTION

The capture of aquatic animals has become part of human culture and behavior since ancient times. Hominins used to inhabit coastal areas or margins of rivers, lakes, and lagoons for shelter and food (Gartside and Kirkegaard, 2009; Prous, 1992), and gradually improved their use of natural resources. Fishing was initially developed for food provision and, later, for medicinal purposes and confection of tools, thus becoming part of the lifestyle and culture of the first hominins.

Hominins primarily fished in estuaries, mangroves, and areas of calm waters such as lakes and rivers (Marean et al., 2007; Stewart, 1994b). Fishes, mollusks, crustaceans, turtles, sirenians, and other aquatic animals were captured, prepared, and conserved using techniques that evolved together with preparation, conservation, and navigation skills (Gartside and Kirkegaard, 2009; Stringer et al., 2008). Besides its initial subsistence purpose, and with the advancement of human civilizations, fishing became an important commercial activity in ancient times. Early it was controlled by elites of Egyptian and Roman cultures through a slavery system, in which slaves performed the captures (Aleem, 1972; Stringer et al., 2008). Regarded as notable traders, Phoenicians also relied on fishing and marine trade as major economic activities. Indeed, one of the most desirable (and costly) products in Phoenician markets, the purple dye, was obtained from purple dye murex gastropods (e.g., *Bolinus brandaris*, *Bolinus trunculus*), which were collected on coastal areas; large mounds of shells indicate that the production of the purple dye was an important sector (Sahrhage and Lundbeck, 1992). Hence, fishery resources were used as a bargaining chip in commercial negotiations. Further, as other aquatic habitats became explored, such as the

Ethnozoology
http://dx.doi.org/10.1016/B978-0-12-809913-1.00008-9

open seas, the capture of new targeted animals, such as cetaceans, followed. Finally, fishing has turned into one of the most important economic activities incorporated by the capitalist system.

Despite its current commercial approach, fishing is also still practiced for subsistence, sport, and recreational purposes. However, commercial fisheries stands out as the major source of several socioeconomic concerns, due to overfishing or the capture of more wildlife than the natural capacity of a population to replace itself through reproduction. Since 1970, overexploitation of aquatic animals has led to a crisis in the fishing industry, with several species becoming commercially extinct, compromising the livelihood of those who directly or indirectly depend on fishing.

Besides the impacts caused by overfishing, several social issues related to ongoing increases in human population and coastal development have also led to aquatic habitat degradation and biodiversity loss. Indeed, aquatic environments face a variety of threats that potentially affect fishing. For instance, pollution by sewage, pesticides, and oil spills; mangrove and riparian forests removal; changes in sediment dynamics in coastal areas; degradation due to tourism; increasing in boat traffic; invasive species; and, under a global perspective, climate change (Reis et al., 2016).

The socio-environmental impacts resulting from overfishing have been widely assessed and discussed, leading to the creation (and further amendments) of agreements, conventions, and policies toward the regulation of fishing activities and the conservation of targeted species and ecosystems. Due to the multifaceted nature of the industry, much comprehensive data is needed to the development of such strategies, thus requiring a multidisciplinary approach involving researchers of several traditional fields, such as fisheries engineering, oceanography, biology, zoology, ecology, anthropology, economy, among others. Nowadays, Ethnozoology has been providing important contributions through understanding the human–animal relationships in fisheries studies.

This chapter discusses the importance of fishery to humans, considering historical aspects of this activity, the development of fishing techniques and strategies, the targeted taxa and their major uses, and the consequent social and ecological implications associated with those practices.

WHEN DID HUMANS START FISHING? A BRIEF HISTORICAL INTRODUCTION

One of the first evidences of fishing dates from 700,000 BC, and consists of a pile of shells found in a midden (an archaeological term used to define refuses of human occupation) in Thailand, thus suggesting the existence of the practice of collecting clams (Pope, 1989). This fact reinforces the idea that, before capturing fishes, hominins were intertidal mollusk collectors (Diegues, 1983). Indeed, kitchen middens had an invaluable importance in determining first fishing activities; as such, waste deposits usually contain animal remains, bones, artifacts, and tools that can be related to hominin activities, including fishing (Prous, 1992). In South Africa, ~164,000 years ago, shellfish exploitation may have been one of the last human diet items added before domestication took place in the end of Pleistocene and a major attractive to human establishment and migration along coastlines (Marean et al., 2007). Although little is known about fisheries in primitive societies, archaeological and ethnological evidence supports that it represented an important food source, before the rise of agriculture (Diegues, 1983; Gartside and Kirkegaard, 2009). Since the first cave paintings (c. 40,000 BC), several lines of evidence indicate that fishes, as well as other aquatic animals, served as food and to a range of other human physical and spiritual necessities and desires (Gartside and Kirkegaard, 2009). Moreover, fishery resources were also acknowledged to be used for commercial, artisanal, and medicinal purposes (Castro et al., 2016;

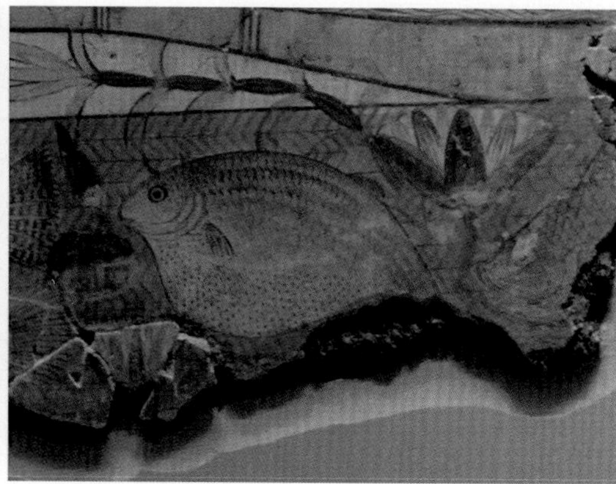

FIGURE 8.1 Fishes depicted in the tomb-chapel of Nebamun, c. 1350 BCE., 18th Dynasty (below, detail of fish in marsh), illustrating the antiquity of the interaction of people and icthyofauna. *Photo credits: Rômulo R.N. Alves (Photographed at the British Museum, London, England).*

El-Deir et al., 2012; Hanazaki and Begossi, 2000; MacKinney, 1946; Pinto et al., 2013, 2015; Rocha et al., 2008).

The prehistoric human-fish relationship dates from the Pleistocene, in which *Homo habilis* captured fishes in lakes and streams in Eastern Africa, with rudimentary or no technology at all (Stewart, 1994b). Archaeological records suggest that specialized fishing technology only arose after 150,000 BC, when *Homo sapiens* developed skills to explore aquatic resources, including the marine realm (Erlandson, 2001). Historical documentation reports the use of spears first at c. 90,000 BC (Yellen et al., 1995), traps at 75,000 BC, hooks at 42,000 BC, and nets at 40,000 BC (Lackey, 2005; Pitcher and Lam, 2015).

Fish processing involved several steps, such as drying, salting, and pickling fish, and was mastered by ancient Egyptians, whose fishing industry dates from over 5000 years ago (Aleem, 1972). Egyptians also improved fish processing, as they produced fish cakes and fillets, using several types of knives, implements, and containers. Fish was exported dry and in this form also comprised a typical part of troops' rations. It is interesting to point out that fishers would eventually join troops on the move to provide freshly caught fish, as ordered, for example, by Ramesses II (19th Dynasty, c. 1290 BC), when his army was sent to Syria (Aleem, 1972).

Fishes became part of the Egyptian culture and were used as symbols and ornaments (Fig. 8.1). At least six different fishes are figured in hieroglyphs, including tilapias, catfishes, and carps. Fish-shaped ornaments, pallets, and glass bottles were commonly manufactured; fishing and fish processing could be found expressed on temples and tombs; and some species, such the elephant fish *Mormyrus*, were considered to be sacred—some embalmed specimens were found in Minya, Upper Egypt (Aleem, 1972). The evidence of ornamental fish ponds in some temples may also reflect the influence of fishery resources on Egyptian culture.

Fishes from the Great Lakes comprised a major food source for the Native Americans by 3000 to 2000 BC. Several fishing gears were present, including spears, hooks and lines, and weirs, and, later (~300 and 200 BC), fishing nets (firstly adapted from hunting nets) (Bogue, 2001). In the medieval period, fishing was practiced within feudal properties, in lakes, lagoons, and in the coast (Hérubel, 1928).

Commercial fisheries were established in the 15th century, after the development of new navigation and preservation technologies, which meant more time spent at the sea and long distance travel, and, hence, more captures and higher fisheries diversity (Diegues, 1983), leading to the establishment of several types of fisheries, gears, and vessels.

FISHING AND FISHERIES DEFINITIONS

The word fishing has its origin from the ancient English word fiscian, which means fishing, capturing, or trying to capture fishes (Harper, 2016). However, the term fishing has been used to refer to the capture of other aquatic animals along with fishes, such as mollusks, crustaceans, mammals, etc. (Alves, 2012; Pinto et al., 2013). Moreover, for many fishers (particularly those involved with small-scale fisheries), those taxa are grouped as "fish." Thus, the expression marine fisheries, as specified by Luna et al. (2004), for example, would comprise wild living resources in coastal and open seas that could be captured. On the other hand, as some aquatic animals are usually considered to be hunted, other terms instead of fishing may apply—in the case of whale hunting, for instance, the term whaling should be used. For this reason, although whaling may be mentioned throughout the text, this chapter did not consider the "fisheries" of hunted animals. It is worth noting that commercial whaling has been banned in most countries, after the 1982 international moratorium, although it still takes place in Norway and Iceland on objections or under reservations (IWC, 2017); moreover, Japan holds a controversial scientific whaling program (Brierley and Clapham, 2016).

Fisheries, which comprise the activities/efforts involved in fishing aquatic animals, can have numerous classifications. One of the broader approaches considers the habitats where the fishery takes place, and, thus, we may refer to it as coastal fisheries (along coastal lines, including riverine/estuarine, reefs, and across other continental shelf ecosystems), marine fisheries (in the open seas), fluvial fisheries (in rivers and streams), and lacustrine fisheries (in lakes, lagoons, and reservoirs) (Figs. 8.2 and 8.3). It should be noted that this chapter primarily focuses on coastal/marine fisheries, as they account for the bulk of captures worldwide (more than 80 million tonnes in 2015) in comparison to inland catches (FAO, 2016).

Fisheries may also be classified according to their scale. Subsistence fisheries comprise those related to food provision for fishers and their families, while commercial fisheries refer to fishing for economic profit. Subsistence fishers can be registered as professional fishers, thus getting professional recognition for their activities; the same applies to commercial fishers, although their registry is mandatory. Recreational and sport fisheries comprise leisure and/or competition activities, with no strict commitment to trade or consumption of the catches.

Artisanal fisheries often refer to small-scale commercial fishing activities (that can partially be for subsistence), which generally take place in domestic waters, employ traditional low-cost fishing techniques/gears and small vessels, and involve few people (Berkes et al., 2001). Those fisheries contrast with large-scale commercial fisheries, which commonly use high-tech fishing gear and large vessels, leading to higher fishing effort and catch rates. Industrial fisheries are conducted in coastal waters or in the open seas, with high crew numbers; however, due to high costs, the number of workers is lower than that of artisanal fisheries, which have accounted for around 90% of the jobs in global fisheries (FAO, 2016).

Fisheries may also be classified (and named) depending on the targeted fishery resource, which mainly correspond to fish, clams/

FIGURE 8.2 Fisheries in freshwater environments. (A) Fisher on the Congo River, Lukolela, Democratic Republic of Congo; (B) A man fishing in the lake (in West Java, Indonesia) using a traditional net; (C) A group of people in Sentarum Lake (West Kalimantan, Indonesia) catching fish with nets; and (D) Fisher catches fish in the upper waters of the Malinau river, East Kalimantan, Indonesia. *Photo credits: (A) Ollivier Girard for Center for International Forestry Research (CIFOR); (B) Ricky Martin for CIFOR, (C) Ramadian Bachtiar for CIFOR, (D) Michael Padmanaba for CIFOR.*

FIGURE 8.3 Coastal fisheries in São Tomé, West Africa. *Photo credit: Luiz Alves Rocha.*

FIGURE 8.4 Capture of mollusks (*Anomalocardia brasiliana*) (A and B) and crabs (*Ucides cordatus*) (C and D) in Northeastern Brazil. *Photo credits: (A) Thelma L.P. Dias; (B–D) José S. Mourão.*

shrimp/lobster/crab (mollusks and crustaceans) (Fig. 8.4), octopus/squid, turtle fisheries, etc. Considering specific fish groups, we may also refer to, for instance, shark, herring, tuna, or swordfish fisheries.

Regarding the amount of time spent at sea, fisheries can be defined as one-day fisheries—when fishers spend a whole day at sea and return, and overnight trips fisheries—when fishers carry out fishing trips lasting several days at sea. Vessel types are of great importance in the latter, as only autonomous and specific boats allow fishers to live aboard for longer periods.

FISHERY RESOURCES AND THEIR USES

Fishery resources have a comprehensive definition and, in general, refer to all animals that predominantly inhabit aquatic habitats and are captured by humans (Gartside and Kirkegaard, 2009). Therefore, invertebrates (e.g., mollusks, crustaceans, echinoderms, and even sponges) and vertebrates (fishes, amphibians, reptiles, mammals) are considered fishery resources, although only fishes and crustaceans (i.e., marine lobsters, shrimps, and crabs) are quantified in fisheries statistics.

Despite several types of fishing resources, fishes (Chondrichthyes, Actinopterygii, and Sarcopterygii) have been by far the most targeted group. Indeed, the higher importance of those animals' commercialization has been driven forward by their use for food supply since the medieval period, being important, for instance, to the economy of several coastal cities in the North Sea and the English Channel (Boyer, 1967). Furthermore, fisheries production has grown throughout decades responding to an increase in demand and in the number of fishers, as well as the improvement of fisheries technologies. In 1950, fishing production comprised 20 million tonnes; in 2014, production has been estimated at 167.2 million tonnes, 87.5% of which (146.3 million tonnes) was destined to human consumption (FAO, 2016).

FISHERY RESOURCES AS FOOD—AND FOR WHAT ELSE?

Generally, a fishery exists for the purpose of providing food to humans and their animals, although fishery resources are also used, to a lesser extent, for medicinal and magic-religious purposes, and as curios or aquarium organisms (Alves et al., 2013a, 2012; Dias et al., 2011; Marques, 1995; Pinto et al., 2015; Rosa et al., 2005).

The medicinal use of fisheries resources is a widespread practice that involves hundreds of species, with fishes comprising a major portion of the animal taxa used as medicine in several health systems of different human cultures (Alves and Alves, 2011; Alves et al., 2013a,b). Their medicinal applications refer to the use of whole animals, body parts, or by-products of dead and, more rarely, live specimens (Alves and Rosa, 2013; El-Deir et al., 2012). Moreover, recent reviews have shown that at least 266 species of marine invertebrates (Alves et al., 2013a) and 24 species of aquatic mammals are used in traditional medicine worldwide (Alves et al., 2013b). In Asian countries, especially China, several fishery resources are also consumed as delicacies—shark fins, seahorses, and sea cucumbers can be used in the preparation of tonic soups (Lack and Sant, 2011).

Several fish, invertebrate, and mammal by-products, whose form and function have an important symbolic interpretation, are used in offerings (which may involve sacrifices) and as magical amulets (Alves et al., 2012; Léo Neto et al., 2009; Leo Neto et al., 2012; Moazami, 2005). Mollusk shells have been used in several cultures (e.g., Africans) during rituals and as symbolic items, or as bargaining chips (Leo Neto et al., 2012). Worldwide, species with a cultural and folkloric appeal have been used in a superstitious context (e.g., as amulets) and/or for medicinal purposes (e.g., as aphrodisiacs). For instance, seahorses, in which males get pregnant, are a symbol of male virility and have been sold in Brazil in bottles with a perfume-based concoction labeled as "seahorse perfume," which serves to attract the opposite sex; moreover, seahorses can be used as amulets for good luck and emotional/financial success (Alves et al., 2012; Rosa et al., 2011). Some of those properties are also attributed to sea stars (Alves et al., 2012).

Further uses have also been reported, such as fish toxins (e.g., pufferfish tetrodotoxin) used in the zombie culture in the Caribbean (Albuquerque et al., 2011; Littlewood and Douyon, 1997) and by Amerindians in hunting and fishing (Léry, 1997); corals and shells used in building cement and bricks; turtle shells, fish bones, spines, and scales used as tools (O'Connor et al., 2011) or sold as handicrafts (Dias et al., 2011; Jennings, 1987; Lunn et al., 2008; Pinto et al., 2013; Rosa et al., 2011; Wood and Wells, 1988); some sponges (Demospongiae) used for house cleaning (Gartside and Kirkegaard, 2009).

Several fishery resources are targeted by the aquarium trade, especially reef species such as corals and fishes (Calado et al., 2003; Monticini, 2010; Rhyne et al., 2012; Wabnitz, 2003; Wood, 2001), such as the clownfish (*Amphiprion ocellaris*) and the palette surgeonfish (*Paracanthurus hepatus*)

FIGURE 8.5 Clownfish (*Amphiprion ocellaris*) (*above*) and palette surgeonfish (*Paracanthurus hepatus*) (*below*), species that inspired the movies *Finding Nemo* and *Finding Dory*, respectively. *Photo credit: Luiz Alves Rocha (A. ocellaris) and Yi-Kai Tea (P. hepatus).*

(Fig. 8.5), well-known due to Pixar's animation *Finding Nemo*, which increased trade rates of these species (Yan, 2016). Sponges, sea anemones, and other organisms can also be found in tanks as pets (Balaji et al., 2009).

Aquatic animals (or parts of them) are also widely used as ornaments and curios, especially sea stars and mollusk shells (Dias et al., 2011), but also fishes, such as seahorses (Vincent et al., 2011). Other species are sought for jewellery confection: for instance, oyster pearls are heavily exploited, attaining high commercial values; red corals (*Corallium rubrum*) are sold in the form of necklaces or earrings (see Chapter 14).

Fisheries resources may also be used for recreational and entertainment purposes. Aquatic species are captured and displayed in aquariums and aquatic parks worldwide (Maple and Perdue, 2013) or may be used as tourist attractions, such as seahorses in Brazil and Papua New Guinea (Cater, 2007; Ternes et al., 2016), and sharks in the Bahamas, South Africa, New Zealand, and Australia (Davis et al., 1997; Garrod and Gössling, 2008; Higham and Lück, 2007). Such tourism-related activities are usually unregulated and not monitored and may raise several conservation and animal welfare concerns. For instance, removing seahorses out of the water and placing them in glass containers for exhibition, as performed by seahorse-watching operators in Brazil, potentially result in stress for the animals (Ternes et al., 2016); additionally, shark-related tourism (e.g., diving) may alter shark behavior, distribution, and even damage reef communities' structure (Burgin and Hardiman, 2015; Topelko and Dearden, 2005). On the other hand, in addition to their educational value, touristic experiences with aquatic animals also support thousands of jobs and local economies, which promotes reduction in captures, as fishers shift their investments toward tourism (Topelko and Dearden, 2005; Vianna et al., 2012).

A BRIEF OVERVIEW OF FISHING GEARS AND TECHNIQUES

Since the first hominins developed manual abilities for manufacturing tools and weapons, fishing became more efficient and widespread. Collection of aquatic organisms improved with the use of several gears such as small hoes, harpoons, traps, hooks, and fishing nets. Later, the construction and improvement of fishing vessels allowed fisheries development, leading to a higher exploitation of aquatic environments.

Manual collection of intertidal mollusks was likely the first fishing technique used to capture aquatic animals. First records of middens suggest that the practice of collecting mollusks by humans dates from 700,000 BC, in Thailand (Pope, 1989); Palaeolithic middens dating back to 300,000 BC were also found in the archaeological site of Terra Amanta, in France (Scarre, 2005). Sponges, crustaceans (lobsters, shrimps, crabs), mollusks (clams, oysters, scallops, whelks), echinoderms (sea cucumbers and sea urchins), as well as turtles and their eggs, are among organisms that are commonly hand-captured, especially by artisanal and subsistence fishers.

The first effective fishing gears also date from the Palaeolithic and comprised spears, arrows, tridents, and harpoons (Bednarik, 2014; Guthrie, 2005; Yellen et al., 1995), which were used for both hunting and fishing. Spears were made of tree branches with a sharp end, which could be improved using sharp stones, shells, or bones (Fig. 8.6). Currently, spears and harpoons (long spear-like gear connected to a line that can be retrieved) have no importance to commercial fisheries, being used for subsistence purposes or by sport divers (harpoons usually shot with guns).

FIGURE 8.6 A Hupa man (Native Americans of California, USA) with his spear (1923). *Source: Edward S. Curtis [Public domain], via Wikimedia Commons.*

Fishing traps date back to c. 75,000 BC (Lackey, 2005; Pedersen, 1995; Pitcher and Lam, 2015); they could be huge, such as dykes and fish weirs, or smaller structures and used in manual catches. Among fishing traps, fish weirs stand out due to the complex technique involved in their construction. They consist of permanent enclosures fixed with wooden posts, strategically positioned and forming a maze during the low tide that direct the fishes to a trap net. Weirs have been built in rivers and estuaries since the Middle Ages, according to the knowledge of the fishers on fish movements and local landscape (Salisbury, 1991) and can be also found in reef ecosystems. In South America (Brazil), weirs have been reported since 1858 (Menezes, 1976) and are locally named fish corrals (Maneschy, 1993; Tavares et al., 2005); in British and Irish estuaries, those traps were used in the preindustrial period (Scearce, 2010).

For several small-scale fishers, traps comprise the main fishing gear, with different types and sizes (Figs. 8.7 and 8.8). Many of those traps are built using wood and vegetal fiber; nonetheless, most of the current traps are made using nylon nets. Fishing traps may be permanent or semipermanent, when built using rocks or wood stakes, or may be removable, such as cages, bottles, baskets, hooks, and sieves, and may use baits depending on the targeted species, which mostly consist of benthic invertebrates (e.g., crabs, lobsters) and demersal fishes.

The use of hooks was improved in the Neolithic (O'Connor et al., 2011; Sahrhage and Lundbeck, 2012), when, later, the first fishing nets were developed (Pringle, 1997). The earliest hooks date from c. 40,000 BC, and were made of bones, found in East Timor, South-eastern Asia

FIGURE 8.7 Fish trap used in the sea Lamu, Kenya MR. *Photo credit: Anthony B. Cunningham.*

FIGURE 8.8 Traditional fish traps, Hà Tây, Vietnam. *Source: Petr Ruzicka from Prague, via Wikimedia Commons.*

(O'Connor et al., 2011). In Brazil, indigenous people used this fishing gear before the 15th century, using a plant called tucum (*Bactris setosa*) to make the lines (Léry, 1997) while hooks were made of plant or fish spines. Later, those materials were replaced with iron, due to the Portuguese influence (Léry, 1997; Silva, 2001). Hand lines and hooks are still used in subsistence and artisanal fisheries, but it has been mostly widespread by recreational fishers. Longlines, which correspond to a set of hooks and lines that can be laid in the water column or on the bottom, are still used by small-scale commercial fisheries, but this fishing technique is widely applied by industrial fisheries, especially for large predator fishes, with lines that are several kilometers long.

In the Bronze Age, hooks of all types and sizes, as well as fishing nets, had already been used in the Middle East (Pitcher and Lam, 2015; Stewart, 1994a). The first nets were sewed with

FIGURE 8.9 Fisher catching fish with a cast net in Sentarum lake, West Kalimantan, Indonesia. *Photo credit: Ramadian Bachtiar for Center for International Forestry Research (CIFOR).*

FIGURE 8.10 Cod end of the trawling net just before discharging fish on deck during stock assessment surveys. *Source: Captain Robert A. Pawlowski, NOAA Corps (Fisheries Collection, Image ID: fish0012) [Public domain], via Wikimedia Commons.*

vegetal fibers, such as cotton (Leite, 1991). As of World War II, fishery had advanced greatly with the popularization of synthetic fishing threads (such as nylon), which were manufactured with petroleum by-products and were more durable than their plant-based predecessors (Diegues, 1983). Artisanal fishers commonly rely on gillnets, cast nets, hand nets, and beach seines for coastal fishing with or without fishing vessels employment (Fig. 8.9). Small-scale fisheries also employ hand-dredges, commonly for the capture of crustaceans and mollusks (e.g., clams). Industrial fisheries employ large fishing nets, especially surrounding nets (e.g., purse seines) and trawl nets (e.g., bottom and mid-water trawls) (Fig. 8.10), both managed by one or two vessels. Boat dredges are also commonly used for benthic fauna exploitation.

The establishment of artificial reefs or use of artificial structures to attract fishes also comprises an ancient technique that is still in use worldwide (Santos et al., 2010). There are reports of the use of artificial reefs to attract tunas, dating to 3000 years ago in the Mediterranean (Purpura, 1990). Earlier, in the Neolithic, African people have already noticed high abundance of fishes near floating or submersed objects (Stone et al., 1991). In Brazil, historical records refer to artificial reefs constructed with branches, bamboo, leaves, and rocks by several indigenous tribes that named this technique as "marambaias," which means, from Tupi, "place of good fishing" (Conceição, 2003). Industrial fisheries, conversely, may rely on fish aggregating devices—structures that comprise natural and/

FIGURE 8.11 Somali fishing fleet. *Photo credit: Luiz Alves Rocha.*

or artificial floating objects that can drift or can be fixed, and under which fishes (such as tuna) naturally aggregate and are captured, commonly using purse-seines (Dagorn et al., 2013; Davies et al., 2014).

Most fishing gears are used aboard fishing vessels, which may be rudimentary wooden structures or high-technology industrial boats, which means that navigation plays an indispensable role in fishing. It is believed that *Homo erectus* probably built the first boats made of wood or skin at c. 800,000 BC, as migrations have been reported even at great distances to the North, to the Mediterranean region, and several thousand miles to the East, facing Asia (Morwood et al., 1998; O'Sullivan et al., 2001). Archaeological data in Wallacea, Indonesia, show that men's history at the sea started in late Palaeolithic (Bednarik, 2014). The earliest evidence of the use of boats for fishing consisted of paddles excavated in Europe dating from the Mesolithic (Sahrhage and Lundbeck, 2012). Nonetheless, evidence of the capture of pelagic fish such as tuna dating back 42,000 years in East Timor suggests high level of maritime skills and technology and deep sea fisheries (O'Connor et al., 2011).

Fisheries type concerning habitats, targeted species, and fishing gears usually determine the fishing vessel type to be used (Nishida et al., 2008). Artisanal and subsistence fisheries generally rely on canoes, rafts, and small wooden boats (Figs. 8.11 and 8.12). "Jangadas," small wooden boats with sails, have already been reported to be used by Greek, Roman, Deutsche, and Gauls (Araujo, 1990; Cascudo, 2002; Mussolini, 1953). In the Philippines, Africa, Brazil, and Central America, most small boats are propelled by paddles and sails, although, in the last decades, several small-scale fishers have motorized their vessels, using onboard or offboard engines. Industrial fisheries use larger motorized vessels for accommodating crew, large seines and/or trawls and have plenty of space for storing and processing the resource captured (Fig. 8.13).

ANIMALS USED IN FISHERIES: AN ANCIENT TRADITIONAL TECHNIQUE STILL USED TODAY

One of the most interesting fishing techniques is the use of animals to help catching fish. This practice dates back to ancient times and

FIGURE 8.12 Fishers and dhows at Pemba, Mozambique. *Photo credit: Anthony B. Cunningham.*

FIGURE 8.13 Crab boat from the North Frisian Islands working in the North Sea. *Source: The original uploader was Jom at German Wikipedia (Joachim Müllerchen, via Wikimedia Commons).*

results from a close relationship between the fisher and the animal helper (that may or may not be trained for that purpose), and still comprises a current approach in several parts of the world, being featured in many ethnobiological/ethnozoological studies (Allen, 2010; Alves et al., 2009; Bezerra et al., 2012; Svanberg et al., 2016).

Several groups can be used for assisting fishing activities, such as birds, mammals, and other fishes (Gabriel et al., 2005; Gudger, 1919b, 1926,

FIGURE 8.14 *Phalacrocorax carbo*, one of the cormorant species used for assisting fishing activities. *Photo credit: Rômulo R.N. Alves.*

1927). Tamed oriental darters (*Anhinga melanogaster*) and cormorants (*Phalacrocorax capillatus; Phalacrocorax carbo*) are good examples (Figs. 8.14 and 8.15) (Gudger, 1926; Svanberg et al., 2016). Cormorant fishing has been a common practice in Japan and in some parts of China; darters have been used in South Asia (Gudger, 1926; Jackson, 1997; Laufer, 1931; Stonor, 1948). In Japan, cormorant fishing dates to before the Heian period (from the 9th to the 12th century); in Gifu, fishers used 12 birds each (record from 1928), a practice that is still currently used (Gabriel et al., 2005). In South America, similar practices were also reported, particularly in pre-Colombian Peru (Birket-Smith, 1966). Despite reports of the use of cormorants for fishing by English, Italian, and French fishers, it has never been a usual practice in Europe (Beike, 2012).

Otters are another example of animal aid in fisheries (Allen, 2010; Gudger, 1927). They have been used for centuries in Asia, Africa, and Europe (Gabriel et al., 2005). In Eurasia, two species seem to have been employed in fishing (Svanberg et al., 2016); in North America, although there is evidence of such practice, it is more limited (Etkin, 2009; Gudger, 1927). The use of trained otters in fisheries dates as far back as the Tang dynasty (618–907), being reported to take place in the Yangtze Chinese region (Svanberg et al., 2016). Currently, otter fishing is still in use as a traditional fishing technique in Asia, particularly in Bangladesh, where they are trained to chase fish toward boats and nets, generally two adult and one young otter (Feeroz et al., 2011); they can be even sold as a fishing aid (Deb, 2015). This practice has been rapidly decreasing due to several socioeconomic and conservation issues, such as fish stock declines, diseases, conflicts over fishing taxes/permissions, and the implementation of mechanical fishing techniques (Feeroz et al., 2011).

FIGURE 8.15 Chinese man with tame fishing cormorant, Yunnan Province, China. *Source: My Hobo Soul, via Wikimedia Commons.*

Fishing with dolphins comprises another traditional technique still currently used, especially in cast-net fisheries. For instance, in Brazil,

FIGURE 8.16 Bottlenose dolphins (*Tursiops truncatus*) interacting with artisanal fishers in Laguna, Santa Catarina State, Brazil. *Photo credit: Carolina Bezamat.*

bottlenose dolphins (*Tursiops truncatus*) help fishers to catch mullets (*Mugil* spp.) in a cooperative system with mutual advantages, as dolphins also obtain fish (Fig. 8.16) (Peterson et al., 2008; Simões-Lopes et al., 1998; Zappes et al., 2011). Other examples of such cooperative fishing interactions include those performed by cast-net fishers and Irrawaddy dolphins (*Orcaella brevirostris*) in Myanmar (Smith et al., 2009), and by aboriginal fishers and bottlenose dolphins (*Tursiops* spp.) in Australia, where both nets and spears can be used (Neil, 2002). Reports describe fishers "calling" the dolphins by splashing on the water or making other types of noise to attract the animals for fishing (Neil, 2002; Smith et al., 2009). However, in Brazil, dolphins initiate and control the fishing. They spontaneously herd fish toward shallow waters where fishers await with their cast nets either on foot or aboard small vessels; fishers, thus, act as a barrier, confusing and spreading the fish, making them easier to capture by the dolphins (Simões-Lopes et al., 1998). And this interaction goes further: fishers are able to recognize dolphins' behaviors that sign the location of fish schools (Simões-Lopes et al., 1998; Zappes et al., 2011) and dolphins reportedly can recognize and show some kind of preference for signaling to a particular fisher (Zappes et al., 2011).

Among domesticated animals, horses and dogs are major assistants in human activities (Alves et al., 2009; Blazina and Kogan, 2016; Clutton-Brock, 1992). In fisheries, horses have a discrete involvement, which mainly involved pulling fishing gears, such as long seine nets, as reported in rivers such as Rio de la Plata and Rio Uruguay, Argentina (Gabriel et al., 2005, 2008). Moreover, horses could be ridden during fishing activities; for instance, the horses of Oost Duinkerke are ridden by shrimp fishers when fishing in the Belgian coast (Boudarel, 1948).

Dogs have been more broadly used in fisheries, especially being trained to chase fish into nets in many different parts of the world (Gabriel et al., 2005), although they can also help landing fish caught in sport fishing lines (Gudger, 1923, 1941) or frighten fish into shallow water where they are captured by fishers, a typical technique employed by Ainu people in Japan and Russia (Weyer, 1959). Similarly, in Tierra del Fuego (Argentina), dogs were used to herd fish toward small beaches where they were captured by nets

FIGURE 8.17 Portuguese water dog, a breed used to chase fish. *Source: Public Domain,* https://en.wikipedia.org/w/index.php?curid=8515709.

FIGURE 8.18 A slender suckerfish (*Echeneis naucrates*) species used to help fishers capture large fishes and sea turtles. *Richard ling, via Wikimedia Commons.*

(Gudger, 1923). In the Portuguese coast, fishers use Portuguese water dogs to chase fish and also to retrieve lost or broken tackle (von Brandt, 1972) (Fig. 8.17). Dogs can also be tamed to capture and bring fish directly to their owners (Gabriel et al., 2005) and are also used in invertebrate fisheries (i.e., crab fisheries) in Malaysia (Burdon, 1951).

Some animals can be used for fishing due to biological characteristics that may improve catches, other than being tamed. Remoras or suckerfish are good examples: they can help fishers capture large fishes and sea turtles, as described from East Africa and the Torres Strait between Australia and Papua New Guinea (De Sola, 1932; Gudger, 1919a,b). For that, a remora is captured alive using hook and line and then has a line (or a ring attached to a line) fastened to its tail. When the targeted fish or turtle is sighted, the remora is released back to the sea; then, when the remora attaches itself to the target, the line is pulled back to the boat, or to shallow waters, where the fish or turtle can be captured or speared. Remoras do attach themselves very strongly and can pull animals weighing 9–10 kg, as reported for the major species used for fishing, the c. 60 cm-long *Echeneis naucrates* (Fig. 8.18) (Gabriel et al., 2005).

FISHING: IMPORTANCE, IMPACTS, AND SOCIO-ENVIRONMENTAL CONCERNS

The importance of fishing in human history is undeniable, even the discovery and exploration of the New World, for instance, can be tied to cod fisheries (Kurlansky, 2011). Today, millions of people worldwide depend directly or indirectly on fisheries as an income source and for subsistence (FAO, 2016). Fishing has become an important global economic activity with the further development of fisheries trade, the manufacturing and use of instruments, boats and fishery resources preservation (Diegues, 1983). The majority of the world fisheries production (c. 87%) have been destined to human consumption (FAO, 2016), fishes comprising 17% of total protein consumption by human populations worldwide (FAO, 2014) and thus representing a valuable source of essential nutrients for several communities (Begossi et al., 2004; Casale, 2011; FAO, 2016; HLPE, 2014; Nishida et al., 2006a).

Besides its significance to the economy and welfare of coastal communities in feeding security and livelihood, fishing has played an important role as part of the cultural identity of coastal communities (FAO, 2007), especially indigenous and traditional fishing groups. Indeed, several cultures have a unique heritage driven by fishing activities. Furthermore, fisheries likely led to the settlement of many human civilizations that exploited coastal localities and later moved further inland, for instance, from Africa to Eurasia (Walter et al., 2000) and within Oceania (O'Connor et al., 2011).

Concomitantly, fisheries also lead to heavy pressures on the targeted species. For instance, more than one-third of the oceans are overexploited (FAO, 2016; Pitcher and Lam, 2015). About 75% of the commercial fish species have been fully or overexploited and some are threatened (FAO, 2014). According to the ONU (2002), most of the fishing areas worldwide have reached their maximum catching potential. Nearly 32% of the commercial fish stocks have been estimated as overfished in 2013 (FAO, 2016). One-third of all groupers (a group that comprises a billion dollar fishery) are now considered threatened due to overfishing (de Mitcheson et al., 2013). It is worth noting that most of the consumed fisheries resources comprise captured wild animals (FAO, 2016); in contrast with bushmeat, wild fishery resources are part of our routine supermarket shopping.

Other fishing techniques, especially netting and trawling, have been shown to affect not only targeted but also nontargeted species. For instance, shrimp and fish trawling produces tonnes of incidental captures (bycatch), which may include other untargeted fish species, turtles, corals, and other animals and are often discarded (Alverson et al., 1994; Beckman, 2013; Eayrs, 2007; Lewison et al., 2004; Nogueira and Alves, 2016). Special concerns are raised to the fact that some fish stocks are being depleted by trawling even before having their structure/composition known or well described. Furthermore, discards often consist of huge amounts of juvenile organisms (Broadhurst, 2000; Kenelly, 2014; White et al., 2006) and, not rarely, threatened species (Lawson, 2017; Moore et al., 2013; Wallace et al., 2013; Zeeberg et al., 2006). In a recent analysis of marine fisheries discards, Zeller et al. (2017) showed that nearly 10% of global total catches were discarded per year during the period of 2000–2014, mostly by industrial fisheries. Along with bycatch, ghost-fisheries (captures by abandoned fishing gears) bring up another challenging issue, rarely assessed in global estimates (Pauly and Zeller, 2016a). Other practices, such as shark finning, which consists of removing sharks fins and releasing the (dead or dying) animal back to the sea has raised conservation concerns and resulted in declines of several species (Clarke et al., 2006, 2013). Overfishing, especially considering top predators (such as sharks and tunas), also gives rise to critical alterations in marine food webs (Pauly et al., 1998; Pauly and Palomares, 2005).

Fisheries resources face other threats that lead to declines in production. For instance, illegal fishing (Agnew et al., 2009; Beddington et al., 2007; Sumaila et al., 2006), degradation, and loss of habitats, such as mangroves, reefs, seagrass beds (Lotze et al., 2006), or, indirectly, climate change (Jackson et al., 2001). Regionally, threats and impacts vary not only spatially and seasonally, but also with the political climate. Currently in the United States, several key legislations related to fisheries are changing, and in Brazil, a change in political powers has significantly decreased fishery regulations (Di Dario et al., 2015; Pinheiro et al., 2015).

Illegal fishing comprises the capture of threatened species and/or those protected by law; captures during the reproductive period; captures of specimens of prohibited sizes; fishing inside no-take areas (Agnew et al., 2011; Ban and Vincent, 2009; Mora et al., 2011). Moreover, fishing may also be considered illegal when it is practiced with the use of forbidden gears, such as lobster fishing using air compressors or fishing nets with mesh sizes under the smaller permitted size (Pinto et al., 2013), using of blast fishing or poisoning compounds (Calado et al., 2014; Slade and Kalangahe, 2015). Worldwide,

c. 20% of captures come from illegal fishing, accounting for 10–23.5 billion dollars annually (WWF, 2017).

Even when conducted legally, fisheries may also be impacted by habitat degradation and/or loss. For instance, pollution by solid wastes or by chemicals have resulted in severe impacts to aquatic animals (Derraik, 2002; Islam and Tanaka, 2004). Habitat destruction affects fish stocks, as degraded habitats may lose their key roles in maintaining targeted populations. Mangroves and reefs, which work as nurseries, feeding and reproductive grounds, and have an important role in keeping the connectivity among other habitats (Mumby and Hastings, 2008; Unsworth et al., 2008), have been considered threatened ecosystems primarily due to their degradation. Coastal development, the rise of touristic activities, development of harbors, industries, and aquaculture facilities (Halpern et al., 2008), comprise other threats to those ecosystems that potentially affect habitats and, consequently, fish stocks (Reis et al., 2016).

Furthermore, climate change has been considered in the habitat degradation context and fishery stocks declines, especially for ecosystems which negatively respond to rising water level and temperature and decreasing pH, such as coral reefs (Jackson et al., 2001). Those changes affect survival (e.g., coral bleaching and death), food web links, reproduction, and migration processes, and consequently, fisheries production (Weatherdon et al., 2016).

Such concerns highlight the need for urgent measures to promote fishery sustainability under local, regional, and global perspectives. The global crisis affecting the fishing sector changes not only the availability and diversity of resources, but has profound impacts on fishers' livelihoods, especially those at the artisanal and subsistence levels (Pauly et al., 2005). Impacts may also affect the cultural heritage of fishing communities, leading small-scale and subsistence fishers to shift their activities toward other "nonartisanal" fishing techniques or even withdraw their involvement with fisheries (Alves and Nishida, 2003; Ruddle, 1993). Moreover, the use of such "new" and often unregulated fishing techniques (commonly employing nonselective fishing gears) seeking faster and more productive outcomes, may reflect an important change in traditional fishing techniques—and also raise new concerns about overfishing in this sector. In Brazil, for example, mangrove crab (*Ucides cordatus*) gatherers have introduced a trap called "tangle-netting" to replace traditional hand-capturing or tamping techniques, which are more effective but slower; tangle netting will potentially contribute to depletion of crab stocks and mangrove and estuary pollution (Alves and Nishida, 2003; Nascimento et al., 2016). Similar trends have been recorded in other localities worldwide, such as in the Asia–Pacific region, where new and more efficient fishing techniques, gear types, and vessels have contributed to the dwindling of traditional community-based marine resource management systems and of the traditional authority that underpins them (Ruddle, 1993).

Some alternative measures to reduce exploitation have been proposed, such as the establishment of aquaculture ventures, incentives to small-scale fisheries, co-management of fishery resources, law enforcement, and punishment of illegal fishing and trade (Agnew et al., 2011; Ban and Vincent, 2009; Mora et al., 2011). However, some measures should be considered with caution, such as aquaculture, since they can impose other threats to the environment, for instance, pollution, deforestation, and introduction of invasive species (Emerson, 1999).

Global guidelines have also been drawn to protect and manage fishery resources and their habitats, such as the Convention on Biological Diversity (CBD) (United Nations, 1993) and the Code of Conduct for Responsible Fisheries (FAO, 1995). Among in situ recommendations, CBD proposed the establishment of protected areas to promote biodiversity conservation (United Nations, 1993). Moreover, the Code of Conduct for Responsible Fisheries (FAO, 1995) proposed that fishery should follow a new strategy worldwide, considering its crisis due to overexploitation and marginalization of millions of artisanal fishers. That document established international principles and rules, based on responsible and sustainable practices toward the conservation, management, and the development of fishery resources. The Code is a comprehensive document and recommends that responsible fisheries should consider the biological, technological, social, and environmental aspects. Furthermore, it points out that fishers' and scientific knowledge should be jointly considered. However, very few of its suggestions are followed.

Protected areas have long been considered a major conservation measure; however, only 3.41% of the oceans are covered by marine protected areas (MPAs), and most of them are not well enforced (Thomas et al., 2014). MPAs should effectively protect vulnerable and highly productive habitats (such as coral reefs and mangroves) and fauna, especially those species that are considered threatened. They should also play an important role in marine connectivity and enhance the abundance and biomass of several species even outside the MPA's boundaries (spill-over effect), thus supporting fisheries in surrounding areas (Forcada et al., 2009; Stobart et al., 2009).

Nonetheless, the establishment of MPA for the sustainability and recovery of fishery resources has become a difficult task since laws are formulated and executed under different government perspectives, thus depending on environmental and development of fisheries policies that may vary from one place to another.

It must be highlighted that effective measures will ideally be based on continuous assessment of fishing stocks. For years, the only available official fisheries data are primarily reported by countries to FAO, which analyses and publishes annual reports in a global context (The State of World Fisheries and Aquaculture—SOFIA). However, the bulk of FAO's analysis comprises data on commercial industrial fisheries catches (primarily on high-valued species of fish, lobster, and shrimp) and aquaculture production. Catches by small-scale, subsistence, and recreational fisheries are often not included or neglected in global assessments, although alternative estimates have improved, especially through fisheries reconstructions (which also include bycatch estimates), thus aggregating data from those underreported fisheries sectors (Zeller et al., 2006; Zeller and Pauly, 2012). The Sea Around Us provided a comprehensive estimate in the form of an atlas of global fisheries, which incorporate catches from small-scale—and other marginalized and problematic—fisheries to official landing data provided by countries (Pauly and Zeller, 2016a). By doing so, catch reconstructions have revealed a disturbing picture of global fisheries catches, which have actually been found to be 53% higher than data reported by FAO (Pauly and Zeller, 2016b).

ETHNOZOOLOGY AND ITS ROLE IN FISHERY STUDIES

Fisheries management has comprised a challenging issue worldwide, particularly due to the limited recognition of the interdependence between fishers and fishery resources (Ommer and Perry, 2011). Such relationship leads to the inherent complexities in the sociocultural universe of fishing communities that use aquatic wildlife for feeding, trade, in medicinal and magic-religious practices, for handicraft manufacturing, etc. The impacts of fishery resources depletion on artisanal and subsistence communities are immensurable, as fishing comprises a vital activity and source of livelihoods for those groups.

Reconciling its social, environmental, and economic dimensions is a major challenge for the development of sustainable fisheries. Moreover, most fishers come from the artisanal small-scale sector, which has very important cultural aspects, although this may have been lost in some communities. Thus, there is a crescent need of an interdisciplinary approach in understanding the role and impacts of fisheries. Such approach should also include the information and knowledge provided by fishers, which can be useful for fisheries assessments as they encompass essential aspects such as how they fish and how they perceive fisheries and its outcomes, how they involve traditions in developing fishing techniques and using fishing gears, how much they produce, and how they manage their activities.

Under this perspective, it is reasonable to acknowledge fishers' traditional knowledge and perceptions as well-recognized tools for understanding fishery processes and impacts (Berkes et al., 1995; Ruddle, 1993, 1994). Indeed, such information have increasingly been incorporated in the development of management strategies for biodiversity uses, especially through ethnozoological studies.

As the relationship between people and aquatic animals is primarily linked to the existence of fishing activities, which virtually affects all coastal civilizations worldwide, ethnozoological studies may fill important gaps that other disciplines alone are not able to. Therefore, the inter- and transdisciplinary nature of such studies may have profound influence in fisheries science. In this perspective, focusing on the importance of ethnobiology in studying marine ecosystems, Narchi et al. (2014) highlighted that its interdisciplinary approach can provide significant contributions in several situations, including (1) make better use of the coastal and marine resources, while contributing to their conservation; (2) contribute to novel management schemes that incorporate fishers' knowledge effectively; (3) facilitate participatory approaches involving indigenous and other coastal populations in learning and decision making for ocean and coastal management; (4) encourage people to think about the link between present, past, and future; and (5) prepare local coastal populations for future changes.

One of the main contributions of ethnozoological studies is that fishers' knowledge potentially supplement information for underreported fisheries (i.e., small-scale, subsistence). Indeed, interviews with fishers are one of the alternatives proposed to assess those fishery sectors (Pauly and Zeller, 2016a). Catch rates estimates, fishing composition, gears used and their impacts are some of the invaluable information those fishers may provide, especially concerning low-valued resources (e.g., several invertebrates, small fishes) and where no ecological studies are available and often neglected in fisheries statistics. In such cases, ethnozoological studies may be the only opportunity to assess trends in distribution, habitat use, and catch rates. For instance, fishers have been able to inform population trends and/or catch rates for several data-poor species like seahorses (Lawson, 2017; Rosa et al., 2005, 2011), and the mangrove crab *U. cordatus* (Linnaeus, 1763) in Brazil (Alves and Nishida, 2003; Capistrano and Lopes, 2012; Jankowsky et al., 2006).

An illustrative example of the relevance of ethnozoological studies is the case of seahorses in Brazil, where they comprised an invaluable tool for seahorse conservation. As most small-scale fisheries, especially on data-poor species, the seahorse fisheries have been underreported in the country. Moreover, almost no information on seahorse trade and uses—and, thus, the purposes of catching—were available before a program of ethnozoological surveys started in the early 2000s (Rosa et al., 2011). Brazil has long been an important exporter of seahorses, but there were large gaps (and overall lack of information) in the assessment of how the seahorse fishery was performed, by whom, how representative were the incomes provided, which were the main species targeted and their uses, and what were the population trends.

The most effective regulation measure in reporting seahorse trade numbers is the

SISCOMEX (Integrated Foreign Trade System). Nonetheless, the system itself only collects export data, as there are specific export quotas for seahorses in Brazil. SISCOMEX reports to the CITES Management and Scientific Authority in the country (the Brazilian Environmental Agency—IBAMA) and potentially is a reliable tool, but it depends on the information in the licences issued by IBAMA. So, no information on actual catches, landings, or domestic trade can be provided by the most effective controlling system in the country—only for exports. Thus, the bulk of information on domestic seahorse catches and trade (live or dried) was provided by those ethnozoological surveys performed along the Brazilian coast. Through the studies, it was observed that different species took part in different trade sectors, comprising a diffuse seahorse trade in the country (Rosa et al., 2011). It was also found that dried seahorses were mainly used in traditional medicines and, to a larger extent than previous thought, in the curio trade. On the other hand, the live seahorse trade was particularly focused in one species (*Hippocampus reidi*), which seems to be the only species exported. However, this raises concerns, as export quotas in the country were defined for two species (*H. reidi* and *Hippocampus erectus*) and through SISCOMEX data, both have been exported, although only *H. reidi* has reportedly been captured for that purpose—thus highlighting misidentification (intentional or not) which could inflate *H. reidi* quotas.

Seahorse uses were also assessed. Surveys performed in fishing communities and markets showed that seahorses are used to treat several diseases, particularly asthma (Alves and Rosa, 2007; Alves et al., 2007; Rosa et al., 2005, 2011). Interviewees were able to provide detailed descriptions of the preparation of seahorse-based remedies, although they could also acknowledge that several illnesses (including asthma) could be treated with other non-threatened species, especially plants. Surveys also demonstrated that seahorses are part of magic-religious rituals, particularly of Afro-Brazilian religions, which are not only performed by fishers (Alves et al., 2012; Leo Neto et al., 2011). Thus, seahorses enter the cultural context of other Brazilian ethnicities and social groups, although to a lesser extent. This shows the complexity of seahorse trade and use in such a large country (Rosa et al., 2011).

But one of the most accomplishments of those studies was to get consistent participation and buy-in from local fishers. They showed a profound knowledge about seahorse ecology and behavior and, thus, could provide important information on fishing captures and trends (Rosa et al., 2005, 2011). Moreover, through collaborative monitoring programs, researchers were able to estimate seahorse catches from artisanal fishers and bycatch from the industrial shrimp fisheries (Rosa et al., 2011). The live trade was mostly originated by the former, and the dried trade, by the latter fisheries type. Moreover, and importantly, it was shown that the income generated by seahorse fishery in Brazil was not very important for fishers.

Finally, we can list some important accomplishments of the ethnozoological studies on seahorses in Brazil. Fishers could provide population trend information, which was alarming. For instance, fishers could report declines of more than 90% in seahorse catches in traditional fisheries locations. Another important result was that fishers were willing to help. They can identify high density areas, impacts, and recognize that declines are true and mostly due to high catch levels. This is especially important since the bulk of the live seahorses traded in the country comes from the wild and, thus, fishers are the very first piece in the trade chain. Where fishers do not capture, there is no trade, and vice versa. Furthermore, fishers know and explore the flagship potential of seahorses and can help in keeping wild populations "safe." For instance, the Ponta do Tubarão Sustainable Development Reserve, a state MPA in NE Brazil where seahorses are protected, has a seahorse logo—interestingly, developed by a fisher in a contest promoted by the fishing community.

All this ethnozoological information was widely used in the proposal for a management plan for seahorses in Brazil, released in 2011. Other studies, though to a lesser extent, have been conducted in the meanwhile, also including an unregulated activity—the use of seahorses as a touristic attraction (Ternes et al., 2016). Moreover, all seahorse species are currently listed in the Brazilian Red List as "vulnerable" and, again, ethnozoological data were fundamental in this context and should continuously be used as a tool for conservation in the country. However, the list is under tremendous political pressure and will likely not be used to enforce fisheries (Di Dario et al., 2015).

As discussed in the case of seahorses, fishers know/make use of behavioral and ecological traits of target species—e.g., migrations, habitat use, depth distribution, etc.—for improving fishing practices and sometimes for conservation. Often referred to as Traditional Ecological Knowledge (TEK) and Local Ecological Knowledge (LEK) (Berkes et al., 2000), fishers' knowledge has been highlighted for its potential contributions to formal data collection, fisheries assessment, and biodiversity conservation (Berkes et al., 1995; Johannes, 1993), as it may assist more flexible management systems, conversely to traditional management (Berkes et al., 2000; Johannes, 1993), especially concerning threatened and overfished species (Gerhardinger et al., 2009). When local and scientific knowledge meet and are complementary assessed, they may comprise powerful tools for wildlife management.

It must be noted that ethnozoological data obtained from stakeholders about fisheries resources are essentially based on their experiences, memories, and perceptions, as well as observed trends noticed by the practice of fishing, diving, and researching (Beaudreau and Levin, 2014). For example, a study conducted at Cook Inlet, Alaska, demonstrated that people show different perceptions about the sustainability of salmon populations, depending on the motivation of the fishing practice (e.g., whether for personal or commercial purposes) and the household income (Loring et al., 2014). For instance, higher-income commercial fishers and respondents more likely perceived salmon fisheries as sustainable, thus suggesting that individual perceptions on resource trends may be related to their socioeconomic status, their interest and relationship with the management system, their perceptions about how resources are distributed (Loring et al., 2014). Moreover, Alves and Nishida (2003) stressed that the socioeconomic context in which fishers are inserted often lead to the development of their economic activities under the subjection of traders that exploit fishers work. Such relationship may result in potential biases.

Those biases in fisheries studies may not only be related to fishers, but also to other stakeholders, even scientists and managers, who may face problems in reconstructing data from collective memory, thus leading to misconceptions about ecological information (Beaudreau and Levin, 2014). This phenomenon was described by Pauly (1995) as the "shifting baseline syndrome," as different generations have different reference points for the same resource, depending on their lifetime, thus leading to different baselines. This can lead to inaccuracies and ineffective stock rebuilding measures and, further, to a gradual accommodation of species losses (Pauly, 1995).

Moreover, some interviewees have reasons to misreport their activities and trends (by under- or overreporting) (Jones et al., 2008), thus leading to bias in their information (Lunn and Dearden, 2006). However, despite such constraints and inaccuracies that may be provided by fishers' perceptions—which are closely related to the context in which they are formed (Daw et al., 2011), fishers' LEK or TEK are considered reliable for research and management purposes, depending on the intended application (Alves and Nishida, 2002; Capistrano and Lopes, 2012;

Mackinson and Nottestad, 1998; Nascimento et al., 2017). For example, if well planned and designed, community-based monitoring and participatory research can be cost-effective and provide reliable scientific data (Carvalho et al., 2009; Holck, 2008), especially when considering small populations or low encounter rates (Coll et al., 2014; Humber et al., 2011, 2017; Rasalato et al., 2010) and highly fished and traded species (Alves and Nishida, 2002; Kay et al., 2012). Furthermore, LEK can provide insights to hypothesis testing about species ecology, shedding light on new scientific information and discoveries (see Chapter 23). However, this can only be accomplished with the adequate application of ethnobiological methods and analyses, which must be in line with the research goals, thus minimizing eventual biases.

Under this perspective, Beaudreau and Levin (2014) stress that the continued development of integrated LEK-science frameworks for understanding environmental change depends on the advancement of tools for aggregating observations in a way that satisfies the demand for quantitative rigor in the ecological sciences but, critically, seeks to avoid inappropriately reductionist treatments of resource users' knowledge. For example, it is important to recognize that knowledge is not simply constructed based on an individual's information environment, but tied inextricably to his or her culture, norms, and values (Berkes et al., 2000).

Hence, fishers' knowledge can be considered an indispensable instrument for the development of fisheries management measures. Fishers permanently interact with the ecosystems in which they are inserted and are capable of perceiving impacts and trends that may go unnoticed by researches, managers, and authorities (Johannes et al., 2000; Silvano and Valbo-Jørgensen, 2008). Furthermore, integrating fishers' (or indigenous) information in the management process makes decision-making more acceptable by those groups, as their participation is effectively inserted in the conservation context. Even critical measures, such as bans or quotas, should become more acceptable when fishers take part in the development of such measures or when they know that their knowledge, perceptions, and opinions are considered (Johannes et al., 2000; Pomeroy and Carlos, 1997). The participation of fishers in the process of management plans implementation has been recognized as of great importance (Berkes et al., 2001; Pomeroy and Berkes, 1997; Pomeroy and Carlos, 1997). Indeed, the comprehension and incorporation of the tight relationship of human societies and fisheries resources in management processes should be acknowledged, improving the understanding of fishing practices characteristics, of how species are exploited, the fishing strategies used, and the socioeconomic reality of fishers (Alves and Nishida, 2003; Silvano and Valbo-Jørgensen, 2008).

Ethnozoological studies also contribute to improve the visibility of groups that actively participate in fishing activities, but are generally marginalized in fisheries assessments and management. Women's participation in fisheries (Fig. 8.4A and B) is a clear example: they are involved in several fishing-related activities, such as capturing, processing, and selling, and also with finance aspects of fisheries (Dye, 1983; Harper et al., 2013; Nascimento et al., 2017; Nishida et al., 2006a; Rocha et al., 2008; Rocha et al., 2012). Nonetheless, women's fishing efforts are often overlooked in fisheries and economic assessments, and in the management and decision-making processes (Harper et al., 2013), which certainly leads to an underestimation of catches and, under a broader perspective, hampers a more complete social-ecological approach of the understanding of human roles in fisheries (Kleiber et al., 2015). This sheds light on the importance and need of interdisciplinary studies, such as those conducted under an ethnobiological approach, in order to demystify the concept that women's participation in fisheries is unimportant, thus promoting opportunities for women to take part in the entire process of fisheries assessments and management.

Several fishing communities face a rapid process of elimination of their original characteristics and culture (Alves and Nishida, 2003; Ruddle, 1993), and traditional fishing techniques, generally less harmful to fishery resources, may also be lost. Under this perspective, ethnozoological studies are important as they value and document fishers' knowledge and perceptions on such processes and their impacts on fish stocks, which may be used in the establishment of particular conservation measures—not only for maintaining fishery resources, but also to help communities keep their traditions and culture. As pointed in previous studies (Alves et al., 2005; Berkes, 1999; Nishida et al., 2006b; Ruddle, 1995), strengthening the cultural values of fishing gives fishers greater visibility, recognition, and political engagement in management processes and improves the dialogue between them and managing agencies (Silvano and Valbo-Jørgensen, 2008).

Fishing activities are characterized by the complexity of the relationships among social and environmental factors, whose understanding is essential for the establishment of management strategies that allow sustainable uses and/or recovery of resources and the subsistence of the people who depend on them. Hence, this activity has roused the interest of several areas of the environmental and social sciences. Ethnozoology may provide an important contribution by evaluating the relationships between people and their catch along human history (Alves and Souto, 2015). Among the ethnozoological subdivisions, ethnoichthyology has been gathering attention from researchers as one of the most important subareas in the field (Alves and Souto, 2011; Lyra-Neves et al., 2015). As pointed out by Alves and Souto (2011), this may reflect the importance of fishes (and other fishery resources) in human culture and economy. Furthermore, most fishery resources generally face loose use and trade controls (or even lack regulations), in comparison to terrestrial game species that are protected by a much more rigorous legislation, which is also much easier to enforce.

References

Agnew, D.J., Pearce, J., Pramod, G., Peatman, T., Watson, R., Beddington, J.R., Pitcher, T.J., 2009. Estimating the worldwide extent of illegal fishing. PLoS One 4, e4570.

Agnew, D.J., Pearce, J., Pramod, G., Peatman, T., Watson, R., Beddington, J.R., Pitcher, T.J., 2011. Estimating the worldwide extent of illegal fishing. In: Hunter III, W. (Ed.), Fisheries Management and Conservation, pp. 27–45.

Albuquerque, U.P., Melo, J.G., Medeiros, M.F.T., Menezes, I.R.A., Moura, G.J.B., El-Deir, A.C.A., Alves, R.R.N., Medeiros, P.M., Araújo, T.A.S., Alves, M.R., 2011. Natural products from ethnodirected studies: revisiting the ethnobiology of the zombie poison. Evidence-based Complementary and Alternative Medicine 2012, 1–19.

Aleem, A.A., 1972. Fishing industry in ancient Egypt. In: Proceedings of the Royal Society of Edinburgh. Section B. Biology, 73, pp. 333–343.

Allen, D., 2010. Otter. (London).

Alverson, D.L., Freeberg, M.H., Murawski, S.A., Pope, J.G., 1994. A Global Assessment of Fisheries by Catch and Discards. FAO.

Alves, R.R.N., 2012. Relationships between fauna and people and the role of ethnozoology in animal conservation. Ethnobiology and Conservation 1, 1–69.

Alves, R.R.N., Alves, H.N., 2011. The faunal drugstore: animal-based remedies used in traditional medicines in Latin America. Journal of Ethnobiology and Ethnomedicine 7 (9), 1–43.

Alves, R.R.N., Nishida, A.K., 2002. A ecdise do caranguejo-uçá, *Ucides cordatus* L. (Decapoda, Brachyura) na visão dos caranguejeiros. Interciencia 27, 110–117.

Alves, R.R.N., Nishida, A.K., 2003. Aspectos socioeconômicos e percepção ambiental dos catadores de caranguejo-uçá *Ucides cordatus* (L. 1763) (Decapoda, Brachyura) do estuário do Rio Mamanguape, Nordeste do Brasil. Interciencia 28, 36–43.

Alves, R.R.N., Rosa, I.L., 2007. Zootherapeutic practices among fishing communities in North and Northeast Brazil: a comparison. Journal of Ethnopharmacology 111, 82–103.

Alves, R.R.N., Rosa, I.L., 2013. Animals in Traditional Folk Medicine: Implications for Conservation. Springer-Verlag, Berlin Heidelberg.

Alves, R.R.N., Souto, W.M.S., 2011. Ethnozoology in Brazil: current status and perspectives. Journal of Ethnobiology and Ethnomedicine 7 (22), 1–18.

Alves, R.R.N., Souto, W.M.S., 2015. Ethnozoology: a brief introduction. Ethnobiology and Conservation 4, 1–13.

Alves, R.R.N., Nishida, A.K., Hernandez, M.I.H., 2005. Environmental perception of gatherers of the crab 'caranguejo-uca' (*Ucides cordatus*, Decapoda, Brachyura) affecting their collection attitudes. Journal of Ethnobiology and Ethnomedicine 1 (10), 1–8.

Alves, R.R.N., Rosa, I.L., Santana, G.G., 2007. The role of animal-derived remedies as complementary medicine in Brazil. BioScience 57, 949–955.

Alves, R.R.N., Mendonça, L.E.T., Confessor, M.V.A., Vieira, W.L.S., Lopez, L.C.S., 2009. Hunting strategies used in the semi-arid region of northeastern Brazil. Journal of Ethnobiology and Ethnomedicine 5, 1–50.

Alves, R.R.N., Rosa, I.L., Léo Neto, N.A., Voeks, R., 2012. Animals for the gods: magical and religious faunal use and trade in Brazil. Human Ecology 40, 751–780.

Alves, R.R.N., Oliveira, T.P.R., Rosa, I.L., Cunningham, A.B., 2013a. Marine invertebrates in traditional medicines. In: Alves, R.R.N., Rosa, I.L. (Eds.), Animals in Traditional Folk Medicine: Implications for Conservation. Springer, Berlim, pp. 263–287.

Alves, R.R.N., Souto, W.M.S., Oliveira, R.E.M.C.C., Barboza, R.R.D., Rosa, I.L., 2013b. Aquatic mammals used in traditional folk medicine: a global analysis. In: Alves, R.R.N., Rosa, I.L. (Eds.), Animals in Traditional Folk Medicine: Implications for Conservation. Springer, Heidelberg, pp. 241–261.

Araujo, N.B.G., 1990. Jangadas. BN Fortaleza.

Balaji, K., Thirumaran, G., Arumugam, R., Kumaraguruvasagam, K.P., Anantharaman, P., 2009. A review on marine ornamental invertebrates. World Applied Sciences Journal 7, 1054–1059.

Ban, N.C., Vincent, A.C.J., 2009. Beyond marine reserves: exploring the approach of selecting areas where fishing is permitted, rather than prohibited. PLoS One 4, e6258.

Beaudreau, A.H., Levin, P.S., 2014. Advancing the use of local ecological knowledge for assessing data-poor species in coastal ecosystems. Ecological Applications 24, 244–256.

Beckman, D., 2013. Marine Environmental Biology and Conservation. Jones & Bartlett Learning, Burlington.

Beddington, J.R., Agnew, D.J., Clark, C.W., 2007. Current problems in the management of marine fisheries. Science 316, 1713–1716.

Bednarik, R.G., 2014. The beginnings of maritime travel. Advances in Anthropology 4, 209–221.

Begossi, A., Hanazaki, N., Ramos, R.M., 2004. Food chain and the reasons for fish food taboos among Amazonian and Atlantic forest fishers (Brazil). Ecological Applications 14, 1334–1343.

Beike, M., 2012. The history of Cormorant fishing in Europe. Vogelwelt 133, 1–21.

Berkes, F., 1999. Sacred Ecology: traditional ecological knowledge and resource management, first ed. Taylor & Francis, Philadelphia, USA.

Berkes, F., Folke, C., Gadgil, M., 1995. Traditional ecological knowledge, biodiversity, resilience and sustainability. In: Perrings, C.A., Mäler, K.G., Folke, C., Holling, C.S., Jansson, B.O. (Eds.), Biodiversity Conservation. Ecology, Economy & Environment, vol. 4, Kluwer Academic Publishers, Dordrecht, pp. 281–299.

Berkes, F., Colding, J., Folke, C., 2000. Rediscovery of traditional ecological knowledge as adaptative management. Ecological Applications 10, 1251–1262.

Berkes, F., Mahon, R., McConney, P., Pollnac, R., Pomeroy, R., 2001. Managing Small-scale Fisheries: Alternative Directions and Methods. International Development Research Centre.

Bezerra, D.M.M., Araujo, H.F.P., Alves, R.R.N., 2012. Captura de aves silvestres no semiárido brasileiro: técnicas cinegéticas e implicações para conservação. Tropical Conservation Science 5, 50–66.

Birket-Smith, K., 1966. Kulturens veje. København.

Blazina, C., Kogan, L.R., 2016. Men and Their Dogs: A New Understanding of Man's Best Friend. Springer International Publishing.

Bogue, M.B., 2001. Fishing the great lakes: an environmental history, 1783–1933. Univ of Wisconsin Press.

Boudarel, N., 1948. Les Richesses de la Mer, Paris.

Boyer, A., 1967. Les Pêches Maritimes. PUF. Colletion Que sais-j, Paris.

Brierley, A.S., Clapham, P.J., 2016. Japan's whaling is unscientific. Nature 529, 283.

Broadhurst, M.K., 2000. Modifications to reduce bycatch in prawn trawls: a review and framework for development. Reviews in Fish Biology and Fisheries 10, 27–60.

Burdon, T.W., 1951. A consideration of the classification of fishing gear and methods. In: Proceedings of the IndoPacific Fisheries Council Sect. II/21, Madras.

Burgin, S., Hardiman, N., 2015. Effects of non-consumptive wildlife-oriented tourism on marine species and prospects for their sustainable management. Journal of Environmental Management 151, 210–220.

Calado, R., Lin, J., Rhyne, A.L., Araújo, R., Narciso, L., 2003. Marine ornamental decapods—popular, pricey, and poorly studied. Journal of Crustacean Biology 23, 963–973.

Calado, R., Leal, M.C., Vaz, M., Brown, C., Rosa, R., Stevenson, T.C., Cooper, C.H., Tissot, B.N., Li, Y., Thornhill, D.J., 2014. Caught in the Act: how the US Lacey Act can hamper the fight against cyanide fishing in tropical coral reefs. Conservation Letters 7, 561–564.

Capistrano, J.F., Lopes, P.F.M.L., 2012. Crab gatherers perceive concrete changes in the life history traits of *Ucides cordatus* (Linnaeus, 1763), but overestimate their past and current catches. Ethnobiology and Conservation 1, 1–21.

Carvalho, A.R., Williams, S., January, M., Sowman, M., 2009. Reliability of community-based data monitoring in the Olifants River estuary (South Africa). Fisheries Research 96, 119–128.

Casale, P., 2011. Sea turtle by-catch in the Mediterranean. Fish and Fisheries 12, 299–316.

Cascudo, L.C., 2002. Geografia dos mitos brasileiros. Global, São Paulo.

Castro, M.S., Martins, I.M., Hanazaki, N., 2016. Trophic relationships between people and resources: fish consumption in an artisanal fishers neighborhood in Southern Brazil. Ethnobiology and Conservation 5, 1–16.

Cater, C., 2007. Perceptions of and interactions with marine environments: diving attractions from great whites to pygmy seahorses. In: Garrod, B., Gössling, S. (Eds.), New Frontiers in Marine Tourism: Diving Experiences, Sustainability, Management. Elsevier, Oxford, pp. 49–64.

Clarke, S.C., McAllister, M.K., Milner-Gulland, E.J., Kirkwood, G.P., Michielsens, C.G.J., Agnew, D.J., Pikitch, E.K., Nakano, H., Shivji, M.S., 2006. Global estimates of shark catches using trade records from commercial markets. Ecology Letters 9, 1115–1126.

Clarke, S.C., Harley, S.J., Hoyle, S.D., Rice, J.S., 2013. Population trends in Pacific Oceanic sharks and the utility of regulations on shark finning. Conservation Biology 27, 197–209.

Clutton-Brock, J., 1992. Horse Power: A History of the Horse and the Donkey in Human Societies. Natural History Museum Publications.

Coll, M., Carreras, M., Ciércoles, C., Cornax, M.J., Gorelli, G., Morote, E., Sáez, R., 2014. Assessing fishing and marine biodiversity changes using fishers' perceptions: the Spanish Mediterranean and Gulf of Cadiz case study. PLoS One 9, e85670.

Conceição, R.N.L., 2003. Ecologia de peixes de recifes artificiais de pneus instalados na costa do estado do Ceará. Universidade Federal de São Carlos.

Dagorn, L., Holland, K.N., Restrepo, V., Moreno, G., 2013. Is it good or bad to fish with FADs? What are the real impacts of the use of drifting FADs on pelagic marine ecosystems? Fish and Fisheries 14, 391–415.

Davies, T.K., Mees, C.C., Milner-Gulland, E.J., 2014. The past, present and future use of drifting fish aggregating devices (FADs) in the Indian Ocean. Marine Policy 45, 163–170.

Davis, D., Banks, S., Birtles, A., Valentine, P., Cuthill, M., 1997. Whale sharks in Ningaloo Marine Park: managing tourism in an Australian marine protected area. Tourism Management 18, 259–271.

Daw, T.M., Robinson, J.A.N., Graham, N.A.J., 2011. Perceptions of trends in Seychelles artisanal trap fisheries: comparing catch monitoring, underwater visual census and fishers' knowledge. Environmental Conservation 38, 75–88.

de Mitcheson, Y.S., Craig, M.T., Bertoncini, A.A., Carpenter, K.E., Cheung, W.W.L., Choat, J.H., Cornish, A.S., Fennessy, S.T., Ferreira, B.P., Heemstra, P.C., 2013. Fishing groupers towards extinction: a global assessment of threats and extinction risks in a billion dollar fishery. Fish and Fisheries 14, 119–136.

De Sola, C.R., 1932. Observations on the use of the sucking-fish or remora, *Echeneis naucrates*, for catching turtles in Cuban and Colombian waters. Copeia 1932, 45–52.

Deb, A.K., 2015. "Something Sacred, Something Secret": traditional ecological knowledge of the artisanal coastal fishers of Bangladesh. Journal of Ethnobiology 35, 536–565.

Derraik, J.G.B., 2002. The pollution of the marine environment by plastic debris: a review. Marine Pollution Bulletin 44, 842–852.

Di Dario, F., Alves, C.B.M., Boos, H., Frédou, F.L., Lessa, R.P.T., Mincarone, M.M., Pinheiro, M.A.A., Polaz, C.N.M., Reis, R.E., Rocha, L.A., 2015. A better way forward for Brazil's fisheries. Science 347, 1079.

Dias, T.L.P., Leo Neto, N.A., Alves, R.R.N., 2011. Molluscs in the marine curio and souvenir trade in NE Brazil: species composition and implications for their conservation and management. Biodiversity and Conservation 20, 2393–2405.

Diegues, A.C., 1983. Pescadores, camponeses e trabalhadores do mar. Ática, São Paulo.

Dye, T., 1983. Fish and fishing on Niuatoputapu. Oceania 53, 242–271.

Eayrs, S., 2007. A Guide to Bycatch Reduction in Tropical Shrimp-Trawl Fisheries. Food and Agricultural Organization of the United Nations, Rome.

El-Deir, A.C.A., Collier, C.A., Almeida Neto, M.S., Silva, K.M.S., Policarpo, I.S., Araújo, T.A.S., Alves, R.R.N., Albuquerque, U.P., Moura, G.J.B., 2012. Ichthyofauna used in traditional medicine in Brazil. Evidence-based Complementary and Alternative Medicine 2012, 1–16:474716.

Emerson, C., 1999. Aquaculture Impacts on the Environment, CSA Guide Discovery Database.

Erlandson, J.M., 2001. The archaeology of aquatic adaptations: paradigms for a new millennium. Journal of Archaeological Research 9, 287–350.

Etkin, N.L., 2009. Foods of Association: Biocultural Perspectives on Foods and Beverages that Mediate Sociability. Tucson.

FAO, Food and Agriculture Organization of the United Nations, 1995. Code of Conduct for Responsible Fisheries. Food and Agriculture Organization, Rome.

FAO, Food and Agriculture Organization of the United Nations, 2007. Small-scale Fisheries. FAO Fisheries and Aquaculture Department.

FAO, Food and Agriculture Organization of the United Nations, 2014. The State of World Fisheries and Aquaculture 2014, Rome.

FAO, Food and Agriculture Organization of the United Nations, 2016. The State of World Fisheries and Aquaculture. Contributing to Food Security and Nutrition for All. Food and Agriculture Organization of the United Nations, Rome.

Feeroz, M.M., Begum, S., Hasan, M.K., 2011. Fishing with otters: a traditional conservation practice in Bangladesh. IUCN Otter Spec. Group Bull. A 28, 14–20.

Forcada, A., Valle, C., Bonhomme, P., Criquet, G., Cadiou, G., Lenfant, P., Sánchez-Lizaso, J.L., 2009. Effects of habitat on spillover from marine protected areas to artisanal fisheries. Marine Ecology Progress Series 379, 197–211.

Gabriel, O., Lange, K., Dahm, E., Wendt, T., 2005. Fish Catching Methods of the World. John Wiley & Sons.

Gabriel, O., Lange, K., Dahm, E., Wendt, T., 2008. Fish Catching Methods of the World. John Wiley & Sons.

Garrod, B., Gössling, S., 2008. New Frontiers in Marine Tourism: Diving Experiences, Sustainability, Management. Routledge.

Gartside, D.F., Kirkegaard, I.R., 2009. A History of Fishing. EOLSS, Paris.

Gerhardinger, L.C., Hostim-Silva, M., Medeiros, R.P., Matarezi, J., Bertoncini, Á., Freitas, M.O., Ferreira, B.P., 2009. Fishers' resource mapping and goliath grouper *Epinephelus itajara* (Serranidae) conservation in Brazil. Neotropical Ichthyology 7, 93–102.

Gudger, E.W., 1919a. On the use of the sucking-fish for catching fish and turtles: studies in Echeneis or remora, II. The American Naturalist 53, 289–311.

Gudger, E.W., 1919b. On the use of the sucking-fish for catching fish and turtles: studies in Echeneis or Remora, III. The American Naturalist 53, 515–525.

Gudger, E.W., 1923. Dogs as fishermen. Natural History 23, 559–568.

Gudger, E.W., 1926. Fishing with the Cormorant. I. The American Naturalist 60, 5–41.

Gudger, E.W., 1927. Fishing with the otter. The American Naturalist 61, 193–225.

Gudger, E.W., 1941. Canine fishermen. Accounts of some dogs that went a-fishing. Natural History Magazine 47, 140–148.

Guthrie, D., 2005. The Nature of Paleolithic Art. University of Chicago Press, Chicaco and London.

Halpern, B.S., Walbridge, S., Selkoe, K.A., Kappel, C.V., Micheli, F., D'agrosa, C., Bruno, J.F., Casey, K.S., Ebert, C., Fox, H.E., 2008. A global map of human impact on marine ecosystems. Science 319, 948–952.

Hanazaki, N., Begossi, A., 2000. Fishing and Niche dimension for food consumption of Caiçaras from Ponta do Almada (Brazil). Human Ecology Review 7, 52–62.

Harper, D., 2016. Online Etymology Dictionary.

Harper, S., Zeller, D., Hauzer, M., Pauly, D., Sumaila, U.R., 2013. Women and fisheries: contribution to food security and local economies. Marine Policy 39, 56–63.

Hérubel, M.A., 1928. L'évolution de la pêche: étude d'économie maritime: ouvrage orné de 9 dessins originaux de L. Haffner. Société d'éditions géographiques, maritimes et coloniales.

Higham, J.E.S., Lück, M., 2007. Marine Wildlife and Tourism Management: Insights from the Natural and Social Sciences. CABI.

HLPE, 2014. High level panel of experts. In: Sustainable Fisheries and Aquaculture for Food Security and Nutrition, Rome.

Holck, M.H., 2008. Participatory forest monitoring: an assessment of the accuracy of simple cost–effective methods. Biodiversity and Conservation 17, 2023–2036.

Humber, F., Godley, B.J., Ramahery, V., Broderick, A.C., 2011. Using community members to assess artisanal fisheries: the marine turtle fishery in Madagascar. Animal Conservation 14, 175–185.

Humber, F., Godley, B.J., Nicolas, T., Raynaud, O., Pichon, F., Broderick, A., 2017. Placing Madagascar's marine turtle populations in a regional context using community-based monitoring. Oryx 51 (3), 542–553.

Islam, M.S., Tanaka, M., 2004. Impacts of pollution on coastal and marine ecosystems including coastal and marine fisheries and approach for management: a review and synthesis. Marine Pollution Bulletin 48, 624–649.

IWC., 2017. International Whaling Commission.

Jackson, C.E., 1997. Fishing with cormorants. Archives of Natural History 24, 189–211.

Jackson, J.B.C., Kirby, M.X., Berger, W.H., Bjorndal, K.A., Botsford, L.W., Bourque, B.J., Bradbury, R.H., Cooke, R., Erlandson, J., Estes, J.A., 2001. Historical overfishing and the recent collapse of coastal ecosystems. Science 293, 629–637.

Jankowsky, M., Pires, J.S.R., Nordi, N., 2006. Contribuição ao manejo participativo do Caranguejo-uçá, *Ucides cordatus* (L., 1763). Cananéia, SP. Boletim do Instituto de Pesca 32, 221–228.

Jennings, M.R., 1987. Impact of the curio trade for San Diego horned lizards (*Phrynosoma coronatum* blainvillii) in the Los Angeles Basin, California: 1885–1930. Journal of Herpetology 21, 356–358.

Johannes, R.E., 1993. Integrating traditional ecological knowledge and manegement with environmental impact assessment. In: Inglis, J.T. (Ed.), Traditional Ecological Knowledge: Concepts and Cases. International Program on Traditional Ecological Knowledge and International Development Research Centre, Ottwa, pp. 33–39.

Johannes, R.E., Freeman, M.M.R., Hamilton, R.J., 2000. Ignore fishers' knowledge and miss the boat. Fish and Fisheries 1, 257–271.

Jones, J.P.G., Andriamarovololona, M.M., Hockley, N., Gibbons, J.M., Milner-Gulland, E.J., 2008. Testing the use of interviews as a tool for monitoring trends in the harvesting of wild species. Journal of Applied Ecology 45, 1205–1212.

Kay, M.C., Lenihan, H.S., Guenther, C.M., Wilson, J.R., Miller, C.J., Shrout, S.W., 2012. Collaborative assessment of California spiny lobster population and fishery responses to a marine reserve network. Ecological Applications 22, 322–335.

Kenelly, S.J., 2014. Solving by-catch problems: successes in developed countries and challenges for protein-poor countries. Journal of the Marine Biological Association of India 56, 19–27.

Kleiber, D., Harris, L.M., Vincent, A.C.J., 2015. Gender and small-scale fisheries: a case for counting women and beyond. Fish and Fisheries 16, 547–562.

Kurlansky, M., 2011. Cod: A Biography of the Fish that Changed the World. Knopf Canada.

Lack, M., Sant, G., 2011. The future of sharks: a review of action and inaction. In: TRAFFIC International and the Pew Environment Group, Washington, DC, USA.

Lackey, R.T., 2005. Fisheries: history, science, and management. In: Lehr, J.H., Keeley, K. (Eds.), Water Encyclopedia: Surface and Agricultural Water. John Wiley & Sons, Inc., New York, pp. 121–129.

Laufer, B., 1931. The Domestication of the Cormorant in China and Japan, Chicago.

Lawson, J.M., 2017. The global search for seahorses in bycatch. Fisheries 42, 34–39.

Leite, A.M., 1991. Manual de Tecnologia da Pesca. Escola Portuguesa de Pesca, Lisboa.

Léo Neto, N.A., Brooks, S.E., Alves, R.R.N., 2009. From Eshu to Obatala: animals used in sacrificial rituals at Candomble "terreiros" in Brazil. Journal of Ethnobiology and Ethnomedicine 5, 1–23.

Leo Neto, N.A., Mourão, J.S., Alves, R.R.N., 2011. "It all begins with the head": initiation rituals and the symbolic conceptions of animals in Candomblé. Journal of Ethnobiology 31, 244–261.

Leo Neto, N.A., Voeks, R.A., Dias, T.L.P., Alves, R.R.N., 2012. Mollusks of Candomble: symbolic and ritualistic importance. Journal of Ethnobiology and Ethnomedicine 8 (10), 1–10.

Léry, J., 1997. Viagem à terra do Brasil. Itatiaia Belo Horizonte.

Lewison, R.L., Crowder, L.B., Read, A.J., Freeman, S.A., 2004. Understanding impacts of fisheries bycatch on marine megafauna. Trends in Ecology & Evolution 19, 598–604.

Littlewood, R., Douyon, C., 1997. Clinical findings in three cases of zombification. The Lancet 350, 1094–1096.

Loring, P.A., Harrison, H.L., Gerlach, S.C., 2014. Local perceptions of the sustainability of Alaska's highly contested Cook Inlet salmon fisheries. Society & Natural Resources 27, 185–199.

Lotze, H.K., Lenihan, H.S., Bourque, B.J., Bradbury, R.H., Cooke, R.G., Kay, M.C., Kidwell, S.M., Kirby, M.X., Peterson, C.H., Jackson, J.B.C., 2006. Depletion, degradation, and recovery potential of estuaries and coastal seas. Science 312, 1806–1809.

Luna, C.Z., Silvestre, G.T., Green, S.J., Carreon III, M.F., White, A.T., 2004. Profiling the status of Philippine marine fisheries: a general introduction and overview. In: DA-BFAR (Department of Agriculture-Bureau of Fisheries, Aquatic Resources) (Ed.), Resources, in Turbulent Seas: The Status of Philippine Marine Fisheries. Coastal Resource Management Project, Cebu City, Philippines, pp. 3–11.

Lunn, K.E., Dearden, P., 2006. Monitoring small-scale marine fisheries: an example from Thailand's Ko Chang archipelago. Fisheries Research 77, 60–71.

Lunn, K.E., Noriega, M.J.V., Vincent, A.C.J., 2008. Souvenirs from the sea: an investigation into the curio trade in echinoderms from Mexico. Traffi Bulletin 22, 19–32.

Lyra-Neves, R.M., Santos, E.M., Medeiros, P.M., Alves, R.R.N., Albuquerque, U.P., 2015. Ethnozoology in Brazil: analysis of the methodological risks in published studies. Brazilian Journal of Biology 75, S184–S191.

MacKinney, L.C., 1946. Animal substances in materia medica. Journal of the History of Medicine and Allied Sciences 1, 149–170.

Mackinson, S., Nottestad, L., 1998. Combining local and scientific knowledge. Reviews in Fish Biology and Fisheries 8, 481–490.

Maneschy, M.C., 1993. Pescadores curralistas no litoral do estado do Pará: evolução e continuidade de uma pesca tradicional. Revista da Sociedade Brasileira da História da Ciência 53–74.

Maple, T.L., Perdue, B.M., 2013. Behavior analysis and training. In: Maple, T.L., Perdue, B.M. (Eds.), Zoo Animal Welfare. Springer, pp. 119–137.

Marean, C.W., Bar-Matthews, M., Bernatchez, J., Fisher, E., Goldberg, P., Herries, A.I.R., Jacobs, Z., Jerardino, A., Karkanas, P., Minichillo, T., 2007. Early human use of marine resources and pigment in South Africa during the Middle Pleistocene. Nature 449, 905–908.

Marques, J.G.W., 1995. Pescando pescadores: etnoecologia abrangente no baixo São Francisco alagoano. NUPAUB-USP, São Paulo, BR.

Menezes, M.F., 1976. Aspectos biológicos da serra, *Scomberomorus maculatus* (Mitchill), capturada por currais-de-pesca. Arquivos de Ciencias do Mar 16, 45–48.

Moazami, M., 2005. Evil animals in the Zoroastrian religion. History of Religions 44, 300–317.

Monticini, P., 2010. The ornamental fish trade: production and commerce of ornamental fish: technical-managerial and legislative aspects. Food and Agriculture Organization of the United Nations.

Moore, J.E., Curtis, K.A., Lewison, R.L., Dillingham, P.W., Cope, J.M., Fordham, S.V., Heppell, S.S., Pardo, S.A., Simpfendorfer, C.A., Tuck, G.N., Zhou, S., 2013. Evaluating sustainability of fisheries bycatch mortality for marine megafauna: a review of conservation reference points for data-limited populations. Environmental Conservation 40, 329–344.

Mora, C., Myers, R.A., Coll, M., Libralato, S., Pitcher, T.J., Sumaila, R.U., Zeller, D., Watson, R., Gaston, K.J., Worm, B., 2011. Management effectiveness of the world's marine fisheries. In: Hunter III, W. (Ed.), Fisheries Management and Conservation, pp. 86–109.

Morwood, M.J., O'Sullivan, P.B., Aziz, F., Raza, A., 1998. Fission-track ages of stone tools and fossils on the east Indonesian island of Flores. Nature 392, 173–176.

Mumby, P.J., Hastings, A., 2008. The impact of ecosystem connectivity on coral reef resilience. Journal of Applied Ecology 45, 854–862.

Mussolini, G., 1953. Aspectos da cultura e da vida social no litoral brasileiro. Revista de Antropologia 1, 81–97.

Narchi, N.E., Cornier, S., Canu, D.M., Aguilar-Rosas, L.E., Bender, M.G., Jacquelin, C., Thiba, M., Moura, G.G.M., De Wit, R., 2014. Marine ethnobiology a rather neglected area, which can provide an important contribution to ocean and coastal management. Ocean & Coastal Management 89, 117–126.

Nascimento, D.M., Alves, A.G.C., Alves, R.R.N., Barboza, R.R.D., Diele, K., Mourão, J.S., 2016. An examination of the techniques used to capture mangrove crabs, *Ucides cordatus*, in the Mamanguape River estuary, northeastern Brazil, with implications for management. Ocean & Coastal Management 130, 50–57.

Nascimento, D.M., Alves, R.R.N., Barboza, R.R.D., Schmidt, A.J., Diele, K., Mourão, J.S., 2017. Commercial relationships between intermediaries and harvesters of the mangrove crab *Ucides cordatus* (Linnaeus, 1763) in the Mamanguape River estuary, Brazil, and their socio-ecological implications. Ecological Economics 131, 44–51.

United Nations, 1993. Vienna Declaration and Programme of Action: The World Conference Human Rights. United Nations, Vienna, Austria.

Neil, D.T., 2002. Cooperative fishing interactions between Aboriginal Australians and dolphins in eastern Australia. Anthrozoos 15, 3–18.

Nishida, A.K., Nordi, N., Alves, R.R.D.N., 2006a. Mollusc gathering in Northeast Brazil: an ethnoecological approach. Human Ecology 34, 133–145.

Nishida, A.K., Nordi, N., Alves, R.R.N., 2006b. The lunar-tide cycle viewed by crustacean and mollusc gatherers in the State of Paraíba, Northeast Brazil and their influence in collection attitudes. Journal of Ethnobiology and Ethnomedicine 2, 1–12.

Nishida, A.K., Nordi, N., Alves, R.R.N., 2008. Embarcações utilizadas por pescadores estuarinos da Paraíba, Nordeste Brasil. Revista de Biologia e Farmácia 3, 1–8.

Nogueira, M., Alves, R.R.N., 2016. Assessing sea turtle bycatch in Northeast Brazil through an ethnozoological approach. Ocean & Coastal Management 133, 37–42.

O'Sullivan, P.B., Morwood, M., Hobbs, D., Suminto, F.A., Situmorang, M., Raza, A., Maas, R., 2001. Archaeological implications of the geology and chronology of the Soa basin, Flores, Indonesia. Geology 29, 607–610.

Ommer, R.E., Perry, R.I., 2011. Introduction. In: Ommer, R.E., Perry, R.I., Cochrane, K., Cury, P. (Eds.), World Fisheries: A Social-Ecological Analysis. Blackwell Publishing Ltd, pp. 3–8.

ONU, Organização das Nações Unidas, 2002. Convenção das Nações Unidas sobre o Direito do Mar: 20° Aniversário (1982–2002). Organização das Nações Unidas.

O'Connor, S., Ono, R., Clarkson, C., 2011. Pelagic fishing at 42,000 years before the present and the maritime skills of modern humans. Science 334, 1117–1121.

Pauly, D., 1995. Anecdotes and the shifting baseline syndrome of fisheries. Trends in Ecology and Evolution 10, 430.

Pauly, D., Palomares, M.L., 2005. Fishing down marine food web: it is far more pervasive than we thought. Bulletin of Marine Science 76, 197–212.

Pauly, D., Zeller, D., 2016a. Global Atlas of Marine Fisheries: A Critical Appraisal of Catches and Ecosystem Impacts. Island Press.

Pauly, D., Zeller, D., 2016b. Catch reconstructions reveal that global marine fisheries catches are higher than reported and declining. Nature Communications 7, 317–320.

Pauly, D., Christensen, V., Dalsgaard, J., Froese, R., Torres, F., 1998. Fishing down marine food webs. Science 279, 860–863.

Pauly, D., Watson, R., Alder, J., 2005. Global trends in world fisheries: impacts on marine ecosystems and food security. Philosophical Transactions of the Royal Society B: Biological Sciences 360, 5–12.

Pedersen, L., 1995. 7000 years of fishing: stationary fishing structures in the Mesolithic and afterwards. In: Fischer, A.E. (Ed.), Man and Sea in the Mesolithic. Coastal Settlement above and Below Present Sea Level. Oxbow Monograph, Oxford, pp. 75–86.

Peterson, D., Hanazaki, N., Simoes-Lopes, P.C., 2008. Natural resource appropriation in cooperative artisanal fishing between fishermen and dolphins (*Tursiops truncatus*) in Laguna, Brazil. Ocean & Coastal Management 51, 469–475.

Pinheiro, H.T., Di Dario, F., Gerhardinger, L.C., Melo, M.R.S., Moura, R.L., Reis, R.E., Vieira, F., Zuanon, J., Rocha, L.A., 2015. Brazilian aquatic biodiversity in peril. Science 350, 1043–1044.

Pinto, M.F., Mourão, J.S., Alves, R.R.N., 2013. Ethnotaxonomical considerations and usage of ichthyofauna in a fishing community in Ceará State, Northeast Brazil. Journal of Ethnobiology and Ethnomedicine 9, 1–11.

Pinto, M.F., Mourão, J.S., Alves, R.R.N., 2015. Use of ichthyofauna by artisanal fishermen at two protected areas along the coast of Northeast Brazil. Journal of Ethnobiology and Ethnomedicine 11, 1–32.

Pitcher, T.J., Lam, M.E., 2015. Fish commoditization and the historical origins of catching fish for profit. Maritime Studies 14, 2.

Pomeroy, R.S., Berkes, F., 1997. Two to tango: the role of government in fisheries co-management. Marine Policy 21, 465–480.

Pomeroy, R.S., Carlos, M.B., 1997. Community-based coastal resource management in the Philippines: a review and evaluation of programs and projects, 1984–1994. Marine Policy 21, 445–464.

Pope, G.G., 1989. Bamboo and human evolution. Natural History 98, 48–57.

Pringle, H., 1997. Ice Age communities may be earliest known net hunters. Science 277, 1203–1204.

Prous, A., 1992. Arqueologia Brasileira. UnB Brasilia.

Purpura, G., 1990. Pesca e stabilimenti antichi per la lavorazione del pesce nella Sicilia occidentale: IV: un bilancio, Atti V Rassegna Archeologia Subacquea. Giardini Naxos 19–21.

Rasalato, E., Maginnity, V., Brunnschweiler, J.M., 2010. Using local ecological knowledge to identify shark river habitats in Fiji (South Pacific). Environmental Conservation 37, 90–97.

Reis, R.E., Albert, J.S., Di Dario, F., Mincarone, M.M., Petry, P., Rocha, L.A., 2016. Fish biodiversity and conservation in South America. Journal of Fish Biology 89, 12–47.

Rhyne, A.L., Tlusty, M.F., Schofield, P.J., Kaufman, L.E.S., Morris Jr., J.A., Bruckner, A.W., 2012. Revealing the appetite of the marine aquarium fish trade: the volume and biodiversity of fish imported into the United States. PLoS One 7, e35808.

Rocha, M.S.P., Mourão, J.S., Souto, W.M.S., Barboza, R.R.D., Alves, R.R.N., 2008. Uso dos recursos pesqueiros no Estuário do Rio Mamanguape, Estado da Paraíba, Brasil. Interciencia 33, 903–909.

Rocha, M.S.P., Santiago, I.M.F.L., Cortez, C.S., Trindade, P.M., Mourão, J.S., 2012. Use of fishing resources by women in the Mamanguape river estuary, Paraíba state, Brazil. Anais da Academia Brasileira de Ciências 84, 1189–1199.

Rosa, I.M.L., Alves, R.R.N., Bonifácio, K.M., Mourão, J.S., Osório, F.M., Oliveira, T.P.R., Nottingham, M.C., 2005. Fishers' knowledge and seahorse conservation in Brazil. Journal of Ethnobiology and Ethnomedicine 1, 1–15.

Rosa, I.L., Oliveira, T.P.R., Osório, F.M., Moraes, L.E., Castro, A.L.C., Barros, G.M.L., Alves, R.R.N., 2011. Fisheries and trade of seahorses in Brazil: historical perspective, current trends, and future directions. Biodiversity and Conservation 20, 1951–1971.

Ruddle, K., 1993. External forces and change in traditional community-based fishery management systems in the Asia-Pacific region. Maritime Anthropological Studies 6, 1–37.

Ruddle, K., 1994. Local knowledge in the folk management of fisheries and coastal marine environments. In: Dyer, C.L., McGoodwin, J.R. (Eds.), Folk Management in the World's Fisheries. University Press of Colorado, Niwot, Colorado.

Ruddle, K., 1995. The role of validated local knowledge in the restoration of fisheries property rights: the example of the New Zealand Maori. In: Hanna, S., Munasinghe, M. (Eds.), Property Rights in a Socialand Ecological Context: Part 2, Case Studies and Design Applications. The Beijer International Institute of Ecological Economics & The World Bank, Stockholm &Washington, DC, pp. 111–119.

Sahrhage, D., Lundbeck, J., 1992. A History of Fishing. Springer-Verlag, Berlin, Germany.

Sahrhage, D., Lundbeck, J., 2012. A History of Fishing. Springer Science & Business Media.

Salisbury, C.R., 1991. Primitive British fish-weirs. In: Good, G.L., Jones, R.H., Ponsford, M.W. (Eds.), Waterfront Archaeology. CBA Research Report, pp. 76–87.

Santos, L.N., Brotto, D.S., Zalmon, I.R., 2010. Fish responses to increasing distance from artificial reefs on the Southeastern Brazilian Coast. Journal of Experimental Marine Biology and Ecology 386, 54–60.

Scarre, C., 2005. The Human Past: World Prehistory & the Development of Human Societies. Thames & Hudson, London.

Scearce, C., 2010. European Fisheries History: Pre-industrial Origins of Overfishing. Proquest Discovery Guides.

Silva, L.G., 2001. A faina, a festa e o rito: uma etnografia histórica sobre as gentes do mar, sécs. XVII ao XIX. Papirus Editora.

Silvano, R.A.M., Valbo-Jørgensen, J., 2008. Beyond fishermen's tales: contributions of fishers' local ecological knowledge to fish ecology and fisheries management. Environment, Development and Sustainability 10, 657–675.

Simões-Lopes, P.C., Fabián, M.E., Menegheti, J.O., 1998. Dolphin interactions with the mullet artisanal fishing on southern Brazil: a qualitative and quantitative approach. Revista Brasileira de Zoologia 15, 709–726.

Slade, L.M., Kalangahe, B., 2015. Dynamite fishing in Tanzania. Marine Pollution Bulletin 101, 491–496.

Smith, B.D., Tun, M.T., Chit, A.M., Win, H., Moe, T., 2009. Catch composition and conservation management of a human–dolphin cooperative cast-net fishery in the Ayeyarwady River, Myanmar. Biological Conservation 142, 1042–1049.

Stewart, H., 1994a. Indian Fishing: Early Methods on the Northwest Coast. University of Washington Press, Seattle.

Stewart, K.M., 1994b. Early hominid utilisation of fish resources and implications for seasonality and behaviour. Journal of Human Evolution 27, 229–245.

Stobart, B., Warwick, R., González, C., Mallol, S., Díaz, D., Reñones, O., Goñi, R., 2009. Long-term and spillover effects of a marine protected area on an exploited fish community. Marine Ecology Progress Series 384, 47–60.

Stone, R.B., McGurrin, J.M., Sprague, L.M., Seaman Jr., W., 1991. Artificial habitats of the world: synopsis and major trends. In: Seaman, W.J., Sprague, I.M. (Eds.), Artificial Habitats for Marine and Freshwater Fisheries. Academic Press, San Diego, California, pp. 31–60.

Stonor, C.R., 1948. Fishing with Indian darter (Anhinga melanogaster) in Assam. Journal of Bombay Natural History Society 47, 746–747.

Stringer, C.B., Finlayson, J.C., Barton, R.N.E., Fernández-Jalvo, Y., Cáceres, I., Sabin, R.C., Rhodes, E.J., Currant, A.P., Rodríguez-Vidal, J., Giles-Pacheco, F., 2008. Neanderthal exploitation of marine mammals in Gibraltar. Proceedings of the National Academy of Sciences 105, 14319–14324.

Sumaila, U.R., Alder, J., Keith, H., 2006. Global scope and economics of illegal fishing. Marine Policy 30, 696–703.

Svanberg, I., Kuusela, T., Cios, S., 2016. Sometimes it is tamed to bring home fish for the kitchen: otter fishing in Northern Europe and beyond. Swedish Dialects and Folk Traditions 138, 79–97.

Tavares, M.C.S., Júnior, I.F., Souza, R.A.L., Brito, C.S.F., 2005. A pesca de curral no Estado do Pará. Boletim Técnico Científico Do Cepnor 5, 115–139.

Ternes, M.L.F., Gerhardinger, L.C., Schiavetti, A., 2016. Seahorses in focus: local ecological knowledge of seahorse-watching operators in a tropical estuary. Journal of Ethnobiology and Ethnomedicine 12, 1–12.

Thomas, H.L., Macsharry, B., Morgan, L., Kingston, N., Moffitt, R., Stanwell-Smith, D., Wood, L., 2014. Evaluating official marine protected area coverage for Aichi Target 11: appraising the data and methods that define our progress. Aquatic Conservation: Marine and Freshwater Ecosystems 24, 8–23.

Topelko, K.N., Dearden, P., 2005. The shark watching industry and its potential contribution to shark conservation. Journal of Ecotourism 4, 108–128.

Unsworth, R.K.F., Salinas De Leon, P., Garrard, S.L., Jompa, J., Smith, D.J., Bell, J.J., 2008. High connectivity of Indo-Pacific seagrass fish assemblages with mangrove and coral reef habitats. Marine Ecology Progress Series 353, 213–224.

Vianna, G.M.S., Meekan, M.G., Pannell, D.J., Marsh, S.P., Meeuwig, J.J., 2012. Socio-economic value and community benefits from shark-diving tourism in Palau: a sustainable use of reef shark populations. Biological Conservation 145, 267–277.

Vincent, A.C.J., Foster, S.J., Koldewey, H.J., 2011. Conservation and management of seahorses and other Syngnathidae. Journal of Fish Biology 78, 1681–1724.

von Brandt, A., 1972. Fish Catching Methods of the World Surrey.

Wabnitz, C., 2003. From ocean to aquarium: the global trade in marine ornamental species. UNEP/Earthprint.

Wallace, B.P., Kot, C.Y., DiMatteo, A.D., Lee, T., Crowder, L.B., Lewison, R.L., 2013. Impacts of fisheries bycatch on marine turtle populations worldwide: toward conservation and research priorities. Ecosphere 4 art40.

Walter, R.C., Buffler, R.T., Bruggemann, J.H., Guillaume, M.M., Berhe, S.M., Negassi, B., Libsekal, Y., Cheng, H., Edwards, R.L., von Cosel, R., Néraudeau, D., Gagnon, M., 2000. Early human occupation of the Red Sea coast of Eritrea during the last interglacial. Nature 405, 65–69.

Weatherdon, L.V., Magnan, A.K., Rogers, A.D., Sumaila, U.R., Cheung, W.W.L., 2016. Observed and projected impacts of climate change on marine fisheries, aquaculture, coastal tourism, and human health: an update. Frontiers in Marine Science 3, 1–21.

Weyer, E., 1959. Primitive Völker Heute. Gütersloh.

White, W.T., Giles, J., Potter, I.C., 2006. Data on the bycatch fishery and reproductive biology of mobulid rays (Myliobatiformes) in Indonesia. Fisheries Research 82, 65–73.

Wood, E.M., 2001. Collection of Coral Reef Fish for Aquaria: Global Trade, Conservation Issues and Management Strategies. Marine Conservation Society, Ross-on-Wye, UK.

Wood, E.M., Wells, S., 1988. The Marine Curio Trade: Conservation Issues: A Report for the Marine Conservation Society. Marine Conservation Society.

WWF, World Wildlife Fund, 2017. World Wildlife Organization, "Overview".

Yan, G., 2016. Saving Nemo–Reducing mortality rates of wild-caught ornamental fish. SPC Live Reef Fish Information Bulletin 21, 3–7.

Yellen, J.E., Brooks, A.S., Cornelissen, E., Mehlman, M.J., Stewart, M., 1995. A middle stone age worked bone industry from Katanda, Upper Semliki Valley, Zaire. Science 268, 553–556.

Zappes, C.A., Andriolo, A., Simões-Lopes, P.C., Di Beneditto, A.P.M., 2011. 'Human-dolphin (*Tursiops truncatus* Montagu, 1821) cooperative fishery'and its influence on cast net fishing activities in Barra de Imbé/Tramandaí, Southern Brazil. Ocean & Coastal Management 54, 427–432.

Zeeberg, J., Corten, A., Graaf, E., 2006. Bycatch and release of pelagic megafauna in industrial trawler fisheries off Northwest Africa. Fisheries Research 78, 186–195.

Zeller, D., Pauly, D., 2012. Reconstruction of marine fisheries catches for key countries and regions (1950–2005). In: Fisheries Centre Research Reports. University of British Columbia.

Zeller, D., Booth, S., Craig, P., Pauly, D., 2006. Reconstruction of coral reef fisheries catches in American Samoa, 1950–2002. Coral Reefs 25, 144–152.

Zeller, D., Cashion, T., Palomares, M., Pauly, D., 2017. Global marine fisheries discards: A synthesis of reconstructed data. Fish and Fisheries , 1–10.

CHAPTER

9

Animal Domestication and Ethnozootechny

Ângelo Giuseppe Chaves Alves[1], Maria Norma Ribeiro[1], Janaina Kelli Gomes Arandas[1], Rômulo Romeu Nóbrega Alves[2]

[1]Universidade Federal Rural de Pernambuco, Recife, Brazil; [2]Universidade Estadual da Paraíba, Campina Grande, Brazil

INTRODUCTION

Humans have historically exercised dominance over other biological species, many of them being fundamental to our survival (Alves, 2016). This dominance was responsible for one of the main milestones in the development of human civilization, namely, the domestication of animals and plants. This process made way for important changes in the lifestyle of human beings, allowing them to abandon their former nomad habits (which were quite common in the beginning of mankind) and pick up sedentary ones, such as settling in certain territories (Alves, 2016).

Naturally, domestication caused changes in the way people interacted with some animal species that began to be important and abundant sources and became easily available. At the same time, this process pushed people farther from other animal species, which would not be susceptible for domestication. As a result, there was a gradual decrease in man's dependence on hunting and fishing when it comes to obtaining animal protein.

Thus, it is safe to say that animal domestication has a great impact on the relations between humans and animals, which makes it a rather relevant theme both in ethnozoology and ethnozootechny. According to Baenninger (1995), the main kinds of interaction between people and animals available in the modern world result from the domestication of other animal species that have played an important role in different aspects of human life, not only due to their functionality but also due to their symbolic importance in many social contexts. In this chapter, we discuss the historical, conceptual, and ethnobiology-related aspects of animal domestication. We will also discourse about the relations between ethnozoology and ethnozootechny, given that the two approaches have been useful for understanding the relationships between humans and other (domestic or wild) animal species along history. In addition, we assume the idea of a continuum, not only between domestic and

Ethnozoology
http://dx.doi.org/10.1016/B978-0-12-809913-1.00009-0

wild animals, in genetic terms, but also between ethnozoology and ethnozootechny, in epistemological terms.

TAMING AND DOMESTICATION

Studies on domestication are common ground for both ethnozoology and ethnozootechny, because it is the turning point in the history of human–animal relationships.

Quite often, people mistake a domestic species for a tamed or trained animal that lives under human dominance. In general, the terms "wild" and "domestic" are complementary and not antagonistic, because they represent the edges of the same process (Dobney and Larson, 2006). The concept or category of "wild" was created when ancient peoples first domesticated plants and animals, before which there was no sense in thinking about such category. The fact is that the dichotomy "wild/domestic" has had profound consequences for human society. Although not every society may consider the wild/domestic distinction, it is an important issue among people who deal with domestic animals in everyday life. For instance, among the Mafulu of New Guinea, wild and domestic pigs are genetically and phenotypically identical (Russel, 2002), and they may interbreed freely in some situations. So, according to the dichotomy wild/domestic, while "wild" is defined as everything that is not domestic, other animals such as pets and totems are difficult to classify. Pets, for instance, can be wild or domestic. Animal domestication has promoted a deep change in human–animal relationships, and it has set the stage for later transformations that include factory farming, genetic engineering, and the transplant of animal parts into human bodies.

The domestication process has been studied from several points of view (Armitage, 1986; Benecke, 1994; Clutton-Brock, 1999). This diversity results from the wide range of human–animal relationships and from the hybrid nature of animal domestication that involves both biological and social components (Clutton-Brock, 1992; Crabtree, 1993).

Most definitions of animal domestication make a distinction between taming and domestication, although some scholars would argue that the boundary is not clear. Taming refers to the relationship between a particular person and a particular animal without long-term effects beyond the lifetime of the animal. On the other hand, domestication is a relationship with a population of animals that often leads to morphological and behavioral changes in that population (Clutton-Brock, 1994; Reitz and Wing, 1999). Hence, taming is a precondition for domestication, but it is not enough. A separate animal population may be herded in ways that do not involve taming, as in ranching (Ingold, 1980). In many hunting societies, people could tame animals such as pet or decoys, but this does not promote significant changes in human–animal relationships nor does it lead to genetic changes in the animal populations (Russel, 2002).

Ducos (1978) states that domestication has both biological and social aspects, and he proposes labeling of animals that are integrated into the human sphere as "domesticated" and those with morphological signs of domestication as "domestic." On the other hand, Ingold (1980) makes the same distinction but uses the opposite terminology. Also, he divides domestication into three processes (taming, herding, and breeding) that do not necessarily cooccur. So, for Ingold (1980), domestication means the social incorporation or appropriation of successive generations of animals by humans.

Digard (1990) conceived domestication as a similar process in all situations, but with deviations in the degree in which it occurs, which varies according to the inherent suitability of the animal species and to the technological and social features of the human society. Perhaps he has the broadest view on domestication, once he states that possession and domination are key features in the process. According to him, there is no point in defining a threshold at which we should consider animals to be fully domesticated. In livestock science, many authors (e.g., Arman, 1979) use the term "semidomesticated" to refer to animals that are not fully domesticated, which includes some fish and bee species that are economically important in many contexts.

HISTORICAL ASPECTS OF DOMESTICATION

Animal domestication began at the end of the Paleolithic era and the beginning of the Neolithic one, around 15,000 BC. This process took place during the Neolithic Revolution, which was a period of transition for human beings, who evolved from living inside caves (when they would obtain food from hunting and harvesting wild vegetables) to becoming farmers/cultivators (Bowman, 1981). The domestication of animals was related to other changes in humans' lifestyle, once they were no longer nomads, but sedentary, which means they settled in and exercised power over territories. When humans became sedentary, the "herd" agency emerged as a new sociality composed by man, animals, and things (Mlekuž, 2013). As a result, the establishment of herds began to represent, from this moment on, more than a mere gathering of animals. It became an ever more complex configuration of the relationship man–animal–environment.

Humans have been domesticating animals for thousands of years, marking one of the great milestones in the development of civilization (Beck and Katcher, 1996) in the first efforts to keep wild animals close. Archeological evidence suggests that distinctly domesticated animals began to emerge approximately 12,000 years ago. However, many of the animals on which humans most depend were domesticated between approximately 9000 and 6000 years before the present time in a region known as the Fertile Crescent, in southwest Asia (Reitz and Wing, 2001). The domestication of animals allowed early human societies to enrich their diets with regular sources of meat, milk, and other animal products (Alves and Souto, 2010). Later, certain domesticated animals provided new sources of muscular energy as pack and mounted animals, or for the traction of ploughs and wagons, thus multiplying man's productive capacity and spatial mobility (Ribeiro, 1998).

The first domesticated animal was probably the wolf (*Canis lupus*) in the Middle Eastern region about 12,000 years ago, which gave rise to dogs (*Canis lupus familiaris*—a subspecies belonging to the same species of wolfs), which greatly aided human hunting activities (Allaby, 2010; Alves et al., 2009; Koster, 2008). Later, other animals were domesticated. Cows, sheep, goats, and pigs all furnished humans with meat and milk, whereas birds, in addition to their meat, provided eggs. Sheep and goats were domesticated in the Middle East between 7000 and 9000 years ago. Cattle were domesticated about 8000 years ago in (modern-day) Iraq and, independently, about 7000 years ago in Pakistan. Pigs were domesticated in the East about 9000 years ago, and horses in southern Russia about 5000 years ago (Allaby, 2010). It is believed that geese were the first farm birds to be domesticated, although it is possible that ducks were domesticated at approximately the same time, as both birds were common in ancient Egypt. The Romans were familiar with Angolan chickens, and the original inhabitants of North America kept domesticated turkeys and may also have domesticated rabbits as a source of food. Animals that contribute to human needs as beasts of burden were also first domesticated in ancient times (although more recently than animals that produce meat and milk). Donkeys and camels were probably the first beasts of burden, after which horses were used in this activity. Mules have similarly been bred for thousands of years. Elephants have been associated with humans for hundreds of years, although they cannot be considered fully domestic animals, once they rarely reproduce in captivity. Llamas have been beasts of burden since remote times in the Andean region of South America (Barsa, 1969).

Civilizations all over the ancient world domesticated animals for several reasons, depending on regional fauna available and which resources that animal species could provide for humans. It is well known that in ancient Egypt, for example, domesticated animals were very important to supply meat, milk, wool, eggs, leather, skins, horns, fat, and manual labor (Fig 9.1). Egyptians domesticated many different types of animals, including cattle, sheep, goats, pigs, geese, and horses. Domesticated animals were also worshipped as manifestations of gods.

THEORIES ON DOMESTICATION

The most convincing evidence of what may have been the first domestication of dogs is found in archeological sites in the Middle East. Research findings indicate that these animals were used for guarding men and for keeping them company when they were hunting, which may explain the unique social bond between man and dog that persists in present time (Dayan, 1994; Muller, 2002; Vonholdt et al., 2010).

FIGURE 9.1 Census of cattle from the tomb of Nebamum, Egypt, about 1400 BC (British Museum). *Photo credits: Rômulo R.N. Alves*

Later on, other species, such as cattle, buffaloes, pigs, horses, rabbits, birds, sheep, and goats, were also domesticated. Among these, the sheep and the goat were the first to be domesticated. Studies show that, during the Neolithic dispersion, the domestic goat was introduced in Europe, including in the Mediterranean Sea's main islands. The domestic goats that live in the continent are physically isolated from their ancestors and quite different from those found in the Mediterranean. This difference is due to the fact that there are large areas in the islands where traditional management practices, geographically separated from the continent, are found. Therefore, they are perceived as interesting models, for they maintain traces of either ancient forms of colonization, historical exchanges, or peculiar rearing systems (Sardina et al., 2006; Vacca et al., 2010; Hughes et al., 2012). Studies done with mitochondrial DNA from goats from the island of Corsica show strong evidence of historical events (from the Neolithic period) that cannot be found in other places (Hughes et al., 2012). The mitochondrial diversity of Corsica's goats is relatively stable since the Middle Age. Besides, no significant changes in the population density or in the genetic diversity were found in the island. The only relevant difference along history was the occurrence of two haplogroups C in the samples of current goats. The findings indicate that the goats that are currently found in Corsica are more similar to those who lived there in the Middle Age than to the ones who can be found today in other places in Europe.

Currently, a growing number of archeological, genetic, and ethnohistorical evidences indicate that neither reproductive isolation nor intentional reproduction were as significant as they were believed to be. The processes of introgression also contributed greatly for the genetic and phenotypic constitution of current populations. Introgression is the exchange of genetic material between populations that are undergoing domestication and those made up of their respective wild ancestors (Giuffra et al., 2000). Cases such as these are clear demonstrations of the integrated action of natural and artificial selection, which contribute to the improvement of our understanding of domestication as a complex biocultural process (Marshall et al., 2014).

In fact, domestication was a result of intentional and nonintentional human actions, and also of the environment modified by man (Larson and Fuller, 2014). According to Zohary et al. (1998), two kinds of selection were complementally at play in the processes of animal domestication: (1) a selection consciously made by humans (selective reproduction) and (2) an unconscious selection caused by the removal of animals from their original wild habitats and their insertion in a new and rather different environment that was designed by man. This environment change automatically led to drastic changes in selection.

The process of domestication, regardless if it is of plants or animals, involves a two-way relationship between humans and other species. Some authors consider it a form of mutualism in which both actors (humans and domesticated species) benefit from the growing mutual dependence. From this perspective, domestication is quite similar to the several mutual relationships found in the natural world (O'Connor, 1997; Zeder, 2012). Moreover, other scholars affirm that domesticated animals somehow manipulate human beings, in a relationship that provided them with a great evolutionary advantage at the expense of man's capacities (Rindos, 1984; Budiansky, 1992; Morey, 1994).

One issue that has been thoroughly debated recently is whether or not all domesticated animals have taken the same path toward domestication, that is, the usual track that is taken from being wild and free to becoming anchored to a partnership with human beings. According to Zeder (2012), there is a wide body of evidence suggesting that, although there are universal attributes in all domesticated animals, the ways toward domestication have varied greatly, once they depend on well-defined biological and cultural parameters, as well as on a set of factors that shape individual cases of animal domestication.

Vigne (2011) has proposed a multistage model to explain the process of domestication, which is conceived as a gradually intensified relationship between humans and animals. According to this model, animal domestication followed a long sequential process that moved from anthropophilia to commensalism, from controlling wild animals to keeping them in captivity, from extensive farming to intensive farming, all of which finally led to the domestication of animals. Even though Zeder (2012) accepts Vigne's (2011) approach, she only recognizes three separate pathways in the relationship between animals and human beings: a "commensal" one, a "prey" one, and a "directed" one (see Fig. 9.2).

The commensal pathway does not begin with the intentional action of man in becoming acquainted with wild animals, be they young or not. Instead, since humans begun to manipulate the surrounding environments, different populations of wild animals may have been attracted by elements of the human niche, which includes the waste of human food and/or smaller animals, which were also attracted by the waste product. From this perspective, animal domestication is seen as a coevolutionary process in which a population responds to a selective pressure by adapting to a new niche that includes other species that are undergoing behavioral transformations. Man-directed selection, on the other hand, which we associate with current domestic populations (for commercial purposes), may be a result of that previous cohabitation that occurred during the commensal phase. Thus, the process was made possible,

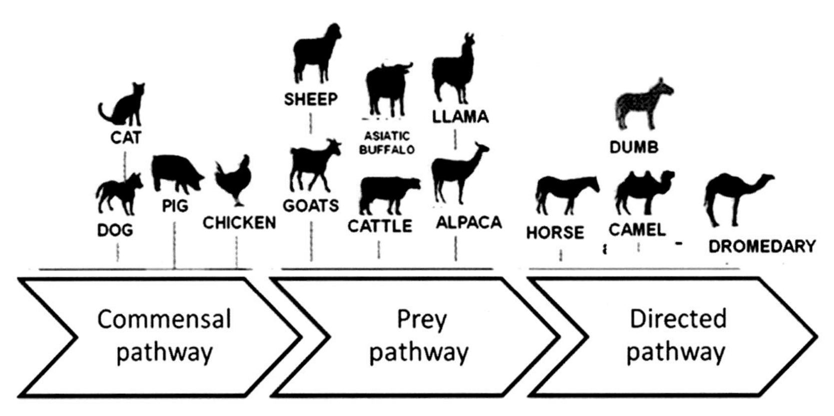

FIGURE 9.2 Pathways for the domestication of some animal species. *Adapted from Zeder, M.A., 2012. The domestication of animals. Journal of Physiological Anthropology 68, 161–190.*

at least at first, without effective human initiative. A textbook example of this is the dog, whose domestication is believed to have begun when wolves were attracted to human encampments by the possibility of feeding on human litter (Coppinger and Coppinger, 2001; Morey, 1994).

Like the commensal pathway, the prey pathway did not begin with human action at all. Man's first aim in this process was not to domesticate animals, but to improve the management of the available resources. The animals that took this pathway were medium- and large-size herbivorous, such as goats and cattle, which were turned into preys. Hence, these animals could have not been attracted by the waste produced by human niches. In this process, it is likely that humans changed their hunting strategies to come closer to their prey. It is believed, then, that the pressure for selecting the animals' features, such as docility, was great. This way, human groups went from hunting to managing herds, which gave them more control over the animals' diet and reproduction (Zeder, 2012). Curiously, this pattern of excessive hunting before domestication suggests that the prey pathway was accidental, and not intentional, like the commensal one (Zohary et al., 1998).

Finally, the directed pathway was the only one in which man had the deliberate aim to domesticate a species. This is due to the fact that humans already owned them and depended on plants and domestic animals. So, even though horses, donkeys, and camels from the Old World have been hunted as prey, they were introduced in the human niche for other reasons, such as for transportation.

The individual progress of the domestic animals and of their human partners along these different pathways varies greatly, as it is shaped by a combination of biological and cultural limitations and opportunities that the animals face when they make a transition to domestication. It probably took a long time for animals who took the commensal pathway to simply become acquainted with human beings and their habitats and to have an active partnership with them (Zeder, 2012).

Animal domestication involved several technical–economic, symbolic, and social aspects associated with the relationships between human societies and animals, which are, then, fundamental in cultural human systems. However, as part of the ecosystem, the relationship between humans and animals also involves ecological approaches, hence they should be considered part of the

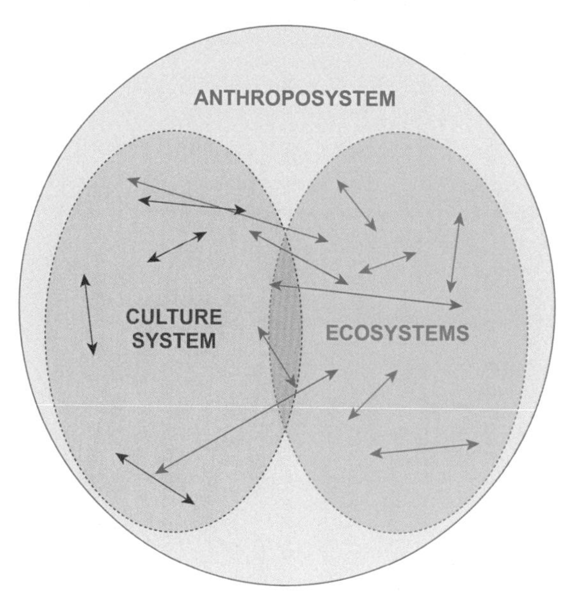

FIGURE 9.3 A scheme representing the anthroposystem, a meta-system made up of the culture system (in red) with its own components (*red arrows*), the ecosystems (in green) in which humans live and the biological interactions between its components (*green arrows*); and the interactions between these two systems (*blue arrows*). *Reproduced from Vigne, J.D., 2011. The origins of animal domestication and husbandry: a major change in the history of humanity and the biosphere. Comptes Rendus Biologies 334, 171–181.*

"anthroposystem," that is, a metasystem that groups cultural and ecological systems and their dynamic interactions along the time (Horard-Herbin and Vigne, 2005; Pascal and Vigne, 2006; Vigne et al., 2009) (see Fig. 9.3).

INTRASPECIFIC DIVERSITY OF LIVESTOCK SPECIES

Animal domestication resulted in a wide intraspecific genetic diversity, which made it possible for humans to use animals in a variety of ways: as food, as tools, as a mere company and for other functions that were established along history. This diversity is especially seen in the breeds that resulted from the combined effect of natural and artificial selection.

The Food and Agriculture Organization of the United Nations (FAO) has organized a Global Databank on Animal Genetic Resources (FAO, 2015) based on data from 182 countries. Up to now, the catalog includes information on 8774 breeds, comprising 19 mammalian species, 17 avian species, and 2 fertile interspecies crosses. The most important of these are represented in Table 9.1.

Humans' interest in domestication, which led to the removal of animals from their natural habitats, promoted environmental changes that allowed for genetic differences that, together with artificial selection, caused a considerable increase in the number and in the extension of genetic variability.

Many breeds were shaped for inhospitable environments and have a leading role in the maintenance of poor human populations in marginal areas, especially in developing countries (FAO, 2007). Intraspecific livestock diversity is an important part of food resources in the world. Among the 8774 breeds cataloged by FAO (2015), 88% are local ones, which means that they are found in only one country. This way, the development of local breeds represents not only a genetic legacy, but also a cultural, historical, and economic one. So, even though there are a few nonintentional aspects in the breeding processes, it is clear that many local races were created and conserved throughout generations by human populations that had little or no contact with technical–scientific institutions at all. These local breeds, then, are related to traditional knowledge and to the search for bottom-up solutions for local issues. In most cases (see Table 9.1), local breeds represent a vast majority of the total number of breeds that are recognized by FAO (2015). Since ethnozoology and ethnozootechny deal with local knowledge and practices associated with animal tenure, it is expected that the conservation of local breeds concerns these two ethnoscientific fields.

Currently, domestic animals remain critical for human development, providing them with

TABLE 9.1 List of Main Domesticated Species Recognized by Food and Agriculture Organization of the United Nations (FAO) and Their Respective Intraspecific Diversity

Species	Scientific Name	Date (Years)	Place of Domestication	Motivation to Domestication		Number of Breeds Recognized by FAO (2015)	
				Ancient	Present	Total Number	Local Breeds
Horse	*Equus caballus*	5,000	West of the steppes in Eurasia (Ukraine, southwest Russia, and West Kazakhstan)	Horsemanship and nutrition	Animal traction and transportation, warfare and combat against drugs and crime, and treatment of illnesses with the contact with animals	905	694 (77%)
Donkey	*Equus asinus europeus*	7,000	Iraq	Carrying heavy objects	Animal traction	174	157 (90%)
	Equus asinus africanus						
Bactrian camel	*Camelus bactrianus*	3,000	Central Asia	Transportation	Transportation, tourism	14	12 (86%)
Dromedary	*Camelus dromedarius*	6,000	Arabia/Asian Southwest	Transportation	Transportation	89	85 (96%)
Pig	*Sus domesticus*	9,000	Asian Southwest and China	Nutrition	Nutrition, fabrication of goods (leather, etc.)	709	543 (77%)
Goat	*Capra hircus*	7,000–9,000	Asian continent/Iran	Religion and nutrition	Nutrition, fabrication of goods (leather, etc.)	681	576 (85%)
Sheep	*Ovis aries*	7,000–9,000	Asian Southwest	Religion and nutrition	Nutrition, fabrication of goods (leather, wool, etc.)	1542	1155 (75%)
Cattle	*Bos indicus*	7,000	Asian South/Pakistan	Nutrition	Nutrition, fabrication of goods (leather, fertilizers, etc.), animal traction and transportation	1408	1019 (72%)
	Bos taurus	8,000	Asian Southwest	Nutrition	Nutrition, fabrication of goods (leather, fertilizers, etc.), animal traction and transportation		

Buffalo	*Bubalus bubalis*	4,000	Asian South	Nutrition	Nutrition, fabrication of goods (leather, etc.), animal traction and transportation	139	123 (88%)
Guinea pig	*Cavia porcellus*	550	Andes	Religion and economy	Use in scientific research	18	17 (94%)
Rabbit	*Oryctolagus cuniculus*	3,000	North America	Nutrition	Nutrition and use in scientific research and in the field of cosmetics	298	236 (79%)
Yak	*Bos grunniens*	10,000	Asia	Nutrition, traction	Nutrition, fur, manure, and transportation	28	28 (100%)
Chicken	*Gallus gallus domesticus*	6,500	Southeast Asia	Nutrition, entertainment (cock fights), and religion	Nutrition (meat and egg production)	1729	1514 (88%)
Ostrich	*Struthio camelus*	1,840	South Africa	Use of the feathers	Production of meat, eggs, and feathers	15	12 (80%)
Turkey	*Meleagris gallopavo*	5,000	Mexico	Nutrition	Nutrition (meat production)	117	92 (79%)
Duck	*Anas platyrhynchos domesticus*	3,500	Old Egypt	Nutrition	Nutrition (meat production)	294	253 (86%)
Muscovy duck	*Cairina moschata*	–	South America Panama	Nutrition	Nutrition	25	24 (96%)
Geese	*Anser anser*	3,500	Old Egypt	Nutrition	Ornamentation and nutrition (meat production)	208	182 (88%)
Pheasant	*Phasianus colchicus*	1,500	Asia	Hunting and nutrition	Ornamentation and nutrition (meat production)	18	18 (100%)
Pigeon	*Columba livia*	4,500	Asia	Nutrition	Messenger, ornamentation, and nutrition (meat production)	73	72 (99%)

Reproduced from FAO, 2015. The second report on the state of the World's animal genetic resources for food and agriculture. In: Scherf, B.D., Pilling, D. (Eds.), FAO Commission on Genetic Resources for Food and Agriculture Assessments. Rome. Available at: http://www.fao.org/3/a-i4787e/index.html; Wiener, G., Han, J.L., Long, R.J., 2003. The Yak, second ed. Regional Office for Asia and the Pacific, Food and Agriculture Organization of the United Nations, RAP Publication, Bangkok; Wing, E., 1986. Domestication of Andean mammals. In: Vuilleumier, F., Monasterio, M. (Eds.), High Altitude Tropical Biogeography. Oxford University Press and the American Museum of Natural History, Oxford, pp. 246–264; Lavallée, D., 1990. La domestication animale en Amerique du Sud - Le point des connaissances. Bulletin de l'Institut Français d'Études Andines 19 (1), 25–44; Crawford, R.D., 1992. Introduction to Europe and diffusion of domesticated turkeys from the America. Archivos de Zootecnia 41 (154), 307–3014; Stahl, P.W., Muse, M.C., Delgado-Espinoza, F., 2006. New evidence for pre-Columbian Muscovy Duck Cairina moschata from Ecuador. IBIS 148, 657–663. http://dx.doi.org/10.1111/j.1474-919X.2006.00564.x; Blay, M., 1991. Cría rentable de patos y gansos. Manual práctico. Editorial: Editorial de Vecchi, Barcelona; Oliveira, D.B., 2005. Aspectos Químicos e Etnomedicinais de Plantas da Dieta de Cervídeos na Reserva Particular do Patrimônio Natural – SESC Pantanal. Tese de Doutorado – Rio de Janeiro - RJ - Universidade Federal do Rio de Janeiro, Brasil; Allaby, M., 2010. Animals: From Mythology to Zoology. Facts On File, Inc, New York; Alves, R.R.N., 2016. Domestication of animals. In: Albuquerque, U.P. (Ed.), Introduction to Ethnobiology. Springer, Switzerland, pp. 221–225.

food, income, transportation, locomotive power, companionship, and entertainment (Figs. 9.4–9.7) (Scanes, 2003). Foods derived from livestock (meat, milk, and eggs) contribute significantly to the dietary intake of energy and nutrients and to the taste and enjoyment of meals (Givens et al., 2004; Hulshof et al., 1999). Livestock is also a dynamic part of agricultural economy, supporting the livelihoods of many families, in particular of those that live in the poorest households in developing countries (Delgado, 1999; FAO, 2009).

Livestock production has evolved fast, signing animals up for slaughter at shorter intervals and in large scale, with important economic and social gains. On the other hand, the increase in production resulted in the worsening of animals' well-being, especially for those who were kept in intensive systems of production (Paranhos da Costa and Cromberg, 1997), hence quite different from how they used to be managed in the beginning of human civilization and from many a number of traditional populations—be they peasants, native, or others—of present day.

Motivated by the tendency to provide more food for the market, humans select animals in

FIGURE 9.4 A Brazilian woman with her animals in Acre, Brazil. *Photo credits: Kate Evans for Center for International Forestry Research (CIFOR).*

FIGURE 9.5 People of Sentarum lake (Indonesia) have their cattle cage floating on the water. *Photo credits: Ramadian Bachtiar for Center for International Forestry Research (CIFOR).*

FIGURE 9.6 Cattle commercialization in Bangladesh. *Photo credits: Terry Sunderland for Center for International Forestry Research (CIFOR).*

FIGURE 9.7 Maasai herdsmen with their domesticated animals, Tanzania, Africa. *Photo credits: Franciany Braga-Pereira.*

a quite intense way. Currently, there is a tendency toward homogenizing and specializing the breeds of domestic animals, which results in the loss of genetic variability and leads the current breeds to risks of extinction at different levels. Consequently, it is adamant that strategies for the conservation of intraspecific genetic diversity of livestock species be sought for, so as to meet the improvement objectives and the conservation objectives simultaneously.

Management of local breeds has been practiced since the beginning of time by local human populations. It relies on knowledge that is built through experience with the herds' genetic attributes, which allows for practices that may contribute to a sustainable use of livestock species' intraspecific biodiversity. This is because the choices made by breeders are often based on thought-out strategies that directly affect the genetic structure of local breeds (see Fig. 9.8).

ETHNOZOOTECHNY OR ETHNOZOOLOGY?

Local zootechnical wisdom or knowledge is the knowledge that rural human populations developed of various techniques used in the management of domestic animals

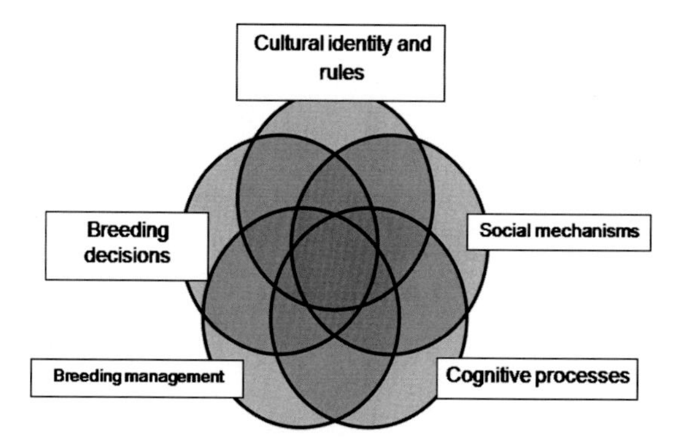

FIGURE 9.8 Elements of local knowledge on animal breeding. *Adapted from Sansthan, L.P.-P., Koehler-Rollefson, I., 2005. Indigenous Breeds, Local Communities: Documenting Animal Breeds and Breeding from a Community Perspective. Lokhit Pashu-Palak Sansthan, Sadri, Rajasthan, India.*

(Baraona, 1987; González, 1993; Vázquez-Varela, 2003). This knowledge has been shaped since the beginning of time, and, naturally, it has suffered transformations along history. Humans' ancient tradition of stock farming and their daily interaction with the herds made it possible for numerous peoples to gather detailed knowledge about domestic species and their needs. Farmers often have traditional ways for selecting animals and are aware of the existence of breeds that have not been documented yet (Ayantunde et al., 2007). In traditional societies, such as those that have small ruminants and other local animals, the objectives of selection are many, once diversity and rusticity are sought for. Farmers select their animals so they will meet different goals, one of which may be to resist environmental changes such as prolonged droughts. Hence, this is a rather different context from the immediate concern of large corporations, which are interested in short-term specialization and productivity.

With a variety of approaches, ethnoscience presents us with viable ways toward the documentation and the appreciation of local knowledge about natural resources (Alves et al., 2010). Two branches of this area that should be noted when we discuss human–animal interaction are ethnozoology and ethnozootechny, which may come into play together.

Roughly, it is possible to say that ethnozoology is concerned with local knowledge about animals, especially (but not only) when it comes to wild species, whereas ethnozootechny is more focused on the interaction between humans and domesticated animals. However, having in mind the complex relationships between the aforementioned categories of animals, as well as those between them and humans, we would rather point out the common aspects of ethnozootechny and ethnozoology, also showing how they complement each other. Biologists may study the intraspecific diversity of domestic animals, whereas zootechnicians, veterinarians, and other professionals from the agrarian sciences may also do research on wild animals. The attempts to understand phenomena such as introgression (Beja-Pereira et al., 2004), for instance, may foster interest about domestic animals' wild ancestors, which can bring zoologists close to the science of domestic animals.

A study on all the aspects that involve local knowledge and its possible connections with the academic knowledge of domestic animals is precisely the focus of ethnozootechny. As a formal academic approach, ethnozootechny was brought about around the mid-20th century so as to integrate social and natural sciences and the farming techniques used with domestic animals (Molénat, 2005). More specifically, it is concerned with the evolution of animal species, as well as with the tools and techniques used by farmers along history and the main causes for the disappearing of certain breeds and its consequences in the herds' loss of genetic diversity (Laurans, 1979).

Ethnozootechny shall be understood, then, as a field of knowledge that is dedicated to the study of practices, knowledge, and beliefs of certain social groups regarding livestock species. Its main focus is on the possible connections, conflicts, and dialogues between the zootechnical knowledge of local human populations and the scientific community (Alves et al., 2010).

CONCLUSION

Due to their interdisciplinary nature, approaches such as ethnozootechny, ethnozoology, and such may offer relevant contributions for the study of human–animal relations, as they include social (history, culture, and economy) and natural (genetic, evolution, ecology) aspects. Many of the difficulties in understanding these themes result from the exaggerated specialization that is so common in the academy. Integrated approaches, then, may allow for a more dynamic observation of the changes that have occurred (and that are occurring now) in the live organisms and in the very sciences that study them. For this to happen, it is essential that history and evolution be taken into consideration.

Prior to the World War II, zootechny encompassed human aspects in a more explicit way. With the development of "modern" large-scale farming aimed at industrialization, this connection was attenuated. This might explain the need for creating the new label "ethnozootechny," in an attempt to reconnect the human aspect to technical–scientific processes. Recently, with the growing appreciation of systems of production that are (socially) fairer and (ecologically) cleaner, the field has become even more susceptible to approaches such as ethnozootechny, agroecology, and such.

Taking a look back in time, we see that the very origin of the term "ethnozootechny" (Mason, 1899) reinforces the integrated view we advocate for here. In his seminal paper, Otis T. Mason, who was an ethnologist and curator of the Smithsonian Institution, presented the "Aboriginal American Zoötechny." According to him, all the human-made transformations ("industries") related to the animal kingdom should be nested under the term "zoötechny," under which the following branches would exist: (1) American Indian zoölogy, or ethnozoölogy in America; (2) exploitive zoötechny—the activities associated with the capture and domestication of animals; (3) elaborative zoötechny—the activities practiced on the animal after capture; (4) ultimate products of zoötechny and their relations with human happiness; (5) social organization and cooperation; (6) the progress of knowledge in zoötechny, including the development of specific language; and (7) religion and the animal kingdom. It would not be fair to say that we seek an uncritical return to an idyllic or underdifferentiated scientific scenario. On the contrary, we argue that we should maintain, at least in part, the capacity of dialoguing and building teams with an open vision to recover an integrative vision. The growing complexity of contemporary socioenvironmental issues (including those related to the human–animal interactions) cannot be sufficiently addressed with the commonly overspecialized scientific vision that is reinforced in many scientific subfields.

In line with the approach pursued by Alves and Albuquerque (2016), we believe that the potential contribution of ethnoscientific approaches, such as ethnozoology, ethnozootechny, and such, would be better achieved if we explored the possibilities of connections and complementarities among them.

References

Alves, R.R.N., 2016. Domestication of animals. In: Albuquerque, U.P. (Ed.), Introduction to Ethnobiology. Springer, Switzerland, pp. 221–225.

Alves, A.G.C., Albuquerque, U.P., 2016. Ethnobiology or ethnoecology? In: Albuquerque, U.P., Alves, R.R.N. (Eds.), Introduction to Ethnobiology. Springer International Publishing, pp. 15–18.

Alves, R.R.N., Souto, W.M.S., 2010. Etnozoologia: conceitos, considerações históricas e importância. In: Alves, R.R.N., Souto, W.M.S., Mourão, J.S. (Eds.), A Etnozoologia no Brasil: Importância, Status atual e Perspectivas, first ed. NUPEEA, Recife, PE, Brazil, pp. 19–40.

Alves, A.G.C., Pires, D.A.F., Ribeiro, M.N., 2010. Conhecimento local e produção animal, Uma perspectiva baseada na Etnozootecnia. Archivos de Zootecnia 59, 45–56.

Alves, R.R.N., Mendonça, L.E.T., Confessor, M.V.A., Vieira, W.L.S., Lopez, L.C.S., 2009. Hunting strategies used in the semi-arid region of northeastern Brazil. Journal of Ethnobiology and Ethnomedicine 5, 1–50.

Allaby, M., 2010. Animals: From Mythology to Zoology. Facts On File, Inc, New York.

Arman, P., 1979. Milk from semi-domesticated ruminants. In: Bourne, G.H. (Ed.). Bourne, G.H. (Ed.), Some Special Aspects of Nutrition, vol. 33. World Rev. Nutr. Diet. Basel, Karger, pp. 198–227.

Armitage, P.L., 1986. Domestication of animals. In: Cole, D.J.A., Brander, G.C. (Eds.), Bioindustrial Ecosystems. Elsevier, Amsterdam, pp. 5–30.

Ayantunde, A.A., Kango, M., Hiernaux, P., Udo, H., 2007. Herders' Perceptions on ruminant livestock breeds and breeding management in southwestern. Nigeria Human Ecology 28, 109–129.

Baenninger, R., 1995. Some consequences of animal domestication for humans. Anthrozoos 8, 69–77.

Baraona, R., 1987. Conocimiento campesino y sujeto social campesino. Revista. Mexicana de Sociologia 49, 167–190.

Barsa, E., 1969. Os animais e a subsistência. Enciclopédia Britânica, São Paulo.

Blay, M., 1991. Cría rentable de patos y gansos. Manual práctico. Editorial: Editorial de Vecchi, Barcelona.

Beja-Pereira, A., England, P.R., Ferrand, N., Jordan, S., Bakhiet, A.O., Abdalla, M.A., Mashkour, M., Jordana, J., Taberlet, P., Luikartvol, G., 2004. African origins of the domestic donkey. Science 304, 1781.

Benecke, N., 1994. Der Mensch und seine Haustiere. Die Geschichte einer jahrtausendealten Beziehung. Theiss-Verlag, Stuttgart, p. 215.

Beck, A.M., Katcher, A.H., 1996. Between Pets and People: The Importance of Animal Companionship. Purdue University Press, West Lafayette, Ind..

Bowman, J.C., 1981. Animais úteis ao homem. Guanabara Koogan, São Paulo.

Budiansky, S., 1992. The Covenant of the Wild: Why Animals Chose Domestication. William Marrow, New York.

Clutton-Brock, J., 1999. A Natural History of Domesticated Mammals. Cambridge University Press, Cambridge.

Clutton-Brock, J., 1994. The unnatural world: behavioural aspects of humans and animals in the process of domestication. In: Manning, A., Serpell, J.A. (Eds.), Animals and Human Society: Changing Perspectives. Routledge, London, pp. 23–35.

Clutton-Brock, J., 1992. The process of domestication. Mammal Review 22, 79–85.

Coppinger, R., Coppinger, L., 2001. Dogs. Schribner, New York.

Crabtree, P.J., 1993. Early animal domestication in the middle east and Europe. In: Schiffer, M.B. (Ed.). Schiffer, M.B. (Ed.), Archaeological Method and Theory, vol. 5. University of Arizona Press, Tucson, pp. 201–245.

Crawford, R.D., 1992. Introduction to Europe and diffusion of domesticated turkeys from the America. Archivos de Zootecnia 41 (154), 307–3014.

Dayan, T., 1994. Early domesticated dogs of the near east. Journal of Archaeological Science 21, 633–640.

Delgado, C.L., 1999. Livestock to 2020: The Next Food Revolution. Intl Food Policy Res Inst.

Digard, J.P., 1990. L'Homme et les animaux domestiques: Anthropologie d'une passion (Man and Domestic Animals: The Anthropology of a Passion). Fayard, Paris.

Dobney, K., Larson, G., 2006. Genetics and animal domestication: new windows on an elusive process. Journal of Zoology 269, 261–271. http://dx.doi.org/10.1111/j.1469-7998.2006.00042.x.

Ducos, P., 1978. Domestication deŽned and methodological approaches to its recognition in faunal assemblages. In: Meadow, R.H., Zeder, M.A. (Eds.), Approaches to Faunal Analysis in the Middle East Peabody Museum Bulletins No. 2. Peabody Museum, Cambridge, pp. 53–56.

FAO, 2009. Animal Genetic Resources a Safety Net for the Future. http://www.fao.org/nr/cgrfa/cthemes/animals/en/.

FAO, 2007. The state of the World's animal genetic resources for food and agriculture–in brief. In: Pilling, D., Rischkowsky, B. (Eds.) Rome.

FAO, 2015. The second report on the state of the World's animal genetic resources for food and agriculture. In: Scherf, B.D., Pilling, D. (Eds.), FAO Commission on Genetic Resources for Food and Agriculture Assessments Rome. Available at: http://www.fao.org/3/a-i4787e/index.html.

Giuffra, E., Kijas, J.M.H., Amarger, V., Carlborg, O., Jeon, J.T., Andersson, L., 2000. The origin of the domestic pig: independent domestication and subsequent introgression. Genetics 154, 1785–1791.

Givens, D.I., Allison, R., Cottrill, B., Blake, J.S., 2004. Enhancing the selenium content of bovine milk through alteration of the form and concentration of selenium in the diet of the dairy cow. Journal of the Science of Food and Agriculture 84, 811–817.

González, J.M., 1993. La sabiduría popular, técnicas y conocimientos científicos tradicionales en Canárias. Centro de la Cultura Popular Canaria, La Laguna.

Horard-Herbin, M.P., Vigne, J.D., 2005. Animaux, environnements et socie' te' s. Errance, Paris.

Hughes, S., Fernández, H., Cucch, i T., Duffraisse, M., Casabianca, F., Istria, D., Pompanon, F., Vigne, J.D., Hänni, C., Taberlet, P., 2012. A dig into the past mitochondrial diversity of Corsican goats reveals the influence of secular herding practices. PLoS One 7 (1), e30272. http://dx.doi.org/10.1371/journal.pone.0030272.

Hulshof, K.F., van Erp-Baart, M.A., Anttolainen, M., Becker, W., Church, S.M., Couet, C., Hermann-Kunz, E., Kesteloot, H., Leth, T., Martins, I., 1999. Intake of fatty acids in Western Europe with emphasis on trans fatty acids: the transfair study. European Journal of Clinical Nutrition 53, 143–157.

Ingold, T., 1980. Hunters, Pastoralists, and Ranchers: Reindeer Economies and Their Transformations. Cambridge University Press, Cambridge.

Koster, J., 2008. The impact of hunting with dogs on wildlife harvests in the Bosawas reserve, Nicaragua. Environmental Conservation 35, 211–220.

Lavallée, D., 1990. La domestication animale en Amerique du Sud - Le point des connaissances. Bulletin de l'Institut Français d'Études Andines 19 (1), 25–44.

Larson, G., Fuller, D.Q., 2014. The evolution of animal domestication. Annual Review of Ecology, Evolution, and Systematics 45, 115–136.

Laurans, R., 1979. L'ethnozootechnie aux confins des sciences de j'homme, de I'écologie et des techniques de I'élevage. Ethnozootechnie 20, 3–12.

Marshall, F.B., Dobney, K., Denham, T., Capriles, J.M., 2014. Evaluating the roles of directed breeding and gene flow in animal domestication. Proceedings of the National Academy of Sciences 111, 6153–6158. http://dx.doi.org/10.1073/pnas.1312984110.

Mason, O.T., 1899. Aboriginal American zootechny. American Anthropologist 1, 45–81.

Mlekuž, D., 2013. The birth of the herd. Society and Animals 21, 150–161.

Molénat, M., 2005. Court historique sur les races en conservation, genèse d'une réalisation au travers de la société. Ethnozootechnie 76, 147–150.

Morey, D., 1994. The early evolution of the domestic dog. American Scientist 82336–82347.

Muller, W., 2002. The first steps of animal domestication. In: Vigne, J.D., Peters, J., Helmer, D. (Eds.). Oxbow Books, Oxford, pp. 34–40.

Oliveira, D.B., 2005. Aspectos Químicos e Etnomedicinais de Plantas da Dieta de Cervídeos na Reserva Particular do Patrimônio Natural – SESC Pantanal. Tese de Doutorado – Rio de Janeiro - RJ - Universidade Federal do Rio de Janeiro, Brasil.

O'Connor, T.P., 1997. Working at relationships: another look at animal domestication. Antiquity 71, 149–156.

Paranhos da Costa, M.J.R., e Cromberg, V.U., 1997. Alguns aspectos a serem considerados para melhorar o bem-estar de animais em sistema de pastejo rotacionado. In: Peixoto, A.M., Moura, J.C., e Faria, V.C., Fundamentos do Pastejo Rotacionado (Eds.). FEALQ, Piracicaba, pp. 273–296.

Pascal, M.O., Vigne, L.J.D., 2006. Invasions biologiques et extinctions. 11 000 ans d'histoire des verte' bre' s en France. Belin, Paris.

Reitz, E., Wing, E.S., 2001. Zooarchaeology. Cambridge manuals in archaeology. Cambridge University Press, Cambridge.

Reitz, E.J., Wing, E.S., 1999. Zooarchaeology. Cambridge University Press, Cambridge.

Ribeiro, D., 1998. O processo civilizatório: etapas da evolução sociocultural. Companhia das Letras, São Paulo.

Rindos, D., 1984. The origins of agriculture: an evolutionary perspective. Academic, Orlando.

Russel, N., 2002. The wild side of animal domestication. Society and Animals 10, 285–302.

Sardina, M.T., Ballester, M., Marmi, J., Finocchiaro, R., van Kaam, J.B., Portolano, B., Folch, J.M., 2006. Phylogenetic analysis of Sicilian goats reveals a new mtDNA lineage. Animal Genetic 37, 376–378.

Stahl, P.W., Muse, M.C., Delgado-Espinoza, F., 2006. New evidence for pre-Columbian Muscovy Duck Cairina moschata from Ecuador. IBIS. 148, 657–663. http://dx.doi.org/10.1111/j.1474-919X.2006.00564.x.

Scanes, C.G., 2003. Biology of Growth of Domestic Animals. Wiley-Blackwell.

Sansthan, L.P.-P., Koehler-Rollefson, I., 2005. Indigenous Breeds, Local Communities: Documenting Animal Breeds and Breeding from a Community Perspective. Lokhit Pashu-Palak Sansthan, Sadri, Rajasthan, India.

Wing, E., 1986. Domestication of Andean mammals. In: Vuilleumier, F., Monasterio, M. (Eds.), High Altitude Tropical Biogeography. Oxford University Press and the American Museum of Natural History, Oxford, pp. 246–264.

Vacca, G.M., Daga, C., Pazzola, M., Carcangiu, V., Dettori, M.L., Cozzi, M.C., 2010. D-loop sequence mitochondrial DNA variability of Sarda goat and other goat breeds and populations reared in the Mediterranean area. Journal of Animal Breeding and Genetics 127, 352–360.

Wiener, G., Han, J.L., Long, R.J., 2003. The Yak, second ed. Regional Office for Asia and the Pacific, Food and Agriculture Organization of the United Nations, RAP Publication, Bangkok.

Vázquez-Varela, J.M., 2003. Introducción á antropoloxía da veterinaria popular en Galicia. Diputación Provincial de Ourense, Ourense, p. 166.

Vigne, J.D., 2011. The origins of animal domestication and husbandry: a major change in the history of humanity and the biosphere. Comptes Rendus Biologies 334, 171–181.

Vigne, J.D., Zazzo, A., Saliège, J.F., Poplin, F., Guilaine, J., Simmons, A., 2009. Pre-Neolithic wild boar management and introduction to Cyprus more than 11,400 years ago. Proceedings of the National Academy of Sciences of the United States of America 106, 16135–16138.

Vonholdt, B.M., Pollinger, J.P., Lohmueller, K.E., Han, E., Parker, H.G., et al., 2010. Genome-wide SNP and haplotype analyses reveal a rich history underlying dog domestication. Nature 464, 898–902.

Zeder, M.A., 2012. The domestication of animals. Journal of Physiological Anthropology 68, 161–190.

Zohary, D., Tchernov, E., Horwitz, L.K., 1998. The role of unconscious selection in the domestication of sheep and goats. Journal of Zoology 245, 129–135.

Further Reading

Digard, J.P., 1994. Relationships between humans and domesticated animals. Interdisciplinary Science Reviews 19, 231–236.

CHAPTER

10

Wild Fauna on the Menu

Rômulo Romeu Nóbrega Alves[1], *Nathalie van Vliet*[2]

[1]Universidade Estadual da Paraíba, Campina Grande, Brazil; [2]Center for International Forestry Research (CIFOR), Bogor, Indonesia

INTRODUCTION

Clearly, one of the most fundamental uses of wild animals is to meet nutritional needs (Reitz and Wing, 2008). People have mainly consumed the muscular and fatty parts of animal carcass, but they also eat visceras (hearts, livers, kidneys, brain, intestines, pancreas, etc.) and also a diversity of subproducts such as honey, eggs, and milk. The use of wild animals as food is prehistoric and started with the history of humankind. Wild meat consumption, for example, has a long history in human evolution (Larsen, 2003), likely going back to the earliest known humanlike ancestor living 5–7 mya (Lee-Thorp et al., 1994; Sillen, 1992). The presence of primitive stone tools beginning at ~2.5 mya in eastern Africa indicates that early humans likely had the capability of cutting and processing meat from animals (De Heinzelin et al., 1999; Shipman, 1986). Earlier 2.5 mya, the archeological evidence of meat consumption is nonexistent, but the commonality of hunting and meat eating by our nearest common ancestor (Larsen, 2003), the chimpanzee, suggests that meat eating has an ancient history, extending before the appearance of a humanlike primate some 6–8 mya (Stanford, 1995).

Archeological researchers have determined that humans have consumed a wide variety of fish, mollusks, birds, mammals, reptiles, and amphibians (Foster and James, 2002; Hamblin, 1985; Jorgenson, 1998; Kyselý, 2008; Masson, 1999; McKillop, 1984, 1985; Pohl, 1981). Although not in possession of a typical carnivore morphology, human beings have succeeded in becoming the most efficient predator in the world (Darimont et al., 2015; Ojasti, 2000), especially in more recent times with the evolution of modern hunting and fishing technologies (Gross, 1975), which have intensified the harvesting of wild meat and other animal-derived products along human history, characterizing two of the oldest and important activities of mankind: fishing and hunting (Alves, 2012). The domestication of animals has been a key element in human history, boosting the use of domestic products as a replacement for harvesting, fishing, and hunting (Cawthorn and Hoffman, 2014).

Whether they are obtained through hunting, fishing, or harvest, wild animal products continue to contribute to the diets of human populations. Worldwide, sources of animal protein embrace species belonging to a huge diversity of genera (Bowman, 1977; Fitter, 1986). While chordates are among the most used by humans,

Ethnozoology
http://dx.doi.org/10.1016/B978-0-12-809913-1.00010-7

invertebrates also play a key role in human diets, particularly insects in terrestrial habitats, and mollusks and crustaceans in aquatic ecosystems. By Food and Agriculture Organization (FAO) estimates, around "one billion people use wild foods in their diet" (Aberoumand, 2009). In general, food security and nontimber forest products, including wild animals, are strongly interlinked in rural communities, especially for the most vulnerable groups (Belcher et al., 2005), even among agricultural communities (Vincetti et al., 2008). Urban communities also rely on wild foods not only in developing landscapes (Nasi et al., 2011; van Vliet et al., 2015) but also in developed countries. Titus et al. (2009) explored the importance of wild game in Alaska, where 80% of the population is urban, and found urban households routinely consuming significant amounts of wild game. In New Zealand, more than 60 species are still in common use, largely because of traditions of Maori groups. In the Wallis Lake catchment, Australia, 88 species are in general use (Gray et al., 2005). In the swamps of Louisiana, large numbers of people still hunt and fish regularly for their own food (Roland, 2006).

In this chapter, our aim is to highlight the importance and diversity of uses of wild animals as sources of food, as well as to discuss the challenges and opportunities for wildlife to remain on the menu.

FRESHWATER AND MARINE ANIMALS USED AS FOOD

In aquatic ecosystems, particularly in coastal regions, where more than 60% of the world's population is concentrated (Vallega, 2013), people rely heavily on invertebrates as a source of protein. In those regions, several communities feed on a diversity of mollusk species and Crustacea. These groups are even more important for human nutrition in estuaries and mangroves. In those areas, edible species of oyster, mussels, cockles, and gastropods are collected extensively for local consumption, usually by the families of local fishermen (Rönnbäck, 1999; Walters et al., 2008). In Brazil, for example, shellfish catching is a largely extractive component of fishing activities concentrated at the periphery of many towns close to estuaries, where at least five main types of mollusks collected by the gatherers in the estuaries: *Crassostrea rhizophorae*, *Mytella guyanensis*, *Anomalocardia brasiliana*, *Tagelus plebeius*, and *Mytella charruana* (Nishida et al., 2006a, 2006b). In the Caribbean and other tropical sites, millions of "Queen conch," *Strombus gigas*, were fished, especially for roasting, as well for the shell, used for decorative purposes (Cattaneo-Vietti, 2016). Different species are consumed in all continents; among the most important are *Ostrea edulis*, *Crassostrea gigas*, *Crassostrea cucullata*, *C. rhizophorae*, *Crassostrea virginica*, *Ostrea lurida*, and *Ostrea chilensis* (Cattaneo-Vietti, 2016).

Cephalopod mollusks, mainly squids and octopuses, are fished and eaten in almost all the warmer parts of the world, notably in Japan (Fitter, 1986). Globally, cephalopod fishing has had an extraordinary development in recent decades. In Eastern Central Atlantic, for example, catch reporting of octopus (*Octopus vulgaris*) started in 1962 to reach 93,000 tons in 1975, but catches have since regularly decreased to reach 9000 tons in 2002 and 8000 tons in 2009. Catches of cuttlefish (*Sepia officinalis*) have fluctuated with an average of 46,000 tons from 1986 to 2004. After 2004 (the peak of the series), catches decreased until 2007 reaching 25,000 ton in 2008. Catches in 2009 were 33,000 tons. Catches of squids (*Loligo vulgaris*) have also decreased since the mid-1990s, and the average catch since 2000 is 8000 tons compared with 23,000 tons for the period 1990–99 (FAO, 2011).

Another important group of invertebrates used as food are Crustacea, particularly those of Malacostraca class, which includes crabs, lobsters, shrimp, crawfish, and prawns. Shrimps are almost certainly the world's most consumed crustaceans of the world, from both the natural environment and aquaculture. In recent decades, a dramatic increase in commercial shrimp farming has been recorded, particularly due to the market demands of Europe, United States, and Japan. Currently, approximately 80% of farmed shrimp is cultured in Asia, including China and Thailand (Kim and Venkatesan, 2015). Another important group are the crabs, commonly harvested by communities living close to Mangroves, where they are the most conspicuous and abundant components of epibenthic macrofauna. In Mangroves from Brazil, for example, Brachyura crabs represent one of the mayor groups in terms of economic relevance (Fig. 10.1). Among the most harvested and commercialized

FIGURE 10.1 Gatherer (above) of the land crab "caranguejo uçá" (*Ucides cordatus*) (below), the most relevant crab species used as food on the coast of Brazil. *Photos: Rômulo R.N. Alves.*

species are *Cardisoma guanhumi*, *Goniopsis cruentata*, *Callinectes* spp., and *Ucides cordatus* (Alves et al., 2005; Alves and Nishida, 2002; Capistrano and Lopes, 2012; Nascimento et al., 2012, 2016, 2017; Nishida et al., 2006b; Nordi et al., 2009). Lobsters also represent another important group used as food by humans, from prehistoric times to the modern era (Spanier et al., 2015). Lobsters are highly valued and sustain some of the most profitable fisheries in all tropical, subtropical, and temperate-cool waters of the world (Briones-Fourzán and Lozano-Álvarez, 2015). In 2012, worldwide landings of marine lobsters were 294,000 metric tons (mt) with a value of around US$2800 million, with an additional 2000 mt produced through aquaculture (FAO, 2014a). Nondecapod crustaceans are not widely consumed, but in some countries, such as Japan, products derived of krill (order Euphausiacea) are commonly used as food (Suzuki and Shibata, 1990). The Japanese Antarctic krill is processed for human consumption as boiled, then frozen krill or peeled krill tail meat is frozen in blocks on board (Nicol et al., 2000).

While less frequently consumed, species from other marine invertebrate groups are also used to feed humans, and these include jellyfish and sea anemones (phylum Cnidaria), nereid polychaete (*Eunice viridis*) (phylum Annelida), sea urchins, sea cucumbers, and occasionally starfish (phylum Echinodermata) (Bieri, 1969; Kim and Venkatesan, 2015). One prominent example is that of the sea urchin, *Paracentrotus lividus*, which gonads are consumed in countries such as France, Spain, Portugal, Italy, and Greece (Boudouresque and Verlaque, 2007; Mamede, 2014).

While a vast number of invertebrate species are edible and are used by humans for food, the importance of vertebrates deserves special emphasis. Fish are among the aquatic vertebrate species most used across the planet (Fig. 10.2). The importance of fishing for mankind has a long history (Alves, 2012). Practically all modern human societies that have ever existed have

FIGURE 10.2 Fishes catch in the Colombia (A), Mozambique (B), and Brazil (C) and (D). *Photo credits: (A) Nathalie van Vliet, (B) Anthony B. Cunningham, (C) and (D) Thelma L. P. Dias.*

to some degree utilized aquatic (ocean, lake, or stream) resources (García-Quijano and Pitchon, 2010). Worldwide, capture fisheries and aquaculture provide 3 billion people with almost 20% of their average per capita intake of animal protein and a further 1.3 billion people with about 15% of their per capita intake (HLPE, 2014). This share can exceed 50% in some countries (Béné et al., 2015). The small-scale fisheries sector is estimated to employ around 90% of the world's fishers, producing almost half of the world's fish and supplying most of the fish consumed in the developing world (FAO, 2016). In West Africa, Asian coastal countries, and many small island states, the proportion of total dietary protein from fish can reach 60% or more (e.g., Gambia, Sierra Leone, Ghana, Cambodia, Bangladesh, Indonesia, Sri Lanka, or the Maldives) (FAO, 2014b). In general, the most important freshwater food fishes belong to the carp (Cyprinidae) and catfish (Siluridae and Ictaluridae) families, but Salmonidade and the eel *Anguilla anguilla* are also important. Nevertheless, most of our fish protein supply comes from the sea, and all maritime nations, including some, such as Poland, with only a tiny coastline, have sea fishing fleets. The most important marine fish families in the Northern hemisphere are the cod (Gadidae), herring (Clupeidae), and mackerel and tunny (Scombridae) families, together with the flatfish (Pleuronectidae and Soleidae) (Fitter, 1986).

Besides fish, humans also explore other aquatic vertebrates, particularly mammals and reptiles. Aquatic mammals, for example, have been utilized by humans for food since ancient times (Colten and Arnold, 1998; Crespo and Hall, 2002; Monks, 2005; Porcasi and Fujita, 2000; Romero et al., 2002; Romero and Creswell, 2005). These animals occupy a variety of habitats, from pelagic continental shelf and coastal marine waters to estuarine, riverine, and lacustrine areas (Vidal, 1993), and in all those ecosystems, these animals are used by humans. It is not surprising, because these animals generally are medium to large sized, thus, being a potential source of large amounts of meat and fat, and other subproducts used for different purposes. In some cases, the potential for marine mammals as source of revenues has served as incentive for people to migrate to coastal areas. In the cold regions of the planet, for instance, the scarcity of animal protein from terrestrial origin and the lack of alternative sources of food explain why human population historically concentrated on the coast, where marine mammals were especially important as source of food (Colten and Arnold, 1998). Large sea mammals, in particular, provide a high caloric return on calories invested in their hunt (i.e., a high benefit/cost ratio), in spite of the relatively high energy expenditure necessary to search for, retrieve, butcher, distribute, cook, and consume them (Smith, 1981; Yesner, 1987). Furthermore, to provide buoyancy and protection against cold, sea mammals (particularly those in high-latitude environments) tend to have large amounts of subcutaneous fat. Fats provide twice the caloric density of proteins or carbohydrates (around 9 kcal/g) and thus provide an excellent source of energy for humans (Speth and Spielmann, 1983). Among societies that inhabit cold regions and are exclusively maritime in exploitation patterns, such as the prehistoric Aleuts or inhabitants of the Fuegian littoral, sea mammals represent an important dietary constituent (Yesner, 1988). Manatees, whales, sea lions, dolphins, and porpoises, among others, are among the aquatic mammals that are used as food in different parts of the world (Hovelsrud et al., 2008; Monks, 2005; Ripple and Perrine, 1999). Peoples' access to these animals is potentiated by the fact that some species are docile and thus easy to capture; others frequently beach on the coast, and others are prey and die in fishing nets (Alves et al., 2013d). The large terrestrial breeding colonies of some pinnipeds have made them highly susceptible to hunting in the distant and near past. As discussed by Rick et al. (2011) due to their large amounts of meat, oil, ivory, and other important raw material and dietary resources, pinnipeds (seals, sea lions, and walruses) and sea otters have been hunted and scavenged by people in the northeastern Pacific for much of the Holocene or earlier. In more recent times, human hunting of pinnipeds and sea otters decimated marine mammal populations, causing the extirpation of local populations of sea otters (*Enhydra lutris*), Guadalupe fur seals (*Arctocephalus townsendi*), and northern elephant seals (*Mirounga angustirostris*) and the extinction of the Steller's sea cow (*Hydrodamalis gigas*) (Ellis, 2003; Scammon, 1968).

REPTILES AND AMPHIBIANS SERVED ON THE MENU

Reptiles are an another important vertebrate group that have served as an important source of protein for human populations around the world (Fig. 10.3 and 10.4) (Alves et al., 2012c; Klemens and Thorbjarnarson, 1995), being mostly consumed in the warmest and wettest tropical and subtropical regions, where the greatest diversity of reptiles is found, not only in number of species but also in body sizes. Of all reptiles, chelonians are the most heavily exploited for human consumption (Alves et al., 2012c; Klemens and Thorbjarnarson, 1995). Almost all species of turtles and terrestrial tortoises, large and small, have served as human food (meat and eggs) at one time or another (Klemens and Thorbjarnarson, 1995). For some human communities in the Amazon

FIGURE 10.3 Examples of reptiles used as food: lizards [*Iguana iguana* (A) and *Tupinambis merianae* (B)]; American crocodile (*Crocodylus acutus*) (C); Giant South American turtle (*Podocnemis expansa*) (D); (E) specimens of *Podocnemis erythrocephala* recently captured by children in Jaú National Park, Amazonas state, Brazil; and (F) the shells of tortoises eaten by Indians from Pará state, northern Brazil. *Photo credits: (A) and (D) Rômulo R.N. Alves. (B) Washington L. S. Vieira. (C) Nathalie van Vliet. (E) and (F) Juarez C. B. Pezzuti.*

region, chelonians have assumed an important role in the diet of local people at a certain time of the year (Alves et al., 2012c; Klemens and Thorbjarnarson, 1995; Pezzuti et al., 2008; Rebêlo et al., 2006; Thorbjarnarson et al., 2000). The seven South American *Podocnemis* river turtles have long been important sources of meat and eggs for local people (Alves and Santana, 2008; Fitter, 1986). *Podocnemis expansa* (the Amazonian turtle), the largest river turtle in South America,

FIGURE 10.4 The leatherback sea turtle (*Dermochelys coriacea*), chelonian in the Gabonese Coast, which eggs are used for food in several locations. *Photo credits: Nathalie van Vliet.*

is one of the most consumed species (Fig. 10.4). The importance of chelonians for human diet is recorded also in Asia, where the soft shell turtles (several species of the family Trionychidae) are commonly consumed, as well as throughout their range North America and Africa (Klein et al., 2007; Klemens and Thorbjarnarson, 1995; van Dijk et al., 2000). Similar to freshwater species, all marine turtle species are used as food sources in all their distribution range (Fig. 10.4) (Frazier, 2005; Humber et al., 2014; Mancini and Koch, 2009; Nogueira and Alves, 2016; Rudrud et al., 2007). While the consumption of crocodilians and their eggs is not as widespread or intense when compared to the exploitation of certain types of turtles (Klemens and Thorbjarnarson, 1995), they have been hunted by humans for their meat and skins (Roth and Merz, 1997). Crocodile meat is considered a delicacy, with a flavor lying between chicken, veal, and fish (Huchzermeyer, 2003), and it is particularly consumed in Australia, South Africa, Thailand, Ethiopia, Cuba, and in regions of the United States (Hoffman and Cawthorn, 2012). Whereas the meat is relished in some societies, it is generally consumed as a by-product of the skin trade (Hoffman, 2008). Snakes and lizards are important items in diet in China and elsewhere in Asia and the Pacific, as well as in Latin America (Alves et al., 2012a; Alves et al., 2012c; Brooks et al., 2010; Fitter, 1986; Klemens and Thorbjarnarson, 1995; Zhou and Jiang, 2004). Some species of lizards are regularly eaten by people. Lizards of the genus *Tupinambis* are hunted for food in Brazil, Argentina, Paraguay, and parts of Bolivia (Fig. 10.4) (Alves et al., 2012a; Fitzgerald, 1994). The consumption of *Iguana iguana* (meat and eggs) is also common in the tropical Americas (Alves et al., 2012c; Klemens and Thorbjarnarson, 1995; Werner, 1991). In North Africa, *Uromastyx* species are considered "fish of the desert" and eaten by nomadic tribes (Grzimek, 2003). The consumption of snakes is not common in several localities, but in Asian countries such as China, Taiwan, Thailand, Indonesia, Vietnam, and Cambodia, these animals are protein sources in many diets (Brooks et al., 2010; Irvine, 1954). Recent information has shown that up to 4000 tons of snake meat is served annually in China, where this animal's meat is commonly served in restaurants in cities such as Foshan, Shanghai, and Yangshuo, particularly appreciated for their medicinal value, taste, and nutritive value (Hoffman and Cawthorn, 2012). Also, in several rural communities in West African regions, snakes are commonly hunted for wild meat, mainly large snakes such as pythons and boas, which move slowly and are easily hunted for their meat (Hoffman and Cawthorn, 2012).

Although amphibians are consumed on a smaller scale than other vertebrates, there are places where these animals constitute an important part of human diet (Fig. 10.5) (Alves, 2012; Warkentin et al., 2009). Among all amphibians, frogs are the most popular and are collected for subsistence or local consumption in many countries in Asia, Africa, and Latin America (Altherr et al., 2011), where they have always been collected locally, comprising an essential source of animal protein (Angulo, 2008; Mohneke et al., 2009, 2010). Mohneke et al. (2009) pointed that at least 32 amphibians (3 Urodela and 29 Anura) are used as food in the world. In Africa, it is probable

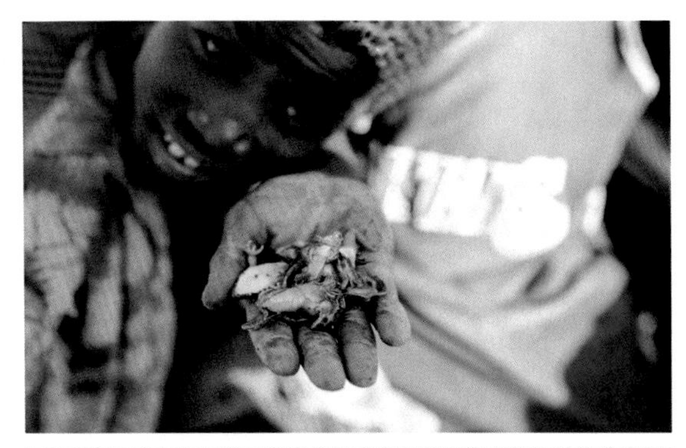

FIGURE 10.5 Children look for frogs for food, Burkina Faso (above), and the species *Leptodactylus vastus* Lutz, 1930 (Northeastern Pepper Frog), amphibian used as food in Brazil. *Photo credits: (A) Ollivier Girard for Center for International Forestry Research (CIFOR). (B) Washington L. S. Vieira.*

that amphibians have always been eaten and used for cultural purposes, with examples occurring in western Cameroon (Gonwouo and Rödel, 2008) and in Gabon (Pauwels et al., 2003). In Asia, Warkentin et al. (2009) summarized some alarming data on the numbers of Asian frogs collected for human consumption. Bangladesh and India, for example, became major exporters of frogs' legs beginning in the 1950s, producing >4000 tons/year over the next 30 years (Niekisch, 1986; Pandian and Marian, 1986). Frogs are also consumed in "haute cuisine" restaurants of countries of the European Union and the United States (Jensen and Camp, 2003; Mohneke et al., 2009, 2010, 2011; Patel, 1993; Warkentin et al., 2009).

FIGURE 10.6 Edible caterpillars used as food in Zambia. *Photo credits: Anthony B. Cunningham.*

Terrestrial Animals Feeding Humans

Terrestrial invertebrates include a huge diversity of animal groups consumed by humans since ancient times. Insects have a very long history of use by humans for food in Africa, Asia, and Latin America (Bodenheimer, 1951; Chakravarthy et al., 2016; DeFoliart, 1997; Meyer-Rochow, 2010; Morris, 2004; Srivastava et al., 2009) and supplement the diets of approximately 2 billion people (Halloran and Vantomme, 2013). There are at least 2037 species of edible insects globally (Jongema, 2015), distributed among different taxonomic groups that have played an important role in the history of human nutrition. Among the main insect groups used for food are the Orthoptera (grasshoppers, locusts, katydids, crickets), Isoptera (termites), Hymenoptera (ants, bees, wasps), Heteroptera (true bugs), Lepidoptera (butterflies, moths), Homoptera (Cicadas and others), and Coleoptera (beetles, weevils) (Fig. 10.6) (Chakravarthy et al., 2016; DeFoliart, 1997; Figueirêdo et al., 2015; Jongema, 2015). The importance of insects as a food source for humans is not surprising, because this is the group with the highest number of species in nature, thereby representing a significant biomass (Meyer-Rochow and Chakravorty, 2013). In addition, the high reproductive rates of many insect species and the lower ecological footprint required to produce an equivalent amount of animal-based protein compared to vertebrates are further attributes to consider (Yen, 2015). Insects not only represent a direct source of food but may also produce important products that complement human diet, such as honey, which has played a key role for human nutrition in many communities across the planet (Toussaint-Samat, 2009). In addition to insects, other terrestrial arthropods also are used as food, such as arachnids (spiders and scorpions) and millipedes (Fig. 10.6) (Enghoff et al., 2014; Meyer-Rochow, 2005)

Birds have always been an important food source for humans. These animals provide eggs and meat that are consumed by people in different countries (Fig. 10.7). Birds may comprise a significant amount of the total biomass obtained through hunting (Alves, 2012; Fernandes-Ferreira et al., 2012; Gonzalez, 2004; González, 1999; Mendonça et al., 2016). Where

FIGURE 10.7 Brazilian bird game species: (A) *Rhea americana* (Linnaeus, 1758); (B) *Nothura boraquira* (Spix, 1825); (C) *Dendrocygna viduata* (Linnaeus, 1766); and (D) *Penelope jacucaca* (Spix, 1825). *Photo credits: (A), (B), and (C) John Philip Medcraft. (D) Dandara Bezerra.*

mammals have become scarcer, birds play an important role in the subsistence of rural families that depend on wildlife for their livelihoods. This trend was observed by Mendonça et al. (2016) in a study on the annual consumption of wild animals by hunter families in the semiarid region of Paraiba (Northeast Brazil). Among the species of birds most consumed for food are Columbidae, Tinamidae, Anatidae, Cracidae, and Phasianidae. Where larger-sized bird species occur, they are the favorite prey for the hunters. Cracids, for example, are traditionally considered the most important birds for subsistence hunting in tropical forests, and they are always present in the diet of all Amazonian rural settlements (Barros et al., 2011; Pierret and Dourojeanni, 1967; Vickers, 1991). In contrast, in semiarid regions from Brazil, species from the families Columbidae and Tinamidae are the most hunted for food, mainly because other larger-sized species are scarce or have been locally depleted by hunting (Albuquerque et al., 2012; Mendonça et al., 2016). The rheas (family Rheidae) (Fig. 10.8), endemic to the Neotropics, have had their wild populations negatively affected by hunting for meat consumption and egg harvesting as subsistence resources (Barbarán, 2004; Barri et al., 2008). Small birds are also captured for food. Their gregarious habit makes them an easy prey that a high number of individuals are usually captured from a group (Bezerra et al., 2012). In the southern part of Pacaya-Samiria National Reserve (Peru) and its surroundings, Gonzalez (2004) registered at least 47 bird species hunted for food in 1996, but undulated tinamous (*Crypturellus undulatus*), anhingas (*Anhinga anhinga*), razor-billed curassows (*Mitu tuberosa*), muscovy ducks (*Cairina moschata*), and olivaceous cormorants (*Phalacrocorax olivaceus*) were the most frequently hunted species. *M. tuberosa* and *C. moschata*, because of their larger size, were the most important species in terms of biomass. A similar situation is observed in the semiarid region of Brazil, where many bird species are used as sources of food, and most of the species hunted belong to the Columbidae and Tinamidae families (Alves et al., 2013a; Bezerra et al., 2011, 2012). Wild bird eggs are an important source of food in some areas (Cott, 1954; González, 1999). Eggs used for food belong mainly to large species and to those nesting colonially; in general, the eggs of such birds are both accessible and palatable. Charadriiformes, Anseriformes, Procellariiformes, and Sphenisciformes are among the most important egg birds utilized (Cott, 1954). Flamingo eggs are eaten in certain areas of the Andes of Bolivia and Chile (Campos, 1986), collection of seabird eggs is a traditional practice in the Caribbean islands that dates back several centuries (Haynes, 1987), and harvesting eggs by indigenous peoples from beach-nesting birds in the Peruvian Amazon is common (Robinson and Redford, 1991). Bird eggs are also an important source of food for local people in many areas of the Peruvian Amazon (Gonzalez, 2004; González, 1999).

The main vertebrate group targeted by hunting activities are mammals (Alves et al., 2016; Barboza et al., 2016; Redford and Robinson, 1987; Robinson and Redford, 1991). These animals comprise the preferred source of food because of their size and the possibility of yielding a greater return for the energy invested in hunting (Albuquerque et al., 2012; Alves et al., 2016; Leopold, 1959; Mesquita and Barreto, 2015; Nasi et al., 2008). Among the most consumed terrestrial mammals are the ungulates, rodents, and primates (Fig. 10.8 and 10.9). As might be expected, hunters focus initially on large animals and continue to hunt them even when their numbers become low (Nasi et al., 2008). It should be highlighted that in areas where there is not an abundance of larger-sized mammals, small mammals may present a similar or greater hunting significance (Alves et al., 2016; Barboza et al., 2016). Dalle Zotte (2014) pointed that the meat of lagomorphs (i.e., rabbits, hares, and pikas) and rodents in particular has long played a vital role in subsistence societies throughout

FIGURE 10.8 Hunting armadillo [*Dasypus kappleri* (A) and *Tayassu pecari* (Link, 1795) (B)], game mammals of the Brazilian Amazon region; (C) small diurnal monkeys in Kisangani, Democratic Republic of Congo; and (D) meat of the deer and *T. pecari* in Guyana. *Photo credits: Juarez C. B. Pezzuti (A). Flávio Barros (B). (C) and (D) Nathalie van Vliet.*

the world where it is considered tasty, nutritious, and often superior to that from conventional livestock (Roth and Merz, 1997). In Africa, for example, the giant rat (*Cricetomys gambianus*) and cane rats or grasscutters (*Thryonomys swinderianus* and *Thryonomys gregorianus*) are commonly hunted for food and highly preferred by consumers (Ntiamoa-Baidu, 1997; Odebode et al., 2011). These species are regarded as pests on many crops and have proven to be well adapted to hunting due to their high reproductive rates (Jeffrey, 1977; Martin, 1983). In the semiarid of northeastern Brazil, small-sized species of rodents (e.g., *Galea spixii*, *Cavia aperea*, and *Kerodon rupestris*) with fast reproduction are still being hunted in a high number of specimens, to compensate for their small biomass (Alves et al., 2016; Barboza et al., 2016). Mammals represent an important source of animal protein in both rural and urban areas (Alvard et al., 1997; Bailey, 2000; Bennett and Robinson, 2000; Newing, 2001; Ojasti, 1997; Schenck et al., 2006; van Vliet, 2011; van Vliet et al., 2014; van Vliet et al., 2015). Even in societies with an agriculture- and livestock-based economy, hunting is important for many families, providing up to 80% of the meat consumed in certain rural communities (Bakarr et al., 2001). For example, the rural population from the Brazilian Amazon alone each year consumes between 9.6 and 23.5 million reptiles,

FIGURE 10.9 Scenes associated with wild meat consumption in tropical countries: (A) slaughtering of the peccaries *Tayassu pecari* in Brazilian Amazon; (B) pirarucu meat in Leticia (Colombia); (C) blue duiker for sale in Maduda, Democratic Republic of Congo; and (D) deer (*Mazama americana*) meat for sale in Guyana. *Photo credits: (A) Juarez C. B. Pezzuti. (B), (C) and (D) Nathalie van Vliet.*

birds, and mammals, representing an estimated total biomass of 67,173–164,692 tons and a yield of 36,392–89,224 tons of wild meat (Peres, 2000). In the moist forests of the Congo Basin in Africa, between 1 and 3 million mt of dressed (slaughtered) wild meat is eaten each year (Fa et al., 2003). The magnitude of the exploitation and consumption of wild meat varies, however, from country to country and is determined primarily by its availability; it is also influenced by governmental controls on hunting and by socioeconomic status and cultural prohibitions. In areas where wildlife still persists, people either collect, hunt, or purchase and eat wild meat for a variety of reasons. Some people depend on wild meat for their supply of animal protein because they have no alternative sources or cannot afford them; others eat wild meat as a matter of preference or as a luxury item/delicacy on special occasions (Alves et al., 2009; Fa et al., 2003; Ntiamoa-Baidu, 1997).

CHALLENGES AND OPPORTUNITIES FOR WILD ANIMALS TO CONTINUE TO FEED HUMANS

Ecological Impacts of Current Harvest Levels

In terrestrial ecosystems, the sustainability of wild animals' extraction for food varies across the regions and across the taxonomic groups: from local extinctions to overpopulation. No simple solution therefore applies to any location and can be generalized to all wild animal groups.

The use of wild mammals in most tropical forest areas poses sustainability concerns hindering the possibility for vulnerable species to remain part of the menu and suggesting the need for innovative sustainable harvest approaches for the most resilient ones. "Defaunation" is often cited as the most evident impact of excessive hunting for tropical terrestrial mammals (Redford, 1992) as it can also have major impacts not just in terms of the extirpation of vulnerable species but also on broad ecological processes (Dirzo et al., 2014; Muller-Landau, 2007). However, smaller and faster-reproducing species tend to be more resilient to hunting than larger species with lower reproductive rates (Cowlishaw et al., 2005; Weinbaum et al., 2013) and may continue to persist in forest–agriculture tropical landscapes despite hunting pressure and habitat degradation. The fact that most animal meat coming from tropical terrestrial ecosystems is derived from those resilient species (e.g., large rodents and small ungulates) represents an unexplored opportunity for the implementation of sustainable use innovative mechanisms. In temperate regions, the situation is almost inversed. Carnivores have been largely reduced, and hunting is the main regulatory factor of most large mammal species (Sinclair et al., 2006). Wildlife management has generally focused on different practices of minimum quota setting (Sinclair and Byrom, 2006). In Europe, where agriculture competes with wildlife habitats, many wildlife populations have increased during the last decades (Calvert and Gauthier, 2005; Côté et al., 2004), and filling the quotas is increasingly challenging given the length of restricted hunting seasons. Despite the very high levels of hunting bags for wild boar *Sus scrofa* in Europe (Keuling et al., 2013), populations are increasing and dispersing into agrarian landscapes causing significant economic costs for farmers (e.g., Frackowiak et al. (2013), Lagos et al. (2012), Saito et al. (2012), Sakurai et al. (2014), Schley et al. (2008)). The high reproduction rates of such species and their adaptability to diverse pressures explain why current harvest rates seem to be insufficient (Keuling et al., 2016). In those contexts, a longer season and increased hunter effort is needed to fill or closely approach the set quota (Loe et al., 2016).

Concerning birds, overexploitation for food and sport purposes is one of the main drivers of extinctions globally and is the second most significant threat to migratory birds (Kirby et al., 2008). In tropical Africa, many hornbills (e.g., the yellow-casqued hornbill *Ceratogymna elata*) are vulnerable to hunting pressure (Whytock et al., 2016), probably because of their low reproductive rates (Trail, 2007). Bird population declines in tropical areas are of concern because of the potential broader implications for ecosystem function, as frugivorous species play an important role in seed dispersal and maintenance of forest structure (Brodie et al., 2009; Poulsen et al., 2013). In Europe, the illegal killing of several bird species is also of major concern (e.g., white-headed Duck ("endangered"), marbled teal ("vulnerable"), and saker falcon (*Falco cherrug*) ("Endangered")). Bird harvest information (for both legal and illegal take) despite being key for assessing the sustainability of exploitation is generally not available, fragmented, or not up-to-date.

In aquatic ecosystems, the reduction of wild fish stocks is a reality, particularly in many tropical marine and freshwaters (Golden et al., 2011, 2016; Molur et al., 2011). Most marine fish stocks are being used at or above their sustainable levels, leaving only around 10.5% of the waters under fished (FAO, 2016). In Europe's seas, overfishing is very high: 50% of fish stocks in the northeast Atlantic Ocean and the Baltic Sea, and over 90% in the Mediterranean and Black Seas are fished above their maximum sustainable yield (FAO, 2016). However, many stocks have been recovering since 2003, largely as a result of better management and significant progress toward fishing at maximum sustainable yield in the EU's northeast Atlantic Ocean and Baltic waters (Cardinale et al., 2013). In tropical marine waters, ocean warming and shifts in net primary production are expected to reduce catch by as much as 30% in the tropics by 2050 relative to recent decades (Cheung et al., 2016), and fish will also probably get smaller by around 20% during this period (Cheung et al., 2013). Freshwater fish have also suffered from extinctions and population declines. However, inland fisheries harvest is often unrecorded or drastically underreported (Lynch et al., 2016), particularly the harvest from small-scale or artisanal fishing (Bartley et al., 2015; Dugan et al., 2010; Hortle, 2007). Besides overfishing, other external factors have also largely contributed to freshwater fish declines. In fact, about 65% of freshwater habitat is classified as moderately or highly threatened by diverse anthropogenic stressors (Vörösmarty et al., 2010). The primary impacts on aquatic ecosystems in developing countries often come from urbanization. For example, besides a few coastal cities, most of the population in Papua New Guinea tends to cluster around rivers and estuaries because of the many resources available from aquatic systems (Sheaves et al., 2016). In developed countries, fishery resources are being affected by destruction and fragmentation of aquatic habitats, sedimentation, and runoff of agrochemicals (Jonsson et al., 2015). For example, in Finland, acidification has already caused the drop-in up to 4400 fish populations (Rask et al., 1995). Another less direct driver of fish declines is climate change and warming. In the Lake Tanganyika, sustained warming during the last hundred years has affected biota by strengthening and shallowing stratification of the water column yielding fish and mollusk declines (Cohen et al., 2016).

Nutrition and Zootherapeutic Benefits Related to Consuming Wild Animals

From a health perspective, the use of wild animals contributes to human nutrition and health in different ways both in rural and urban areas. In some situations, the consumption of the wild meat is often contributed to the purported medicinal benefits derived from the animal parts (Alves et al., 2012b; Alves et al., 2013e; Klemens and Thorbjarnarson, 1995; Castro et al., 2016), and this enduring connection between medicinal therapy and food animals goes well beyond the understanding that adequate nutrition sustains a person's health (Alves et al., 2013b). The animals most frequently used for traditional medicine are vertebrates, mostly harvested through hunting or fishing. There is a high degree of overlap between medicinal and alimentary uses of those wild animals, and the difference between food and medicine may not always be clear (Alves et al., 2013b; Alves et al., 2013c). Fishes illustrate this situation well. The meat of these animals, besides being consumed as an important source of food, is also considered important for human health. In fact, fishes are considered an excellent and low-fat source of protein and provide many health benefits to consumers of their meat (Bouzan et al., 2005; Daviglus et al., 2002; El-Deir et al., 2012).

Communities that depend heavily on wild animals tend to enjoy comparatively high standards of health. Studies of the daily protein intake provided by wild meat and wild fish have shown that international standard nutrient recommendations are met in a substantial proportion of cases (Dufour, 1991; Murrieta et al., 2008; Nardoto et al., 2011; Piperata et al., 2011; Sirén and Machoa, 2008). While many studies tend to emphasize the importance of wild meat as a source of protein, its contribution to micronutrient intake and fat has also widely been proven for Neotropical and African wild meat species (Golden et al., 2011; Sarti et al., 2015; Sirén and Machoa, 2008). Freshwater fish and seafood are also known for their beneficial effects on human health. Several studies have repeatedly linked fish consumption, especially those species whose content in omega-3 polyunsaturated fatty acids (PUFAs) is high, with healthier hearts (Abeywardena and Patten, 2011; Davidson et al., 2011; Delgado-Lista et al., 2012). The benefits of fish and seafood consumption on health are also due to the content of high-quality protein, vitamins, as well as other essential nutrients (Domingo, 2016). The role of fish in mitigating mental disorders, such as depression and dementia, is also receiving increased attention from scientists (Domingo, 2016). Insects are also increasingly known for their nutritional value and increasingly mentioned for their totally unexplored potential for future use. When compared to conventional production animals, insects are suggested to be an interesting protein source because they have a high reproductive capacity, high nutritional quality, very low water and land utilization, and high feed conversion efficiency (Dossey, 2013; Oonincx, 2015; Shockley and Dossey, 2014).

ZOOTHERAPEUTIC USES AND SANITARY CONCERNS

The nutritional or therapeutic benefits of wild animals also depend on the quality of the environment from which they are harvested, and the cleaning, transportation, and conservation/cooking practices. In Central Africa, several specimens of wild meat were found to be contaminated by toxic elements, such as nickel, chromium, and polycyclic aromatic hydrocarbons (Abdul et al., 2014; Igene et al., 2015). Fish species have also proved to be harmful when they accumulate contaminants such as methyl mercury and dioxins present in their habitats. For example, in the Great Lakes, persistent bioaccumulative toxicants such as methyl mercury and polychlorinated biphenyls have resulted in over 20 years of fish consumption advisories to reduce fish consumption (Dellinger and Ripley, 2016).

Besides the potential for contamination with toxic elements, the consumption of wild animals (for food or medicinal purposes) may also generate health risks, even in most natural and unintervened environments, due to the variety of zoonosis that may spread to humans (van Vliet et al., 2017). The likelihood of an association between wild meat consumption and the spread of zoonotic diseases has been largely publicized in recent years, especially in Africa, where there have been a number of virulent outbreaks of viral hemorrhagic fevers. This outbreak was given extensive coverage in the conservation media, and there were calls for the wild meat trade to be banned throughout its range (Osofsky, 2014;

Young, 2014) based on the argument that lowering the consumption in wild animals (especially bats and primates) could be the key to preventing epidemics (Pooley et al., 2015). The main risk of zoonotic transfer of viruses is the immediate butchery level, in handling fresh meat, not in consuming it. However, there are a wide range of potentially dangerous parasites and bacteria that may be transmitted by the fecal–oral route or through the consumption of rough wild meat material. The most abundant parasites in African wild meat species are *Trichuris* sp., *Ancylostoma* sp., and *Strongyloides fuelleborni*. *Trichuris* sp. frequently transmitted from simians to humans (Kurpiers et al., 2016). In Latin America, the most frequent parasites and bacteria are *Toxoplasma gondii* (Hamilton et al., 2014), *Echinococcus vogeli* (Mayor et al., 2015), *Capillaria hepatica* (Soares et al., 2011), *Brucella suis* (Lord and Lord, 1991), and *Mycobacterium leprae* (Frota et al., 2012). Despite their prevalence in wild meat species and the observed cases of severe complications due to gastrointestinal infections, these silent parasites and bacteria remain rather shadowed behind the more mediatic zoonotic outbreaks. Fish also carry a diversity of zoonoses. Fishborne nematode infections in humans are common in countries where people have a traditional custom of consuming live, raw, smoked, lightly cooked, or marinated fish and/or squid (Eiras et al., 2016). Several species of fishborne nematodes are recognized as causative agents for human diseases (Cross and Belizario, 2007). Three species and one genus of bacteria known to be fishborne zoonoses are also commonly found: *Mycobacterium* spp., *Clostridium botulinum*, *Streptococcus iniae*, and *Vibrio vulnificus* (Gauthier, 2015). In temperate freshwaters, zoonoses are also common. For example, some fish species (*Perca fluviatilis*, *Esox lucius*, *Lota lota*) found in lakes from Italy, France, and Switzerland have been recently found to be infected by *Diphyllobothrium latum* (Cestoda: Diphyllobothriidae), a common fishborne zoonosis.

The Place of Wild Animal Foods in a World Dominated by Domestic Sources of Meat

With the initial domestication of animals, the original diets of early humans changed slowly, but more rapidly with advancing technology after the Industrial Revolution (Bharucha and Pretty, 2010). Because domestication developed out of food gathering, many farmers continually blurred the distinction between the cultivated and the uncultivated (Mazhar and Buckles, 2007). Many wild species are actually still found within the fields themselves. The harvesting of wild fish species from paddy fields in Thailand (Halwart, 2008), Bangladesh (Mazhar and Buckles, 2007), and Cambodia (Guttman, 1999) is a good example of the latter. However, over the last 15 years, industrial domestic sources of meat have largely overtaken wild animal foods in most urban areas of the world as well as in most rural areas connected to markets (Popkin, 1993, 1994). The first most recent nutritional transition, lasting about 60 years, involved the increase in dietary diversity, meat, and lightly processed products. The second, most recent transition involved poorer populations which nutrition became restricted to cheap, highly processed, and high calorie foods, whereas wealthier populations could enjoy dietary diversity including increasingly expensive fruits and vegetables (Hawkes, 2006). The urban centers of several middle-income Asian countries including Thailand, China, and India went through both transitions in the quick succession of 40 years (Popkin, 2003). Recent studies continue to evidence nutritional transitions from wild meat and fish to industrial meats and processed foods in many remote areas such as the small indigenous communities from the Amazon (Byron, 2003; Nardoto et al., 2011; van Vliet et al., 2015), pigmy hunter-gatherer communities in Central Africa (Dounias and Froment, 2011), communities in India (Rathore, 2009; Samson and Pretty, 2006), and the Inuit communities from the Arctic (Huet et al., 2012; Samson and

Pretty, 2006). Global trends indicate that more people will come to depend solely on processed and cultivated foods (Johns and Maundu, 2006), thus marginalizing wild foods. While for many indigenous communities, wild foods still outweigh modern store-bought items in terms of nutrient content, their gradual replacement by store-bought produce already causes significant negative impacts on nutritional security at household and community levels (Loring and Gerlach, 2009; Samson and Pretty, 2006) and health risks at individual level (Nardoto et al., 2011; Sarti et al., 2015; Silva and Padez, 2010). Research has also pointed the declines in local ecological knowledge as communities rely increasingly on processed foods and move away from land-based livelihoods (Pilgrim et al., 2008). Wild food species are declining anyway in many agricultural landscapes (Bharucha and Pretty, 2010) due to the homogenization of agricultural landscapes increasingly limiting the availability and use of wild foods of nutritional importance (Pretty, 2002; Scoones et al., 1992). Their continued availability depends on the maintenance of synergies between farming and wild biodiversity (Pretty, 2008).

As such, although wild animals continue to have a fundamental role as a source of food for several human populations, domestic species currently represent the main animals used for global meat production (Cawthorn and Hoffman, 2014). In this context, domestic mammals and birds are the main groups involved, producing meat, eggs, and milk, which as a whole represent the main protein sources used by humans and generate a powerful global market. In the last two decades, the growth in livestock production reflects the global trend of consumers increasingly moving away from the consumption of beef toward poultry, pork, eggs, and dairy products (FAO, 2014). Indeed, in 2012, cattle produced 63 million tons of meat, sheep 8.5 million tons, and goats 5.3 million tons, but these species were far outranked by pigs (109 million tons) and poultry (105 million tons; (FAOSTAT, 2014)). In Latin America, poultry production nearly duplicated in one decade (2001–11), bovine meat also increased but only by one-third over this same period (Rodríguez et al., 2016). In China, for example, consumption of animal products increased by nearly 40% between 1989 and 1997 (Frazão et al., 2008). With a projected world population of nearly 10 billion people by 2050, an unprecedented increase in demand for animal protein including meat, eggs, milk, and other animal products is inevitable (National Academy of Sciences, 2015), and production will have to increase by more than 200 million tons to reach the target of 470 million tons, according to FAO estimates (FAO, 2009).

To meet the burgeoning demand for meat and animal protein, large-scale domestic meat production systems will need to be put in place. However, there are also obvious costs associated with large-scale production systems and these need to be acknowledged and carefully analyzed to maximize access to animal protein while minimizing the environmental and health costs. While the land sparing theory suggests that the intensification of production systems might help reduce further grabbing over natural areas (Green et al., 2005), the land conversion necessary to produce enough meat to cover for the increasing demand is so immense that additional land will inevitably have to be withdrawn from currently forested or wild areas (Alexandratos and Bruinsma, 2012; Foley et al., 2011; Newbold et al., 2014; Ramankutty et al., 2002) with tremendous consequences for biodiversity and the availability of wild foods (Alkemade et al., 2013; Newbold et al., 2014). At global level, animal production is estimated to contribute about 14.5% of all GHG emissions (Flachowsky and Kamphues, 2012), and thus any significant increase in agroindustrial production would impact significantly on global warming. In addition to the environmental costs, large-scale industrial systems are also a matter of concern in relation to nutrition and health issues. Highly intensified, industrial-style production (often in periurban settings) is increasingly common (Thornton, 2010) and could escalate threats to public health and the environment in many tropical forest regions (Leibler et al., 2009; Liverani et al., 2013). Indeed, disease spillovers are frequent in intensive farming, particularly the industrial poultry (Graham et al., 2008; Leibler et al., 2009; Mennerat et al., 2010). Despite the undeniable achievements in the control and eradication of some zoonosis in industrialized countries (Perry et al., 2013), the reality is different in tropical regions where veterinary assistance is more rarely available and where the distribution of the products involves long commodity chains with unhygienic conditions.

CONCLUSIONS: THE FUTURE OF WILD ANIMAL FOODS

Beyond 2015, governments around the world will be faced with the need to comply with the new internationally validated sustainable development goals (SDGs) (UN General Assembly, 2015), which place food security in the center of the debate for poverty eradication and sustainable development (UNDP, 2014). Whereas past approaches conceptualized food security as a production issue, which should be addressed through increasing the amount of food produced, the current approach focuses on the accessibility of nutritious food, as well as its availability (Manning et al., 2013; Olukoshi, 2013).

In the context of the *new food equation* (NFE), national and international governing bodies increasingly acknowledge the multifunctional character of food systems, which are deeply linked to public health, biodiversity conservation, and natural resources management (Morgan and Sonnino, 2010). Within this NFE for sustainable food strategies, some key central concepts are the balance between *localization of food chains* (by dovetailing local production and consumption of seasonal foods), *globalization* (by promoting use of fairly traded produce from other countries), *food culture* (by taking into account hedonistic and sociocultural values of food) (Morgan, 2010), and biodiverse diets (Burlingame et al., 2006; Englberger et al., 2010; Flyman and Afolayan, 2006; Johns and Eyzaguirre, 2006; Ruel, 2003).

The environmental, health, and cultural trade offs generated by totally replacing wild foods by domestic and processed sources of meat are immense and deserve close attention. Therefore, within this new concept of sustainable food systems, there is scope that wild foods will continue to be in the menus and may well remain legitimate components of diversified diets. A clear prerequisite is to bring the contribution of wild animal foods into national statistics so that the food security, health, and ecological sustainability issues can be openly discussed. We believe this is a necessary precursor for a more penetrating debate on the feasibility of the various options for sustainable use to contribute to the welfare and nutrition of range state populations, in ways that are economically sound and compliant with standards of public health and that take a much broader view of sustainable food strategies.

References

Abdul, I.W., Amoamah, M.O., Abdallah, A., 2014. Determinants of polycyclic aromatic hydrocarbons in smoked bushmeat. International Journal of Nutrition and Food Sciences 3, 1–6.

Aberoumand, A., 2009. Nutritional evaluation of edible Portulaca oleracia as plant food. Food Analytical Methods 2, 204–207.

Abeywardena, Y.M., Patten, G.S., 2011. Role of ω3 longchain polyunsaturated fatty acids in reducing cardio-metabolic risk factors. Endocrine, metabolic & immune disorders-drug targets (formerly current drug targets-immune. Endocrine & Metabolic Disorders) 11, 232–246.

Albuquerque, Araújo, E., Lima, A., Souto, A., Bezerra, B., Freire, E.M.X., Sampaio, E., Casas, F.L., Moura, G., Pereira, G., Melo, J.G., Alves, M., Rodal, M., Schiel, M., Neves, R.L., Alves, R.R.N., Azevedo-Júnior, S., Telino Júnior, W., 2012. Caatinga revisited: ecology and conservation of an important seasonal dry forest. Scientific World Journal 2012.

Alexandratos, N., Bruinsma, J., 2012. World Agriculture towards 2030/2050: The 2012 Revision. ESA Working Paper. ESA Working paper, Rome, FAO.

Alkemade, R., Reid, R.S., van den Berg, M., de Leeuw, J., Jeuken, M., 2013. Assessing the impacts of livestock production on biodiversity in rangeland ecosystems. Proceedings of the National Academy of Sciences 110, 20900–20905.

Altherr, S., Goyenechea, A., Schubert, D.J., 2011. Canapés to Extinction: The International Trade in Frogs' Legs and its Ecological Impact, a Report by Pro Wildlife. Defenders of Wildlife and Animal Welfare Institute. Pro Wildlife, Munich (Germany), Washington, D.C. (USA), p. 33.

Alvard, M.S., Robinson, J.G., Redford, K.H., Kaplan, H., 1997. The sustainability of subsistence hunting in the Neotropics. Conservation Biology 11, 977–982.

Alves, R.R.N., Nishida, A., Hernandez, M., 2005. Environmental perception of gatherers of the crab 'caranguejo-uca' (*Ucides cordatus*, Decapoda, Brachyura) affecting their collection attitudes. Journal of Ethnobiology and Ethnomedicine 1, 1–8.

Alves, R.R.N., Nishida, A., 2002. A ecdise do caranguejo-uçá, *Ucides cordatus* L. (Decapoda, Brachyura) na visão dos caranguejeiros. Interciencia 27, 110–117.

Alves, R.R.N., 2012. Relationships between fauna and people and the role of ethnozoology in animal conservation. Ethnobiology and Conservation 1, 1–69.

Alves, R.R.N., Feijó, A., Barboza, R.R.D., Souto, W.M.S., Fernandes-Ferreira, H., Cordeiro-Estrela, P., Langguth, A., 2016. Game mammals of the Caatinga biome. Ethnobiology and Conservation 5, 1–51.

Alves, R.R.N., Leite, R.C., Souto, W.M.S., Bezerra, D.M.M., Loures-Ribeiro, A., 2013a. Ethno-ornithology and conservation of wild birds in the semi-arid Caatinga of northeastern Brazil. Journal of Ethnobiology and Ethnomedicine 9, 1–12.

Alves, R.R.N., Mendonça, L.E.T., Confessor, M.V.A., Vieira, W.L.S., Lopez, L.C.S., 2009. Hunting strategies used in the semi-arid region of northeastern Brazil. Journal of Ethnobiology and Ethnomedicine 5, 1–50.

Alves, R.R.N., Oliveira, T.P.R., Rosa, I.L., 2013b. Wild animals used as food medicine in Brazil. Evidence-based Complementary and Alternative Medicine 2013.

Alves, R.R.N., Pereira Filho, G.A., Silva Vieira, K., Souto, W.M.S., Mendonças, L.E.T., Montenegro, P.F.G.P., Almeida, W.O., Vieira, W.L.S., 2012a. A zoological catalogue of hunted reptiles in the semiarid region of Brazil. Journal of Ethnobiology and Ethnomedicine 8, 27.

Alves, R.R.N., Rosa, I.L., Léo Neto, N.A., Voeks, R., 2012b. Animals for the gods: magical and religious faunal use and trade in Brazil. Human Ecology 40, 751–780.

Alves, R.R.N., Santana, G.G., 2008. Use and commercialization of Podocnemis expansa (Schweiger 1812) (Testudines: Podocnemididae) for medicinal purposes in two communities in North of Brazil. Journal of Ethnobiology and Ethnomedicine 4, 1–6.

Alves, R.R.N., Souto, W.M.S., Barboza, R.R.D., Bezerra, D.M.M., 2013c. Primates in traditional folk medicine: world overview. In: Alves, R.R.N., Rosa, I.L. (Eds.), Animals in Traditional Folk Medicine: Implications for Conservation. Springer, Berlin, pp. 135–170.

Alves, R.R.N., Souto, W.M.S., Oliveira, R.E.M.C.C., Barboza, R.R.D., Rosa, I.L., 2013d. Aquatic mammals used in traditional folk medicine: a global analysis. In: Alves, R.R.N., Rosa, I.L. (Eds.), Animals in Traditional Folk Medicine: Implications for Conservation. Springer, Heidelberg, pp. 241–261.

Alves, R.R.N., Vieira, K.S., Santana, G.G., Vieira, W.L.S., Almeida, W.O., Souto, W.M.S., Montenegro, P.F.G.P., Pezzuti, J.C.B., 2012c. A review on human attitudes towards reptiles in Brazil. Environmental Monitoring and Assessment 184, 6877–6901.

Alves, R.R.N., Vieira, W.L.S., Santana, G.G., Vieira, K.S., Montenegro, P.F.G.P., 2013e. Herpetofauna used in traditional folk medicine: conservation implications. In: Alves, R.R.N., Rosa, I.L. (Eds.), Animals in Traditional Folk Medicine: Implications for Conservation. Springer-Verlag, Berlin Heidelberg, pp. 109–133.

Angulo, A., 2008. Consumption of Andean frogs of the genus Telmatobius in Cusco, Peru: recommendations for their conservation. Traffic Bulletin 21, 95–97.

Bailey, N., 2000. Global and historical perspectives on market hunting: implications for the African bushmeat crisis. Sustainable Development and Conservation Biology - University of Maryland and Bushmeat Crisis Task Force Silver Spring, Maryland 1–44.

Bakarr, M.I., Fonseca, G., Mittermeier, R.A., Rylands, A.B., Painemilla, K.W., 2001. Hunting and Bushmeat Utilisation in the African Rainforest: Perspectives towards a Blueprint for Conservation Action. Conservation International, Washington, DC.

Barbarán, F.R., 2004. Usos mágicos, medicinales y rituales de la fauna en la Puna del Noroeste Argentino y Sur de Bolivia. Contribuciones al manejo de vida silvestre en Latinoamérica 1, 01–26.

Barboza, R.R.D., Lopes, S.F., Souto, W.M.S., Fernandes-Ferreira, H., Alves, R.R.N., 2016. The role of game mammals as bushmeat in the Caatinga, northeast Brazil. Ecology and Society 21, 1–11.

Barri, F.R., Martella, M.B., Navarro, J.L., 2008. Effects of hunting, egg harvest and livestock grazing intensities on density and reproductive success of lesser rhea Rhea pennata pennata in Patagonia: implications for conservation. Oryx 42, 607–610.

Barros, F.B., Pereira, H.M., Vicente, L., 2011. Use and knowledge of the razor-billed curassow pauxi tuberosa (spix, 1825) (galliformes, Cracidae) by a riverine community of the oriental amazonia, Brazil. Journal of Ethnobiology and Ethnomedicine 7 (1), 1–11.

Bartley, T.J., Braid, H.E., McCann, K.S., Lester, N.P., Shuter, B.J., Hanner, R.H., 2015. DNA barcoding increases resolution and changes structure in Canadian boreal shield lake food webs. DNA Barcodes 3, 30–43.

Belcher, B., Ruíz-Pérez, M., Achdiawan, R., 2005. Global patterns and trends in the use and management of commercial NTFPs: implications for livelihoods and conservation. World Development 33, 1435–1452.

Béné, C., Barange, M., Subasinghe, R.P., Pinstrup-Andersen, P., Merino, G., Hemre, G., Williams, M., 2015. Feeding 9 billion by 2050–Putting fish back on the menu. Food Security 7, 261–274.

Bennett, E.L., Robinson, J.G., 2000. Hunting of wildlife in tropical forests. The World Bank Environment Department Papers 1–42.

Bezerra, D.M.M., Araujo, H.F.P., Alves, R.R.N., 2012. Captura de aves silvestres no semiárido brasileiro: técnicas cinegéticas e implicações para conservação. Tropical Conservation Science 5, 50–66.

Bezerra, D.M.M.S.Q., Araujo, H.F.P., Alves, R.R.N., 2011. The Use of Wild Birds by Rural Communities in the Semi-arid Region of Rio Grande Do Norte State, Brazil. Bioremediation, Biodiversity and Bioavailability 5 (1), 117–120.

Bharucha, Z., Pretty, J., 2010. The roles and values of wild foods in agricultural systems. Philosophical Transactions of the Royal Society of London B: Biological Sciences 365, 2913–2926.

Bieri, R., 1969. Invertebrates as food. Encyclopedia of Marine Resources 324–325.

Bodenheimer, F.S., 1951. Insects as Human Food. Springer.

Boudouresque, C.F., Verlaque, M., 2007. Ecology of Paracentrotus lividus. In: Lawrence, J.M. (Ed.), Developments in Aquaculture and Fisheries Science. Elsevier, Amsterdam, pp. 243–285.

Bouzan, C., Cohen, J.T., Connor, W.E., Kris-Etherton, P.M., Gray, G.M., König, A., Lawrence, R.S., Savitz, D.A., Teutsch, S.M., 2005. A quantitative analysis of fish consumption and stroke risk. American Journal of Preventive Medicine 29, 347–352.

Bowman, J.C., 1977. Animals for Man. Edward Arnold., London.

Briones-Fourzán, P., Lozano-Álvarez, E., 2015. Lobsters: ocean icons in changing times. ICES Journal of Marine Science 72, i1–i6.

Brodie, J.F., Helmy, O.E., Brockelman, W.Y., Maron, J.L., 2009. Bushmeat poaching reduces the seed dispersal and population growth rate of a mammal-dispersed tree. Ecological Applications 19, 854–863.

Brooks, S.E., Allison, E.H., Gill, J.A., Reynolds, J.D., 2010. Snake prices and crocodile appetites: aquatic wildlife supply and demand on Tonle Sap Lake, Cambodia. Biological Conservation 143, 2127–2135.

Burlingame, B., Charrondiere, R., Halwart, M., 2006. Basic human nutrition requirements and dietary diversity in rice-based aquatic ecosystems. Journal of Food Composition and Analysis 19, 770.

Byron, E.M., 2003. Market Integration and Health: The Impact of Markets and Acculturation on the Self-perceived Morbidity, Diet, and Nutritional Status of the Tsimane'Amerindians of Lowland Bolivia. University of Florida.

Calvert, A.M., Gauthier, G., 2005. Effects of exceptional conservation measures on survival and seasonal hunting mortality in greater snow geese. Journal of Applied Ecology 42, 442–452.

Campos, L.C., 1986. Distribution, Human Impact, and Conservation of Flamingos in the High Andes of Bolivia. University of Florida, Gainesville.

Capistrano, J.F., Lopes, P.F.M.L., 2012. Crab gatherers perceive concrete changes in the life history traits of *Ucides cordatus* (Linnaeus, 1763), but overestimate their past and current catches. Ethnobiology And Conservation 1, 1–21.

Cardinale, M., Dörner, H., Abella, A., Andersen, J.L., Casey, J., Döring, R., Kirkegaard, E., Motova, A., Anderson, J., Simmonds, E.J., 2013. Rebuilding EU fish stocks and fisheries, a process under way? Marine Policy 39, 43–52.

Castro, M.S., Martins, I.M., Hanazaki, N., 2016. Trophic relationships between people and resources: fish consumption in an artisanal fishers neighborhood in Southern Brazil. Ethnobiology and Conservation 5, 1–16.

Cattaneo-Vietti, R., 2016. Mand and Shells: Molluscs in the History. Bentham Science Publishers Ltd.

Cawthorn, D.M., Hoffman, L.C., 2014. The role of traditional and non-traditional meat animals in feeding a growing and evolving world. Animal Frontiers 4, 6–12.

Chakravarthy, A.K., Jayasimha, G.T., Rachana, R.R., Rohini, G., 2016. Insects as human food. In: Chakravarthy, A.K. (Ed.), Economic and Ecological Significance of Arthropods in Diversified Ecosystems. Springer, pp. 133–146.

Cheung, W.W.L., Jones, M.C., Lam, V.W.Y., Miller, D.D., Ota, Y., Teh, L., Sumaila, U.R., 2016. Transform High Seas Management to Build Climate Resilience in Marine Seafood Supply. Fish and Fisheries.

Cheung, W.W.L., Sarmiento, J.L., Dunne, J., Frölicher, T.L., Lam, V.W.Y., Palomares, M.L.D., Watson, R., Pauly, D., 2013. Shrinking of fishes exacerbates impacts of global ocean changes on marine ecosystems. Nature Climate Change 3, 254–258.

Cohen, A.S., Gergurich, E.L., Kraemer, B.M., McGlue, M.M., McIntyre, P.B., Russell, J.M., Simmons, J.D., Swarzenski, P.W., 2016. Climate warming reduces fish production and benthic habitat in Lake Tanganyika, one of the most biodiverse freshwater ecosystems. Proceedings of the National Academy of Sciences 113, 9563–9568.

Colten, R.H., Arnold, J.E., 1998. Prehistoric marine mammal hunting on California's northern channel islands. American Antiquity 63, 679–701.

Côté, S.D., Rooney, T.P., Tremblay, J.P., Dussault, C., Waller, D.M., 2004. Ecological impacts of deer overabundance. Annual Review of Ecology, Evolution, and Systematics 35, 113–147.

Cott, H.B., 1954. The exploitation of wild birds for their eggs. Ibis 96, 129–149.

Cowlishaw, G., Mendelson, S., Rowcliffe, J., 2005. Evidence for post-depletion sustainability in a mature bushmeat market. Journal of Applied Ecology 42, 460–468.

Crespo, E.A., Hall, M.A., 2002. Interactions between aquatic mammals and humans in the context of ecosystem management. In: Evans, P.G.H., Raga, J.A. (Eds.), Marine Mammals: Biology and Conservation. Kluwer Academic/Plenum Publishers, New York, pp. 463–490.

Cross, J.H., Belizario, V., 2007. Capillariasis. In: Murrell, D.K., Fried, B. (Eds.), Food-borne Parasitic Zoonoses: Fish and Plant-borne Parasites. Springer, US, pp. 209–234.

Dalle Zotte, A., 2014. Rabbit farming for meat purposes. Animal Frontiers 4, 62–67.

Darimont, C.T., Fox, C.H., Bryan, H.M., Reimchen, T.E., 2015. The unique ecology of human predators. Science 349, 858–860.

Davidson, M.H., Kling, D., Maki, K.C., 2011. Novel developments in omega-3 fatty acid-based strategies. Current Opinion in Lipidology 22, 437–444.

Daviglus, M., Sheeshka, J., Murkin, E., 2002. Health benefits from eating fish. Comments on Toxicology 8, 345–374.

De Heinzelin, J., Clark, J.D., White, T., Hart, W., Renne, P., WoldeGabriel, G., Beyene, Y., Vrba, E., 1999. Environment and behavior of 2.5-million-year-old Bouri hominids. Science 284, 625–629.

DeFoliart, G.R., 1997. An overview of the role of edible insects in preserving biodiversity. Ecology of Food and Nutrition 36, 109–132.

Delgado-Lista, J., Perez-Martinez, P., Lopez-Miranda, J., Perez-Jimenez, F., 2012. Long chain omega-3 fatty acids and cardiovascular disease: a systematic review. British Journal of Nutrition 107, S201–S213.

Dellinger, M.J., Ripley, M.P., 2016. Mercury risks versus nutritional benefits of tribal commercial fish harvests in the Upper Laurentian Great Lakes. Human and Ecological Risk Assessment: An International Journal 22, 1036–1049.

Dirzo, R., Young, H.S., Galetti, M., Ceballos, G., Isaac, N.J.B., Collen, B., 2014. Defaunation in the Anthropocene. Science 401, 345.

Domingo, J.L., 2016. Nutrients and chemical pollutants in fish and shellfish. Balancing health benefits and risks of regular fish consumption. Critical Reviews in Food Science and Nutrition 56, 979–988.

Dossey, A.T., 2013. Why insects should be in your diet. Scientist 27, 22–23.

Dounias, E., Froment, A., 2011. From foraging to farming among present-day forest hunter-gatherers: consequences on diet and health. International Forestry Review 13, 294–304.

Dufour, D.L., 1991. Diet and nutritional status of Ameridians: a review of the literature. Cadernos de Saúde Pública 7, 481–502.

Dugan, P.J., Barlow, C., Agostinho, A.A., Baran, E., Cada, G.F., Chen, D., Cowx, I.G., Ferguson, J.W., Jutagate, T., Mallen-Cooper, M., 2010. Fish migration, dams, and loss of ecosystem services in the Mekong basin. Ambio 39, 344–348.

Eiras, J.C., Pavanelli, G.C., Takemoto, R.M., Yamaguchi, M.U., Karkling, L.C., Nawa, Y., 2016. Potential risk of fish-borne nematode infections in humans in Brazil–Current status based on a literature review. Food and Waterborne Parasitology 5, 1–6.

El-Deir, A.C.A., Collier, C.A., Almeida Neto, M.S., Silva, K.M.S., Policarpo, I.S., Araújo, T.A.S., Alves, R.R.N., Albuquerque, U.P., Moura, G.J.B., 2012. Ichthyofauna used in traditional medicine in Brazil. Evidence-based Complementary and Alternative Medicine 2012 (ID 474716), 1–16.

Ellis, R., 2003. The Empty Ocean. Island Press/Shearwater Books, Washington.

Enghoff, H., Manno, N., Tchibozo, S., List, M., Schwarzinger, B., Schoefberger, W., Schwarzinger, C., Paoletti, M.G., 2014. Millipedes as food for humans: their nutritional and possible antimalarial value—a first report. Evidence-based Complementary and Alternative Medicine 2014, 1–10.

Englberger, L., Kuhnlein, H.V., Lorens, A., Pedrus, P., Albert, K., Currie, J., Pretrick, M., Jim, R., Kaufer, L., 2010. Pohnpei, FSM case study in a global health project documents its local food resources and successfully promotes local food for health. Pacific health dialog 16, 121–128.

Fa, J.E., Currie, D., Meeuwig, J., 2003. Bushmeat and food security in the Congo Basin: linkages between wildlife and people's future. Environmental Conservation 30, 71–78.

FAO, Food and Agriculture Organization, 2009. Global Agriculture towards 2050. Italy, FAO, Rome.

FAO, Food and Agriculture Organization, 2014a. FAO Yearbook. Fishery and Aquaculture Statistics 2012. FAO, Rome.

FAO, Food and Agriculture Organization, 2014b. The State of World Fisheries and Aquaculture 2014, Rome.

FAO, Food and Agriculture Organization, 2011. Review of the State of World Marine Fishery Resources. Fisheries and Aquaculture Technical Paper, Rome.

FAO, Food and Agriculture Organization, 2014. Statistical Year Book 2014. Latin America and the Caribbean Food and Agriculture. Food and Agriculture Organization of the United Nations Regional Office for the Latin America and the Caribbean, Santiago do Chile.

FAO, Food and Agriculture Organization, 2016. The State of World Fisheries and Aquaculture. Contributing to Food Security and Nutrition for All. Food and Agriculture Organization of the United Nations, Rome.

FAOSTAT, 2014. Food and Agriculture Organization Statistics Database.

Fernandes-Ferreira, H., Mendonça, S.V., Albano, C., Ferreira, F.S., Alves, R.R.N., 2012. Hunting, use and conservation of birds in Northeast Brazil. Biodiversity and Conservation 21 (1), 221–244.

Figueirêdo, R.E.C.R., Vasconcellos, A., Policarpo, I.S., Alves, R.R.N., 2015. Edible and medicinal termites: a global overview. Journal of Ethnobiology and Ethnomedicine 11, 1–7.

Fitter, R.S.R., 1986. Wildlife for Man: How and Why We Should Conserve Our Species. Collins, London, London.

Fitzgerald, L.A., 1994. Tupinambis lizards and people: a sustainable use approach to conservation and development. Conservation Biology 8, 12–15.

Flachowsky, G., Kamphues, J., 2012. Carbon footprints for food of animal origin: what are the most preferable criteria to measure animal yields? Animals 2, 108–126.

Flyman, M.V., Afolayan, A.J., 2006. The suitability of wild vegetables for alleviating human dietary deficiencies. South African Journal of Botany 72, 492–497.

Foley, J.A., Ramankutty, N., Brauman, K.A., Cassidy, E.S., Gerber, J.S., Johnston, M., Mueller, N.D., O'Connell, C., Ray, D.K., West, P.C., 2011. Solutions for a cultivated planet. Nature 478, 337–342.

Foster, M.S., James, S.R., 2002. Dogs, deer, or guanacos: zoomorphic figurines from Pueblo Grande, central Arizona. Journal of Field Archaeology 29, 165–176.

Frackowiak, W., Gorczyca, S., Merta, D., Wojciuch-Ploskonka, M., 2013. Factors affecting the level of damage by wild boar in farmland in north-eastern Poland. Pest Management Science 69, 362–366.

Frazão, E., Meade, B., Regmi, A., 2008. Converging patterns in global food consumption and food delivery systems. Amber Waves 6, 22–29.

Frazier, J., 2005. Traditional and cultural use of marine turtles. In: Third Meeting of the Signatory States, Bangkok.

Frota, C.C., Lima, L.N.C., Rocha, A.S., Suffys, P.N., Rolim, B.N., Rodrigues, L.C., Barreto, M.L., Kendall, C., Kerr, L.R.S., 2012. Mycobacterium leprae in six-banded (Euphractus sexcinctus) and nine-banded armadillos (Dasypus novemcinctus) in Northeast Brazil. Memórias do Instituto Oswaldo Cruz 107, 209–213.

García-Quijano, C.G., Pitchon, A., 2010. Aquatic Ethnobiology. In: Stepp, J.R. (Ed.), Ethnobiology. Encyclopedia of Life Support Systems (EOLSS). UNESCO, Oxford, UK.

Gauthier, D.T., 2015. Bacterial zoonoses of fishes: a review and appraisal of evidence for linkages between fish and human infections. The Veterinary Journal 203, 27–35.

Golden, C., Allison, E.H., Cheung, W.W.L., Dey, M.M., Halpern, B.S., McCauley, D.J., Smith, M., Vaitla, B., Zeller, D., Myers, S.S., 2016. Fall in fish catch threatens human health. Nature 534, 317–320.

Golden, C.D., Fernald, L.C.H., Brashares, J.S., Rasolofoniaina, B.J.R., Kremen, C., 2011. Benefits of wildlife consumption to child nutrition in a biodiversity hotspot. Proceedings of the National Academy of Sciences 108, 19653–19656.

Gonwouo, L.N., Rödel, M.O., 2008. The importance of frogs to the livelihood of the Bakossi people around Mount Manengouba, Cameroon, with special consideration of the Hairy Frog, Trichobatrachus robustus. Salamandra 44, 23–34.

Gonzalez, J.A., 2004. Human use and conservation of economically important birds in seasonally flooded forests of the northeastern Peruvian Amazon. In: Silvius, K., Bodmer, R., Fragoso, J. (Eds.), People in Nature: Wildlife Conservation in South and Central America. Columbia University Press, New York, pp. 344–361.

González, J.A., 1999. Effects of harvesting of waterbirds and their eggs by native people in the northeastern Peruvian Amazon. Waterbirds 217–224.

Graham, J.P., Leibler, J.H., Price, L.B., Otte, J.M., Pfeiffer, D.U., Tiensin, T., Silbergeld, E.K., 2008. The animal-human interface and infectious disease in industrial food animal production: rethinking biosecurity and biocontainment. Public Health Reports 123, 282–299.

Gray, M.C., Altman, J.C., Halasz, N., 2005. The Economic Value of Wild Resources to the Indigenous Community of the Wallis Lake Catchment. Australia National University, Center for Aboriginal Economic Policy Research Discussion Paper 272, Canberra.

Green, R.E., Cornell, S.J., Scharlemann, J.P.W., Balmford, A., 2005. Farming and the fate of wild nature. Science 307, 550–555.

Gross, D.R., 1975. Protein capture and cultural development in the Amazon basin. American Anthropologist 77, 526–549.

Grzimek, B., 2003. Grzimek's Animal Life Encyclopedia – Reptiles. Thomson – Gale.

Guttman, H., 1999. Rice field fisheries: a resource for Cambodia. Naga, The ICLARM Quarterly 22, 11–15.

Halloran, A., Vantomme, P., 2013. The Contribution of Insects to Food Security, Livelihoods and the Environment. Food and Agriculture Organization of the United Nations, Rome.

Halwart, M., 2008. Biodiversity, nutrition and livelihoods in aquatic rice-based ecosystems. Biodiversity 9, 36–40.

Hamblin, N.L., 1985. The Role of Marine Resources in the Maya Economy: A Case Study from Cozumel, Mexico, Prehistoric Lowland Maya Environment and Subsistence Economy. Harvard university, Cambridge, pp. 159–173. The Peabody museum of american archaeology and ethnology.

Hamilton, C.M., Katzer, F., Innes, E.A., Kelly, P.J., 2014. Seroprevalence of Toxoplasma gondii in small ruminants from four Caribbean islands. Parasites & Vectors 7, 1–4.

Hawkes, C., 2006. Uneven dietary development: linking the policies and processes of globalization with the nutrition transition, obesity and diet-related chronic diseases. Globalization and Health 2, 1–18.

Haynes, A.M., 1987. Human exploitation of seabirds in Jamaica. Biological Conservation 41, 99–124.

HLPE, High Level Panel of Experts, 2014. Sustainable Fisheries and Aquaculture for Food Security and Nutrition. (Rome).

Hoffman, L.C., 2008. The yield and nutritional value of meat from African ungulates, camelidae, rodents, ratites and reptiles. Meat Science 80, 94–100.

Hoffman, L.C., Cawthorn, D.M., 2012. What is the role and contribution of meat from wildlife in providing high quality protein for consumption. Animal Frontiers 2, 40–53.

Hortle, K.G., 2007. Consumption and the yield of fish and other aquatic animals from the Lower Mekong Basin. MRC Technical Paper 16, 1–88.

Hovelsrud, G.K., McKenna, M., Huntington, H.P., 2008. Marine mammal harvests and other interactions with humans. Ecological Applications 18.

Huchzermeyer, F.W., 2003. Crocodiles: Biology, Husbandry and Diseases. CABI Publishing, Wallingford.

Huet, C., Rosol, R., Egeland, G.M., 2012. The prevalence of food insecurity is high and the diet quality poor in Inuit communities. The Journal of Nutrition 142, 541–547.

Humber, F., Godley, B.J., Broderick, A.C., 2014. So excellent a fishe: a global overview of legal marine turtle fisheries. Diversity and Distributions 20, 579–590.

Igene, J., Okoro, K., Ebabhamiegbebho, P., Evivie, S., 2015. A Study Assessing Some Metal Elements Contamination Levels in Grasscutter (Thryonomys Swinderianus Temminck) Meat.

Irvine, F.R., 1954. Snakes as food for man. British Journal of Herpetology 1, 183–189.

Jeffrey, S., 1977. How Liberia uses wildlife. Oryx 14, 168–173.

Jensen, J.B., Camp, C.D., 2003. Human exploitation of amphibians: direct and indirect impacts. In: Semlitsch, R.D. (Ed.), Amphibian Conservation. Smithsonian Books, Washington DC, pp. 199–213.

Johns, T., Eyzaguirre, P.B., 2006. Linking biodiversity, diet and health in policy and practice. Proceedings of the Nutrition Society 65, 182–189.

Johns, T., Maundu, P., 2006. Forest biodiversity, nutrition and population health in market-oriented food systems. Unasylva 57, 34–40.

Jongema, Y., 2015. List of Edible Insect Species of the World.

Jonsson, M., Hedström, P., Stenroth, K., Hotchkiss, E.R., Vasconcelos, F.R., Karlsson, J., Byström, P., 2015. Climate change modifies the size structure of assemblages of emerging aquatic insects. Freshwater Biology 60, 78–88.

Jorgenson, J.P., 1998. The Impact of Hunting on Wildlife in the Maya Forest of Mexico, Timber, Tourists and Temples: Conservation and Development in the Maya Forests of Belize, Guatemala and Mexico. Island Press, Washington D.C, pp. 179–194.

Keuling, O., Baubet, E., Duscher, A., Ebert, C., Fischer, C., Monaco, A., Podgórski, T., Prevot, C., Ronnenberg, K., Sodeikat, G., 2013. Mortality rates of wild boar Sus scrofa L. in central Europe. European Journal of Wildlife Research 59, 805–814.

Keuling, O., Strauß, E., Siebert, U., 2016. Regulating wild boar populations is "somebody else's problem"!-Human dimension in wild boar management. Science of the Total Environment 554, 311–319.

Kim, S., Venkatesan, J., 2015. Introduction to seafood science. In: Kim, S. (Ed.), Seafood Science: Advances in Chemistry, Technology and Applications. CRC Press, Boca Raton, pp. 1–13.

Kirby, J.S., Stattersfield, A.J., Butchart, S.H.M., Evans, M.I., Grimmett, R.F.A., Jones, V.R., O'Sullivan, J., Tucker, G.M., Newton, I., 2008. Key conservation issues for migratory land and waterbird species on the world's major flyways. Bird Conservation International 18, S49–S73.

Klein, G., Andreoletti, O., Budka, H., Buncic, S., Colin, P., Collins, J.D., De Koeijer, A., Griffin, J., Havelaar, A., Hope, J., 2007. Public health risks involved in the human consumption of reptile meat scientific opinion of the panel on biological hazards. The EFSA Journal 578, 1–55.

Klemens, M.W., Thorbjarnarson, J.B., 1995. Reptiles as a food resource. Biodiversity and Conservation 4, 281–298.

Kurpiers, L.A., Schulte-Herbrüggen, B., Ejotre, I., Reeder, D.M., 2016. Bushmeat and emerging infectious diseases: lessons from Africa. In: Angelici, F.M. (Ed.), Problematic Wildlife: A Cross-disciplinary Approach. Springer, Switzerland, pp. 507–551.

Kyselý, R., 2008. Frogs as a part of the Eneolithic diet. Archaeozoological records from the Czech Republic (Kutná Hora-Denemark site, Rivnác culture). Journal of Archaeological Science 35, 143–157.

Lagos, L., Picos, J., Valero, E., 2012. Temporal pattern of wild ungulate-related traffic accidents in northwest Spain. European Journal of Wildlife Research 58, 661–668.

Larsen, C.S., 2003. Animal source foods and human health during evolution. The Journal of Nutrition 133, 3893–3897.

Lee-Thorp, J.A., van der Merwe, N.J., Brain, C.K., 1994. Diet of Australopithecus robustus at Swartkrans from stable carbon isotopic analysis. Journal of Human Evolution 27, 361–372.

Leibler, J.H., Otte, J., Roland-Holst, D., Pfeiffer, D.U., Magalhaes, R.S., Rushton, J., Graham, J.P., Silbergeld, E.K., 2009. Industrial food animal production and global health risks: exploring the ecosystems and economics of avian influenza. Ecohealth 6, 58–70.

Leopold, A.S., 1959. Wildlife of Mexico: The Game Birds and Mammals. Univ of California Press, Berkeley and Los Angeles.

Liverani, M., Waage, J., Barnett, T., Pfeiffer, D.U., Rushton, J., Rudge, J.W., Loevinsohn, M.E., Scoones, I., Smith, R.D., Cooper, B.S., 2013. Understanding and managing zoonotic risk in the new livestock industries. Environmental Health Perspectives 121, 873.

Loe, L.E., Rivrud, I.M., Meisingset, E.L., Bøe, S., Hamnes, M., Veiberg, V., Mysterud, A., 2016. Timing of the hunting season as a tool to redistribute harvest of migratory deer across the landscape. European Journal of Wildlife Research 62, 315–323.

Lord, V.R., Lord, R.D., 1991. Brucella suis infections in collared peccaries in Venezuela. Journal of Wildlife Diseases 27, 477–481.

Loring, P.A., Gerlach, S.C., 2009. Food, culture, and human health in Alaska: an integrative health approach to food security. Environmental Science & Policy 12, 466–478.

Lynch, A.J., Myers, B.J.E., Chu, C., Eby, L.A., Falke, J.A., Kovach, R.P., Krabbenhoft, T.J., Kwak, T.J., Lyons, J., Paukert, C.P., 2016. Climate change effects on North American inland fish populations and assemblages. Fisheries 41, 346–361.

Mamede, N.M.G., 2014. Relações ecológicas entre ouriços-do-mar e os seus predadores e presas no sudoeste de Portugal Continental. Universidade de Lisboa.

Mancini, A., Koch, V., 2009. Sea turtle consumption and black market trade in Baja California Sur, Mexico. Endangered Species Research 7, 1–10.

Manning, R., Scott, C.H., Haddad, L., 2013. Whose goals count? Lessons for setting the next development goals. IDS Bulletin 44, 1–9.

Martin, G.H.G., 1983. Bushmeat in Nigeria as a natural resource with environmental implications. Environmental Conservation 10, 125–132.

Masson, M.A., 1999. Animal resource manipulation in ritual and domestic contexts at Postclassic Maya Communities. World Archaeology 31, 93–120.

Mayor, P., Baquedano, L.E., Sanchez, E., Aramburu, J., Gomez-Puerta, L.A., Mamani, V.J., Gavidia, C.M., 2015. Polycystic Echinococcosis in Pacas, Amazon Region, Peru. Emerging Infectious Diseases 21, 456.

Mazhar, F., Buckles, D., 2007. Food Sovereignty and Uncultivated Biodiversity in South Asia: Essays on the Poverty of Food Policy and the Wealth of the Social Landscape. Academic Foundation.

McKillop, H., 1984. Prehistoric Maya reliance on marine resources: analysis of a midden from Moho Cay, Belize. Journal of Field Archaeology 11, 25–35.

McKillop, H.I., 1985. Prehistoric exploitation of the manatee in the Maya and circum-Caribbean areas. World Archaeology 16, 337–353.

Mendonça, L.E.T., Vasconcellos, A., Souto, C.M., Oliveira, T.P.R., Alves, R.R.N., 2016. Bushmeat consumption and its implications for wildlife conservation in the semi-arid region of Brazil. Regional Environmental Change 16, 1649–1657.

Mennerat, A., Nilsen, F., Ebert, D., Skorping, A., 2010. Intensive farming: evolutionary implications for parasites and pathogens. Evolutionary Biology 37, 59–67.

Mesquita, G.P., Barreto, G.P., 2015. Evaluation of mammals hunting in indigenous and rural localities in Eastern Brazilian Amazon. Ethnobiology And Conservation 4, 1–14.

Meyer-Rochow, V., 2010. Entomophagy and its impact on world cultures: the need for a multidisciplinary approach. Edible Forest Insects 23–36.

Meyer-Rochow, V.B., 2005. Traditional food insects and spiders in several ethnic groups of northeast India, Papua New Guinea, Australia and New Zealand. In: Paoletti, M.G. (Ed.), Ecological Implications of Minilivestock: Potential of Insects, Rodents, Frogs and Snails for Sustainable Development. Science Publishers Inc, Enfield, NH, pp. 389–413.

Meyer-Rochow, V.B., Chakravorty, J., 2013. Notes on entomophagy and entomotherapy generally and information on the situation in India in particular. Applied Entomology and Zoology 48, 105–112.

Mohneke, M., Onadeko, A.B., Hirschfeld, M., Rödel, M.O., 2010. Fried and dried: amphibians in local and regional food markets in West Africa. Traffic Bulletin 22, 117–128.

Mohneke, M., Onadeko, A.B., Rödel, M.O., 2009. Exploitation of frogs – a review with a focus on West Africa. Salamandra 45, 193–202.

Mohneke, M., Onadeko, A.B., Rödel, M.O., 2011. Medicinal and dietary uses of amphibians in Burkina Faso. African Journal of Herpetology 60, 78–83.

Molur, S., Smith, K.G., Daniel, B.A., Darwall, W.R.T., 2011. The status and distribution of freshwater biodiversity in the Western Ghats, India. IUCN, International Union for Conservation of Nature and Natural Resources, pp. 1–115.

Monks, G.G., 2005. The Exploitation and Cultural Importance of Sea Mammals. Oxbow Books.

Morgan, K., 2010. Local and green, global and fair: the ethical foodscape and the politics of care. Environment and Planning a 42, 1852–1867.

Morgan, K., Sonnino, R., 2010. The urban foodscape: world cities and the new food equation. Cambridge Journal of Regions, Economy and Society 3, 209–224.

Morris, B., 2004. Insects and Human Life. Berg Publishers.

Muller-Landau, H.C., 2007. Predicting the long-term effects of hunting on plant species composition and diversity in tropical forests. Biotropica 39, 372–384.

Murrieta, R.S.S., Bakri, M.S., Adams, C., Oliveira, P.S.S., Strumpf, R., 2008. Consumo alimentar e ecologia de populações ribeirinhas em dois ecossistemas amazônicos: um estudo comparativo. Revista de Nutrição 21, 123s–133s.

Nardoto, G.B., Murrieta, R.S.S., Prates, L.E.G., Adams, C., Garavello, M.E.P.E., Schor, T., Moraes, A., Rinaldi, F.D., Gragnani, J.G., Moura, E.A.F., 2011. Frozen chicken for wild fish: nutritional transition in the Brazilian Amazon region determined by carbon and nitrogen stable isotope ratios in fingernails. American Journal of Human Biology 23, 642–650.

Nascimento, D.M., Alves, A.G.C., Alves, R.R.N., Barboza, R.R.D., Diele, K., Mourão, J.S., 2016. An examination of the techniques used to capture mangrove crabs, *Ucides cordatus*, in the Mamanguape River estuary, northeastern Brazil, with implications for management. Ocean & Coastal Management 130, 50–57.

Nascimento, D.M., Alves, R.R.N., Barboza, R.R.D., Schmidt, A.J., Diele, K., Mourão, J.S., 2017. Commercial relationships between intermediaries and harvesters of the mangrove crab *Ucides cordatus* (Linnaeus, 1763) in the Mamanguape River estuary, Brazil, and their socio-ecological implications. Ecological Economics 131, 44–51.

Nascimento, D.M., Ferreira, E.N., Bezerra, D.M., Rocha, P.D., Alves, R.R.N., Mourão, J.S., 2012. Capture techniques' use of Caranguejo-uçá crabs (*Ucides cordatus*) in Paraíba state (northeastern Brazil) and its socio-environmental implications. Anais da Academia Brasileira de Ciências 84, 1051–1064.

Nasi, R., Brown, D., Wilkie, D., Bennett, E., Tutin, C., van Tol, G., Christophersen, T., 2008. Conservation and Use of Wildlife-based Resources: The Bushmeat Crisis, Technical Series N 33. Secretariat of the Convention on Biological Diversity, Montreal, and Center for International Forestry Research (CIFOR), Bogor. 50 Pages.

Nasi, R., Taber, A., van Vliet, N., 2011. Empty forests, empty stomachs? Bushmeat and livelihoods in the Congo and Amazon Basins. International Forestry Review 13, 355–368.

National Academy of Sciences, 2015. Global Considerations for Animal Agriculture Research, Critical Role of Animal Science Research in Food Security and Sustainability. The National Academies Press, Washington, D.C.

Newbold, T., Hudson, L.N., Phillips, H.R.P., Hill, S.L.L., Contu, S., Lysenko, I., Blandon, A., Butchart, S.H.M., Booth, H.L., Day, J., 2014. A global model of the response of tropical and sub-tropical forest biodiversity to anthropogenic pressures. Proceedings of the Royal Society of London B: Biological Sciences 281, 20141371.

Newing, H., 2001. Bushmeat hunting and management: implications of duiker ecology and interspecific competition. Biodiversity and Conservation 10, 99–118.

Nicol, S., Forster, I., Spence, J., 2000. Products derived from krill. In: Everson, I. (Ed.), Krill: Biology, Ecology and Fisheries. Blackwell Science, pp. 262–283.

Niekisch, M., 1986. The international trade in frogs' legs. Traffic Bulletin 8, 7–10.

Nishida, A.K., Nordi, N., Alves, R.R.N., 2006a. Mollusc gathering in northeast Brazil: an ethnoecological approach. Human Ecology 34, 133–145.

Nishida, A.K., Nordi, N., Alves, R.R.N., 2006b. The lunar-tide cycle viewed by crustacean and mollusc gatherers in the State of Paraíba, Northeast Brazil and their influence in collection attitudes. Journal of Ethnobiology and Ethnomedicine 2, 1–12.

Nogueira, M., Alves, R.R.N., 2016. Assessing sea turtle bycatch in Northeast Brazil through an ethnozoological approach. Ocean & Coastal Management 133, 37–42.

Nordi, N., Nishida, A.K., Alves, R.R.N., 2009. Effectiveness of two gathering techniques for *Ucides cordatus* in northeast Brazil: implications for the sustainability of mangrove ecosystems. Human Ecology 37, 121–127.

Ntiamoa-Baidu, Y., 1997. Wildlife and Food Security in Africa. Food & Agriculture Organization of the UN (FAO).

Odebode, A.V., Awe, F., Famuyide, O.O., Adebayo, O., Ojo, O.B., Daniel, G., 2011. Households' consumption patterns of grasscutter (Thryonomys swinderianus) meat within Ibadan Metropolis, Oyo state, Nigeria. Continental Journal of Food Science and Technology 5, 49–57.

Ojasti, J., 1997. Wildlife Utilization in Latin America: Current Situation and Prospects for Sustainable Management. Food and Agriculture Organization of the United Nations, Rome.

Ojasti, J., 2000. Manejo de Fauna Silvestre Neotropical. In: Dallmeier, F. (Ed.), SI/MAB Series. Smithsonian Institution/MAB Biodiversity Program, Washington D.C, p. 290.

Olukoshi, A., 2013. The MDGs, empowerment and accountability in Africa: retrospect and prospects. IDS Bulletin 44, 14–21.

Ooninex, D.G.A.B., 2015. Insects as Food and Feed: Nutrient Composition and Environmental Impact. Wageningen University.

Osofsky, S., 2014. How to Keep Viruses in the Wild from Finding Humans. CNN.

Pandian, T.J., Marian, M.P., 1986. Production and utilization of frogs: an ecological view. Proceedings: Animal Sciences 95, 289–301.

Patel, T., 1993. French may eat Indonesia out of frogs. New Scientist 1868, 7.

Pauwels, O.S.G., Rodel, M.O., Toham, A.K., 2003. Leptopelis notatus (Anura: Hyperoliidae) in the Massif du Chaillu, Gabon: from ethnic wars to soccer. Hamadryad 27, 271–273.

Peres, C.A., 2000. Effects of subsistence hunting on vertebrate community structure in Amazonian forests. Conservation Biology 14, 240–253.

Perry, B.D., Grace, D., Sones, K., 2013. Current drivers and future directions of global livestock disease dynamics. Proceedings of the National Academy of Sciences 110, 20871–20877.

Pezzuti, J.C.B., Silva, D.F., Rebelo, G.H., Lima, J.P., 2008. A captura de quelônios no Parque Nacional do Jaú, Amazonas. In: Silva, F.P.C., Gomes-Silva, D.A., Melo, J.S., Nascimento, V.M. (Eds.), Coletânea de Textos: manejo e monitoramento de fauna silvestre em florestas tropicais Belém, pp. 150–156.

Pierret, P.V., Dourojeanni, M.J., 1967. Importancia de la Caza para Alimentación Humana en el Curso Inferior del Río Ucayali. Perú. Revista Forestal del Perú 1, 10–21.

Pilgrim, S.E., Cullen, L.C., Smith, D.J., Pretty, J., 2008. Ecological knowledge is lost in wealthier communities and countries. Environmental Science & Technology 42, 1004–1009.

Piperata, B.A., Spence, J.E., Da-Gloria, P., Hubbe, M., 2011. The nutrition transition in Amazonia: rapid economic change and its impact on growth and development in Ribeirinhos. American Journal of Physical Anthropology 146, 1–13.

Pohl, M., 1981. Ritual continuity and transformation in Mesoamerica: reconstructing the ancient Maya cuch ritual. American Antiquity 46, 513–529.

Pooley, S., Fa, J.E., Nasi, R., 2015. No conservation silver lining to Ebola. Conservation Biology 29, 965–967.

Popkin, B.M., 1993. Nutritional patterns and transitions. Population and Development Review 19, 138–157.

Popkin, B.M., 1994. The nutrition transition in low-income countries: an emerging crisis. Nutrition Reviews 52, 285–298.

Popkin, B.M., 2003. The nutrition transition in the developing world. Development Policy Review 21, 581–597.

Porcasi, J.F., Fujita, H., 2000. The dolphin hunters: a specialized prehistoric maritime adaptation in the southern California Channel Islands and Baja California. American Antiquity 65, 543–566.

Poulsen, J.R., Clark, C.J., Palmer, T.M., 2013. Ecological erosion of an Afrotropical forest and potential consequences for tree recruitment and forest biomass. Biological Conservation 163, 122–130.

Pretty, J., 2008. Agricultural sustainability: concepts, principles and evidence. Philosophical Transactions of the Royal Society of London B: Biological Sciences 363, 447–465.

Pretty, J.N., 2002. Agri-culture: Reconnecting People, Land, and Nature. Routledge.

Ramankutty, N., Foley, J.A., Norman, J., McSweeney, K., 2002. The global distribution of cultivable lands: current patterns and sensitivity to possible climate change. Global Ecology and Biogeography 11, 377–392.

Rask, M., Mannio, J., Forsius, M., Posch, M., Vuorinen, P.J., 1995. How many fish populations in Finland are affected by acid precipitation? Environmental Biology of Fishes 42, 51–63.

Rathore, M., 2009. Nutrient content of important fruit trees from arid zone of Rajasthan. Journal of Horticulture and Forestry 1, 103–108.

Rebêlo, G.H., Pezzuti, J.C.B., Lugli, L., Moreira, G., 2006. Pesca Artesanal de Quelônios no Pesca Artesanal de Quelônios no Pesca Artesanal de Quelônios no Parque Nacional do Jaú (AM) arque Nacional do Jaú (AM). Boletim Do Museu Paraense Emílio Göeldi 1.

Redford, K.H., 1992. The empty forest. BioScience 42, 412–422.

Redford, K.H., Robinson, J.G., 1987. The game of choice: patterns of indian and colonist hunting in the Neotropics. American Anthropologist 89, 650–667.

Reitz, E.J., Wing, E.S., 2008. Zooarchaeology. Cambridge Univ Pr.

Rick, T.C., Braje, T.J., DeLong, R.L., 2011. People, pinnipeds, and sea otters of the Northeast Pacific, human impacts on seals, sea lions, and sea otters: integrating archaeology and ecology in the Northeast Pacific. Univ of California Pr 1–17.

Ripple, J., Perrine, D., 1999. Manatees and Dugongs of the World. Voyageur Press.

Robinson, J.G., Redford, K.H., 1991. Neotropical Wildlife Use and Conservation. University of Chicago Press, Chicago.

Rodríguez, D.I., Anríquez, G., Riveros, J.L., 2016. Food security and livestock: the case of Latin America and the Caribbean. Ciencia e Investigación Agraria 43, 5–15.

Roland, G., 2006. Atchafalaya Houseboat: My Years in the Louisiana Swamp. Louisiana State University Press, Baton Rouge, Louisiana.

Romero, A., Baker, R., Cresswell, J.E., Singh, A., McKie, A., Manna, M., 2002. Environmental history of marine mammal exploitation in Trinidad and Tobago, WI, and its ecological impact. Environment and History 8, 255–274.

Romero, A., Creswell, J., 2005. In the land of the mermaid: how culture, not ecology, influenced marine mammal exploitation in the Southeastern Caribbean. In: Romero, A., West, S.E. (Eds.), Environmental Issues in Latin America and the Caribbean. Springer, Dordrecht, pp. 3–30.

Rönnbäck, P., 1999. The ecological basis for economic value of seafood production supported by mangrove ecosystems. Ecological Economics 29, 235–252.

Roth, H.H., Merz, G., 1997. Wildlife Resources: A Global Account of Economic Use. Springer Science & Business Media.

Rudrud, R.W., Kroeker, J.W., Leslie, H.Y., Finney, S.S., 2007. The sea turtle wars: culture, war and sea turtles in the Republic of the Marshall Islands. SPC Traditional Marine Resource Management and Knowledge Information Bulletin 21, 3–29.

Ruel, M.T., 2003. Operationalizing dietary diversity: a review of measurement issues and research priorities. The Journal of Nutrition 133, 3911S–3926S.

Saito, M., Koike, F., Momose, H., Mihira, T., Uematsu, S., Ohtani, T., Sekiyama, K., 2012. Forecasting the range expansion of a recolonising wild boar Sus scrofa population. Wildlife Biology 18, 383–392.

Sakurai, R., Jacobson, S.K., Ueda, G., 2014. Public perceptions of significant wildlife in Hyogo, Japan. Human Dimensions of Wildlife 19, 88–95.

Samson, C., Pretty, J., 2006. Environmental and health benefits of hunting lifestyles and diets for the Innu of Labrador. Food Policy 31, 528–553.

Sarti, F.M., Adams, C., Morsello, C., Van Vliet, N., Schor, T., Yagüe, B., Tellez, L., Quiceno-Mesa, M.P., Cruz, D., 2015. Beyond protein intake: bushmeat as source of micronutrients in the Amazon. Ecology and Society 20, 1–15.

Scammon, C.M., 1968. The Marine Mammals of the Northwestern Coast of North America. Originally Published in 1874. Dover, New, York.

Schenck, M., Nsame Effa, E., Starkey, M., Wilkie, D., Abernethy, K., Telfer, P., Godoy, R., Treves, A., 2006. Why people eat bushmeat: results from two-choice, taste tests in Gabon, central Africa. Human Ecology 34, 433–445.

Schley, L., Dufrêne, M., Krier, A., Frantz, A.C., 2008. Patterns of crop damage by wild boar (Sus scrofa) in Luxembourg over a 10-year period. European Journal of Wildlife Research 54, 589–599.

Scoones, I., Melnyk, M., Pretty, J.N., 1992. The Hidden Harvest: Wild Foods and Agricultural Systems. A Literature Review and Annotated Bibliography. Sustainable Agriculture Programme. International Institute for Environment and Development, London.

Sheaves, M., Baker, R., Abrantes, K.G., Connolly, R.M., 2016. Fish biomass in tropical estuaries: substantial variation in food web structure, sources of nutrition and ecosystem-supporting processes. Estuaries and Coasts 1–14.

Shipman, P., 1986. Scavenging or hunting in early hominids: theoretical framework and tests. American Anthropologist 88, 27–43.

Shockley, M., Dossey, A.T., 2014. Insects for human consumption. In: Morales-Ramos, J., Rojas, G., Shapiro-Ilan, D.I. (Eds.), Mass Production of Beneficial Organisms: Invertebrates and Entomopathogens. Elsevier, Oxford, pp. 617–652.

Sillen, A., 1992. Strontium-calcium ratios (Sr/Ca) of Australopithecus robustus and associated fauna from Swartkrans. Journal of Human Evolution 23, 495–516.

Silva, H., Padez, C., 2010. Body size and obesity patterns in Caboclo populations from Pará, Amazonia, Brazil. Annals of Human Biology 37, 218–230.

Sinclair, A.R.E., Byrom, A.E., 2006. Understanding ecosystem dynamics for conservation of biota. Journal of Animal Ecology 75, 64–79.

Sinclair, A.R.E., Fryxell, J.M., Caughley, G., 2006. Wildlife Ecology and Management. Blackwell Scientific Publications, Malden.

Sirén, A., Machoa, J., 2008. Fish, wildlife, and human nutrition in tropical forests: a fat gap? Interciencia 33, 186.

Smith, N.J.H., 1981. Caimans, Capybaras, otters, manatees, and man in amazonia. Biological Conservation 19, 177–187.

Soares, M.C.P., Nunes, H.M., Silveira, F.A.A., Alves, M.M., Souza, A.J.S., 2011. Capillaria hepatica (Bancroft, 1893) (Nematoda) entre populações indígenas e mamíferos silvestres no noroeste do Estado do Mato Grosso, Brasil, 2000. Revista Pan-Amazônica de Saúde 2, 35–40.

Spanier, E., Lavalli, K.L., Goldstein, J.S., Groeneveld, J.C., Jordaan, G.L., Jones, C.M., Phillips, B.F., Bianchini, M.L., Kibler, R.D., Díaz, D., 2015. A concise review of lobster utilization by worldwide human populations from prehistory to the modern era. ICES Journal of Marine Science 72, i7–i21.

Speth, J.D., Spielmann, K.A., 1983. Energy source, protein metabolism, and hunter-gatherer subsistence strategies. Journal of Anthropological Archaeology 2, 1–31.

Srivastava, S.K., Babu, N., Pandey, H., 2009. Traditional insect bioprospecting-As human food and medicine. Indian Journal of Traditional Knowledge 8, 485–494.

Stanford, C.B., 1995. Chimpanzee hunting behavior and human evolution. American Scientist 83, 256–261.

Suzuki, T., Shibata, N., 1990. The utilization of Antarctic krill for human food. Food Reviews International 6, 119–147.

Thorbjarnarson, J.B., Lagueux, C.J., Bolze, D., Klemens, M.W., Meylan, A.B., 2000. Human use of turtle: a worldwide perspective. In: Klemens, M.W. (Ed.), Turtle Conservation. Smithsonian Institution Press, Washington and London, pp. 33–84.

Thornton, P.K., 2010. Livestock production: recent trends, future prospects. Philosophical Transactions of the Royal Society B: Biological Sciences 365, 2853–2867.

Titus, K., Haynes, T.L., Paragi, T.F., 2009. The importance of Moose, Caribou, deer and small game in the diet of Alaskans. In: Watson, R.T., Fuller, M., Pokras, M., Hunt, W.G. (Eds.), Ingestion of Lead from Spent Ammunition: Implications for Wildlife and Humans. The Peregrine Fund, Boise, Idaho, USA, pp. 137–143.

Toussaint-Samat, M., 2009. A History of Food. John Wiley & Sons.

Trail, P.W., 2007. African hornbills: keystone species threatened by habitat loss, hunting and international trade. Ostrich-journal of African Ornithology 78, 609–613.

UN General Assembly, 2015. Transforming Our World: The 2030 Agenda for Sustainable Development. A/70/L.1. Seventieth Session, Agenda Items 15 and 116.

UNDP, 2014. Human Development Report. UNDP, New York, USA.

Vallega, A., 2013. Fundamentals of Integrated Coastal Management. Springer Science & Business Media.

van Dijk, P.P., Stuart, B.L., Rhodin, A.G.J., 2000. Asian Turtle Trade. Chelonian Research Foundation, Lunenburg, Massachusetts.

van Vliet, N., 2011. Livelihood Alternatives for the Unsustainable Use of Bushmeat, p. 46 Report prepared for the CBD Bushmeat Liaison Group. Technical Series. Secretariat of the Convention on Biological Diversity, Montreal.

van Vliet, N., Mesa, M.P.Q., Cruz-Antia, D., Aquino, L.J.N., Moreno, J., Nasi, R., 2014. The uncovered volumes of bushmeat commercialized in the Amazonian trifrontier between Colombia, Peru & Brazil. Ethnobiology and Conservation 3, 1–11.

van Vliet, N., Moreno, J., Gómez, J., Zhou, W., Fa, J.E., Golden, C., Alves, R.R.N., Nasi, R., 2017. Bushmeat and human health: assessing the evidence in tropical and subtropical forests. Ethnobiology and Conservation 6, 1–45.

van Vliet, N., Quiceno-Mesa, M.P., Cruz-Antia, D., Tellez, L., Martins, C., Haiden, E., Oliveira, M.R., Adams, C., Morsello, C., Valencia, L., 2015. From fish and bushmeat to chicken nuggets: the nutrition transition in a continuum from rural to urban settings in the Tri frontier Amazon region. Ethnobiology and Conservation 4.

Vickers, W.T., 1991. Hunting yields and game composition over ten years in an Amazonian village. In: Robinson, J.G., Redford, K.H. (Eds.), Neotropical Wildlife Use and Conservation. University of Chicago Press, Chicago, pp. 53–81.

Vidal, O., 1993. Aquatic mammal conservation in Latin America: problems and perspectives. Conservation Biology 788–795.

Vincetti, B., Eyzaguirre, P., Johns, T., 2008. The nutritional role of forest plant foods for rural communities. In: Coler, C.J.P. (Ed.), Human Health and Forests: A Global Overview of Issues, Practice and Policy. Earthscan, London, UK, pp. 63–96.

Vörösmarty, C.J., McIntyre, P.B., Gessner, M.O., Dudgeon, D., Prusevich, A., Green, P., Glidden, S., Bunn, S.E., Sullivan, C.A., Liermann, C.R., 2010. Global threats to human water security and river biodiversity. Nature 467, 555–561.

Walters, B.B., Rönnbäck, P., Kovacs, J.M., Crona, B., Hussain, S.A., Badola, R., Primavera, J.H., Barbier, E., Dahdouh-Guebas, F., 2008. Ethnobiology, socio-economics and management of mangrove forests: a review. Aquatic Botany 89, 220–236.

Warkentin, I.G., Bickford, D., Sodhi, N.S., Bradshaw, C.J.A., 2009. Eating frogs to extinction. Conservation Biology 23, 1056–1059.

Weinbaum, K.Z., Brashares, J.S., Golden, C.D., Getz, W.M., 2013. Searching for sustainability: are assessments of wildlife harvests behind the times? Ecology Letters 16, 99–111.

Werner, D.I., 1991. The rational use of green iguanas. In: Robinson, J.G., Redford, K.H. (Eds.), Neotropical Wildlife Use and Conservation. The University of Chicago Press, Chicago, pp. 181–201.

Whytock, R.C., Buij, R., Virani, M.Z., Morgan, B.J., 2016. Do large birds experience previously undetected levels of hunting pressure in the forests of Central and West Africa? Oryx 50, 76–83.

Yen, A.L., 2015. Can edible insects help alleviate the bushmeat crisis? Journal of Insects as Food and Feed 1, 169–170.

Yesner, D.R., 1987. Life in the "Garden of Eden": constraints of marine diets for human societies. In: Harris, M., Ross, E. (Eds.), Food and Evolution. Temple University Press, Philadelphia, pp. 285–310.

Yesner, D.R., 1988. Effects of prehistoric human exploitation on Aleutian sea mammal populations. Arctic Anthropology 28–43.

Young, R., 2014. Take Bushmeat off the Menu before Humans Are Served Another Ebola.

Zhou, Z., Jiang, Z., 2004. International trade status and crisis for snake species in China. Conservation Biology 18, 1386–1394.

CHAPTER

11

Insects as Human Food

Arnold van Huis

Wageningen University, Wageningen, The Netherlands

INTRODUCTION

The eating of insects is not very common in Western countries, while in the tropics insects are often a regular part of the diet. By consulting the scientific literature on insects as food and feed worldwide, Jongema (2015) listed more than 2000 different arthropod species. They belong to the following groups: Coleoptera (beetles, often the larvae) (31%), Lepidoptera (caterpillars) (17%), Hymenoptera (wasps, bees, and ants) (15%), Orthoptera (crickets, grasshoppers, and locusts) (14%), Hemiptera (true bugs) (11%), Isoptera (termites) (3%), Odonata (dragonflies), Diptera (flies), and others (9%).

Why is the eating of insects largely limited to tropical countries? There are several reasons. First, edible insect species are available throughout the year, while in temperatezones during the winter period they are in a resting stage, called diapause or quiescence. Second, insect species in tropical zones are larger probably because insects respire through a complex network of tubes (called the tracheal system) that delivers oxygen-containing air to every cell of the body. With higher temperature the diffusion of oxygen is faster, shown by the gigantic insect species during the Permian era (about 290 million to 250 million years ago); besides oxygen levels were much higher than at present (Harrison et al., 2010). Third, tropical insect species are often aggregated facilitating harvesting, such as caterpillars of *Imbrasia* spp. or locust swarms. Also the future queens and kings of termites (the reproductives) swarm at the first rains after the dry season (their nuptial flights). Four, people are closer into contact with nature, because the weather permits people to live outdoors, although urbanization in tropical countries is countering this effect. In Africa this will increase from 40% in 2014 to 56% in 2050 and in Asia during the same period from 48% to 64% (UN, 2014). Five, Westerners do not value insects very much and often regard them as dirty, disgusting, and dangerous (Looy et al., 2014). The main argument however is that it is easier to harvest a meal of insects from nature in warmer regions than in temperate zones.

Another explanation why insects were never domesticated in the West was given by DeFoliart (1999). He stated that historically insects were

Ethnozoology
http://dx.doi.org/10.1016/B978-0-12-809913-1.00011-9

not competitive as food items. Agriculture originated mainly in the fertile crescent of the Middle East, where crops such as wheat, barley, and several legumes were first grown and animals such as sheep, goats, pigs, and cattle were first domesticated (Diamond, 1997). Insects in the Middle East were of minor importance.

A number of authors have given an overview of the eating of insects in different continents in particular (Bergier, 1941; Bodenheimer, 1951), and DeFoliart (2012) in his online bibliography. Evans et al. (2015) explain the term "entomophagy" in a historical context. They indicated that the term was mainly used by people who do not eat insects, denoting a peculiar eating habit from other cultures. An example is the title of the book of Bergier (1941) "*Entomophagous People and Edible Insects*" (translated from French). A historical overview of edible insects was also given by Costa-Neto and Dunkel (2016).

An overview of the practices of harvesting and eating insects in different parts of the world will be given. For early hominids, it was probably an important part of the diet. I will indicate the difficulty of the taxonomic identification of species as often vernacular names are used. Some examples are given of how important insects are in the diet and pharmacopeia of people. There are taboos which prevent people from eating insects. I will give examples of ingenious harvesting practices showing people's intricate knowledge about the biology and ecology of insect species. This knowledge about eating insects now seems to benefit Western countries as it is increasingly realized that using insects either as human food or animal feed offers considerable environmental benefits.

HISTORY

McGrew (2001) showed that in early hominid diets, insects were the only kind of invertebrate food eaten by primates across species from tree shrews to humans as they are nutritious and abundantly present. Four of the main five insect groups eaten (Coleoptera, Lepidoptera, Hymenoptera, Orthoptera) (McGrew, 2001) were the same as the grouping made by Jongema (2015), who, however, listed the fifth group Isoptera (termites) in the sixth place. Backwell and d'Errico (2001) gave evidence that foraging for termites as food was practised by early hominids for nearly a million of years in South Africa. They did so by studying the wear patterns on bones used as a tool by *Australopithecus robustus* to dig termites from mounds.

Madsen and Kirkman (1988) concluded from ethnographic, ethnohistoric, and archeological data that insects may have been major components of meals in the Great Basin in the United States. They studied the lakeside cave at the margin of the Great Salt Lake. Apparently during plagues, probably occurring at least once every 2 years, the grasshopper *Melanoplus sanguinipes* (Orthoptera: *Acrididae*) had flown or was blown into the lake, resulting in salted and sun-dried grasshoppers, tens of kilometers along the beach. The densities were estimated to range between 1800 and 24,000 per meter. Deposits, such as coprolites in the lakeside cave, suggested that 4500 years ago, these grasshoppers were collected on the beach, winnowed in the cave, and consumed. Because of the salt, the grasshoppers were naturally preserved, and as such could be eaten for several years. Prehistoric human coprolites with chitinous exoskeletons of insects were also found in other places of the United States (Arizona, Arkansas, Colorado, Texas, Kentucky, Missouri, Nevada, Utah) and in Mexico and Peru (Brothwell and Brothwell, 1998, pp. 67–72; Reinhard and Bryant, 1992). However, these may not have been all dietary components, as there is a possibility that they were remains of coprophagous insects or that insects were eaten accidentally. However, the amount present in coprolites suggest that they were eaten as food.

Silk was produced in China around the fourth or the fifth century BC (Vainker, 2004). However, it is not sure whether at the time the cocoons were already eaten. Mitsuhashi (2008) mentions that in the Shanxi province in China, cocoons of the wild silkworm, *Theophila religiosae* (Lepidoptera: *Bombycidae*), were found 2000 to 2500 years ago with large holes in them suggesting that they were eaten. It is likely because the cocoons are put in hot water to prevent the moth from escaping and ruining the silk, and so the cooked pupae could have been eaten. So, the silkworm served two purposes: sericulture and food. Such as double production purpose may also be the case for honey, which is often eaten together with the bee larvae (Ghosh et al., 2016; Hocking and Matsumura, 1960).

According to Sutton (1995), anthropologists seem often unaware of the important role of insects in food history, and the reasons are the Western bias against insects (DeFoliart, 2012), the generally low "visibility" of insects in archeology (Madsen and Kirkman, 1988), and the overemphasis on mammals and hunters (Kornfeld et al., 1996).

NOMENCLATURE OF EDIBLE INSECTS

One of the difficulties in knowing which insects we are dealing with are the vernacular names. For example, ants of the genus *Atta* and *Acromyrmex* (Hymenoptera: *Formicidae*) in Peruvian forests have up to 12 different names depending on the locality, their instar, or the castes (Dourojeanni, 1965). A very clear example about the confusion caused by common names is the term "white ants" which is very often used for termites, while ants (Hymenoptera) are from a complete different order than termites (Isoptera). This is also exemplified by the book *Soul of the white ant* (Marais, 1973), which actually deals with termites. In New Zealand, The Maori called the larvae of a large hawkmoth *Agrius* (=*Sphinx*) *convolvuli* (Lepidoptera: *Sphingidae*) "Anuhe" (Miller, 1952). This is a foliar pest of the sweet potato and has 20 other synonyms. The same name was, however, used for the edible larvae of the ghost moth *Aenetus* (=*Charagia*) *virescens* (Lepidoptera: *Hepialidae*) which lives for the first 5 to 6 years in tree trunks.

Yen et al. (1997) did a study in Central Australia on the use of vernacular names of insects with the aboriginal group Anangu, who speak Pitjantjatjara. They have no general name for insects or invertebrates. Edible grubs are called "Maku." However, it depends on which part of the host plant they are found. Grubs from the river red gum, *Eucalyptus camaldulensis*, have different "Maku" names depending on whether they are derived from the roots or the trunk or the branches. Moreover, there are at least 24 other Maku names depending from which host plant they are harvested. In Australia the term "witchetty grub" or "bardi grub" can denote a large, white, wood-eating larvae of particularly the cossid moth *Endoxyla leucomochla* (Lepidoptera: *Cossidae*), which feeds on the roots of the witchetty bush. However, the term may also apply to larvae of other cossid moths, ghost moths (*Hepialidae*), and longhorn beetles (*Cerambycidae*) (Wikipedia, 2016).

Costa Neto and Ramos-Elorduy (2006) list a considerable number of edible insect species from Brazil, but they are also confronted with the difficulty of converting vernacular names to scientific ones.

In the Sahelian region many grasshopper species are eaten (Van Huis, 2003). De Groot (1995) found that by showing pinned preserved grasshopper specimens to subsistence millet farmers in Niger that women knew more grasshopper species by their vernacular names than men because they had to harvest and cook them. The Mofu-Gudur is a Sahelian ethnic group in the extreme north of Cameroon. They could identify 65 different grasshopper species with local names: large ones 15 (of which 13 are edible); small ones 41 (36 edible); not classified 9; mantids 8 (5 edible) (Barreteau, 1999). This means that almost all species are considered edible. If not, the name often reveals it, for example, *Humbe tenuicornis* (Orthoptera: *Acrididae*) meaning "the one that pierces the neck." The names reflect the following: (1) whether "they are males or females (the nutritious females are more appreciated); (2) their life cycle, distinguishing larvae from adults; (3) their host plant or host tree; (4) characteristics, for example, from a culinary point of view (fat, meager, or toxic such as *Zonocerus variegatus* (Orthoptera: *Pyrgomorphidae*) although this species is eaten as well), color, form, brightness (light from white millet, or reddish from red millet); (5) environment, for example, desert or riverine, or dry and rainy season; (6) others, for example, a species that when eaten or even touched will cause swollen breasts and cut milk delivery. In central Mexico, the Pjiekakjoo name the invertebrates according to physical characteristics such as similitude with animals, fruits, and tools; properties when cooked; and hardness, ethology, and ecological traits (Aldasoro Maya and Gómez, 2016).

FIGURE 11.1 Edible insects with the bamboo caterpillar, *Omphisa fuscidentalis* (Lepidoptera: *Crambidae*), in the front, Tlat Dong Makkhai Market near Vientiane, Lao People's Democratic Republic. *Photo: Arnold van Huis.*

IMPORTANCE OF EDIBLE INSECTS IN DIETS

Most insect species that are collected from nature are seasonally available as they depend on the host plant or host tree. That is why on the diverse vegetation in the mountainous regions of the Lao People's Democratic Republic (Lao PDR) a broader variety of insects are available than on the Vientiane plain (Nonaka et al., 2008). In the northeast of Lao PDR, especially in the rainy season of June and July many species of insects can be harvested, such as termites, crickets, and beetles, while in the dry season only bamboo caterpillars (Figs. 11.1 and 11.2) and dragonflies are available (Hanboonsong and Durst, 2014).

FIGURE 11.2 Vegetables packed with the bamboo caterpillar, *Omphisa fuscidentalis* (Lepidoptera: *Crambidae*), Tlat Dong Makkhai Market near Vientiane, Lao People's Democratic Republic. *Photo: Arnold van Huis.*

Some insects species in Lao PDR are available throughout the year, and these are often aquatic ones, such as the giant water bug, *Lethocerus indicus* (Hemiptera: *Belostomatidae*), (Figs. 11.3 and 11.4) or the water scorpion *Laccothrephes* sp. (Hemiptera: *Nepidae*) (Nonaka et al., 2008). The Asian weaver ant *Oecophylla*

FIGURE 11.3 Overview of edible insects for sale with the giant waterbug, *Lethocerus indicus* (Hemiptera: *Belostomatidae*), on the left-hand side, Tlat Dong Makkhai Market near Vientiane, Lao People's Democratic Republic. *Photo: Arnold van Huis.*

FIGURE 11.4 Giant waterbug, *Lethocerus indicus* (Hemiptera: *Belostomatidae*), Klong Toey Market, Bangkok, Thailand. *Photo: Arnold van Huis.*

FIGURE 11.5 Pupae of the weaver ant, *Oecophylla smaragdina* (Hymenoptera: *Formicidae*) on a heap of ice, Klong Toey Market, Bangkok, Thailand. *Photo: Arnold van Huis.*

smaragdina (Hymenoptera: *Formicidae*) is one of the most favored edible insect species in the Lao PDR and Thailand and what is eaten is the queen brood (large larvae and pupae) (Fig. 11.5). The queen brood is collected from February to April, and erroneously called "ant eggs" indicating that they are not aware that the larvae and pupae develop into future queens (Van Itterbeeck et al., 2014).

In the Sahelian region certain grasshopper species are not eaten during the rainy season as they have a bad taste, but are appreciated during the cool part of the dry season when the harmattan is blowing (Seignobos et al., 1996).

It is difficult to indicate how important insects are as food in tropical regions, because the food source is seasonal. In Africa, in some parts of the Central African Republic it was estimated to be 15% of the meat diet (Roulon-Doko, 1998) and in some parts of the Democratic Republic of Congo 10% (Gomez et al., 1961); see also Van Huis (2003). In the Delta region in Nigeria, 29% of the respondents consumed termites, 24% palm weevils, 18% crickets, and 11% grasshoppers (Okore et al., 2014). A survey among student of tertiary institutions in southwestern Nigeria revealed that 58% consumed termites, 36% *Anapleptes trifaciata* (Coleoptera: *Scarabeidae*) and 33% the palm weevil *Rhynchophorus phoenicis* (Coleoptera: *Curculionidae*) (Lawal et al., 2010).

At least 32 Amerindian groups in the Amazon base use terrestrial invertebrates as food (Paoletti et al., 2000a,b). For example, the Guajibo during the rainy season (July/August) derive over 60% of their animal protein from insects, especially grasshoppers and the palm weevil, *Rhynchophorous palmarum* (Coleoptera: *Curculionidae*) (Paoletti et al., 2000a,b). The most important insects eaten by the Tukanoan Indians in the northwest Amazon are palm weevils, ants (*Atta* spp.), termites (*Syntermes* sp.), and noctuid and saturniid caterpillars, providing up to 12% of the crude protein derived from animal foods in men's diets and 26% in women's diets during one season of the year (Dufour, 1987).

TOTEMS AND TABOOS

Indigenous people respect sacred forests, taboos, totems and ancestral spirits, and in this way are their own custodians and legislators of environmental management (Meyer-Rochow, 2009). Offending those is believed to cause illness, death, drought, and disappearance. For example, the Kaingang in the Amazon associate ants with spirits of their ancestors and therefore do not kill ants (Posey, 2002). However, education, Christianity, population growth, colonial policies, infrastructure development, and entrepreneurship are factors that make disappear the fear-of-nature in rural areas and erode sustainable environmental management practices (Dzerefos et al., 2013). For the Kalanga people of the Zaka district in Masvingo, Zimbabwe, edible insects are a source of food and income for the community. A rain-making ceremony is essential to bring vital water for the development of the edible stinkbug, *Encosternum delegorguei* (Hemiptera: *Tessaratomidae*), edible caterpillars, and wild fruits (Risiro et al., 2013). After a harvesting ceremony, the community start to collect the edible insects, and this ensures that insects are harvested only when mature and in abundance. This system protects the system from overharvesting and extinction of the insects. A particular clan is left to oversee the harvesting of the insects, thus avoiding overexploitation.

At certain times of the year the grasshopper *Ruspola differens* (Orthoptera: *Tettigoniidae*), locally called "*nsenene*" is found in large numbers in Uganda and the Baganda consider it a great delicacy. Roscoe (1965, pp. 144–145) mentions that before anyone may eat the first meal of the season, a male member of the grasshopper clan must jump over his wife, or have sexual intercourse with her; otherwise some members of the family would fall ill. The ceremony took place in order that other clans might eat freely of the grasshopper.

A lot of indigenous societies have totem groups. These are groups within tribes, which regard an object, often an animal, as their ancestor (Bodenheimer, 1951; 11, pp. 76–82). Therefore, the eating of the totem animal is forbidden to all members of the totem clan, or only to a very moderate consumption during ceremonies. So, in Australia there are totems for several insect species such as the witchetty grub totem of the Arunta tribe. Once a year a ceremony would take place in order to increase the abundance of the grubs, and only after the ceremony other totem groups are allowed to eat the insect.

The African palm weevil *R. phoenicis* invades palms that have been damaged by other insects such as the African rhinoceros beetle *Oryctes monoceros* (Coleoptera: *Scarabaeidae*) or by human activity such as tapping palm wine. Cut trees also serve as breeding sites. Children in Kwara State of Nigeria are discouraged to eat the larvae (Fasoranti and Ajiboye, 1993). One of the reasons mentioned is that children may get drunk (referring to the palm wine). This may be not just an excuse, as Dounias (2003) mentions that the larvae have a palm wine taste. However, there is also an economic reason (Fasoranti and Ajiboye, 1993): as the larvae are very tasty, it may encourage children to cut the trees depriving the community of palm oil, palm kernels, and palm wine. The same authors mention that members of the Ire clan (the majority are blacksmiths) of the Yoruba tribe in Nigeria do not eat crickets, because they worship the Iron god Ogun, and this god does not accept animals that have no blood.

The larvae of the silkworm *Anaphe infracta* (Lepidoptera: *Thaumetopoeidae*) feed on several trees in central, eastern, and southern Africa. However, the larvae are not so much appreciated as food. It is when the larvae form the communal cocoon and pupate that they are most appreciated. It is even possible to collect the cocoon and store it at home for future use. Both larvae and pupae from the cocoon are eaten. However, in Zambia, with certain ethnic groups it is believed that pregnant women will not be able to deliver the child when they eat the insect. The explanation is that the child is confined to the womb as the larvae to the cocoon (Silow, 1976, p. 122).

Christianity or maybe other more Western beliefs may have an influence of what people can and cannot eat. For example, Ndlovu (2015) mentions "*Insects and animals are used by all Zimbabwean cultures as food and it is from these organisms that the people derive meat foods. Christianity and tradition have prescribed some of these as uneatable and as a result some people may starve amidst some insects or animals that they are prohibited to eat.*" Silow (1976, p. 212) mentions from Zambia that because Europeans reacted with disgust when offered caterpillars as food, many villagers came to believe that real Christians do not eat caterpillars. Those that came from mission schools as teachers or public servants often adopted an attitude of aversion toward the larvae, saying that they are food for villagers and laborers.

HARVESTING PRACTICES

The harvesting practices indicate the knowledge the indigenous population has with the biology, behavior, and ecology of the insects. Very often insects are collected by hand. For example, the best part of the day to collect grasshoppers, is early in the morning when it is fairly easy to catch the cold-blooded animals. I will give a number of different practices used for a number of important species.

Using Light or Sound

The most favored termite caste as food is the reproductive one. The swarming termites have their nuptial flight after the first rains following the dry season. After the flight, they shed their wings. The most common way to collect those that emerge during the night is to have a light source above a receptacle of water. They are attracted by the light, fall into the water and are then scooped out. However, it is also possible to make them emerge by pounding rhythmically on the soil with stones and sticks to simulate heavy rainfall and sometimes simultaneously water is poured over the hill (Dounias, 2016).

The edible grasshopper *Ruspolia differens* (Orthoptera: *Tettigonioidea*) is also collected using light sources. Each year there are accidents in Kampala, the capital of Uganda on the road, when women and children, collect those from street lights during the period when they swarm (Owen, 1973, p. 132). However, they are also commercially collected in mass by using huge lamps shining in the sky. They are attracted by the light, hit iron sheets which reflect the light, and fall in drums placed underneath (Agea et al., 2008). The author observed in Lao PDR that a normal light trap (lamp in front of an iron sheet and a bucket with water underneath) was used to collect insects during the night. We were told that the harvest was used for feeding chickens, but that selected ones were reserved for human consumption.

In Cameroon and the Central African Republic, to check whether larvae of the palm weevil *Rhynchophorus* spp. are ready to be harvested, women put their ears to the trunk of the palm tree and decide based on the sound of nibbling larvae (Chesquière, 1947; Muafor et al., 2015). In the Democratic Republic of Congo, sound from whistles made from grass is used by children to attract edible crickets (Malaisse, 1997, p. 240 citing Centner, 1963).

Harvesting From Nesting Structures

Apart from using a light source, the reproductives of termites can be collected directly from the nest (Van Huis, 2003). Holes or trenches are dug near the termite hill; the termites are then attracted by light or fire and then swept into the dug structures. Even tents are built over the holes from where the termites emerge. Depending on the ethnic group, soldiers of termites are sometimes eaten. They are often collected by children, who insert grass stems or reeds into holes of a broken part of the termite hill. The soldiers bite into the stem or reed and are then stripped into a bowl. This is done by Amerindians in the Amazon when collecting soldiers of the termite *Syntermes* spp. (Paoletti et al., 2000a,b). Also in Africa, this is common practice to extract the soldiers (Malaisse, 1997, p. 228).

In Columbia, the future queens of the ant *Atta laevigata* (Hymenoptera: *Formicidae*) have their nuptial flight in April and May. These are collected and considered not only delicacies but are also used as aphrodisiac (Granados et al., 2013). However, the collection from the nest always causes injuries, as the worker and soldier ants defend their nests. Women of the San of the Central Kalahari in the Republic of Botswana locate the nest ant *Camponotus* sp. (Hymenoptera: *Formicidae*) by looking for drained sand (Nonaka, 1996). Once a nest is found, it is poked with a digging stick and the ground around tapped by hand. The emerging ants are collected, but this has to be done quick as the ants bite. Even a few ants are carried home as they are used for seasoning, adding an acidic flavoring to their food.

Queen brood of weaver ants in Lao PDR is produced and harvested from February to April. To collect the brood, use is made of a 4–6 m bamboo stick with sharpened tip and a basket hanging behind the tip. People pierce the nest with the stick and shake it such that the brood falls into the basket (Van Itterbeeck et al., 2014). In the mountainous areas of the central region of Japan, the wasp *Vespula flavipes* (Hymenoptera: *Vespidae*) is a seasonal delicacy (Nonaka, 2010). Collectors attract the wasps by offering them small pieces of meat with tiny ribbons attached to them. When the wasps fly back to the nest, the ribbons enable the collectors to follow the wasp. Once the nest is found, it can be harvested. Often only a part of the nest is harvested in order to allow the colony to recover and ensure future harvests.

Another example in which nests are used to harvest concerns honey pot ants (Hymenoptera: *Formicidae*). Workers of the ants feed nectar to other workers, called repletes, which hang in chambers and have crops swollen with honey. The repletes supply the colony with honey in times of scarcity. They are eaten as a snack by native Americans of the American Southwest (*Myrmecocystus* spp.) and by the aboriginals in Australia (*Camponotus inflatus*) (Conway, 1994). The harvesting in Australia is mainly done by women, who know where the nests are and how the ants get the honey, but seem unaware of how the repletes develop.

From Tanzania (Harris, 1940) and Zambia (Silow, 1976) it is known that the larvae of the silkworm *Anaphe panda* (Lepidoptera: *Thaumetopoeidae*) are eaten. These caterpillars construct communal nests in the branches of trees. In the Kakamega forest in Kenya, sleeve nets are used to decrease mortality by predators and parasitoids (Mbahin et al., 2010).

Exploring Hibernation or Aestivation Sites

In Sudan, the sorghum bug, *Agonescelis versicolor* (Hemiptera: *Pentatomidae*) is not only a pest of sorghum but is also used to extract an edible oil from it (Delmet, 1973). The insect is collected during the dry season when they aggregate in mountain cracks. The edible stinkbug *Encosternum delegorguei* (Hemiptera: *Tessaratomidae*) is eaten by several ethnic groups in South Africa (Dzerefos et al., 2013). The harvest from trees in woodland and plantations is facilitated when the insects aggregate into football-size clusters during the winter season. However, to ward off predators the stinkbugs produce noxious defense chemicals, which stain the skin and affect vision. So, protective gear is worn and harvesting is nocturnal when the insect is immobilized by cold.

Bogong moths *Agrotis infusa* (Lepidoptera: *Noctuidae*) aestivate in rock crevices in the snowy mountains and Victoria Alps in the southeast of Australia after having migrated from up to 1000km from the inland plains of eastern Australia (Green et al., 2001). The aboriginals, during annual ceremonies, used to smoke them out of the crevices and collect them on bark sheets, nets, or skins (Flood, 1980). The large quantities of moths and the ease of gathering made them a reliable food source. They were carefully cooked, winnowed, and eaten, or preserved by pounding them into cakes (Rigby, 2011). Whereas the consumption of caterpillars is quite common, this is one of the few examples where the adults of Lepidoptera are eaten.

Facilitating Collection

Besides harvesting from nature, the indigenous population also knows ways to encourage the availability of edible insect populations, also called semidomestication. Several examples are given by Van Itterbeeck et al. (2014). For example, providing egg-laying sites of reed and grasses for aquatic Hemiptera in lakes of Mexico, manipulating the habitat to increase edible caterpillars in Africa, and cutting palm trees deliberately to encourage palm weevils to lay their eggs. In the latter case, in the Venezuelan Amazon they make cuts in the trunk of the palm tree to encourage oviposition by palm weevils (*Rhynchophorus* spp.) (Choo et al., 2009). They discern whether the larvae are ready to harvest by counting the days or lunar cycles or examining the color of sawdust expelled from the entry holes larvae create as they tunnel through the palm trunks, and they also know the difference in the development times of the two species, *R. palmarum* and *R. barbirostris*.

A simple kind of rearing to have the edible insect in close proximity to where people live is practised in Africa, where branches with young caterpillars are cut from trees and transported and placed on the same tree species, but near the village (Malaisse, 1997, pp. 207–208).

Exploring Certain Habitats

Harvesting edible insects requires often an intimate knowledge about their biology and ecology. In Australia, the insects that are most commonly eaten by aboriginals are "edible grubs," which are coleopterous or lepidopterous larvae that feed in trunks, branches, or roots of plants. These are locally called bardi or witjuti grubs and there are at least 25 plant species that harbor these insects (Yen, 2005). Australian aboriginals also know about "bush coconuts" or "bloodwood apples" which contain larvae that can be eaten (Yen et al., 2016). The larvae is a scale insect *Cystococcus* sp. (Hemiptera: *Eriococcidae*) which induces galls (outgrowth of plant tissue caused by an insect laying an egg in it) on bloodwoods (*Myrtaceae*) (Semple et al., 2015). Aboriginals eat the larva and the white coconut-like flesh of the inner gall (Gullan and Cranston, 2005).

In the Democratic Republic of Congo, Latham (2005) listed edible caterpillar species to be found on about 50 plant species (Latham, 2005). For African forests in general the biology and ecology of the insect is known for caterpillars, termites, and a number of other insect species (Malaisse, 1997, pp. 199–241). The latter also provides information about harvesting practices and nutritional value of species.

GENDER PARTICIPATION IN COLLECTION AND MARKETING

Women seem to be more involved in the collection and consumption of edible insects than men in Africa (Niassy et al., 2016). Reasons could be: more vulnerability to malnutrition; more dependence on local natural resources; and insects as the only protein and fat sources available.

Tukanoan Indians in the northwest Amazon spend a lot of time in collecting ant and termite soldiers and women do it more than men or children (Dufour, 1987). Men collect insect species that require felling and splitting of trees to harvest the larvae. Men, women, and older children are mostly responsible for collecting caterpillars and palm weevils and the alates of the ant *Atta* sp. and the termite *Syntermes* sp. Concerning the consumption, in the period that insects (in particular ants and termites) are most abundant, they provide 12% of the animal protein in men's diet and 26% in women's diet. Van Huis (2003) also indicates from Africa that insects are often collected by women and children and by the latter when the catch is small or difficult (e.g., cicadas and crickets).

The collection of the edible grasshopper, *R. differens*, a delicacy in Uganda, is an activity for women, who collect the grasshoppers for their husbands, who would in turn buy a traditional dress for the women in Buganda (Agea et al., 2008). Children help their mothers. When collected in large-scale using electricity, lamps, iron sheets, and drums, men tend to dominate. Large-volume trading chains and long-distance trade are lucrative and these are mostly activities of the men.

The collection and processing of the mopane caterpillar, *Imbrasia belina*, (Lepidoptera: *Saturniidae*) in southern Africa is mainly done by poor rural women (Ghazoul, 2006). Women also are the main sellers of mopane caterpillars in market stalls and by the roadside (Kozanayi and Frost, 2002). They sell the caterpillars in small volumes whereas men tend to dominate large-volume trade. In Limpopo, South Africa, female-headed households are more likely to participate in the mopane caterpillar markets, probably because these are resource constrained, lacking access to productive assets (land, labor, capital) which limits their agricultural production capabilities (Baiyegunhi and Oppong, 2016).

FIGURE 11.6 Caterpillars for sale on the market in the Democratic Republic of Congo. *Photo: Giulio Napolitano, FAO.*

On the contrary, in the Central African Republic it is only the men (mainly school boys) who are responsible for the collection of caterpillars, while the sale is only done by women (also school girls), partly (about 25%) by women fruit and vegetable sellers (Mbetid-Bessane, 2005) (Fig. 11.6).

HABITAT DESTRUCTION

Edible caterpillars *Eucheira socialis* (Lepidoptera: *Pieridae*) occur in mountainous regions of Mexico and live on *Arbutus* sp. (*Ericacaeae*) trees, where about 200 of them construct a silken, bag-like nest. However, the trees have been cut to produce firewood, and the nests can hardly be found anymore; the Pjiekakajoo community has only the memory of how, many years ago, the hills were white due to the quantity of nests (Aldasoro Maya and Gómez, 2016).

Since prehistoric times eggs of aquatic Hemiptera of the family *Corixidae* are harvested from bundles of reeds which people keep at the bottom of Mexican lakes. However, drying up of the lakes, improper cultivation techniques, and pollution endangers the harvest (Ramos-Elorduy, 2006).

However, when an edible insect species is in high demand, causing a surge in prices, than overexploitation is a threat. One clear example is the "escamoles" of *Liometopum* sp. (Hymenoptera: *Formicidae*) that are harvested because of their abundance and popularity (Ramos-Elorduy, 2006). They were first collected only by trained men, who retrieved only the reproductive caste in such a way that the nests could be exploited in later seasons. However, now untrained people harvest also the workers, and this leaves the nest in peril. From Australia, Yen (2009b) mentions that even an increased demand by ecotourism and restaurant markets may threaten edible insects such as honey ants and witchuti and bardi grubs.

AGRICULTURAL PESTS AS FOOD

Grasshopper species are often agricultural pests. However, there are several examples in which grasshoppers can be physically controlled by handpicking the insect as food (Fig. 11.7). Rice field grasshoppers of the genus *Oxya* spp. (Orthoptera: *Acrididae*) have been traditional food in most of Asia. In Korea, they were used as side dishes in meals and as a drinking snack. However, insecticide use during the 1960s and 1970s greatly reduced grasshopper populations (Pemberton, 2003). When in the 1980s the government put less emphasis on the countryside, farmers, especially in some highland areas, stopped using insecticides and consequently grasshoppers as food experienced a revival.

In the 1970s, Thailand suffered a major plague of the Bombay locust *Nomadacris succincta* (Orthoptera: *Acrididae*) (Hanboonsong, 2010).

FIGURE 11.7 The red locust, *Nomadacris septemfasciata* (Orthoptera: *Acrididae*), a pest of graminaceous crops, in Malagasy Markets (Betioky and around). *Photo: Annie Monard, FAO.*

FIGURE 11.8 Ready-to-eat fried and seasoned grasshopper in plastic bags, Klong Toey Market, Bangkok, Thailand. *Photo: Arnold van Huis.*

Aerial spraying did not succeed in controlling the pest. From 1978 to 1981 a campaign to urge people to eat the locust was started, as this was an old practice from the past. The publicity campaign promoted the use of locusts for deep-fried snacks, as a ground-up ingredient in crackers, or as a cooking sauce (Fig. 11.8). The campaign was successful. The Bombay locust became a popular food and is no longer considered a major pest by most farmers.

In Assam, India, a root feeding pest, the beetle *Lepidiota mansueta* (Coleoptera: *Scarabaeidae*) has since 2005 become an extremely serious pest of many field crops on Majuli river island in the Brahmaputra River (Bhattacharyya et al., 2015). On this largest mid-river deltaic island of the world, the crop area is 30,000 ha of which 70% is affected. The crops concerned are potato, sugarcane, *Colocasia* sp., and green gram with damages up to 50%. With assistance of the Assam Agricultural University, dishes have been developed such as roasted beetle fry with tomato and plain roasted beetle and beetle curry, which have become popular food (Borah, 2016).

In central Mexico, the grasshopper *Sphenarium purpurascens* (Orthoptera: *Pyrgomophidae*) is a pest of corn, bean, and alfalfa in Mexico. The pest is often controlled by applying organophosphorus insecticides. However, the grasshoppers are also captured and eaten as food. Cerritos and Cano-Santana (2008) showed that manual harvesting reduces the density of the grasshopper and mentioned as advantages: the harvested product can be used as food, no costs of pesticide use, and less environmental contamination. Cerritos Flores et al. (2015) calculated that the estimated biomass of this insect in Mexico would be 350,000 tons per year, generating a gross income of US$ 350 million (10–55 individuals/m^2 over > 1 million ha).

MEDICINAL USES

Insects are not only eaten for food but often used for medicinal purposes. In Meyer-Rochow (2017) examples can be found of using insects against different ailments. For other examples in a number of countries in different continents see Table 11.1.

However, it is sometimes difficult to make a distinction between superstition and real medicinal uses. The "Doctrine of Signatures" may play a role, for example, hairy caterpillars being used to cure baldness (Van Huis, 2002). Meyer-Rochow (1979) mentions that in Kwantung province in China water beetles are used as urine-inhibitor while the dung beetle is used against diarrhea in Thailand. However, one has to be careful to judge quickly if it is superstition. For example, in East Africa it is a widespread custom by adolescent girls to let water beetles of the families *Gyrinidae* and *Dytiscidae* bite in their nipples in order to stimulate breast growth (Van Huis, 2002). One explanation is that girls want to transfer the properties of these breast-shaped beetles to themselves. However, Kutalek and Kassa (2005) showed that gyrinids produce, among other substances, norsesquiterpenes, and that the dytiscids also possess prothoracic defensive glands, which produce, among other substances, hormone-like steroids.

In China it is popular to use caterpillars infected with fungi as medicine (Yen, 2015). The entomophagous fungus *Cordyceps* (Hypocreales: *Clavicipitaceae*) parasitizes larvae, pupae, or adults of insects. Of the 300 species reported so far, no other species is considered as medicinally important and costly as *Cordyceps sinensis* (Arora, 2015). It is a native of high Himalayan mountains in Tibet, Nepal, India, and Bhutan, at an altitude ranging from 3000 to 5000 m, and commonly known as "yartsa-gunbu" in Tibet and as "Keera ghas" or "Keera jhar" (insect herb) in Indian mountains. In nature, it is parasitic on the larvae of a small moth *Hepialus armoricanus* (Lepidoptera: *Hepialidae*). It is world's most efficient and expensive (Liang, 2011) medicinal mushroom and considered as a traditional Chinese medicine having multiple medicinal and pharmacological properties and also used to treat respiratory and immune disorders; pulmonary diseases; renal, liver, and cardiovascular diseases; hyposexuality; and hyperlipidemia with among others as bioactive ingredients: cordycepin, adenosine, and ergosterol (Arora, 2015). It is advertised as combating fatigue (Liang, 2011) and the success of Chinese women athletes,

TABLE 11.1 Articles Dealing With Arthropods as Medicine in Different Continents

Continent	Country and Reference
Africa	Cameroon (Tamesse et al., 2016), Mali (Lehmana et al., 2007), Nigeria (Lawal and Banjo, 2007)
Asia	China (Huang, 1998; Read, 1940; Wang, 2014), India (Chakravorty et al., 2011), Malaysia (Chung et al., 2001), South Korea (Pemberton, 1999; Pemberton, 2005)
Australia	(Cherry, 1991)
Latin America	Brazil (Costa-Neto, 2002; Posey, 2002), Mexico (Ramos-Elorduy de Conconi, 1988)

achieving world records at the 1500 and 3000 m in 1993 was attributed to using this fungus (Steinkraus and Whitfield, 1994).

The weaver ant *Oecophylla smaragdina* is a delicacy of certain ethnic groups living in the forests of Kerala, India. They make oil using hot extraction of crushed worker ants to treat inflammation of joints and skin infections (Vidhu and Evans, 2015).

Chitin is known to have immunological properties, among others in India (Chakravorty et al., 2011). Chitin is the most abundant polysaccharide in the world after cellulose, and it is a material of the exoskeleton of insects, parasites, and fungi. Chitin and chitin derivatives can stimulate innate immune cells and may have beneficial effects on inflammatory responses like those in asthma and other lung disorders (Lee et al., 2008).

An overview of insects used as aphrodisiac in different parts of the world is given by Motte-Florac (2016). Often social insects serve that purpose, for example, gravid new queens of *Atta laevigata* (Hymenoptera: *Formicidae*) in Colombia and Venezuela are given as dowry; in China, powder of the ant *Polyrhachis vicina* (Hymenoptera: *Formicidae*) is a very popular traditional medicine, in particular to "rejuvenate older people"; the weaver ant *Oecophylla longinoda* is used in Cameroon; the queen of termites is often also used as such in Africa (Van Huis, 2002) but also in the Amazon (Posey, 2002).

FROM HARVESTING AND SEMIDOMESTICATION TO FARMING

If we want to promote the eating of insects, it is no option to promote harvesting from nature, as this resource is limited. Edible insects from nature are already endangered by overexploitation, habitat change, and pollution. The other problem with harvesting is that the resource is seasonal and can only be made available throughout the year by preservation practices. It is also not possible to ensure food safety, as edible insects harvested from crops may be contaminated with pesticides, and rearing under controlled conditions obviates such problems. Another way of making insects more available is semidomestication and Van Itterbeeck and Van Huis (2012) give examples for grasshoppers, aquatic Hemiptera, and palm weevils.

Small-scale rearing of crickets appears to be very successful as demonstrated in Thailand where 20,000 farms rear 7500 tons a year (Hanboonsong et al., 2013). It has been shown that small-scale farming of palm weevils is possible: for example, in Thailand (Hanboonsong et al., 2013), Lao PDR (Hanboonsong and Durst, 2014), Cameroon (Muafor et al., 2015), and Venezuela (Cerda et al., 2001).

Rearing has been attempted for the mopane caterpillar but problems due to disease spreading through a captive population remain unresolved; it was technically feasible but was not cost-effective (Ghazoul, 2006).

There are ways of making insects more palatable by rearing them on specific food plants. Such an example is given for caterpillars. When a caterpillar has several host trees, one among which is *Julbernardia paniculata*, they are transferred to this tree and left there for some time to develop and make them tastier as indicated by gourmets (Silow, 1976, p. 210).

This means that industrial rearing of species that can be harvested the whole year, such as crickets, should be attempted and a number of projects aiming to achieve this have been initiated.

CONCLUSIONS

The information available on the practice of eating insects from all over the world is rather limited. What is available are overviews on the practice of insects as food in standard works (Bergier, 1941; Bodenheimer, 1951; DeFoliart, 2012). From some countries, there is a lot of information available due to the activities of some scientists. Examples are Dr. J. Ramos-Elorduy Blásquez from Mexico, who produced an impressive list of publications on the topic (Pino Moreno, 2016), and Latham (2005) and Malaisse and Parent (1980) who both published on edible caterpillar species from the Democratic Republic of Congo. However, a lot of information is also hidden in books dealing with anthropological studies. With the Westernization of tropical societies this information is getting increasingly difficult to get and may get lost. This was already apparent when I studied the traditional use of arthropods in Africa where I mention that my informants often had to consult parents, grandparents, and village elders in order to obtain correct information (Van Huis, 1996). Several authors have indicated that a more concerted effort is necessary to collect this information before it disappears, such as from the aboriginals in Australia (Yen, 2010), the Amerindians in the Amazon (Paoletti et al., 2000a,b), or several ethnic groups in India (Chakravorty et al., 2011). Also, Costa-Neto (2015) mentions from Latin America that many species have not been collected or identified. He also blames the Western bias against the eating of insects: the negative perception of insects as food by Westerners "contaminates" indigenous cultures: people used to eat arthropods, now considering it only for poor and backward people. DeFoliart (1999) and Yen (2009a) feel that acculturation to Western lifestyles tends to cause a reduction in the use of insects as food. Looy et al. (2014) dwell a bit more on the insect cuisine being a threat to the psychological and cultural identity of Westerners.

There is an increasing interest in the West of using insects as food and feed. The 2013 FAO book *Edible Insects: Future Prospects for Feed and Food Security* (Van Huis et al., 2013) was downloaded more than 7 million times. The conference "Insects to feed the world" was attended by 450 participants from 45 countries (Van Huis and Vantomme, 2014). There is an exponential increase in scientific articles on insects as food and feed (Van Huis, 2015). Consulting of the "Web of Science," using the word "edible insects," indicated that only in 2016, 51 articles on edible insects were produced compared to 18 during the 5 years from 2005 to 2009. So instead of looking down at the practice of eating insects as a primitive habit, Westerners now are keen to learn about this food habit from indigenous societies. We now realize that the eating of insects as a protein source has many advantages compared to our conventional production animals. Those advantages can be divided in several categories which have been summarized in Van Huis (2016): nutritonal value being similar or better; less greenhouse gases; less ammonia emissions; less land area needed; low feed conversion efficiency; and insects being able to convert low-value organic side streams into high-value protein products.

Therefore, it is recommended that much more effort will be spent on collecting information about which arthropod species are harvested, preserved, prepared, and marketed. And this should be done before this information is lost. We also see that in some countries like in Thailand there is reappreciation of insects as food and that they are increasingly reared to satisfy the increasing demand (Durst and Hanboonsong, 2015). Also in other tropical countries there is renewed interest. It is necessary to be aware of the Western cultural bias. The challenge is now to collect and evaluate traditional practices.

References

Agea, J.G., Biryomumaisho, D., Buyinza, M., Nabanoga, G.N., 2008. Commercialization of *Ruspolia nitidula* (nsenene grasshoppers) in Central Uganda. African Journal of Food Agriculture and Development 8, 319–332.

Aldasoro Maya, E.M., Gómez, B., 2016. Insects and other invertebrates in the Pjiekakjoo (Tlahuica) culture in Mexico state, Mexico. Journal of Insects As Food and Feed 2, 43–52.

Arora, R.K., 2015. Cordyceps sinensis (Berk.) Sacc.-an entomophagous medicinal fungus - a review. International Journal of Recent Advances in Multidisciplinary Research 2, 161–170.

Backwell, L.R., d'Errico, F., 2001. Evidence of termite foraging by Swartkrans early hominids. Proceedings of the National Academy of Sciences 98, 1358–1363.

Baiyegunhi, L.J.S., Oppong, B.B., 2016. Commercialisation of mopane worm (*Imbrasia belina*) in rural households in Limpopo Province, South Africa. Forest Policy and Economics 62, 141–148.

Barreteau, D., 1999. Les Mofu-Gudur et leurs criquets. In: Baroin, C., Boutrais, J. (Eds.), L'homme et l'animal dans le bassin du lac Tchad, Actes du colloque du reseau Mega-Tchad, Orleans 15–17 octobre 1997. Editions IRD (Institut de Recherche pour le Developpement), Collection Colloques et Seminaires, no. 00/354. Université Nanterre, Paris, pp. 133–169.

Bergier, E., 1941. Peuples entomophages et insectes comestibles: etude sur les moeurs de l'homme et de l'insecte. Imprimerie Rulliere Freres, Avignon.

Bhattacharyya, B., Pujari, D., Bhuyan, U., Handique, G., Baruah, A.A.L.H., Dutta, S.K., Tanaka, S., 2015. Seasonal life cycle and biology of *Lepidiota mansueta* (Coleoptera: Scarabaeidae): a serious root-feeding pest in India. Applied Entomology and Zoology 50, 435–442.

Bodenheimer, F.S., 1951. Insects as Human Food; a Chapter of the Ecology of Man. Dr. W. Junk, Publishers, The Hague. 352 pp.

Borah, A., 2016. The Majuli beetle turns from pest to delicacy. In: 2016 India Climate Dialogue of 10 August, New Delhi.

Brothwell, D., Brothwell, P., 1998. Food in Antiquity: A Survey of the Diet of Early People. John Hopkins University Press, Baltimore.

Cerda, H., Martinez, R., Briceno, N., Pizzoferrato, L., Manzi, P., Tommaseo Ponzetta, M., Marin, O., Paoletti, M.G., 2001. Palm worm: (*Rhynchophorous palmarum*) traditional food in Amazonas, Venezuela - nutritional composition, small scale production and tourist palatibility. Ecology of Food and Nutrition 40, 13–32.

Cerritos Flores, R., Ponce-Reyes, R., Rojas-García, F., 2015. Exploiting a pest insect species *Sphenarium purpurascens* for human consumption: ecological, social, and economic repercussions. Journal of Insects as Food and Feed 1, 75–84.

Cerritos, R., Cano-Santana, Z., 2008. Harvesting grasshoppers *Sphenarium purpurascens* in Mexico for human consumption: a comparison with insecticidal control for managing pest outbreaks. Crop Protection 27, 473–480.

Chakravorty, J., Ghosh, S., Meyer-Rochow, V.B., 2011. Practices of entomophagy and entomotherapy by members of the Nyishi and Galo tribes, two ethnic groups of the state of Arunachal Pradesh (North-East India). Journal of Ethnobiology and Ethnomedicine 7, 5. http://dx.doi.org/10.1186/1746-4269-7-5.

Cherry, R., 1991. Use of insects by Australian aborigines. American Entomologist 32, 8–13.

Chesquière, J., 1947. Les insectes palmicoles comestibles (Appendice II). In: Lepesme, P., Bourgogne, J., Cairashi, E., Paulian, R., Villiers, A. (Eds.), Les Insectes des Palmiers. Paul Lechevalier, Paris, pp. 791–793.

Choo, J., Zent, E.L., Simpson, B.B., 2009. The importance of traditional ecological knowledge for palm-weevil cultivation in the Venezuelan Amazon. Journal of Ethnobiology 29, 113–128.

Chung, A.Y.C., Khen, C.V., Unchi, S., Tingek, S., Wong, A., 2001. A survey on traditional uses of insects and insect products as medicine in Sabah. Malaysian Naturalist 55, 24–29.

Conway, J.R., 1994. Honey ants. American Entomologist 40, 229–234.

Costa-Neto, E.M., 2002. The use of insects in folk medicine in the State of Bahia, northeastern Brazil, with notes on insects reported elsewhere in Brazilian folk medicine. Human Ecology 30, 245–263.

Costa-Neto, E.M., 2015. Anthropo-entomophagy in Latin America: an overview of the importance of edible insects to local communities. Journal of Insects as Food and Feed 1, 17–23.

Costa-Neto, E.M., Dunkel, F.V., 2016. Insects as food: history, culture and modern uses around the world (Chapter 2). In: Dossey, A.T., Morales-Ramos, J.A., Guadalupe Rojas, M. (Eds.), Insects as Sustainable Food Ingredients: Production, Processing and Food. Academic Press, Amsterdam, pp. 29–60.

Costa Neto, E.M., Ramos-Elorduy, J., 2006. Los insectos comestibles de Brasil: Etnicidad, diversidad e importancia en la alimentación. Boletín Sociedad Entomológica Aragonesa 38, 423–442.

De Groot, A.A., 1995. La protection des vegetaux dans les cultures de subsistance: le cas du mil au Niger de l' Ouest. Département de Protection des Végétaux (Centre Agrhymet), Niamey.

DeFoliart, G., 2012. The Human Use of Insects as a Food Resource: A Bibliographic Account in Progress. http://labs.russell.wisc.edu/insectsasfood/the-human-use-of-insects-as-a-food-resource/.

DeFoliart, G.R., 1999. Insects as food: why the western attitude is important. Annual Review of Entomology 41, 21–50.

Delmet, C., 1973. Extraction d'huile comestible d'un insecte (*Agnoscelis versicolor*) au djebel Guli (Soudan). Études Rurales 52, 138–140.

Diamond, J., 1997. Guns, Germs and Steel – the Fates of Human Societies. W.W. Norton & Company, New York.

Dounias, E., 2003. l' Exploitation meconnue d'une ressource connue: la collecte des larves comestibles de charancons dans les palmiers-raphia au sud du Cameroun.

Dounias, E., 2016. Des moissons éphémères. L'art de collecter et de consommer les termites sous les tropiques. In: Le Gall, P., Motte-Florac, E. (Eds.), Savoureux insectes: d'aliment traditionelles à l'innovation gastronomique. Presses universitaires de Rennes/Institut de Recherche pour le développement, pp. 273–339.

Dourojeanni, M.J., 1965. Denominaciones vernaculares de insectos y algunos otros invertebrados en la Selva del Peru. Revista Peruana de Entomologia 8, 131–137.

Dufour, D.L., 1987. Insects as food: a case study from the Northwest Amazon. American Anthropologist, New Series 89, 383–397.

Durst, P.B., Hanboonsong, Y., 2015. Small-scale production of edible insects for enhanced food security and rural livelihoods: experience from Thailand and Lao People's Democratic Republic. Journal of Insects As Food and Feed 1, 25–31.

Dzerefos, C.M., Witkowski, E.T.F., Toms, R., 2013. Comparative ethnoentomology of edible stinkbugs in southern Africa and sustainable management considerations. Journal of Ethnobiology and Ethnomedicine 9, 1–12.

Evans, J., Alemu, M.H., Flore, R., Frøst, M.B., Halloran, A., Jensen, A.B., Maciel-Vergara, G., Meyer-Rochow, V.B., Münke-Svendsen, C., Olsen, S.B., Payne, C., Roos, N., Rozin, P., Tan, H.S.G., Huis, A.v., Vantomme, P., Eilenberg, J., 2015. 'Entomophagy': an evolving terminology in need of review. Journal of Insects As Food and Feed 1, 293–305.

Fasoranti, J.O., Ajiboye, D.O., 1993. Some edible insects of Kwara state, Nigeria. American Entomologist 39, 113–116.

Flood, J., 1980. Of moths and men (Chapter 6). In: The Moth Hunters: Aboriginal Prehistory of the Australian Alps. Australian Institute of Aboriginals Studies, Canberra.

Ghazoul, J., 2006. Mopani Woodlands and the Mopane Worm: Enhancing Rural Livelihoods and Resource Sustainability. Final Technical Report. DFID, London.

Ghosh, S., Jung, C., Meyer-Rochow, V.B., 2016. Nutritional value and chemical composition of larvae, pupae, and adults of worker honey bee, *Apis mellifera ligustica* as a sustainable food source. Journal of Asia-Pacific Entomology 19, 487–495.

Gomez, P.A., Halut, R., Collin, A., 1961. Production de proteines animales au Congo. Bulletin Agricole du Congo 52, 689–700.

Granados, C.C., Acevedo, C.D., Guzman, C.L.E., 2013. Tostado y harina de la hormiga santandereana *'Atta laevigata'*. Biotecnología en el Sector Agropecuario y Agroindustrial 11, 68–74.

Green, K., Broome, L., Heinze, D., Johnston, S., 2001. Long distance transport of arsenic by migrating bogong moths from agricultural lowlands to mountain ecosystems. The Victorian Naturalist 118, 112–116.

Gullan, P.J., Cranston, P.S., 2005. Insects as food: an outline of entomology. In: The Insects. Blackwell Publishing, Oxford, pp. 10–20.

Hanboonsong, Y., 2010. Edible insects and associated food habits in Thailand. In: Durst, P.B., Johnson, D.V., Leslie, R.L., Shono, K.E. (Eds.), Forest insects as Food: Humans Bite Back. FAO Regional Office for Asia and the Pacific, Bangkok, Thailand, pp. 173–182.

Hanboonsong, Y., Durst, P.B., 2014. Edible Insects in Lao PDR: Building on Tradition to Enhance Food Security. Food and Agriculture Organization of the United Nations. Regional Office for Asia and The Pacific, Bangkok. Rap Publication 2014/12.

Hanboonsong, Y., Jamjanya, T., Durst, P.B., 2013. Six-legged Livestock: Edible Insect Farming, Collection and Marketing in Thailand. Food and Agriculture Organization of the United Nations, Regional Office for Asia and the Pacific, Bangkok.

Harris, W.V., 1940. Some notes on insects as food. Tanganyika Notes and Records 9, 45–48.

Harrison, J.F., Kaiser, A., VandenBrooks, J.M., 2010. Atmospheric oxygen level and the evolution of insect body size. Proceedings of the Royal Society B 277, 1937–1946.

Hocking, B., Matsumura, F., 1960. Bee brood as food. Bee World 41, 113–120.

Huang, Z., 1998. Insects in Chinese Culture: Insect as Medicine. http://zacharyhuang.com/pubs/insect.html.

Jongema, Y., 2015. List of Edible Insect Species of the World. Laboratory of Entomology, Wageningen University, The Netherlands. http://www.wur.nl/en/Expertise-Services/Chair-groups/Plant-Sciences/Laboratory-of-Entomology/Edible-insects/Worldwide-species-list.htm.

Kornfeld, M., Boyle, K.V., Clark, G.A., Enloe, J.G., Kozlowski, J., Mora, R., Straus, L.G., Tabarev, A.V., Zilhão, J., 1996. The big-game focus: reinterpreting the archaeological record of cantabrian upper paleolithic economy [and comments and reply]. Current Anthropology 37, 629–657.

Kozanayi, W., Frost, P., 2002. Marketing of Mopane Worm in Southern Zimbabwe. Mopane Worm Market Survey: Southern Zimbabwe. Institute of Environmental Studies, University of Zimbabwe, pp. 1–30.

Kutalek, R., Kassa, A., 2005. The use of gyrinids and dytiscids for stimulating breast growth in East Africa. Journal of Ethnobiology 25, 115–128.

Latham, P., 2005. Edible Caterpillars and Their Food Plants in Bas-Congo Province, Democratic Republic of Congo. https://www.researchgate.net/publication/309121521_Edible_caterpillars_and_their_food_plants_of_Bas-Congo_province_DR_Congo_2016.

Lawal, O.A., Banjo, A.D., 2007. Survey for the usage of arthropods in traditional medicine in southwestern Nigeria. Journal of Entomology 4, 104–112.

Lawal, O.A., Banjo, A.D., Fasunwon, B.T., Osipitan, A.A., Omogunloye, O.A., 2010. Entomophagy among tertiary Institutions in southwestern Nigeria. World Journal of Zoology 5, 47–52.

Lee, C.G., Silva, C.A.D., Lee, J.-Y., Hartl, D., Elias, J.A., 2008. Chitin regulation of immune responses: an old molecule with new roles. Current Opinion in Immunology 20, 684–689.

Lehmana, A.D., Dunkel, F.V., Klein, R.A., Ouattara, S., Diallo, D., Gamby, K.T., N'Diaye, M., 2007. Insect management products from Malian traditional medicine—establishing systematic criteria for their identification. Journal of Ethnopharmacology 110, 235–249.

Liang, Y., 2011. Making gold: commodification and consumption of the medicinal fungus chong-cao in Guangdong and Hong Kong. Hong Kong Anthropologist 5, 1–17.

Looy, H., Dunkel, F.V., Wood, J.R., 2014. How then shall we eat? Insect-eating attitudes and sustainable foodways. Agriculture and Human Values 31, 131–141.

Madsen, D.B., Kirkman, J.E., 1988. Hunting hoppers. American Antiquity 53, 593–604.

Malaisse, 1997. Se nourir en foret claire africaine: approche ecologique et nutritionnelle. Les Presses Agronomiques de Gembloux. 384 pp.

Malaisse, F., Parent, G., 1980. Les chenilles comestibles du Shaba meridional. Naturalistes Belges 61, 2–24.

Marais, E., 1973. The Soul of the White Ant. First published in 1937 under 'Die Siel van die Mier'.Cox & Wyman Ltd., London. 163 pp. http://www.myrmecofourmis.com/sites/default/files/public/uploads/document/1/the-soul-of-the-white-ant_eugene-marais_ebook_18053.pdf.

Mbahin, N., Raina, S.K., Kioko, E.N., Mueke, J.M., 2010. Use of sleeve nets to improve survival of the Boisduval silkworm, Anaphe panda, in the Kakamega Forest of western Kenya. Journal of Insect Science 10. http://dx.doi.org/10.1673/031.010.0601.

Mbetid-Bessane, E., 2005. Commercialization of edible caterpillars in Central African Republic. Tropicultura 23, 3–5.

McGrew, W.C., 2001. The other faunivory: primate insectivory and early human diet. In: Stanford, C.B., Bunn, H.T. (Eds.), Meat-eating and Human Evolution. Oxford University Press, Oxford, pp. 160–178.

Meyer-Rochow, V.B., 2009. Food taboos: their origins and purposes. Journal of Ethnobiology and Ethnomedicine 5, 18. http://dx.doi.org/10.1186/1746-4269-5-18.

Meyer-Rochow, V.B., 2017. Therapeutic arthropods and other, largely terrestrial, folk-medicinally important invertebrates: a comparative survey and review. Journal of Ethnobiology and Ethnomedicine 13, 9. http://dx.doi.org/10.1186/s13002-017-0136-0.

Meyer-Rochow, V.B., 1979. The diverse uses of insects in traditional societies. Ethnomedicine 5, 287–300.

Miller, D., 1952. The insect people of the Maori. The Journal of the Polynesian Society 61 61 pp.

Mitsuhashi, J., 2008. Entomophagy: human consumption of insects. In: Capinera, J.L. (Ed.), Encyclopedia of Entomology. Springer, Dordrecht, pp. 1341–1343.

Motte-Florac, E., 2016. Des insectes aphrodisiaques au menu. In: Le Gall, P., Motte-Florac, E. (Eds.), Savoureux insectes: d'aliment traditionelles à l'innovation gastronomique. Presses universitaires de Rennes/Institut de Recherche pour le développement, pp. 149–196.

Muafor, F.J., Gnetegha, A.A., Gall, P.L., Levang, P., 2015. Exploitation, Trade and Farming of Palm Weevil Grubs in Cameroon. Center for International Forestry Research (CIFOR), Working Paper 178, Bogor, Indonesia.

Ndlovu, S., 2015. Names as indigenous knowledge for making meat edible and/or inedible: implications on food security in Zimbabwe. In: Felecan, O. (Ed.), Actele Conferinței Internaționale de Onomastică. Ediția a III-a: Conventional/unconventional in onomastics, pp. 751–762.

Niassy, S., Affognon, H.D., Fiaboe, K.K.M., Akutse, K.S., Tanga, C.M., Ekesi, S., 2016. Some key elements on entomophagy in Africa: culture, gender and belief. Journal of Insects As Food and Feed 2, 139–144.

Nonaka, K., 1996. Ethnoentomology of the central Kalahari san. African Study Monographs, Suppl. 22, 29–46.

Nonaka, K., 2010. Cultural and commercial roles of edible wasps in Japan. In: Durst, P.B., Johnson, D.V., Leslie, R.L., Shono, K.E. (Eds.), Forest insects as Food: Humans Bite Back, Proceedings of a Workshop on Asia-Pacific Resources and Their Potential for Development, FAO, Regional Office for Asia and the Pacific, Bangkok, 19–21 February, 2008, pp. 123–130.

Nonaka, K., Sivilay, S., Boulidam, S., 2008. The Biodiversity of Insects in Vientiane. National Agriculture and Forestry Institute (NAFRI) and Research Institute for Hamanity and Nature (RIHN), Kansai Art Printing, Nara, Japan.

Okore, O., Avaoja, D., Nwana, I., 2014. Edible insects of the Niger Delta area in Nigeria. Journal of Natural Sciences Research 4, 1–9.

Owen, D.F., 1973. Man's Environmental Predicament: An Introduction to Human Ecology in Tropical Africa. University Press, Oxford.

Paoletti, M.G., Buscardo, E., Dufour, D.L., 2000a. Edible invertebrates among Amazonian Indians: a critical review of disappearing knowledge. Environment, Development and Sustainability 2, 195–225.

Paoletti, M.G., Dufour, D.L., Cerda, H., Torres, F., Pizzoferrato, L., Pimentel, D., 2000b. The importance of leaf- and litter-feeding invertebrates as sources of animal protein for the Amazonian Amerindians. Proceedings of the Royal Society of London. Series B, Biological Sciences 267, 2247–2252.

Pemberton, R.W., 1999. Insects and other arthropods used as drugs in Korean traditional medicine. Journal of Ethnopharmacology 65, 207–216.

Pemberton, R., 2003. Persistance and change in traditional uses of insects in contemporary East Asian cultures. In: Motte-Florac, E., Thomas, J.M.C. (Eds.), Les insectes dans la tradition oral. Peeters-SELAF, Paris, pp. 139–154.

Pemberton, R.W., 2005. Contemporary use of insects and other arthropods in traditional Korean medicine (hambang) in South Korea and elsewhere. In: Paoletti, M.G. (Ed.), Ecological Implications of Minilivestock: Potential of Insects, Rodents, Frogs and Snails. Science Publishers, Inc., Enfield, New Hampshire, pp. 459–474.

Pino Moreno, J.M., 2016. The career of Dr J. Ramos-Elorduy Blásquez. Journal of Insects As Food and Feed 2, 3–14.

Posey, D.A., 2002. Insects, foods, medicines in folklore in Amazonia. In: Motte-Florac, E., Thomas, J.M.C. (Eds.), Les insectes dans la tradition oral. Peeters-SELAF, Paris, pp. 221–237.

Ramos-Elorduy de Conconi, J., Pino Moreno, C.J.M., 1988. The utilization of insects in the emperical medicine of ancient Mexicans. Journal of Ethnobiology 8, 195–202.

Ramos-Elorduy, J., 2006. Threatened edible insects in Hidalgo, Mexico and some measures to preserve them. Journal of Ethnobiology and Ethnomedicine 2, 51. http://dx.doi.org/10.1186/1746-4269-2-51.

Read, B.E., 1940. Insect used in Chinese medicine. Journal of the North China Asiatic Society 71, 22–32.

Reinhard, K.J., Bryant, V.M., 1992. Coprolite analysis: a biological perspective on archaeology. Archaeological Method and Theory 4, 245–288.

Rigby, K., 2011. Getting a Taste for the Bogong Moth. Australian Humanities Review 50 http://australianhumanitiesreview.org/2011/05/01/getting-a-taste-for-the-bogong-moth/.

Risiro, J., Doreen, T.T., Basikiti, A., 2013. Indigenous knowledge systems and environmental management: a case study of Zaka district, Masvingo province, Zimbabwe. International Journal of Academic Research in Progressive Education and Development 2, 19–39.

Roscoe, J., 1965. The Baganda: An Account of Their Native Customs and Beliefs. Frank Cass & Co., Ltd., London.

Roulon-Doko, P., 1998. Chasse, cueillette et cultures chez les Gbaya de Centrafrique. L'Harmattan, Paris. 540 pp.

Seignobos, C., Jean-Philippe, D., Henri-Pierre, A., 1996. Les Mofu et leurs insectes. Journal d'agriculture traditionnelle et de botanique appliquée 38, 125–187.

Semple, T.L., Gullan, P.J., Hodgson, C.J., Hardy, N.B., Cook, L.G., 2015. Systematic review of the Australian 'bush coconut' genus Cystococcus (Hemiptera: Eriococcidae) uncovers a new species from Queensland. Invertebrate Systematics 29, 287–312.

Silow, C.A., 1976. Edible and Other Insects of Mid-western Zambia, Studies in Ethno-entomology II. Institutionen foer Allmaen och Jaemfoerande Etnografi vid Uppsala Unoversiteit. Occasional Papers V, 223 pp.

Steinkraus, D.C., Whitfield, J.B., 1994. Chinese caterpillar fungus and world record runners. American Entomologist 40, 235–239.

Sutton, M.Q., 1995. Archaeological aspects of insect use. Journal of Archaeological Method and Theory 2, 253–298.

Tamesse, J.L., Kekeunou, S., Tchatchouang, L.J., Ndegue, O.L.M., Aissatou, L.M., Tombouck, D., Youssa, B., 2016. Insects as food, traditional medicine and cultural rites in the west and south regions of Cameroon. Journal of Insects As Food and Feed 2, 153–160.

UN, 2014. World Urbanization Prospects: The 2014 Revision, Highlights. United Nations, Department of Economic and Social Affairs. Population Division (ST/ESA/SER.A/352).

Vainker, S.J., 2004. Chinese Silk: A Cultural History. Rutgers University Press, New Brunswick.

Van Huis, A., 1996. The traditional use of arthropods in Sub Saharan Africa. In: Proceedings of the Section Experimental and Applied Entomology of the Netherlands Entomological Society (N.E.V.), Amsterdam 7, pp. 3–20.

Van Huis, A., 2002. Medical and stimulating properties ascribed to arthropods and their products in sub-Saharan Africa. In: Motte-Florac, E., Thomas, J.M.C. (Eds.), Les insectes dans la tradition orale - Insects in oral literature and tradition. Peeters-SELAF, Paris, pp. 367–382.

Van Huis, A., 2003. Insects as food in sub-Saharan Africa. Insect Science and Its Application 23, 163–185.

Van Huis, A., 2015. Edible insects contributing to food security? Agriculture & Food Security 4, 1–9.

Van Huis, A., 2016. Edible insects are the future? Proceedings of the Nutrition Society 75, 294–305.

Van Huis, A., Itterbeeck, J.V., Klunder, H., Mertens, E., Halloran, A., Muir, G., Vantomme, P., 2013. Edible Insects: Future Prospects for Food and Feed Security. FAO, Rome. Forestry Paper 171.

Van Huis, A., Vantomme, P., 2014. Conference report: insects to feed the world. Food Chain 4, 184–192.

Van Itterbeeck, J., Sivongxay, N., Praxaysombath, B., Van Huis, A., 2014. Indigenous knowledge of the edible weaver ant *Oecophylla smaragdina* Fabricius (Hymenoptera: Formicidae) from the Vientiane plain, Lao PDR. Ethnobiology Letters 5, 4–12.

Van Itterbeeck, J., Van Huis, A., 2012. Environmental manipulation for edible insect procurement: a historical perspective. Journal of Ethnobiology and Ethnomedicine 8, 1–19.

Vidhu, V.V., Evans, D.A., 2015. Ethnoentomological values of *Oecophylla smaragdina* (Fabricius). Current Science 109, 572–579.

Wang, D., 2014. The R&D of Insects as Food and Medicine in China. Northwest A&F University, China.

Wikipedia, 2016. Witchetty Grub. https://en.wikipedia.org/wiki/Witchetty_grub.

Yen, A., Flavel, M., Bilney, C., Brown, L., Butler, S., Crossing, K., Jois, M., Napaltjarri, Y., Napaltjarri, Y., West, P., Wright, B., 2016. The bush coconut (scale insect gall) as food at Kiwirrkurra, Western Australia. Journal of Insects As Food and Feed 2, 293–299.

Yen, A.L., 2005. Insects and other invertebrate foods of the Australian aborigines. In: Paoletti, M.G. (Ed.), Ecological Implications of Minilivestock: Potential of Insects, Rodents, Frogs and Snails. Science Publishers, Inc., Enfield, New Hampshire, pp. 367–388.

Yen, A.L., 2009a. Edible insects: traditional knowledge or western phobia? (Special Issue: trends on the edible insects in Korea and Abroad.). Entomological Research 39, 289–298.

Yen, A.L., 2009b. Entomophagy and insect conservation: some thoughts for digestion. Journal of Insect Conservation 13, 667–670.

Yen, A.L., 2010. Edible insects and other invertebrates in Australia: future prospects. In: Durst, P.B., Johnson, D.V., Leslie, R.L., Shono, K.E. (Eds.), Forest insects as Food: Humans Bite Back. FAO, Regional Office for Asia and the Pacific, Bangkok, Thailand, pp. 65–84.

Yen, A.L., 2015. Conservation of Lepidoptera used as human food and medicine. Current Opinion in Insect Science 12, 102–108.

Yen, A.L., Gillen, J., Gillespie, R., Vanderwal, R., The Mutitjulu Community, 1997. A preliminary assessment of Anangu knowledge of central Australian invertebrates. Memoirs of the Museum of Victoria 56, 631–634.

CHAPTER

12

Current Levels, Recent Historical Trends, and Drivers of Wildmeat Trade in the Amazon Tri-Frontier Region Between Colombia, Peru, and Brazil

Nathalie van Vliet[1], *Jessica Moreno*[2], *Juanita Gomez*[2], *Laurane L'haridon*[2], *Lindon Neves de Aquino*[3], *François Sandrin*[2], *Liliana Vanegas*[2], *Robert Nasi*[1]

[1]Center for International Forestry Research (CIFOR), Bogor, Indonesia; [2]Fundación SI, Bogotá, Colombia; [3]Universidad Federal do Amazonas, Benjamin Constant, Brazil

INTRODUCTION

Despite the various social and economic drivers that are rapidly pulling rural communities in the Amazon away from the use of forest products (Sills et al., 2011), wildmeat is still a fundamental source of protein, fat, and micronutrients for many people (Siren and Machoa, 2008; Sarti et al., 2015; Morsello et al., 2015). Rural wildmeat consumption in the Amazon has been estimated around 150,000 tons/year, or about 63 kg per person per annum (Nasi et al., 2011). While urban wildmeat trade and consumption in the Amazon were considered insignificant for many years (Rushton et al., 2005), studies demonstrate that urban consumption of wildlife is still widespread in Amazonia's towns (Parry et al., 2014; van Vliet et al., 2014). Indeed, there are a number of large well-known urban markets where wild animals are sold for human consumption. The Belen market in Peru, for example, supplies wildmeat to Iquitos, the largest city in the Peruvian rainforest (Rushton et al., 2005), where large volumes of wildmeat are sold regularly (Bodmer and Lozano, 2001; Claggett, 1998). Other significant urban wildmeat markets in the Amazon region exist in towns like Pompeya, Ecuador (WCS, 2007), Abaetetuba in Pará, Brazil (Baía et al., 2010), and in the Amazon tri-frontier towns of Leticia, Tabatinga, Benjamin Constant, and Caballococha (van Vliet et al., 2014).

Long-term wildmeat market datasets have been proposed as valuable sources of information

Ethnozoology
http://dx.doi.org/10.1016/B978-0-12-809913-1.00012-0

to complement ecological data on population trends, for the evaluation of sustainability (Crookes et al., 2005; Brashares et al., 2004). Besides, understanding the local social and ecological context is also necessary to interpret monitoring data (Coad et al., 2013). We suggest that the wildmeat trade system may be viewed as a socioecological system (van Vliet et al., 2015c) where both exogenous and internal drivers need to be understood in order to interpret the changes observed in volumes and species composition. In the Amazon, the monitoring of wildmeat trade remains a challenge, probably because buying and selling wildlife is illegal in all Amazonian countries and the sale of wildmeat occurs in secret selling points (Sampaio, 2003; Bodmer et al., 2004). As such, besides the market of Iquitos, no other markets in the Amazon have been monitored over time. This constitutes a serious impediment for managers and policy makers who are left without a clear understanding of the trends and the underlying drivers of wildmeat trade and their implications for policy action.

In this chapter, we consider the wildmeat trade system as a socioecological system and use a combination of qualitative and quantitative methods, based on participatory approaches, to understand the levels, trends, and drivers of change in wildmeat sales in urban centers from the Amazon tri-frontier region between Peru, Brazil, and Colombia. In methodological terms, our participatory approach is novel as it considers the stakeholders of the trade as part of the data collection and scenario building process. We use our results to discuss the possible future of wildmeat trade in the region and discuss management options to mitigate those impacts.

WILDMEAT TRADE IN THE TRI-FRONTIER REGION BETWEEN BRAZIL, COLOMBIA, AND PERU

Description of the Social and Ecological Context

Our study was conducted in the tri-frontier region, in four towns—Leticia in Colombia (37,832 people), Caballococha in Peru (7,885 people), Tabatinga (52,272 people) and Benjamin Constant (33,411 people) in Brazil (Fig. 12.1). These towns originated from different migration waves of indigenous (mostly Ticuna and Yagua in Peru; Ticuna, Cocama, Yagua, and Uitoto in Colombia; Ticuna in Brazil) European colonists (*colonos*) or people of European and American Indian ancestry (*mestizo*) (Domínguez, 1985; DANE, 2007; IBGE, 2010; INEI, 2008, 2010, 2011; Suárez-Mutis et al., 2010).

The region is still largely forested, divisible into eight main habitat types, though largely composed of terra firme forest (dry unflooded areas), varzea forest (regularly flooded by white waters), and marshy forests (flooded by blackwaters in certain seasons) (Coelho, 2005) (4°12′54″S 69°56′28″O, Fig. 12.1). On the Colombian side *terra firme* forests predominate up to Puerto Nariño, whereas along the Brazilian and Peruvian sides marshy and varzea forests are more common. Climate in the region is typically equatorial. Average temperature is 26.2°C with a mean relative humidity of >86%. Rainfall is high, with an annual average of 3270 mm and a monthly mean of 266 mm (Domínguez, 1985). Seasonal flooding is characteristic of rivers in the region, where large tracts of rainforest are inundated during high rainfall periods. The lowest flood stage (around 445 cm) occurs between July and September (Rudas and Prieto, 2005), while the highest stage (1686 cm) is typical in May (Domínguez, 1985). Lowest rainfall is typical for August, increasing during September with a sharper rise between January and April.

The local economy is mainly based on small slash-and-burn cultivation (*chagras*) and some trade. Agricultural food production, often protein poor, is complemented by hunting and fishing (Eden, 1990). Tourism provides some alternative income, but drug trafficking (particularly along the Peruvian border) and the illegal extraction of cedar (*Cedrela* spp.) in Brazil and Peru contribute to the region's alternative source of wealth (Riaño, 2003; Zarate, 2008).

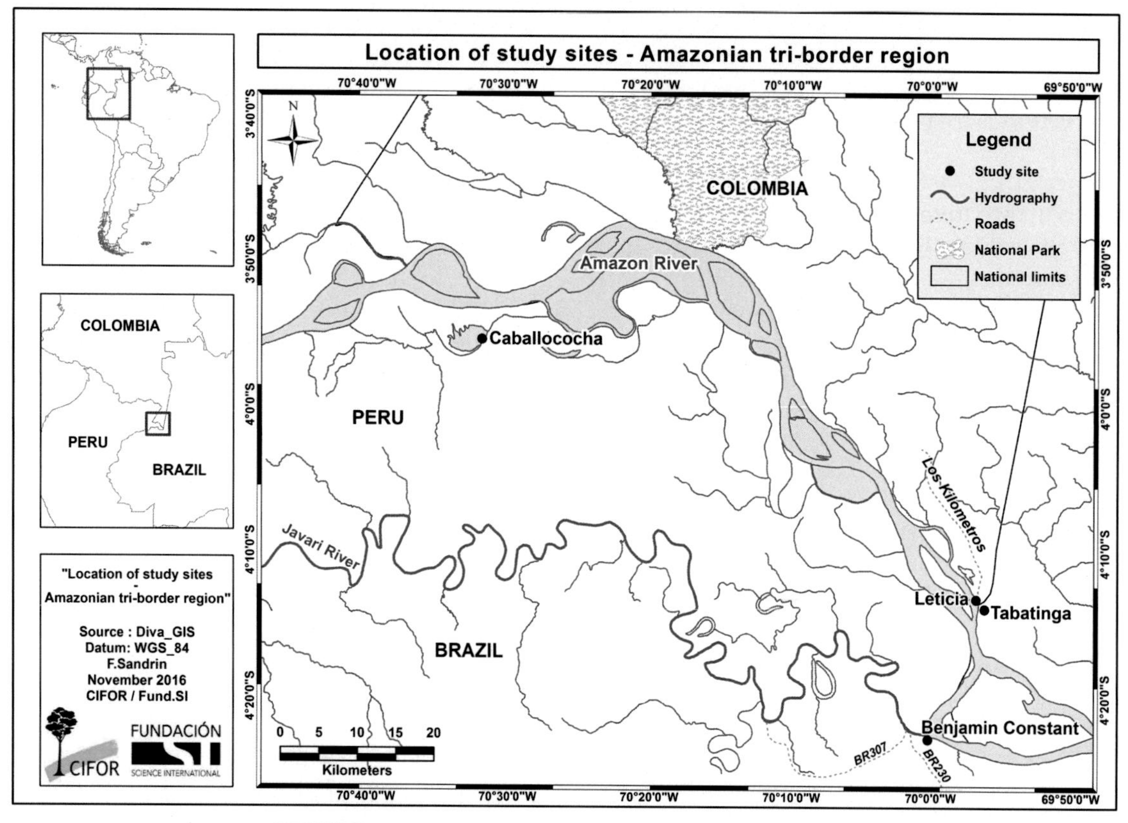

FIGURE 12.1 Study area, the Amazonian tri-frontier towns.

Urbanization, shifting cultivation, and ranching is expanding along the Colombian border, coca and cash crops in Peru, and logging and oil palm plantations on the Brazilian sector (Fig. 12.2).

FIGURE 12.2 Lively harbor life in Leticia. *Photo: Alvaro Moreno.*

Approach Used to Understand the Wildmeat Trade System

To understand the structure of the wildmeat trade system, we developed a combination of participatory methodologies covering a period of 3 consecutive years that were compatible with the fact that wildmeat trade occurs mainly in hidden channels. We spent 3–4 months in 2012 exclusively observing the market, discussing with consumers, identifying and approaching the traders through informal discussions, sharing meals, and traveling around to potential source areas. This investment of time was crucial to gain the confidence of different stakeholders and gather qualitative information on the sale points, the typology of stakeholders in the chain (from the hunters to the urban consumers), the main trade routes, and the means of transportation so as to define the overall catchment area and the relationships among stakeholders. This description was published in van Vliet et al. (2015a).

Since 2012, we spent 4 more years developing a collaborative relationship with the different stakeholders in the chain, involving them as researchers in our work and building confidence and mutual interest. In December 2015, it became possible to invite the different stakeholders from the three countries to meet and discuss a joint vision of their activity, exchange on the drivers of change and describe the changes observed in the last 20 years. For this, we organized a workshop in Leticia inviting both hunters and traders from the different towns in the tri-frontier. Thirteen hunters participating in wildmeat trade (out of 91 in the region) and 7 traders (out of 29 in the region), representative from the three countries, attended our meeting. The workshop was divided into two sessions. The first session aimed at characterizing the different components of the wildmeat trade socioecological system. The facilitators invited the participants to describe the different stakeholders involved in the wildmeat trade, the species used, the hunting grounds, and the internal and external factors influencing the system. This session was meant to build a common vision of the system and complement the information described in van Vliet et al. (2015a). The second session used a historical trend analysis where participants were first invited to describe the trends in a number of variables that characterize the trade (volumes sold, most traded species, prices, number of stakeholders involved) over time since 1995. The time period was chosen so as to cover a period of time that most participants could recall. Then they listed, in chronological order, the major events that have led to the present state of the wildmeat trade system. To facilitate the discussion, the facilitator invited the participants to describe the changes observed for each component of the socioecological system over time (changes in habitat, climate changes, changes in infrastructure, changes in human population, changes in urbanization etc.). The technique was valuable in promoting discussion of major external and local factors that have led to the present situation (Sayer et al., 2007). It also provided the basis for valuable discussion of possible future scenarios. The discussions provided insights into origins of the present situation and the drivers of change that could lead to both desirable and undesirable future outcomes.

Structure of the Wildmeat Trade Socioecological System

The wildmeat trade system in towns from the tri-frontier region can be understood as a socioecological system as shown in Fig. 12.3.

The ecological system is composed of three main components: (1) the hunting grounds in private or indigenous land; (2) the habitats where wildmeat species are found (both secondary and primary forests in terra firme, varzea, and marshy forests); and (3) the species hunted and sold [mostly *Cuniculus paca* (paca), *Tapirus terrestris* (tapir), *Pecari tajacu* (collared peccary), and *Tayassu pecari* (white-lipped peccary)].

The social system is composed of the traders in markets, the hunters that provide them with meat (and traders provide them with inputs such as ammunitions or gasoline), the restaurants and other clients who purchase the meat, as well as the local authorities who enforce the prohibition to sell meat. The relationship among these stakeholders and their resources are described in detail in van Vliet et al. (2015a). In Peru, wildmeat trade is closely linked to coca activities either because the main clients are coca workers (in the case of Caballococha) or because wildmeat follows the same routes as drugs (from Caballococha to Brazilian towns along the Amazon river).

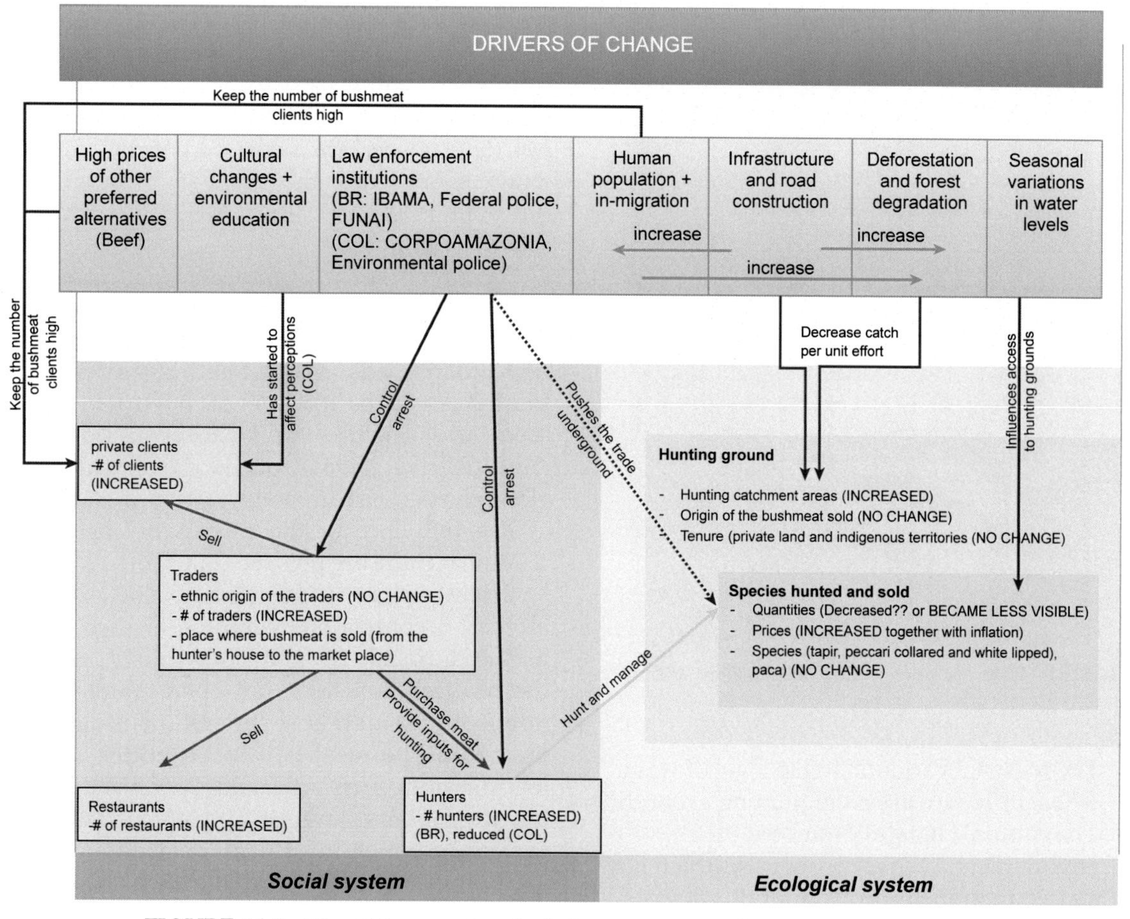

FIGURE 12.3 The wildmeat socioecological system in towns from the tri-frontier region.

Enforcement in Brazil targets the hunters through the FUNAI office (National Indian Foundation), which controls the illegal access by colono and caboclo hunters to indigenous territories and IBAMA (Instituto Brasileiro do Meio Ambiente e dos Recursos Naturais Renováveis), which enforces the trade in urban areas. In Colombia, CORPOAMAZONIA is the regional environmental agency in charge of enforcing the illegality of the trade. Although commercial hunting in theory could be legal under law 611/2000 and Decree 4688/2005, it is not possible to obtain a trading license due to current legal bottlenecks (van Vliet et al., 2015d) (Fig. 12.4).

The main drivers of change in the system in order of importance as perceived by the users are as follows:

1. The levels of law enforcement, which influence the way wildmeat is sold and the risks taken by the hunters and traders;
2. The high prices of other preferred meats, such as beef, which maintain wildmeat at very competitive prices;

FIGURE 12.4 Salting peccary meat being sold at the market. *Photo: François Sandrin.*

3. The seasonal variation in the level of water, which influence access to hunting grounds;
4. The cultural changes promoted by environmental school programs, which have started to influence consumption choices by the younger generation, particularly on the Colombian border.

Historical Changes Observed in the Wildmeat Trade System Since 1995

The changes observed in wildmeat trade since 1995 are presented in Table 12.1.

Species: The list of species sold has not changed over the years.

Number of stakeholders: The number of stakeholders involved in the commodity chain has increased, particularly in Brazil. In Peru, the sale of wildmeat was associated with coca production activities because the *cocaleros* (workers in coca leaf recollection) were the mayor consumers of wildmeat. With the anticoca program in Peru, the sale of wildmeat has also dropped.

Volumes: The volumes of wildmeat sold have not reduced despite increased control and law enforcement (since the opening of the FUNAI office in Brazil and the CORPOAMAZONIA office in Colombia).

Catchment area: Twenty years ago, the wildmeat used to come mainly from the Javari and the Amazon rivers. Since the construction of peri-urban roads, many hunting grounds are located along the road. As such, current catchment areas include the Javari river sides, the Amazon river sides, Atacuari river side, and along the peri-urban roads.

The influence of the drivers of change on wildmeat trade are presented in Table 12.2.

Effect of law enforcement: The effect of increased control has been the change in the way wildmeat is sold: previously sold at the hunters' houses in an open manner, it is now sold by market sellers who bear most of the risks. The meat is sold in hidden fridges in Colombia. It is still sold openly in Brazil, but the traders have an efficient information system in which informants, placed in strategic points of the town, may call by phone in case the authorities are seen to approach the market place.

Prices of other preferred alternatives: With inflation, the prices of other preferred foods have increased and maintained wildmeat at competitive prices. While wildmeat prices have also increased with inflation (due to the increased costs of ammunitions and fuel), they have remained lower than those of other highly preferred meats.

Demography: Human population increase and in-migration have increased demand for food and driven the conversion of forest land into farms, infrastructure, and increased urban areas, reducing habitat for wildlife and increasing the efforts that hunters must invest to hunt the same amount as in previous years.

TABLE 12.1 Changes in Wildmeat Trade Variables Since 1995 as Perceived by Local Stakeholders

	Country	1995	2000	2005	2010	2015	Observations
Quantities sold	*Brazil*	XXXXX	XXXXX	XXXX	XXXX	XXXX	Quantities are not necessarily reduced but they have found ways to sell more underground.
	Peru	XXXXX	XXXXX	XXXXX	XXXXX	XX (decline in 2014 after coca eradication campaign)	
	Colombia	XXXXX	XXXXX	XXXXX	XXXXX	XXXX	
Species sold	*Brazil and Colombia*	Tapir, peccary (white-lipped and collared), paca					NO CHANGE
	Peru	Tapir, peccary (white-lipped and collared), paca)				no tapir anymore	NA
Bushmeat prices	*Brazil*	2.5 reals	X	XX	XXX	15 reals	Increase with inflation
	Peru	X	XX	XXX	8 soles	10–12 soles	
	Colombia	300 COP	X	XX	XXX	12–13 000 COP	
Origin of the bushmeat traded in urban areas	*Brazil*	Javari, Asentamento, Cajari			Javari, Br307, Asentamento, Cajari		No change, except that Br307 road opened more access to hunting grounds between Benjamin Constant and Atalaia do Norte
	Peru	Tierra, Amarilla, and Atacuari rivers					
	Colombia	Caballococha, indigenous communities along the Amazon, peri-urban area of los Kilometros					
Catch per unit effort	*all*	XXXX	XXXX	XXX	XX	XX	Longer distances are necessary to find prey
Places where bushmeat is sold	*Brazil*	Sold at hunters' houses	Sold openly in markets, with a very efficient information network to avoid confiscations				Trade occurs increasingly underground
	Colombia	Sold at hunters' houses	Sold openly in markets		Sold undergound, only to known consumers		

Continued

TABLE 12.1 Changes in Wildmeat Trade Variables Since 1995 as Perceived by Local Stakeholders—cont'd

	Country	1995	2000	2005	2010	2015	Observations
Number of clients	*Brazil and Colombia*	Increased with increased urban population and the high price of beef					Cultural changes are not seen as influencing the demand for bushmeat
	Peru	No information			Coca eradication campaign drastically reduced the number of clients who were all coca workers		
	Colombia	XXXX	XXXX	XXXX	XXX		Demand might decrease in Colombia due to environmental campaigns that influence the choices of the younger generation
Number of hunters	*Brazil*	XX	XXX	XXXX	XXXXX	XXXXXXX	Differences among countries in the number of hunters involved in the trade
	Peru				XXX	X	
	Colombia	XXXX	XXX	XX	X	X	
Colombia Number of traders in urban markets	*Brazil*	n=5	–	–	–	n=40	Increased number of traders but particularly in Brazil
	Caballococha	–	–	–	n=5	n=1	
	Colombia	n=2	–	–	–	n=5	

The number of X represents a qualitative measure of each variable and it is trends over time.

TABLE 12.2 Changes in the Drivers of Change and Their Influence on Wildmeat Trade Since 1995 as Perceived by Local Stakeholders

Driver	Country	1995	2000	2005	2010	2015	Comments
Control and law enforcement	*Brazil*	Demarcation Valle do Javari and creation of a National Indian Foundation (FUNAI) office					Increased enforcement has influenced the way bushmeat is sold and the risks incurred by hunters and traders. Law enforcement has potentially slightly reduced biomass sold but has not influenced prices
	Peru	Very little or no law enforcement, only sporadic environmental campaigns					
	Colombia		CORPOAMAZONIA		Bushmeat trade becomes punishable by penal court		
Inflation	*all*	X	XX	XXX	XXXX	XXXXX	Prices of ammunition and fuel have increased with inflation, and this has had consequences in bushmeat prices
Deforestation, timber extraction, agriculture	*all*	X	XX	XXX	XXXX	XXXXX	Deforestation and forest degradation has reduced habitat and food for wildlife increasing effort for hunting
Human population	*Brazil*	X	XX	In-migration along the road	XXXX	XXXXX	Increased urban and rural population has increased bushmeat demand in urban areas, but also increased hunting pressure in rural areas
	Colombia	X	XX	XXX	XXXX	XXXXX	
Road construction	*Brazil*			BR 307 built between Benjamin Constant and Atalaia do Norte			New rural settlements were created along the newly built road increasing agricultural activities and reducing wildlife habitat
Environmental education campaigns in schools	*Brazil*				X	XX	Environmental education campaigns influence the way the youth value conservation and start being opposed to the use of natural resources
	Peru	None					
	Colombia			X	XX	XXXX	

The number of X represents a qualitative measure of each variable and it is trends over time.

FIGURE 12.5 Participants to the workshop from Brazil, Colombia, and Peru. *Photo: Nathalie van Vliet.*

Environmental school programs: Environmental school programs seem to be influencing consumption patterns of the younger generation, which now demises wildmeats for chicken or industrial meats (Fig. 12.5).

CONTEMPORARY WILDMEAT TRADE VOLUMES IN THE TRI-FRONTIER REGION BETWEEN BRAZIL, COLOMBIA, AND PERU

Participatory Quantitative Assessment of Contemporary Wildmeat Trade

Based on preliminary work on the wildmeat trade in the region (see van Vliet et al., 2014 for more information on the methodology), we replicated a participatory monitoring system in 2013, 2014, 2015, and 2016. We trained eight market traders (out of N=29 traders and sampling equal to 27%) to act as research assistants for the collection of information on trade in the towns of Leticia (Colombia), Tabatinga (Brazil), Benjamin Constant (Brazil), and Caballococha (Peru). Monitors were chosen according to three main criteria: (1) good geographical representation of the sample; (2) regular presence in the market (traders that sold wildmeat occasionally were not selected); (3) a willingness and trust to participate in the project. The eight monitors selected are actually the only ones that sell wildmeat on a daily basis. Thus, the sum of what they sell represents a minimum of the total wildmeat sales in the towns from the tri-frontier.

In 2015, due to the insecurity situation created by the anticoca campaign in Peru, wildmeat traders did no longer sell in the markets and it was impossible for the researchers to monitor any underground trade still going on in Caballococha. Traders from Caballococha reported that the trade in wildmeat was very insignificant during the anticoca campaign. As a result, the data for 2015 only reports wildmeat traded in towns from Brazil and Colombia and trade in Peru was considered null.

To ensure the quality of self-reported data, we chose to limit the monitoring period to a short representative period, rather than having a long-term and nonterminal monitoring system which would have probably created biases due to research fatigue. As such, market sellers were asked to record data for 10 consecutive days, during the low-level water (September) and 10 consecutive days during the high-level waters (May). In addition, to ensure greater accuracy, the researchers visited market sellers on a daily basis during the monitoring period in case the traders wanted to double-check methodological matters before filling in the questionnaires. For all records of wildmeat carcasses sold, the traders registered information on species, condition (fresh, salted, smoked), biomass, and price in a notebook specifically designed by the research team for data collection. In 2015 and 2016, we actually changed the paper monitoring notebooks by tablets and the monitoring process was made using a questionnaire developed using a phone App called Kobocollect. This methodology proved to be very practical for the users, as it allowed entering data in a very user-friendly manner, saving time as well (van Vliet et al., 2017) (Fig. 12.6).

FIGURE 12.6 Indigenous hunter selling fresh tapir meat in Brazil. *Photo: Lindon Neves de Aquino.*

FIGURE 12.7 Salted tapir meat traded at the market. *Photo: Nathalie van Vliet.*

Current Wildmeat Trade Levels

The biomass sold during the 20-day monitoring period per year as registered by the eight traders participating in this study varied along the years. The amount sold during the study period was equivalent to 5.9 tons in 2013, 4.3 tons in 2014, 1.2 tons in 2015 (only for Brazil and Colombia), and 2.5 tons in 2016. Based on biomass amounts traded in 2013, 2014, 2015, and 2016, extrapolated to the whole year, the minimum biomass traded in markets from the catchment area varied from 21.6 tons/year to 106.2 tons/year, and the maximum calculated as if all the 29 traders were equally active varies from 78.3 tons/year to 385 tons/year. Biomass sold per trader was higher during the high-water period (differences between the two periods were nonsignificant in 2013; Kruskal Wallis test, $P=.94$), 2015 (Kruskal Wallis test, $P=.13$), and 2016 (Kruskal Wallis test, $P=.89$), but significant in 2014 (Kruskal Wallis, $P<.001$) (Figs. 12.3 and 12.7).

During the 4-year monitoring period, traders in Brazil contributed to 65% of the total biomass sold. Biomass per trader was significantly lower in Colombia than in Peru (Kruskal Wallis test, $P<.02$) and Brazil (Kruskal Wallis test, $P<.001$). Wildmeat volumes steadily decreased in Colombia and Brazil from 2013 to 2015, and then increased again in 2016. In Peru, volumes increased from 2013 to 2014 and then dropped to 0 in 2015 to recover in 2016 to a value higher than that of 2014. The variations between the years were higher in Brazil and Peru than in Colombia (Fig. 12.8).

Species Sold and Prices

Market traders sold at least 18 wildmeat species in 2013, 14 species in 2014, 11 species

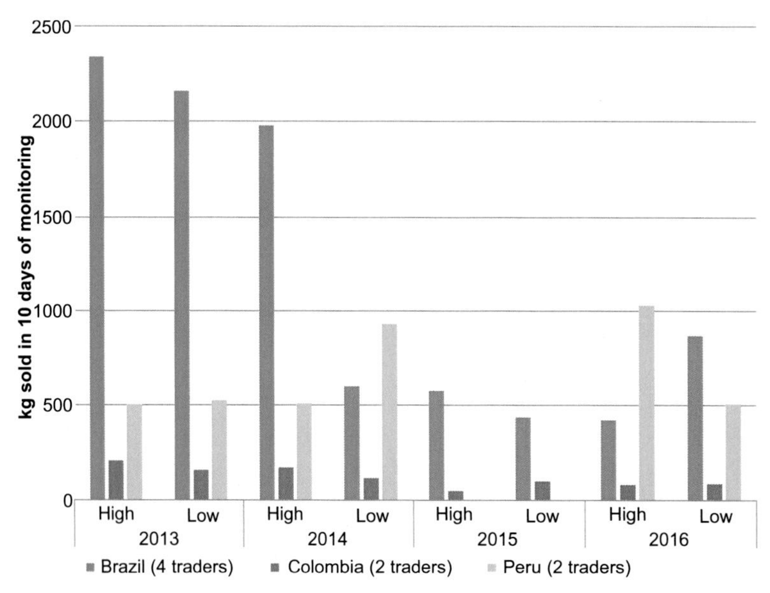

FIGURE 12.8 Biomass traded in 2013, 2014, 2015, and 2016, during low and high level water seasons in Colombia, Peru, and Brazil (10 monitoring days per season each year).

in 2015, and 9 in 2016. In the 4 years, mammals contributed 86.6% of the total biomass, whereas birds and reptiles accounted for 2.4% and 10.9%, respectively. In terms of biomass, the most commercialized species were *Cuniculus paca* (paca), *Tapirus terrestris* (tapir), *Pecari tajacu* (collared peccary), *Mazama* sps. (deer), and *Tayassu pecari* (white-lipped peccary) (Fig. 12.4), but the contribution of each species to the total biomass varied considerably between the 4 years. Indeed, while paca was the most important species in 2013 (22% of the total biomass sold), it only contributed to 13% of the biomass in 2015. On the contrary, while tapir contributed to only 25% of the total biomass in 2013, it represented as much as 68% in 2015 (Fig. 12.9).

Wildmeat is sold fresh (32%), salted (35%), chilled/frozen (17%), alive (7.1%), and smoked (8.9%). In Leticia, wildmeat is almost exclusively sold fresh (92%), whereas in Caballococha it is mostly salted (31.2%), smoked (30%), and fresh (30%) and for the case of Tabatinga and Benjamin Constant it is sold mostly salted (40.2%), fresh (25%), and chilled/frozen (23%) (Fig. 12.10).

The average price for fresh meat from the most consumed species in the market place was US$ 4.7/kg. Given enforcement, market sellers are unable to increase prices when wildmeat becomes scarcer (e.g., dry season). Smoked or salted wildmeat was about 20% cheaper than fresh wildmeat. The main clients of wildmeat traders in markets were *colono* or *mestizo* families, restaurant owners, and public authorities. Coca workers were the main clients of market sellers in Caballococha. Wildmeat was significantly cheaper in Peru, than in Colombia or Brazil (Kruskal Wallis test, $P<.001$). Prices for the whole region (in USD equivalent for each of the monitoring periods) decreased steadily from 2013 to 2015 and then increased again in 2016 (Table 12.3). However, this trend is probably due to the increased value of the USD from 2013 to 2015. Indeed, when prices are compared within each

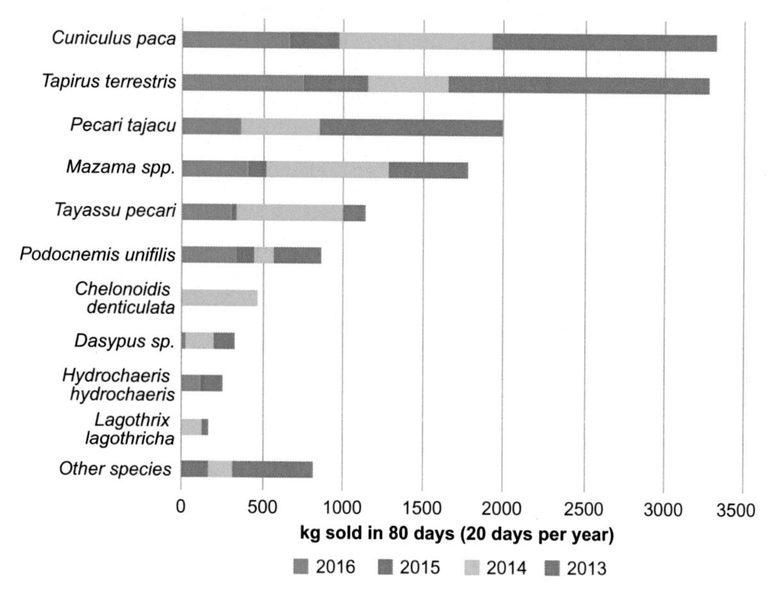

FIGURE 12.9 Species and biomass traded in 2013, 2014, 2015, and 2016 (20 monitoring days per year).

FIGURE 12.10 Fish and wildmeat market. *Photo: Nathalie van Vliet.*

country over the study period using the local currency, a steady and significant increase in prices was observed for Colombia (Kruskal Wallis test, $P = .002$) and Brazil (Kruskal Wallis test, $P < .001$), although this increase is lower than that of national inflation.

WILL THE DEMISE OF WILDMEAT TAKE PLACE IN THE TRI-FRONTIER BETWEEN PERU, BRAZIL, AND COLOMBIA?

What Is the Effect of Urbanization?

Our study suggests that urban wildmeat trade in the tri-frontier region is maintained over time by the increased in-migration to the area and the high prices of other preferred meats (such as beef). Urbanization of forest wilderness is particularly important in the Brazilian Amazon, where eight million city-dwellers are poor (IBGE, 2010). As shown by Parry et al. (2010), remote headwaters are abandoned through out-migration to local urban centers, motivated by poor access to public services, transport, and trade. These results support the conclusions from Parry and Peres (2015) that urbanization of forested wilderness continue to drive demand for wildlife as food in Amazonia. Despite increased access to cheap chicken and processed foods in

TABLE 12.3 Average Sale Prices per Kilogram of Fresh Wildmeat (USD/kg) for the Five Most Traded Species in Terms of Biomass, During the Four Monitoring Years

Most Traded Species (In Terms of Biomass)	Average Selling Price (USD)											
	Brazil				Colombia				Peru			
	2013	2014	2015	2016	2013	2014	2015	2016	2013	2014	2015	2016
Cuniculus paca	5.39	5.0	3.96	4.43	6.82	7.3	5.28	4.41	4.24	4.32	–	3.99
Tapirus terrestris	5.45	5.28	4.31	4.49	–	–	–	–	3.34	3.04	–	2.58
Pecari tajacu	5.48	–	–		6.0	6.0	–	–	4.91	4.05	–	3.60
Mazama spp.	5.46	5.42	3.96	4.68	7.0	5.86	4.2	3.6	3.84	3.69	–	3.76
Tayassu pecari	5.71	5.51	4.71		6.0	–	–	–	4.26	3.77	–	3.27

the region (van Vliet et al., 2015b), demand for wildmeat remains high as an occasional luxury meal, given the cultural attachment to traditional and tasty meats (Morsello et al., 2015). Our study shows once again that the demise of wildmeat is not taking place in urban Amazon. Only consumed sporadically, it remains part of the menu and demand is not bound to disappear in the next generation. There are a number of factors that might curve the demand in the future though: the effect of environmental education on consumption choices and the availability, prices, and food preferences of the younger generations.

Is Law Enforcement Helping to Reduce the Trade?

Wildmeat trade is greatly affected at times and in places where law enforcement efforts are put in place, but not necessarily in terms of volumes traded. During our study period from 2013 to 2016, the volumes traded varied particularly according to the presence/absence of law enforcement patrols. For example, between 2014 and 2015, patrols increased in all countries at the same time: in Brazil due to security increase during the football world cup, in Colombia due to presidential elections, and in Peru due to a severe anticoca campaign, which resulted in a drop of wildmeat sales in the region.

However, according to the stakeholders, law enforcement mostly influences the ways and places where wildmeat is sold. The sporadic enforcement activities clearly lead to more underground channels, but not to the reduction of the trade. As also demonstrated by Parry and Peres (2015), blanket bans and heavy punishments are poorly enforced and offer limited deterrent to illegal behavior. Instead, they incentivize the development of diverse strategies to avoid confiscations, such as trade via cell phones, informants posted at strategic positions around the market to inform traders if the environmental police is around, etc. Despite law enforcement, the number of stakeholders in the trade chain has increased over the last 20 years.

Is the Level of Trade Ecologically Unsustainable?

Whether the trade is sustainable or not, requires a species by species assessment based not only on trade data, but also on population surveys over time. In this study we only analyze one side of the coin by providing information on trade levels and species sold. Currently, the mean volume of wildmeat traded in markets from the tri-frontier towns may reach up to 385 tons per year for an urban population of 146,555 habitants (2.6 kg/hab/year). These results are about three times higher than those described by Baia et al. (2010) in Abaetetuba, Brazil (130,000 habitants): 6 tons in 17 days, equivalent to about 128 tons per year (0.98 kg/hab/year). Our results show that wildmeat trade in Amazonian towns is not totally insignificant as compared to those reported for Central African countries, as was thought before (Rushton et al., 2005). We also show that the volumes traded vary from year to year according to several drivers that need better long-term understanding in order to find appropriate solutions to unsustainable use.

The list of species sold has not changed and the source areas have not changed either in the last decades, indicating some stability in the system. However, the hunters perceive a reduction in catch per unit effort, particularly due to habitat degradation and transformation into farmland. Overall, two of the most commercialized species in biomass terms are listed as "Least Concern" by IUCN red list (collared peccary and paca) (IUCN, 2016), but tapir and the white-lipped peccary, listed as "Vulnerable," are also among the most commercialized. The yellow-footed tortoise and yellow-spotted turtle are among the most frequently traded species and are both classified as "Vulnerable" by the IUCN red list. The sustainability of the trade therefore requires close attention. Indeed, with the increased number of stakeholders involved and decreased suitable wildlife habitat, it is important to monitor whether hunting is becoming unsustainable in some areas for the most vulnerable species.

References

Baía Jr., P.C., Guimarães, D.A., Le Pendu, Y., 2010. Non-legalized commerce in game meat in the Brazilian Amazon: a case study. Revista de Biología Tropical 58 (3), 1079–1088.

Bodmer, R.E., Lozano, E.P., 2001. Rural development and sustainable wildlife use in Peru. Conservation Biology 15 (4), 1163–1170.

Bodmer, R.E., Lozano, E.P., Fang, T.G., 2004. Economic analysis of wildlife use in the Peruvian Amazon. In: Silvins, K.M., Bodmer, R., Fragoso, J.M.V. (Eds.), People and Nature: Wildlife Conservation in South and Central America. Columbia University, New York, USA, pp. 191–207.

Brashares, J.S., Arcese, P., Sam, M.K., Coppolillo, P.B., Sinclair, A.R.E., Balmford, A., 2004. Wildmeat hunting, wildlife declines, and fish supply in West Africa. Science 306, 1180–1183.

Claggett, P.R., 1998. The spatial extent and composition of wildlife harvests among three villages in the Peruvian Amazon. In: 1998 meeting of the Latin American Studies Association. The Palmer House Hilton Hotel, Chicago, Illinois.

Crookes, D.J., Ankudey, N., Milner-Gulland, E.J., 2005. The value of a long-term wildmeat market dataset as an indicator of system dynamics. Environmental Conservation 32 (4), 333–339.

Coad, L., Schleicher, J., Milner-Gulland, E.J., Marthews, T.R., Starkey, M., Manica, A., Balmford, A., Mbombe, W., Dio Bineni, T.R., Abernethy, K.A., 2013. Social and ecological change over a decade in a village hunting system, Central Gabon. Conservation Biology 27, 270–280. http://dx.doi.org/10.1111/cobi.12012.

Coelho, M.R., Fidalgo, E.C.C., Araújo, F.O., Santos, H.G., Mendonça-Santos, M.L., Pérez, D.V., Moreira, .F.M.S., 2005. Levantamento pedológico de uma área-piloto relacionada ao projeto BiosBrasil (Conservation and sustainable management of below-ground biodiversity: phase I) município de Benjamin Constant (AM): janela 6. Boletim de Pesquisa e Desenvolvimento 68. Embrapa, Rio de Janeiro, Brazil. [Online] URL: http://ainfo.cnptia.embrapa.br/digital/bitstream/item/89680/1/bpd68-2005-area-piloto-proj-bios-br.pdf.

DANE (Departamento Administrativo Nacional de Estadística), 2007. Censo General 2005. [online] URL: http://www.dane.gov.co/index.php?option=com_content&view=article&id=307&Itemid=124.

Domínguez, C., 1985. Amazonia Colombiana: Visión General. Biblioteca Banco Popular, Bogotá, Colombia.

Eden, M.J., 1990. Ecology and Land Management in Amazonia. Belhaven Press, London, U.K.

IBGE (Instituto Brasileiro de Geografia e Estatística), 2010. Joint Statistical Publication by BRIC Countries. Brazil, Russia, India, China. Gerência de Editoração/Centro de Documentação e Disseminação de Informações - CDDI/IBGE, Rio de Janeiro, Brazil.

INEI (Instituto Nacional de Estadística e Informática), 2008. Perú: Crecimiento y distribución de la población, 2007. Fondo de Población de las Naciones Unidas, Lima, Perú.

INEI (Instituto Nacional de Estadística e Informática), 2010. Perú: Análisis Etnosociodemográfico de las Comunidades Nativas de la Amazonía, 1993 y 2007. Dirección Técnica de Demografía e Indicadores Sociales – Fondo de población de las Naciones Unidas, Lima, Perú.

INEI (Instituto Nacional de Estadística e Informática), 2011. Perú: anuario de estadísticas ambientales – 2011. INEI, Lima, Perú. [Online] URL: http://www.inei.gob.pe/BiblioINEIPub/BancoPub/Est/Lib0978/index.html.

IUCN (International Union for Conservation of Nature), 2016. IUCN Red List of Threatened Species. Version 2013.2. Cambridge, United Kingdom [Online] URL: www.iucnredlist.org.

Morsello, C., Yagüe, B., Beltreschi, L., van Vliet, N., Adams, C., Schor, T., Quiceno-Mesa, M.P., Cruz, D., 2015. Cultural attitudes are stronger predictors of bushmeat consumption and preference than economic factors among urban Amazonians from Brazil and Colombia. Ecology and Society 20 (4), 21.

Nasi, R., Taber, A., van Vliet, N., 2011. Empty forests, empty stomachs? Wildmeat and livelihoods in the Congo and Amazon Basins. International Forestry Review 13 (3), 355–368.

Parry, L., Day, B., Amaral, S., Peres, C., 2010. Drivers of rural exodus from Amazonian headwaters. Population and Environment 32, 137–176.

Parry, L., Barlow, J., Pereira, H., 2014. Wildlife harvest and consumption in Amazonia's urbanized wilderness. Conservation Letters 7 (6), 565–574.

Parry, L., Peres, C.A., 2015. Evaluating the use of local ecological knowledge to monitor hunted tropical forest wildlife over large spatial scales. Ecology and Society 20 (3), 15.

Riaño, E., 2003. Organizando su espacio, construyendo su territorio. Transformaciones de los asentamientos Ticuna en la ribera del Amazonas colombiano. Unibiblos, Bogotá, Colombia.

Rudas, A., Prieto, A., 2005. Flora of the Amacayacu Amazonas National Natural Park, Colombia. Missouri Botanical Garden Press.

Rushton, J., Viscarra, R., Viscarra, C., Basset, F., Baptista, R., Brown, D., 2005. How Important Is Wildmeat Consumption in South America: Now and in the Future? Odi Wildlife Policy Briefing (1) ODI, London, United Kingdom. [Online] URL: http://www.odi.org.uk/resources/download/2418.pdf.

Sampaio, P.A.M., 2003. Comércio ilegal de carne de animais silvestres em quatro feiras livres do estuario amazônico, Estado do Pará-Brasil. Trabalho de Conclusão de Curso, Universidade Federal do Pará, Centro de Ciências Biológicas, Belém, Pará, Brazil.

Sarti, F.M., Adams, C., Morsello, C., Van Vliet, N., Schor, T., Yagüe, B., Tellez, L., Quiceno-Mesa, M., Cruz, D., 2015. Beyond protein intake: bushmeat as source of micronutrients in the Amazon. Ecology and Society 20 (4), 22.

Sayer, J., Campbell, B., Petheram, L., Aldrich, M., Perez, M.R., Endamana, D., Dongmo, Z.L.N., Defo, L., Mariki, S., Doggart, N., Burgess, N., 2007. Assessing environment and development outcomes in conservation landscapes. Biodiversity and Conservation 16, 2677–2694.

Sills, E., Shanley, P., Paumgarten, F., de Beer, J., Pierce, A., 2011. Evolving perspectives on non-timber forest products. In: Shackelton, S., Shackelton, C., Shanley, P. (Eds.), Non-timber Forest Products in the Global Context. Springer-Verlag Berlin Heidelberg, Germany, pp. 23–51.

Sirén, A.H., Machoa, J.D., 2008. Fish, wildlife, and human nutrition in tropical forests: a fat gap? Interciencia 33 (3), 186–193.

Suárez-Mutis, M., Mora, C., Pérez, L., Peiter, P., 2010. Interacciones transfronterizas y salud en la frontera Brasil-Colombia-Perú. Revista Mundo Amazónico 1, 243–266.

van Vliet, N., Quiceno-Mesa, M.P., Cruz-Antia, D., Neves de Aquino, L.J., Moreno, J., Nasi, R., 2014. The uncovered volumes of bushmeat commercialized in the Amazonian trifrontier between Colombia, Peru & Brazil. Ethnobio Conserv 3, 7. http://dx.doi.org/10.15451/ec2014-11-3.7-1-11.

van Vliet, N., Quiceno, M.P., Cruz, D., Neves de Aquino, L.J., Yagüe, B., Schor, T., Hernandez, S., Nasi, R., 2015a. Wildmeat networks link the forest to urban areas in the trifrontier region between Brazil, Colombia, and Peru. Ecology and Society 20 (3), 21.

van Vliet, N., Quiceno-Mesa, M.P., Cruz-Antia, D., Tellez, L., Martins, C., Haiden, E., de Oliveira, M.R., Adams, C., Morsello, C., Valencia, L., Bonilla, T., Yagüe, B., Nasi, R., 2015b. From fish and wildmeat to chicken nuggets: the nutrition transition in a continuum from rural to urban settings in the Colombian Amazon region. Ethnobiology and Conservation 4 (6), 1–12.

van Vliet, N., Fa, J., Nasi, R., 2015c. Managing hunting under uncertainty: from one-off ecological indicators to resilience approaches in assessing the sustainability of bushmeat hunting. Ecology and Society 20 (3), 7. http://dx.doi.org/10.5751/ES-07669-200307.

van Vliet, N., Gomez, J., Quiceno-Mesa, M.P., Escobar, J.F., Andrade, G., Vanegas, L.A., Nasi, R., 2015d. Sustainable wildlife management and legal commercial use of bushmeat in Colombia: the resource remains at the cross-road. International Forestry Review 17 (4).

van Vliet, N., Sandrin, F., Vanegas, L., Ĺharidon, L., Fa, J.E., Nasi, R., 2017. High-tech participatory monitoring in aid of adaptive hunting management in the Amazon. Unasylva 249, FAO.

WCS (Wildlife Conservation Society), 2007. El tráfico de carne silvestre en el Parque Nacional Yasuní: Caracterización de un mercado creciente en la Amazonía norte del Ecuador. Programa Ecuador Boletín No. 2 WCS, Quito, Ecuador. [Online] URL: http://s3.amazonaws.com/WCSResources/file_20110823_035823_ecu_pub_ProgramaEcuadorBoletin2_2007_oAuB.pdf.

Zarate, C., 2008. Fronteras en la globalización: localidad, biodiversidad y comercio en la Amazonia. Observatorio Andino. Universidad Javeriana, Fundación Konrad Adenauer. V. 1, Bogotá, Colombia.

Further Reading

Ojasti, J., 2000. Manejo de fauna neotropical. Instituto de Zoología Tropical, Caracas, Venezuela.

CHAPTER

13

Animals and Human Health: Where Do They Meet?*

Rômulo Romeu Nóbrega Alves[1], *Iamara da Silva Policarpo*[2]

[1]**Universidade Estadual da Paraíba, Campina Grande, Brazil;** [2]**Universidade Federal da Paraíba, João Pessoa, Brazil**

INTRODUCTION

Humans share the planet with a bewildering variety of other animals and plants, forming an intricate web of interactions. Disturbances that negatively impact the environment or the health of biological organisms will affect the harmonious functioning of their interactions; thus, human health depends on the health of the organisms with which we interact and the environment in which we all live.

The interrelationships between society and nature and the importance of environmental health to human health have become widely acknowledged (Alves and Rosa, 2007; WHO, 2005), drawing attention to the fact that biodiversity losses will have direct or secondary effects on human well-being. Thus, human health must not be considered in isolation, for it depends on the quality of the environment in which we live—for people to be healthy, the environment must be healthy (Alves and Rosa, 2007). The recognition that human, animal, and environmental health are linked generated the "One Health" concept, which is defined by the One Health Commission (2010) as "the collaborative effort of multiple disciplines to obtain optimal health for people, animals, and our environment." Mi et al. (2016) emphasized that one health seeks to understand the interactions between humans, animals, and environmental factors, and their impacts on health.

Extremely close connections have existed between humans and animals throughout history (Alves, 2012), and we likewise share hundreds of illnesses. Cross-species transmissions and the emergence and eventual evolution of a plethora of infectious pathogens have been observed ever since the establishment of human–animal interfaces (Reperant et al., 2012), with links between human and animal health having profound effects on almost every aspect of our lives.

*This chapter is a revised and updated version of the section "The role of fauna in human health," extracted from an article by Alves, R.R.N., 2012. "Relationships Between Fauna and People and the Role of Ethnozoology in Animal Conservation. Ethnobiology and Conservation 1," 1–69.

Ethnozoology
http://dx.doi.org/10.1016/B978-0-12-809913-1.00013-2

The links between animals and human health have been substantiated throughout the history of mankind, from causes to cure of human diseases. Seven main aspects should be highlighted when considering this connection: (1) animals as the cause/disseminator of diseases for humans and vice versa; (2) animals as sentinels of human health; (3) the use of animals in traditional medicine systems; (4) animal-assisted therapy; (5) biotherapy; (6) animals as a source of drugs; and (7) use of animals in medical research. These aspects will be briefly discussed in this chapter.

Animals as Cause/Disseminator of Disease in Humans and Vice Versa

Since ancient times, human beings have related the appearance of certain diseases and epidemics to the presence or influence of animals that are considered to presage bad omens, diseases, and death (Ávila-Pires, 1989). This is not surprising, considering that the natural world has a strong influence on the transmission of disease to humans from animals and vice versa, and the perception of more primitive societies therefore certainly reflects daily experiences. As pointed by Wolfe et al. (2007), human hunter/gatherer populations currently suffer, and presumably have suffered for millions of years, from infectious diseases similar or identical to diseases of other wild primate populations.

When an infectious agent responsible for a human disease is also capable of infecting other species, these species may act as reservoirs or vectors for the disease (European Commission, 2011). Arthropods, for example, transmit hundreds of different known infectious and parasitic agents to humans and animals around the world. These vectors include almost all forms of blood-sucking arthropods: mosquitoes, ticks, mites, biting midges, sand flies, kissing bugs, bed bugs, black flies, lice, fleas, and deer and horse flies (Seymour, 1984).

Diseases and infections that are naturally transmitted between animals and humans are known as zoonoses (Bell et al., 1988; Krauss, 2003) and have been known to affect human health throughout history (Kruse et al., 2004). Such diseases have an important impact on public health and economy, and on the conservation of wildlife (Cleaveland et al., 2001). Exposure of humans to zoonoses occurs in different ways, from well-known or well-understood direct transmission routes, such as bites and rabies, to less obvious pathways, the risk factors or potential exposure routes of which are difficult to recognize and are interlinked in a relationship network between human beings, animals, and the environment (Friend, 2006). The most frequent sources of zoonose transmission comprise food and contaminated water, vector insect bites and scratches, or bites from infected animals (Chomel, 2002).

Zoonotic diseases account for approximately 75% of emerging infectious diseases (Chomel et al., 2007; Taylor et al., 2001). A literature search showed that more than 800 human pathogens are zoonotic (Taylor et al., 2001; Woolhouse and Gowtage-Sequeria, 2005). According to Weiss (2001), some of these pathogens may cause serious diseases in wild animals but, in some cases, the animals act as reservoirs, without showing any clinical symptoms (Williams et al., 2002). As mentioned above, zoonoses can be transmitted by direct contact with infected animals, dead or alive, which are used by humans in several ways, including consumption as food or as pets. The consumption of animal products as food or in traditional medicine, for example, facilitates the transmission of serious and widespread zoonoses, such as tuberculosis or rabies (De Smet, 1991; Schnurrenberger and Hubbert, 1981; Still, 2003). Another example deserving mention is avian influenza (Influenza A) viruses; these are responsible for highly contagious acute illness in humans, pigs, horses, marine mammals, and birds, occasionally resulting in devastating epidemics and pandemics (Bengis et al., 2004).

Wild animals constitute an important but poorly known reservoir of emerging infectious diseases, most of which are of zoonotic concern (Pérez, 2009). The trade in wildlife for food consumption, medicines, and as pets, among other uses, involves the capture and sale of billions of animals of incredibly wide varieties of species (Alves et al., 2010a; Alves et al., 2013a; Pérez, 2009; Roldán-Clarà et al., 2014). Wildlife commercialization, both legal and illegal, is considered a significant driver of zoonotic diseases—leading to the introduction of zoonoses and/or foreign diseases that may impact domestic animals and/or native wildlife species (Karesh et al., 2005; Rostal et al., 2012). Hunting and the consumption of bushmeat are important routes for the introduction and transmission of zoonotic diseases (Van Vliet et al., 2017). Any wildlife species harvested for bushmeat could be a potential source of zoonotic diseases that could be transferred during hunting, butchering, or preparation (Karesh and Noble, 2009; Wolfe et al., 2000). Armadillos, for example, are widely hunted as food resources and for medicinal uses, but are natural reservoirs of etiological agents of several zoonotic diseases that affect humans—such as leprosy, trichinosis, coccidioidomycosis (valley fever), Chagas disease, and typhus (Silva et al., 2005). More than 100 occurrences of pulmonary mycosis, for example, were recorded in 40 municipalities in Piauí state in northeastern Brazil (Alves et al., 2016). The exotic pet trade deals with an increasing range of wild animal species, from invertebrates to mammals (Pérez, 2009), and it is believed that epidemics such as SARS (severe acute respiratory syndrome), monkey pox, and avian influenza H5N1 emerged from wildlife markets (Brown, 2004; Burgos and Burgos, 2007; Check, 2004; Karesh et al., 2007; Sleeman, 2006; Warwick et al., 2011).

There is a rising threat from emerging infectious diseases spreading to people and other animals, fueled by human activities ranging from the handling of bushmeat and the trade in exotic animals to the destruction of wild habitat (Lilley et al., 1997; Patz et al., 2000; Walsh et al., 1993). Despite warnings of the potential significance for human disease of changing patterns in their relationship with animals and the natural world, scientists have continued to treat human and animal health as largely independent disciplines, while historians have also neglected this important aspect of human disease (Hardy, 2003). In this sense, it is crucial that the interdependence between animal and humans be considered in the development of new public health practices.

Animal as Sentinels of Human Health

As discussed previously, animals suffer a similar spectrum of disease as humans (Bell et al., 1988; Krauss, 2003) and, therefore, may be sensitive indicators of environmental hazards and provide an early warning system for public health intervention (Reif, 2011). The concept that disease occurrence in nonhuman animal populations (wild and domestic) can serve as a sentinel warning of an environmental threat to human health has a long history (Rabinowitz et al., 2005).

Animals have served in numerous cases as "sentinels" of environmental threats near the living or working environments (Van der Schalie et al., 1999), and humans can, in return, sometimes serve as sentinels for animal health. The potential for animals to serve as sentinels for humans (or vice versa) depends on the type of linkages and contacts between specific animal populations and neighboring humans (Rabinowitz and Conti, 2013). Terrestrial wildlife, companion animals, food production animals, and aquatic animal populations can be monitored as sentinels for environmental impacts caused by pathogens, contaminants, and/or land-use changes (Rabinowitz and Conti, 2013).

Several historical examples illustrate animals' usefulness as predictors of human illness (Rabinowitz and Conti, 2013; Reif, 2011). In the 1870s, fattened cattle experienced high mortality at a stock show in London's Smithfield Market associated with a dense industrial fog—a precursor to the air pollution episodes typified by the infamous London Fog of 1952, during which thousands of residents died (Glickman et al., 1991). In the 1950s, recognition of neurobehavioral disturbances in the cat population of Minamata, Japan, preceded a severe episode of neurological disease among local residents caused by consumption of seafood contaminated with methylmercury (Tsuchiya, 1992). Sediments, shellfish, and fish in Minamata Bay became contaminated with mercuric chloride as the result of effluent discharges from a chemical plant. The ataxic "dancing cats of Minamata" were a warning sign. Unfortunately, it was not recognized in time to prevent the human epidemic (Reif, 2011). In 1962, it was cases of lead poisoning in cattle and horses living in the vicinity of a smelter that alerted the Minnesota State Health Department to conduct surveillance for lead exposure in local human populations (Hammond and Aronson, 1964). Another classic example of this is the historic use of canaries by miners to detect the presence of toxic gases in coal mines (Burrell and Seibert, 1916). Dying crows and other birds signaled the appearance and spread of the West Nile virus infection in the Western hemisphere. As the disease spread, monitoring of dead crows was used as a sentinel system for early warning of human disease risk (Julian et al., 2002).

Animal sentinels may potentially be used to address a range of surveillance questions including: (1) detection of a known pathogen in a new area; (2) detection of changes in the prevalence or incidence of a pathogen or disease over time; (3) determining the rates and direction of pathogen spread; (4) testing specific hypotheses about the ecology of a pathogen; and (5) evaluating the efficacy of potential disease control interventions (McCluskey, 2003). Appropriate use of animal sentinels can facilitate the early detection and identification of outbreaks, which is of critical importance both for the success of control and prevention efforts (Chomel, 2003; Kahn, 2006) and for reducing the magnitude of subsequent outbreaks (Ferguson et al., 2005). However, the potential of animal sentinel surveillance can only be fully realized if information sourced from animal populations is acted upon. For example, an Ebola outbreak in central Africa was the result after insufficient preventive health measures were taken despite warnings of an imminent human outbreak being provided from monitoring of Ebola deaths in primate sentinels (Rouquet et al., 2005).

Studies of the effects of environmental exposure on domestic and wild animals can corroborate or inform epidemiologic studies in humans (Reif, 2011). Currently, however, physicians assessing environmental health risks to patients do not routinely include animal sentinel data in their clinical assessments. Public health practitioners are unlikely to respond to mortality events in animals that are not clearly due to West Nile virus or other known zoonoses, such as rabies (Rabinowitz et al., 2005). Reasons for the underuse of animal sentinel data by human health professionals may include limited understanding of the relationships existing between animal, human, and ecosystem health; insufficient knowledge of veterinary medicine; and few institutional protocols to incorporate animal data into public health surveillance (Stephen and Ribble, 2001). Both human and animal health professionals have gained an increasing awareness that disease events in animal populations may have direct relevance to human health (Scotch et al., 2009).

Traditional Medicine

It is known that at the dawn of recorded history humans often ate, or wore on their person, some portion of an animal that was thought to have a healing or protective influence (MacKinney, 1946); this highlights the intertwining of the origin of the medicinal use of faunal elements with their use as food. In the same context, Chemas (2010) remarked that the treatment of illnesses using animal-based remedies is an extremely old practice, the most remote antecedent of which is a carnivore diet, closely followed by the ritual ingestion of deceased persons (e.g., close relatives, warriors) as a means of absorbing their virtues (e.g., courage, virility), and subsequently by a true medicinal use inseparable from magical-religious elements. These observations are in line with the view of nature as providing many things for humankind, including tools for the first attempts at therapeutic intervention (Nakanishi, 1999).

Although plants and plant-derived materials make up the majority of the ingredients used in most traditional medical systems worldwide, whole animals, animal parts, and animal-derived products also constitute important elements of the materia medica (Alakbarli, 2006; Alves et al., 2013b; Alves and Rosa, 2005; Moquin-Tandon, 1861; Scarpa, 1981; Stephenson, 1832; Unnikrishnan, 2004). Products derived from medicinal animals are directly used in the confection of popular remedies and magical items such as charms, amulets, and talismans that are widely sought after in traditional medicinal practices (Alves and Rosa, 2013a; Anyinam, 1995).

The antiquity in the use of medicinal animals and its persistency through times are a testimony to the importance of those therapeutic resources to mankind (Alves et al., 2013b). In modern societies, zootherapy constitutes an important alternative to the many other known therapies practiced worldwide (Alves and Rosa, 2013a). Wild and domestic animals and their by-products (e.g., hooves, skins, bones, feathers, tusks) form important ingredients in the preparation of curative, protective, and preventive medicine (Adeola, 1992; Alves and Alves, 2011; Alves et al., 2012; Ashwell and Walston, 2008; Martinez, 2013; Whiting et al., 2011; Williams et al., 2013).

Many cultures still employ traditional medicine incorporating animal-derived remedies. Probably the most famous of these are the Chinese, who use animals to treat a variety of ailments. Although less known and less frequently studied, Latin America and Africa both have a long tradition of using their equally varied and rich fauna, including many endangered species, to treat all kinds of ailments (Alves and Rosa, 2013b). Zootherapeutic practices are also found in Europe (Ceríaco, 2013; Quave et al., 2010; Voultsiadou, 2010).

Mammals, fish, reptiles, birds, mollusks, and insects, including many threatened species, are prominently used in traditional medicine (Figs. 13.1 and 13.2) (Alves et al., 2010b; Alves et al., 2008; Alves et al., 2013g; Ferreira et al., 2012, 2013; Williams et al., 2013), substantiating the importance of taking into account their harvesting in the context of animal conservation. Many marine (Alves and Dias, 2010; Alves et al., 2013c) and terrestrial invertebrates (Costa-Neto, 2005; Figueirêdo et al., 2015; Kritsky, 1987; Pemberton, 1999) make up part of the therapeutic arsenal of popular medicine (Figs. 13.3–13.5).

Articles and review texts have revealed the high numbers of animal species used in traditional medicinal practices throughout the world (Table 13.1). Researchers have reported more than 1500 animal species that have some medicinal use in traditional Chinese medicine (Yinfeng et al., 1997). In Latin America, at least

FIGURE 13.1 Medicinal animal-derived products (crocodile skulls, antelope horns, and a diversity of carnivore and nonhuman primate skulls) for sale in Benin, West Africa. *Photo credits: Anthony B. Cunningham.*

FIGURE 13.2 The tegu lizard (*Salvator merianae*) and boa snake (*Boa constrictor*), reptiles species often used in Brazilian traditional medicine. *Photo credits: John Philip Medcraft.*

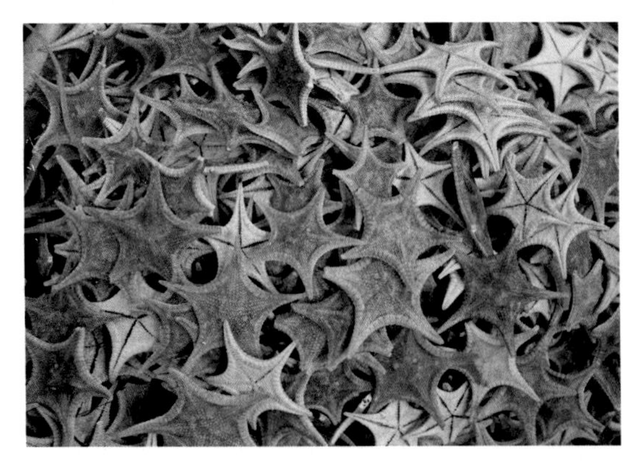

FIGURE 13.3 Dried starfish for sale in the traditional Chinese medicine market in Chengdu, Sichuan. *Photo credits: Anthony B. Cunningham.*

584 animal species have been reported as being used in traditional medicinal practices (Alves and Alves, 2011). Worldwide, at least 284 reptiles and 47 amphibians (Alves et al., 2013g), 110

FIGURE 13.4 The centipedes (*Scolopendra subspinipes*), known as wugong in Chinese, widely sold in traditional Chinese medicine markets. *Photo credits: Anthony B. Cunningham.*

primates (Alves et al., 2013e), 108 mammalian carnivores (Alves et al., 2013d), and 266 marine invertebrates (Alves et al., 2013c) are used in popular medicines. Research in 25 African countries has shown that at least 354 bird species are used there in traditional curing practices (Williams et al., 2013) (Table 13.1). Some groups, such as seahorses (*Hippocampus* spp.) are widely employed for medicinal purposes (Fig. 13.6). Rosa et al. (2013) reported that of 48 species recognized as valid by Project Seahorse (2016), 20 species were cited in the literature as having medicinal uses. It is important to point out that the number of animal species used in popular medicinal practices must certainly be larger than what has so far been recorded, as in spite of studies focusing on this theme, many regions have not been closely examined—indicating that further studies will be indispensable to increasing our understanding of the links between the traditional uses of animals and conservation biology, the sustainable management of natural resources, public health policies, and biological prospection (Alves and Rosa, 2006).

The world's animals and plants—including a number of species used in traditional medicine—face threats ranging from habitat loss to the global wildlife trade (Alves and Rosa,

FIGURE 13.5 Examples of raw materials derived from medicinal animals sold in Brazilian cities. *Photo credits: Rômulo R.N. Alves.*

2013a). There has been an increasing demand for traditional medicines (Alves and Rosa, 2007; Robinson and Zhang, 2011), and the link between traditional medicine and the loss of certain species has become apparent (Alves et al., 2007; Call, 2006). This trend bears important implications for the conservation of the many species of flora and fauna, on which traditional remedies are based (Alves and Rosa, 2013a; Lee, 1999). Unfortunately, whereas the use of traditional remedies used to be a localized practice, the globalization of commerce in combination with the increased popularity of natural approaches to health worldwide has created a level of demand that threatens the survival of many vulnerable species of wildlife (IFAW, 2011). The case of vertebrates threatened by trade for traditional medicine, including rhinos, tigers, bears, pangolins, turtles, seahorses, monkeys, tigers, rhinoceros, and bears is well-known (Fig. 13.7).

Medicinal use of fauna represents an additional pressure for many species, and has been indicated as an important cause of population decline. Thus, not only should the use of these animals in popular medicine be considered, but also their exploitation by the pharmaceutical industry (Marques, 1997). As Shaw (2009) points out, any pharmaceutical scientist who is involved in contemporary natural product research has to get involved in, or at the very least become familiar with, global issues of species conservation and/or biodiversity.

TABLE 13.1 Richness of Animal Species Used in Traditional Folk Medicine According to Literature

Animal Group	Number of Medicinal Species	Geographic Coverage	References
Herpetofauna	331	Worldwide	Alves et al. (2013g)
Primates	110	Worldwide	Alves et al. (2013e)
Carnivorous mammals	108	Worldwide	Alves et al. (2013d)
Aquatic mammals	24	Worldwide	Alves et al. (2013f)
Seahorses	20	Worldwide	Rosa et al. (2013)
Marine invertebrates	266	Worldwide	Alves et al. (2013c)
Birds	354	Africa (25 countries)	Williams et al. (2013)
All taxons combined	584	Latin America	Alves and Alves (2011)
All taxons combined	54	Portugal	Ceríaco (2013)
All taxons combined	109	India	Mahawar and Jaroli (2008)
All taxons combined	137	Nigeria	Soewu (2013)
All vertebrate groups	147	South Africa	Whiting et al. (2013)
Mammals	87	Benin (West Africa)	Djagoun et al. (2013)

BIOTHERAPY

The use of live organisms for treating human and animal illnesses is known as biotherapy (Grassberger et al., 2013). As pointed out by those authors, this is an ancient practice, although it has now attracted the interest of many clinicians, biologists, biochemists, and patient advocates and has emerged as a rapidly advancing multidisciplinary field of medicine. Some of the principal animals used in biotherapy are maggots, leeches, bees, parasitic worms, and fish (Table 13.2), giving rise to the fields of maggot therapy, hirudotherapy, apitherapy, helminth therapy, and ichthyotherapy, which will all be briefly discussed below.

Hirudotherapy

The use of medicinal leeches for curative purposes is one of the oldest known practices in medicine (Gileva and Mumcuoglu, 2013) and is called hirudotherapy or leech therapy. Although more than 650 species of leeches have been described, only a few are used for medicinal (therapeutic) purposes. The European medicinal leech, *Hirudo medicinalis*, is one of the most extensively studied annelids and the most frequently used in modern medical practices (Gileva and Mumcuoglu, 2013). Similar results have also been obtained in some cases with *Hirudo verbana* and *Hirudo michaelseni* (Van Wingerden and Oosthuizen, 1997; Whitaker et al., 2012). Medicinal leeches apparently reduce blood coagulation and relieve venous pressure resulting from blood pooling (especially after plastic surgery) and stimulate blood circulation (Godfrey, 1997). Leeches are now applied to treat a wide array of diseases or conditions, including reconstructive plastic surgery; venous, cardiovascular, neurological, gynecological, osteomuscular, and ophthalmologic disorders; periodontal and oral mucosal diseases; and cancer (Gileva and Mumcuoglu, 2013; Gilyova, 2005; Scott, 2002). Specific applications of hirudotherapy can vary in different countries or regions. Hirudotherapy is officially

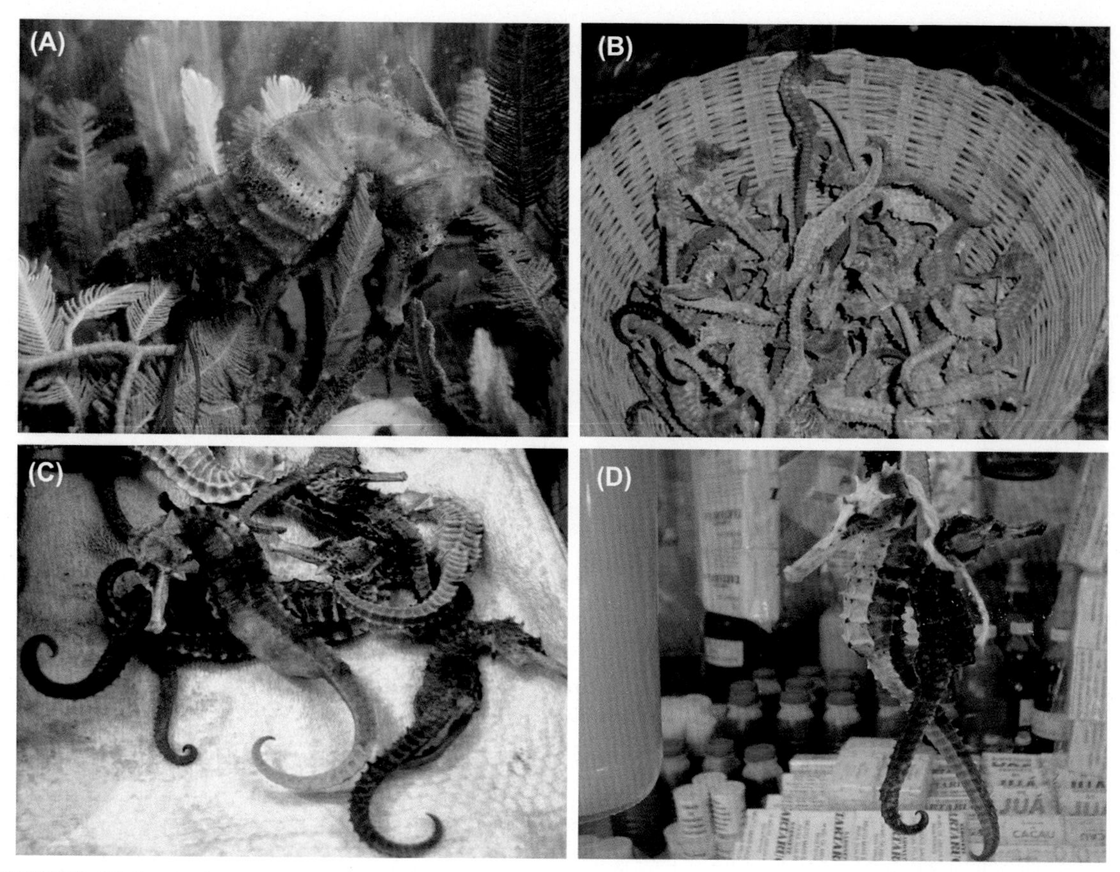

FIGURE 13.6 The longsnout seahorse, *Hippocampus reidi* (A), species commonly used for medicinal purposes in Brazil, traded in the dried form. Dried seahorse specimens for sale (B–D) in Brazilian cities. *Photo credits: (A) Thelma L.P. Dias, (B and D) Rômulo R.N. Alves, (C) Ierecê L. Rosa.*

recognized as an alternative therapy for numerous internal diseases including osteoarthritis, phlebitis, hypertension, and glaucoma in some Eastern Europe, Russian, and Asian countries, while American and European practitioners emphasize the value of leeches in microvascular and reconstructive surgery for both pediatric and adult populations (Gilyova, 2005).

Maggot Therapy

One of the most interesting applications of insects as therapeutic agents is maggot therapy, which involves the treatment of superficial or deep wounds with the help of blowfly larvae (Costa-Neto, 2005). This method was accidentally discovered during the World War I and was widely used during the 1930s and 1940s, being indicated for infected wounds that are difficult to heal, such as osteomyelitis, abscesses, burns, wounds on diabetic patients, pressure ulcers, traumatic lesions, tumors, and untreatable gangrene (Martini and Sherman, 2003). The advantages of maggot therapy, also called larval therapy, maggot debridement therapy, and biosurgery, include its profound efficacy in debriding necrotic tissue and its relative safety and simplicity; it is frequently

FIGURE 13.7 Tigers and rhinos, examples of endangered animals impacted due to use and traffic of their parts for traditional medicines. *Photo credits: Rômulo R.N. Alves.*

utilized when conventional medical treatments and surgeries are not capable of deterring progressive tissue destruction (Sherman et al., 2013). The therapy treatment consists of applying live, sterile, laboratory raised, blowfly larvae (Diptera: *Calliphoridae*) to lesions or chronic and/or infected wounds to remove necrotic/infected tissues and promote healing (Martini and Sherman, 2003). The larvae can be applied to a wound in a direct (free-range) or indirect (contained) manner (Gunjegaonkar et al., 2016). Blowfly larvae, *Lucilia sericata*, are frequently used, although other species have also been employed, such as *Lucilia cuprina*, *Phormia regina*, and *Calliphora vicina* (Sherman et al., 2000). Maggots like those of *L. sericata* feed on dead tissue where gangrene-causing bacteria thrive. As they eat, they also secrete allantoin, a chemical that inhibits bacterial growth (Costa-Neto, 2005). Medicinal maggots have three actions: they clean wounds by dissolving the dead (necrotic) infected tissue, disinfect the wound by killing bacteria, and stimulate wound healing (Sherman et al., 2000). The larva debride the necrotic tissue using their oral suction apparatus, which liberates digestive enzymes and dissolves the infected tissue; the wound is disinfected through the secretion of antibacterial substances liberated by the larva that activate macrophages, induce wound healing, and stimulate healthy tissue growth (Dallavecchia et al., 2011).

Use of maggot therapy declined with the introduction of modern antibiotics and improvements in surgical debridement, although the use of maggot therapy has been returning due to growing antibiotic resistance and potential adverse effects associated with antibiotic use (Gunjegaonkar et al., 2016). Initially, maggot therapy was used for treating chronic wounds in humans, but was later used with animals as a likewise safe and effective method of curing serious wounds (Munir et al., 2016). Numerous researchers have shown that larval therapy is much more effective at debriding wounds than conventional treatments (Dallavecchia et al., 2011; Sherman et al., 1986). Maggot therapy is commonly used in the United States, Israel, and Europe to treat various types of infected wounds, such as those on the feet of diabetics, postoperative infections, ulcer scabs, and leg ulcers (Ratcliffe et al., 2011; Sherman et al., 2000).

Apitherapy—Bee Venom Therapy

Bee venom therapy is a biotherapeutic medical treatment that uses bee venom to treat numerous illnesses (Kim, 2013). Bee venom has a long history of use as a folk remedy for treating diseases such as arthritis, angiocardiopathy, back pain, musculoskeletal pain, cancerous

TABLE 13.2 Animals Species Used in Biotherapy

Species	References
HIRUDOTHERAPY	
Hirudo medicinalis (Linnaeus, 1758)	Gileva and Mumcuoglu (2013)
Hirudo verbana (Carena, 1820)	Van Wingerden and Oosthuizen (1997) and Whitaker et al. (2012)
Hirudo michaelseni (Augener, 1936)	Van Wingerden and Oosthuizen (1997) and Whitaker et al. (2012)
Poecilobdella granulosa (Savigny, 1822)	Lone et al. (2011)
MAGGOT THERAPY	
Calliphora vicina (Robineau-Desvoidy, 1830)	Teich and Myers (1986)
Chrysomya rufifacies (Macquart, 1842)	Teich and Myers (1986)
Lucilia caesar (Linnaeus, 1758)	Baer (1931) and McLellan (1932)
Lucilia cuprina (Wiedemann, 1830)	Fine and Alexander (1934)
Lucilia illustris (Meigen, 1826)	Lerclercq (1990)
Lucilia sericata (Meigen, 1826)	Baer (1931)
Phormia regina (Meigen, 1826)	Baer (1931), Horn et al. (1976), Reames et al. (1988), and Robinson (1933)
Protophormia terraenovae (Robineau-Desvoidy, 1830)	Lerclercq (1990)
Wohlfahrtia nuba (Wiedemann, 1830)	Grantham-Hill (1933)
Musca domestica (Linnaeus, 1758)	Grantham-Hill (1933)
APITHERAPY	
Apis melifera (Linnaeus, 1758)	Kim (2013)
ICHTHYOTHERAPY	
Garra rufa (Heckel, 1843)	Grassberguer and Sherman (2013)
HELMINTH THERAPY	
Ascaris lumbricoides (Linnaeus, 1758)	Correale and Farez (2007, 2011)
Enterobius vermicularis (Linnaeus, 1758)	Correale and Farez (2007, 2011)
Hymenolepis diminuta (Rudolphi, 1819)	Correale and Farez (2007, 2011)
Hymenolepis nana (Siebold, 1852)	Correale and Farez (2007, 2011)
Strongyloides stercoralis (Bavay, 1876)	Correale and Farez (2007, 2011)
Trichuris suis (Schrank, 1788)	Fleming et al. (2011)
Necator americanus (Stiles, 1902)	Wolff and Broadhurst (2012)
Trichuris trichiura (Linnaeus, 1771)	Wolff and Broadhurst (2012)

tumors, multiple sclerosis, and skin diseases, and it is known to promote wound healing (Alqutub et al., 2011; Beck, 1935; Cherniack, 2010; Roy et al., 2015). The use of bee venom is quite ancient, but has attracted growing interest now due to its positive therapeutic results (Moreira, 2012). Bee venom is a colloidal substance that can be dialyzed through membranes and absorbed through the skin (Shimpi et al., 2016); it is composed of approximately 18 active compounds including enzymes, bioamines, and peptides with important biological effects (Yasui, 2012). Bee venom has been reported to have both a central analgesic mechanism and peripheral analgesic action due to its anti-inflammatory action (Shin et al., 2012).

Bee venom can be applied by direct stings from bees (apipuncture or bee sting therapy) or through injections of a venom extract (bee venom therapy)—both requiring the experience of a qualified health professional (Lucache et al., 2015). Bee venom therapy involves the use of the venom produced by *Apis melifera*, marketed under the names of Apitoxina and Apitox. The venom is applied intradermally, never intravenously, and used for treating various autoimmune diseases, neurological disturbances, and chronic and inflammatory illnesses (Kim, 2013). Apipuncture is a method used by health professionals or licensed practitioners of acupuncture, with the venom being injected by holding a live bee (with forceps) on the affected area of the patient and allowing it to sting that person. Before initiating this type of treatment, however, the patient must be examined for allergic reactions to the venom (Kim, 2013; Yasui, 2012). Bee venom therapy stimulates the immunological system through the hypothalamus, pituitary gland, and suprarenal glands, inducing the body to produce its own curative substances. The efficiency of bee venom therapy has been evaluated in both laboratory and clinical experiments with humans (Yasui, 2012); however, this curative mechanism is not yet well understood, but a series of chemical compounds acting together in the body have been identified (Lucache et al., 2015).

Ichthyotherapy

The term "ichthyotherapy" was proposed in 2006 (Grassberguer and Sherman, 2013) and readily adopted, being defined as an alternative therapy for treating skin diseases with the so-called "doctor fish of Kangal," *Garra rufa* (Fig. 13.8) (Heckel, 1843) (Grassberguer and Sherman, 2013), a small, freshwater cyprinid fish native to the Middle East (Froese and Pauly, 2016). The use of this species is directed toward treating skin problems such as psoriasis, eczema, dermatitis, acne, calluses, and hardness (Ozcelik and Akyol, 2011). Several underlying mechanisms have been suggested for the observed efficacy of ichthyotherapy. One obvious mechanism is physical contact with the fish, which feeds on desquamating skin, leading to a rapid reduction of superficial skin scales (Grassberguer and Sherman, 2013). *G. rufa* are toothless fish that consume the dead skin cells of people that they come into direct contact with in the water, without affecting healthy skin (Cabral and Carneiro, 2014). The

FIGURE 13.8 "Doctor fish of Kangal," *Garra rufa*, fish species used in ichthyotherapy. *Photo credits: Tacyana P.R. Oliveira.*

underlying mechanisms of ichthyotherapy are not yet totally understood, but the most visible effect of exposure to this fish is the removal of excess skin layers, although the dramatic observed reductions in the inflammatory component, especially among patients with psoriasis, suggest additional mechanisms (possibly molecular). More complete biochemical studies will need to be undertaken to identify and characterize the properties of *G. rufa* in that context (Grassberguer and Sherman, 2013).

Helminth Therapy

The therapeutic uses of helminths (parasitic worms) have been tested in laboratory trials as new approaches to treating a variety of allergic and autoimmune illnesses (Khan and Fallon, 2013). This type of therapy is called helminth therapy and consists of the inoculation of the patient with specific parasitic intestinal nematodes (helminths). Diseases such as ulcerative colitis, multiple sclerosis, rheumatoid arthritis, Crohn disease, celiac disease, and autism are among the health problems potentially treatable with helminths. A number of such organisms are currently being investigated for their use in therapeutic treatments (see Table 13.2), including *Trichuris suis* (Fleming et al., 2011), *Necator americanus* (Elliot et al., 2013), *Trichuris trichiura* (Correale and Farez, 2011); *Hymenolepis diminuta, Ascaris lumbricoides, Strongyloides stercoralis, Enterobius vermicularis,* and *Hymenolepis nana* (Correale and Farez, 2011; Leonardi-Bee et al., 2006).

It is appropriate to note that helminth therapies will probably be increasingly used in developed societies where epidemics of inflammatory disorders are most prevalent, and thus in people never previously exposed to helminths, while the desirable protective effects of helminth infections of humans in field studies have been reported for people in endemic areas (Scrivener et al., 2001). The theory that helminth infection protects against autoimmune diseases can be tested by comparing the prevalence of those diseases in highly helminth-exposed and less- or nonexposed human populations. There is strong evidence that helminth exposure results in changes to the immune system that decrease the risk of developing immune disorders, thereby preventing the onset of immune-mediated diseases that have become common in developed countries (Elliot et al., 2013).

Animal-Assisted Therapy

Another way in which animals can be used to ameliorate human health conditions is through human involvement with living creatures as a form of therapy (Alves et al., 2009). For centuries people have noted that animals can have positive influences on human health and functioning (Nimer and Lundahl, 2007). Florence Nightingale suggested in the 19th century that a bird might be a primary source of pleasure for people confined to their rooms due to medical problems (McConnell, 2002). The use of animals as therapy has intensified now and has received many names, such as animal-assisted therapy, pet therapy, animal-assisted activities, pet-facilitated therapy, pet-assisted therapy, animal-facilitated therapy, animal-assisted interventions, and animal visitation (Connor and Miller, 2000; Fine, 2010; Hooker et al., 2002; Kruger et al., 2006). All of these practices have the common focus of utilizing animals as facilitators for patient recovery and for establishing positive therapies, especially for patients with special needs, children with cognitive or emotional disturbances, and older people (Oliva, 2010).

Both domestic and domesticated animals have found medicinal uses as co-therapists (Silveira, 1998). Among the animals most commonly used for therapeutic purposes are dogs, cats, horses, dolphins, small tame animals such as rabbits and gerbils, and aquarium fish. Typically, reptiles are frowned upon in therapeutic settings because of their high risk of carrying disease or causing injuries to clients, and because of the difficulty of providing those animals with proper care and safe environments. Farm animals are often useful therapy animals, with the most common being llamas and pot-bellied pigs (Chandler, 2012).

Animal-assisted activities can occur in a variety of settings in which people interact with (talk to, pet, groom) companion animals while the animal's handler is present. Intense attachments can rapidly develop between people and pets during these encounters (Reperant et al., 2012), and researchers have shown that animal-assisted therapies provide numerous positive effects to patient health and can stimulate the development of diverse abilities such as learning, language acquisition, motricity, among others (Oliva, 2010). Contact with animals has proven to be an efficient method for stimulating and helping individuals with mobility problems, mental disabilities, or behavioral problems (Alves and Rosa, 2013b). Animal-assisted therapy can be used, for example, as part of a patient's physical therapy treatment plan, to decrease anxiety in psychiatric patients and agitation in older adults with dementia (Barker and Dawson, 1998; Batson, 1998). Examples include gradually increasing the number of brush strokes on a dog in order to exercise an impaired hand, and eliciting a relaxation response through horseback riding activities with children with spastic cerebral palsy (McGibbon et al., 1998). Children with autism spectrum disorders were more likely to respond appropriately in therapy sessions involving live dogs than with a stuffed toy dog or a ball (Martin and Farnum, 2002). Some nursing home residents were found to show lower cortisol levels during dog visits than human visits. During and after animal visits, hospitalized patients used fewer analgesics, reporting less pain and less depression, and heart failure patients showed decreased anxiety and epinephrine levels (Beck, 2000; Cole et al., 2007).

Fauna as Source of Drugs

Throughout human history, people have used various natural materials to cure their illnesses and improve their health (Alves and Rosa, 2007). Wildlife not only contributes to traditional medicine but also modern medicine, with natural extracts being used by pharmaceutical companies as raw material for the manufacture of drugs (Rose et al., 2012; Sifuna, 2012). The drugs we use to maintain human health are still predominantly derived from plant and animal species (Fitter, 1986). Historically, while the use of plants as medicines has been extensively recognized, studied, and reviewed, studies on the use of fauna as a source of drugs have only been produced now, and have demonstrated the enormous potential of fauna as a source of natural products and drugs (Alves and Albuquerque, 2013).

Several studies have shown that natural animal resources are highly promising in the search for new products of medicinal or pharmaceutical interest (Alves and Albuquerque, 2013; Chivian, 2002; Dossey, 2010; Fusetani, 2000). This potential is perhaps even greater for animals than for plants, considering that the number of animal species is several times greater (Alves and Albuquerque, 2012). For example, Trowell (2003) points out that there are at least 16 times as many insect species as there are plant species, yet plant chemistry has been studied 7000 times more than insect chemistry, based on a comparison of the amount of research undertaken per species.

Marine animals also represent an exceptional source of bioactive natural products, many of which exhibit structural features not found in terrestrial natural products (Faulkner, 1998; Ireland et al., 1993; Seedhouse, 2010). Drugs have been derived from sharks, sting rays, corals, sea anemones, mollusks, annelids, sponges, sea squirts, sea cucumbers, and horseshoe crabs (Alves and Albuquerque, 2013; Fitter, 1986). Shark liver contains lipids that enhance human resistance to cancer, and the horseshoe crab *Limulus* not only has a serum that isolates tumor cells and white blood cells from the whole blood of cancer patients, but is also the source of substances used to detect bacterial toxins in human body fluids (Fitter, 1986). Invertebrates are proving to yield increasing numbers of antibiotic agents, blood coagulants and anticoagulants, and neuromuscular, as well as anticancer compounds (Alves and Albuquerque, 2013; Fitter, 1986; Myers, 1979; Seedhouse, 2010). Approximately 2500 new metabolites were reported from a variety of marine organisms during the decade of 1977–87 (Ireland et al., 1993). Already, more than 15,000 natural products have been discovered, and this number continues to grow. While bioprospecting and deep-ocean exploration are in their infancy, the novel biology of the organisms discovered to date and their potential for revolutionizing the medical realm means that scientific interest will be increasingly focused on realizing the potential that exists in the deep ocean. And, inevitably, as a growing body of science reaffirms that deep-sea biodiversity holds major promise for the treatment of human diseases, exploration will surely venture ever deeper in search of untapped resources (Fusetani, 2000; Seedhouse, 2010).

Various terrestrial vertebrates, particularly amphibians and reptiles, have been of great interest in pharmacological studies (Alves and Albuquerque, 2013). One excellent example of successful drug development from a component of snake venom (*Bothrops jararaca*, Wied, 1824) is that of the inhibitors of angiotensin-converting enzyme. This enzyme is responsible for converting an inactive precursor into the locally active hormone angiotensin, which causes blood vessels to constrict and hence raises blood pressure (Bisset, 1991). Another good example is the work initially conducted by Daly during the 1960s on the skin secretions of dendrobatid frogs from Ecuador, and of other "poison dart" frog species in Central and South America. This work has led to the identification of a number of alkaloid toxins that bind to multiple receptors in the membranes of nerve and muscle cells (Chivian, 2002). The anticarcinogenic activities of Indian monocellate cobra and Russell's viper venom were studied in carcinoma, sarcoma, and leukemia models. Under in vivo experiments, it was found that the sublethal doses of the Indian *Elapidae* (monocellate cobra) and *Viperidae* (Russell's viper) venom caused cytotoxicity in Ehrlich ascites carcinoma (EAC) cells; it increased the life span of EAC-bearing mice and reinforced its antioxidant system (Debnath et al., 2007). Similarly, antitumor activity of *Hydrophidae* (*Lapemis curtus*) venom was also established against EAC in Swiss albino mice in vivo and HeLa and Hep2 tumor cell cultures in vitro (Karthikeyan et al., 2008). Ferreira et al. (2010) evaluated the anti-inflammatory activity of topically administered *Tupinambis merianae* fat in animal models (male and female Swiss mice—*Mus musculus*). In this first experimental test of the antiinflammatory activity of *T. merianae* fat in in vivo models, the authors found that it had significant topical anti-inflammatory activity and reduced inflammation edema in mouse ears caused by croton oil (single and multiple applications), phenol, and arachidonic acid. Similarly, numerous animals produce substances with antimicrobial activities that are effective as defense mechanisms against infections by microorganisms and have synergetic effects when associated with antibiotics (Coutinho et al., 2004)—and researchers have demonstrated positive results in relation to these activities in diverse species (Table 13.3).

There has been increasing attention paid to animals, both vertebrates and invertebrates, as a source for new medicines (Chivian, 2002). Animals have been methodically tested by pharmaceutical companies as sources of drugs for modern medical science (Kunin and Lawton, 1996), and the current percentage of animal sources for producing essential medicines is

TABLE 13.3 Examples of Animal Species With Pharmacological Activities

Species	Activity	References
Macoma birmanica (Philipi, 1949)	Antibacterial	Adhya et al. (2009)
Trionyx sinensis (Wiegmann, 1935)	Antibacterial	Thammasirirak et al. (2006)
Amyda cartilaginea (Boddaert, 1770)	Antibacterial	Thammasirirak et al. (2006)
Chelonia mydas (Linnaeus, 1758)	Antibacterial	Thammasirirak et al. (2006)
Nasutitermes corniger (Motschulsky, 1855)	Modulation of the antibiotic activity	Coutinho et al. (2009, 2010) and Chaves et al. (2014)
Atta sexdens rubropilosa (Forel, 1908)	Antifungal	Masaro et al. (2001)
Squalus acanthias (Linnaeus, 1758)	Antibacterial	Donia and Hamann (2003)
Leptodactylus macrosternum (Miranda-Ribeiro, 1926)	Antibacterial	Cabral et al. (2013)
Leptodactylus vastus (Adolf Lutz, 1930)	Antibacterial	Cabral et al. (2013)
Pseudocanthotermes spiniger	Antibacterial and antifungal	Lamberty et al. (2001)
Gallus gallus domesticus (Linnaeus, 1758)	Modulation of the antibiotic activity	Coutinho et al. (2014)
Rhinella jimi (Stevaux, 2002)	Modulation of the antibiotic activity	Sales et al. (2015)
Rhynocoris marginatus (Fabricius, 1794)	Antibacterial	Sahayaraj et al. (2006)
Catamirus brevipennis (Servile)	Antibacterial	Sahayaraj et al. (2006)
Tropidurus hispidus (Spix, 1825)	Modulation of the antibiotic activity; antiinflammatory	Santos et al. (2012, 2015)
Tropidurus semitaeniatus (Spix, 1825)	Modulation of the antibiotic activity	Santos et al. (2012)
Spilotes pullatus (Linnaeus, 1758)	Modulation of the antibiotic activity	Oliveira et al. (2014)
Dinoponera australis (Roger, 1861)	Antibacterial	Cologna et al. (2005)
Dinoponera quadriceps (Santschi, 1921)	Antibacterial	Cardoso et al. (2010)
Boa constrictor (Linnaeus, 1758)	Modulatory of the antibiotic activity; antiinflammatory	Ferreira et al. (2011, 2014)
Crotalus durissus (Lineu, 1758)	Antiinflammatory	Ferreira et al. (2014)
Iguana iguana (Linnaeus, 1758)	Antiinflammatory	Ferreira et al. (2014)
Euphractus sexcinctus (Linnaeus, 1758)	Antiinflammatory	Ferreira et al. (2014)
Tupinambis merianae (Linnaeus, 1758)	Antiinflammatory	Ferreira et al. (2010)
Naja atra (Cantor, 1842)	Antiinflammatory	Zhu et al. (2016)

fairly significant. Of the 252 chemicals selected as essential by the World Health Organization, 11.1% are derived from plants and 8.7% from animals (Marques, 1997). Of the 150 prescription drugs currently in use in the United States of America, 27 have animal origin (World Resources Institute, 2000). Although the potential of faunal biodiversity is well known, a careful strategy is required if species are to be exploited sustainably. One of the main conservation concerns about the exploitation of fauna in the search for bioactive compounds is the possible overharvesting of target organisms (Alves and Albuquerque, 2013). Some taxa with known pharmacological potential are especially susceptible to overexploitation; for example, marine species such as cone shells and mollusks have been overharvested as sources of clinical neuropharmaceuticals (Sukarmi and Sabdono, 2011). Harvesting of reef organisms for the discovery and development of pharmaceuticals is causing increased concern, since it has been perceived by many as unsustainable and a threat to conservation (Hunt and Vincent, 2006; Sukarmi and Sabdono, 2011).

Bioprospecting continues to generate considerable debate (Alves and Albuquerque, 2013), as critics dispute the idea that the commodification of nature will contribute to conservation (Simpson, 1997), or that natural products have a future in the discovery of new drugs (Firn, 2003). Regardless of the perspective adopted, as highlighted by Barrett and Lybbert (2000), the need to conserve precious biodiversity is clear, especially as we begin to appreciate the magnitude of the spiritual, social, and economic services it provides.

Animals in Biomedical Research

In addition to their use in traditional medicine, in biotherapies, and as sources of medicinal drugs, animals are essential to research projects—with both direct and indirect implications for human health. Animal experimentation in the context of scientific research has contributed greatly to the development of medical science and technology, including the development of prophylactic measures and treatments for diseases that affect humans (Chorilli et al., 2009).

Although research using animals intensified during the last century, this technique is known to have been employed since ancient times (Alves, 2012), and animal research has formed the basis for much of the progress in understanding and treating human (and animal) diseases (Schacter, 2006). There are records of experimentation with animals reaching back to ancient Rome, but not until the Renaissance did scholars begin to seriously study how the body works. Leonardo da Vinci (1452–1519) and other artists and anatomists made early anatomical investigations of muscle and bone structure, and William Harvey (1578–1657) discovered the circulation of blood through his experiments with live deer. Much of the live animal experimentation during this period, both in England and France, was based on the view of the French philosopher René Descartes (1596–1650) that animals were incapable of feeling pain. The 19th century French physiologist, Claude Bernard (1813–78) and his teacher, François Magendie (1783–1855), conducted wide-ranging animal experiments, including surgery, the use of drugs, and the removal of body parts from many species (Bishop and Nolen, 2001).

From ancient to modern times, the use of animals in research has become one of the most important ways to better understand aspects of the anatomy, biochemistry, genetics, nutrition and physiology of humans. Additionally, knowledge of the transmission mechanisms and treatment of human diseases are associated with such research. According to Bishop and Nolen (2001), many, if not most, of the spectacular innovations in medical understanding and treatment of today's human maladies have been based on research using animals.

Animals are used, for example, to develop new surgical techniques; test the efficacy and possible side effects of new drugs; determine the preventative and curative virtues of new medicines against diseases; test the safety of new chemicals used in the food industry; and check the quality of new batches of drugs and medicines (Bowman, 1977; Fitter, 1986). The discovery of antibiotics, analgesics, anesthetics, and antidepressants; the success of organ transplant development; catheterization; cardiac pacemaker and several other surgical techniques; practically all research protocols about safety, toxicity, effectiveness, and quality control of new drugs—all these pass through the use of laboratory animals. Other examples of scientific contributions arising from studies of animals are the discovery of insulin, the development of vaccines against several diseases, and serum production (Fagundes and Taha, 2004). Many drugs used by humans are directly produced from animals, for example, hormones used to overcome problems of fertility in humans are derived from cattle; insulin used to keep diabetics alive comes mainly from the pancreases of cattle and pigs. In addition, many vaccines are produced on animal tissue and on chicken eggs. Measles vaccine, for example, is produced on canine kidney tissue, as well as on eggs (Bowman, 1977).

Approximately 35 million animals are used worldwide in research each year, including dogs, monkeys, and cats, although 90% are laboratory rats, mice, and birds (Bishop and Nolen, 2001). Nonhuman primate species, because of their similarity to humans, are among the principal groups of animals used in biomedical research (Carlsson et al., 2004; Fitter, 1986), which are associated with significant contribution to advances in human health and disease control (Fitter, 1986). A wide variety of nonhuman primate species are used in these studies, involving at least 56 different extant species or subspecies (Carlsson et al., 2004), notably the Asian rhesus monkey *Macaca mulatta* and the African green or vervet monkey *Cercopithecus aethiops* (Carlsson et al., 2004; Fitter, 1986). These mammals are important and often essential for research in HIV/AIDS, malaria, cardiovascular diseases, cancers, and hepatitis, and also for the production and testing of drugs and vaccines. Similarly, armadillos have been used in medical research since the mid-1800s (Sharma et al., 2013). Studies involving these mammals have contributed to our knowledge of various infectious diseases, including syphilis and Chagas disease (Sharma et al., 2013; Wicher et al., 1983). However, the nine-banded armadillo *Dasypus novemcinctus* Linnaeus, 1758 has been most exploited as a model for leprosy (Peña et al., 2008; Scollard, 2008; Sharma et al., 2013) The armadillo is the only animal model in which protection against dissemination of leprosy bacilli or progress of nerve damage can be evaluated. Bacterins of heat-killed *Mycobacterium leprae* or viable BCG have been shown to protect armadillos against *M. leprae* challenge or enhance their immunity to the organism (Kirchheimer et al., 1978). Armadillos can also be used for testing new diagnostic candidates because they are the only host in which the true status of infection can be determined, and the long incubation period of leprosy in humans can cause confounding results (Sharma et al., 2013). Another animal group used in medical research experiments are marsupials (Jurgelski, 1984). The opossum, *Didelphis virginiana*, for example, is used in endocrinological, embryological, anatomical, psychiatric, and neurological research (Wiedorn, 1954).

A dynamic tension exists between support for scientific enquiry, mostly to alleviate human disease and public concern about animal suffering (Bishop and Nolen, 2001). This discussion is not new, however, and as early as the 16th century philosophers were debating the morality of animal experimentation, with their arguments centering on whether animals felt pain and the moral status of animals as living, sentient creatures (Schacter, 2006). Although the discovery of anesthetics and their use in animal experiments might have been expected to somewhat quiet this issue, revulsion at the use and potential misuse of animals for human betterment sustains a significant activism opposed to any use of animals in research (Schacter, 2006).

CONCLUSIONS

Humans are animals that live in association with thousands of other animal species and share the same environment and a wide diversity of diseases that can be mutually transmitted. If, on one hand, animals are vectors of human diseases, they are also indispensable for their treatments and cures. Products derived from animals are fundamental ingredients of both traditional remedies and modern drugs; live animals can alert us to approaching epidemics and are protagonists and agents in many therapies, and are fundamental to research efforts that seek to understand human illnesses and test modern medicines. It is therefore clear that human health depends on the health of both animals and our environment, and that health strategies must always take this intricate interdependence into consideration.

References

Adeola, M.O., 1992. Importance of wild animals and their parts in the culture, religious festivals, and traditional medicine, of Nigeria. Environmental Conservation 19, 125–134.

Adhya, M., Singha, B., Chatterjee, B.P., 2009. Purification and characterization of an N-acetylglucosamine specific lectin from marine bivalve *Macoma birmanica*. Fish & Shellfish Immunology 27, 1–8.

Alakbarli, F., 2006. Medical Manuscripts of Azerbaijan. Heydar Aliyev Foundation, Baku.

Alqutub, A.N., Masoodi, I., Alsayari, K., Alomair, A., 2011. Bee sting therapy-induced hepatotoxicity: a case report. World Journal of Hepatology 3, 268–270.

Alves, R.R.N., 2012. Relationships between fauna and people and the role of ethnozoology in animal conservation. Ethnobiology and Conservation 1, 1–69.

Alves, R.R.N., Albuquerque, U.P., 2012. Ethnobiology and conservation: why do we need a new journal? Ethnobiology and Conservation 1, 1–3.

Alves, R.R.N., Albuquerque, U.P., 2013. Animals as a source of drugs: bioprospecting and biodiversity conservation. In: Alves, R.R.N., Rosa, I.L. (Eds.), Animals in Traditional Folk Medicine: Implications for Conservation. Springer Heidelberg, pp. 67–89.

Alves, R.R.N., Alves, H.N., 2011. The faunal drugstore: animal-based remedies used in traditional medicines in Latin America. Journal of Ethnobiology and Ethnomedicine 7, 1–43.

Alves, R.R.N., Dias, T.L.P., 2010. Usos de invertebrados na medicina popular no Brasil e suas implicações para conservação. Tropical Conservation Science 3, 159–174.

Alves, R.R.N., Rosa, I.L., 2005. Why study the use of animal products in traditional medicines? Journal of Ethnobiology and Ethnomedicine 1, 1–5.

Alves, R.R.N., Rosa, I.L., 2006. From cnidarians to mammals: the use of animals as remedies in fishing communities in NE Brazil. Journal of Ethnopharmacology 107, 259–276.

Alves, R.R.N., Rosa, I.L., 2007. Biodiversity, traditional medicine and public health: where do they meet? Journal of Ethnobiology and Ethnomedicine 3, 1–9.

Alves, R.R.N., Rosa, I.L., 2013a. Animals in Traditional Folk Medicine: Implications for Conservation. Springer-Verlag, Berlin Heidelberg.

Alves, R.R.N., Rosa, I.L., 2013b. Introduction: toward a plural approach to the study of medicinal animals. In: Alves, R.R.N., Rosa, I.L. (Eds.), Animals in Traditional Folk Medicine: Implications for Conservation. Springer-Verlag, Berlin Heidelberg.

Alves, R.R.N., Rosa, I.L., Santana, G.G., 2007. The role of animal-derived remedies as complementary medicine in Brazil. BioScience 57, 949–955.

Alves, R.R.N., Vieira, W.L.S., Santana, G.G., 2008. Reptiles used in traditional folk medicine: conservation implications. Biodiversity and Conservation 17, 2037–2049.

Alves, R.R.N., Oliveira, M.G.G., Barboza, R.R.D., Singh, R., Lopez, L.C.S., 2009. Medicinal animals as therapeutic alternative in a semi-arid region of Northeastern Brazil. Forsch Komplementärmedizin/Research in Complementary Medicine 16, 305–312.

Alves, R.R.N., Barboza, R.R.D., Souto, W.M.S., 2010a. A global overview of canids used in traditional medicines. Biodiversity and Conservation 19, 1513–1522.

Alves, R.R.N., Souto, W.M.S., Barboza, R.R.D., 2010b. Primates in traditional folk medicine: a world overview. Mammal Review 40, 155–180.

Alves, R.R.N., Rosa, I.L., Léo Neto, N.A., Voeks, R., 2012. Animals for the Gods: magical and religious faunal use and trade in Brazil. Human Ecology 40, 751–780.

Alves, R.R.N., Lima, J.R.F., Araújo, H.F., 2013a. The live bird trade in Brazil and its conservation implications: an overview. Bird Conservation International 23, 53–65.

Alves, R.R.N., Medeiros, M.F.T., Albuquerque, U.P., Rosa, I.L., 2013b. From past to present: medicinal animals in a historical perspective. In: Alves, R.R.N., Rosa, I.L. (Eds.), Animals in Traditional Folk Medicine: Implications for Conservation. Springer, pp. 11–23.

Alves, R.R.N., Oliveira, T.P.R., Rosa, I.L., Cunningham, A.B., 2013c. Marine invertebrates in traditional medicines. In: Alves, R.R.N., Rosa, I.L. (Eds.), Animals in Traditional Folk Medicine: Implications for Conservation. Springer, Berlin, pp. 263–287.

Alves, R.R.N., Pinto, L.C.L., Barboza, R.R.D., Souto, W.M.S., Oliveira, R.E.M.C.C., Vieira, W.L.S., 2013d. A global overview of carnivores used in traditional medicines. In: Alves, R.R.N., Rosa, I.L. (Eds.), Animals in Traditional Folk Medicine: Implications for Conservation. Springer, pp. 171–206.

Alves, R.R.N., Souto, W.M.S., Barboza, R.R.D., Bezerra, D.M.M., 2013e. Primates in traditional folk medicine: world overview. In: Alves, R.R.N., Rosa, I.L. (Eds.), Animals in Traditional Folk Medicine: Implications for Conservation. Springer, Berlin, pp. 135–170.

Alves, R.R.N., Souto, W.M.S., Oliveira, R.E.M.C.C., Barboza, R.R.D., Rosa, I.L., 2013f. Aquatic mammals used in traditional folk medicine: a global analysis. In: Alves, R.R.N., Rosa, I.L. (Eds.), Animals in Traditional Folk Medicine: Implications for Conservation. Springer, Heidelberg, pp. 241–261.

Alves, R.R.N., Vieira, W.L.S., Santana, G.G., Vieira, K.S., Montenegro, P.F.G.P., 2013g. Herpetofauna used in traditional folk medicine: conservation implications. In: Alves, R.R.N., Rosa, I.L. (Eds.), Animals in Traditional Folk Medicine: Implications for Conservation. Springer-Verlag, Berlin Heidelberg, pp. 109–133.

Alves, R.R.N., Melo, M.F., Ferreira, F.S., Trovão, D.M.B.M., Dias, T.L.P., Oliveira, J.V., Lucena, R.F.P., Barboza, R.R.D., 2016. Healing with animals in a semiarid northeastern area of Brazil. Environment, Development and Sustainability 18, 1733–1747.

Anyinam, C., 1995. Ecology and ethnomedicine: exploring links between current environmental crisis and indigenous medical practices. Social Science & Medicine 40, 321–329.

Ashwell, D., Walston, N., 2008. An Overview of the Use and Trade of Plants and Animals in Traditional Medicine Systems in Cambodia, first ed. TRAFFIC Southeast Asia, Greater Mekong Programme, Ha Noi, Vietnam.

Ávila-Pires, F.D., 1989. Zoonoses: hospedeiros e reservatórios. Cadernos de Saúde Pública 5, 82–97.

Baer, W.S., 1931. The treatment of chronic osteomyelitis with the maggot (larva of the blow fly). The Journal of Bone & Joint Surgery American 13, 438–475.

Barker, S.B., Dawson, K.S., 1998. The effects of animal-assisted therapy on anxiety ratings of hospitalized psychiatric patients. Psychiatric Services 49, 797–801.

Barrett, C.B., Lybbert, T.J., 2000. Is bioprospecting a viable strategy for conserving tropical ecosystems? Ecological Economics 34, 293–300.

Batson, K., 1998. The effect of a therapy dog on socialization and physiological indicators of stress in persons diagnosed. In: Wilson, C.C., Turner, D.C. (Eds.), Companion Animals in Human Health. Sage, Thousand Oaks, p. 203.

Beck, B.F., 1935. Bee Venom Therapy: Bee Venom, Its Nature, and Its Effect on Arthritic and Rheumatoid Conditions. D. Appleton-Century Company.

Beck, A., 2000. The use of animals to benefit humans, animal-assisted therapy. In: Fine, A.H. (Ed.), The Handbook on Animal Assisted Therapy: Theoretical Foundations and Guidelines for Practice. Academic Press, San Diego.

Bell, J.C., Palmer, S.R., Payne, J.M., 1988. The Zoonoses (Infections Transmitted from Animals to Man). Arnold, London.

Bengis, R.G., Leighton, F.A., Fischer, J.R., Artois, M., Mörner, T., Tate, C.M., 2004. The role of wildlife in emerging and re-emerging zoonoses. Revue Scientifique et Technique, Office International Epizooties 23, 497–511.

Bishop, L.J., Nolen, A.L., 2001. Animals in research and education: ethical issues. Kennedy Institute of Ethics Journal 11, 91–113.

Bisset, N.G., 1991. One man's poison, another man's medicine? Journal of Ethnopharmacology 32, 71–81.

Bowman, J.C., 1977. Animals for Man. Edward Arnold., London.

Brown, C., 2004. Emerging zoonoses and pathogens of public significance—an overview. Revue Scientifique et Technique, Office International Epizooties 23, 435–442.

Burgos, S., Burgos, S.A., 2007. Influence of exotic bird and wildlife trade on avian influenza transmission dynamics: animal-human interface. International Journal of Poultry Science 6, 535–538.

Burrell, G.A., Seibert, F.M., 1916. Gases Found in Coal Mines. Miners' Circ. 14. Bur. Mines, Dep. Inter., Washington, DC.

Cabral, H., Carneiro, J., 2014. O papel da ictioterapia no tratamento da psoríase: relato de caso. Revista Portuguesa de Medicina Geral e Familiar 30, 402–405.

Cabral, M.E.S., Dias, D.Q., Sales, D.L., Oliveira, O.P., Teles, D.A., Sousa, J.G.G., Coutinho, H.D.M., Costa, J.G.M., Kerntopf, M.R., Alves, R.R.N., 2013. Evaluations of the antimicrobial activities and chemical compositions of body fat from the amphibians *Leptodactylus macrosternum* Miranda-Ribeiro (1926) and *Leptodactylus vastus* Adolf Lutz (1930) in Northeastern Brazil. Evidence-Based Complementary and Alternative Medicine 2013, 1–7.

Call, E., 2006. Mending the Web of Life: Chinese Medicine and Species Conservation. International Fund for Animal Welfare, Massachusetts.

Cardoso, J.S., Gonçalves, J.M., Fagundes, E.N., Costa Neto, E.M., Uetanabaro, A.P.T., 2010. Tratamento de infecções cutâneas com *Dinoponera quadriceps*: atividade antimicrobiana comprovada. In: Costa-Neto, E.M., Alves, R.R.N. (Eds.), Zooterapia: Os Animais na Medicina Popular Brasileira. Nupeea, Recife, pp. 141–157.

Carlsson, H.E., Schapiro, S.J., Farah, I., Hau, J., 2004. Use of primates in research: a global overview. American Journal of Primatology 63, 225–237.

Ceríaco, L.M.P., 2013. A review of fauna used in zootherapeutic remedies in Portugal: historical origins, current uses, and implications for conservation. In: Alves, R.R.N., Rosa, I.L. (Eds.), Animals in Traditional Folk Medicine. Springer-Verlag, Berlin-Heidelberg, pp. 317–345.

Chandler, C.K., 2012. Animal Assisted Therapy in Counseling. Routledge, New York.

Chaves, T.P., Clementino, E.L.C., Felismino, D.C., Alves, R.R.N., Vasconcellos, A., Coutinho, H.D.M., Medeiros, A.C.D., 2014. Antibiotic resistance modulation by natural products obtained from *Nasutitermes corniger* (Motschulsky, 1855) and its nest. Saudi Journal of Biological Sciences 22, 404–408.

Check, E., 2004. Health concerns prompt US review of exotic-pet trade. Nature 427, 277.

Chemas, R.C., 2010. A zooterapia no âmbito da medicina civilizada. I. Organoterapia humana e animal stricto sensu. In: Costa-Neto, E.M., Alves, R.R.N. (Eds.), Zooterapia: Os Animais na Medicina Popular Brasileira, first ed. NUPEEA, Recife, PE, Brazil, pp. 75–102.

Cherniack, E.P., 2010. Bugs as drugs, part 1: insects: the "new" alternative medicine for the 21st century. Alternative Medicine Review 15, 124–135.

Chivian, E., 2002. Biodiversity: Its Importance to Human Health. Harvard Medical School, Boston.

Chomel, B.B., 2002. Zoonosis bacterianas de aparición reciente. Revista Panamericana de Salud Publica 11, 50–55.

Chomel, B.B., 2003. Control and prevention of emerging zoonoses. Journal of Veterinary Medical Education 30, 145–147.

Chomel, B.B., Belotto, A., Meslin, F.X., 2007. Wildlife, exotic pets, and emerging zoonoses. Emerging Infectious Diseases 13, 6–11.

Chorilli, M., Michelin, D.C., Salgado, H.R.N., 2009. Animais de laboratório: o camundongo. Revista de Ciências Farmacêuticas Básica e Aplicada 28, 11–23.

Cleaveland, S., Laurenson, M.K., Taylor, L.H., 2001. Diseases of humans and their domestic mammals: pathogen characteristics, host range and the risk of emergence. Philosophical Transactions of the Royal Society of London. Series B: Biological Sciences 356, 991–999.

Cole, K.M., Gawlinski, A., Steers, N., Kotlerman, J., 2007. Animal-assisted therapy in patients hospitalized with heart failure. American Journal of Critical Care 16, 575–585.

Cologna, C.T., Barbosa, D.B., Santana, F.A., Rodovalho, C.M., Oliveira, L.A., Brandeburgo, M.A.M., 2005. Estudo da peçonha de Dinoponera australis—Roger, 1861 (Hymenoptera, Ponerinae), V Encontro interno de Iniciação científica. Convênio CNPq/UFU, Uberlândia.

Connor, K., Miller, J., 2000. Animal-assisted therapy: an in-depth look. Dimensions of Critical Care Nursing 19, 20–26.

Correale, J., Farez, M., 2007. Association between parasite infection and immune responses in multiple sclerosis. Annals of Neurology 61, 97–108.

Correale, J., Farez, M.F., 2011. The impact of parasite infections on the course of multiple sclerosis. Journal of Neuroimmunology 233, 6–11.

Costa-Neto, E.M., 2005. Entomotherapy, or the medicinal use of insects. Journal of Ethnobiology 25, 93–114.

Coutinho, H.D.M., Bezerra, D.A.C., Lôbo, K., Barbosa, I.J.F., 2004. Atividade antimicrobiana de produtos naturais. Revista Conceitos 77, 77–85.

Coutinho, H., Vasconcellos, A., Lima, M., Almeida-Filho, G., Alves, R.R.N., 2009. Termite usage associated with antibiotic therapy: enhancement of aminoglycoside antibiotic activity by natural products of *Nasutitermes corniger* (Motschulsky 1855). BMC Complementary and Alternative Medicine 9, 35.

Coutinho, H.D.M., Vasconcellos, A., Freire-Pessôa, H.L., Gadelha, C.A., Gadelha, T.S., Almeida-Filho, G.G., 2010. Natural products from the termite *Nasutitermes corniger* lowers aminoglycoside minimum inhibitory concentrations. Pharmacognosy Magazine 6, 1–4.

Coutinho, H., Aquino, P., Leite, J., Leandro, L., Figueredo, F., Matias, E., Guedes, T., 2014. Modulação da atividade antibacteriana do tecido adiposo da *Gallus gallus domesticus* (Linnaeus, 1758)/Modulatory antibacterial activity of body fat from *Gallus gallus domesticus* (Linnaeus 1758). Comunicata Scientiae 5, 380.

Dallavecchia, D.L., Proença, B.N., Aguiar Coelho, V.M., 2011. Biotherapy: an efficient alternative for the treatment of skin lesions. Revista de Pesquisa: Cuidado é Fundamental Online 3, 2071–2087.

De Smet, P.A.G.M., 1991. Is there any danger in using traditional remedies? Journal of Ethnopharmacology 32, 43–50.

Debnath, A., Chatterjee, U., Das, M., Vedasiromoni, J.R., Gomes, A., 2007. Venom of Indian monocellate cobra and Russell's viper show anticancer activity in experimental models. Journal of Ethnopharmacology 111, 681–684.

Djagoun, C.A.M.S., Akpona, H.A., Mensah, G.A., Nuttman, C., Sinsin, B., 2013. Wild mammals trade for zootherapeutic and mythic purposes in Benin (West Africa): capitalizing species involved, provision sources, and implications for conservation. In: Alves, R.R.N., Rosa, I.L. (Eds.), Animals in Traditional Folk Medicine: Implications for Conservation. Springer-Verlag, Berlin Heidelberg, pp. 367–381.

Donia, M., Hamann, M.T., 2003. Marine natural products and their potential applications as anti-infective agents. The Lancet Infectious Diseases 3, 338–348.

Dossey, A.T., 2010. Insects and their chemical weaponry: new potential for drug discovery. Natural Product Reports 27, 1737–1757.

Elliot, D.E., Pritchard, D.I., Weinstock, J.V., 2013. Helminth therapy. In: Grassberger, M., Sherman, R.A., Gileva, O.S., Kim, C.M.H., Mumcuoglu, K.Y. (Eds.), Biotherapy—History, Principles and Practice: A Practical Guide to the Diagnosis and Treatment of Disease Using Living Organisms. Springer, Dordrecht, The Netherlands, pp. 177–190.

European Commission, 2011. Biodiversity and Health.

Fagundes, D.J., Taha, M.O., 2004. Modelo animal de doença: critérios de escolha e espécies de animais de uso corrente. Acta Circurgica Brasileira 19, 59–65.

Faulkner, D.J., 1998. Marine natural products. Natural Product Reports 15, 113–158.

Ferguson, N.M., Cummings, D.A.T., Cauchemez, S., Fraser, C., Riley, S., Meeyai, A., Iamsirithaworn, S., Burke, D.S., 2005. Strategies for containing an emerging influenza pandemic in Southeast Asia. Nature 437, 209–214.

Ferreira, F.S., Brito, S.V., Saraiva, R.A., Araruna, M.K.A., Menezes, I.R.A., Costa, J.G.M., Coutinho, H.D.M., Almeida, W.O., Alves, R.R.N., 2010. Topical anti-inflammatory activity of body fat from the lizard *Tupinambis merianae*. Journal of Ethnopharmacology 130, 514–520.

Ferreira, F.S., Silva, N.L.G., Matias, E.F.F., Brito, S.V., Oliveira, F.G., Costa, J.G.M., Coutinho, H.D.M., Almeida, W.O., Alves, R.R.N., 2011. Potentiation of aminoglycoside antibiotic activity using the body fat from the snake *Boa constrictor*. Revista Brasileira de Farmacognosia 21, 503–509.

Ferreira, F.S., Albuquerque, U.P., Coutinho, H.D.M., Almeida, W.O., Alves, R.R.N., 2012. The trade in medicinal animals in northeastern Brazil. Evidence-based Complementary and Alternative Medicine 2012, 1–20.

Ferreira, F.S., Fernandes-Ferreira, H., Leo Neto, N., Brito, S.V., Alves, R.R.N., 2013. The trade of medicinal animals in Brazil: current status and perspectives. Biodiversity and Conservation 22, 839–870.

Ferreira, F.S., Brito, S.V., Sales, D.L., Menezes, I.R.A., Coutinho, H.D.M., Souza, E.P., Almeida, W.O., Alves, R.R.N., 2014. Anti-inflammatory potential of zootherapeutics derived from animals used in Brazilian traditional medicine. Pharmaceutical Biology 52, 1403–1410.

Figueirêdo, R.E.C.R., Vasconcellos, A., Policarpo, I.S., Alves, R.R.N., 2015. Edible and medicinal termites: a global overview. Journal of Ethnobiology and Ethnomedicine 11, 1–7.

Fine, A.H., 2010. Handbook on Animal-assisted Therapy: Theoretical Foundations and Guidelines for Practice. Academic Press.

Fine, A., Alexander, H., 1934. Maggot therapy—technique and clinical application. Journal of Bone and Joint Surgery 16, 572–582.

Firn, R.D., 2003. Bioprospecting–why is it so unrewarding? Biodiversity and Conservation 12, 207–216.

Fitter, R.S.R., 1986. Wildlife for Man: How and Why We Should Conserve Our Species. Collins, London, London.

Fleming, J.O., Isaak, A., Lee, J.E., Luzzio, C.C., Carrithers, M.D., Cook, T.D., Field, A.S., Boland, J., Fabry, Z., 2011. Probiotic helminth administration in relapsing–remitting multiple sclerosis: a phase 1 study. Multiple Sclerosis Journal 17, 743–754.

Friend, M., 2006. Disease Emergence and Resurgence: The Wildlife-human Connection. US Department of the Interior, US Geological Survey.

Froese, R., Pauly, D., 2016. Garra rufus.

Fusetani, N., 2000. Drugs from the Sea. S Karger Pub.

Gileva, O.S., Mumcuoglu, K.Y., 2013. Hirudotherapy. In: Grassberger, M., Sherman, R.A., Gileva, O.S., Kim, C.M.H., Mumcuoglu, K.Y. (Eds.), Biotherapy-History, Principles and Practice: A Practical Guide to the Diagnosis and Treatment of Disease Using Living Organisms. Springer, Dordrecht, Heidelberg, New York, London, pp. 31–76.

Gilyova, O.S., 2005. Modern hirudotherapy–a review. The BeTER LeTTER 2, 1–3.

Glickman, L.T., Fairbrother, A., Guarino, A.M., Bergman, H.L., Buck, W.B., 1991. Animals as Sentinels of Environmental Health Hazards. National Research Council, Committee on Animals as Monitors of Environmental Hazards, Washington, DC.

Godfrey, K., 1997. Use of leeches and leech saliva in clinical practice. Nursing Times 93, 62–63.

Grantham-Hill, C., 1933. Preliminary note on the treatment of infected wounds with the larva of *Wohlfartia nuba*. Transactions of the Royal Society of Tropical Medicine and Hygiene 27, 93–98.

Grassberger, M., Sherman, R.A., Gileva, O.S., Kim, C.M.H., Mumcuoglu, K.Y., 2013. Biotherapy–History, Principles and Practice: A Practical Guide to the Diagnosis and Treatment of Disease Using Living Organism. Springer.

Grassberguer, M., Sherman, R.A., 2013. Ichthyotherapy. In: Grassberger, M., Sherman, R.A., Gileva, O.S., Kim, C.M.H., Mumcuoglu, K.Y. (Eds.), Biotherapy—History, Principles and Practice: A Practical Guide to the Diagnosis and Treatment of Disease Using Living Organisms. Springer Science+Business Media, Dordrecht, The Netherlands, pp. 147–176.

Gunjegaonkar, S.M., Kshirsagar, S.S., Bayas, J.P., 2016. Biosurgical therapy-overview. International Journal of Pharmacological Research 6, 133–137.

Hammond, P.B., Aronson, A.L., 1964. Lead poisoning in cattle and horses in the vicinity of a smelter. Annals of the New York Academy of Sciences 111, 595–611.

Hardy, A., 2003. Animals, disease, and man: making connections. Perspectives in Biology and Medicine 46, 200–215.

Hooker, S.D., Freeman, L.H., Stewart, P., 2002. Pet therapy research: a historical review. Holistic Nursing Practice 17, 17–23.

Horn, K.L., Cobb, A.H., Gates, G.A., 1976. Maggot therapy for subacute mastoiditis. Archives of Otolaryngology 102, 377–379.

Hunt, B., Vincent, A.C.J., 2006. Scale and sustainability of marine bioprospecting for pharmaceuticals. AMBIO: A Journal of the Human Environment 35, 57–64.

IFAW, I.F.f.A.W., 2011. Traditional Medicine.

Ireland, C.M., Copp, B.R., Foster, M.P., McDonald, L.A., Radisky, D.C., Swersey, J.C., 1993. Biomedical potential of marine natural products. Marine Biotechnology 1, 1–43.

Julian, K.G., Eidson, M., Kipp, A.M., Weiss, E., Petersen, L.R., Miller, J.R., Hinten, S.R., Marfin, A.A., 2002. Early season crow mortality as a sentinel for West Nile virus disease in humans, northeastern United States. Vector Borne and Zoonotic Diseases 2, 145–155.

Jurgelski, W., 1984. The marsupial as a biomedical model. International Journal of Toxicology 3, 343–355.

Kahn, L.H., 2006. Confronting zoonoses, linking human and veterinary medicine. Emerging Infectious Diseases 12, 556–561.

Karesh, W.B., Noble, E., 2009. The bushmeat trade: increased opportunities for transmission of zoonotic disease. Mount Sinai Journal of Medicine: A Journal of Translational and Personalized Medicine 76, 429–434.

Karesh, W.B., Cook, R.A., Bennett, E.L., Newcomb, J., 2005. Wildlife trade and global disease emergence. Emerging Infectious Diseases 11, 1000–1002.

Karesh, W.B., Cook, R.A., Gilbert, M., Newcomb, J., 2007. Implications of wildlife trade on the movement of avian influenza and other infectious diseases. Journal of Wildlife Diseases 43, S55–S59.

Karthikeyan, R., Karthigayan, S., Sri Balasubashini, M., Somasundaram, S.T., Balasubramanian, T., 2008. Inhibition of Hep2 and HeLa cell proliferation in vitro and EAC tumor growth in vivo by *Lapemis curtus* (Shaw 1802) venom. Toxicon 51, 157–161.

Khan, A.R., Fallon, P.G., 2013. Helminth therapies: translating the unknown unknowns to known knowns. International Journal for Parasitology 43, 293–299.

Kim, C.M.H., 2013. Apitherapy–bee venom therapy. In: Grassberger, M., Sherman, R.A., Gileva, O.S., Kim, C.M.H., Mumcuoglu, K.Y. (Eds.), Biotherapy—History, Principles and Practice: A Practical Guide to the Diagnosis and Treatment of Disease Using Living Organisms. Springer Science+Business Media, Dordrecht, The Netherlands, pp. 77–112.

Kirchheimer, W.F., Sanchez, R.M., Shannon, E.J., 1978. Effect of specific vaccine on cell-mediated immunity of armadillos against *M. leprae*. International Journal of Leprosy and Other Mycobacterial Diseases 46, 353–357.

Krauss, H., 2003. Zoonoses: infectious diseases transmissible from animals to humans. American Society for Microbiology.

Kritsky, G., 1987. Insects in traditional Chinese medicines. Proceedings of the Indiana Academy of Science 96, 289–291.

Kruger, K.A., Serpell, J.A., Fine, A.H., 2006. Animal-assisted interventions in mental health: definitions and theoretical foundations. In: Fine, A.H. (Ed.), Handbook on Animal-Assisted Therapy: Theoretical Foundations and Guidelines for Practice. Academic Press, California, pp. 21–38.

Kruse, H., Kirkemo, A.M., Handeland, K., 2004. Wildlife as source of zoonotic infections. Emerging Infectious Diseases 10, 2067–2072.

Kunin, W.E., Lawton, J.H., 1996. Does biodiversity matter? Evaluating the case for conserving species. In: Gaston, K.J. (Ed.), Biodiversity. A Biology of Numbers and Difference Blackwell Science. Oxford, UK, pp. 367–387.

Lamberty, M., Zachary, D., Lanot, R., Bordereau, C., Robert, A., Hoffmann, J.A., Bulet, P., 2001. Insect immunity constitutive expression of a cysteine-rich antifungal and a linear antibacterial peptide in a termite insect. Journal of Biological Chemistry 276, 4085–4092.

Lee, S.K.H., 1999. Trade in traditional medicine using endangered species: an international context. In: 2nd Australian Symposium on Traditional Medicine and Wildlife Conservation. Australia, Melbourne.

Leonardi-Bee, J., Pritchard, D., Britton, J., 2006. Asthma and current intestinal parasite infection: systematic review and meta-analysis. American Journal of Respiratory and Critical Care Medicine 174, 514–523.

Lerclercq, M., 1990. Utilisation de larves de dipteres-Maggot Therapy-en medecine: historique et actualite. Bulletin et Annales de la Societe Royale Belge d'Entomologie 126, 41–50.

Lilley, B., Lammie, P., Dickerson, J., Eberhard, M., 1997. An increase in hookworm infection temporally associated with ecologic change. Emerging Infectious Diseases 3, 391–393.

Lone, A.H., Ahmad, T., Anwar, M., Habib, S., Sofi, G., Imam, H., 2011. Leech therapy-a holistic approach of treatment in unani (greeko-arab) medicine. Ancient Science of Life 31, 31–35.

Lucache, B., Albu, E., Mungiu, O.C., 2015. Acupuncture and bee venom therapy in the chronic low back pain: a short review. African Journal of Traditional, Complementary and Alternative Medicines 12, 76–80.

MacKinney, L.C., 1946. Animal substances in materia medica. Journal of the History of Medicine and Allied Sciences 1, 149–170.

Mahawar, M.M., Jaroli, D.P., 2008. Traditional zootherapeutic studies in India: a review. Journal of Ethnobiology and Ethnomedicine 4, 17.

Marques, J.G.W., 1997. Fauna medicinal: Recurso do ambiente ou ameaça à biodiversidade. Mutum 1, 4.

Martin, F., Farnum, J., 2002. Animal-assisted therapy for children with pervasive developmental disorders. Western Journal of Nursing Research 24, 657–670.

Martinez, G.J., 2013. Use of fauna in the traditional medicine of native Toba (qom) from the Argentine Gran Chaco region: an ethnozoological and conservationist approach. Ethnobiology and Conservation 2, 1–43.

Martini, R.K., Sherman, R.A., 2003. Terapia de desbridamento com larvas. Journal Brasileiro de Medicina 85, 82–85.

Masaro Jr., A.L., Della Luccia, T.M.C., Barbosa, L.C.A., Maffia, L.A., Morandi, M.A.B., 2001. Inhibition of the germination of *Botrytis cinerea* Pers. Fr. conidia by extracts of the mandibular gland of *Atta sexdens* rubropilosa forel (Hymenoptera: Formicidae). Neotropical Entomology 30, 403–406.

McCluskey, B.J., 2003. Use of sentinel herds in monitoring and surveillance systems. In: Salman (Ed.), Animal Disease Surveillance and Survey Systems: Methods Applications. Iowa State Press, Iowa, pp. 119–133.

McConnell, E., 2002. Myths & facts…about animal-assisted therapy. Nursing 32, 76.

McGibbon, N.H., Andrade, C.K., Widener, G., Cintas, H.L., 1998. Effect of an equine-movement therapy program on gait, energy expenditure, and motor function in children with spastic cerebral palsy: a pilot study. Developmental Medicine & Child Neurology 40, 754–762.

McLellan, N.W., 1932. The maggot treatment of osteomyelitis. Canadian Medical Association Journal 27, 256–260.

Mi, E., Mi, E., Jeggo, M., 2016. Where to now for one health and ecohealth? EcoHealth 13, 12–17.

Moquin-Tandon, A., 1861. Elements of Medical Zoology. Baillière.

Moreira, D.R., 2012. Apiterapia no tratamento de patologias. Revista F@pciência 9, 21–29.

Munir, T., Malik, M.F., Hashim, M., Qureshi, M.A., Naseem, S., 2016. Therapeutic applications of blowfly maggots: a review. Journal of Entomology and Zoology Studies 4 (5), 33–36.

Myers, N., 1979. The Sinking Ark: A New Look at the Problem of Disappearing Species. Pergamon Press, New York.

Nakanishi, K., 1999. An historical perspective of natural products chemistry. In: Comprehensive Natural Products Chemistry: Isoprenoids Including Carotenoids and Steroids.

Nimer, J., Lundahl, B., 2007. Animal-assisted therapy: a meta-analysis. Anthrozoos 20, 225–238.

Oliva, V.N.L.S., 2010. Terapia assistida por Animais. In: Costa- Neto, E.M., Alves, R.R.N. (Eds.), Zooterapia: Os Animais na Medicina Popular Brasileira. Nuppea, Recife, pp. 127–140.

Oliveira, O.P., Sales D.L., Dias, D.Q., Cabral, M.E.S, Araújo Filho, J.A., Teles, D.A., Sousa, J.G.G., Ribeiro, S.C., Freitas, F.R.D., Coutinho, H.D.M., Kerntopf, M.R., Costa, J.G.M., Alves, R.R.N., Almeida, W.O., 2014. Antimicrobial activity and chemical composition of fixed oil extracted from the body fat of the snake *Spilotes pullatus*. Pharmaceutical Biology 52, 740–744.

One Health, 2010. One Health Commission.

Ozcelik, S., Akyol, M., 2011. Kangal hot spring with fish (Kangal fishy Health spa) & Psoriasis. La Presse Thermale et Climatique 148, 141–147.

Patz, J.A., Graczyk, T.K., Geller, N., Vittor, A.Y., 2000. Effects of environmental change on emerging parasitic diseases. International Journal for Parasitology 30, 1395–1405.

Pemberton, R.W., 1999. Insects and other arthropods used as drugs in Korean traditional medicine. Journal of Ethnopharmacology 65, 207–216.

Peña, M.T., Adams, J.E., Adams, L.B., Gillis, T.P., Williams, D.L., Spencer, J.S., Krahenbuhl, J.L., Truman, R.W., 2008. Expression and characterization of recombinant interferon gamma (IFN-γ) from the nine-banded armadillo (*Dasypus novemcinctus*) and its effect on *Mycobacterium leprae*-infected macrophages. Cytokine 43, 124–131.

Pérez, J.M., 2009. Parasites, pests, and pets in a global world: new perspectives and challenges. Journal of Exotic Pet Medicine 18, 248–253.

Project Seahorse, 2016. Why Seahorses? Essential Facts About Seahorses.

Quave, C.L., Lohani, U., Verde, A., Fajardo, J., Rivera, D., Obón, C., Valdes, A., Pieroni, A., 2010. A comparative assessment of zootherapeutic remedies from selected areas in Albania, Italy, Spain and Nepal. Journal of Ethnobiology 30, 92–125.

Rabinowitz, P., Conti, L., 2013. Links among human health, animal health, and ecosystem health. Annual Review of Public Health 34, 189–204.

Rabinowitz, P.M., Gordon, Z., Holmes, R., Taylor, B., Wilcox, M., Chudnov, D., Nadkarni, P., Dein, F.J., 2005. Animals as sentinels of human environmental health hazards: an evidence-based analysis. EcoHealth 2, 26–37.

Ratcliffe, N.A., Mello, C.B., Garcia, E.S., Butt, T.M., Azambuja, P., 2011. Insect natural products and processes: new treatments for human disease. Insect Biochemistry and Molecular Biology 41, 747–769.

Reames, M.K., Christensen, C., Luce, E.A., 1988. The use of maggots in wound debridement. Annals of Plastic Surgery 21, 388–391.

Reif, J.S., 2011. Animal sentinels for environmental and public health. Public Health Reports 126, 50.

Reperant, L.A., Cornaglia, G., Osterhaus, A.D.M.E., 2012. The importance of understanding the human–animal interface: from early hominins to global citizens. In: Mackenzie, J.S., Jeggo, M., Daszak, M., Richt, J.A. (Eds.), One Health: The Human-Animal-Environment Interfaces in Emerging Infectious Diseases. Springer, Heidelberg, New York, Dordrecht, London, pp. 49–81.

Robinson, W., 1933. The use of blowfly larvae in the treatment of infected wounds. Annals of the Entomological Society of America 26, 270–276.

Robinson, M.M., Zhang, X., 2011. The World Medicines Situation 2011–Traditional Medicines: Global Situation, Issues and Challenges. World Health Organization, Geneva.

Roldán-Clarà, B., Lopez-Medellín, X., Espejel, I., Arellano, E., 2014. Literature review of the use of birds as pets in Latin-America, with a detailed perspective on Mexico. Ethnobiology and Conservation 3, 1–18.

Rosa, I.L., Defavari, G.R., Alves, R.R.N., Oliveira, T.P.R., 2013. Seahorses in traditional medicines: a global overview. In: Alves, R.R.N., Rosa, I.L. (Eds.), Animals in Traditional Folk Medicine. Springer, Berlin, pp. 207–240.

Rose, J., Quave, C.L., Islam, G., 2012. The four-sided triangle of ethics in bioprospecting: pharmaceutical business, international politics, socio-environmental responsibility and the importance of local stakeholders. Ethnobiology and Conservation 1, 1–25.

Rostal, M.K., Olival, K.J., Loh, E.H., Karesh, W.B., 2012. Wildlife: the need to better understand the linkages. In: Mackenzie, J.S., Jeggo, M., Daszak, P., Richt, J.A. (Eds.), One Health: The Human-Animal-Environment Interfaces in Emerging Infectious Diseases. Springer, Berlin Heidelberg, pp. 101–125.

Rouquet, P., Froment, J.M., Bermejo, M., Kilbourn, A., Karesh, W., Reed, P., Kumulungui, B., 2005. Wild animal mortality monitoring and human Ebola outbreaks, Gabon and Republic of Congo, 2001–2003. Emerging Infectious Diseases 11, 283–290.

Roy, S., Saha, S., Pal, P., 2015. Insect natural products as potential source for alternative medicines—a review. World Scientific News 19, 80–94.

Sahayaraj, K., Borgio, J.F., Muthukumar, S., Anandh, G.P., 2006. Antibacterial activity of *Rhynocoris marginatus* (Fab.) and *Catamirus brevipennis* (Servile)(Hemiptera: reduviidae) venomS against human pathogens. Journal of Venomous Animals and Toxins Including Tropical Diseases 12, 487–496.

Sales, D.L., Oliveira, O.P., Cabral, M.E., Dias, D.Q., Kerntopf, M.R., Coutinho, H.D.M., Costa, J.G.M., Freitas, F.R.D., Ferreira, F.S., Alves, R.R.N., Almeida, W.O., 2015. Chemical identification and evaluation of the antimicrobial activity of fixed oil extracted from *Rhinella jimi*. Pharmaceutical Biology 53, 98–103.

Santos, I.J.M., Coutinho, H.D.M., Ferreira Matias, E.F., Costa, J.G.M., Alves, R.R.N., Almeida, W.O., 2012. Antimicrobial activity of natural products from the skins of the semiarid living lizards *Ameiva ameiva* (Linnaeus, 1758) and *Tropidurus hispidus* (Spix, 1825). Journal of Arid Environments 76, 138–141.

Santos, I.J.M., Leite, G.O., Costa, J.G.M., Alves, R.R.N., Campos, A.R., Menezes, I.R.A., Freita, F.R.V., Nunes, M.J.H., Almeida, W.O., 2015. Topical anti-inflammatory activity of oil from *Tropidurus hispidus* (Spix, 1825). Evidence-Based Complementary and Alternative Medicine 2015.

Scarpa, A., 1981. Pre-scientific medicines: their extent and value. Social Science & Medicine 15A, 317–326.

Schacter, B.Z., 2006. The New Medicines: How Drugs Are Created, Approved, Marketed, and Sold. Greenwood Publishing Group.

Schnurrenberger, P.R., Hubbert, W.T., 1981. An Outline of the Zoonoses. Iowa State University Press, Ames, IA.

Scollard, D.M., 2008. The biology of nerve injury in leprosy. Leprosy Review 79, 242–253.

Scotch, M., Odofin, L., Rabinowitz, P., 2009. Linkages between animal and human health sentinel data. BMC Veterinary Research 5, 15.

Scott, K., 2002. Is hirudin a potential therapeutic agent for arthritis? Annals of the Rheumatic Diseases 61, 561–562.

Scrivener, S., Yemaneberhan, H., Zebenigus, M., Tilahun, D., Girma, S., Ali, S., McElroy, P., Custovic, A., Woodcock, A., Pritchard, D., 2001. Independent effects of intestinal parasite infection and domestic allergen exposure on risk of wheeze in Ethiopia: a nested case-control study. The Lancet 358, 1493–1499.

Seedhouse, E., 2010. Ocean Outpost: The Future of Humans Living Underwater. Springer.

Seymour, C., 1984. Arthropod vectors of human disease in the United States. Clinical Microbiology Newsletter 6, 85–89.

Sharma, R., Lahiri, R., Scollard, D.M., Pena, M., Williams, D.L., Adams, L.B., Figarola, J., Truman, R.W., 2013. The armadillo: a model for the neuropathy of leprosy and potentially other neurodegenerative diseases. Disease Models and Mechanisms 6, 19–24.

Shaw, C., 2009. Advancing drug discovery with reptile and amphibian venom peptides: venom-based medicines. The Biochemical Society 31, 34–37.

Sherman, R.A., Wyle, F., Vulpe, M., 1986. Maggot therapy for treating pressure ulcers in spinal cord injury patients. The Journal of Spinal Cord Medicine 18, 71–74.

Sherman, R.A., Hall, M.J.R., Thomas, S., 2000. Medicinal maggots: an ancient remedy for some contemporary afflictions. Annual Review of Entomology 45, 55–81.

Sherman, R.A., Mumcuoglu, K.Y., Grassberger, M., Tantawi, T.I., 2013. Maggot therapy. In: Grassberger, M., Sherman, R.A., Gileva, O.S., Kim, C.M.H., Mumcuoglu, K.Y. (Eds.), Biotherapy-History, Principles and Practice: A Practical Guide to the Diagnosis and Treatment of Disease Using Living Organisms. Springer Science+Business Media, Dordrecht, The Netherlands, pp. 5–29.

Shimpi, R., Chaudhari, P., Deshmukh, R., Devare, S., Bagad, Y., Bhurat, M., 2016. A review: phamacotherapeutics of bee venom. World Journal of Pharmacy and Pharmaceutical Sciences 5, 656–667.

Shin, B.C., Kong, J.C., Park, T.Y., Yang, C.Y., Kang, K.W., Choi, S., 2012. Bee venom acupuncture for chronic low back pain: a randomised, sham-controlled, triple-blind clinical trial. European Journal of Integrative Medicine 4, e271–e280.

Sifuna, N., 2012. The future of traditional customary uses of wildlife in modern Africa: a case study of Kenya and Botswana. Journal of Biophysical Chemistry 2, 31–38.

Silva, E., Rosa, P., Arruda, M., Rúbio, E., 2005. Determination of duffy phenotype of red blood cells in *Dasypus novemcinctus* and *Cabassous* sp. Brazilian Journal of Biology 65, 555–557.

Silveira, N., 1998. Gatos, a emoção de lidar. Editora Leo Christiano, São Paulo, Brasil.

Simpson, R.D., 1997. Biodiversity prospecting: shopping the wilds is not the key to conservation. Resources 126, 12–15.

Sleeman, J., 2006. Wildlife zoonoses for the veterinary practitioner. Journal of Exotic Pet Medicine 15, 25–32.

Soewu, D.A., 2013. Zootherapy and biodiversity conservation in Nigeria. In: Alves, R.R.N., Rosa, I.L. (Eds.), Animals in Traditional Folk Medicine: Implications for Conservation. Springer-Verlag, Berlin Heidelberg, pp. 347–365.

Stephen, C., Ribble, C., 2001. Death, disease and deformity; using outbreaks in animals as sentinels for emerging environmental health risks. Global Change & Human Health 2, 108–117.

Stephenson, J., 1832. Medical Zoology, and Mineralogy; or Illustrations and Descriptions of the Animals and Minerals Employed in Medicine, and of the Preparations Derived from Them: Including Also an Account of Animal and Mineral Poisons. John Wilson.

Still, J., 2003. Use of animal products in traditional Chinese medicine: environmental impact and health hazards. Complementary Therapies in Medicine 11, 118–122.

Sukarmi, S., Sabdono, A., 2011. Ethical perspectives of sustainable use of reef's invertebrates as a source of marine natural products. Journal of Coastal Development 11, 97–103.

Taylor, L.H., Latham, S.M., Mark, E.J., 2001. Risk factors for human disease emergence. Philosophical Transactions of the Royal Society of London. Series B: Biological Sciences 356, 983–989.

Teich, S., Myers, R.A., 1986. Maggot therapy for severe skin infections. Southern Medical Journal 79, 1153–1155.

Thammasirirak, S., Ponkham, P., Preecharram, S., Khanchanuan, R., Phonyothee, P., Daduang, S., Srisomsap, C., Araki, T., Svasti, J., 2006. Purification, characterization and comparison of reptile lysozymes. Comparative Biochemistry and Physiology Part C: Toxicology & Pharmacology 143, 209–217.

Trowell, S., 2003. Drugs from bugs: the promise of pharmaceutical entomology. Futurist 37, 17–19.

Tsuchiya, K., 1992. Historical perspectives in occupational medicine. The discovery of the causal agent of minamata disease. American Journal of Industrial Medicine 21, 275–280.

Unnikrishnan, E., 2004. Materia Medica of the Local Health Traditions of Payyannur Centre for Development Studies.

Van der Schalie, W.H., Gardner Jr., H.S., Bantle, J.A., De Rosa, C.T., Finch, R.A., Reif, J.S., Reuter, R.H., Backer, L.C., Burger, J., Folmar, L.C., 1999. Animals as sentinels of human health hazards of environmental chemicals. Environmental Health Perspectives 107, 309–315.

Van Vliet, N., Moreno, J., Gómez, J., Zhou, W., Fa, J.E., Golden, C., Alves, R.R.N., Nasi, R., 2017. Bushmeat and human health: assessing the evidence in tropical and subtropical forests. Ethnobiology and Conservation 6, 1–45.

Van Wingerden, J.J., Oosthuizen, J.H., 1997. Use of the local leech *Hirudo michaelseni* in reconstructive plastic and hand surgery. South African Journal of Surgery (Suid-Afrikaanse tydskrif vir chirurgie) 35, 29–31.

Voultsiadou, E., 2010. Therapeutic properties and uses of marine invertebrates in the ancient Greek world and early Byzantium. Journal of Ethnopharmacology 130, 237–247.

Walsh, J.F., Molyneux, D.H., Birley, M.H., 1993. Deforestation: effects on vector-borne disease. Parasitology 106, S55–S75.

Warwick, C., Lindley, S., Steedman, C., 2011. How to handle pets: a guide to the complexities of enforcing animal welfare and disease control. Environmental Health News 8 (18–19).

Weiss, R.A., 2001. The Leeuwenhoek lecture 2001. Animal origins of human infectious disease. Philosophical Transactions of the Royal Society of London. Series B: Biological Sciences 356, 957–977.

Whitaker, I.S., Oboumarzouk, O., Rozen, W.M., Naderi, N., Balasubramanian, S.P., Azzopardi, E.A., Kon, M., 2012. The efficacy of medicinal leeches in plastic and reconstructive surgery: a systematic review of 277 reported clinical cases. Microsurgery 32, 240–250.

Whiting, M.J., Williams, V.L., Hibbitts, T.J., 2011. Animals traded for traditional medicine at the Faraday market in South Africa: species diversity and conservation implications. Journal of Zoology 284, 84–96.

Whiting, M.J., Williams, V.L., Hibbitts, T.J., 2013. Animals traded for traditional medicine at the Faraday market in South Africa: species diversity and conservation implications. In: Alves, R.R.N., Rosa, I.L. (Eds.), Animals in Traditional Folk Medicine: Implications for Conservation. Springer-Verlag, Berlin Heidelberg, pp. 421–473.

WHO, 2005. Ecosystems and Human Well-Being : Health Synthesis : A Report of the Millennium Ecosystem Assessment, first ed. WHO, Geneva, Switzerland.

Wicher, K., Kalinka, C., Walsh, G.P., 1983. Attempt to infect the nine-banded armadillo with *Treponema pallidum*. International Archives of Allergy and Immunology 70, 285–287.

Wiedorn, W.S., 1954. A new experimental animal for psychiatric research: the opossum, *Didelphis virginiana*. Science 119, 360–361.

Williams, E.S., Yuill, T., Artois, M., Fischer, J., Haigh, S.A., 2002. Emerging infectious diseases in wildlife. Revue Scientifique et Technique, Office International des Epizooties 21, 139–158.

Williams, V.L., Cunningham, A.B., Bruyns, R.K., Kemp, A.C., 2013. Birds of a feather: quantitative assessments of the diversity and levels of threat to birds used in African traditional medicine. In: Alves, R.R.N., Rosa, I.L. (Eds.), Animals in Traditional Folk Medicine: Implications for Conservation. Springer, pp. 383–420.

Wolfe, N.D., Eitel, M.N., Gockowski, J., Muchaal, P.K., Nolte, C., Prosser, A.T., Torimiro, J.N., Weise, S.F., Burke, D.S., 2000. Deforestation, hunting and the ecology of microbial emergence. Global Change & Human Health 1, 10–25.

Wolfe, N.D., Dunavan, C.P., Diamond, J., 2007. Origins of major human infectious diseases. Nature 477, 279–283.

Wolff, M.J., Broadhurst, M.J., 2012. Helminthic therapy: improving mucosal barrier function. Trends in Parasitology 28, 187–194.

Woolhouse, M.E., Gowtage-Sequeria, S., 2005. Host range and emerging and reemerging pathogens. Emerging Infections Diseases 11, 1842–1847.

World Resources Institute, W.R.R, 2000. People and Ecosystems the Fraying Web of Life. World Resources Institute, Washington, DC, p. 389.

Yasui, A.M., 2012. Avaliação da toxicidade e da resposta cutânea local induzidas por doses diluídas de veneno de abelha em cães. Universidade Federal Rural do Rio de Janeiro, Seropédica.

Yinfeng, G., Xueying, Z., Yan, C., Di, W., Sung, W., 1997. Sustainability of Wildlife Use in Traditional Chinese Medicine, Conserving China's Biodiversity: Reports of the Biodiversity Working Group (BWG). China Council for International Cooperation on Environment and Development, pp. 190–220.

Zhu, Q., Huang, J., Wang, S., Qin, Z., Lin, F., 2016. Cobrotoxin extracted from *Naja atra* venom relieves arthritis symptoms through anti-inflammation and immunosuppression effects in rat arthritis model. Journal of Ethnopharmacology 1–9.

CHAPTER

14

Use and Commercialization of Animals as Decoration*

Rômulo Romeu Nóbrega Alves[1], *Ellori Laíse Silva Mota*[1,2], *Thelma Lúcia Pereira Dias*[1]

[1]**Universidade Estadual da Paraíba, Campina Grande, Brazil;** [2]**Universidade Federal da Paraíba, João Pessoa, Brazil**

INTRODUCTION

The most common use of nonhuman animals by humans is to meet their nutritional needs (Reitz and Wing, 2008). However, apart from being a major source of food, wildlife has always been an important source of raw material, such as hide, fur, skin, hair, feathers, ivory, antlers, horns, bones, and a large variety of substances used for medicinal and spiritual/religious purposes (Alves and Rosa, 2013; Alves et al., 2012; Roth and Merz, 1997). The use of animal by-products inappropriate for alimentary consumption illustrates a human strategy to optimize the use of an obtained resource.

Early humans lived alongside the animals they hunted, eating their meat, using their hides, and putting their bones and tusks to use as building materials, or fashioning them into ornaments or weapons (Pedersen, 2004). The contribution of animals to this aspect of human welfare is extensive and varied (Fitter, 1986). Animal products, such as horn, shell, bone, ivory, and antler, have been used by humans as ornaments and decorative materials, as personal accessories, as decoration or curios (items of interest or curiosity), and as tools (Dias et al., 2011; Pedersen, 2004). Such uses have fostered a growing trade that is closely tied to the curio trade, which involves the sale of dried animals or animal parts for the primary purpose of use as ornamentation and/or decoration.

Animal parts, such as bones, teeth, whole skeletons, shells, antlers, feathers, fur, and claws, among others, are used for the production of decorative objects and the manufacture of handmade jewelry (Nijman and Nekaris, 2014). Species that naturally possess morphological adornments, such as horns and varied sharp or colored protuberances (e.g., feathers, skins,

*This chapter is a revised and updated version of the section "Animal as ornaments, decoration, tools and others purposes," extracted from article "Alves, R.R.N., 2012. Relationships between fauna and people and the role of ethnozoology in animal conservation. Ethnobiology and Conservation 1, 1–69."

Ethnozoology
http://dx.doi.org/10.1016/B978-0-12-809913-1.00014-4

and shells) or thorns, are often more attractive to the decorative trade. For example, lizards (Jennings, 1987), birds (use of feathers) (Nijman and Nekaris, 2014), marine mollusk shells (Dias et al., 2011), whole turtles or shell parts (Ceballos and Fitzgerald, 2004), and hard corals (Wood and Wells, 1988; Wood et al., 2012), among others, are particularly common components of the trade in wild animals for decorative purposes.

For purposes of definition, the use of animals for decoration is considered the use of the whole animal as an ornament, as a part of a decorative piece or part of the decoration of a decorative object. Large markets that trade animals and/or their parts have been expanding throughout the world and have involved legally protected and globally endangered wild species (Nijman and Nekaris, 2014; Nijman et al., 2015). Species included in the IUCN red list, listed by CITES or even protected by current local laws are commonly documented as being involved in wildlife trade for decorative purposes (Nijman and Nekaris, 2014). This is the case of the Hawksbill sea turtle, *Eretmochelys imbricata*, which is critically endangered (IUCN, 2016) and listed in Appendix I of CITES (Nijman and Nekaris, 2014), but is still exploited for various purposes, including ornamentation, tools, and food (Lam et al., 2012). Depending on the level of knowledge of the trader, or even the strictness of the laws directed to control such illegal activity, trade can take place clandestinely, or be freely offered to buyers.

One of the issues to be considered in documenting trade in wild species for decorative purposes is the origin of the exploited individuals. Many endemic species from certain regions or countries, or with very restricted distributions, are often marketed on different continents. This means that whole animals or parts of animals are crossing international borders freely, thereby fueling the trade in decorative and souvenir products throughout the world. For example, an endemic gastropod of the northeast coast of Brazil, *Voluta ebraea* (Dias et al., 2011) possesses a limited distribution along a narrow coastline (from the state of Pará to southern state of Bahia), but its shells, obtained from individuals of several Brazilian states, are sold on websites in Italy, Switzerland, Greece, and Portugal, among other countries.

The impact of the removal of individuals from these populations on the ecosystem is unknown. The high demand of this trade can lead to the decline of entire populations in certain areas, leading fishermen to explore other locations to meet demand, which, in the medium term, results in a progressive decline in the population stocks of target species (Jennings, 1987).

Although widely known and widespread, the use and trade of animals for decorative and ornamental purposes is still poorly quantified around the world. Trade in these products depends on agreements for these data to be shared; however, many of the disclosures are inconsistent and divergent (Wells and Barzdo, 1991; Wood et al., 2012).

This chapter presents an overview of the use of animals for decoration and as ornaments based on information obtained in the literature. In addition to listing the major animal taxa involved in decorative trade, we report their main uses and discuss ecological concerns arising from population declines of selected taxa and the implications for conservation. Special focus is given to marine fauna, which includes several species used for decorative purposes and for making ornaments, and for which the volume of information available regarding these uses is greater than that for terrestrial animals.

To assess the use of marine animal by-products for ornaments, decoration, or curios, we compiled qualitative and quantitative data from a comprehensive literature review of journal articles, books, book chapters, published technical reports, and databases, such as the Red list of the International Union for Conservation of Nature (www.iucnredlist.org) and the International Convention on Trade in Wild Fauna and Flora (www.cites.org). The literature consulted was compiled from searches using Google Scholar and from the websites of scientific journals.

THE MAIN TAXA AND DERIVED PRODUCTS USED FOR DECORATION

A wide variety of wildlife products has been used for decoration and ornamental purposes, including ivory, coral, turtle, and mollusk shells; reptile and other animals skins; and feathers (Dias et al., 2011; Oldfield, 2001; Pedersen, 2004). Wild animal-derived decorations and ornaments are often high-value luxury products. They may be bought for decoration or fashion or used as part of traditional wear or for luck and prestige. Some ornaments are an integral part of local tradition—for example, combs made from *bekko* (tortoise shell) are part of traditional Japanese wedding dress—but others are merely a response to the demands of fashion (Bagniewska and MacDonald, 2010).

Terrestrial Taxa

Birds are one of the main groups of terrestrial animals to have been exploited as a source of products for use as trappings or decoration (Collar et al., 2007; Kothari, 2007). Initially, feathers were regarded as a status symbol, and have been a very significant component of tribal attire in all parts of the world. Native Americans wore feathers as headdresses or jewelry and, in some cases, accorded them talismanic powers. Dubin and Jones (1999) pointed out that Native Americans believed that feathers embody the life energy of the animal, with the eagle being the most valued due to it frequenting close to the heavens. It was believed that an eagle feather imparts power that protects the wearer and aids warriors and hunters by conferring the eagles hunting stealth. The symbolic/decorative use of feathers has also been recorded elsewhere, such as in Africa where feathered headdresses or necklaces are worn by shamans or "witch doctors," possibly together with animal teeth and bird beaks. In South America, whole capes were covered in bright feathers, such as the scarlet feathers of the ibis. Even in Scotland, the number of eagle feathers worn on a bonnet signified rank and, although it is an old custom, is still acknowledged. Only the clan chief can wear three eagle feathers, while a chieftain can wear two (Pedersen, 2004).

Throughout the world, people have used bird plumes to decorate themselves, particularly in the form of headdresses. Styles vary widely, as do the reasons for creating headdresses and for preferring certain bird feathers over others (Biebuyck and Van den Abbeele, 1984). The plumes used and the manner in which they are worn may, for instance, mark the wearer's achievements, as in the coup complex of the Plains Indians. Alternatively, they may indicate initiatory status and knowledge as in, for example, the Baktaman of New Guinea. In other societies feathers may indicate social standing and wealth, and may even be worn to enhance the beauty of the rich, as in Edwardian England where the demands of "society belles" for bird of paradise plumes brought some of these majestic creatures to the verge of extinction in parts of Melanesia. Elsewhere, plumed headdresses may serve as regimental insignia to clearly distinguish warriors in battle as, for example, among the Zulu peoples of the last century. They may also be intended to confuse an enemy as to the number of opponents, such as the ostrich feather rings of the Masai. Other societies have included the plumes of certain birds in their headgear in the belief that this will effect some sympathetic transfer of the birds' qualities to them; for instance, that an owl's plumes might improve night vision (Sillitoe, 1988).

The practice of wearing feathers in hair or as headgear is ancient and widespread among peoples of several continents. Volpi (2016) notes that among bodily adornment and clothing decoration, feathers have occupied a privileged position in various cultures. From primitive people to women of the European court, ornamental plumage has a long history of popularity (Doughty, 1975). For example, the ostrich (*Struthio camelus*) was esteemed for its plumage by the Romans and Greeks. Remains of hunting scenes in Roman villas included ostriches among game animals captured or killed, and dignitaries wore their feathers as head ornaments (Doughty, 1975).

Feathers have been systematically used for ornamental purposes in Western clothing since at least the 13th century. From the 15th century until the late 18th century men from ruling classes wore hats decorated with feathers of all kinds, but especially the tail feathers of the peacock, pheasant, and cockerel (Yardwood, 1988). During the 16th century, women began to use feathers in headdresses, hats, and fans, and adornments made from feathers began to be worn by both the aristocracy and the haute bourgeoisie in Europe (Mertens, 2014). This practice can be attributed to the expansion of Spanish fashion during the Golden Age (1521–1643); the diversity and abundance of feathers and birds brought into the market by voyages of discovery in the Far East increased trade with the East (Colomer and Amalia, 2014; Mertens, 2014). In the late 1700s, women of the French court were already adept at using feathers as a material, and fashionable accessory was in high demand. Marie Antoinette, the last queen of France, was one of the pioneers in using headdress with feather plumes (Fig. 14.1), thus stimulating the fashion of wearing bird feathers in women's hats (Doughty, 1975). This practice became widespread in Europe throughout the 18th and 19th centuries, especially in cities like London and Paris, and spread across the Atlantic to the United States where it was received by an eagerly accepting audience of urban women (Daniel, 2014). Particularly in the 19th century, native and imported feathers were used in ladies' headdresses, muffs, hoods, etc. (Cumming et al., 2010).

FIGURE 14.1 Portrait of Marie Antoinette (1755–93) wearing a headdress ornamented with feathers and pearls. *Source: Jean-Baptiste Gautier Dagoty (1740–1786) | Date=ca.1775* | http://commons.wikimedia.org/wiki/File:Marie-Antoinette,_1775.

Throughout Europe and the United States, not only feathers but also entire birds were used primarily for headdresses and brooches (Arnold, 2011). The demand for feathers for ornamental purposes greatly increased during this period, leading to the exploitation of a greater diversity of birds. As a consequence, the hunting of birds for these purposes intensified (Schindler, 2001). In many places, such as New Guinea, Central America, and South America, men were attracted to bird hunting, with the aim of providing feathers to fashion centers, especially Paris and London (Doughty, 1975). Feathers and exotic birds stuffed into wooden trunks lined with paper or tin sheets arrived in Europe. Upon their arrival, whole birds that had been prepared in their countries of origin were emptied and stuffed with tow or were opened up, flattened, and their hides tanned. London was the main market for feathers, hosting fashion sales at dockyards where bird feathers from all over the world were on sale (Breton, 1914).

The exploited species were from all parts of the world, and encompassed a wide variety of bird groups. In New Guinea, birds of paradise were hunted, whereas in India it was bird-kings, pheasants, and herons, and in Russia owls, jays, and handles. In the swamps and rainforests from Guatemala to Brazil, herons and parrots were sought. In Africa, ostriches were exploited, while in the southern steppes of South America it was rheas (Schindler, 2001). Exploited birds encompassed a great richness of species, as pointed out by Hornaday (1913), who compiled a list of dozens of birds that were involved in the feather markets of the time (Table 14.1). Demand was constantly growing as a result of the growth of the bourgeoisie, who had sufficient financial means to indulge in the dynamic fashion world (Schindler, 2001). The practice of hunting of wild birds to harvest their feathers, especially the more decorative plumes which were sold for use as ornamentation, such as aigrettes in millinery, was known as "plume hunting" (McIver, 2003). Driven by the dictates of fashion, hunting for bird plumes drove some species to the brink of extinction (Daniel, 2014).

Today feathers are again popular in jewelry, most commonly hanging in pendants, earrings, or necklaces. They are often dyed, because most birds having naturally exotically colored plumage are now protected by conservation treaties, although this does not always prevent them from being exploited. Most feathers currently used for decoration or ornamentation come from domestic fowl, such as ducks, or from game birds (Pedersen, 2004). Along with other animal products, feathers are frequently marketed as components of objects sold as crafts. In Brazil, for example, there is widespread trade in handicrafts containing parts of wild animals, although prohibited by law. In 2007, the Brazilian Institute for the Environment and Renewable Natural Resources (IBAMA), for example, carried out an operation, called "Sad Fashion," which had the purpose of curbing the practice of selling stuffed animals, ornaments, handmade accoutrements, and products made from the parts of wild animals. Among the seized products were bird feathers, butterflies, starfish, reptile skins, jaguar or other feline skins, and objects made from marine fish and invertebrates. During the operation, 8567 pieces of handicrafts made by using parts of illegal Brazilian wildlife were seized.

Different mammal products (e.g., horn, leather, and teeth) are also used as trappings and for decoration (Pedersen, 2004). Ivory was probably the first organic material used for human ornamentation (Kunz, 1916). It is dentine from the teeth or tusks of mammals and is most commonly thought of as derived from elephants, but teeth or tusks from walruses, hippopotamuses, marine whales, and dolphins, as well as swines, such as wild boar and warthogs, have also been used (Pedersen, 2004). Ivory has been used as a

TABLE 14.1 Examples of Birds Exploited by Feather Markets Throughout the 18th and 19th Centuries

American Egret, Snowy Egret, Scarlet Ibis, Green Ibis, Herons (generally), Marabou Stork, Pelicans (all species), Bustard, Greater Bird of Paradise, Lesser Bird of Paradise, Red Bird of Paradise, Twelve-Wired Bird of Paradise, Black Bird of Paradise, Rifle Bird of Paradise, Jobi Bird of Paradise, King Bird of Paradise, Magnificent Bird of Paradise, Impeyan Pheasant, Tragopan Pheasant, Argus Pheasant, Silver Pheasant, Golden Pheasant, Jungle Cock, Peacock, Condor, Vultures (generally), Eagles (generally), Hawks (generally), Crowned Pigeon (two species), Choncas, Pitta, Magpie, Touracou (or Plantain-Eater), Velvet Birds, Grives, Mannikin, Green Parrot, Dominos (Sooty Tern), Garnet Tanager, Grebe, Green Merle, Horphang, Rhea, Sixplet, Starling, Tetras, Emerald-Breasted Hummingbird, Blue-Throated Hummingbird, Amethyst Hummingbird, Resplendent Trogon (several species), Cock-of-the-Rock, Macaw, Toucan, Emu, Sun-Bird, Owl, Kingfisher, Jabiru Stork, Albatross, Tern (all species), Gull (all species)

Reproduced from Hornaday, W.T., 1913. Our Vanishing Wild Life: Its Extermination and Preservation. Charles Scribner's Sons.

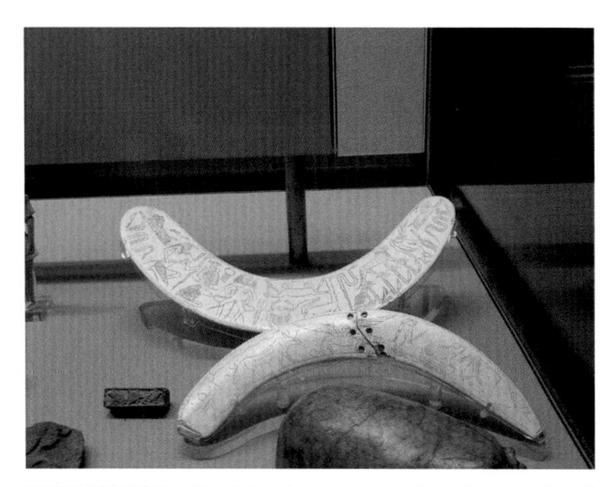

FIGURE 14.2 Semicircular ivory wands—decorated with fearsome deities—housed in the British Museum, London, England (Late Middle Kingdom, about 1759 BC). *Photo credits: Rômulo R.N. Alves.*

FIGURE 14.3 Handbag made from the skin of a dwarf crocodile from West Africa. Dated 20th century. Item displayed in the Natural History Museum, London, England. *Photo credits: Rômulo R.N. Alves.*

decorative material since the dawn of civilization, and sculptures made of ivory are known from more than 30,000 years ago (Conard, 2003). Much ivory work survives from ancient Egyptian, Greek, and Roman art, especially in sculpture, bas-relief, and inlay (Erickson and Gross, 2005) (Fig. 14.2). These authors pointed out that objects made of ivory epitomized valor and prestige before the advent of modern weaponry, when hunting elephants was very dangerous. For centuries people have viewed elephant ivory as a valuable commodity, and used it for carvings, jewelry, and other artifacts (Naylor, 2005). The Trojans wore buckles and pins fashioned from elephant tusks, and adorned their war chariots with pieces of ivory. Two thousand years ago, the Romans honored illustrious men with handsomely chiseled writing tablets and scepters carved from ivory. From prehistoric times to the present day, ivory has been sought as a luxury item of multiple uses. Thus, the killing of elephants to satisfy the demand for ivory has been considered the major factor in the reduction of their populations throughout most of human history (Kingdon and Pagel, 1997).

Skins, furs, feathers, and fibers from many mammal, reptile, bird, and fish species are traded internationally to make clothing, boots and shoes, bags, and other items (Figs. 14.3 and 14.4) (Oldfield, 2001). Such animal-based materials have been used throughout the globe to fulfill human desire for adornment (Wilcox, 1951). Fur, for example, has been an important material for apparel and accessories since prehistoric times (Schwebke and Krohn, 1970), when it was considered a luxury item and a status symbol. During the Middle Ages, many expensive furs from animals like leopard, mink, chinchilla, beaver, marten, and sable, among others, were consumed by the higher classes, whereas commoners, who could not afford luxurious furs, used less expensive fur from cats, dogs, rabbits, and sheep (Wilcox, 1951). In today's fashion, animal fur continues to be an important material for apparel and accessories (Lee, 2014; Strege, 2014). Lee (2014) noted that big-name designers, such as Yves Saint Laurent, Christian Dior, and Karl Lagerfeld, use fur to create garments that boost the level of fashion excitement. Leather from various animals, such as cattle, buffalo, calf, ox, lamb, sheep, goat, and deer, are used in more expensive soft leather apparels (Plannthin, 2016). Hides of species of snakes, iguanas, leopards, tigers, jaguars, lizards, elephants, and crocodiles have, at

FIGURE 14.4 Film actress and opera singer Geraldine Farrar, standing facing right, wearing a leopard skin robe and feather headdress (about 1917). *Source = Hartsook Studios. | Author = Unknown | Permission = {{PD-US}}.*

different times in history, been considered more beautiful and desirable than other hides (Ewing, 1981; Plannthin, 2016; Wilcox, 1951). Fur has not only been used for lining garments and as collars, cuffs and trims, but also for bedspreads, pillows, hats, gloves, and the lining of shoes (Ewing, 1981). The hunting of animals for their hides (for leather) or fur has played a very important role in the exploitation of fauna throughout history, leading some species to near extinction. The fashion for spotted cat fur came to an abrupt halt because these animals were placed on the endangered species list of the Endangered Species Act of 1973, along with some species of seal and otter (Ewing, 1981). Now, the use of animal-based materials for fashion products has been strongly criticized by animal rights advocates (Lee, 2014; Olson and Goodnight, 1994; Sneddon et al., 2010; Summers et al., 2006).

Marine Taxa

Marine animals also provide products that are used and traded as curiosities, souvenirs, crafts, jewelry (Dias et al., 2011; Grey et al., 2005; Wood and Wells, 1995) or as decorative, utilitarian or nonutilitarian artifacts, and even as contemplative items. These curiosities and souvenirs are generally manufactured using intact dead marine animals or their parts (Grey et al., 2005). Worldwide, the marine curio and souvenir market encompasses about 5000 species of mollusks (bivalves and gastropods), 40 species of corals, and unknown numbers of sponges and echinoderms (Wood and Wells, 1988). In the last two decades efforts to document and quantify the marine species used in this trade have been undertaken in several parts of the world, especially for the most widely used species of fish (Gössling et al., 2004; Grey et al., 2005; Vincent, 1996), sea turtles (Tröeng and Drews, 2004), mollusks (Dias et al., 2011; Gössling et al., 2004; Salamanca and Pajaro, 1996; Wells, 1981, 1989; Wood and Wells, 1995; Wood and Wells, 1988), corals (USCRTF, 2000), and echinoderms (Lunn et al., 2008) (Fig. 14.5) (Table 14.2).

Mollusk shells comprise the majority of marine curio trade items in terms of the number of species involved and the volume commercialized (Wood and Wells, 1988); in the Philippines alone approximately 1000 species of mollusks are traded and sold (Wood and Wells, 1995). Shell collecting was an extremely popular hobby in the 19th century, yet ornamental/curio/souvenir shell trade has intensified with the advent of internet shopping (Gössling et al., 2004), while the development of tourism (especially in tropical countries) has increased the market for souvenirs, with a consequent pressure on marine resources. In Brazil, for example, a total of 126 species (41 bivalves and 85 gastropods) have been found being sold individually as decorative pieces or incorporated into utilitarian objects (Dias et al., 2011). Fig. 14.6 illustrates the use of shells and starfish for decorative purpose.

FIGURE 14.5 Decorative use of sponges and crustaceans. Dried individual tube and massive sponges sold for decoration (A) and (B), Red lobster (*Panulirus argus*) bottled as a decorative object (C), and mangrove uçá crab (*Ucides cordatus*) as an ash-tray ornament (D). *Photos credits: Whitney Rose Petrey at* http://maritimeculture.blogspot.com.br/2012/02/tarpon-springs.html *(A) and (B); Nivaldo A. Léo-Neto (C); Thelma L. P. Dias (D).*

In some parts of the world, trade in wild species for decorative purposes targets tourists, but it can also comprise predominantly local buyers (Nijman and Nekaris, 2014). The level of exploitation of some species is alarming. For example, Mexico alone (Lunn et al., 2008) recorded a stock of approximately 40,000 starfish and 8,600 sea urchins in stores selling handicrafts and decorative products in the year 2004. Nijman et al. (2015) reported that at least 42,000 shells were seized by Indonesian authorities between 2005 and 2013, including more than 3,000 shells of *Nautilus pompilius*, 32,000 of *Cassis cornuta*, and 2,000 of *Tridacna gigas*. According to these authors, at least two-thirds of these shells were destined for China and the United States. Trade in coral for ornamentation purposes, although still in existence since the 1950s, has gradually been replaced by trade in live specimens for use in aquariums (Wood et al., 2012). According to Bentley (1998), in 1990 alone Indonesia exported almost one million coral pieces, mainly to the United States and Japan. In the Pacific Region, the taxa *Fungia* spp., *Pocillopora* spp., *Porites* spp., and *Acropora* spp. were the most exploited until the 1990s (Wood et al., 2012). Precious corals (e.g., *Corallium*, *Paracorallium*) are mainly used in the production of jewelry such as rings, necklaces, pendants, earrings, and carved objects such as statues (Tsounis et al., 2010).

In addition to shells, pearls represent another mollusk product with high ornamental value (Cariño and Monteforte, 2009;

TABLE 14.2 Main Species of Marine Animals Used as Ornaments and Decorations

Phylum/Groups	Specific Taxa	Main Uses
PORIFERA		
Sponges	*Spongia* spp., *Hippospongia communis*	Dried animals for decoration
CNIDARIA		
Stony corals	*Meandrina braziliensis*, *Acropora* spp., *Fungia* spp.	Whole animal as decorative/utilitarian piece such as washbasin, as ornaments in several objects, and handicrafts
Gorgonians	*Cirrhipathes* spp., *Antipathes* spp., *Corallium* spp.	Whole animal as decorative piece, as ornaments in several objects, jewelry, and handicrafts
MOLLUSCA		
Gastropods	*Cassis cornuta*, *Cassis tuberosa*, *Charonia* spp., *Ovula* spp., *Cypraea* spp., *Conus* spp., *Lambis* spp., *Oliva* spp., *Tonna* spp. *Turbo* spp., *Strombus* spp., *Pleuroploca trapezium*, Volutidae spp., *Haliotis* spp.	Whole animal as decorative piece, as ornaments in several objects, jewelry, and handicrafts. Some species are used as tools such as knives, spoons, and trumpets
Bivalves	*Tridacna* spp., *Hippopus hippopus*, Pectinidae spp., *Pinctada margaritifera*, *Pinctada maxima*, *Atrina vexilum*	Whole animal as decorative/utilitarian piece such as washbasin, as ornaments in several objects, jewelry, and handicrafts. Some species are used as tools for cleaning and shaping claypots
Cephalopods	*Nautilus pompilius*	Whole animal or the sectioned shell for decoration
ARTHROPODA		
Crustaceans	*Ucides cordatus*, *Panulirus* spp.	Whole dried animal for decoration and as ornaments, wet inside decorative bottles or as part of decorative objects
ECHINODERMS		
Sea stars	*Oreaster* spp., *Pisaster* spp.	Whole animal as decorative piece, as ornaments in several objects, jewelry, and handicrafts (shell mobiles, necklaces)
Sea urchins	*Echinometra* spp., *Tripneustes* spp.	Whole animal as decorative piece, as ornaments in several objects, jewelry, and handicrafts (toys, candle holders)
CHORDATA		
Sea turtles	*Chelonia mydas*, *Eretmochelys imbricata*, *Caretta caretta*	Whole animal as decoration or tortoise shell or *bekko* as ornaments in several objects, jewelry, and handicrafts
Sharks	*Isurus oxyrinchus*, *Galeocerdo cuvier*, *Sphyrna* spp.	Whole taxidermized animal, wet in jars or the jaws as decoration, or as ornaments in jewelry and handicrafts
Osteichthyes	*Megalops atlanticus*, *Lutjanus griseus*, *Diodon* spp., *Hippocampus* spp.	Whole taxidermized or dried animal as decoration, or as ornaments in jewelry and handicrafts
Mammals	*Dugong dugon*	Teeth and bones as ornaments in jewelry and handicrafts, and decoration

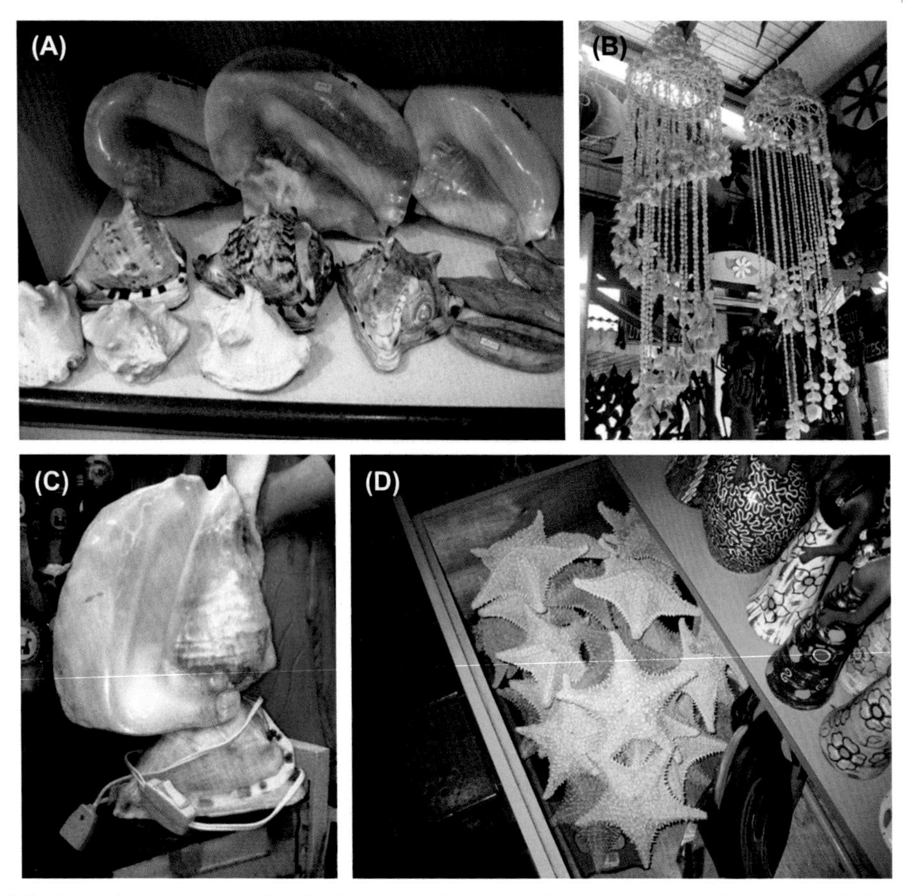

FIGURE 14.6 Some decorative use of mollusks and starfish: Individual gastropod shells (*Lobatus goliath*, *Cassis tuberosa*, and *Lobatus costatus*) sold for decoration (A), Ceiling lamp made of gastropod shells (B), Table lamp produced with shells of the large-sized gastropods *Lobatus goliath* and *Cassis tuberosa* (C), and individual starfish (*Oreaster reticulatus*) sold for decoration (D). *Photos credits: Thelma L. P. Dias (A), (C) and (D); Nivaldo A. Léo-Neto (B).*

Cattaneo-Vietti, 2016). Pearls are produced by mollusks as a self-defense response against external mechanical stimuli caused by irregular solid particles or foreign organisms that lodge in the animal and cannot be expelled, provoking irritation to exposed body tissue (Cariño and Monteforte, 2009). Pearls are the oldest known gems and have fascinated people since ancient times (Cariño and Monteforte, 2009). They have been used for human adornment for a long period of time (Charpentier et al., 2012; Strack, 2006). Their importance in decoration has given pearls a venerated position (Farn, 2013), being considered a talisman of fecundity and a symbol of perfection and absolute purity by many cultures (Cattaneo-Vietti, 2016).

Pearls have long been a symbol of power, and many people of significant social status have adorned themselves with pearl jewels (Cartier, 2014), including emperors and empresses, kings and queens, generals, nobles, and priests (Kunz and Stevenson, 1908). This is well exemplified by Assyrian and Persian bas-reliefs, which show how sovereigns and great personages of those countries adorned themselves profusely with pearls. They wore them not only in their jewelry, but also on their garments and even in their beards. The coins of Persian kings also bear testimony to the use of this gem in ancient Persia, since the sovereigns are represented wearing tiaras ornamented with triple rows of pearls (Kunz and Stevenson, 1908). The ornamental value of pearls is also well documented for the 15th century, when they were worn by persons of rank and fashion (Kunz and Stevenson, 1908). For example, "a remarkable 1483 portrait of Margaret, wife of James III of Scotland, which is now preserved at Hampton Court, shows her wearing such wonderful pearl ornaments that she might well be called Margaret from her decorations" (Kunz and Stevenson, 1908) (Fig. 14.7).

Most pearl-producing mollusks are bivalves, and the most famous pearls are produced by various species of pearl oysters of the Red Sea and Western Indian Ocean, but also in Australia and Central America (Cattaneo-Vietti, 2016). Species valued for their pearls are also found in Sri Lanka (*Pinctada fucata*), the Persian Gulf (*Pinctada margaritifera* var *erythraensis*), Venezuela (*Pinctada imbricata* and *Pteria colymbus*), and Sharks Bay, Australia (*Pinctada maxima*). The species commercially exploited for both their shells and their pearls are those of the gulfs of California and Panama (*Pinctada mazatlanica* and *Pteria sterna*), the Red Sea (*P. m. erythraensis*), and the South and Central Pacific (*P. maxima* and *Pinctada margaritifera cummingi*) (Cariño and Monteforte, 2006).

FIGURE 14.7 Queen Margaret of Scotland (1456–86) wearing pearl ornaments. *Source: Hugo van der Goes [Public domain], via Wikimedia Commons.*

Marine fish have also been exploited as marine curios, in particular groups such as sharks and seahorses (Grey et al., 2005; Rosa et al., 2011; Vincent, 1996). Each year, several hundred thousand seahorses are captured for souvenirs. Dried seahorses are popular curios in Asia, Latin America, and parts of Europe and are fashioned into earrings, brooches, and key chains (Rosa et al., 2011; Vincent et al., 2011).

Sharks and turtles have been heavily exploited for use as decorative pieces; as part of ornamental objects; and as part of hair adornments, bracelets, rings, and various pieces of jewelry. Seizures undertaken between 2000 and 2008 involving countries in East Asia recorded at least 9180 products extracted from sea turtles (or *bekko*). Of these, 2062 involved the entire taxidermized animal, 1161 were handcrafted, and 798 were shell plates (Lam et al., 2012). These authors recorded 10,080 *bekko* products for sale in 58 stores in Japan. These products included a wide range of objects for personal decoration, such as necklaces, tie clips, eyeglass frames, bracelets, and brooches. The range of refined *bekko* products sold in Japan reflects a tradition of craft activities involving marine turtles of over 30 years. *Bekko* trade in Asia exploits at least five species of sea turtles (*Chelonia mydas*, *Eretmochelys imbricata*, *Caretta caretta*, *Lepidochelys olivacea*, and *Dermochelys coriacea*); however, this number may be an underestimate since the manufacture of some handicrafts precludes species identification (Lam et al., 2012; Stiles, 2008).

Trade in shark products is well-known and even well quantified based on the sale of meat and fins (Chen and Phipps, 2002; Graham, 2004). However, the decorative use of sharks (mostly body parts) is still poorly documented and quantified. Records of the use of teeth, jaws, and other parts of the body by some traditional populations around the world are known, and are attributed to the presence and growth of tourism (Rose, 1996). In 2009, in Bangladesh alone, about 200 kg of teeth and 1 ton of cartilaginous shark spines were documented as raw material for the production of decorative products intended for sale (Hoq et al., 2011). According to Grey et al. (2005), in the period between 1997 and 2001 most of the shark pieces sold as marine curiosities were imported from the United States, such as teeth (66%) or worked jewelry (24%), of which 10% were imported as pieces of jaws, skulls, trophies, or sculptures (Fig. 14.8). The value of decorative pieces from sharks, such as jaws and teeth, can vary according to the species from which it was

FIGURE 14.8 Shark jaws and juvenile sharks bottled for sale as decorative pieces (A) and (C), Dried seahorse hanging over a dried gorgonian coral (B). *Photos: Jillian Morris at* http://shark-girl.blogspot.com.br/2013/ *(A) and (C); Thelma L. P. Dias (B).*

extracted (Grey et al., 2005). For example, the jaw of the great white shark (*Carcharodon carcharias*) can cost around USD 10,000 and a single tooth can be sold for up to USD 100 (Gosling, 2003). In Bali, Nijman and Nekaris (2014) recorded at least 380 dried stingray tails (*Neotrygon kuhlii* and *Pastinachus sephen*) for sale as decorative objects, as well as hand-carved daggers made of dugong (*Dugong dugon*) bones. Fig. 14.8 shows some shark products and seahorses sold as decoration.

CONCLUSIONS

The use of animals for decoration or as part of decorative pieces, as well as for personal use, has a long history in various human cultures. Some products of animal origin have great financial value and have been considered glamorous by both historical and contemporary societies. However, the growing demand for animal products for decorative or ornamental purposes has caused an increase in the collection of some species of mollusks and corals, as well as the hunting of wild animals, to supply this ornamental curio trade. This situation is even more serious when one considers that control over the exploitation of animals or their parts for decorative purposes has never been established. As discussed previously, feathers, which are popular in jewelry, most commonly hanging in pendant earrings or necklaces, are often extracted from endangered birds or rare species that are trafficked illegally.

Shark teeth used to decorate collars are obtained from individuals captured by nonselective fishing methods or by predatory fisheries, placing natural populations at risk and disrupting ecosystem stability. Mollusks, through their shells and pearls, are the most widely used marine taxa for decorative purposes, either as a whole animal or as decorative objects. Sea turtle populations have been decimated as a result of exploitation for the consumption of their meat and the use of parts of their shell for the manufacture of a great diversity of decorative pieces.

All these activities center around humans, who participate in all stages of the chain of exploitation, from the capture to the manufacture and trade of animals and their products. It is important to understand the distinction between cultural use and trade so that conservation measures can benefit the animal species involved and the human populations that depend on these products. In this context, ethnozoological research is of vital importance because it can acquire knowledge that can support efficient measures of control, management, and conservation of the fauna used for decorative purposes.

References

Alves, R.R.N., Rosa, I.L., 2013. Animals in Traditional Folk Medicine: Implications for Conservation. Springer-Verlag, Berlin Heidelberg.

Alves, R.R.N., Rosa, I.L., Léo Neto, N.A., Voeks, R., 2012. Animals for the gods: magical and religious faunal use and trade in Brazil. Human Ecology 40, 751–780.

Arnold, L.J.M., 2011. Postcards from Nowhere. University of Sydney.

Bagniewska, J., MacDonald, D., 2010. Executive Summary of the Report Animal Welfare, International Development, Biodiversity Conservation – the Road to Peaceful Coexistence Wildlife Conservation Research Unit University of Oxford.

Bentley, N., 1998. An overview of the exploitation, trade and management of corals in Indonesia. Traffic Bulletin 17, 67–78.

Biebuyck, D.P., Van den Abbeele, N., 1984. The Power of Headdresses: A Cross-cultural Study of Forms and Functions. Leopold III Foundation for Exploration & Nature Conservation, Brussels.

Breton, A., 1914. "L' industrie des plumes [feather's industry]. In: Tissandier, G. (Ed.), La Nature; Revue des Sciences et leurs applications aux Arts et a l' Industrie. Masson et Cie, Paris, pp. 279–281.

Cariño, M., Monteforte, M., 2006. Une historie mondiale des perles et des nacres: pêche, culture et commerce. Ed. L'Harmattan, Col, Maritimes, Paris.

Cariño, M., Monteforte, M., 2009. An environmental history of nacre and pearls: fisheries, cultivation and commerce. Global Environment 3, 48–71.

Cartier, L.E.H., 2014. Sustainability and Traceability in Marine Cultured Pearl Production. University of Basel.

Cattaneo-Vietti, R., 2016. Mand and Shells: Molluscs in the History. Bentham Science Publishers Ltd.

Ceballos, C.P., Fitzgerald, L.A., 2004. The trade in native and exotic turtles in Texas. Wildlife Society Bulletin 32, 881–891.
Charpentier, V., Phillips, C.S., Méry, S., 2012. Pearl fishing in the ancient world: 7500 BP. Arabian Archaeology and Epigraphy 23, 1–6.
Chen, V.Y., Phipps, M.J., 2002. Management and Trade of Whale Shark in Taiwan. A Traffic East Asia Report. Traffic East Asia, Taipei.
Collar, N.J., Long, A.J., Jaime, P.R.G., 2007. Birds and People: Bonds in a Timeless Journey. BirdLife International.
Colomer, J.L., Amalia, D., 2014. Vestir a La Española En Las Cortes Europeas (Siglos XVI Y XVII). [Spanish Fashion at the Courts of Early Modern Europe]. Centro de Estudios Europa Hispanica (CEEH), Madrid.
Conard, N.J., 2003. Palaeolithic ivory sculptures from southwestern Germany and the origins of figurative art. Nature 30, 37.
Cumming, V., Cunnington, C.W., Cunnington, P.E., 2010. The Dictionary of Fashion History. Berg, Oxford.
Daniel, J., 2014. The Plume Hunters. WINC, pp. 18–23.
Dias, T.L.P., Leo Neto, N.A., Alves, R.R.N., 2011. Molluscs in the marine curio and souvenir trade in NE Brazil: species composition and implications for their conservation and management. Biodiversity and Conservation 20, 2393–2405.
Doughty, R.W., 1975. Feather Fashions and Bird Preservation: A Study in Nature Protection. Univ of California Press.
Dubin, L.S., Jones, P., 1999. North American Indian Jewelry and Adornment: From Prehistory to the Present. Abradale Press.
Erickson, L.G., Gross, T.S., 2005. Exotic materials in the decorative arts: keys to the identification of ivory. California Homes 38–39.
Ewing, E., 1981. Fur in Dress. B.T. Batsford Ltd, London.
Farn, A.E., 2013. Pearls: Natural, Cultured and Imitation. Elsevier.
Fitter, R.S.R., 1986. Wildlife for Man: How and Why We Should Conserve Our Species. Collins, London, London.
Gosling, M., 2003. Great White Sharks under Attack by Curio Pirates. Cape Times Online. http://www.capetimes.co.za.
Gössling, S., Kunkel, T., Schumacher, K., Zilger, M., 2004. Use of molluscs, fish, and other marine taxa by tourism in Zanzibar, Tanzania. Biodiversity and Conservation 13, 2623–2639.
Graham, R.T., 2004. Global whale shark tourism: a 'golden goose' of sustainable and lucrative income. Shark News 16, 8–9.
Grey, M., Blais, M.A., Vincent, A.C.J., 2005. Magnitude and trends of marine fish curio imports to the USA. Oryx 39, 413–420.
Hoq, M.E., Yousuf Haroon, A.K., Hussain, M.G., 2011. Shark Fisheries in the Bay of Bengal, Bangladesh: Status and Potentialities, Support to Sustainable Management of the BOBLME Project. Bangladesh Fisheries Research Institute (BFRI), Bangladesh, p. 76.
Hornaday, W.T., 1913. Our Vanishing Wild Life: Its Extermination and Preservation. Charles Scribner's Sons.
IUCN, 2016. Red List of Threatened Species. Version 2016.3.
Jennings, M.R., 1987. Impact of the curio trade for San Diego horned lizards (*Phrynosoma coronatum blainvillii*) in the Los Angeles Basin, California: 1885–1930. Journal of Herpetology 21, 356–358.
Kingdon, J., Pagel, M., 1997. The Kingdon Field Guide to African Mammals. Academic Press London.
Kothari, A., 2007. Birds in Our Lives. Universities Press.
Kunz, G.F., 1916. Ivory and the Elephant in Art. Doubleday, Page and Co., New York.
Kunz, G.F., Stevenson, C.H., 1908. The Book of the Pearl: The History, Art, Science, and Industry of the Queen of Gems. The Century co.
Lam, T., Ling, X., Takahashi, S., Burgess, E.A., 2012. Market Forces: An Examination of Marine Turtle Trade in China and Japan. TRAFFIC East Asia.
Lee, M., 2014. The Effects of Product Information on Consumer Attitudes and Purchase Intentions of Fashion Products Made of fur, Leather, and Wool. Iowa State University.
Lunn, K.E., Noriega, M.J.V., Vincent, A.C.J., 2008. Souvenirs from the sea: an investigation into the curio trade in echinoderms from Mexico. Traffic Bulletin 22, 19–32.
McIver, S.B., 2003. Death in the Everglades: The Murder of Guy Bradley, America's First Martyr to Environmentalism. University Press of Florida, Gainesville, FL.
Mertens, W., 2014. Feather fans: renaissance and high bloom in western fashion, 1830–1930. In: Swan, J., Debo, K., Dirix, E., O'Neill, A., K., V.G., Marschner, J. (Eds.), Birds of Paradise: Plumes and Feathers in Fashion. Lannoo, Antwerp, pp. 6–29.
Naylor, R.T., 2005. The underworld of ivory. Crime, Law and Social Change 42, 261–295.
Nijman, V., Nekaris, K.A.I., 2014. Trade in wildlife in Bali, Indonesia, for medicinal and decorative purposes. Traffic Bulletin 26, 31–33.
Nijman, V., Spaan, D., Nekaris, K.A.I., 2015. Large-scale trade in legally protected marine mollusc shells from Java and Bali, Indonesia. PLoS One 10, e0140593.
Oldfield, S., 2001. The Trade in Wildlife: Regulation for Conservation. Earthscan Publications Ltd, London.
Olson, K.M., Goodnight, G.T., 1994. Entanglements of consumption, cruelty, privacy, and fashion: the social controversy over fur. Quarterly Journal of Speech 80, 249–276.
Pedersen, M.C., 2004. Gem and Ornamental Materials of Organic Origin. Elsevier Butterworth-Heinemann.
Plannthin, D.K., 2016. Animal ethics and welfare in the fashion and lifestyle industries. In: Muthu, S.S., Gardetti, M.A. (Eds.), Green Fashion. Springer, pp. 49–122.
Reitz, E.J., Wing, E.S., 2008. Zooarchaeology. Cambridge Univ Pr.

Rosa, I.L., Oliveira, T.P.R., Osório, F.M., Moraes, L.E., Castro, A.L.C., Barros, G.M.L., Alves, R.R.N., 2011. Fisheries and trade of seahorses in Brazil: historical perspective, current trends, and future directions. Biodiversity and Conservation 20, 1951–1971.

Rose, D.A., 1996. An Overview of World Trade in Sharks and Other Cartilaginous Fishes. TRAFFIC International, Cambridge, UK.

Roth, H.H., Merz, G., 1997. Wildlife Resources: a Global Account of Economic Use. Springer Science & Business Media.

Salamanca, A.M., Pajaro, M.G., 1996. The utilization of seashells in the Philippines. Traffic Bulletin 16, 61–72.

Schindler, H., 2001. Plumas como enfeites da moda. História, Ciências. Saúde–manguinhos 8, 1089–1108.

Schwebke, W.P., Krohn, B.M., 1970. How to Sew Leather, Suede, fur. The Bruce Publishing Company, New York.

Sillitoe, P., 1988. From head-dresses to head-messages: the art of self-decoration in the highlands of Papua New Guinea. Man 23, 298–318.

Sneddon, J., Lee, J.A., Soutar, G.N., 2010. An exploration of ethical consumers' response to'animal friendly' apparel labelling. Journal of Research for Consumers 18, 1–10.

Stiles, D., 2008. An Assessment of the Marine Turtle Products Trade in Viet Nam. TRAFFIC Southeast Asia, Petaling Jaya, Selangor, Malaysia.

Strack, E., 2006. Pearls. Rühle-Diebener-Verlag, Stuttgart, Germany.

Strege, G., 2014. Fur as fashion in America. Fashion, Style & Popular Culture 1, 413–432.

Summers, T.A., Belleau, B.D., Xu, Y., 2006. Predicting purchase intention of a controversial luxury apparel product. Journal of Fashion Marketing and Management: An International Journal 10, 405–419.

Tröeng, S., Drews, C., 2004. Money Talks: Economic Aspects of Marine Turtle Use and Conservation. WWF International, Switzerland.

Tsounis, G., Rossi, S., Grigg, R.W., Santangelo, G., Bramanti, L., Gili, J.M., 2010. The exploitation and conservation of precious corals. Oceanography and Marine Biology: An Annual Review 48, 161–212.

USCRTF, United States Coral Reef Task Force, 2000. International Trade in Coral and Coral Reef Species: The Role of the United States. USCRTF—United States Coral Reef Task Force, Washington.

Vincent, A.C.J., 1996. The International Trade in Seahorses. Traffic International, Cambridge.

Vincent, A.C.J., Foster, S.J., Koldewey, H.J., 2011. Conservation and management of seahorses and other Syngnathidae. Journal of Fish Biology 78, 1681–1724.

Volpi, M.C., 2016. The exotic west: the circuit of carioca featherwork in the nineteenth century. Fashion Theory 20, 127–151.

Wells, S.E., 1981. International trade in ornamental corals and shells. In: Fourth International Coral Reef Symposium, pp. 323–330.

Wells, S.M., 1989. Impacts of the precious shell harvest and trade: conservation of rare or fragile resources. In: Caddy, J. (Ed.), Marine Invertebrate Fisheries: Their Assessment and Management. Wiley-Interscience, New York, pp. 443–454.

Wells, S.M., Barzdo, J.G., 1991. International trade in marine species: is CITES a useful control mechanism? Coastal Management 19, 135–154.

Wilcox, R.T., 1951. The Mode in Furs. Charles Scribner's Sons, New York.

Wood, E., Wells, S., 1995. The shell trade: a case for sustainable utilization. In: Kay, E.A. (Ed.), The Conservation Biology of Molluscs. IUCN Species Survival Commission, Cambridge.

Wood, E.M., Wells, S., 1988. The Marine Curio Trade: Conservation Issues: A Report for the Marine Conservation Society. Marine Conservation Society.

Wood, W., Malsch, K., Miller, J., 2012. International trade in hard corals: review of management, sustainability and trends. In: Proceedings of the 12th International Coral Reef Symposium, Cairns, Australia.

Yardwood, D., 1988. The Encyclopedia of World Costume. B.T. Batsford, London.

CHAPTER

15

The Role of Animals in Human Culture*

Rômulo Romeu Nóbrega Alves, Raynner Rilke Duarte Barboza

Universidade Estadual da Paraíba, Campina Grande, Brazil

INTRODUCTION

People share the world with a variety of other animals and often establish relationships with them, some of which began in the very distant past. Animals are valuable to people not only because they provide utilitarian and economic benefits but also because they have been incorporated into our sense of place and are enshrined in long-standing cultural practices (Allaby, 2010; Alves and Souto, 2015; Shepard, 1996). Animals have long drawn the attention of humankind by capturing our imagination and influencing our thoughts, dreams, and fears (Alves and Souto, 2015; Gross and Vallely, 2013; Hogue, 2003), and giving rise to myths, legends, and fables, which direct human attitudes toward them.

Any understanding of an animal, and what that particular animal mean to us, will be informed by, and be inseparable from, our knowledge of its cultural representation (Forscher, 2007). The most widespread way of representing animals has been to assign them figurative, cultural, social, or political meaning, that is, to depict them as metaphors or allegories for human beings or human traits. Such cultural meanings are constructed from a set of salient traits with which a particular kind of animal can be associated (Palmeri, 2006). The cultural importance of animals has been reflected in art, literature, symbolism, music, mythology, and religion, among other important human representations (Kalof and Resl, 2007; Kothari, 2007; Malamud et al., 2007; Resl, 2007; Senior, 2009). Klingender (1971) highlighted that, during all periods of time in human history, animals have been used in art and literature to symbolize religious, social, and political beliefs, and artists have found constant inspiration in the grace and beauty of animal forms.

Considering the importance of animals to people's lives, it is not surprising to find animals playing a host of roles in oral and written traditions throughout human history, ranging from folk to holy writings of the world's most prominent religions (Hogue, 2003). In this chapter we discuss the significance of animals in the main human cultural manifestations, which have perpetuated from the most primitive to the most contemporary societies. It should be emphasized that, in many cases, these cultural manifestations of animals are closely interwoven with

*This chapter is a revised and updated version of the section "Animals in symbolism, myth, religion, magic and art" from the paper: "Alves, R.R.N., 2012. Relationships between fauna and people and the role of ethnozoology in animal conservation. Ethnobiology and Conservation 1, 1–69."

Ethnozoology
http://dx.doi.org/10.1016/B978-0-12-809913-1.00015-6

their utilitarian value, and so understanding all aspects of value is fundamental to understanding human attitudes and perceptions directed toward animals.

ANIMALS IN RELIGIOUS PRACTICES

Among the most relevant cultural features, religious beliefs and practices have long influenced human perception and use of natural resources (Alves and Rosa, 2008; Alves et al., 2012; Berkes, 2001; Tomalin, 2004). Animals have, and continue to play, an important role in religious practices worldwide (Alves et al., 2009a; Berkes, 2001; Leo Neto and Alves, 2010; Léo Neto et al., 2009; Leo Neto et al., 2011; Leo Neto et al., 2012; Lugira, 2009; McNeely, 2001; Nikoloudis, 2001; Tomalin, 2004), and their relationships with humans have encompassed a supernatural dimension since ancient times (Alves and Souto, 2010).

Animals have always played a prominent role in the history of religions. On the basis of the ethnographic record and knowledge of ancient religions, specialists have concluded that animals were considered to be both ancestors and protectors and played a central role in the magic-religious behavior of traditional societies. It has also been suggested that the cult of animals—termed "totemism"—was the earliest form of religion (Alinei, 2000). As Allaby (2010) reports, all human cultures produce myths and these frequently include totemic animals and animal gods. Totemic ideas and practices reveal animals playing key roles in the construction of the identities of individuals, clans, and ethnic groups (Olupona, 1993). In some places, such as in some African and South American countries, for example, totemic animals play key roles in the religious and cultural practices of local people, who consider these animals attached to their own origin, ancestors, and place, and thus maintain close relationships with them (Alves et al., 2010; Olupona, 1993).

Over millennia, anthropomorphic goddesses and gods slowly replaced animal deities. The more archaic divinities, however, accompanied their more human successors, often as mascots or alternate forms. Athena, for example, was pictured with an owl, Zeus with an eagle; Odin was accompanied by ravens and by wolves; Mary, mother of Christ, was often shown with a dove. The monkey god, Hanuman, who fought alongside the hero, Rama, in the epic Ramayana, is now perhaps the most popular figure in the Hindu pantheon. There is something of an archaic mother goddess in the "wicked witch" of Halloween, pictured with a faithful spider at her side (Sax, 2002).

Figures that blend human and animal features became common with the transition from hunting and gathering to agriculture. The gods and goddesses of ancient Egypt often had a human torso and the head of an animal—crocodile, hippopotamus, ram, baboon, jackal, cat, falcon, or ibis (Fig. 15.1). Similarly, the Greeks had their centaurs and satyrs, whereas the Hebrews had cherubim and seraphim. Garuda, the carrier of the Hindu god Vishnu, had the torso and face of a man but the wings and beak of an eagle. Yet another fantastic animal was al-Borak, the steed on which Mohammed made his flight to Heaven; she had the body of a horse and the face of a woman, and she could see the dead (Sax, 2002).

In the first millennium BC, animal cults became an increasing prominent feature of Egyptian religious activity. The Egyptians considered some animals as gods or their intercessors. Such animals were so important that some of them were bred in sacred temples, mummified after death, and often buried with their owners. Among the most numerous are the mummies of cats (associated with the goddess Bastet) and ibises (representative of the god Thoth), but this practice also involved other animals such as bulls, snakes, falcons, crocodiles, fish, and baboons (Fig. 15.2).

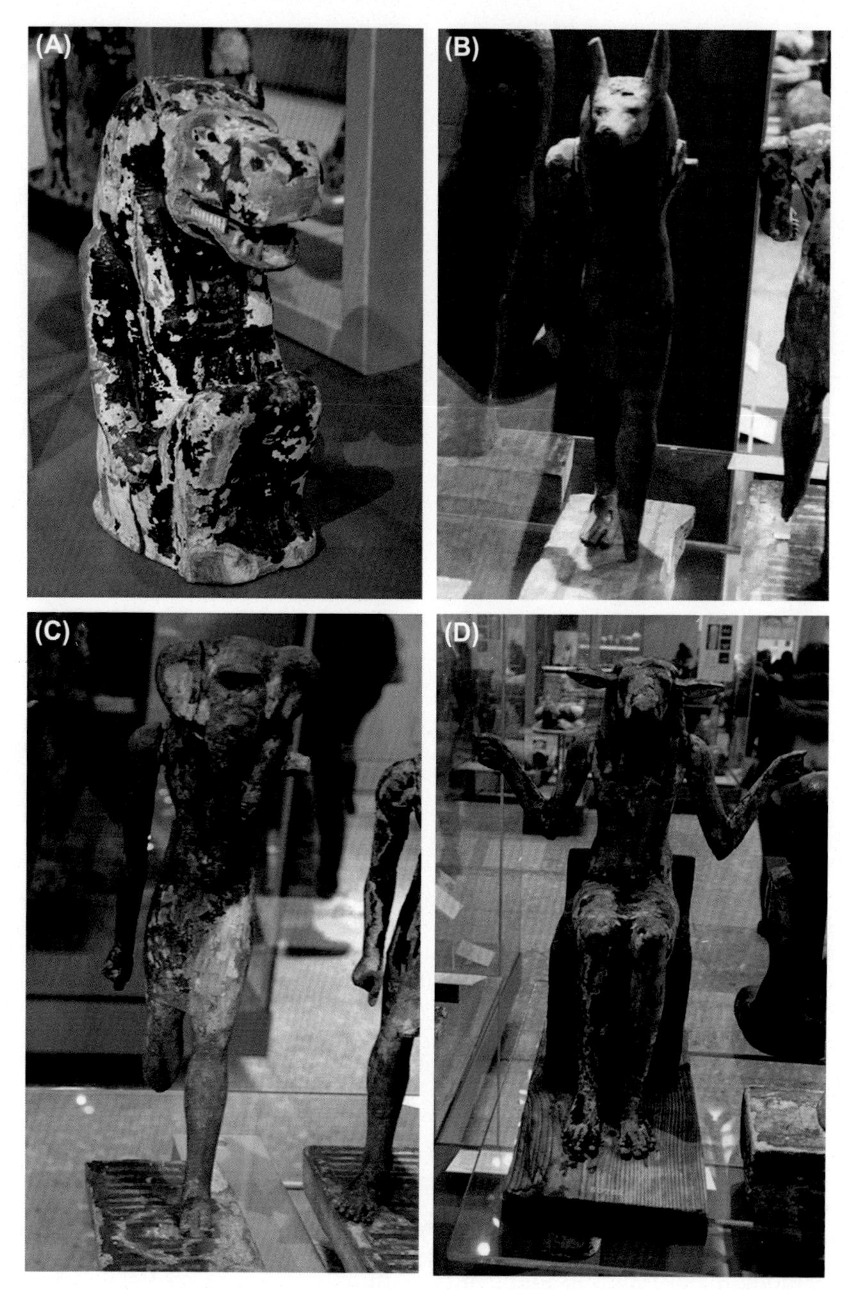

FIGURE 15.1 Statues of deities (with human torso and animal head) of ancient Egypt exhibited at the British Museum, London. (A) Hippopotamus-headed deity, late 18th Dynasty, about 1295 BC; (B) wooden figure of a jackal-headed deity, 19th or 20th Dynasty, about 1295–1070 BC; (C) wooden figure of a baboon-headed god, 19th or 20th Dynasty, about 1295–1070 BC; and (D) ram-headed deity, late 18th Dynasty, about 1295 BC. *Photo credits: Rômulo R.N. Alves.*

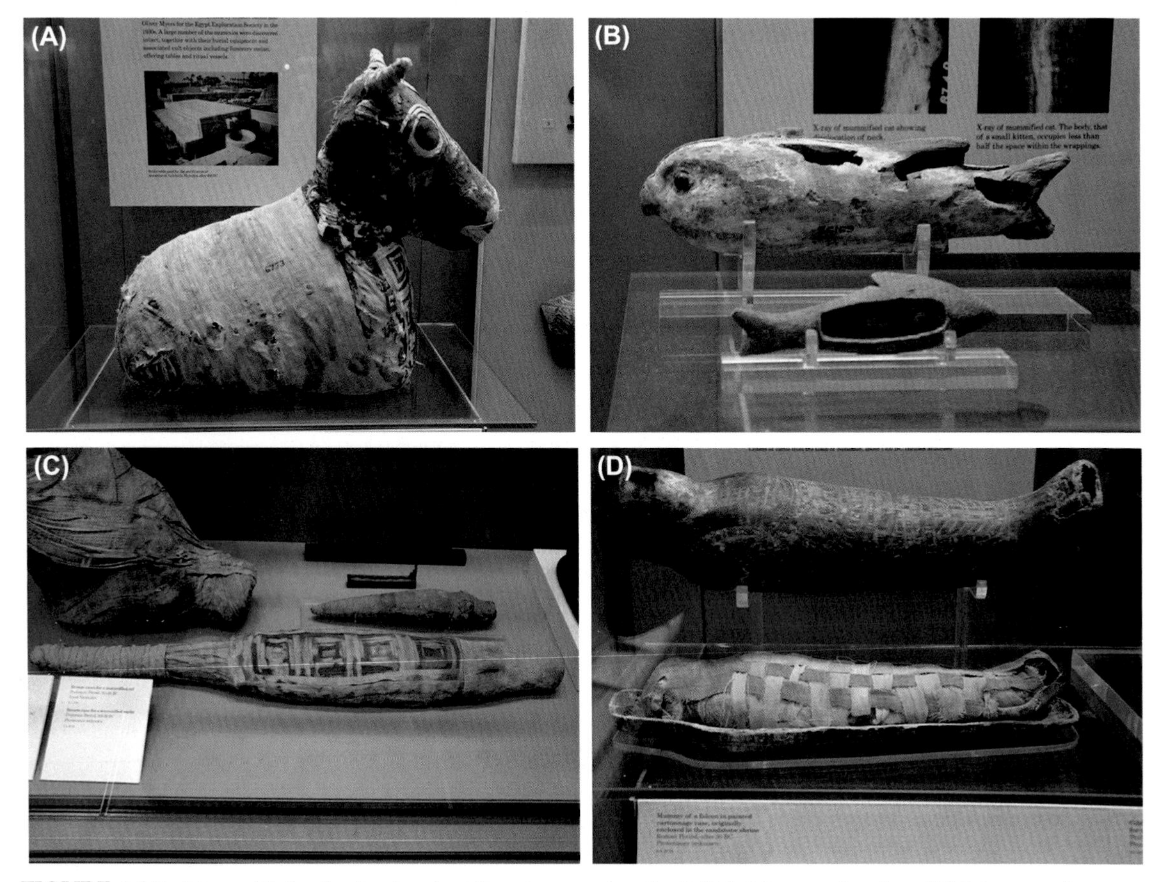

FIGURE 15.2 Mummified animals of ancient Egypt exposed at the British Museum, London. (A) Mummy of a young bull, Roman period after 30 BC, from Thebes; (B) falcon-headed wooden coffin for a mummy of a fish, Roman period after 30 BC, from Saqqara, and mummy of a fish in painted wooden coffin, Ptolemaic or Roman period, after 305 BC; (C) mummies of small crocodiles, Roman period after 30 BC, from El-Hiba; (D) mummy of a falcon in painted cartonnage case, originally enclosed in the sandstone shrine, Roman period after 30 BC. *Photo credits: Rômulo R.N. Alves.*

At a greater or lesser extent, the magic-religious importance of animals has persisted until the present day and varies greatly among different cultures and their religious practices. Belief in the spiritual value of nature is most often associated with indigenous people (Basso, 1996; Berkes, 2008; Callicott, 1994; Cruikshank, 2005) but is common among major religions and even among people with no religious affiliation at all (Haggan, 2011). From ancient times to the present, many species were, and remain, associated with religious practices; animals considered as sacred, zoomorphical, or anthropozoomorphical gods continue to be worshiped and involved in religious rituals. Products from dead animals are frequently used to decorate altars and religious temples in a variety of countries of the world (Adeola, 1992; Alves et al., 2009a; Chan, 2006; Leo Neto and Alves, 2010; Léo Neto et al., 2009; Leo Neto et al., 2011; Leo Neto et al., 2012). Animals serve in sacrificial offerings, their body parts are integral parts of magical amulets and nasal ablutions, and their form and function take on profound symbolic significance—good or evil, powerful or meek, clever or dull (Léo Neto et al., 2009; Moazami, 2005).

The release of captive wild animals for spiritual or religious purposes is commonly practiced in many parts of eastern and southern Asia (Gilbert et al., 2012). Known as merit release, "prayer release," or by the Chinese name of Feng Sheng, the activity is rooted in Buddhist teachings, although the followers of other belief systems also participate to a lesser degree (Ahmed, 1997; Chan, 2006; Liu et al., 2012; Severinghaus and Chi, 1999). Within Buddhism, the liberation of animals from captivity is believed to be the most powerful means of attaining spiritual merit (Zangpo, 2005). Practitioners believe that relieving, suffering, and prolonging life will enhance personal karma, promoting advancement to a state of enlightenment. However, in many places, this act of benevolence has given rise to a commercial trade through which wildlife (primarily birds, fish, and turtles) are captured solely for the purpose of merit release (Chan, 2006; McClure and Chaiyaphun, 1971; Schoppe, 2008).The ethics of supplying animals specifically for the purpose of release has been questioned on the grounds of animal welfare, ecological impact, and health (Shiu and Stokes, 2008). Furthermore, the release of nonindigenous species can lead to the establishment of self-sustaining populations, with potentially deleterious effects on local ecology. The red-eared slider, *Trachemys scripta* (Schoepff, 1792), for example, a species that is commonly sold for release purposes (Goh and O'Riordan, 2007), has become the most abundant freshwater chelonian in many parts of Asia (Ramsay et al., 2007).

Further evidence of the importance of animals in religious practices is the architecture and ornamentation of religious temples, many of which depict either real or imaginary animals, as can be observed in historic churches all over the world, but especially in Europe (Figs. 15.3–15.5), testifying that animals occupied an important position at the time. The external architecture of many churches incorporates animal figures, as also does the internal decoration, as recorded by Garcia (2012), for the Sé Cathedral, São Paulo, Brazil, where sculptures of animals of the Brazilian fauna, such as armadillos, egrets, lizards and toucans, are found inside the building (Fig. 15.5). Similarly, many historical churches and cathedrals in Europe have representations of animals (real or imaginary) (Fig. 15.4), often reflecting their symbolic role in religious practices. Wentersdorf (1983–84) pointed out that there is a profusion of wood and stone carving in churches and cathedrals, as exemplified by the misericords with their depiction of biblical and fabulous, episodes or (more commonly) emblematic images, which include humans, birds, beasts, and monstrous creatures, representing activities, virtues, and vices.

Among the most common architectural elements in churches and cathedrals (and other Gothic buildings) are gargoyles (Fig. 15.6). In most cases they function in water drainage, being positioned in the protruding part of roof rails. Many gargoyles were shaped in the form of animals, real or imaginary, and others combining animals and people. It is believed that these creatures were placed in medieval cathedrals to alert Catholic believers that evil never sleeps and the faithful should always be in a state of alertness and continuous vigilance, even in holy places.

ANIMALS IN MYTHOLOGY

Around the world, and throughout history, people have created stories, legends, and myths, which have been incorporated into different cultural activities of mankind. Myths address questions that humans have always posed about their origins, their environments, their ultimate destinies, and the meanings of their lives (Stookey, 2004). Unsurprisingly, animals play key roles in many myths that, in part, reflect the consequences of differing everyday life experiences of people and that are passed from generation to generation. Many myths have a great influence on people's perception of fauna (Alves

FIGURE 15.3 Internal architecture of St. Peter's Basilica (Vatican City, Italy) with faunistic decorative elements (dog, lion, ram, and pigeon). *Photo credits: Rômulo R.N. Alves.*

et al., 2017) and represent a powerful determinant of how animals are treated in a given society (Preece, 2002).

Animals, real or imaginary, capture human imagination in many ways and figure prominently in narratives throughout myth tradition (Haenn et al., 2014; Stookey, 2004; Ulric, 1996). The significance of mythology in human life is well illustrated by religious traditions, many of which are guided by myths and legends, including those that deal with the origin of gods and the world of humans (Alves et al., 2012). Virtually all manner of beliefs have involved animals (Bowman, 1977), and they have, and continue to play, important roles in religious practices worldwide (Alves et al., 2012; McNeely, 2001). The cult of animals, which considers manifestations or incarnations of deities, called Zoölatry, has existed among all cultures, but reached its greatest development among the Egyptians, who adored a vast Pantheon of deified bulls, rams, cats, mice, ibises, sparrows, hawks, crocodiles, and a multitude of mongrel creations of the imagination (Evans, 1896).

In almost all cultural traditions there are animals associated with creation myths, narratives

FIGURE 15.4 Fabulous animals decorating the interior of St. Peter's Basilica (Vatican City, Italy). *Photo credits: Rômulo R.N. Alves.*

that address the question of how the universe or the earth and its inhabitants first came into existence (Stookey, 2004). Turtles, for example, have an important role in mythologies around the world (Garfield, 1986). This author pointed out that turtles were a sacred symbol of the world's creation for many human groups in both the Americas and Asia. For example, several North American Indian tribes, including the Iroquois, Seri, and Mandans, believed that the earth rested on the back of a giant turtle (Jobes, 1962). In the stories of Africa's Khoisan people, a mantis is the first living creature to appear on the earth and the creator of other living beings, including people (Stookey, 2004). Nonhuman primates represent another group of animals with important significance in mythology; because of their physical similarity to humans, these animals have inspired a number of myths about human origins (Alves et al., 2016). Mayan creation myths, for example, describe the existence of earlier worlds destroyed by various cataclysms, and the societies of one earlier world were transformed into monkeys and survived (Alves et al., 2017). These animals were, and remain, associated with religious and magical practices, and they have been portrayed with human qualities or as gods in paintings or other artistic products (Alves et al., 2013; Mittermeier et al., 2007). For instance, from the beginning of Egyptian history until at least the beginning of the Christian period, baboons held a very consistent and important role in ancient Egyptian religious beliefs (Fig. 15.1) and were represented as both demon and protector (Alves et al., 2016).

Myths in which animals are considered demons, tricksters, or heroes are also common (Stookey, 2004). The fox, for example, appears in the folklore of many cultures as a figure of cunning or trickery. In Europe, in the Middle Ages and Renaissance, foxes were associated with wiliness and fraudulent behavior and were sometimes burned as symbols of the Devil (Benton, 1997). Similarly, in Dogon mythology (Africa), the fox represents disobedience and disorder within the world and is considered the enemy of water, fertility, and civilization. A mythical fox named Yurugu is said to have been condemned by Amma (the creator god) to search the world for a lost twin (Griaule, 2005). In some societies, however, the fox can be either a good or a bad omen depending on how it is portrayed in the particular tale, as occurs in Chinese mythology (Kang, 2006). The Moche people of ancient Peru worshiped animals and often depicted the fox in their art (Berrin and Larco Museum, 1997).

Among the various categories of myths surrounding animals are those involving

FIGURE 15.5 Faunistic elements decorating the external architecture of the Sacred Family Cathedral (Sagrada Familia), Barcelona, Catalonia, Spain ((A) frog, (B) lizard, (C) snake, and (D) horse); and animals from the Brazilian fauna ((E) armadillo and (F) egret) as part of the inside decoration of the Sé Cathedral, São Paulo, Brazil. *Photo credits: Rômulo R.N. Alves (A–D) and Douglas Nascimento (E and F).*

FIGURE 15.6 Gargoyles in the form of animals as architectural elements of St. Vitus Cathedral at Prague Castle, Prague, Czech Republic. *Photo credits: Rômulo R.N. Alves.*

transmutation or changes between human and animal states (Alves et al., 2017). Many Native American myths, for example, describe a time when animal people and human beings were not clearly distinguished from one another. In these mythical narratives, animals and human beings assume each other's identities through metamorphosis or by the magical powers of shape-shifting (Stookey, 2004). For instance, among the Matsigenka Indians of the Peruvian Amazon, it is believed that humans were the first to inhabit the earth and they were slowly transformed into different animal species, starting with primates. At a festival celebration, the first shaman, called Yavireri, managed to transform two groups of people into woolly monkeys and spider monkeys (Shepard, 2002).

Animals sometimes appear in myths and legends as symbols of certain characteristics they are believed to represent. Some primates are considered strong and intelligent animals, as in parts of Africa where, for instance, some tribes hunt chimpanzees because their mythology holds that they will absorb their intelligence by eating them (Alves et al., 2017). On the other hand, in the stories of people who do not live closely with other primates, monkeys and apes are seen as unintelligent, primitive, and violent. It is important to point out that the perception of primates, as positive or negative, can change over time. An example is in Japan, where the perception of monkeys has changed over time and people have found multiple and often contradictory meanings to the presence of monkeys (Mito and Sprague, 2013).

The mythologies also include numerous imaginary or fabulous animals (Borgès and Guerrero, 1957; Sperber, 1996; Stookey, 2004). These are represented by a variety of monsters and creatures combining features of different animals such as the minotaur (man and bull), the centaur (man and horse), the sphinx (woman, bird, and lion in the Greek version), the griffin (eagle and lion), the hippogriff (horse and griffin), and the barometz (lamb and plant). Sometimes they possess unique features such as the phoenix (immortal and periodic bird), the cerberus (a three-headed dog), the catoblepas (with its deadly gaze but with a head so heavy it can only stare at the ground), and the dahu (legs shorter on one side). Of course hybrids and monsters can freely combine, thus the dragon, the Hydra of Lerna, and the Chimaera (Sperber, 1996).

ANIMALS IN ART AND LITERATURE

The influence of animals in cultural and artistic manifestations is richly illustrated in history. There are many types of stories, fables, folk tales, legends, myths, and proverbs that have been generated from relationships between humans and other animals and that also have been passed from generation to generation through oral traditions (Alves, 2012). Even before humans could write, they drew pictures of animals on cave walls (Allen, 1983); these cave paintings, mostly representing animal figures, can be considered the first expressions of human art. In addition to cave walls, early on people also drew pictures of animals on rocks and pottery. These animals existed as symbols capable of transmitting cultural meaning both for those who drew them as well as for those who came long later (Spears et al., 1996). Such artistic renditions conveyed to the observer the cultural significance that the animals represented for the civilization. For example, the early cave paintings at Lascaux and Altamira portray a sense of awe for the animal (Baky, 1980). The figurative representation of wild animals is, of course, a process of great antiquity, as the many examples of rock art and cave painting around the world demonstrate (Beardsworth and Bryman, 2001). Chauvet, in southeast France, is the most ancient; some of its vivid panels of horses, lions, and rhinoceros are as much as 31,000 years old (Chauvet et al., 1996).

FIGURE 15.7 Still Life of Dead Game with a Heron and Spaniel 1977 (William Gouw Ferguson). Painting portrayed at Scottish National Gallery, Edinburgh, Scotland. *Photo credits: Rômulo R.N. Alves.*

Animals have been depicted in art throughout history in many different ways, and they still serve as a source of inspiration for numerous different types of artists. Painters and sculptors often portray animals in their works. For example, many paintings show animals or activities involving animals, such as hunting and fishing (Fig. 15.7). Since the 15th century, and especially in the 16th and 17th centuries, many artists painted local and exotic wild animals portrayed in rural and urban European landscapes (Fig. 15.8). Domestic animals, especially those used as pets, such as birds, dogs, and cats, were also commonly portrayed by these artists (Fig. 15.9). This is not surprising, because people naturally become attached to their pets and may include them alongside images of themselves or surround themselves with representations of the animals they love. In addition to "real" animals, fabulous creatures also have been imagined and recorded in objects, architecture, and art in most human societies. Imaginary beasts repeatedly have the attributes of several animals—combining their best (or worst) features to create a terrifying monster or "super" animal (Fig. 15.10). Among the various artists who portrayed fauna in their works, the famous painters Vincent van Gogh, Eugène Delacroix, and Franz Marc are remarkable, to mention just a few (Fig. 15.11).

FIGURE 15.8 Wild and domestic animals depicted in paintings exhibited at National Gallery London. Above, Surprised! 1891—Henri Rousseau and below, The Large Dort, about 1650. *Photo credits: Rômulo R.N. Alves.*

As human cultures evolved, representations of wild creatures, in literary, artistic, and symbolic terms, became ever more elaborate and sophisticated (Beardsworth and Bryman, 2001).

FIGURE 15.9 Pets portrayed by artists in paintings exposed at the National Gallery London. Lady Cockburn and Her Three Eldest Sons, 1773—Sir Joshua Reynolds (on the left) and Woman with a Cat, about 1880–82—Edouard Manet (on the right). *Photo credits: Rômulo R.N. Alves.*

FIGURE 15.10 Paintings exposed at the Modern Art National Museum, Rome, Italy. Creta 1931–32, Alberto Savinio (on the left) and Una and the Lion 1860, William Bell Scott (on the right). *Photo credits: Rômulo R.N. Alves.*

FIGURE 15.11 Two Crabs, 1889 (Vincent van Gogh). Painting exposed at the National Gallery London, England. *Photo credits: Rômulo R.N. Alves.*

Through time, animals have appeared in paintings, literature, music, dance, sculptures, carvings, and prints. Testimonies to the cultural importance of animals are abundant in museums of natural history and art (e.g., British Museum, Louvre, Natural History Museum Vienna) (Fig. 15.12), in books, in historical and modern buildings that portray animals in their architecture, and in squares and public places of different cities in the world, where animal sculptures are so often easy to find (Fig. 15.12–15.17).

In music, animal sounds have been a source of inspiration for composers, as well as a direct source of music (Clark and Rehding, 2001; Turner and Freedman, 2004). As recently as the 1500s, some people believed that the origin of music had its roots in animal song and other natural sounds (Turner and Freedman, 2004). Links between animals and music include the documentation of songs of various animals in musical notation (Clark and Rehding, 2001; Clark, 1879; Mathews, 1910) and theories of the development of melodious bird songs from the harmonic structure of single-note sounds (Clark, 1879). Many musical compositions have described animals in their lyrics, often with their skills glorified or their features attributed to humans.

The earliest musical instruments have been found to derive from a common source: hunting implements. Loud instruments (percussion instruments, reeds, trumpets) were used to call or repulse prey and to signal among hunters. Some literacy sources from ancient Greece and China, as well as iconographic material from Egypt and Mexico, provide later descriptions of the music/hunt association (Lawergren, 1988). Even today, some cultures possess examples of their music imitating natural sounds. In some instances, this feature is related to shamanistic beliefs or practices (Deschênes, 2002; Hoppál, 2006), but it may also serve as entertainment (game) or for practical functions (luring animals in hunt) (Alves et al., 2009b; Bezerra et al., 2012; Deschênes, 2002; Fernandes-Ferreira et al., 2012; Kothari, 2007).

As with music, animal movements have inspired dance in many human cultures (Alves, 2012). Imitative animal dances were a basic theme of different cultures (Royce, 2007). Among some indigenous peoples of North America, for instance, the dance movements imitated the movements and rhythms of the animals, and the dancers wore the skins or others parts of the animals to assume their spirit and appease it for hunting, killing, and eating them (Kassing, 2007; Royce, 2007). Similarly, dances in which behaviors of apes and other animals are imitated by humans have been documented—for example, among the aboriginal peoples of Australia and Africa—and such dances could have existed early in mankind's evolutionary history when *Homo erectus* acquired the cognitive capacity for deliberate imitation (Francis, 1991). As mentioned before, animals were important as gods in Egypt, and according to Kassing (2007), some of their dances are presumed to have imitated animal movements because many gods had animal heads and the Egyptians were well acquainted with animal dances. Also, Greek religion is populated with sacred animals and beginning with the early Greeks, animal dances were a predominant theme and are cited throughout Greek literature and history. Pig, boar, bear, lion, and fish

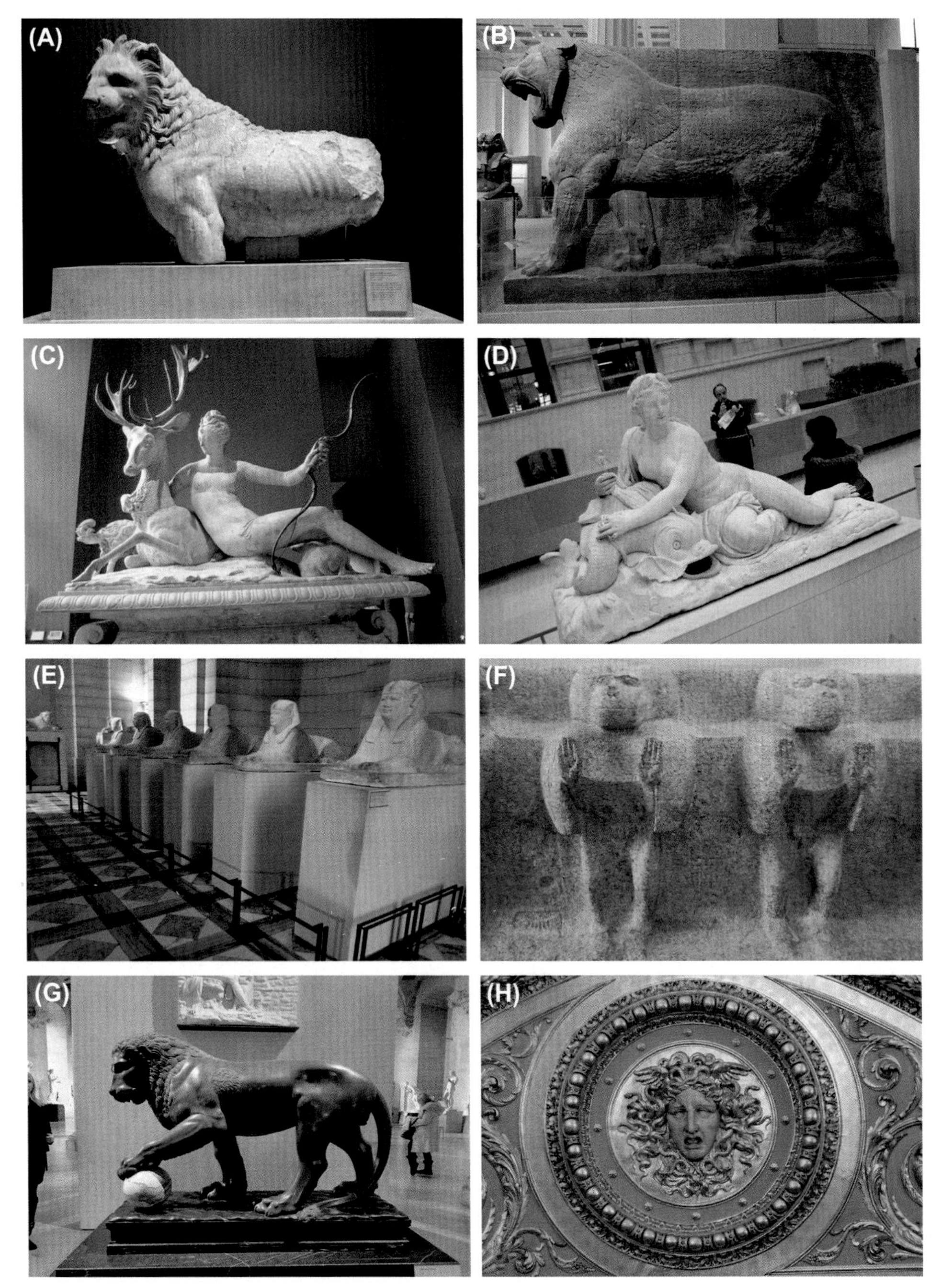

FIGURE 15.12 Anthropozoomorphical and animal representations in sculptures and statues exhibited in the British Museum, London, England (A and B), and Louvre Museum, Paris, France (C–G). (A) Lion from the Mausoleum at Halicarnassus (modern-day Bodrum), (B) colossal guardian lion, (C) Greek goddess Artemis (Latin: Diana) with a deer, (D) mythical animal and Amphitrite, the wife of the Greek god of the sea, Poseidon (Roman: Neptune), (E) Egyptian sphinxes, (F) baboons, (G) lion sculpture, and (H) medusa (with snakes in place of hair). *Photo credits: (A, B, E–H) Rômulo R.N. Alves, (C and D) José S. Mourão.*

FIGURE 15.13 Examples of zoomorphic sculptures and statues in some cities of the world: (A and B)–Paris, France; (C) La Plata, Argentina; (D and F) Budapest, Hungary; and (E) Versailles, France. *Photo credits: (A and B) José S. Mourão; (C–F) Rômulo R.N. Alves.*

dances honored deities by imitating their movements (Kassing, 2007).

Animals also hold an important place in written literature. Examples are the writings of Shakespeare, which contain over 70 abstractions appearing in animal form, and over 4000 allusions are made to animals in his writings (Yoder, 1947). The 19th and 20th centuries produced such

FIGURE 15.14 Equestrian statues in some cities of the Europe: (A) Edinburgh, Scotland; (B) London, England; (C) Vienna, Austria; (D) Budapest, Hungary; (E) Prague, Czech Republic; and (F) Rome, Italy. *Photo credits: Rômulo R.N. Alves.*

FIGURE 15.15 Lion statues in London, England (above), Budapest, Hungary (below left), and Edinburgh, Scotland (below right). *Photo credits: Rômulo R.N. Alves.*

writers as Melville (speculating about a whale's feelings), Dickinson (who wrote over 300 poems featuring animals), Hemingway (who wrote about the fear experienced by a fish before it is about to be caught), Twain, and Steinbeck (Allen, 1983). The point of view of the animal, as depicted by these authors, served as a precursor to the anthropomorphic (the attribution of human characteristics to nonhuman entities) use of animals in many settings (Spears et al., 1996). Additionally, writers such as Nietzsche, Kafka, Max Ernst, D.H. Lawrence, and Darwin advanced philosophies that caused traditional boundaries between the animal world and that of human beings to appear insignificant while simultaneously expressing the idea that anthropomorphism exists to serve human psychological needs (Norris, 1985).

ANIMALS IN SYMBOLISM

Because of the close relationship between humans and other animals throughout history, all cultures have used animals to reflect the nature of humanity and to symbolize societal and individual human characteristics (Merrill, 1990). The use of animal symbolism in cognitive and expressive behavior is probably as old as human consciousness (Lawrence, 1997). Evidence for this includes humankind's earliest artistic and ritualistic representations, which were created in the likeness of various living creatures who inhabited the same environment (Lawrence, 1997). According to Walens (1987) the "symbolization of animals is an essential feature in reflections about the nature of humanity, of the characteristics of individuals and

FIGURE 15.16 Mythical animal sculptures and statues decorating public spaces in some cities: (A and B) London, England; (C) Rome, Italy; (D) Paris, France; and (E and F) Vienna, Austria. *Photo credits: Rômulo R.N. Alves.*

their societies, of the surrounding world and its forces, and of the cosmos as a whole."

The power of human affinity for animals is manifest not only in direct interactions between people and animals but also through the process of using animals to symbolize many aspects of life (Lawrence, 1997). The cultural importance of animals has resulted in their widespread use

FIGURE 15.17 Representation of humans with animals in some cities of Europe (A) Barcelona, Spain; (B and D) London, England; and (C) Budapest, Hungary. *Photo credits: Rômulo R.N. Alves.*

as symbols, icons, and representations throughout history. The morals of Aesop's fables, as colorfully portrayed by animals, have continued to be relevant to modern man (Spears et al., 1996). Societies have handed down morals through the continued use of animals to symbolically communicate important cultural values (Chesterton, 1912).

Examples of the use of animals as symbols are also found in religion. In the Bible, a donkey served to warn Balaam of God's will. The Bible also tells the story of a crafty, talking serpent who skillfully managed to deceive Adam and Eve in the Garden of Eden (Spears et al., 1996). Fish are commonly cited in both the Old and New Testaments of the Bible, although no particular species is named. The biblical fish category also includes marine mammals. According to the Book of Jonah, a so-called "great fish" swallowed the prophet Jonah, keeping him in its belly for 3 days before vomiting him up; other passages refer to it as a whale (Twining, 2004). Both foxes and jackals were well known by the ancient Hebrews, and their cunning was proverbial among them and is still recognized by many human cultures worldwide (Finkel, 1993). The Behemoth, a beast mentioned in the book of Job (Old Testament), is suggested to have identities ranging from a mythological creature to an elephant, hippopotamus, rhinoceros, or buffalo

(Metzger and Coogan, 2004). Some Young Earth creationists argue that the description of the Behemoth is actually a description of a dinosaur. Metaphorically it has come to represent any extremely large or powerful entity. Another ancient creature is the Leviathan, which is mentioned at least six times in the Hebrew Bible, and represents a development of the earlier Canaanite sea monster described as a servant of the sea god Yammu in the Baal Cycle discovered in the ruins of Ugarit (Heider, 1999). Some Jewish sources describe the Leviathan as a dragon who lives over the Sources of the Deep and who, along with the male land monster Behemoth, will be served up to the righteous at the end of time (Heider, 1999). The Leviathan of the Middle Ages was used as an image of Satan, endangering both God's creatures—by attempting to eat them—and God's creation—by threatening it with upheaval in the waters of Chaos (Labriola, 1982).

The use of animal imagery to symbolize morality was a practice that derived immediately from the Middle Ages and flourished throughout the Renaissance. It involved the use of symbols taken from pagan literature, the Bible, treatises by naturalists, and folklore and became commonplace in both didactic and secular literature.

Animals provide multidimensional symbols that can be used to transmit cultural meanings (Morgado, 1993). Robin (1977) argued that, from the earliest days of civilization, animals have been used symbolically to portray a wide range of human qualities, including the orderliness of bees, the treachery of the hyena, the hypocrisy of the crocodile (crying "crocodile tears"), the cunning of the fox, the friendship/loyalty of the dog, the inconstancy of the chameleon, and the peace of the turtle dove. As such, animal symbols are defined culturally and belong to the collective memory of members of society. They form the basis for "an inexhaustible repository that novelists, poets, artists, dramatists, film makers, and even advertisers draw on, either consciously or intuitively, when they wish to evoke an immediate yet profound response" (Rowland, 1973). Even early humans saw animals as exemplifying human traits (Spears et al., 1996).

Animals, both real and mythical, are used as symbols by nations and monarchies (Collar et al., 2007; Sleeboom, 2002). Most countries choose an animal for a national symbol that is closely related to the country. For example, the Andean Condor plays an important role in the folklore and mythology of the South American Andean Region and is a national symbol of Argentina, Bolivia, Chile, Colombia, Ecuador, and Peru (Herrmann et al., 2010; Reid, 1957). The giraffe is a national symbol in Africa, and to many people the large herds of ungulates on the savannah symbolize the African environment and its history (Kaltenborn et al., 2006). In other cases, animals are chosen because they represent some coveted attribute such courage or strength. For example, the eagle of the United States is a symbol of strength and bravery (Fig. 15.18) (Ajala, 1991) and, as a vehicle of communicating patriotism, the eagle is rivaled only by the flag, the Declaration of Independence, the Constitution, the National Anthem, and the Pledge of Allegiance (Center and Walsh, 1981). In some form or another, eagles are also a symbol for Armenia, Albania, Austria, Egypt, Germany, Mexico, Panama, Philippines, Poland, Romania, Russia, and Serbia (Lawrence, 1990). Large carnivores such as tigers and lions are also popular as national symbols. The Royal Bengal Tiger, for example, is the national animal of both India and Bangladesh (Sexton, 2011). Likewise, the lion is a royal symbol of power and is used throughout Africa. In the Dande village in Zimbabwe, southern Africa, the royal ancestors are often referred to as mhondoro (lion). It is believed that at the demise of a chief, his spirit would travel to the bush and "enter the body of a lion" (Lan, 1985). The mhondoro spirit rules over the extent of territory that he has either conquered during his reign or that had once been passed on to him from a former ruler (Lan, 1985). These "spirit provinces," as the territories are called, constitute segregated

FIGURE 15.18 Examples of animals used as national symbols (Eagle—United States and Giraffe—Tanzania). *Photo credits: Anthony B. Cunningham (Eagle) and Franciany Braga-Pereira (Giraffe).*

regions that are demarcated by rivers (Olupona, 1993). The mhondoro protectors of these lands manifest themselves as lions during celebrations.

In contrast to the above, nonhuman spirits can manifest themselves as hyenas and are referred to as shave (jnashave) (Lan, 1985). The mashave spirit may pass witchcraft to unsuspecting men and women. The lion and the hyena, then, represent two opposing forces that control the universe, a good and benevolent force versus an evil and destructive force. Whereas the former symbolizes the royal ancestors, the latter consists of the destructive witches. Based on this concept and Dande's thought and practice, it is taboo to kill lions (Lan, 1985). The symbolic opposition between royalty and witchcraft, as represented in the images of strong and agile animals such as lions and birds, runs through royal and nonroyal cult objects in Africa (Olupona, 1993).

Among the totems of countries or regions, there are also imaginary animals, such as the unicorn, for example, which, along with the lion, are symbols of the United Kingdom (Fig. 15.19). In England the use of the lion as a national symbol dates back to Richard I (a.k.a. Richard the Lionheart, AD 1157–99), whose military exploits in the Crusades and death in battle fashioned the prototype brave, courageous English lionheart (Hand, 2002). The symbolism of animals is particularly obvious in sport. The English see lions as brave, proud animals and use a symbol portraying three of them on the front of the jersey of their national football team. Australia is, of course, famous for its kangaroos, and unsurprisingly their national rugby team is more commonly known as the Wallabies, a kind of small kangaroo. Similarly, the South African team is known as the Springboks, a type of African antelope, and New Zealanders, whether or not in the context of sport, are commonly known as Kiwis; a kiwi is a native New Zealand bird that cannot fly (BBC, 2007). In Brazil, Straube (2010) analyzed the iconography of birds in national football and

FIGURE 15.19 The unicorn and the lion are symbols of the United Kingdom. The image shows these animals represented in hall of the Royal Palace, Edinburgh Castle, Scotland. *Photo credits: Rômulo R.N. Alves.*

FIGURE 15.20 The vulture is the mascot of Brazil's most popular football team, Flamengo. *Photo Source:* http://www.baixakis.com.br/papel-de-parede-flamengo-para-baixar/flamengo/.

recorded 51 official shields that possessed one or more bird species. For example, the most popular team in Brazil (Flamengo) is symbolized by a vulture (Fig. 15.20).

FINAL REMARKS

Since the earliest days of recorded history, the influence of animals on the lives of humans has been remarkable. Paintings on the walls of caves demonstrated early on the fascination humans had with fauna and represents the earliest known prehistoric figurative art. Since then, animals have been represented in a wide variety of cultural expressions of humankind worldwide, which has been intensifying over time, thus reflecting the complexity of the multifaceted relationship between humans and nonhuman animals. Every human culture has a set of legends, myths, fables, and folk tales that are transmitted between generations through oral, visual, and written communication. As expected, not all relationships between people and animals are harmonious, with antagonism frequently being the case when animals are considered as competitors or harmful, sometimes reflecting real situations, but on other occasions reflecting diverse cultural representations, such as folklore and religious traditions. In many places, for instance, animal species are considered deities and worshiped, whereas in others they may be associated with negative attributes.

Worldwide, complex cultural practices associated with fauna influence how animals are viewed, used, and treated across human cultures. Therefore, understanding the cultural role played by animals is essential to understand the relationships humans have with them, and thus it represents an important field of ethnozoological research.

References

Adeola, M.O., 1992. Importance of wild animals and their parts in the culture, religious festivals, and traditional medicine, of Nigeria. Environmental Conservation 19, 125–134.

Ahmed, A., 1997. Live Bird Trade in Northern India. TRAFFIC-India.

Ajala, V.O., 1991. The image of corporate symbol. Africa Media Review 5, 61–74.

Alinei, M., 2000. A stratigraphic and structural approach to the study of magico-religious motivations. Južnoslovenski Filolog 56, 75–92.

Allaby, M., 2010. Animals: From Mythology to Zoology. Facts on File, Inc., New York.

Allen, M., 1983. Animals in American Literature. University of Illinois Press.
Alves, R.R.N., 2012. Relationships between fauna and people and the role of ethnozoology in animal conservation. Ethnobiology and Conservation 1, 1–69.
Alves, R.R.N., Léo Neto, N.A., Santana, G.G., Vieira, W.L.S., Almeida, W.O., 2009a. Reptiles used for medicinal and magic religious purposes in Brazil. Applied Herpetology 6, 257–274.
Alves, R.R.N., Mendonça, L.E.T., Confessor, M.V.A., Vieira, W.L.S., Lopez, L.C.S., 2009b. Hunting strategies used in the semi-arid region of northeastern Brazil. Journal of Ethnobiology and Ethnomedicine 5, 1–50.
Alves, R.R.N., Rosa, I.L., 2008. Use of Tucuxi dolphin *Sotalia fluviatilis* for medicinal and magic/religious purposes in North of Brazil. Human Ecology 36, 443–447.
Alves, R.R.N., Rosa, I.L., Léo Neto, N.A., Voeks, R., 2012. Animals for the gods: magical and religious faunal use and trade in Brazil. Human Ecology 40, 751–780.
Alves, R.R.N., Souto, W.M.S., 2010. Etnozoologia: conceitos, considerações históricas e importância. In: Alves, R.R.N., Souto, W.M.S., Mourão, J.S. (Eds.), A Etnozoologia no Brasil: Importância, Status atual e Perspectivas, first ed. NUPEEA, Recife, PE, Brazil, pp. 19–40.
Alves, R.R.N., Souto, W.M.S., 2015. Ethnozoology: a brief introduction. Ethnobiology and Conservation 4, 1–13.
Alves, R.R.N., Souto, W.M.S., Barboza, R.R.D., 2010. Primates in traditional folk medicine: a world overview. Mammal Review 40, 155–180.
Alves, R.R.N., Souto, W.M.S., Barboza, R.R.D., 2016. The role of nonhuman primates in religious and folk medicine beliefs. In: Waller, M.T. (Ed.), Ethnoprimatology: Primate Conservation in the 21st Century. Springer, Switzerland, pp. 117–135.
Alves, R.R.N., Souto, W.M.S., Barboza, R.R.D., 2017. Primates in mythology. In: Fuentes, A. (Ed.), The International Encyclopedia of Primatology. John Wiley & Sons, Inc.
Alves, R.R.N., Souto, W.M.S., Barboza, R.R.D., Bezerra, D.M.M., 2013. Primates in traditional folk medicine: world overview. In: Alves, R.R.N., Rosa, I.L. (Eds.), Animals in Traditional Folk Medicine: Implications for Conservation. Springer, Berlin, pp. 135–170.
Baky, J., 1980. Human and Animals. The HW Wilson Company, New York.
Basso, K.H., 1996. Wisdom Sits in Places: Landscape and Language Among the Western Apache. University of New Mexico Press.
BBC, W.S., 2007. National Symbols.
Beardsworth, A., Bryman, A., 2001. The wild animal in late modernity: the case of the Disneyization of zoos. Tourist Studies 1, 83–104.
Benton, J.R., 1997. Holy Terrors: Gargoyles on Medieval Buildings. Abbeville Press, New York, NY.
Berkes, F., 2001. Religious traditions and biodiversity. Encyclopedia of Biodiversity 5, 109–120.
Berkes, F., 2008. Sacred Ecology. Routledge, London and New York.
Berrin, K., Larco Museum, 1997. The Spirit of Ancient Peru: Treasures from the Museo Arqueológico Rafael Larco Herrera. Thames and Hudson, New York.
Bezerra, D.M.M., Araujo, H.F.P., Alves, R.R.N., 2012. Captura de aves silvestres no semiárido brasileiro: técnicas cinegéticas e implicações para conservação. Tropical Conservation Science 5, 50–66.
Borgès, J.L., Guerrero, M., 1957. Manual de zoologia fantastica. Fundo de Cultura Económica, México.
Bowman, J.C., 1977. Animals for Man. Edward Arnold., London.
Callicott, J.B., 1994. Earth's Insights: A Multicultural Survey of Ecological Ethics from the Mediterranean Basin to the Australian Outback. University of California Press, Berkeley CA.
Center, A.H., Walsh, F.E., 1981. Public Relations Practices: Case Studies. Prentice-Hall, Inc., New Jersey.
Chan, S.W., 2006. Religious Release of Birds in Hong Kong. University of Hong Kong.
Chauvet, J.M., Deschamps, E.B., Hillaire, C., Bahn, P., 1996. Dawn of Art: The Chauvet Cave: the Oldest Known Paintings in the World. HN Abrams.
Chesterton, G.K., 1912. Aesop's Fables, Introduction. Crown Publishers.
Clark, S., Rehding, A., 2001. Introduction. In: Clark, S., Rehding, A. (Eds.), Music Theory and Natural Order from the Renaissance to the Early Twentieth Century. Cambridge University Press, Cambridge, MA, pp. 1–13.
Clark, X., 1879. Animal music, its nature and origin. The American Naturalist 13, 209–223.
Collar, N.J., Long, A.J., Jaime, P.R.G., 2007. Birds and People: Bonds in a Timeless Journey. BirdLife International.
Cruikshank, J., 2005. Do Glaciers Listen? Local Knowledge, Colonial Encounters, and Social Imagination. UBC Press, Vancouver BC.
Deschênes, B., 2002. "Inuit Throat-Singing". Musical traditions. The Magazine for Traditional Music Throughout the World.
Evans, E.P., 1896. Animal Symbolism in Ecclesiastical Architecture. W. Heinemann, London.
Fernandes-Ferreira, H., Mendonça, S.V., Albano, C., Ferreira, F.S., Alves, R.R.N., 2012. Hunting, use and conservation of birds in Northeast Brazil. Biodiversity and Conservation 221–244.
Finkel, A., 1993. The Essence of the Holy Days: Insights from the Jewish Sages. J. Aronson, Northvale, NJ.
Forscher, H., 2007. Animals in the Landscape: An Analysis of the Role of the Animal Image in Representations of Identity in Selected Australian Feature Films from 1971 to 2001. Bond University.

Francis, S.T., 1991. The origins of dance: the perspective of primate evolution. Dance Chronicle 14, 203–220.
Garcia, G., 2012. A fauna brasileira escondida na Catedral da Sé.
Garfield, E., 1986. The turtle: A most ancient mystery. 1. Its role in art, literature, and mythology. Current Contents 293–297.
Gilbert, M., Sokha, C., Joyner, P.H., Thomson, R.L., Poole, C., 2012. Characterizing the trade of wild birds for merit release in Phnom Penh, Cambodia and associated risks to health and ecology. Biological Conservation 153, 10–16.
Goh, T.Y., O'Riordan, R.M., 2007. Are tortoises and freshwater turtles still traded illegally as pets in Singapore? Oryx 41, 97–100.
Griaule, G., 2005. Dogon Religion, Encyclopedia of Religion, second ed. MacMillan, Detroit, pp. 2390–2392.
Gross, A., Vallely, A., 2013. Animals and the Human Imagination: A Companion to Animal Studies. Columbia University Press, New York.
Haenn, N., Schmook, B., Reyes, Y.M., Calmé, S., 2014. A cultural consensus regarding the king vulture?: preliminary findings and their application to Mexican conservation. Ethnobiology and Conservation 3, 1–15.
Haggan, N., 2011. You don't know what you've got 'Til it's Gone': the case for spiritual values in marine ecosystem management. In: Ommer, R.E., Perry, R.I., Cochrane, K., Cury, P. (Eds.), World Fisheries: A Social-ecological Analysis. Wiley Online Library, Oxford, UK, pp. 224–246.
Hand, D., 2002. Football, Cultural Identities and the Media: A Research Agenda.
Heider, G.C., 1999. Tannîn, Dictionary of Deities and Demons in the Bible. Wm. B. Eerdmans Publishing, Grand Rapids.
Herrmann, T., Costina, M.I., Costina, A.M., 2010. Roost sites and communal behavior of Andean condors in Chile. Geographical Review 100, 246–262.
Hogue, J., 2003. Cultural entomology. In: Resh, V.H., Carde, R.T. (Eds.), Encyclopedia of Insects. Academic Press, New York, pp. 273–281.
Hoppál, M., 2006. Music of shamanic healing. In: Kilger, G. (Ed.), Macht Musik. Musik als Glück und Nutzen für das Leben. Wienand Verlag, Köln.
Jobes, G., 1962. Dictionary of Mythology, Folklore and Symbols. Scarecrow Press.
Kalof, L., Resl, B., 2007. A Cultural History of Animals: In Antiquity. Berg.
Kaltenborn, B.P., Bjerke, T., Nyahongo, J.W., Williams, D.R., 2006. Animal preferences and acceptability of wildlife management actions around Serengeti National Park, Tanzania. Biodiversity and Conservation 15, 4633–4649.
Kang, X., 2006. The Cult of the Fox: Power, Gender, and Popular Religion in Late Imperial and Modern China. Columbia University Press.
Kassing, G., 2007. History of Dance: An Interactive Arts Approach. Human Kinetics.
Klingender, F., 1971. Animals in Art and Thought to the End of the Middle Ages. Routledge.
Kothari, A., 2007. Birds in Our Lives. Universities Press.
Labriola, A.C., 1982. The medieval view of history in paradise lost. In: Mulryan, J. (Ed.), Milton and the Middle Ages. Bucknell University Press, pp. 115–134.
Lan, D., 1985. Guns & Rain: Guerrillas & Spirit Mediums in Zimbabwe. Univ. of California Press.
Lawergren, B., 1988. The origin of musical instruments and sounds. Anthropos 83, 31–45.
Lawrence, E.A., 1990. Symbol of a nation: the bald eagle in American culture. Journal of American Culture 13, 63–69.
Lawrence, E.A., 1997. Hunting the Wren: Transformation of Bird to Symbol: A Study in Human-animal Relationships. Univ Tennessee Press.
Leo Neto, N.A., Alves, R.R.N., 2010. A Natureza sagrada do Candomblé: análise da construção mística acerca da Natureza em terreiros de candomblé no Nordeste de Brasil. Interciencia 35, 568–574.
Léo Neto, N.A., Brooks, S.E., Alves, R.R.N., 2009. From Eshu to Obatala: animals used in sacrificial rituals at Candomble "terreiros" in Brazil. Journal of Ethnobiology and Ethnomedicine 5, 1–23.
Leo Neto, N.A., Mourão, J.S., Alves, R.R.N., 2011. "It all begins with the head": initiation rituals and the symbolic conceptions of animals in Candomblé. Journal of Ethnobiology 31, 244–261.
Leo Neto, N.A., Voeks, R.A., Dias, T.L.P., Alves, R.R.N., 2012. Mollusks of Candomble: symbolic and ritualistic importance. Journal of Ethnobiology and Ethnomedicine 8 (10), 1–10.
Liu, X., Mcgarrity, M.E., Li, Y., 2012. The influence of traditional Buddhist wildlife release on biological invasions. Conservation Letters 5, 107–114.
Lugira, A.M., 2009. African Traditional Religion. Chelsea House Pub, New York.
Malamud, R., Kalof, L., Pohl-Resl, B., 2007. A Cultural History of Animals in the Modern Age. Berg.
Mathews, F.S., 1910. Field Book of Wild Birds and Their Music. The Knickerbocker Press, New York.
McClure, H.E., Chaiyaphun, S., 1971. The sale of birds at the Bangkok "Sunday Market", Thailand. Natural History Bulletin of the Siam Society 24, 41–78.
McNeely, J.A., 2001. Religions, traditions and biodiversity. COMPAS Magazine 20–22.
Merrill, B.R., 1990. 'Behold the lamb of god': the Savior's use of animals as symbols. In: Top, B.L., Van Orden, B.A. (Eds.), Lord of the Gospels: The 1990 Sperry Symposium on the New Testament, pp. 129–147.
Metzger, B.M., Coogan, M.D., 2004. The Oxford Guide to People and Places of the Bible. Oxford University Press.
Mito, Y., Sprague, D.S., 2013. The Japanese and Japanese Monkeys: Dissonant Neighbors Seeking Accommodation in a Shared Habitat, the Macaque Connection. Springer, pp. 33–51.

Mittermeier, R.A., Ratsimbazafy, J., Rylands, A.B., Williamson, L., Oates, J.F., Mbora, D., Ganzhorn, J.U., Rodríguez-Luna, E., Palacios, E., Heymann, E.W., 2007. Primates in peril: the world's 25 most endangered primates, 2006–2008. Primate Conservation 22, 2–40.

Moazami, M., 2005. Evil animals in the Zoroastrian religion. History of Religions 44, 300–317.

Morgado, M.A., 1993. Animal trademark emblems on fashion apparel: a semiotic interpretation. Clothing and Textiles Research Journal 11, 31–38.

Nikoloudis, S., 2001. Animal sacrifice in the Mycenaean world. Journal of Prehistoric Religion 15, 32–38.

Norris, M., 1985. Beasts of the Modern Imagination: Darwin, Nietzsche, Kafka, Ernst, and Lawrence. Johns Hopkins University Press, Baltimore, MD.

Olupona, J.K., 1993. Some notes on animal symbolism in African religion and culture. Anthropology and Humanism 18, 3–12.

Palmeri, F., 2006. Humans and Other Animals in Eighteenth-Century British Culture: Representations, Hybridity, Ethics. Ashgate Publishing Company, Miami, USA.

Preece, R., 2002. Awe for the Tiger, Love for the Lamb: A Chronicle of Sensibility to Animals. UBC Press.

Ramsay, N.F., Kaye, P., Ng, A., O'Riordan, R.M., Chou, L.M., 2007. The red-eared slider (*Trachemys scripta elegans*) in Asia: a review. In: Gherardi, F. (Ed.), Biological Invaders in Inland Waters: Profiles, Distribution, and Threats. Springer, Dordrecht, The Netherlands.

Reid, J.T., 1957. An aspect of symbolic nationalism in Spanish America:(aspirations and emblems). Hispania 40, 73–75.

Resl, B., 2007. A Cultural History of Animals in the Medieval Age. Berg, Oxford.

Robin, P.A., 1977. Animal Lore in English Literature. Folcroft Press, London, UK.

Rowland, B., 1973. Animals with Human Faces. University of Tennessee Press, Knoxville, TN.

Royce, A.P., 2007. Dance at the dawn of time. In: Kassing, G. (Ed.), History of Dance: An Interactive Arts Approach. Human Kinetics, pp. 25–38.

Sax, B., 2002. The Mythological Zoo: An Encyclopedia of Animals in World Myth, Legend and Literature. ABC-CLIO, Inc., Santa Barbara.

Schoppe, S., 2008. Science in CITES: The Biology and Ecology of the Southeast Asian Box Turtle and its Uses and Trade in Malaysia. Petaling Jaya, Selangor, Malaysia.

Senior, M., 2009. A Cultural History of Animals in the Age of Enlightenment. Berg Publishers.

Severinghaus, L.L., Chi, L., 1999. Prayer animal release in Taiwan. Biological Conservation 89, 301–304.

Sexton, C., 2011. The Bengal Tiger. Pilot Books.

Shepard, G.H., 2002. Primates in Matsigenka subsistence and world view. In: Fuentes, A., Wolfe, L. (Eds.), Primates Face to Face: Conservation Implications of Human-nonhuman Primate Interconnections. Cambridge University Press, Cambridge, pp. 101–136.

Shepard, P., 1996. The Others: How Animals Made Us Human. Island Press, Washington, DC.

Shiu, H., Stokes, L., 2008. Buddhist animal release practices: historic, environmental, public health and economic concerns. Contemporary Buddhism 9, 181–196.

Sleeboom, M., 2002. The power of national symbols: the credibility of a dragon's efficacy. Nations and Nationalism 8, 299–313.

Spears, N.E., Mowen, J.C., Chakraborty, G., 1996. Symbolic role of animals in print advertising: content analysis and conceptual development. Journal of Business Research 37, 87–95.

Sperber, D., 1996. Why are perfect animals, hybrids, and monsters food for symbolic thought? Method & Theory in the Study of Religion 8, 143–169.

Stookey, L.L., 2004. Thematic Guide to World Mythology. Greenwood Publishing Group.

Straube, F.C., 2010. As aves nos símbolos do futebol brasileiro: Escudos. Atualidades Ornitológicas 158, 33–48.

Tomalin, E., 2004. Bio-divinity and biodiversity: perspectives on religion and environmental conservation in India. Numen 51, 265–295.

Turner, K., Freedman, B., 2004. Music and environmental studies. The Journal of Environmental Education 36, 45–52.

Twining, L., 2004. Symbols and Emblems of Early and Mediaeval Christian Art. Kessinger Publishing, USA.

Ulric, K., 1996. Mythical animals: a reference pathfinder. Collection Building 15, 32–35.

Walens, S., 1987. Animals. In: Eliade, M. (Ed.), The Encyclopedia of Religion. Macmillan Press, New York, pp. 291–296.

Yoder, A.E., 1947. Animal Analogy in Shakespeare's Character Portrayal. King's Crown Press.

Zangpo, S., 2005. Releasing Life: An Ancient Buddhist Practice in the Modern World. The Corporate Body of the Buddha Educational Foundation, Taiwan.

CHAPTER

16

Fauna at Home: Animals as Pets

Rômulo Romeu Nóbrega Alves[1], *Luiz Alves Rocha*[2]

[1]Universidade Estadual da Paraíba, Campina Grande, Brazil; [2]California Academy of Sciences, San Francisco, CA, United States

INTRODUCTION

The world is shared by humans and a diversity of other animals, implying the occurrence of a variety of interactions between them (Alves, 2012). Humans used products derived from fauna for numerous purposes since ancient times. In addition, humans have made use of and established a variety of interactions with live animals for work, companionship, agriculture (livestock), research (laboratory animals), etc. Among these, the use of animals as pets is the most intimate human–animal relationships (Franklin, 1999). The roots of this interaction are very ancient, but it clearly peaked when domestication became a part of human culture (Bueno, 2009). Considering that domestication is invariably associated with some form of captive or controlled breeding, it is possible that the species we now classify as "domestic" were simply those that bred most readily as pets within the hunter-gatherer milieu (Serpell, 1990). Whatever the case, Digard (1990) takes perhaps the broadest view of domestication by proposing a definition that includes not only pets, but also animals captured from the wild for human use and that are not necessarily bred in captivity.

Humans have a long history of associations with animals, dating back to prehistoric hunting (Honess and Wolfensohn, 2010). Primitive peoples found human–animal relationships to be important to their survival and pet keeping, as one of them, was common in hunter-gatherer societies (National Institutes of Health, 1987). Fitter (1986) suggested that the whole array of domesticated animals may well owe their origin to some Cro-Magnon or Neanderthal child or old person adopting an orphaned wolf cub. The first domesticated animal was probably the wolf, which gave rise to dogs that have been living intimately with *Homo sapiens* for at least 30,000–32,000 years. By the early Neolithic age (c. 8000 BC), the burial of dogs—a symbolic act reserved predominantly for kin—was already common among hunter-gatherers (Losey et al., 2013).

Evidence reveals that the keeping of pets has been practiced continuously throughout human history (Serpell, 2011). Ancient cultures from throughout the world are known to have caught, kept, and bred animals as pets (Carrete and Tella, 2008; Collar et al., 2007), such as Egyptian, Greek, and Roman civilizations; dogs and cats were kept as pets in imperial homes

Ethnozoology
http://dx.doi.org/10.1016/B978-0-12-809913-1.00016-8

of both China and Japan (Serpell and McCune, 2012). This practice has persisted and spread throughout history, and currently involves a wide variety of animals throughout the world (Alves, 2012), in cultural, social, economic, and ecological aspects of human life.

The maintenance of a pet is motivated by several factors and represents a form of dynamic mutual interaction between people and animals that has implications for both. It is, therefore, a practice of extreme significance in the lives of a large number of people and which, of course, has aroused the attention of many academic areas, such as anthropology, sociology, philosophy, ethology, veterinary medicine, biology, animal sciences, and ethnozoology. Here, we will briefly discuss the importance of the interactions between people and pets and the implications of these relationships. For the purpose of this chapter, pets will be considered "an animal kept within a domestic setting, where its main purpose is for personal interest, entertainment, or companionship" (Bush et al., 2014).

ANIMALS KEPT AS PETS

Given that the practice of keeping pets is strongly associated with the practice of animal domestication, it is not surprising that the most popular pet species are domesticated animals (Waller, 2016). In addition to dogs and cats, rabbits, guinea pigs, hamsters, and budgerigars are among the most common pets kept throughout the world (Fig. 16.1). These animals have been subjected to selective breeding pressures, including goal-oriented breeding by humans, for over hundreds or thousands of years, and thus have undergone significant genetic changes from their wild ancestors (Pierce, 2016). Humans have been selecting animals for their abilities and physical and behavioral characteristics for thousands of years, which has produced a great diversity of breeds as exemplified by dogs and cats. The result has been a boost to the popularity of these animals as pets around the world. For example, more than half of all American households share their home with at least one dog or cat, with populations of these pets in the United States having risen from approximately 61 and 75 million in 2001, to 76 and 85 million in 2012, respectively (Pet Food Institute, 2013). Currently there are an estimated 8.5 million dogs and 7.5 million cats in the United Kingdom (Pet Food Manufacturer's Association, 2016), and in Japan, an estimated 9.9 million cats and almost 10 million dogs were part of households in 2015 (Japanese Pet Food Association, 2016).

The practice of keeping pets, however, is not limited to domesticated animals. The list of species used as pets also includes wild animals, termed by some authors as "exotic" pets. In this context, exotic does not refer to a species being nonnative to the place where they are found, but instead is used as an adjective to describe any object in society that is unique, dangerous, or exciting (Mitchell and Tully, 2009). These authors also point out that there are other adjectives commonly used to describe exotic pets, including nontraditional or nondomestic, which encompass a wide variety of taxa from invertebrates to mammals.

Invertebrates (terrestrial and aquatic) have become increasingly popular as pets (Mitchell and Tully, 2009), as evidenced by the increasing global trade in these animals (Kumschick et al., 2016). Mitchell and Tully (2009) emphasize that there are numerous explanations for this trend, including the "awe factor," the relative inexpensiveness of the animals, and their low-maintenance and husbandry requirements. Cnidarians, gastropods, arachnids, myriapods, insects, crustaceans, and echinoderms are among the invertebrate groups common to the captive pet trade and/or aquaculture (Kumschick et al., 2016; Mitchell and Tully, 2009).

Live ornamental aquatic animals, which include aquatic invertebrates and fishes (Fig. 16.2), have experienced noticeable increase in interest from hobbyists now and have driven an ornamental pet trade that involves numerous countries (Calado et al., 2003; Ng et al., 2016; Wabnitz, 2003).

FIGURE 16.1 Some of the most popular pets in the world: dogs (A and B), cats (C), budgerigars (D), guinea pigs (E), and hamsters (F). *Photo credits: (A) Luiz A. Rocha, (B) Tacyana P.R. Oliveira, (C) Thelma L.P. Dias, (D) Jerry Tiller, via Wikimedia Commons, (E) Sandos at the English language Wikipedia, (F) TetraHydroCannabinol, via Wikimedia Commons.*

For example, Wabnitz (2003) pointed out that a total of 140 species of stony coral (scleractinians) and more than 500 species of invertebrates (other than corals) are traded as marine ornamentals. Calado et al. (2003) indicated that at least 128 species of decapod crustaceans are marketed for aquarium purposes. In a study conducted in Singapore, Ng et al. (2016) recorded 59 species of freshwater bivalves and gastropods marketed for the same purpose. Nonetheless, fishes represent the animal group with the largest number of species involved in the ornamental pet trade (Mitchell and Tully, 2009; Rhyne et al., 2012; Wabnitz, 2003). Wabnitz (2003) indicated that at least 1471 species of fish are traded worldwide, with the best estimate of annual global trade ranging between 20 and 24 million individuals. Damselfishes (Pomacentridae) make up almost half of the trade, with species of angelfishes (Pomacanthidae), surgeonfishes (Acanthuridae), wrasses (Labridae), gobies (Gobiidae), and butterflyfishes (Chaetodontidae) accounting for approximately another 25%–30%. The most traded species are the blue-green

FIGURE 16.2 Examples of common ornamental fishes. (A) *Zebrasoma flavescens*, (B) *Centropyge flavissima*, (C) *Dascyllus aruanus*, (D) *Pseudanthias squamippinis*, (E) *Pseudanthias squamippinis*, (F) *Labroides dimidiatus*. *Photo credit: Luiz A. Rocha.*

damselfish (*Chromis viridis*), the clown anemonefish (*Amphiprion ocellaris*), the whitetail dascyllus (*Dascyllus aruanus*), the sapphire devil (*Chrysiptera cyanea*), and the threespot dascyllus (*Dascyllus trimaculatus*) (Wabnitz, 2003). Seahorses (genus *Hippocampus*) are also among the most popular fishes for use as aquarium pets (Rosa et al., 2005, 2011; Vincent, 1996; Vincent et al., 2011), with most of the 48 species recognized as valid by Project Seahorse (2016) being marketed for this purpose. Some of these fishes (Fig. 16.3) can reach values of thousands of dollars due to their beauty and rarity in the trade.

Birds represent another group commonly used as pets (Alves et al., 2013a,b; Anderson, 2010; Roldán-Clarà et al., 2014). Of all the known bird species of the world (n = 9856 species), 3649 species (37%) are utilized as pets (Butchart, 2008). As a country, Brazil illustrates this situation well. Bird diversity is exceptionally high in Brazil, with 1919 avian species (Piacentini et al., 2015), and bird-keeping activities are common throughout the country, both in rural and urban areas (Fig. 16.4) (Alves et al., 2010). Birds represent the most traded group among all animals involved in wild animal

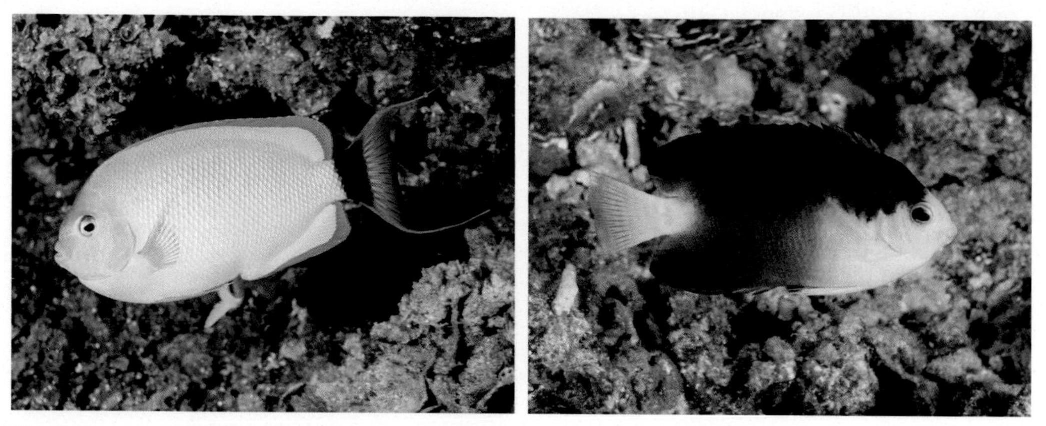

FIGURE 16.3 *Genicanthus personatus* (*left*) and *Centropyge nahacki* (*right*) are among the most expensive marine aquarium fishes. *Photo credit: Luiz A. Rocha.*

FIGURE 16.4 Wild birds used as pets in Brazil. (A) *Amazona aestiva*, (B) *Eupsittula cactorum*, (C) *Icterus jamacaii*, and (D) *Patagioenas picazuro*. *Photo credits: (A–C) John Philip Medcraft and (D) Rômulo R.N. Alves.*

traffic in Brazil (Renctas, 2001), with more than 400 species being marketed for use as pets (Alves et al., 2013a,b). The high species richness of birds kept as pets is not surprising given that their beautiful appearance and singing abilities (or both) have attracted people since ancient times (Carrete and Tella, 2008; Collar et al., 2007; Grier, 2010; Tidemann and Gosler, 2010). Presently, keeping birds as pets in Brazil represents a practice that crosses class lines, as well as racial and ethnic barriers.

Most of the bird species kept as pets belong to families of the orders passeriformes and psittaciformes. Species of the family Psittacidae (Fig. 16.5), for example, which includes parrots, parakeets, macaws, and cockatoos, are among the world's most popular pet birds due to their capacity to imitate human voices, as well as their intelligence, beauty, and docility—being second only to dogs and cats in overall popularity (Fernandes-Ferreira et al., 2012; Fitzgerald, 1989; Hardie and Gaski, 1989; Hemley, 1994; Roldán-Clarà et al., 2014). Another very popular family is Emberizidae, which are favored as cage-birds since they are extremely hardy, granivorous, of a small size allowing large numbers of them to be kept together in small cages, with great singing ability, and colorful plumage (Alves et al., 2013a,b; Alves et al., 2010; Frisch and Frisch, 1981; Licarião et al., 2013; Sick, 1997).

FIGURE 16.5 Children and their pet birds (psittaciformes) in northern (*above*) and northeastern regions of Brazil. *Photo credits: Neil Palmer/CIAT for Center for International Forestry Research and Rômulo R.N. Alves, respectively.*

In the past few decades, amphibians and reptiles (collectively known as herpetofauna) have gained popularity among breeders of animal pets (Fig. 16.6). Among amphibians, frogs, toads, and salamanders are the most common groups kept as pets (Mitchell and Tully, 2009; Wombwell, 2014), fueling an amphibian pet trade that involves a great diversity of species (Magalhães and São-Pedro, 2012; Natusch and Lyons, 2012; Prestridge et al., 2011; Wombwell, 2014). Based on data from the IUCN Red List (2015), Auliya et al. (2016) pointed out that more than 290 amphibian species are targeted for the international pet trade and consumption as food, emphasizing that species from South and Central America and Madagascar are predominantly collected for the pet trade. Species of the family Dendrobatidae (poison dart frogs) of Central and South America are extremely popular in the European herpetological pet trade (Gorzula, 1996).

In the case of reptiles, their use as pets has expanded globally, generating trade that involves high values at the international level

FIGURE 16.6 Reptiles kept as pets in Brazil. (A) *Pantherophis guttatus*, (B) *Boa constrictor*, (C) *Eublepharis macularius*, (D) *Stigmochelys pardalis*. *Photos: Courtesy of keepers.*

and which has been increasing over the last 30 years (Alves et al., 2012a,b,c; Romagosa, 2015). As a consequence, reptiles currently represent the third most species-rich vertebrate class after fishes and birds in the international pet trade (Bush et al., 2014). The United States, the United Kingdom, and countries of the European Union are the primary consumers of reptiles and amphibians as pets (Auliya, 2003; Auliyai et al., 2016; Hoover, 1998). In the United States, the number of individuals imported of some of these groups has more than doubled since the early 1970s. For example, about 320,000 lizards and snakes were imported per year in the early 1970s, and more than one million were imported per year in the early 2000s (Romagosa, 2015). Auliyai et al. (2016) reviewed data on the global reptile pet trade and found that the European Union plays a significant role, with over 20 million individuals imported over a 10-year period. Japan also stands out (Auliya, 2003), as it is one of the largest importers of live reptiles, having imported >317,000 specimens of reptiles from 52 countries during 2013.

According to Hoover (1998), the wide commercialization of reptiles is due to their relative abundance, the wide variety of species that can be acquired, the development of improved techniques for raising them in captivity, the growth in the number of restrictions

FIGURE 16.7 Examples of wild mammals used as pets in Brazil (*left*) and Indonesia (*right*). *Photo credits: Hugo Fernandes-Ferreira and Hari Priyadi for Center for International Forestry Research (CIFOR), respectively.*

concerning the commercialization of other species of animals, and mainly because they generally require less care than mammals or birds. In fact, today the demand for reptiles for domestic breeding is intense, and a wide variety of species are available to potential breeders on online sites for the purchase, sale, or exchange of wild animals.

In addition to domestic mammals, several wild species are also raised as pets, including marsupials, ferrets, deer, squirrels, hedgehogs, and primates (Fig. 16.7). This last group exemplifies the situation well since thousands of primates are hosted as pets in people's homes in virtually all areas where they co-occur with humans, including both urban and rural settings (Alves et al., 2016; Duarte-Quiroga and Estrada, 2003; Soulsbury et al., 2009). Local species are typically kept as pets in countries with native primate populations (Soulsbury et al., 2009). For example, in Mexico City, Duarte-Quiroga and Estrada (2003) found that 82% of pet primates belonged to three native species and were sourced from the wild. This has also been observed in other countries such as Brazil and Indonesia, where many native primates are kept as pets (Alves et al., 2016; Wright, 2005). Species of primates are also kept as pets in many countries where they do not occur naturally, in which cases the specimens are derived from captive breeding, from surplus zoo stock or from illegal wildlife trade. Soulsbury et al. (2009) pointed out that there are no accurate figures on the number of primates kept as pets because most countries do not have a recognized body recording the number or species of privately owned primates. Previous research, however, indicates that the numbers are significant in some countries. Japan, for instance, imported 16,000 primates for the pet trade pre-CITES (Convention on International Trade in Endangered Species of Wild Fauna and Flora) (Matsubayashi et al., 1986). Post-CITES, this number dropped, and between 1981 and 1983 about 7% (569 of 8553) of primates imported into Japan were for the pet trade (Matsubayashi et al., 1986). In the United States, the number of primates kept as pets is as high as 15,000 (Born Free USA, 2005). In countries with wild primate populations, ownership of primates is widespread but difficult to quantify (Fuentes, 2006) because animals are often kept clandestinely, with many obtained from the illegal trade.

There are still cases of people who maintain exotic animals that are unpredictable and incapable of being domesticated or tamed, including lions, tigers, cougars, wolves, bears, alligators, and venomous snakes and other reptiles (Animal Protection Institute, 2005). Of course keeping these animals requires a large financial investment, so it is generally people with high incomes that choose these animals as pets.

WHY DO HUMANS KEEP PETS?

Initially the keeping of animals was predominantly utilitarian (Honess and Wolfensohn, 2010). Dogs, which are considered the first domesticated species, have long been recognized as excellent hunting assistants and certainly the initial motivation for humans to keep them as pets is linked to this practice. Several other animals (e.g., cattle, sheep, pigs, goats, and poultry) were later domesticated for their utility for meat, eggs, and other products (Alves, 2016). It is significant to note that the keeping of animals as pets did not merely come as a result of the domestication of animals for utilitarian purposes (Hirschman, 1994), such as for food and protection, but appears to have resulted from a human desire for companionship with other species (see Messent and Serpell, 1981). From the earliest days of European exploration, native peoples were recorded as engaged in pet keeping. For example, from the early 1500s onward, explorers and missionaries described native inhabitants of the Americas as keeping a wide variety of companion animals: tapirs, wolves, raccoons, bears, monkeys, peccaries, moose, mice, rats, squirrels, and birds (Working Party Council for Science and Society, 1988). This situation persists to this day in several hunter-gatherer societies, which keep few or no domestic animals for utilitarian purposes but commonly keep a wide variety of wild animals as pets (Hirschman, 1994). For example, South American natives keep agoutis, pacas, parrots, rodents, sloths, and monkeys (Savishinsky, 1986). In some cases young mammals are suckled by the women of the tribe, cared for like children and, upon their death, are buried near the house or hammock of their keeper (Serpell, 1986). Similar situations were recorded in historical documents, which reveal that native peoples frequently captured young animals of various kinds and reared them as favorites, and perhaps even sold them or presented them as curiosities to colonizers and visiting naturalists (Alves, 2012). Other examples include the Australian Aborigines, who keep tamed wallabies, possums, dingoes, bandicoots, and cassowaries, and the Inuit of the Arctic regions, who keep pet bear cubs, foxes, porcupines, birds, wolves, and baby seals (Serpell, 1986). Thus, the lives of humans and other animal species are, and have been, closely intertwined in ways that surpass utilitarian necessity and approximate interpersonal bonding (Hirschman, 1994).

It is clear that the reasons for keeping a pet have changed and diversified throughout history. Dogs illustrate this situation well, originally being used for hunting or tracking prey, participating in wars by protecting troops, serving as guards, pulling sleds, serving as food, etc. (Bernard and Demaret, 1996). Initially there was no emotional link as there is today, so much so that in the 17th century, when guard dogs and shepherds reached an advanced age, which prevented them from performing their functions satisfactorily, they were sacrificed by hanging or drowning (Bernard and Demaret, 1996). Later, dogs came to be maintained for reasons of status, aesthetics and companionship, and often came to be considered "family members."

It should be noted that in many situations, the choice of a pet is motivated by multiple reasons. A hunter, for example, can keep a dog because they are excellent hunting assistants (Alves et al., 2009; Koster, 2008; Koster and Tankersley, 2012) or for protection against wild animals. However, if dogs, by instinct of predation, attack domestic livestock, they are often slaughtered. Also, in both urban and rural areas, dogs are raised in homes as sentinels to warn of potential dangers and ward off thieves. Dogs are also used as assistants for people with physical disabilities (service dogs). Even with dogs being kept for practical reasons, emotional bonds are generated during the coexistence with their owners. People of different cultures and social circumstances choose to have pets for different reasons (Archer, 1997). One such reason is the potential benefit to human health. Interestingly, one of the earliest recorded uses of animals in health care is by the founder of modern nursing. In 1860, Florence Nightingale (Nightingale, 1969) observed that "A small pet is often an excellent companion for the sick, for long chronic cases especially." She suggested the use of a caged bird as the only pleasure an invalid who had been confined for years to the same room might enjoy (Jorgenson, 1997). Nowadays, researchers have been investigating the contribution of pets to the quality of life of their owners (Beck and Katcher, 1984, 1996; Eshuis et al., 2016), and their positive effects on the physical, psychological, and social aspects of people's lives (Batson, 1998; Beetz et al., 2011; Raina et al., 1999; Wilson and Turner, 1997).

Choosing an animal to be kept as a pet is also influenced by the particular characteristics of the species chosen and the profile of its future owners. Animal attributes considered for choosing a pet are, for example, beauty, intelligence, singing ability (for birds), utility value, or simply its rarity or behavior. Some people choose their pets simply by personal fascination with a type of animal (Herzog, 2014). Obviously, empathy and fascination with a species can be strongly influenced by cultural factors (Herzog, 2014), from stories in the media to portrayals of animals in literature and film (Waller, 2016). Ghirlanda et al. (2014), for example, showed that the release of movies featuring dogs is often associated with an increase in the popularity of featured breeds, which may have an influence on the choice of pet dogs. The global dissemination of wildlife films—even those with apparently pro-welfare messages such as *Finding Nemo* and *Rio* (featuring a clownfish [*A. ocellaris* (Fig. 16.8)] and blue macaws [*Anodorhynchus hyacinthinus*], respectively)—coincide with peaks in demand for featured species (Yong et al., 2011). According to Yan (2016), in the 5 years since *Finding Nemo* first aired, rising demand in the marine aquarium

FIGURE 16.8 *Ampiprion ocellaris*, fish species whose popularity as a pet increased after the release of the animated film *Finding Nemo*. *Photo credit: Luiz A. Rocha.*

trade has forced some clownfish populations to decline by as much as 75%. Predicting this situation in the release of the sequel *Finding Dory* (featuring the species *Paracanthurus hepatus*, perciformes: Acanthuridae; Fig. 16.9), aquarium hobbyists and animal protection and conservation groups undertook a campaign urging consumers not to buy fish like "Dory," which is much harder to keep in aquaria than "Nemo" and only recommended it for expert aquarists. Another example of media influence on pet selection was recorded in Indonesia, where the offering of owls in bird markets was very limited until the release of the *Harry Potter series* in the early 2000s, when their popularity as pets increased in that country (Nijman and Nekaris, 2017). Internet companies that source pets and provide increased access to information are driving demand for yet more novel pets, especially exotics (Bush et al., 2014; Magalhães and São-Pedro, 2012; Wu, 2007).

Religious beliefs and practices also have long influenced human perceptions of animals and their uses (Alves et al., 2012a,b,c). Waller (2016) points out that religious principles, traditions, and practices lead diverse peoples to embrace certain animals yet shun others, thereby exerting a direct influence on the selection of a pet. In Western cultures, for example, snakes are wrapped in legends, myths, and beliefs—perhaps the best known of these tales being the loss of paradise by Adam and Eve due to the guile of a snake (Leeming, 2003), generating a negative perception of snakes, which are usually associated with evil (Alves et al., 2012a,b,c).

It should be noted that a species popularity as a pet is closely linked to culture. For example, Waller (2016) describes how ferrets are loved in Israel, but hated in New Zealand. This author also provides the example of rabbits, which are one of the most popular companion animals today in Japan, embodying *kawaii* (cute) and generally not considered food by the Japanese, but the same rabbit serves both as companion and food in Europe and North America. Similarly, dogs are worshiped in many countries of the world and rarely used for food purposes, but are widely consumed as food in countries such as Vietnam, Taiwan, and China (Waller, 2016).

In addition to requiring dedication and a time commitment on the part of the owner, the maintenance of a pet has a financial cost, which includes expenses of food, veterinary care, equipment, water, and electricity, among others. Such costs vary according to the animal. Data from Associação Brasileira da Indústria

FIGURE 16.9 *Paracanthurus hepatus*, a species of pet fish whose demand increased due to the movie *Finding Dory*. *Photo credit: Yi-Kai Tea.*

de Produtos para Animais de Estimação (ABINPET; Brazilian Association of the Pet Products Industry) estimate that the average monthly cost of keeping dogs in Brazil ranges from US$ 66 to US$ 124, depending on the size of the animal. The overall monthly spending on cats is US$ 36. The research also calculated the monthly costs for fish, rodents, birds and reptiles as US$ 20, US$ 16, US$ 2, and US$ 4, respectively (ABINPET, 2016). Thus, a respondent's income was certainly an important factor in influencing their choice of a pet.

The circumstances that lead people to adopt animals vary across the globe, as well as across regions within individual countries and between urban and rural areas. Even in urban areas, the place where a person resides, whether in an apartment or a house, influences the decision of having a pet and if so, what type of pet. In rural areas, and among hunting communities, it is common to adopt orphaned pups of wild animals as pets (e.g., wild felines raccoons, deer, primates etc.) (Fig. 16.10), after the parents are slaughtered during hunting (Alves et al., 2016; Barboza et al., 2016). This situation seems to have been perpetuated throughout history since it was probably the offspring of animals that were hunted and killed that humans used to start breeding and domesticating (Alves and Souto, 2010). Other examples illustrate the diversity of situations that motivate the choice of a pet. In some regions of Nigeria, for example, chimpanzee babies make more desirable pets than monkeys because of their more human-like appearance (Nyanganji et al., 2011). In Brazil, one of the reasons people maintain jabuti tortoises, *Chelonoidis carbonaria*, is the popular belief that their presence helps to prevent illnesses such as bronchitis and asthma (Alves et al., 2012a,b,c).

PROBLEMS ASSOCIATED WITH KEEPING PETS

Pet keeping represents a form of symbiosis (Wilson, 1975), or a prolonged and intimate relationship between animals of different species (Archer, 1997). Such relationships are not always beneficial to the groups directly involved and may even indirectly harm other animal groups. Potential problems associated with pet breeding

FIGURE 16.10 Specimen of wild cat *Leopardus* sp. kept as pet in Brazil. *Photo credit: Raynner R.D. Barboza.*

influence aspects of public health, economics, ecology, conservation, and animal welfare.

As discussed above, one of the motivations for having a pet are the potential health benefits to its owners. Paradoxically, there are a number of zoonotic diseases that can be mutually transmitted between pets and their owners. Pickering et al. (2008) explains that transmission may be direct or indirect through contact; aerosols; bites or scratches; from contamination of the environment, food, or water; or via disease-carrying vectors. These authors also point out that in many cases the owners of pets are not aware of the risks associated with changes in the behavioral and/or physical characteristics of their pets. Additionally, many owners do not have an adequate understanding of disease transmission, methods of preventing transmission, and appropriate maintenance facilities for their pets, especially for nontraditional ones.

From an ecological perspective, the capture of wild animals for use as pets has a direct impact, with it being one of the reasons for population declines of many species. This practice is also linked to other activities, such as trade in wild animals and biological invasions, which have obvious conservation implications. The domestic and international pet trade constitutes one of the most representative forms of commercialization of wild fauna (Alves et al., 2013a,b; Bush et al., 2014; Franke and Telecky, 2001; Roldán-Clarà et al., 2014). According to Baker et al. (2013), one-fifth of wildlife trade nowadays is driven by demand for pets or animals for use in entertainment. Such trade involves an extensive richness of taxa, especially vertebrates. For example, Bush et al. (2014) state that the exotic pet trade involved at least 585 bird, 485 reptile, and 113 mammal species. In addition, more than 290 amphibian species are targeted for the international pet trade and consumption purposes (Auliya et al., 2016).

The capture of animals from the wild for pet trade has been identified as the cause of population decline of many species. In addition to direct capture, many species end up dying during some stage of the marketing or trafficking chain. The illegal trade of wild birds for pets in Brazil illustrates this situation well (Alves et al., 2013b), with the majority of wild birds commercialized as pets dying before ever reaching their final destinations. To try to circumvent surveillance and enforcement, traffickers transport the animals in extremely limited and overcrowded spaces and do not supply them with water or food. Under these conditions, birds often fight, mutilating and killing each another in the process. It has been estimated that at least three animals are lost for each animal product traded; among live animals this figure is even higher—only 1 out of 10 animals captured survives (Redford, 1992). The main causes of mortality are (1) wounded animals dying after escaping and (2) damaged animals being discarded because they are considered "below standard" (Redford, 1992). Another example involves poison arrow frogs (family Dendrobatidae), which are one of the most sought-after groups of amphibians in the international pet trade (Lötters and Ulber, 2007). From 2000 to 2014, more than 180,000 Dendrobatidae were reported as legally traded by exporting countries on a global scale, but many steps within the trade chain are poorly documented (Auliya et al., 2016), such as mortality rates from the point of harvest to the exporter and transport mortalities from the exporter to importation destinations (see Ashley et al., 2014; Brown et al., 2011; Wombwell, 2014). This situation highlights another important issue associated with the use of pets: animal welfare. Animals may be subject to maltreatment not only during the capturing and marketing process but also during their lifetime in captivity where maintenance conditions are, in many cases, inappropriate.

Another serious ecological problem associated with the pet trade is the introduction of nonindigenous species. The lack of knowledge and irresponsibility of many pet keepers lead to the introduction of exotic animals (Romagosa, 2015),

FIGURE 16.11 The Indo-Pacific red lionfish (*Pterois volitans*) photographed in Belize. This is an invasive tropical fish that has spread throughout new marine environments through the aquarium trade. *Photo credit: Luiz A. Rocha.*

a well-known problem in animal conservation. This is more serious considering that in many places trade in live animals is largely unregulated (Smith et al., 1999), and the number of introductions of nonnative species by this route continues to increase, and is expected, along with their associated impacts, to be of continued concern (Goss and Cumming, 2013; Kraus, 2009). The number of biological invasions associated with the practice of keeping exotic pets is very high. For example, Krysko et al. (2011) reported that 84% of the 137 nonindigenous reptile and amphibian species in Florida, USA, were attributed to introduction via the pet trade (although, the vast majority of these were reptiles, with only three amphibian species having become established). At a global level, Kraus (2003) analyzed the published records of introduced exotic reptiles and amphibians and found that the pet trade represented one of the main routes of introduction (34% of introductions), involving 72 species, of which 36 established exotic populations in new environments, mainly lizards (37%), turtles (25%), and frogs (22%).

The introduction of nonnative species through the wildlife and pet trade may be the result of accidental escapes or intentional release (Goss and Cumming, 2013). Following introduction, a fraction of nonnative species may become established and spread more aggressively than others; these "invasive" species are known to cause serious ecological (Crowl et al., 2008; Mack et al., 2000) and economic (Pimentel et al., 2005) impacts. An illustrative example is the lionfish, *Pterois volitans* (Fig. 16.11). Native to the Indian and Pacific oceans; this species is considered a voracious predator and poses a threat to reef ecosystems outside its natural area of occurrence. The species was introduced in Florida, USA, from aquarium specimens. The first record was made in 1985, but records became frequent in the 1990s. In the Caribbean, where there are no natural predators, the species has rapidly proliferated and represents one of the greatest threats to local marine biodiversity. Without control, native fish species end up having their populations drastically reduced, causing serious ecological problems in these environments. The species was also recorded on the Brazilian coast, raising concern about the potential impacts that it may have on the marine ecosystems there (Ferreira et al., 2015; Hixon et al., 2016; Palmer et al., 2016).

Domestic pets can also have an impact on wildlife. In this case the main documented problems are related to "feral animals" (from Latin *fera*, meaning "a wild beast"), a term used to refer to an animal living in the wild, but descended from domesticated individuals that escaped from a domestic or captive situation and are living more or less as wild animals. Studies identify feral dogs and cats as representing a serious threat to wild species (Loss et al., 2013; Medina et al., 2011; Young et al., 2011). Cats, for example, have been listed among the 100 worst non-native invasive species in the world (Medina et al., 2011). These authors point out that major impacts of feral cats on native vertebrates have been reported from at least 120 different islands and on at least 175 vertebrates (25 reptiles, 123 birds, and 27 mammals). They also suggest that feral cats on islands are responsible for at least 14% of global bird, mammal, and reptile extinctions and are the main threat to almost 8% of critically endangered birds, mammals, and reptiles. Similarly, impacts of feral dogs on wildlife have raised serious conservation concerns in many parts of the world (Hughes and Macdonald, 2013; Sepúlveda et al., 2014; Young et al., 2011). In addition to directly preying on wild animals, both feral dogs and cats also compete with native predators and disseminate diseases among wildlife.

FINAL REMARKS

Keeping animals as pets represents one of the earliest ethnozoological interactions recorded in human history. The relationship has intensified with animal domestication and has perpetuated throughout the world until the present day. Various species, both domesticated and wild, are kept as pets by humans, with mammals and birds being among the most popular pets raised in the world. Nowadays, however, wild animal breeding has become increasingly popular. The choice of a species for a pet depends on a number of factors including the profile of the keeper, the beauty and behavior of the animal, its utility value (companion, hunting, pet therapy, etc.), maintenance costs, and even the media. Despite a variety of positive points associated with keeping pets, there are also many potential problems, including transmission of zoonoses, invasions of exotic species, illegal trafficking of wildlife, and animal welfare. Thus, keeping animals as pets represents a practice of extreme relevance to humans, which involves issues concerning public health, ethics, economy, ecology, conservation—and of course, ethnozoology.

References

ABINPET, Associação Brasileira da Indústria de Produtos para Animais de Estimação, 2016. Custo médio mensal de manutenção de animais de estimação.

Alves, R.R.N., 2012. Relationships between fauna and people and the role of ethnozoology in animal conservation. Ethnobiology and Conservation 1, 1–69.

Alves, R.R.N., 2016. Domestication of animals. In: Albuquerque, U.P. (Ed.), Introduction to Ethnobiology. Springer, Switzerland, pp. 221–225.

Alves, R.R.N., Souto, W.M.S., 2010. Etnozoologia: conceitos, considerações históricas e importância. In: Alves, R.R.N., Souto, W.M.S., Mourão, J.S. (Eds.), A Etnozoologia no Brasil: Importância, Status atual e Perspectivas, first ed. NUPEEA, Recife, PE, Brazil, pp. 19–40.

Alves, R.R.N., Mendonça, L.E.T., Confessor, M.V.A., Vieira, W.L.S., Lopez, L.C.S., 2009. Hunting strategies used in the semi-arid region of northeastern Brazil. Journal of Ethnobiology and Ethnomedicine 5, 1–50.

Alves, R.R.N., Nogueira, E., Araujo, H., Brooks, S., 2010. Bird-keeping in the Caatinga, NE Brazil. Human Ecology 38, 147–156.

Alves, R.R.N., Pereira Filho, G.A., Silva Vieira, K., Souto, W.M.S., Mendonças, L.E.T., Montenegro, P.F.G.P., Almeida, W.O., Vieira, W.L.S., 2012a. A zoological catalogue of hunted reptiles in the semiarid region of Brazil. Journal of Ethnobiology and Ethnomedicine 8 (27), 1–29.

Alves, R.R.N., Rosa, I.L., Léo Neto, N.A., Voeks, R., 2012b. Animals for the gods: magical and religious faunal use and trade in Brazil. Human Ecology 40, 751–780.

Alves, R.R.N., Vieira, K.S., Santana, G.G., Vieira, W.L.S., Almeida, W.O., Souto, W.M.S., Montenegro, P.F.G.P., Pezzuti, J.C.B., 2012c. A review on human attitudes towards reptiles in Brazil. Environmental Monitoring and Assessment 184, 6877–6901.

Alves, R.R.N., Leite, R.C., Souto, W.M.S., Bezerra, D.M.M., Loures-Ribeiro, A., 2013a. Ethno-ornithology and conservation of wild birds in the semi-arid Caatinga of northeastern Brazil. Journal of Ethnobiology and Ethnomedicine 9, 1–12.

Alves, R.R.N., Lima, J.R.F., Araújo, H.F., 2013b. The live bird trade in Brazil and its conservation implications: an overview. Bird Conservation International 23, 53–65.

Alves, R.R.N., Feijó, A., Barboza, R.R.D., Souto, W.M.S., Fernandes-Ferreira, H., Cordeiro-Estrela, P., Langguth, A., 2016. Game mammals of the Caatinga biome. Ethnobiology and Conservation 5, 1–51.

Anderson, P.K., 2010. Human–bird Interactions. In: Duncan, I.J.H., Hawkins, P. (Eds.), The Welfare of Domestic Fowl and Other Captive Birds. Springer, Dordrecht, pp. 17–51.

Animal Protection Institute, 2005. A Life Sentence, the Sad and Dangerous Realities of Exotic Animals in Private Hands in the U.S. Animal Protection Institute, Sacramento, CA.

Archer, J., 1997. Why do people love their pets? Evolution and Human Behavior 18, 237–259.

Ashley, S., Brown, S., Ledford, J., Martin, J., Nash, A.E., Terry, A., Tristan, T., Warwick, C., 2014. Morbidity and mortality of invertebrates, amphibians, reptiles, and mammals at a major exotic companion animal wholesaler. Journal of Applied Animal Welfare Science 17, 308–321.

Auliya, M., 2003. Hot trade in cool creatures: a review of the live reptile trade in the European Union in the 1990s with a focus on Germany. TRAFFIC Europe, Brussels, Belgium.

Auliya, M., García-Moreno, J., Schmidt, B.R., Schmeller, D.S., Hoogmoed, M.S., Fisher, M.C., Pasmans, F., Henle, K., Bickford, D., Martel, A., 2016. The global amphibian trade flows through Europe: the need for enforcing and improving legislation. Biodiversity and Conservation 25 (13), 2581–2595.

Auliyai, M., Altherr, S., Ariano-Sanchez, D., Baard, E.H., Brown, C., Brown, R.M., Cantu, J.C., Gentile, G., Gildenhuys, P., Henningheim, E., Hintzmann, J., Kanari, K., Krvavac, M., Lettink, M., Lippert, J., Luiselli, L., Nilson, G., Nguyen, T.Q., Nijman, V., Parham, J.F., Pasachnik, S.A., Pedrono, M., Rauhaus, A., Córdova, D.R., Sanchez, M.E., Schepp, U., van Schingen, M., Schneeweiss, N., Segniagbeto, G.H., Somaweera, R., Sy, E.Y., Türkozan, O., Vinke, S., Vinke, T., Vyas, R., Williamson, S., Ziegler, T., 2016. Trade in live reptiles, its impact on wild populations, and the role of the European market. Biological Conservation 204, 103–119.

Baker, S.E., Cain, R., Van Kesteren, F., Zommers, Z.A., D'cruze, N., Macdonald, D.W., 2013. Rough trade: animal welfare in the global wildlife trade. BioScience 63, 928–938.

Barboza, R.R.D., Lopes, S.F., Souto, W.M.S., Fernandes-Ferreira, H., Alves, R.R.N., 2016. The role of game mammals as bushmeat in the Caatinga, northeast Brazil. Ecology and Society 21, 1–11.

Batson, K., 1998. The effect of a therapy dog on socialization and physiological indicators of stress in persons diagnosed. In: Wilson, C.C., Turner, D.C. (Eds.), Companion Animals in Human Health. Sage, Thousand Oaks, p. 203.

Beck, A.M., Katcher, A.H., 1984. A new look at pet-facilitated therapy. Journal of the American Veterinary Medical Association 184, 414–421.

Beck, A.M., Katcher, A.H., 1996. Between Pets and People: The Importance of Animal Companionship. Purdue Univ Press.

Beetz, A., Kotrschal, K., Turner, D.C., Hediger, K., Uvnäs-Moberg, K., Julius, H., 2011. The effect of a real dog, toy dog and friendly person on insecurely attached children during a stressful task: an exploratory study. Anthrozoos 24, 349–368.

Bernard, P., Demaret, A., 1996. Why Have Pets? Present and Permanent Reasons.

Born Free USA, Summer 2005. Monkey Business. WLT, p. 24.

Brown, J.L., Twomey, E., Amezquita, A., Souza, M.B.D., Caldwell, J.P., Loetters, S., May, R.V.O.N., Melo-Sampaio, P.R., Mejia-Vargas, D., Perez-Pena, P., 2011. A taxonomic revision of the neotropical poison frog genus Ranitomeya (Amphibia: Dendrobatidae). Zootaxa 3083, 1–120.

Bueno, F.G.C., 2009. Homem que não ladra, cão que não morde: a Comunicação Interespécies marcando a Cultura pós-moderna. Universidade de Sorocaba, Sorocaba, SP.

Bush, E.R., Baker, S.E., Macdonald, D.W., 2014. Global trade in exotic pets 2006–2012. Conservation Biology 28, 663–676.

Butchart, S.H.M., 2008. Red List Indices to measure the sustainability of species use and impacts of invasive alien species. Bird Conservation International 18, S245–S262.

Calado, R., Lin, J., Rhyne, A.L., Araújo, R., Narciso, L., 2003. Marine ornamental decapods—popular, pricey, and poorly studied. Journal of Crustacean Biology 23, 963–973.

Carrete, M., Tella, J.L., 2008. Wild-bird trade and exotic invasions: a new link of conservation concern? Frontiers in Ecology and the Environment 6, 207–211.

Collar, N.J., Long, A.J., Jaime, P.R.G., 2007. Birds and People: Bonds in a Timeless Journey. BirdLife International.

Crowl, T.A., Crist, T.O., Parmenter, R.R., Belovsky, G., Lugo, A.E., 2008. The spread of invasive species and infectious disease as drivers of ecosystem change. Frontiers in Ecology and the Environment 6, 238–246.

Digard, J.P., 1990. L'homme et les animaux domestiques: anthropologie d'une passion. Fayard, Paris.

Duarte-Quiroga, A., Estrada, A., 2003. Primates as pets in Mexico city: an assessment of the species involved, source of origin, and general aspects of treatment. American Journal of Primatology 61, 53–60.

Eshuis, J., Enders-Slegers, M.J., Verheggen, T., 2016. Anthrozoology in The Netherlands: connecting science and practice: situating human-animal engagement within cultures. In: Pręgowski, M.P. (Ed.), Companion Animals in Everyday Life. Springer, New York, pp. 27–41.

Fernandes-Ferreira, H., Mendonça, S.V., Albano, C., Ferreira, F.S., Alves, R.R.N., 2012. Hunting, use and conservation of birds in Northeast Brazil. Biodiversity and Conservation 21 (1), 221–244.

Ferreira, C.E.L., Luiz, O.J., Floeter, S.R., Lucena, M.B., Barbosa, M.C., Rocha, C.R., Rocha, L.A., 2015. First record of invasive lionfish (*Pterois volitans*) for the Brazilian coast. PLoS One 10, e0123002.

Fitter, R.S.R., 1986. Wildlife for Man: How and Why We Should Conserve Our Species. Collins, London.

Fitzgerald, S., 1989. International Wildlife Trade: Whose Business Is it? World Wildlife Fund.

Franke, J., Telecky, T.M., 2001. Reptiles as Pets: An Examination of the Trade in Live Reptiles in the United States. Humane Society of the United States, Washington, DC.

Franklin, A., 1999. Animals & Modern Cultures – a Sociology of Human-Animal Relations in Modernity. Sage, London.

Frisch, J.D., Frisch, S., 1981. Aves Brasileiras. In: Volume 1. Editora Dalgas-Ecoltec Ecologia Técnica e Comércio.

Fuentes, A., 2006. Human-nonhuman primate interconnections and their relevance to anthropology. Ecological and Environmental Anthropology 2, 1–11.

Ghirlanda, S., Acerbi, A., Herzog, H., 2014. Dog movie stars and dog breed popularity: a case study in media influence on choice. PLoS One 9, e106565.

Gorzula, S., 1996. The trade in dendrobatid frogs from 1987 to 1993. Herpetological Review 27, 116–122.

Goss, J.R., Cumming, G.S., 2013. Networks of wildlife translocations in developing countries: an emerging conservation issue? Frontiers in Ecology and the Environment 11, 243–250.

Grier, K.C., 2010. Pets in America: a history. UNC Press Books.

Hardie, L.C., Gaski, A.L., 1989. Wildlife Trade Education Kit. 132.

Hemley, G., 1994. International Wildlife Trade: A CITES Sourcebook. WWF/Island Press, Washington.

Herzog, H.A., 2014. Biology, culture, and the origins of pet-keeping. Animal Behavior and Cognition 1, 296–308.

Hirschman, E.C., 1994. Consumers and their animal companions. Journal of Consumer Research 20, 616–632.

Hixon, M.A., Green, S.J., Albins, M.A., Akins, J.L., Morris Jr., J.A., 2016. Lionfish: a major marine invasion. Marine Ecology Progress Series 558, 161–165.

Honess, P.E., Wolfensohn, S.E., 2010. Welfare of exotic animals in captivity. In: Tynes, V.V. (Ed.), Behavior of Exotic Pets. Willey-Blackwell, pp. 215–223.

Hoover, C., 1998. The US role in the international live reptile trade: amazon tree boas to Zululand dwarf chameleons. TRAFFIC, North America.

Hughes, J., Macdonald, D.W., 2013. A review of the interactions between free-roaming domestic dogs and wildlife. Biological Conservation 157, 341–351.

Japanese Pet Food Association, 2016. In: National Survey of Keeping Conditions of Cats and Dogs (In Japanese).

Jorgenson, J., 1997. Therapeutic use of companion animals in health care. Journal of Nursing Scholarship 29, 249–254.

Koster, J., 2008. The impact of hunting with dogs on wildlife harvests in the Bosawas Reserve, Nicaragua. Environmental Conservation 35, 211–220.

Koster, J.M., Tankersley, K.B., 2012. Heterogeneity of hunting ability and nutritional status among domestic dogs in lowland Nicaragua. Proceedings of the National Academy of Sciences 109, E463–E470.

Kraus, F., 2003. Invasion pathways for terrestrial vertebrates. In: Carlton, J., Ruiz, G., Mack, R. (Eds.), Invasive Species: Vectors and Management Strategies, Island Press, Washington, DC, pp. 68–92.

Kraus, F., 2009. Alien Reptiles and Amphibians: A Scientific Compendium and Analysis. Springer-Verlag, United Kingdom.

Krysko, K.L., Burgess, J.P., Rochford, M.R., Gillette, C.R., Cueva, D., Enge, K.M., Somma, L.A., Stabile, J.L., Smith, D.C., Wasilewski, J.A., 2011. Verified non-indigenous amphibians and reptiles in Florida from 1863 through 2010: outlining the invasion process and identifying invasion pathways and stages. Zootaxa 1–64.

Kumschick, S., Devenish, A., Kenis, M., Rabitsch, W., Richardson, D.M., Wilson, J.R.U., 2016. Intentionally introduced terrestrial invertebrates: patterns, risks, and options for management. Biological Invasions 18, 1077–1088.

Leeming, D.A., 2003. From Olympus to Camelot: The World of European Mythology. Oxford University Press, USA.

Licarião, M.R., Bezerra, D.M.M., Alves, R.R.N., 2013. Wild birds as pets in Campina Grande, Paraíba state, Brazil: an ethnozoological approach. Anais da Academia Brasileira de Ciências 85, 201–213.

Losey, R.J., Garvie-Lok, S., Leonard, J.A., Katzenberg, M.A., Germonpré, M., Nomokonova, T., Sablin, M.V., Goriunova, O.I., Berdnikova, N.E., Savel'ev, N.A., 2013. Burying dogs in ancient Cis-Baikal, Siberia: temporal trends and relationships with human diet and subsistence practices. PLoS One 8, e63740.

Loss, S.R., Will, T., Marra, P.P., 2013. The impact of free-ranging domestic cats on wildlife of the United States. Nature Communications 4, 1396.

Lötters, S., Ulber, T., 2007. In: Chimaira (Ed.), Poison Frogs: Biology, Species & Captive Care.

Mack, R.N., Simberloff, D., Mark Lonsdale, W., Evans, H., Clout, M., Bazzaz, F.A., 2000. Biotic invasions: causes, epidemiology, global consequences, and control. Ecological Applications 10, 689–710.

Magalhães, A.L.B., São-Pedro, V.A., 2012. Illegal trade on non-native amphibians and reptiles in southeast Brazil: the status of e-commerce. Phyllomedusa: Journal of Herpetology 11, 155–160.

Matsubayashi, K., Gotoh, S., Suzuki, J., 1986. Changes in import of non-human primates after ratification of CITES (Washington Convention) in Japan. Primates 27, 125–135.

Medina, F.M., Bonnaud, E., Vidal, E., Tershy, B.R., Zavaleta, E.S., Josh Donlan, C., Keitt, B.S., Corre, M., Horwath, S.V., Nogales, M., 2011. A global review of the impacts of invasive cats on island endangered vertebrates. Global Change Biology 17, 3503–3510.

Messent, P.R., Serpell, J.A., 1981. An historical and biological view of the pet-owner bond. In: Fogle, B. (Ed.), Interrelations Between People and Pets. Charles C. Thomas, Springfield, IL, pp. 5–22.

Mitchell, M., Tully Jr., T.N., 2009. Manual of Exotic Pet Practice. Elsevier Health Sciences, St. Louis, Missouri.

National Institutes of Health, 1987. The Health Benefits of Pets (1988-216-107). US Government Printing Office, Washington, DC.

Natusch, D.J.D., Lyons, J.A., 2012. Exploited for pets: the harvest and trade of amphibians and reptiles from Indonesian New Guinea. Biodiversity and Conservation 21, 2899–2911.

Ng, T.H., Tan, S.K., Wong, W.H., Meier, R., Chan, S.Y., Tan, H.H., Yeo, D.C.J., 2016. Molluscs for sale: assessment of freshwater gastropods and bivalves in the ornamental pet trade. PLoS One 11, e0161130.

Nightingale, F., 1969. Notes on Nursing. Dover Publications, New York.

Nijman, V., Nekaris, K.A.I., 2017. The Harry Potter effect: the rise in trade of owls as pets in Java and Bali, Indonesia. Global Ecology and Conservation 11, 84–94.

Nyanganji, G., Fowler, A., McNamara, A., Sommer, V., 2011. Monkeys and apes as animals and humans: ethnoprimatology in Nigeria's Taraba region. In: Sommer, V., Ross, C. (Eds.), Primates of Gashaka: Socioecology and Conservation in Nigeria's Biodiversity Hotspot. Springer, New York, USA, pp. 101–134.

Palmer, G., Hogan, J.D., Sterba-Boatwright, B.D., Overath, R.D., 2016. Invasive lionfish *Pterois volitans* reduce the density but not the genetic diversity of a native reef fish. Marine Ecology Progress Series 558, 223–234.

Pet Food Institute, 2013. Cat and Dog Population.

Pet Food Manufacturer's Association, 2016. Pet Population 2016.

Piacentini, V.Q., Aleixo, A., Agne, C.E., Mauricio, G.N., Pacheco, J.F., Bravo, G.A., Brito, G.R.R., Naka, L.N., Olmos, F., Posso, S., Silveira, L.F., Betini, G.N., Carrano, E., Franz, I., Lees, A.C., Lima, L.M., Pioli, D., Schunck, F., Amaral, F.R., Bencke, G.A., Cohn-Haft, M., Figueiredo, L.F.A., Straube, F.C., Cesari, E., 2015. Annotated checklist of the birds of Brazil by the Brazilian Ornithological Records Committee/Lista comentada das aves do Brasil pelo Comitê Brasileiro de Registros Ornitológicos. Revista Brasileira de Ornitologia-Brazilian Journal of Ornithology 23, 90–298.

Pickering, L.K., Marano, N., Bocchini, J.A., Angulo, F.J., 2008. Exposure to nontraditional pets at home and to animals in public settings: risks to children. Pediatrics 122, 876–886.

Pierce, J., 2016. Run, Spot, Run: The Ethics of Keeping Pets. University of Chicago Press.

Pimentel, D., Zuniga, R., Morrison, D., 2005. Update on the environmental and economic costs associated with alien-invasive species in the United States. Ecological Economics 52, 273–288.

Prestridge, H.L., Fitzgerald, L.A., Hibbitts, T.J., 2011. Trade in non-native amphibians and reptiles in Texas: lessons for better monitoring and implications for species introduction. Herpetological Conservation Biology 6, 324–339.

Raina, P., Waltner-Toews, D., Bonnett, B., Woodward, C., Abernathy, T., 1999. Influence of companion animals on the physical and psychological health of older people: an analysis of a one-year longitudinal study. Journal of the American Geriatrics Society 47, 323–329.

Redford, K.H., 1992. The empty forest. BioScience 42, 412–422.

Renctas, 2001. 1º relatório nacional sobre o tráfico de fauna silvestre. (Brasília).

Rhyne, A.L., Tlusty, M.F., Schofield, P.J., Kaufman, L.E.S., Morris Jr., J.A., Bruckner, A.W., 2012. Revealing the appetite of the marine aquarium fish trade: the volume and biodiversity of fish imported into the United States. PLoS One 7, e35808.

Roldán-Clarà, B., Lopez-Medellín, X., Espejel, I., Arellano, E., 2014. Literature review of the use of birds as pets in Latin-America, with a detailed perspective on Mexico. Ethnobiology and Conservation 3, 1–18.

Romagosa, C.M., 2015. Contribution of the live animal trade to biological invasions. In: Canning-Clode, J. (Ed.), Biological Invasions in Changing Ecosystems. DE GRUYTER OPEN, pp. 116–134.

Rosa, I.M.L., Alves, R.R.N., Bonifácio, K.M., Mourão, J.S., Osório, F.M., Oliveira, T.P.R., Nottingham, M.C., 2005. Fishers' knowledge and seahorse conservation in Brazil. Journal of Ethnobiology and Ethnomedicine 1, 1–15.

Rosa, I.L., Oliveira, T.P.R., Osório, F.M., Moraes, L.E., Castro, A.L.C., Barros, G.M.L., Alves, R.R.N., 2011. Fisheries and trade of seahorses in Brazil: historical perspective, current trends, and future directions. Biodiversity and Conservation 20, 1951–1971.

Savishinsky, J.S., 1986. Pet ideas: the domestication of animals, human behavior, and human emotions. In: Matcher, A., Beck, A.M. (Eds.), New Perspectives in Our Lives with Companion Animals. University of Pennsylvania Press, Philadelphia, pp. 112–131.

Seahorse, P., 2016. Why Seahorses? Essential Facts about Seahorses.

Sepúlveda, M.A., Singer, R.S., Silva-Rodríguez, E., Stowhas, P., Pelican, K., 2014. Domestic dogs in rural communities around protected areas: conservation problem or conflict solution? PLoS One 9, e86152.

Serpell, J., 1986. In the Company of Animals: A Study of Human-Animal Relationships.

Serpell, J., 1990. Pet-keeping and animal domestication: a reappraisal. In: CluttonBrock, J. (Ed.), The Walking Larder: Patterns of Domestication, Pastoralism, and Predation. Taylor & Francis, London, pp. 10–21.

Serpell, J.A., 2011. Historical and cultural perspectives on human-pet interactions. In: McCardle, P., McCune, S., Griffin, J., Esposito, L., Freund, L. (Eds.), Animals in Our Lives: Human-Animal Interaction in Family, Community, and Therapeutic Settings. Paul H. Brookes Publishing Co, Baltimore, MD, pp. 11–22.

Serpell, J., McCune, S., 2012. Livro de bolso do WALTHAM® sobre interações entre humanos e animais. Beyond Design Solutions Ltd.

Sick, H., 1997. Ornitologia Brasileira. Nova Fronteira, Rio de Janeiro.

Smith, C.S., Lonsdale, W.M., Fortune, J., 1999. When to ignore advice: invasion predictions and decision theory. Biological Invasions 1, 89–96.

Soulsbury, C.D., Iossa, G., Kennell, S., Harris, S., 2009. The welfare and suitability of primates kept as pets. Journal of Applied Animal Welfare Science 12, 1–20.

Tidemann, S., Gosler, A., 2010. Ethno-ornithology: Birds, Indigenous People, Culture and Society, first ed. Earthscan/James & James.

Vincent, A.C.J., 1996. The International Trade in Seahorses. Traffic International, Cambridge.

Vincent, A.C.J., Foster, S.J., Koldewey, H.J., 2011. Conservation and management of seahorses and other Syngnathidae. Journal of Fish Biology 78, 1681–1724.

Wabnitz, C., 2003. From Ocean to Aquarium: The Global Trade in Marine Ornamental Species. UNEP/Earthprint.

Waller, S., 2016. Companion animals and nuisance species: adventures in the exotic, the wild, the illegal, and cross-cultural comfort zones. In: Pregowski, M.P. (Ed.), Companion Animals in Everyday Life. Springer, pp. 13–25.

Wilson, E.O., 1975. Sociobiology: The New Synthesis. Harvard University Press, Cambridge, MA.

Wilson, C.C., Turner, D.C., 1997. Companion Animals in Human Health. Sage Publications.

Wombwell, E.L., 2014. Emerging Infectious Disease and the Trade in Amphibians. University of Kent.

Working Party Council for Science, Society, 1988. Companion Animals in Society. Oxford University Press, Oxford.

Wright, J., 2005. The Primate Trade in Indonesia: A Rural Perspective. University of Manchester, Manchester, UK.

Wu, J., 2007. World without borders: wildlife trade on the Chinese-language internet. Traffic Bulletin 21, 75–84.

Yan, G., 2016. Saving Nemo–Reducing mortality rates of wild-caught ornamental fish. SPC Live Reef Fish Information Bulletin 21, 3–7.

Yong, D.L., Fam, S.D., Lum, S., 2011. Reel conservation: can big screen animations save tropical biodiversity. Tropical Conservation Science 4, 244–253.

Young, J.K., Olson, K.A., Reading, R.P., Amgalanbaatar, S., Berger, J., 2011. Is wildlife going to the dogs? Impacts of feral and free-roaming dogs on wildlife populations. BioScience 61, 125–132.

CHAPTER

17

What About the Unusual Soldiers? Animals Used in War

Rômulo Romeu Nóbrega Alves, Raynner Rilke Duarte Barboza
Universidade Estadual da Paraíba, Campina Grande, Brazil

INTRODUCTION

Human use of other animals has a long history. Since ancient times, live animals have performed several activities in a wide range of sectors including agriculture, transportation, entertainment, military and police activities (Alves, 2012, 2016; Alves and Souto, 2015). Therefore, as pointed out by Cunningham (1995), "Animals have participated in the development of practically every aspect of human civilization—in war and peace, in work and at play, in love (companion animals) and hate (varmints), in life (biological experiments) and in death (pet cemeteries)."

Human conflicts have been documented for a long time, ranging from clashes between small tribes to the great wars, which mark the history of humankind. Among the main causes of conflicts in the past, and continuing into the present, are disputes over territories and boundaries, as well as the natural resources they contain. In battling enemies, a wide variety of strategies have been used to attack and defend, and which have varied according to location and historical period. Even elements derived from biodiversity have made up the armory used by humans in such conflicts. Products derived from plants and animals, such as branches, bones, fur, skins, and leather, have been widely used, for example, in the manufacture of weapons and armor. Bows (and arrows), for instance, are considered one of the most ancient and widely used weapons throughout human history. Their manufacture employed materials available in nature, such as raw hide, animal tendons, flax, hemp, silk, or sinew. Arrows were made of straight or straightened, strong, but light, wood. These weapons have appeared worldwide in virtually every military force from antiquity to at least the end of the 16th century (Tucker, 2015). Animal by-products were also components of warfare strategies, such as wearing animal skins as camouflage, for example (Tucker, 2015).

As pointed out by Kistler (2011), "The history of our world has been shaped by wars, and our wars have been shaped by animals." Domesticated or wild animals were, and remain, often used key elements for human military purposes (Lawrence, 1991), playing major roles as beasts of burden, messengers, weapons, protectors, mascots, and friends. The importance

Ethnozoology
http://dx.doi.org/10.1016/B978-0-12-809913-1.00017-X

of animals to military activities has been significant. Estimates indicate that over 16 million animals served during the World War I, and by 1916 alone the warring nations had raised 103 cavalry divisions with over a million horses (Gardiner, 2006; van Vliet, 2007). The diversity of animals, and their roles in battle, have been extensive, ranging from being simple military mascots to "field warriors."

Nowadays, the use of animals in wars and on battlefields has been decreasing with the growth of mechanization and new technologies. Nevertheless, there are several species that continue to be used in military and police activities in various parts of the world, especially dogs and horses, which are used in military transportation, patrol, persecution, and battling narcotics, among other uses. In this chapter, we will briefly discuss the role of animals in military activities, focusing on the major animal groups involved and their respective roles in the warfare throughout the history of mankind.

BRIEF HISTORY

The territory and the natural resources shared by animals are among the main reasons for them to fight each other, including humans. Conflict among human groups has a long, continuous history marked by local, national, or international fighting and war (Kanyamibwa, 1998). It is not surprising, therefore, that military activities are documented in the earliest written records. Keeley (1997) illustrated this well by highlighting some of these records, as follows: "The earliest Egyptian hieroglyphs record the victories of Egypt's first pharaohs, the Scorpion King and 'Narmer.' The first secular literature or history recorded in cuneiform recounts the adventures of the Sumerian warrior–king Gilgamesh. The earliest written parts of the Books of Moses, the "Jstrand" (called so because in its passages the name given God is Yahweh or, corruptly, Jehovah), culminate in the brutal Hebrew conquest of Canaan. The earliest annals of the Chinese, Greeks, and Romans are concerned with wars and warrior kings. Most Mayan hieroglyphic texts are devoted to the genealogies, biographies, and military exploits of Mayan kings. The folklore and legends of preliterate cultures, the epic oral traditions that are the precursors to history, are equally bellicose."

Likewise, records of the use of animals in warfare and other combat-related activities are ancient. Fauna had played several roles in the armies of the ancient world, functioning as baggage and transport animals, as well as carrying soldiers into combat (Cooper, 2000; Keeley, 1997; Willekes, 2016). There is an abundance of remote documents attesting to the military use of animals, such as dogs, horses, elephants, and camels (both Dromedary and Bactrian) (Sloane, 1955; Willekes, 2016).

The use of elephants (*Elephas maximus*) in war, for example, has been documented since the time of the Indus Valley Civilizations, c.2000 BC (Clutton-Brock, 1984). War elephants were used on battle fronts to pursue the enemy, destroy their alignment strategies, and incite panic (Whitney and Smith, 1911). They were used in many battles and in turn came to influence the campaigns of Alexander the Great and became especially important in the armies of the Hellenistic east (Willekes, 2016). One of most famous examples of the use of elephants in war involves the general Hannibal Barca, who commanded an armed force that included well-trained war elephants for striking toward the Pyrenees in northern Italy (Figs. 17.1 and 17.2) (Cooper, 2000; Willekes, 2016). In 280 BC, Pyrrhus, king of Epirus, used elephants against the Romans when his army invaded Italy, more than 50 years before Hannibal's more famous exploit (Cooper, 2000). The bellicose relevance of elephants inspired the creation of the term "Elephantry" to refer to military squads with elephant-mounted soldiers prepared to fight.

The horse was the most widely used animal throughout the recorded history of warfare. The

FIGURE 17.1 War elephants depicted with Hannibal Barca crossing the Rhône. Made in 1878. *From Henri Motte (1846–1922) (scan book) [Public domain], via Wikimedia Commons.*

FIGURE 17.2 Engraving of the Battle of Zama by Cornelis Cort, 1567. *From Cornelis Cort [Public domain], via Wikimedia Commons.*

most primitive indication of the use of horses in warfare probably dates from 4000 and 3000 BC in Eurasia. An ancient Sumerian engraving (from 2500 BC) portrays some type of equine pulling wagons in a war battle. In the year 1600 BC, elaborate harnesses and illustrations of sturdy chariots made chariot warfare well-known and famous in all ancient Near East; as a matter of fact, the earliest document written for training horses for war was a guidebook for chariot horses formulated around 1350 BC (Eglan, 2015). Another mammal with ancient use in human conflicts was the camel, which was used by armies of the Arabian Peninsula and the Near East as baggage animals and in combat. For example, Arabian camel cavalry mounted on dromedaries are depicted in a series of reliefs from the Assyrian palace at Nineveh c.645 (Willekes, 2016). The use of dogs in theaters of war extends back many thousands of years to the very beginning of recorded history, for wall-drawings and bas-reliefs found among the tombs of Egypt, Greece, and Assyria clearly show that the Egyptians, Assyrians, and Grecians made use of dogs in repelling the enemy (Sloane, 1955).

In addition to the aforementioned animals, several others have a long history of military use, including other mammals, birds, snakes, and insects, which have been used for various purposes depending on their physical or behavioral particularities. As pointed out by Kistler (2011), "Before the 20th century, no army could forgo the use of vast numbers of animals in its ranks."

FAUNA USED IN WAR AND THEIR FUNCTIONS

From the past to the present, a wide diversity of animals has been used in human warfare. The species chosen depended on the function it was to perform. For example, animals that are used directly on the battlefield tended to be large, so terrestrial vertebrates are predominant among the animals used for direct combat. On the other hand, birds, by their ability to fly, were efficient as messengers. In general, vertebrates, and mainly mammals, stand out among animals used in wars, with dogs, horses, mules, elephants, oxen, and camels being the main participants. Birds and reptiles also have representatives among war animals, as well as several invertebrates, such as worms, fleas, and bees.

Animals have served as vital resources before, during, and after wars. In preparation for battle,

FIGURE 17.3 Statue of Alexander the Great and his horse Bucephalus, located in front of Edinburgh's City Chambers (by John Steell). *From Stefan Schäfer, Lich.* http://creativecommons.org/licenses/by-sa/3.0.

for example, equids such as mules, donkeys, asses, and horses played key roles in transporting armaments and soldiers and their supplies (Gorzoni, 2010; Lawrence, 1991; Willekes, 2016). In addition to transportation, horses were the most important animals for combat, both in chariot forces and cavalry (Willekes, 2016), but also assumed other functions such as hauling heavy guns, moving supplies, and carrying the wounded (Coatesworth, 2016). The military versatility of horses placed them among the most emblematic of the "animals of war," and made them the very symbol of conquest (Lawrence, 1991). Historical military records evidenced the importance of horses in battles, expressing the respect and admiration of the military for these animals throughout history. One of the best known examples is the case of Bucephalus, the horse of Alexander the Great (Fig. 17.3), which was memorialized after death by his grief-stricken owner, who ordered a city built over the horse's grave (Lawrence, 1991). Napoleon I of France also loved his famous war mount Marengo (Fig. 17.4). Thus, not only horsemen and great generals gained fame, but their horses as well, thereby making them figures of history. Not surprisingly, a great number of equestrian statues depicting horses mounted by soldiers or generals are common in ancient museums and castles (Fig. 17.5). As a result of the regular involvement of horses in wars throughout history, thousands of these animals were killed until World War II. During the World War I, for example, there were appalling losses among pack animals and artillery traction (Gorzoni, 2010), and an estimated 8 million horses and countless mules and donkeys died (Coatesworth, 2016).

FIGURE 17.4 Napoleon crossing the Alps mounted on his horse (possibly Marengo). Painted by Jacques-Louis David. *Reproduced from Jacques-Louis David [Public domain], via Wikimedia Commons.*

Llamas also participated in military conflicts, generally assuming the same functions of equids. During Incan warfare, for example, these animals were used as personal mounts for warriors and as transportation of necessary supplies and battlefield apparatuses (Gorzoni, 2010).

Dogs have also had a prominent military role. Kistler (2011) pointed out that dogs helped humans fight in conflicts long before

FIGURE 17.5 The Cavalry Man – Royal Scots Dragoon Guards Museum – Edinburgh Castle. *Photo credits: Rômulo R.N. Alves.*

organized warfare began. Historically, certain dogs were bred for fighting in battle, and they functioned as lethal weapons of war (Varner and Varner, 1983). War dogs were used by many civilizations worldwide, for instance, Egyptians, Persians, Baganda, Greeks, Britons, Alans, Slavs, and the Romans (Forster, 1941; Greenhalgh, 1973). According to accounts of the Greeks and Romans, dogs were much appreciated as sentinels or guards, though they were also often taken into battle (Forster, 1941). The earliest use of war dogs in combat, documented in classical sources, was by Alyattes of Lydia against the Cimmerians around 600 BC. The Lydian dogs slaughtered many intruders and pursued others (Forster, 1941).

All major nations involved in war used dogs in numerous roles that are as varied as the different breeds. The most famous dog breed used in battle was the Molossus dog from the Molossia region of Epirus, recognized by the Romans as the strongest and most resilient dog known to man (De Prisco and Johnson, 1990). During late antiquity, Attila the Hun often used large mastiff or molosser-type breeds, strapped with armor or spiked collars, in his campaigns (Greenhalgh, 1973). Other cultures used armor-plated dogs to protect caravans or strike opponents. The Spanish conquistadors used to train aggressive dogs to kill prisoners (Varner and Varner, 1983). Later on, Frederick the Great used hounds as weapons guards and messengers during the Seven Years' War with Russia. In France, Napoleon used dogs during his crusades, and until 1770 dogs were guardians of naval installations. In the United States, the first official use of dogs for military purposes was during the Seminole Wars (Greenhalgh, 1973), but they were also used during the American Civil War, to send messages, defend, and guard prisoners (Eglan, 2015). In the beginning of the 1930s, the Soviet Union used large dogs for antitank purposes. Early on this strategy involved fitting hounds with tilt-rod mines and having them run beneath enemy tanks, at which time the mines would be automatically detonated (Eglan, 2015).

War dogs were much appreciated for their endurance and capabilities of facilitating logistics, communication, and transportation of loads over different terrains and landscapes and in varying climatic conditions. The Belgian army used trained dogs to pull their weapons and supplies on wheeled carriages or, reportedly, even the wounded in carts (Willmott, 2003). The Soviet Red Army also used hounds to drag wounded men to aid stations during World War II (Willmott, 2003). A famous Yorkshire terrier (4 lb) named Smoky was used to carry a telegraph wire through a 4–8-inch diameter, 70-ft long pipe to ensure point-to-point communication between army units (Putney, 2003).

Dogs have also been traditionally used to locate mines, and in many cases to track deserters, fugitives, and enemy troops, coinciding somewhat with the duties of a scout dog (Eglan, 2015). As sentries, dogs were used to protect base camps or other main areas at night and sometimes during the day; this is considered one of the earliest military-related uses of dogs. During the Cold War, the American military used sentry dog teams outside of nuclear weapon storage areas (Rogak, 2011).

In the World War II, trained Russian suicide dogs carrying primed bombs strapped on their backs hurled themselves under enemy tanks, thereby blowing them up. Similarly, the Japanese employed "commando" dogs who pulled carts containing time bombs (Cooper, 2000; Dempewolff, 1943). Dogs also served as messengers, sentries, guards, trackers, pack animals carrying food and ammunition, and as mine detectors (Fig. 17.6). They have also been used as pest destroyers, such as killing rats in trenches. Rescue dogs have saved many lives, especially when victims were buried under rubble. Kistler (2011) states that rescue dogs learned to hunt for wounded men in trenches, in fox-holes, behind trees, in mud, and in barbed wire. Wounded men would often crawl deep into holes to protect themselves from further harm, yet this hid them from medics. The dogs scented on fresh blood and sometimes smelled for breath to find unconscious men who could not move or cry out. Red Cross dogs also carried water bottles and small medical supplies so that lightly wounded men could patch themselves up well enough to return to the rear for help (Rogers, 2005; Sanderson, 1997).

FIGURE 17.6 US Marine "Raiders" and their dogs, which are used for scouting and running messages, starting off for the jungle frontlines in Bougainville, late 1943. Reproduced from https://commons.wikimedia.org/wiki/File%3AMarine-war-dogs.jpg.

Dogs in modern military roles are often referred to as Military Working Dogs (MWD)—United Kingdom or police dogs or K-9s—United States. Their roles are nearly as varied as they were for ancient dogs, including law enforcement, drug and explosives detection, search and rescue, and as sentries, trackers, scouts, and mascots (Fig. 17.7) (Eglan, 2015). However, they are still used in frontline formations. For instance, in 2011, 600 US military dogs took part actively during the Iraq and Afghanistan conflicts (Harris, 2011).

As mentioned previously, elephants have taken part in war since early times (Kistler, 2011;

FIGURE 17.7 A Belgian shepherd Malinois atop an M2A3 Bradley fighting vehicle in Iraq in 2007. *From US Air Force photo by Staff Sgt. Stacy L. Pearsall, Wikimedia commons, public domain.*

Moore, 1986). Kistler (2011) highlights that of "all the creatures used by military forces throughout history, only elephants have been regularly trained to mangle enemies in personal combat. From time to time dogs sprang upon the enemy, but pachyderms have crushed foes for thousands of years." Moore (1986) emphasizes that elephants were not only used to stampede cavalry and to trample down infantry, but also were used to carry fighting machines, protected by armor, with steel blades fastened to their tusks, and saddled with towers of howdahs containing several men from which missiles of various kinds could be thrown. On occasion the elephants of rival forces would have a duel a la mort, the rest of the forces halting to contemplate the contest. Alexander the Great was one of the precursors of the use of these animals in war, and this tradition passed to other warriors; among them was Pharaoh Ptolemy II, who used them in fighting battles. Other warriors have used elephants in the battlefields: Darius III, King of Persia, and General Cargill Anibal, who benefited from elephants in all phases of his campaign against Rome (Gorzoni, 2010). Elephants continued to be used for transportation and logistics in warfare right up to the World War II, when they were used to transport guns and supplies, and to assist in engineering projects such as road and bridge building (Fig. 17.8) (Williams, 1950).

Endowed with great size and strength, elephants were true "tanks" of war, being often covered with metal plates on the head and flanks, in addition to swords fixed on the trunks. When put against enemies, they were difficult to stop (Gorzoni, 2010). As expected, the use of elephants as weapons was followed by the introduction of antielephant tactics and techniques to deter, deflect, and destroy them (Chrystal, 2015). Despite their strength, elephants were easily frightened; so a strategy adopted to nullify the military power of these animals was to seek ways to daunt them. For this, other animals were frequently used. Romans, for instance, came to the battle equipped with ox-drawn wagons with long spikes to impale the beasts. The war pig was a particularly popular weapon against the elephant (Chrystal, 2015), according to several ancient sources. As Grant (1978) describes, wild boars were used in the battle of 272 BC, when the Romans fought against the Tarantines. During the confrontation, the Romans released several boars, which frightened the Tarantino army pachyderms in such a way that the animals turned against their own troop due to a strong desperate behavior caused by the herd of wild boars (Gorzoni, 2010). In his natural history records, Pliny the Elder writes that "Elephants are frightened by the tiniest sound of a pig squealing; and when wounded and frightened, they always give ground" (Rackham et al., 2014). Aelianus (1959) also recorded that elephants were scared by swine squeals (and cattle with horns), and described how the Romans used many squealing swine (and cattle) to ward off the war elephants of Pyrrhus in 275 BC. In his book *History of the Wars*, Procopius (2007) recorded that during the 16th century, the defenders of the city of Edessa strategically hung up a squealing swine to ward off the Khosrau's single siege elephant. Historical accounts of incendiary pigs were chronicled by the military writer Polyaenus (Polyaenus, 1994) and by Aelianus (Scholfield, 1971). Both writers stated that Antigonus II Gonatas' siege of Megara (266 BC) was encroached when the Megarians soaked some pigs with combustible oil or resin, set them alight, and released them in the

FIGURE 17.8 A military elephant in World War I pulls heavy equipment in Sheffield, England. *From Mattes [Public domain], via Wikimedia Commons.*

direction of the opponent's assembled war elephants. The elephants exhibited a panic behavior due the flaring and the squealing sounds of the pigs, resulting in a massacre of their own troops by crushing them to death.

Resistant and well-adapted to arid environments, camels have been legendary animals throughout history for their service in war in desert environments, where armies also used them as freight animals instead of horses and mules (Fig. 17.9) (Cooper, 2000; Lawrence, 1991; Willekes, 2016). Camels can cover unusually long distances hauling quantities of goods far more efficiently than any other beast of burden. These qualities made camels highly appreciated in warfare and long distance trade within their original areas of distribution and beyond (Bartosiewicz, 2014). Camels gained more prestige and strategic importance in military operations in their natural habitat (Bartosiewicz, 2014), and camel cavalries were used in wars throughout all of Africa and along the Middle East, and remain in use by India's contemporary Border Security Force. They were first known to have been used as animals of war in Turkey some 2600 years ago. The conquests of the Achaemenid king Cyrus the Great (559–530) would have been impossible without the logistic support of one humped camels. During the period of the Ottoman Empire (AD 1299–1923) camels were used extensively and were dubbed "Military Heavy Transport Vehicles." Up to 60,000 camels were used at one time by the Ottoman military (Yilmaz et al., 2011). At the beginning of World War I (1914) the number of female *Camelus dromedarius* imported annually into Turkey was still in the region of 7000–8000 heads (Bulliet, 1975).

Pigeons represent the most prominent birds in the history of warfare. These intelligent animals were used for communication and photographic espionage. Due to their outstanding flight capability and remarkable homing ability, trained pigeons can efficiently reach a stipulated location and return to their starting point. It is not surprising, therefore, that they were used as messengers during war, a vital role for winning battles, especially in the past, when communication technology was nonexistent or undeveloped. These birds were responsible for carrying notifications through a small tube attached to the leg. In other cases, they were trained to carry small cameras placed under their wings, with the aim of strategically logging pictures of their enemies (Fig. 17.10) (Gorzoni, 2010). Messenger pigeons have been among the most effective war animals, with their use as military aids beginning

FIGURE 17.9 Ottoman camel corps at Beersheba during the First Suez Offensive of World War I, 1915. *Reproduced from American Colony Jerusalem/Wikimedia Commons.*

FIGURE 17.10 A pigeon fitted with a German unmanned camera (probably aerial reconnaissance in World War I). *Reproduced from Bundesarchiv, Bild 183-R01996, via Wikimedia Commons.*

at least 2000 years ago and continuing through World War II and the Korean Conflict (Lawrence, 1991). There are several historical events (some dramatic) that highlight the relevance of pigeons in war. An example occurred in the Battle of Verdum during World War I (1914–18), when major Frenchman Sylvain Raynal and his troops were isolated among German soldiers, and so he sent messages with distress calls through a pigeon. The animal arrived at its destination, fulfilling its mission, but then dying because of close contact with clouds of gases and smoke. Due to its accomplishment, the animal was honored in the Fort Vaux esplanade, having inscribed in its lapide the last message of the commander Raynal (Gorzoni, 2010). In the same war, a pigeon saved the crew of a seaplane that fell on the high seas and was out of communication. The military pigeon was sent with a message indicating the coordinates of the vessel. The animal is now embalmed and exhibited in the officers' room of the British Air Force (Gorzoni, 2010). At the Battle of the Argonne Forest, an American army pigeon, Cher Ami (Fig. 17.11), became a special hero when he flew 25 miles in 25 min to deliver an urgent message that saved many lives. During his flight, Cher Ami was struck by bullets that destroyed his left eye, penetrated his breastbone, and shattered one leg. But the pigeon managed to reach his destination with the message still attached to the torn ligaments of his leg. Cher Ami lived and ultimately received the Croix de Guerre for bravery (Baynes and Wister, 1925; Stevens, 1988).

FIGURE 17.11 Stuffed war pigeon Cher Ami on display at the Smithsonian Institute. *From Smithsonian.* https://www.si.edu/encyclopedia_Si/nmah/cherami.htm.

Due to their important role in war communications, pigeons also became targets of enemy forces (Gorzoni, 2010; Snyders, 2015). One of the strategies adopted by soldiers was the use of hawks and falcons to intercept messenger pigeons and their message. The use of these birds of prey as message trackers was recorded in the Hundred Years' War, when France fought against England. The Germans used the hawks and falcons in World War II, when 16,500 English pigeons were parachuted to serve the French commanders. Some German officers trained in falconry (art of training hawks and falcons) made field inroads in search of the group of pigeons. Once found, they would release the birds of prey whenever a pigeon was released, which killed their prey in midair, so that when it fell to the ground the falconers were able to retrieve and intercepted the message (Gorzoni, 2010). Falcons were also trained to keep airstrips cleared of smaller birds (Lawrence, 1991).

In addition to pigeons and hawks and falcons, other birds were used in war as messengers, and as indicators of the presence of poison gas or the approach of the enemy. In World War II, the British made use of sea gulls, conditioned by the release of great amounts of bread, to warn them of the presence of submarines (Lawrence, 1991). The territorial behavior and vocalizations of some bird species were used to detect intruders and ambushes (Lawrence, 1991). During World War II, canaries were used by the Allies as indicators of poison gas in tunnels dug in the vicinity of the enemy border (Gorzoni, 2010). In addition to canaries, other animals were widely employed in World War I to detect the presence of noxious substances in the trenches, including mice and cats (Lawrence, 1991). Garden slugs, acutely sensitive to the presence of mustard gas in the air, proved to be ideal for this task. By monitoring their reactions, it was possible to determine the proportion of gas in the air and to sound the signal for putting on masks when the concentration reached the danger point for people (Lawrence, 1991).

As expected, terrestrial animals have been used more in human warfare because the majority of conflicts were fought on land (Kistler, 2011). Dolphins and sea lions are among the representatives of the aquatic fauna used in warfare (Fig. 17.12). For example, during the Vietnam War, dolphins were used to detect underwater swimmers, but, since information about those operations is classified, the exact tasks required of these animals are not generally known (Lawrence, 1991). Unconfirmed suspicions indicate that, in addition to detecting mines, tagging submarines with listening devices, and attaching bombs to ships, dolphins may have been trained to attack and even kill North Vietnamese divers (Sleeper, 1990). Sea lions were trained to detect the noise of propellers from enemy submarines during the World War I. Whenever they found a submarine, the animals were rewarded with fish. However, because they could not distinguish the noise of an allied or enemy submarine, the use of these animals was interrupted (Gorzoni, 2010). A number of species have been used in bomb detection roles, on both land and sea. These include more widely known examples such as dogs, a number of rodent species (i.e., rats and ferrets), and aquatic animals including dolphins, sea lions, and orcas (Salter, 2015).

FIGURE 17.12 Bottlenose dolphin (*Tursiops truncatus*) used by US Navy Marine Mammal Program (NMMP) on mine clearance operations, with locator beacon. *Reproduced from https://en.wikipedia.org/wiki/File:NMMP_dolphin_with_locator.jpeg.*

Serpents were also used in wars. One of the earliest reports of military use of these reptiles occurred in a naval battle against King Eumenes of Pergamon in 184 BC. Hannibal of Carthage had clay pots filled with venomous snakes and instructed his sailors to throw them onto the decks of enemy ships (Rothschild, 1964). The best known case involving the use of snakes in wars occurred during the Vietnam War, when Vietnamese soldiers built tunnels in which they placed poisonous snakes to bite the enemy as they passed (Gorzoni, 2010).

During the World War II, monkeys, dogs, cats, rabbits, guinea pigs, rats, and pigeons were used to test substances used in chemical warfare. Swine and dogs were completely blown-up in trial atom bomb explosions, and countless animals were conscripted for measuring the effects of radiation burns (Baynes and Wister, 1925; Cooper, 2000; Dempewolff, 1943). Plans were made in World War II to bombard the Japanese with bats equipped with incendiary bombs (Lawrence, 1991).

In spite of the importance of vertebrates in wars, invertebrates also have been used, mainly to destabilize enemy armies. In this scenario, several groups of invertebrates, mainly arthropods, became especially useful as biological weapons (Lockwood, 2012; Peterson, 1995). Lockwood (2012) stressed that "The use of insects as weapons of war, tools of terrorism, and instruments of torture extends from the opportunism of prehistorical assaults to the calculated tactics of modern, asymmetric conflict." Vectors of various diseases, insects were used to transmit illness and cause casualties in enemy armies; for this, the animals must be in contact with the enemies to infect them. Such strategies have been documented for a long time, including the practice of launching biological material into a besieged city, such as that recorded in the 14th century when the French launched, from a long distance, rotting livestock into the castle at Thun l'Ev'eque (Geissler and van Courtland Moon, 1999). Another example occurred in 1343, when Mongol forces under the leadership of Janibeg bombarded the Genoese seaport of Kaffa on the Crimean peninsula (Croddy, 2002; Geissler, 1986; Mayor, 2003; Robertson and Robertson, 1995). The siege lasted for 3 years, until the Mongol camping site was destroyed by bubonic plague. But before giving up his attack, the Mongol khan catapulted corpses inside the town (Geissler, 1986). Lockwood (2012) points out that in the early 20th century the relationship between insects and disease was exploited, and vectors were mass produced as microguided missiles loaded with pathogens.

The term Entomological Warfare (EW) has been used in literature in reference to a specific kind of biological warfare that uses mainly insects in a direct strike or as vectors of a biological agent (Larsen and Smith, 2005; Lockwood, 1987, 2008). In essence, EW occurs in three variations. According to Lockwood (2008), one type comprises infecting insects with a pathogen and then dispersing them over target zones (Endicott and Hagermann, 1998). The insects then act as a main vector, infecting any living being they might bite. Another type of EW is a straight insect attack on crops; in this case, the animal may not be infected with any risk pathogen but instead represents a threat to agriculture (Endicott and Hagermann, 1998). The last method encompasses the utilization of uninfected insects, such as bees and wasps, to strategically attack the enemy (Lockwood, 2008), which is discussed further below.

During World War I, German infiltrators were blamed for trying to spread plague bacteria in Russia in 1915 and 1916. For this effort to have been effective, the dissemination of the disease would probably have required the human flea, *Pulex irritans* L., as a vector. Also, the Germans were accused of having injected mule and horses with glanders and livestock animals with anthrax. Among the Allied Powers of World War II, the Canadian Army commanded a groundbreaking effort of vector-borne warfare (Gage and Kosoy, 2005). Later the Japanese became committed to developing the plague flea as a weapon, while France also pursued EW programs. Like the Germans, the French carried out tests with the Colorado potato beetle (*Lepinotarsa decemlineata*), intended to target the enemy's food sources (Croddy and Wirtz, 2005). Japan benefited from entomological warfare over China in many ways during World War II (Croddy and Wirtz, 2005). The so-called Unit 731, Japan's notorious biological warfare unit commanded by Lt. General Shirō Ishii, used fleas and flies infected with plague or covered with cholera to contaminate the population in China (Gage and Kosoy, 2005). The Japanese army released the insects from low-flying planes and by dropping missiles and bombs full of the insects and disease (Croddy and Wirtz, 2005).

With regard to the Cold War (1947–91), the Soviet Union explored the method, developed the technique, and tested an EW program as a key part of an anticrop and antianimal Biological Warfare scheme. The Soviets used ticks to transmit animal pathogens, such as foot and mouth disease. They also used avian ticks to transmit *Chlamydophila psittaci* to chickens. Furthermore, the nation claimed to have elaborated a dynamic mass insect-breeding facility, capable of producing thousands or millions of parasitic insects per day (Lockwood, 2012). The United States also explored the potential of EW during the Cold

War, and army intelligence developed strategies for an entomological warfare facility capable of outputting 100 million yellow fever–infected mosquitoes per month (Hay, 1999).

The United States has also used EW in campaigns of noncombat situations. During 1990, the nation subsidized a $6.5 million program intended to develop and breed caterpillars (Lockwood, 2008). The animals were to be released in Peru on coca fields as part of a program called American War on Drugs (Lockwood, 2008). During 2002, the US entomological antidrug effort at Fort Detrick aimed to discover an insect vector for a virus that attacks the opium poppy (Lockwood, 2012).

One of the groups of invertebrates most used in war has been bees, which have a history of being used as weapons in different warfare situations. For long periods of time, besieged Europeans made use of bees to repulse intruders. In the year 1908, the Danish and Norwegians assailed Chester, England. When the Scandinavians began to tunnel under the city fortifications, the English threw beehives into the subterranean passage and ended the incursion (Free, 1982). The inhabitants of Gussing, Hungary, protected their hometown in 1289 against an Austrian army by dumping bees on the enemy (Ambrose, 1974). During World War I, "At the Battle of Tanga in German East Africa, the enemy cunningly hid wires and canes among the grasses and bushes through which the British were advancing. These when trodden upon lifted the lids from hives of wild bees, which poured out and caused terrible distress to the British troops" (Baynes and Wister, 1925).

Other invertebrates that have been used in military activities include spiders, scorpions, and worms. One example was recorded in about AD 198, when the Parthian city of Hatra (near Mosul, Iraq) repulsed the Roman army led by Septimius Severus by hurling clay pots filled with live scorpions at them (Eneh, 2012). Other examples include the use of the glow worms, which were carried as an aid in guiding tanks during World War I, and spiders, which were kept in "farms" during World War II where they were put to work producing the extra-fine yet strong silken strands needed for making highly precise telescopic gun and bomb sights (Lawrence, 1991).

In addition to all the roles played by wildlife in wars, as discussed earlier, animals of many kinds have also been adopted officially and unofficially as pets and mascots by the armed forces. Among the most common military mascots are dogs and horses, but mascots ranged from hens to donkeys, badgers to sheep, and even eagles, lions, and bears had their place in camp and on the battlefield (van Vliet, 2007). One of the most well-known wild animal military mascots is Wojtek, the bear of the 22nd Transport Company of the Polish Army Service Corps, who saw action at Monte Cassino in 1944 (Fig. 17.13). van Vliet (2007) stresses that mascots brought loyalty

FIGURE 17.13 Soldier interacting with Wojtek, a mascot bear of the 22nd Transport Artillery Company (Army Service Corps, 2nd Polish Corps), World War II. *From Imperial War Museum [Public domain], via Wikimedia Commons.*

and enthusiasm, while the act of nurturing the animals offset boredom for soldiers in camp. The relevance of animals as mascots is reflected in the many honors they have received, including medals for bravery, postwar retirement benefits, and military burials with honors.

The mechanization and new technologies of modern times have reduced the usefulness of animals in war (Kistler, 2011), but has not put a complete stop to their conscription. Equids, such as horses and mules, for example, continue to have a role in the US military of the 21st century. During the Soviet War in Afghanistan, for example, the United States made use of a great number of donkeys and mules to transport armaments and provisions over Afghanistan's rocky landscape to the mujahedeen (Bearden and Risen, 2004). The use of mules by US forces has continued in the Afghanistan War between 2001 and 2014. Moreover, several species continue to be used in military and police activities in various countries around the world, mainly dogs and horses, which are used for transportation, rescue, patrol, persecution, and battling narcotics, among other activities. Many countries still preserve small groups or units of mounted riders for patrolling and inspection, and military horse units are frequently used for formal and educational purposes.

Finally, it is worth noting that wars do not only involve animal species that have been "recruited" to act directly in conflicts. Military activities cause serious environmental impacts (Brauer, 2009; Gleditsch, 1998; Gurses, 2012; Jarrett, 2003; Machlis and Hanson, 2011), which naturally affect all of the fauna in the regions where fighting occurs. These impacts date to the earliest military histories and have increased over time with the scale and technology of warfare (Hanson et al., 2009). In the modern era, biological, chemical, and nuclear weapons have expanded the potential for catastrophic, landscape-level change and long-term contamination (Dudley and Woodford, 2002; Glasstone and Dolan, 1977).

FINAL REMARKS

Throughout history, animals have played significant roles in various aspects of human life, including the various conflicts and wars that have been features of all of humankind's history. The list of animals involved in military activities is diverse, ranging from worms to large mammals, both domestic and wild. The roles various animals played in warfare varied according to the species involved. In many cases the use of native animals from the regions where the battles occurred was preferred, while in other scenarios exotic animals were imported to be incorporated into armies.

Animals were recruited to participate in virtually every stage of warfare, from logistical support and transport of combatants, weapons, and supplies to the treatment of the wounded. Animals were decisive in many conquests, especially in the earliest recorded wars, when technology and mechanization were poorly developed. In addition to being used to attack enemies, animal species were also involved in defense strategies. Millions of animals were killed in human wars, and in some cases, they became symbols and mascots, gaining military recognition and honors for the deeds accomplished during the conflicts.

References

Aelianus, 1959. Trans. A.F. Scholfield. On the Characteristics of Animals, vol. III. William Heinemann Ltd., London.

Alves, R.R.N., 2012. Relationships between fauna and people and the role of ethnozoology in animal conservation. Ethnobiology and Conservation 1, 1–69.

Alves, R.R.N., 2016. Domestication of animals. In: Introduction to Ethnobiology. Springer, Switzerland, Albuquerque, UP, pp. 221–225.

Alves, R.R.N., Souto, W.M.S., 2015. Ethnozoology: a brief introduction. Ethnobiology and Conservation 4, 1–13.

Ambrose, J.T., December 1974. Insects in warfare. Army 33–38.

Bartosiewicz, L., 2014. Camels in the front line. Anthropozoologica 49, 297–302.

Baynes, E.H., Wister, O., 1925. Animal Heroes of the Great War. The Macmillan Company.

Bearden, M., Risen, J., 2004. The Main Enemy: the Inside Story of the CIA's Final Showdown with the KGB. Presidio Press.

Brauer, J., 2009. War and Nature: the Environmental Consequences of War in a Globalized World. Rowman & Littlefield.

Bulliet, R.W., 1975. Camel and the Wheel. Harvard University Press, Harvard.

Chrystal, P., 2015. Roman Military Disasters: Dark Days & Lost Legions. Pen and Sword.

Clutton-Brock, J., 1984. The master of game: the animals and rituals of medieval venery. Biologist 31, 167–171.

Coatesworth, J., 2016. Understanding the Horse's Skin and Coat. The Crowood Press.

Cooper, J., 2000. Animals in War. Corgi Books, London.

Croddy, E., 2002. Chemical and Biological Warfare: a Comprehensive Survey for the Concerned Citizen. Springer, New York.

Croddy, E., Wirtz, J.J., 2005. Weapons of Mass Destruction: an Encyclopedia of Worldwide Policy, Technology, and History. ABC-CLIO.

Cunningham, P.F., 1995. Topics awaiting study: investigable questions on animal issues. Society & Animals 3, 89–106.

De Prisco, A., Johnson, J.B., 1990. The Mini-atlas of Dog Breeds. TFH Publications, United States.

Dempewolff, R.F., 1943. Animal Reveille. Doubleday, Doran, Incorporated.

Dudley, J.P., Woodford, M.H., 2002. Bioweapons, biodiversity, and ecocide: potential effects of biological weapons on biological diversity. BioScience 52, 583–592.

Eglan, J., 2015. Beasts of War: the Militarization of Animals. lulu.com.

Endicott, S., Hagermann, E., 1998. The United States and Biological Warfare: Secrets from the Early Cold War and Korea. Indiana University Press, USA.

Eneh, O.C., 2012. Biological weapons-agents for life and environmental destruction. Research Journal of Environmental Toxicology 6, 65–87.

Forster, E.S., 1941. Dogs in ancient warfare. Greece and Rome 10, 114–117.

Free, J.B., 1982. Bees and Mankind. George Allen & Unwin, London.

Gage, K.L., Kosoy, M.Y., 2005. Natural history of plague: perspectives from more than a century of research. Annual Review of Entomology 50, 505–528.

Gardiner, J., 2006. The Animals' War: Animals in Wartime from the First World War to the Present Day. Portrait.

Geissler, E., 1986. Biological and Toxin Weapons Today. Oxford University Press, USA.

Geissler, E., van Courtland Moon, J.E., 1999. Biological and Toxin Weapons: Research, Development and use from the Middle Ages to 1945. SIPRI Biological and Chemical Warfare Studies. Oxford University Press, New York.

Glasstone, S., Dolan, P.J., 1977. The Effects of Nuclear Weapons, third ed. U.S. Government Printing Office, Washington, DC.

Gleditsch, N.P., 1998. Armed conflict and the environment: a critique of the literature. Journal of Peace Research 35, 381–400.

Gorzoni, P., 2010. Animais nas Guerras: A Força do Exército dos Bichos nas Grandes Batalhas. Matrix.

Grant, M., 1978. History of Rome. Weidenfeld & Nicolson.

Greenhalgh, P.A.L., 1973. Early Greek warfare: Horsemen and chariots in the Homeric and Archaic ages. Cambridge University Press.

Gurses, M., 2012. Environmental consequences of civil war: evidence from the Kurdish conflict in Turkey. Civil Wars 14, 254–271.

Hanson, T., Brooks, T.M., Fonseca, G.A.B., Hoffmann, M., Lamoreux, J.F., Machlis, G., Mittermeier, C.G., Mittermeier, R.A., Pilgrim, J.D., 2009. Warfare in biodiversity hotspots. Conservation Biology 23, 578–587.

Harris, G., 2011. A Bin Laden Hunter on Four Legs. The New York Times Company.

Hay, A., 1999. A magic sword or a big itch: an historical look at the United States biological weapons programme. Medicine, Conflict and Survival 15, 215–234.

Jarrett, R., 2003. The environment: collateral victim and tool of war. BioScience 53, 880–882.

Kanyamibwa, S., 1998. Impact of war on conservation: Rwandan environment and wildlife in agony. Biodiversity and Conservation 7, 1399–1406.

Keeley, L.H., 1997. War Before Civilization: the Myth of the Peaceful Savage. Oxford University Press.

Kistler, J.M., 2011. Animals in the Military: from Hannibal's Elephants to the Dolphins of the US Navy. ABC-CLIO.

Larsen, J.A., Smith, J.M., 2005. Historical Dictionary of Arms Control and Disarmament. Scarecrow Press, Lanham, MD.

Lawrence, E.A., 1991. Animals in war: history and implications for the future. Anthrozoos 4, 145–153.

Lockwood, J.A., 1987. Entomological warfare: history of the use of insects as weapons of war. Bulletin of the Entomological Society of America 33, 76–82.

Lockwood, J.A., 2008. Six-legged Soldiers: Using Insects as Weapons of War. Oxford University Press.

Lockwood, J.A., 2012. Insects as weapons of war, terror, and torture. Annual Review of Entomology 57, 205–227.

Machlis, G.E., Hanson, T., 2011. Warfare Ecology. Springer, pp. 33–40.

Mayor, A., 2003. Greek Fire, Poison Arrows, and Scorpion Bombs: Biological & Chemical Warfare in the Ancient World. Overlook Duckworth, New York.

Moore, J., 1986. Elephants in war. Canadian Veterinary Record 27, 312–313.

Peterson, R.K.D., 1995. Insects, disease, and military history. American Entomologist 41, 147–161.

Polyaenus VI, L., 1994. Polyaenus' Stratagems of War. Ares Pub.
Procopius, 2007. History of the Wars, Books I and II. Echo Library.
Putney, W.W., 2003. Always Faithful: a Memoir of the Marine Dogs of WWII. Potomac Books, United States.
Rackham, H., Jones, W.H.S., Eichhholz, D.E.T., 2014. Pliny's Natural History. Loeb Classical Library, London.
Robertson, A.G., Robertson, L.J., 1995. From asps to allegations: biological warfare in history. Military Medicine 160, 369–373.
Rogak, L., 2011. The Dogs of War: the Courage, Love, and Loyalty of Military Working Dogs. St. Martin's Griffin, United States.
Rogers, K.M., 2005. First Friend: a History of Dogs and Humans. St. Martin's Press, New York.
Rothschild, J.H., 1964. Tomorrow's Weapons, Chemical and Biological. McGraw-Hill, New York.
Salter, C., 2015. Animals and war: anthropocentrism and technoscience. NanoEthics 9, 11–21.
Sanderson, J., 1997. War Dog Heroes: True Stories of Dog Courage in Wartime. Scholastic Inc., New York.
Scholfield, A.F., 1971. Aelian, on Animals. 2.
Sleeper, B., 1990. Uncle Sam wants you, cetaceans. Animals 123, 23–27.
Sloane, C.F., 1955. Dogs in war, police work and on patrol. The Journal of Criminal Law, Criminology, and Police Science 46, 385.
Snyders, H., 2015. 'More than just human heroes' the role of the pigeon in the first world war. Scientia Militaria: South African Journal of Military Studies 43, 133–150.
Stevens, P.D., 1988. Real Animal Heroes. Sharp and Dunnigan, Chico, CA.
Tucker, S.C., 2015. Instruments of War: Weapons and Technologies that have Changed History. ABC-CLIO.
van Vliet, D., 2007. Animals and war. In: Heathcote-James, E. (Ed.), Psychic Pets – How Animal Intuition and Perception has Changed Human Lives. John Blake, London, England.
Varner, J.G., Varner, J.J., 1983. Dogs of the Conquest. University of Oklahoma Press Norman, OK.
Whitney, W.D., Smith, B.E., 1911. Elephantry, the Century Dictionary and Cyclopedia. Century Company, p. 2257.
Willekes, C., 2016. Animals in war. In: Phang, S.E., Spence, I., Kelly, D., Londey, P. (Eds.), Conflict in Ancient Greece and Rome: the Definitive Political, Social, and Military Encyclopedia. ABC-CLIO, pp. 100–101.
Williams, J.H., 1950. Elephant Bill. Hart-Davis, London.
Willmott, H.P., 2003. First World War. Dorling Kindersley Publishers, London, United Kingdom.
Yilmaz, O., Ertugrul, M., Wilson, R.T., 2011. The domestic livestock resources of Turkey: Camel. Journal of Camel Practice and Research 18, 21–24.

CHAPTER

18

The Ethnozoological Role of Working Animals in Traction and Transport*

Rômulo Romeu Nóbrega Alves
Universidade Estadual da Paraíba, Campina Grande, Brazil

INTRODUCTION

The domestication of animals transformed the way humans interact with the animal world, resulting in intensified exploitation of fauna by people. The principal motivation for the domestication of most species was their potential to provide food products for humans (Alves, 2016; Diamond, 2002; Muller, 2002; Russell, 2002). Resources derived from domestic fauna have been widely used in different ways by humans to fulfill several needs (Alves, 2016). These resources, or products, have been classified into two categories, which were mentioned without any precise definition by Sherratt (1981), then defined by Greenfield (2005) and Greenfield et al. (1988): primary products, which can be extracted only upon the death of the animal, such as meat, hide, bones, etc. (Bökönyi, 1974; Sherratt, 1983); and secondary products, which can be exploited during the lifetime of the animal, such as milk, strength, fleece, manure, etc. (Greenfield et al., 1988).

The diversity of uses of domestic animals illustrates the many roles they play for human societies. When alive, domestic animals provide various products (milk, eggs, fleece, manure, etc.) and are used as companions, experimental subjects, sources of entertainment, and providers of muscle power. In the case of muscle power, animals are kept and trained by humans to perform tasks in a wide range of sectors including agriculture, transportation, construction, tourism, and mining, as well as for cultural and domestic services; such animals are classified as "working animals."

The relationships between people and working animals date from the beginning of civilization. Several terms have been used to refer to working animals kept mainly to do muscular work, including traction animals and draft animals; but for the purposes of this chapter, all of these categories will hereafter be considered, and referred to, as working animals. Working animals have been used in agriculture and transport for thousands of years, illustrating well the

*This chapter is a revised and updated version of the section "Animal traction and transport" from the paper: "Alves, R.R.N., 2012. Relationships between fauna and people and the role of ethnozoology in animal conservation. Ethnobiology and Conservation 1, 1–69."

Ethnozoology
http://dx.doi.org/10.1016/B978-0-12-809913-1.00018-1

convergence of the domestication of plants and animals to potentiate food production around the world. The use of these animals, however, is not restricted to agriculture and rural areas, but also extends to cities, where they are used to transport materials, cargo, or people. In this chapter, the role of working animals that perform transport and traction activities will be discussed briefly, highlighting the ethnozoological role of these animals throughout their shared history with humans.

A BRIEF HISTORY OF THE USE OF WORKING ANIMALS FOR TRACTION AND TRANSPORT

The use of working animals is an ancient practice that has persisted into the present (Guthiga et al., 2007). Their use originated when humans realized that in addition to providing meat and other by-products, many animals possessed great muscle power that could be harnessed as a source of force to transport loads and people, and to pull plows, sledges, carts, and even war chariots (Davis, 1987). There is also evidence that the use of animal power is associated with the complex aspects of technological innovations, such as the invention of the wheel, of vehicles, and of harnessing equipment. These events represent important historical milestones in human history and were instrumental in exploiting new geographical areas and in increasing food production.

Although the exploitation of domestic animals for traction lies at the foundation of historic and modern strategies throughout the Old World, little is known about its origins (Greenfield, 1988). Most of the evidence comes from two geographically distinct sources: the Mesopotamian Uruk-Early Dynastic and Egyptian Old Kingdom periods (late fourth and early third millennia BC) and the Eneolithic and Bronze Age of Central Europe (3300–1000 BC) (Fig. 18.1). The early Mesopotamian/Egyptian evidence is largely iconographic (cylinder seals, temple, and tomb friezes), although many later historic cuneiform and hieroglyphic texts document secondary-product exploitation (Green, 1980; Sherratt, 1981; Simoons, 1971).

Some of the earliest examples for the use of traction animals can be seen in Sumerian pictograms and cylinder seals dating from the end of the fourth millennium BC (Davis, 1987; Greenfield, 1988). In the royal tombs of Ur (early third millennium BC), for example, the ox was harnessed to the king's chariot, and to the chariots taken as booty and used as tribute by Assyrian kings in the first millennium BC (Aynard, 1972). Paleopathological evaluation of bones of cattle and horses suggests that these animals were used as draft animals and/or for carrying cargo (Marković et al., 2014; Sherratt, 1981, 1983). In Central Europe, there is considerable artifactual evidence implying the use of domestic animals for traction by the beginning of the post-Neolithic period (late fourth and early third millennia BC), including ceramic sieves and changes in vessel type; remains of wool, wagons, and wagon parts in burials and waterlogged contexts; wagon-shaped figurines and vessels; plow marks in paleosols; terracotta

FIGURE 18.1 Painted wooden figurine showing a peasant plowing with a pair of oxen (c.2025–1850 BC, 11th–12th Dynasty) illustrating the antiquity of the use of traction animals in agriculture. *Photo credits: Rômulo R.N. Alves (Photographed at the British Museum, London, England).*

models of plows; and wooden plows in waterlogged contexts (Greenfield, 1988).

The archeological evidence suggests that cattle were probably the first animals used for traction and that the earliest forms of harness for traction and power were developed using cattle, and subsequently transferred to and modified for other species. The horse is another example of an animal used for traction since ancient times. It is considered to be the animal on which specific riding techniques were first developed, though cattle and other animals had previously been used for carrying people. From about 3000 BC onward there is increasing evidence that, apart from cattle and horses, reindeer, onagers (wild asses), asses, Bactrian camels, dromedaries, buffaloes, yaks, and elephants were used for traction and transport (Bowman, 1977).

Working animals have always played a very important role in all aspects of people's lives throughout human history, until the advent of the industrial revolution and the exploitation of fossil fuels; a period in which the development of all civilizations was strongly dependent on the muscle power of humans and animals. It is worth noting that throughout the developing world, working animals are still a vitally important part of highly appropriate and effective systems of food production and transportation, as discussed in the next section.

WORKING ANIMALS AND THEIR USES

The use of working animals is declining in many parts of the world as a result of increased mechanization. Spugnoli and Dainelli (2013) emphasized that animal traction has been almost completely eliminated from most industrialized countries, apart from exceptional or ideological uses. However, this trend is not universal and in some places there has even been a return to the use of working animals as a source of power, even when there is access to mechanization. This persistence is associated with cultural, economic, and ecological factors. Although it is most commonly used in developing countries, many developed countries are seeing a resurgence of interest in the use of animal power (Bradbury, 2002). In continental Europe, for example, farmers are increasingly viewing horse-powered systems utilizing the latest horse-drawn equipment as a modern, ecologically and economically sound alternative for use in organic operations (Herold et al., 2009).

Despite the advances of mechanization, the importance of working animals as a source of power is likely to continue in the foreseeable future, and their uses are spread worldwide. They have been particularly important in Asia (FAO, 2007), but relatively unimportant in sub-Saharan Africa, where their usage is restricted by heavy soils and the presence of *tsetse* flies. Nonetheless, animal traction is of great importance in parts of Africa. In Gambia, for example, 73.4 percent of crop fields are cultivated using animal power. In Latin America, the Caribbean, and the Near and Middle East, animal power is, again, vital to the livelihoods of many small-scale farmers (FAO, 2007).

From an ecological point of view, animal power is considered sustainable and environmentally friendly. Fuller and Aye (2012) stress that, on a global scale, human and animal power combined is the largest single contributor of renewable energy, and represents the most important energy sources for farmers in developing countries. Given this scenario, growing interest in environmentally safe techniques, as practised in organic and biodynamic agriculture, has renewed interest in the use of animal power in place of machine power (Fuller and Aye, 2012; Heinberg, 2006).

Economic factors also influence the persistence of the use of working animals in agricultural practices, such as the cost and maintenance of agricultural tractors, which is prohibitive to small farmers. Fuel prices, for example, increase production costs when machines are used by farmers. Thus, it is not surprising that many farmers rely on working animals to perform activities associated with agricultural production. This situation becomes even more relevant when we consider that many farmers who use working animals have low incomes. According to IFAD (2011), two billion people live and work on small farms in developing countries, and most of these smallholder farmers live on less than $2 per day. Among these workers, the use of working animals remains popular and may even increase.

In many parts of the world, working animals are an affordable, appropriate, and sustainable technology (Starkey, 1994). In some areas, the use of animals and mechanical tractors are used in a complementary way. For example, tractors can be used to plow the land before planting, while animal plows are used during weeding. There are also cases where motor power is still inappropriate. For example, in mountainous terrain, in paddy fields, and in areas without roads or which are short of fossil and other fuels, the use of animals for traction and transport is dominant and is likely to remain so (Bowman, 1977).

Around the world, millions of working animals have been used in agricultural activities, and in the transport of production, cargo, and people (Figs. 18.2–18.4). Pearson (2005) pointed out that it is impossible to obtain precise information on the number of animals used for work

FIGURE 18.2 Horse-drawn carriages in Vienna, Austria (A and B), Rome, Italy (C) and Prague, Czech Republic (D), used in sightseeing tours and wedding ceremonies. *Photo credits: Rômulo R.N. Alves.*

purposes worldwide because while most countries maintain statistics on livestock numbers, for ruminants they do not distinguish their use for draft from their use for beef or milk. Previous global estimates of the population of working animals indicated 300 (Wilson, 2003) and 400 million (Barwell and Ayre, 1982). Other estimates showed that in the developing world in 1994, 51% of the 921 million cattle, 35% of the 135 million buffalo, 65% of 43 million horses, 87% of 43 million donkeys, 70% of 14 million mules, and 15% of 20 million camels were used for work (Pearson, 1999). Figures now indicate that 200–250 million working animals are used throughout the world (Starkey, 2010).

It is believed that the earliest working animals were bovines (*Bos taurus*), with camels, donkeys, and horses coming later for use in

FIGURE 18.3 Working animals used for the transport of water in Africa (A) and Brazil (B); for pulling plows (C) and carts (D) in Brazil; (E) A child mounted on a buffalo on Marajo Island, north Brazil, where they are used for transport and hauling; and (F) Camels used for transport in Marrakesh, Morocco. *Photo credits: (A) Ollivier Girard for Center for International Forestry Research (CIFOR); (B) John Philip Medcraft; (C) Angelo Giuseppe Chaves Alves; (D and E) Rômulo R.N. Alves and (F) Ierecê Lucena Rosa.*

FIGURE 18.4 Donkey carts loaded with wood (A and B) and used for transporting rubber (C) and bricks (D) in Africa. *Photo credits: (A and D) Ollivier Girard for Center for International Forestry Research (CIFOR); (B and C) Daniel Tiveau for CIFOR.*

carrying people and goods, and for traction (Davis, 1987). Other working animals have been used throughout history, including mules, asses, buffalo, yaks, llamas, alpacas, reindeer, elephants, elk, moose, sheep, goats, and dogs (Goe and McDowell, 1980). According to Starkey (2000), cattle are the major work animals worldwide. He also points out that it is common to use male animals because they are stronger than females and cattle herds always produce a surplus of males. Additionally, castrated animals are more docile than intact males (Starkey, 2000). The strategy of castrating working animals also extends to other species of traction animals, such as mules, donkeys, and horses.

The choice of animal to be used depends on the region and purpose of use. In arid and semiarid regions of the world, for example, cattle, donkeys, mules, horses, and camels are common working animals. Yaks are important pack animals in the high mountain ranges of Asia, where sheep and goats are also sometimes used for this purpose (FAO, 2007). In the humid tropics, water buffaloes are used in great frequency (Pearson, 2005). In Asia, for example, buffaloes are particularly suited to working in swampy conditions (FAO, 2007). Similarly, on the island of Marajó in the north region of Brazil, these animals are everywhere and are used in agriculture and as a means of transportation for the population, including the local police (Fig. 18.3E).

Activities performed by working animals comprise pulling implements for cultivation and harvesting, including sledges, and subsequently wheeled vehicles for both peaceful and military purposes; carrying packs, bricks, and other construction materials; providing tillage, weeding, and rubbish collection; carrying water and riders; serving tourism and ceremonies (such as weddings and festivals); and for driving machinery such as threshing, milking, and irrigation equipment (Bowman, 1977; Pritchard et al., 2005; Ramaswamy, 1998) (Figs. 18.2–18.4). Working animals are also used in the timber industry and to power stationary equipment, such as water pumps, sugarcane crushers, and grinding mills. Less widespread is their use in the movement of materials for small-scale building projects and road, dam, and reservoir construction within rural areas (Pearson, 1999, 2005).

Working animals are multipurpose and play a vital role in both urban and rural transportation (Bradbury, 2002). In eastern and southern Africa, they are vital to rural economies and help to relieve the burdens of transport on rural households (Denis, 1999). In the Brazilian semiarid region, working animals are widely used on small rural properties for several activities related to rural life, including soil preparation, planting, harvesting, and transport of produce. For example, donkeys are used to transport water, people, and carts, and to pull plows (Figs. 18.3 and 18.4). These practices also apply to several other animals, such as oxen, buffaloes, and horses.

In several Latin American countries, donkeys, mules, and horses were, for a long time, the best and most efficient means of transporting cargo over long distances, while the ox cart was used for the same purpose, but for short journeys. In Brazil, for example, from the 17th century to the beginning of the 20th century, donkey troops were fundamental to the country's economic development. The drivers of horse or mule troops, called "tropeiros," crossed extensive geographical areas transporting cattle and goods. These troops brought raw materials and agricultural products to the more developed centers, and returned with fabrics, perfumes, dishes, and other materials. The routes of these troops could take several weeks and involved several regions of the country. Small settlements began to appear along the troops' routes where the tropeiros would stop to exchange merchandise and let the cattle graze. In these settlements, commerce naturally developed to attend to the troops, while at the same time the tropeiros took and brought merchandise for them. The tropeiros and their animals thus provided an important contribution to the development of the regions they passed through and were responsible for the economic and cultural integration of many remote regions of colonial Brazil and the appearing towns and cities. The animals used by the tropeiros included horses, donkeys, and mules, with the latter three being more commonly used in the transport of cargo. These animals possessed a high degree of resistance and were able to overcome even the most rugged terrain with firmness and persistence, and thus were appropriate transport, particularly for merchandise.

Animals are also used to carry water and wood, drag logs, and to transport people. In Africa, animals provide a more affordable and efficient way of moving goods over short distances, and offer an efficient alternative to human methods of transporting goods such as head loading (Anderson and Denis, 1994). Both rural and urban businesses and entrepreneurs in South Africa utilize animal-drawn carts for the transport of goods, with the profitability of animal power often exceeding that of motorized transport, depending on factors such as distance traveled and loading time (Starkey, 1995). Animal transport also has the advantage of being able to operate on low-quality tracks and paths, and does not require imported fuel. Arriaga and Jordan (1999) list the income-generating transport activities of equids owned by poor, small farm owners in Mexico, supplementary to their role in agricultural cultivation, which included rental for the purposes of transport, transporting harvest, distributing manure and fertilizers to fields, and other transport activities such as carrying building and domestic materials.

It is important to remember that, in addition to furnishing a source of power, these same animals provide milk, fuel, wool, hair, off-spring, and by-products, such as hides, horns, hooves, and meat, at the end of their working lives (Goe and McDowell, 1980). In fact, in some cases, such animals are even kept alive for milk production after their productive life.

THE WELFARE OF WORKING ANIMALS

Although some authors point out that the relationship between domestic animals and humans has been mutualistic, with humans using several animals as a workforce or for their products, while in turn providing shelter, food, general care, and protection against predators (Heffner, 1999; Jordão et al., 2011), there is an increasingly debated concern about the welfare of working animals. According to FAO (2014), animal welfare is recognized as a central component of responsible animal husbandry and therefore should be a core part of all working animal systems. The welfare of working animals, as well as being a common good per se, has a direct impact on their health and their capacity to carry out tasks and jobs, which in turn affects the livelihoods of their owners and owners' families. As pointed out by Fuller and Aye (2012), many factors can reduce the output of working animals significantly, including stress, malnutrition, a poor fitting harness, etc. The impact of poor nutrition, for example, is significant because thin, underfed, or sick animals will not be able to work efficiently. As an example, Pearson (1991) points out that output of oxen and buffalo can decline by as much as 50% in these situations.

The most commonly identified causes of poor welfare of working animals include ill-treatment and the working of animals that are unfit or immature, poor management practices (including overwork), poor nutrition, and inadequate basic health care (including foot and hoof care) and inappropriate harness and other equipment (e.g., cattle yokes used on equids). Other factors contributing to poor animal welfare include ignorance on behalf of the animal owner, lack of veterinary care, and in some cases, traditional practices (e.g., the practice of burning (thermocautery) horses' legs) (Rahman and Kahn, 2014). It should be remembered that there are some procedures adopted during the training and preparation of working animals that are questioned by animal welfare activists. One such procedure is castration, which has been used as a way to make animals more docile and powerful. This practice deprives an animal of its reproductive life and is often performed without proper veterinary care and without adopting procedures to minimize pain when the animal is being neutered.

In the last few years several examples of abuse of working animals have been recorded, including stress and excessive tension caused by intense overload, injuries, and the use of sharp instruments, such as spikes and whips to force animals to work beyond their capabilities. Animals are also deprived of food (and/or given bad quality food), water, and appropriate rest for long periods of time, as well as behavioral freedom (Jordão et al., 2011), and in some cases, animals that are sick or in advanced pregnancy are nonetheless forced to work (Ramaswamy, 1998).

Animals who can no longer work are often sent to a slaughterhouse, even when they could have continued to live without working for many years. There are some cases, however, when they continue to be bred for the production of milk and fertilizer. In addition, there are situations when animals are abandoned after being considered disabled, which is a common occurrence with donkeys and mules. The case of donkeys in the Brazilian northeast region is very illustrative. Historically exploited by locals, donkeys have become a symbol of the Brazilian northeast region, because they are strong animals mainly used for transporting people and cargo. With the popularization of the use of motor vehicles, tractors, and motorcycles, even in rural areas of the region, donkeys have lost popularity as pack animals. These animals, once widely used as a means of transport and as pack animals, are now devoid of commercial value and are abandoned. As a result, they frequently begin to reproduce indiscriminately, and are now found in large numbers in the region. Abandoned, they arrive at thoroughfares and highways causing accidents, and thus have come to be considered a problem by some residents of the region. The administrators of highways and prefectures collect these animals, but do not have a destination for them. Among the proposals to solve the problem, local authorities have even suggested their slaughter to provide meat for school lunches, prisons, or for export to China. However, these proposals did not gain popular support and were considered disrespectful to the culture of the northeast region, for which the animal is a symbol. Nongovernmental animal protection organizations have also been opposed to these proposals.

FINAL REMARKS

The domestication of animals and plants has had great importance in the history of human civilization, boosting supply of food and others products. One of the oldest ways to increase productivity was through the use of animal power in agricultural activities. From this scenario emerged one of the most important forms of interaction between humans and domesticated animals; their use as workers. This not only enabled increased agricultural production, but also served as a source of valuable products throughout their useful life (e.g., meat, milk, fiber, fertilizer). In this sense, working animals certainly represent one of the most pronounced forms of exploitation of wildlife by humans, considering that they provide products and services during their lives and provide other products when slaughtered.

Over the centuries, working animals have been used for work in cultivation to agricultural production. The mechanization of agriculture led to a reduction in the use of working animals, especially in developed countries. Nonetheless, many people today still rely on animal power to complement human labor in agriculture and transport, and working animals remain vital to a large portion of the world's population, especially in developing countries. The importance of working animals is not restricted to rural areas, and their use is common in many cities in the world, such as to pull carts (for freight) and carriages (to transport people to tourist attractions). They are also fundamental in transporting products from rural areas to be marketed in cities. Regardless of where they are used, activity with working animals involves a large number of people in activities of great social, economic, cultural, and environmental importance.

Throughout their lives, working animals are subject to a diversity of situations during their training and the activities they perform, which have raised concerns about their well-being. Despite the services that these animals provide, there remain many situations in which they suffer ill-treatment, are poorly fed, or even abandoned when their work potential is reduced. Considering that the quality of life of working animals depends on human care, the need to adopt good animal welfare practices, including veterinary care, provision of food and other living conditions appropriate to the needs and nature of the animals, is evident.

References

Alves, R.R.N., 2016. Domestication of animals. In: Albuquerque, U.P. (Ed.), Introduction to Ethnobiology. Springer, Switzerland, pp. 221–225.

Anderson, M., Denis, R., 1994. Improving animal based transport: options, approaches, issues and impact. In: Starkey, P., Mwenya, E., Stares, J. (Eds.), Improving Animal Traction Technology. Proceedings of the First Workshop of the Animal Traction Network for Eastern and Southern Africa. Technical Centre for Agricultural and Rural Cooperation, The Netherlands.

Arriaga, C., Jordan, L.V.B., 1999. Economics of Draught Animal Ownership in Smallholder Campesino (Peasant) Hill Slope Agricultural Production Systems in the State of Mexico, Draught Animal News. Centre for Tropical Veterinary Medicine, University of Edinburgh, Edinburgh.

Aynard, J.M., 1972. Animals in Mesopotamia. In: Brodrick, A.H. (Ed.), Animals in Archaeology. Praeger, New York, pp. 42–68.

Barwell, L., Ayre, M., 1982. The Harnessing of Draught Animals. Intermediate Technology Publications, London.

Bökönyi, S., 1974. The History of Domestic Mammals in Central and Eastern Europe. Akademiai Kiado, Budapest.

Bowman, J.C., 1977. Animals for Man. Edward Arnold., London.

Bradbury, H., 2002. The Gharry Horses of Gonder Project Management Interventions to Improve the Welfare of the Gharry Horses of Gonder Ethiopia. University of Wales, Bancor, UK.

Davis, S.J.M., 1987. The Archaeology of Animals. Routledge, London.

Denis, R., 1999. Meeting the challenges of animal based transport. In: Starkey, P., Kaumbutho, P. (Eds.), Meeting the Challenges of Animal Traction. A Resource Book of the Animal Traction Network for Eastern and Southern Africa (ATNESA), Harare, Zimbabwe. Intermediate Technology Publications, London.

Diamond, J., 2002. Evolution, consequences and future of plant and animal domestication. Nature 418, 700–707.

FAO, Food and Agriculture Organization of the United Nations, 2007. The state of the World's animal genetic resources for food and agriculture. In: Rischkowsky, B., Pilling, D. (Eds.). Food and Agriculture Organization, Rome.

FAO, Food and Agriculture Organization of the United Nations, 2014. The Role, Impact and Welfare of Working (Traction and Transport) Animals, Animal Production and Health Report. No. 5, Rome.

Fuller, R.J., Aye, L.U., 2012. Human and animal power–The forgotten renewables. Renewable Energy 48, 326–332.

Goe, M.R., McDowell, R.E., 1980. Animal Traction: Guidelines for Utilization. Cornell International Agriculture, New York.

Green, M.W., 1980. Animal husbandry at Uruk in the archaic period. Journal of Near Eastern Studies 39, 1–35.

Greenfield, H.J., 1988. The origins of milk and wool production in the old world: a zooarchaeological perspective from the central Balkans [and comments]. Current Anthropology 29, 745–748.

Greenfield, H.J., 2005. A reconsideration of the secondary products revolution in south-eastern Europe: on the origins and use of domestic animals for milk, wool, and traction in the central Balkans. In: Mulville, J., Outram, A. (Eds.), Proceedings of the 9th Conference of the International Council of Archaeozoology, Durham, pp. 14–31.

Greenfield, H.J., Chapman, J., Clason, A.T., Gilbert, A.S., Hesse, B., Milisauskas, S., 1988. The origins of milk and wool production in the old World: a zooarchaeological perspective from the central Balkans [and comments]. Journal of Consumer Research 29, 573–593.

Guthiga, P.M., Karugia, J.T., Nyikal, R.A., 2007. Does use of draft animal power increase economic efficiency of smallholder farms in Kenya? Renewable Agriculture and Food Systems 22, 290–296.

Heinberg, R., 2006. Fifty million farmers. Energy Bulletin. Post Carbon Institute.

Heffner, H.E., 1999. The symbiotic nature of animal research. Perspectives in Biology and Medicine 43, 128–139.

Herold, P., Schlechter, P., Scharnhölz, R., 2009. Modern Use of Horses Inorganic Farming. European Draught Horse Federation.

IFAD, International Fund for Agricultural Development, 2011. Food Prices: Smallholder Farmers Can Be Part of the Solution. International Fund for Agricultural Development.

Jordão, L.R., Faleiros, R.R., Aquino Neto, H.M., 2011. Animais de trabalho e aspectos éticos envolvidos: revisão crítica. Acta Veterinaria Brasilica 5, 33–40.

Marković, N., Stevanović, O., Nešić, V., Marinković, D., Krstić, N., Nedeljković, D., Radmanović, D., Janeczek, M., 2014. Palaeopathological study of cattle and horse bone remains of the ancient roman city of Sirmium (Pannonia/Serbia). Revue de Médecine Vétérinaire 165, 77–88.

Muller, W., 2002. The First Steps of Animal Domestication. Oxbow Books, Oxford.

Pearson, A., 1991. Animal power: matching beast and burden. Appropriate Technology 18, 11–14.

Pearson, R.A., 1999. Work animal power. In: Payne, W.J.A., Wilson, R.T. (Eds.), An Introduction to Animal Husbandry in the Tropics, fifth ed. Blackwell Science, Oxford, pp. 782–798.

Pearson, R.A., 2005. Contributions to Society: Draught and Transport, Encyclopedia of Animal Science. Marcel Dekker Inc., USA, pp. 248–250.

Pritchard, J.C., Lindberg, A.C., Main, D.C.J., Whay, H.R., 2005. Assessment of the welfare of working horses, mules and donkeys, using health and behaviour parameters. Preventive Veterinary Medicine 69, 265–283.

Rahman, A., Kahn, S., 2014. Discussion Paper on the Future Role of the OIE With Respect to the Welfare of Working Animals. OIE Working Group on Animal Welfare.

Ramaswamy, N.S., 1998. Technology and management of working animal systems for increasing the productivity and welfare with special focus on donkeys. In: 3rd International Colloquium on Working Equines, Mexico.

Russell, N., 2002. The wild side of animal domestication. Society & Animals 10, 286–302.

Sherratt, A., 1981. Plough and pastoralism: aspects of the secondary products revolution. In: Hodder, I., Isaac, G., Hammond, N. (Eds.), Patterns of the Past: Studies in Honour of David Clarke. Cambridge University Press, Cambridge, pp. 261–305.

Sherratt, A., 1983. The secondary exploitation of animals in the old World. World Archaeology 15, 90–104.

Simoons, F.J., 1971. The antiquity of dairying in Asia and Africa. Geographical Review 61, 431–439.

Spugnoli, P., Dainelli, R., 2013. Environmental comparison of draught animal and tractor power. Sustainability Science 8, 61–72.

Starkey, P., 1994. A worldwide view of animal traction highlighting some key issues in eastern and southern Africa – improving animal traction technology. In: Starkey, P., Mwenya, E., Stares, J. (Eds.), First Workshop of the Animal Traction Network for Eastern and Southern Africa. Technical Centre for Agricultural and Rural Cooperation, Netherlands.

Starkey, P., 1995. Portraying animal traction in South Africa: empowering rural communities. South African Network of Animal Traction, Development Bank of South Africa.

Starkey, P., 2000. The history of working animals in Africa. In: Blench, R.M., MacDonald, K. (Eds.), The Origins and Development of African Livestock: Archaeology, Genetics, Linguistics and Ethnography. University College Press, London, UK.

Starkey, P., 2010. Sector analysis and policy branch (AGAL). Livestock for Traction: World Trends, Key Issues and Policy Implications, Background Paper Prepared for Livestock Information. Food and Agriculture Organisation (FAO), Rome, Italy, p. 43.

Wilson, R.T., 2003. The environmental ecology of oxen used for draught power. Agriculture, Ecosystems & Environment 97, 21–37.

CHAPTER

19

Wildlife Attractions: Zoos and Aquariums

Rômulo Romeu Nóbrega Alves[1], *Walter Lechner*[2]

[1]Universidade Estadual da Paraíba, Campina Grande, Brazil; [2]University of Vienna, Vienna, Austria

INTRODUCTION

Historically, humans have obtained dominance over numerous animal species, many of them being fundamental to their survival. People have been using animal-derived products mainly for food, as well as for clothing, tools, medicinal, and magic-religious purposes. Besides using animal-derived products, humans have always interacted with a lot of living animals. Those were kept at home as pets, kept and bred at farms, or observed in their natural habitats, in zoos, ponds, and aquariums. This evidences the human fascination for wildlife from time immemorial.

Domestication allowed close contact with animal species that have been kept as pets until present days. In modern times, besides keeping domesticated mammals, such as dogs and cats, and birds, wild animals have become increasingly common as pets. However, to keep wild, exotic, and sometimes dangerous species, it was necessary to maintain them in collections. This practice was recorded since ancient times, has been perpetuated throughout history, and has been evolved into zoos and aquariums. Nowadays those institutions are spread across the globe (Alves, 2012). Precursors were ancient menageries, where our forefathers showed animals for private and public purposes. The term "menagerie" dates to the early 16th century and refers to a "collection of wild animals kept in captivity." Menagerie is derived from the French word *ménage*, "to manage," and *rie*, a place (Veltre, 1996). The name Menagerie was originally applied to a place for domestic animals, with reference to their nurture and training, passed to mean "any collection of animals" (Rennie, 1831).

The admiration and fascination for wild animals is not restricted to those terrestrial species that are often kept in the menageries and zoos. In spite of the greater difficulties inherent in maintaining aquatic animals, especially larger ones, human interest in keeping them in captivity is also old (Alves, 2012). Robinson (1996) pointed that aquariums developed somewhat later than zoos, after the glass tank was perfected. When large saltwater tanks/aquariums began to be established in the late 19th century, they frequently were settled on the edge of the sea, because of a fixation with marine wonders

Ethnozoology
http://dx.doi.org/10.1016/B978-0-12-809913-1.00019-3

and proximity to sea life. Later they were also located inland (Robinson, 1996).

Nowadays a combination of zoos and aquariums at the same place is common. Those institutions present a great diversity of terrestrial and aquatic animals to the public and attract millions of visitors worldwide. Zoos and aquariums present an important and growing part of entertainment and leisure industry. But zoos and aquariums fulfill also functions of research, education, and conservation. All those aspects are shortly discussed in this chapter. It also contains a brief history of zoos and aquariums, which are the most important places for people living in urban areas to interact with wildlife—what demonstrates their importance from an ethnozoological perspective. As pointed by Wagoner and Jensen (2010), zoos represent one of the primary points of engagement between live animals, biological science, and publics of all ages.

BRIEF HISTORY OF ZOOS AND AQUARIUMS

The keeping of wild animals by humans has a very long history (Morris, 2011). Well before the establishment of zoos or aquariums, humans kept animals in captivity for diverse purposes, chiefly as creatures of worship (Zuckerman, 1979). Some animals were kept for companionship, such as cats and dogs; others were kept as symbols of wealth because powerful animals conferred superior status on their owners. Some species were believed to be incarnations of either gods or dead ancestors, and others were used as sacrifices.

The ancient Egyptians probably were the first in human history to establish kinds of zoos. Archeological excavations revealed that already around 3500 BC in the city Hierakonpolis, ancient capital of Egypt, animal species such as elephants, hippos, wildcats, baboons, and others were kept in captivity (Rose, 2010).

By the 3rd century BC, Romans had introduced violent uses of animals into their gladiatorial contests and triumphal processions (Zuckerman, 1979). Gladiators fought each other or fought against wild animals that were kept in cages beneath the stadium (Hopkins and Beard, 2011). A contest with a wild, half-starved animal was also a method of public execution. The games held in the Colosseum were hugely popular. The opening spectacle of the Colosseum games lasted for 100 days, and by its end, more than 9000 animals and 10,000 condemned prisoners had been killed (Allaby, 2010). The Roman public had an insatiable appetite for such events, but well-to-do citizens also kept pets and small menageries. Some people even kept pet lions, usually with their claws and teeth removed (Allaby, 2010).

Due to aquatic life, Chinese emperors kept and bred carps in vases and other tanks already many centuries BC, ancient Egyptians and Sumerians also kept fishes in their homes not only for nutritional but also for recreational purposes (Brunner, 2003). Plinius (AD 77) reported that wealthy Romans kept morays in special tanks in their properties. Keeping and breeding fishes was quite popular in the Roman Empire starting in the 1st century BC (Schmölke and Nikulina, 2008). The first book about ornamental (gold)fish breeding was probably written in the late 16th century in China (Zhang, 1596). The first report of fishes kept in glass bowls was maybe given in 1665 by the British aristocrat Samuel Pepys in his famous diaries, which have been published later in numerous editions (e.g., Latham and Mattews, 1970).

In Europe, aquariums became very popular in the middle of the 19th century. The first "real aquarium" was presented to the public in the Great Exhibition in London 1851. As pointed out by Clary and Wandersee (2005), in the 1850s and early 1860s, several issues of the English Punch magazine, a popular magazine containing cartoons, revealed illustrations attesting to the popularity of aquariums in the home.

The English naturalist and writer Philip Henry Gosse wrote two books, which became very popular, *A naturalist's rambles on the Devonshire coast* and *The aquarium: an unveiling of the wonders of the deep sea* (Gosse, 1853, 1854). Those books were the beginning of the first aquarium boom in England.

Only little later in Leipzig, Germany, the biologist Emil Adolf Roßmäßler published two articles in a very popular magazine and finally wrote a book about aquariums (Roßmäßler, 1854, 1856, 1857), which started the aquarium boom in the German-speaking countries.

Home aquariums led to public aquariums, and in 1853, the first public aquarium was opened in Regent's Park, London. Public aquariums soon followed in other English locations, as well as in Germany, France, and the United States. The aquarium concept, using self-contained and self-sustaining systems, was originated in England during the 1850s. Prior to this development, there were ornamental fish ponds, these being open systems supplied with water from nearby seas, lakes, or streams (Kisling, 2001).

From those times on, not only aristocrats and wealthy people were keeping fishes and other aquatic life in their homes, but this hobby became also popular for normal people. Huge advancements in aquarium techniques, related not only to tanks, but also to heating, filtration, and lighting, made it possible to keep not only tough species, but also sensitive ones from tropical regions and coral reefs. Nowadays freshwater and saltwater aquariums are widespread and popular in homes around the world.

From the very earliest times, rulers and wealthy aristocrats have assembled collections of wild or exotic animals and maintained them in parks or gardens adjacent to their palaces. The animals were not kept in captivity so that natural historians might study them, but for entertainment, and the animals themselves were usually chosen because, although exotic, they were not so violent as to be difficult or dangerous to manage (Allaby, 2010). These private collections or "menageries" of animals were maintained for entertainment or to impress visitors, and became status symbols for their princely owners (Allaby, 2010; Anderson, 1995). One such collection was the Versailles menagerie opened in 1665, when Louis XIV arranged a botanical garden and an enclosure for lions and elephants around his house in a pattern that is said to have inspired Bentham's Panopticon prison of the late 18th century (Mullan and Marvin, 1998). In 1804, in keeping with the democratic order of postrevolutionary France, the royal animals were moved to Paris and a zoological garden at *Jardin des Plantes* was opened to the public (Anderson, 1995).

In the Middle Ages, most European monarchs and many nobles owned menageries (Allaby, 2010). An early example is that of the Emperor Charlemagne in the 8th century. He had three menageries, at Aachen, Nijmegen, and Ingelheim, located in present-day Netherlands and Germany. They housed the first elephants seen in Europe since the Roman Empire, in addition to other animals such as bears, camels, lions, monkeys, and rare birds (Loisel, 1912). Charlemagne received exotic animals for his collection as gifts from rulers of Africa and Asia (James, 1966). In the 18th century, traveling collections of animals were popular in Europe and in America as a form of popular entertainment and were also known as menageries. In 1716, in Boston, a lion became the first wild animal to be exhibited in America. Other animals quickly followed. In America these exhibitions were combined with circus shows based on clowns, acrobats, and other human as well as animal performers. A single ticket allowed entrance to both the circus and the menagerie, and the menagerie was as popular and important as the circus (Allaby, 2010).

The modern zoo has its origins in the royal menageries of ages past, where animals were kept for the purposes of exhibition and entertainment for the privileged few (Bennett, 1829). The earliest important Western European menageries after the Middle Ages began in the

Italian, Portuguese, and Spanish courts. The most important transitions to the modern zoo took place in Vienna, Berlin, and Paris. The oldest still-existing zoo, the Vienna Zoo in Austria, evolved from the Imperial Menagerie at the Schönbrunn Palace in Vienna, an aristocratic menagerie founded by the Habsburg monarchy (Strehlow, 2000). The zoo was constructed by Adrian van Stekhoven in 1752 next to Schloss Schönbrunn at the order of the Holy Roman Emperor Francis I, husband of Maria Theresia, to serve as an imperial menagerie (Morris, 2011; Strehlow, 2000). It was centered around a pavilion intended for use for imperial breakfasts (Fig. 19.1). The central pavilion and the menagerie building were built by Jean Nicolas Jadot de Ville-Issey (Baratay and Hardouin-Fugier, 2004). A small zoo had already existed on the premises since 1540, but the complex was only opened to the public in 1779 (Kunze, 2000).

ZOOS AND AQUARIUMS: TOURISM AND ENTERTAINMENT

Today most larger cities have their zoos and/or aquariums (Figs. 19.2–19.4); the number of zoos worldwide is nearly innumerable (Bell, 2001). Those institutions are maintained by public authorities or private institutions and foundations, and in Manaus, Brazil, there is even a zoo run by the military, the Zoológico do Centro de Instrução de Guerra na Selva, called CIGS zoo.

Nearly 500 public zoos and aquariums exist in the United States, and more than 700 zoos and aquariums exist only in the German-speaking countries of Europe (Germany, Switzerland, and Austria) (Petzold and Sorge, 2012). The largest zoo of the world is the San Diego Zoo Safari Park with 728 ha; the zoo housing most different animal species is the Berlin Zoo and Aquarium in Germany with nearly 1500 species. The biggest aquarium was for many years the Georgia Aquarium in Atlanta with a 30 million liter tank, recently topped by the Chinese Chimelong Ocean Kingdom on the island Hengqin with nearly 49 million liters.

Hundreds of millions of people visit those institutions every year, and many zoos are large attractions not only for tourists, but also for the local population. Some of the aquariums also present shows with trained dolphins and other animals, which are crowd pullers, nevertheless discussed ambivalently because of animal welfare.

FIGURE 19.1 Central pavilion of the Schönbrunn Zoo, Vienna, Austria, where the Imperial family used to go to watch wild animals and for imperial breakfasts. *Photo credits: Rômulo R.N. Alves.*

ZOOS AND AQUARIUMS AS SCIENTIFIC AND EDUCATIONAL INSTITUTIONS

Modern zoos and aquariums exist to educate. There is recreation, there is research, but maybe most important is education of young people. A zoo helps children understand that taking care of our environment has a significant impact on animal life and welfare. The educational value of zoos for children and adolescents is enormous (Jensen, 2011). For many people in our modern world, zoos and aquariums are also a rare opportunity to get in real close contact to wildlife and nature, apart from the daily TVs and computers.

FIGURE 19.2 Some examples of animals kept in zoos. (A, B, and E) Zoo Schönbrunn, Vienna, Austria; (C) Buenos Aires Zoo; (D) Bronx Zoo, New York, United States; and (F) Lujan Zoo, Buenos Aires, Argentina. *Photo credits: (A–C and E) Rômulo Romeu Nóbrega Alves, (D) Itamar Barbosa, and (F) Christine Eloy.*

There has always been science in zoos, and aquariums have always been used for science (Marliave and Newman, 1995; Schwemmer and Thompson, 1995). Many scientists kept and keep aquatic creatures and plants in tanks for their studies (Figs. 19.5–19.9).

The British chemist Joseph Priestley, who discovered the oxygen, studied water plants in tanks already in the second half of the 18th century. In Messina, Sicily, the French marine biologist Jeanne Villepreux-Power studied marine life, including the paper nautilus, *Argonauta argo*,

FIGURE 19.3 Big aquariums with large tropical fishes are popular in modern zoos. This Amazon tank is in the Zoo Berlin, Germany. *Photo credits: Walter Lechner.*

FIGURE 19.5 Many aquariums and zoos do not only present exotic wildlife but also many native animals. This tank with coldwater fishes is from the Vienna zoo in Austria. *Photo credits: Walter Lechner.*

FIGURE 19.4 Due to modern and advanced techniques, it is possible to keep corals and other sensitive animals in aquariums. This is a rather small coral reef tank in the aquarium of the Vienna zoo. *Photo credits: Walter Lechner.*

FIGURE 19.6 "Tunnel-style" aquariums, where people can pass "below water," are very popular nowadays. This is an Amazon fish tank in the aquarium of the Vienna zoo. *Photo credits: Egon Heiss.*

in tanks and aquariums in the 1830s (Swaby, 2015). The British zoologist Anna Thynne maintained corals and sponges in marine aquariums at her home to study them and was probably the first to build a stable sustained marine aquarium (Stott, 2003). In the early 20th century, Austrian Nobel Prize winner Konrad Lorenz studied the territorial behavior of sticklebacks in tanks, and nowadays aquariums are common in science not only to keep study species but also to conduct behavioral or chemical studies. Not only aquariums in science institutions are used for scientific purposes, but zoos and public aquariums are also important places to conduct science. Countless students and scientists do their studies in those institutions, mainly performing behavioral studies on animals or working in animal conservation studies.

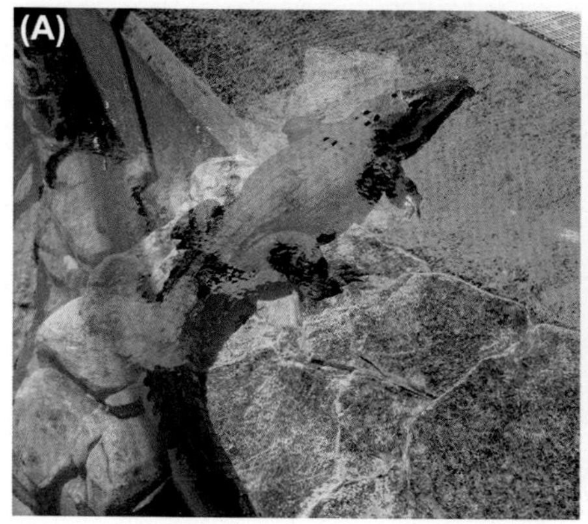

FIGURE 19.7 A pond (A) and a tank (B) with huge saltwater crocodiles (*Crocodylus porusus*) in the public aquarium "Crocosaurus Cove" in Darwin, Australia (B). *Photo credits: Walter Lechner.*

THE ROLE OF ZOOS AND AQUARIUMS IN ANIMAL CONSERVATION

Nowadays zoos and aquariums play an important role in conservation of endangered species (Hayward, 2011; Hoffmann et al., 2010). The question if some species could be preserved in zoos and aquariums after extinction in the wild has already been discussed for many decades (e.g., Conway, 1969; Olney et al., 1993; Soulé et al., 1986; Tudge, 1992).

Modern zoos are organized, and many of those institutions contribute to preservation of some species in international breeding programs (Barongi et al., 2015; Penning et al., 2009). The World Association of Zoos and Aquariums (WAZA) is the umbrella organization of most programs (http://www.waza.org/en/site/conservation). It brings together the European Endangered Species Programme (EEP), the species survival plans of the American Zoo and Aquarium Association (AZA), the Australasian Species Management Program (ASMP) of the Australasian Regional Association of Zoological Parks and Aquaria (ARAZPA), the African Preservation Program (APP) of the Pan-African Association of Zoological Gardens and Aquaria (PAAZA), the activities of the Japanese Association of Zoos and Aquariums (JAZA), the South Asian Zoo Association for Regional Cooperation (SAZARC), and the South East Asian Zoo Association (SEAZA). Actually 22 national or regional zoo and aquarium organizations and 213 zoos and aquariums in 46 nations take part in those activities. According to WAZA and the IUCN Red List of Threatened Species, there are actually 33 species known as extinct in the wild, but kept in captivity, and 31 of these species are actively bred in zoos, aquariums, and similar institutions, preventing their

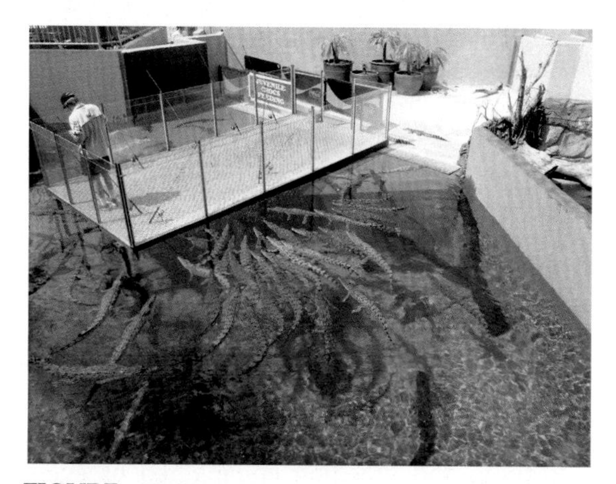

FIGURE 19.8 As attraction visitors can feed small crocodiles with fish rods at Crocosaurus Cove in Darwin, Australia. *Photo credits: Walter Lechner.*

FIGURE 19.9 Beautiful biotope tanks with plants (A and B) and rainbow fishes in the Territory Wildlife Park near Darwin, Northern Territory, Australia (A). *Photo credits: Walter Lechner.*

outright extinction. Nearly 20 species more have already been extinct in nature, but resettled from populations in captivity.

Famous examples for resettlement of species after extinction in the wild are the California condor (Toone and Wallace, 1994) and Przewalski's horse. Several European and North American zoos are cooperated in a breeding program of Przewalski's horse.

But more than 50 animal species, such as mollusks, crustaceans, insects, fishes, amphibians, reptilians, birds, and mammals, have already been resettled in their natural habitats after being extinct in the wild.

Zoos and aquariums are also very active in keeping and breeding endangered or even extinct in the wild fish species, and for those activities they often cooperate with private aquarists. Those groups mainly work with small freshwater fish species, which can also be easily kept at home by specialist aquarists. For example, there are groups taking care of splitfins (Goodeidae), pupfish (a group of the family Cyprinodontidae), African cichlids of Lake Victoria (Olney et al., 1993), and Australian and Papua New Guinean Rainbow fishes (Melanotaeniidae).

Conservation, education, research, and recreation are the main goals of zoos and aquariums. The importance of zoos and aquariums for the first of those four goals is by far greatest, and worldwide local zoos and aquariums are in some way included in most nature conservation and protection programs.

THE CONTROVERSIAL DEBATE ON THE ROLE OF ZOOS AND AQUARIUMS

As mentioned previously, modern zoos and aquariums fulfill four broad functions: entertainment, research, education, and conservation. All these justifications have been criticized by scholars in disciplines ranging from anthropology to zoology (Jamieson, 2015), and in recent years, zoos and aquariums have come under intense criticism. Counterarguments mainly refer ethics, animal rights, and animal welfare (Bostock, 2003; Jamieson, 2015). For example, anti-zoo/aquarium scholars challenge that the conditions of maintenance of zoo animals are adequate for a "good life" and they also doubt the value of scientific research conducted in these institutions (Young, 2016). Others argue that entertainment for humans should not be at the expense of animals that have to be caged. A further argument against zoos is that using captive animals for learning about fauna is not a good way to educate as the animals are perceived as enslaved. And also the prospect of saving endangered species with captive animals is not ideal (Mullan and Marvin, 1998; Munro, 2004). This scenery evidences that the zoos and aquariums represent one of more controversial issues in human–animal interactions, rethinking the social, cultural, economical, and environmental importance of those institutions. As pointed by Wickins-Dražilová (2006), the continuing existence of zoos (and aquariums) and their positive functions (education, recreation, conservation, and science) can be ethically justified only if zoos guarantee the welfare of their animals.

FINAL REMARKS

People have been capturing wild animals for different purposes, and zoos and aquariums originated from those habits—institutions, which are under the eldest in human history. In the beginning, the motivation for keeping collections of wild animals was personal pleasure and to show off wealth and status. But zoos and aquariums continued to evolve and their role as cultural institutions has changed in many ways. Nowadays, reasons for keeping animals in zoos and aquariums include recreational, educational, research, and conservation purposes. Nevertheless, nowadays the importance of those institutions is discussed ambivalently, questioning ethic and animal welfare topics.

Irrespective of this controversy, zoos and aquariums are spread across the world. They directly involve a large number of people who work there as animal keepers, veterinarians, researchers, etc. By the way, more than 700 million people visit zoos and aquariums globally each year (www.waza.org/en/site/zoos-aquariums). There are no exact numbers about how many zoos and aquariums are in the world, because some countries do not license their zoos and so official records do not exist. Nevertheless, it is estimated that there are more than 2800 zoos and aquariums in the world. Those numbers show the ethnozoological importance of those old institutions. They are the first places for interactions of persons in urban areas with wild animals.

References

Allaby, M., 2010. Animals: From Mythology to Zoology. Facts On File, Inc., New York.

Alves, R.R.N., 2012. Relationships between fauna and people and the role of ethnozoology in animal conservation. Ethnobiology and Conservation 1, 1–69.

Anderson, K., 1995. Culture and nature at the Adelaide zoo: at the frontiers of 'human' geography. Transactions of the Institute of British Geographers 20, 275–294.

Baratay, E., Hardouin-Fugier, E., 2004. Zoo: A History of Zoological Gardens in the West. Reaktion Books.

Barongi, R., Fisken, F.A., Parker, M., Gusset, M., 2015. Committing to Conservation: The World Zoo and Aquarium Conservation Strategy.

Bell, C.E., 2001. Encyclopedia of the World's Zoos. Fitzroy Dearborn Publishers, London and Chicago.

Bennett, E.T., 1829. The Tower Menagerie: Comprising the Natural History of the Animals Contained in that Establishment; with Anecdotes of Their Characters and History. Jennings, London.

Bostock, S.S.C., 2003. Zoos and Animal Rights. Routledge.

Brunner, B., 2003. Wie das Meer nach Hause kam. Die Erfindung des Aquariums. Transit Buchverlag, Berlin.

Clary, R.M., Wandersee, J.H., 2005. Through the looking glass: the history of aquarium views and their potential to improve learning in science classrooms. Science & Education 14, 579–596.

Conway, W.G., 1969. Zoos: their changing roles. Science 163, 48.

Gosse, P.H., 1853. A Naturalist's Rambles on the Devonshire Coast. John Van Vorst, London.

Gosse, P.H., 1854. The Aquarium: An Unveiling of the Wonders of the Deep Sea. John Van Vorst, London.

Hayward, M.W., 2011. Using the IUCN Red List to determine effective conservation strategies. Biodiversity and Conservation 20, 2563–2573.

Hoffmann, M., Hilton-Taylor, C., Angulo, A., Böhm, M., Brooks, T.M., Butchart, S.H.M., Carpenter, K.E., Chanson, J., Collen, B., Cox, N.A., Darwall, W.R.T., Dulvy, N.K., Harrison, L.R., Katariya, V., Pollock, C.M., Quader, S., Richman, N.I., Rodrigues, A.S.L., Tognelli, M.F., Vié, J.-C., Aguiar, J.M., Allen, D.J., Allen, G.R., Amori, G., Ananjeva, N.B., Andreone, F., Andrew, P., Ortiz, A.L.A., Baillie, J.E.M., Baldi, R., Bell, B.D., Biju, S.D., Bird, J.P., Black-Decima, P., Blanc, J.J., Bolaños, F., Bolivar, G.W., Burfield, I.J., Burton, J.A., Capper, D.R., Castro, F., Catullo, G., Cavanagh, R.D., Channing, A., Chao, N.L., Chenery, A.M., Chiozza, F., Clausnitzer, V., Collar, N.J., Collett, L.C., Collette, B.B., Fernandez, C.F.C., Craig, M.T.,

Crosby, M.J., Cumberlidge, N., Cuttelod, A., Derocher, A.E., Diesmos, A.C., Donaldson, J.S., Duckworth, J.W., Dutson, G., Dutta, S.K., Emslie, R.H., Farjon, A., Fowler, S., Freyhof, J., Garshelis, D.L., Gerlach, J., Gower, D.J., Grant, T.D., Hammerson, G.A., Harris, R.B., Heaney, L.R., Hedges, S.B., Hero, J.-M., Hughes, B., Hussain, S.A., Icochea, M.J., Inger, R.F., Ishii, N., Iskandar, D.T., Jenkins, R.K.B., Kaneko, Y., Kottelat, M., Kovacs, K.M., Kuzmin, S.L., La Marca, E., Lamoreux, J.F., Lau, M.W.N., Lavilla, E.O., Leus, K., Lewison, R.L., Lichtenstein, G., Livingstone, S.R., Lukoschek, V., Mallon, D.P., McGowan, P.J.K., McIvor, A., Moehlman, P.D., Molur, S., Alonso, A.M., Musick, J.A., Nowell, K., Nussbaum, R.A., Olech, W., Orlov, N.L., Papenfuss, T.J., Parra-Olea, G., Perrin, W.F., Polidoro, B.A., Pourkazemi, M., Racey, P.A., Ragle, J.S., Ram, M., Rathbun, G., Reynolds, R.P., Rhodin, A.G.J., Richards, S.J., Rodríguez, L.O., Ron, S.R., Rondinini, C., Rylands, A.B., Sadovy de Mitcheson, Y., Sanciangco, J.C., Sanders, K.L., Santos-Barrera, G., Schipper, J., Self-Sullivan, C., Shi, Y., Shoemaker, A., Short, F.T., Sillero-Zubiri, C., Silvano, D.L., Smith, K.G., Smith, A.T., Snoeks, J., Stattersfield, A.J., Symes, A.J., Taber, A.B., Talukdar, B.K., Temple, H.J., Timmins, R., Tobias, J.A., Tsytsulina, K., Tweddle, D., Ubeda, C., Valenti, S.V., Paul van Dijk, P., Veiga, L.M., Veloso, A., Wege, D.C., Wilkinson, M., Williamson, E.A., Xie, F., Young, B.E., Akçakaya, H.R., Bennun, L., Blackburn, T.M., Boitani, L., Dublin, H.T., Fonseca, G.A.B., Gascon, C., Lacher, T.E., Mace, G.M., Mainka, S.A., McNeely, J.A., Mittermeier, R.A., Reid, G.M., Rodriguez, J.P., Rosenberg, A.A., Samways, M.J., Smart, J., Stein, B.A., Stuart, S.N., 2010. The impact of conservation on the status of the World's vertebrates. Science 330, 1503–1509.

Hopkins, K., Beard, M., 2011. The Colosseum. Profile Books.

James, F., 1966. Zoos of the World: The Story of Animals in Captivity. Aldus Book, London.

Jamieson, D., 2015. Against zoos. In: Pojman, L.P., Pojman, P., McShane, K. (Eds.), Environmental Ethics: Readings in Theory and Application. Nelson Education, pp. 121–128.

Jensen, E., 2011. Learning about Animals, Science and Conservation at the Zoo: Large-Scale Survey-Based Evaluation of the Educational Impact of the ZSL London Zoo Formal Learning Programme. University of Warwick and Zoological Society of London, London.

Kisling, V.N., 2001. Zoo and Aquarium History: Ancient Animal Collections to Zoological Gardens. CRC Press, London.

Kunze, G., 2000. Tiergarten Schönbrunn. LW Werbe- und Verlags GmbH, Vienna.

Latham, R., Mattews, W., 1970. The Diary of Samuel Pepys – A New and Complete Transcription. vol. 6. Bell & Hyman, London, p. 1665.

Loisel, G., 1912. Histoire des menageries de Pantiquite a nos jours. Octave Doin et Fils, Paris.

Marliave, J.B., Newman, M.A., 1995. The role of aquariums in promoting scientific research. In: Schwemmer, C.M. (Ed.), The Ark Evolving: Zoos and Aquariums in Transition. Front Royal. Smithsonian Institution Conservation and Research Center, Washington.

Morris, D., 2011. A short history of zoos. In: Rees, P.A. (Ed.), An Introduction to Zoo Biology and Management. Wiley-Blackwell, pp. 31–48.

Mullan, B., Marvin, G., 1998. Zoo Culture. Univ of Illinois Pr.

Munro, L., 2004. Animals,' nature' and human interests. In: White, R. (Ed.), Controversies in Environmental Sociology. Cambridge University Press, Cambridge, pp. 61–76.

Olney, P.J., Mace, G., Feistner, A., 1993. Creative Conservation: Interactive Management of Wild and Captive Animals. Springer, Dordrecht.

Penning, M., Reid, G., Koldewey, H., Dick, G., Andrews, B., Arai, K., Garratt, P., Gendron, S., Lange, J., Tanner, K., 2009. Turning the Tide: A Global Aquarium Strategy for Conservation and Sustainability. World Association of Zoos and Aquariums Executive Office, Bern.

Petzold, D., Sorge, S., 2012. Abenteuer Zoo, second ed. L. Stocker Verlag, Graz.

Plinius, G.S.M., AD 77. Naturalis Historia, Rome.

Rennie, J., 1831. The Menageries: Quadrupeds Described and Drawn from Living Subjects. Charles Knight, London.

Robinson, M., 1996. Foreword. In: Hoage, R.F., Deiss, W.A. (Eds.), New Worlds, New Animals: From Menagerie to Zoological Park in the Nineteenth Century. Johns Hopkins University Press, Baltimore, MD/London.

Rose, M., 2010. World's first zoo – Hierakonpolis. Egypt Archaeology Magazine 63, 25.

Roßmäßler, E.A., 1854. Der Ocean auf dem Tische. Die Gartenlaube 2, 392.

Roßmäßler, E.A., 1856. Der See im Glase. Die Gartenlaube 4, 252–256.

Roßmäßler, E.A., 1857. Das Süßwasser-Aquarium. Eine Anleitung zur Herstellung und Pflege desselben. Mendelssohn, Leipzig.

Schmölke, U., Nikulina, E.A., 2008. Fischhaltung im antiken Rom und ihr Ansehenswandel im Licht der politischen Situation. Schriften des naturwissenschaftlichen Vereins für Schleswig-Holstein 70, 36–55.

Schwemmer, C.M., Thompson, S., 1995. A short history of scientific research in zoological gardens. In: Schwemmer, C.M. (Ed.), The Ark Evolving : Zoos and Aquariums in Transition. Front Royal: Smithsonian Institution Conservation and Research Center, Washington.

Soulé, M., Gilpin, M., Conway, W., Foose, T., 1986. The millenium ark: how long a voyage, how many staterooms, how many passengers? Zoo Biology 5, 101–113.

Stott, R., 2003. Theatres of Glass: The Woman Who Brought the Sea to the City. Short Books Ltd., London.

Strehlow, H., 2000. Zoological gardens of Western Europe. In: Kisling, V.N. (Ed.), Zoo and Aquarium History: Ancient Animal Collections to Zoological Gardens. CRC Press, pp. 75–116.

Swaby, R., 2015. Headstrong: 52 Women Who Changed Science-and the World. Broadway Books, New York.

Toone, W.D., Wallace, M.P., 1994. The extinction in the wild and reintroduction of the California condor (*Gymnogyps californianus*). In: Olney, P.J.S., Mace, G.M., Feistner, A.T.C. (Eds.), Creative Conservation: Interactive Management of Wild and Captive Animals. Springer, Dordrecht, pp. 411–419.

Tudge, C., 1992. Last Animals at the Zoo: How Mass Extinction Can Be Stopped. Island Press.

Veltre, T., 1996. Menageries, metaphors, and meanings. In: Hoage, R.J., William, A.D. (Eds.), New Worlds, New Animals: From Menagerie to Zoological Park in the Nineteenth Century. John Hopkins University Press, Baltimore, pp. 19–29.

Wagoner, B., Jensen, E., 2010. Science learning at the zoo: evaluating children's developing understanding of animals and their habitats. Psychology & Society 3, 65–76.

Wickins-Dražilová, D., 2006. Zoo animal welfare. Journal of Agricultural and Environmental Ethics 19, 27–36.

Young, C.C., 2016. Zoos and aquariums. In: Montgomery, G.M., Largent, M.A. (Eds.), A Companion to the History of American Science. Wiley-Blackwell, pp. 553–565.

Zhang, Q., 1596. Book of Vermilion Fish. China. (Translated from simplified Chinese: 张谦德, 朱砂鱼谱).

Zuckerman, L., 1979. Great Zoos of the World: Their Origins and Significance. Weidenfeld and Nicolson., London.

CHAPTER

20

From Roman Arenas to Movie Screens: Animals in Entertainment and Sport

Rômulo Romeu Nóbrega Alves, Raynner Rilke Duarte Barboza

Universidade Estadual da Paraíba, Campina Grande, Brazil

INTRODUCTION

Human lives are intimately connected to animals. As shown in the different chapters of this book, animals play several roles in human societies, and humans depend directly or indirectly on animals in various ways. Reflecting their role of dominance over other animals, humans have exploited them as resources in many ways, among which are the activities of sport and entertainment (Alves, 2012; Alves and Souto, 2015; Bowman, 1977).

Human entertainment is inherent in the various types of interactions they have with animals, and was most likely already present in some of the oldest human activities involving animals, such as hunting and fishing. The feeling of pleasure in the practice of these activities surely provided additional stimulus, even if the main reason for the activities was to obtain animal resources (Alves et al., 2009; Barboza et al., 2016; Fernandes-Ferreira et al., 2012; Young et al., 2016). However, in some cases the aspects of leisure and entertainment represent the main motivation for an activity, as is the case with sport hunting and fishing (Freire et al., 2016; Ihde et al., 2011; Lindsey et al., 2007). Bowman (1977) corroborates this idea by highlighting that "Some modern forms of sport involving animals probably derive from the days when humans needed to hunt to obtain their food."

The uses of animals as objects of human entertainment are diverse and have been recorded in early written records. Ancient documents from Assyria, Babylonia, and Egypt speak of walled menageries and "paradise parks" stocked with wild animals (Kalof, 2011). The wealthy and dominant, particularly sovereigns, collected wildlife, sometimes undertaking excursions into isolated regions for that purpose. They were normally maintained to entertain and amaze others, as an attribute of their supremacy, and a representation of the geographical extent of their dominion. Eventually, the fauna kept in such imperial menageries became what is considered the start of zoological gardens (Allaby, 2010).

Ancient Rome's capture, transport, and display of wild "beasts" for public amusement is also amply documented (Allaby, 2010; Lucius, 2015). Public games were a major part of Roman culture and involved animals on a

Ethnozoology
http://dx.doi.org/10.1016/B978-0-12-809913-1.00020-X

FIGURE 20.1 External and internal view of Coliseum of Rome (Italy), where animal games took place in ancient Rome. *Photo credits: Tacyana P.R. Oliveira.*

massive scale (Coley, 2010). Entertainment contests that employed the presence of animals were known as *venationes* (plural of venatio), a derivation of the Latin words, *venator* and *venation* (hunter and hunt), cognates of the verb *venor*, -ari, and the noun vena (blood vessel) (Kyle, 1998). The most famous of the places where *venationes* took place were the gigantic Flavian Amphitheater and the Coliseum of Rome (Fig. 20.1) (Bomgardner, 2013). Prior to the Coliseum, there was a variety of venues in which the people of Rome viewed the spectacle of a wild-beast hunt (Kidd, 2012). Entertainment shows pitting ferocious beasts against gladiators first took place in open areas in Rome, such as forums or circuses, but these places were quite dangerous because they put the lives of spectators at risk of injury by thrown weapons or by escaped animals (Adkins and Adkins, 1994). In the earlier times before the Coliseum, such *venationes* often took place in the Circus Maximus, and originally included only animals considered Italian (e.g., bears, boars, deer, and bulls) before progressing to the use of exotic animals brought in during the later period of the Republic (Kidd, 2012).

In the coliseum, many animals were used in *venationes* of lush demonstrations and exhibitions of tamed and executed animals. The majority of specimens used are known due to historical and iconographic fonts, and their occurrence in the amphitheater confirmed by archeological data. Examples of literary sources were emphasized by Cassius Dio, Pliny, and Historia Augusta (coliseo; coliseum). Countless animals, fierce and timid, carnivores and herbivores, from elephants to ostriches, were killed in spectacles in Rome (Kyle, 1998). For example, just during the inaugural games of the amphitheater (in 80 CE), over 9000 wild animals were slaughtered, while in 107 CE the Emperor Trajan celebrated his military conquest of Dacia with 120 days of celebrations involving the killing of more than 11,000 animals (Shelton, 2007). In addition to horses used in circuses and amphitheaters, animals imported from various parts of the world were put on exposition to the public, not only to be part of the show, but also for another purpose: to play an active role in human games, whether as a hunter or prey (Coley, 2010). Even though these great shows gradually became sporadic due to the fall of the Western Roman Empire, other practices of violent entertainment encompassing nonhuman animals persisted into the Medieval Age and the Renaissance, including cockfighting, bullbaiting, bearbaiting, and several other animal blood sports, which would naturally result in the violent death of the participants involved (Kiser, 2011).

Throughout recorded history, humans continued to exploit millions of animals for entertainment or sport. People sought to watch them directly in natural environments (e.g., bird watching, safaris) or in local circuses, arenas, zoos, aquatic parks, and racetracks. Animals became attractions of television channels and films and were represented in documentaries, movies, and cartoons. Therefore, there is an enormous human investment of time, energy, and money in animal-related leisure and recreation (Bryant, 1993), evidencing its importance to a wide variety of aspects of human life. In this chapter, we briefly discuss the several ways humans have obtained pleasure and entertainment from animals, focusing on the major animal groups involved and the human activities derived from these interactions.

USING ANIMALS AS ENTERTAINMENT

Humans are part of nature, the very place where they acquire the resources needed for their survival. Natural environments also offer the opportunity for humans to find pleasure, a crucial aspect to their well-being. As pointed out by several authors (Arlinghaus et al., 2007; Franklin, 1999; Schwab, 2004), the most natural source of pleasure is nature, and in this regard animals stand out as having been exploited in different manners by humans. In rural areas, interactions with animals result in varied moments of leisure and entertainment for people, while in urban areas, people seek contact with wild animals in zoos, oceanariums, and aquariums, or travel in search of contact with wild animals in natural areas that allow the observation of wildlife (Shackley, 1996). A large part of the worldwide human population keeps domestic or wild animals as pets, which form part of their daily lives, including their leisure activities. This diversity of interactions with fauna is recognized as being important to human well-being by providing experiences that, among others, involve social, psychological, aesthetic, physical, culinary, scientific, philosophic, and spiritual components, all of which can be a source of pleasure (Arlinghaus et al., 2007). Some of the key interactions with animals that provide leisure, entertainment, and pleasure are discussed briefly below.

Sport Hunting and Fishing

The motivations that lead humans to hunt or fish are diverse, ranging from the acquisition of a food resource to fulfilling cultural beliefs. Among these motivations is the entertainment associated with the activity of hunting and fishing, something that has been well documented since ancient times when kings and nobles undertook hunts as a sport and pastime. They hunted lions and large wild animals to show their power and wealth. In the Old Kingdom of ancient Egypt, pharaohs and other "dignitaries hunted large animals for recreation: the peasants hunted smaller animals—geese, ducks and quail—to supplement their meagre diets" (Usman, 2009). Allsen (2006) states that for elites across Eurasia, hunting served as entertainment on many levels; it was something to do and something to talk about, at times quite incessantly. Furthermore, hunting was also a spectator sport, with great audiences present to watch events involving game (Allsen, 2006). While the act of hunting for recreation has changed over the course of history, the sport has always been popular for the status it imparts on the hunter. Historically, it was widely believed that a trophy could not be bought, but had to be earned with knowledge, skill, and experience in the name of sportsmanship (Brower, 2005). Currently, sport hunting (or trophy hunting) continues to be practiced by representatives of the nobility (in addition to other followers), who engage in this sport in their respective countries or in Africa (Figs. 20.2 and 20.3). In countries such as England, for example, hunting—especially for

FIGURE 20.2 Trophy hunter with prize hippopotamus (Zambia, May 2010). *From Lord Mountbatten, via Wikimedia Commons.*

FIGURE 20.3 Hunter carrying a huge bear's head on his back in Kodiak National Wildlife Refuge (Alaska, USA), an important recreational sport place. *From William A. Troyer [Public domain], via Wikimedia Commons.*

fox—remains a fun activity for the traditional nobility. Sport hunting is also very popular in countries of North America, particularly upon the opening of the hunting seasons of several species of vertebrates. A report by the International Fund for Animal Welfare (IFAW, 2016) estimated that as many as 1.7 million hunting trophies could have been traded between nations from 2004 to 2014. At least 200,000 trophies of threatened taxa, or an average of 20,000 trophies per year, were traded between nations during the same period.

Similarly, recreational fishing has a long history and has received increasing attention recently (Fig. 20.4) (Arlinghaus et al., 2007). One of the earliest documents showing an ancient angling technique as a recreational activity, and not motivated by personal or commercial consumption, was derived from a painting of 3290 BC, which shows a noble Egyptian citizen holding a fish as a trophy (Pitcher and Hollingworth, 2002). The Greeks and the ancient Romans regarded the practice of recreational fishing as an activity employed only by peasants, slaves, and children and was therefore a practice to be avoided by nobles and scholars of society (Pitcher and Hollingworth, 2002). In Europe, fishing as a source of protein and for recreation greatly increased in popularity and was practiced by many people in the 19th and 20th centuries (Arlinghaus et al., 2007). Currently, hunting and fishing as sport are disseminated activities that have attracted the attention of several researchers worldwide (Arlinghaus et al., 2007).

Watching Animals

Animals have always fascinated human beings and the simple practice of observing them provides a source of satisfaction for many people. Dedicating part of one's time to watching animals, whether in their natural environment or in captivity, is an old practice and has been perpetuated to the present day. Wildlife interaction (observing, touching, feeding, photographing, or otherwise experiencing wild animals), therefore, occurs in a wide variety of settings throughout the world (Orams, 2002), and several studies reveal that it is becoming increasingly popular (Arlinghaus et al., 2007; Garrod and Gössling, 2008; Hoppe, 2005;

FIGURE 20.4 Sport fishing on the coast of Paraiba State, Northeast Brazil. *Photo credits: Benigno Veloso.*

FIGURE 20.5 People watching a handler feed sea lions in the Schönbrunn Zoo, Vienna, Austria. *Photo credits: Rômulo R.N. Alves.*

Steven et al., 2015a,b). Interactions with animals occur during activities such as hunting; fishing; and visits to aquariums, zoos, and museums (Fig. 20.5). Furthermore, during the 20th century the reduction in rail and bus fares and the expansion of car ownership meant that increasingly more people were able to see animals in the wild (Franklin, 1999). This perspective contributed to the development of activities aimed at the observation of wildlife, such as "bird watching," "whale watching," and safaris, which have gained much popularity. As an example, research conducted in the United States published in 2011 revealed that nearly one-third of the US population enjoyed wildlife watching. The study points out that practices such as observing, feeding, or photographing wildlife was enjoyed by 71.8 million people 16 years old and older in 2011. Of this group, 22.5 million people took trips away from home for the purpose of enjoying wildlife, while 68.6 million stayed within a mile of home to participate in wildlife watching activities (U.S. Census Bureau, 2011).

Bird watching is an ancient hobby, which originated in Europe and spread throughout the world (Fig. 20.6) (Green and Jones, 2010; Huxley, 1916; Hvenegaard et al., 1989; Urf, 2004). It attracts millions of fans who are

FIGURE 20.6 Bird watchers observing the white-tailed lapwing at Caerlaverock, Scotland. *Reproduced from MPF, via Wikimedia Commons.*

FIGURE 20.7 Tourists and locals enjoying whale watching on the coast of Bar Harbor, Maine, USA. *Reproduced from NightThree at English Wikipedia, via Wikimedia Commons.*

attracted by the beauty and behavior of birds, and who drive the economies of several countries (Hvenegaard et al., 1989; Steven et al., 2015a,b). Similarly, watching aquatic mammals arouses the interest of people (Fig. 20.7) (Hoyt and Hvenegaard, 2002). The industry involved with watching whales in their natural environment originated in Baja, California/Mexico, and Hawaii and dates from the 1950s (Tilt, 1987). Since then, whale watching has spread and proliferated throughout the world, by sea, land, and air, involving all 83 species of whales, dolphins, and porpoises (Hoyt, 2001), and expanded to its greatest popularity since the mid-1980s (Higham and Lück, 2007). According to Hoyt (2001), tourism for dolphin and whale watching has grown considerably in the last few decades and is present in at least 87 countries. In addition, this industry stimulates trade of more than 1 billion dollars annually, representing the largest and most important economic activity of many communities in several countries worldwide (Hoyt and Hvenegaard, 2002). For example, in Australia and New Zealand, tourism activities involving cetacean observation became quite significant in the late 1980s (Orams, 1999). Similarly countries such as Indonesia, Hong Kong, Tonga, and the Solomon Islands have benefited from these activities, obtaining great economic benefit through the industry of whale and dolphin observation. In South America, Brazil stands out in observation tourism with the dolphin *Sotalia fluviatilis* considered one of the main tourist attractions of some beaches of the northeast region (Albuquerque and Souto, 2013). Although wildlife watching typically focuses greater attention on large vertebrates, some smaller underwater inhabitants have also served as attractions, such as seahorses in Papua, New Guinea (Cater, 2007), and Brazil (Ternes et al., 2016).

The observation attention that some animal groups receive from humans can lead to other sources of animal entertainment, such as visits to environments that possess great faunal diversity, such as African savannas or coral reefs, and participating in activities such as safaris or dives, in terrestrial and marine environments, respectively. Safaris have become a popular tourist attraction, especially in African countries that possess great species richness of large vertebrates (Fig. 20.8). Having originated as elite colonialist hunting trips, safaris have undergone changes during the 20th century, evolving into journeys for drawing, filming, counting, and observing animals in nature (Hoppe, 2005). On safaris, people spend much of their time simply

FIGURE 20.8 Tourists watching a leopard resting on a tree during a safari in Serengeti National Park, Tanzania, Africa. *From Yathin S Krishnappa, via Wikimedia Commons.*

observing the animals and the ecological interactions among them. Events of reproduction and predation, for example, arouse special attention in people.

Many people are not satisfied with just observing wild animals and enjoy close encounters with them, and so seek to get as close as possible to, or even interact directly with them, such as making contact and/or feeding. The desire to dive with sharks (Fig. 20.9), for example, has led to a growing industry within the diving world that involves attracting sharks with bait. Such experiences are available with either significant protection (cages or chain mail suits, in the case of sharks) or with no protection, where only guides wear protective equipment and safety is based on understanding the behavior of the animals (Cater, 2007). Feeding, swimming, and other interactions (especially touching and observing) with dolphins has become the main attraction at many locations (Fig. 20.10) (Hoyt, 2001; Orams, 2002; Sá Alves et al., 2012). At several sites of the Brazilian Amazon, for example,

FIGURE 20.9 People inside a shark-proof cage with sharks swimming outside, Hawaii, USA. *From Kalanz from Honolulu, Hawaii (Sharks outside cage), via Wikimedia Commons.*

both feeding and swimming with dolphins are promoted as commercial tourist activities (Alves et al., 2013; Sá Alves et al., 2012).

Orams (2002) emphasizes that interacting with wildlife by feeding has become one of the main forms of direct contact provided to wildlife admirers in the animal's natural environment. This author further points out that such practices can result in psychological, social, and economic benefits that are experienced on the human side of the interaction and, in a limited number of cases, benefits to wildlife as well. However, several negative impacts associated with such practices are also highlighted, including the alteration of natural behavior and population patterns; dependency and habituation; health-disease-injury and the occurrence of intra- and interspecific aggression during such interactions. Perrine (1989), for example, reports that some divers have had fingers, earlobes, lips, faces, and even arms injured or even bitten off when feeding larger fish, such as moray eels, barracuda, groupers, and sharks.

Promoting Animal Fights

Some leisure and entertainment activities involve competitive fighting to the death between animals or between animals and humans (Bowman, 1977). The promotion of

FIGURE 20.10 Feeding of the river dolphin, *Inia geoffrensis*, in the state of Amazonas, Brazil. *Photo credits: Walter Lechner (left) and Tacyana P.R. Oliveira (right).*

animal fights involves both domestic and wild animals. Bullfighting is perhaps the most well-known spectator "sport" involving the killing of animals for entertainment (https://www.league.org.uk/bullfighting-and-bull-running). This practice is a popular activity in Spain, Portugal, Southern France, and many Latin American countries. The type of bullfighting that takes place, particularly in Spain, is usually fatal for either the matador or the bull. The goal of bullfighting is for the matador to stab the bull, particularly in the head and between the horns. These shows attract thousands of spectators to the arenas where these fights take place. Another form of entertainment involving bulls is the "running of the bulls," a practice which involves letting bulls or cows loose to run with, and in front of, jeering crowds along a designated route to their final destination, often the bull ring. The Spanish city of Pamplona hosts the world's most famous bullfight (during San Firmino Festival) (Fig. 20.11), while others, less famous, are held in Portugal, Mexico, Peru, and Colombia (Felizola, 2011).

FIGURE 20.11 Bullfight during the Fiesta de San Fermín in Pamplona, Spain. *From Iñaki Larrea derivative work: Stegop, via Wikimedia Commons.* http://creativecommons.org/licenses/by-sa/2.0.

In Brazil, oxen are involved in a variety of popular festivals, many of which have been the subject of much controversy due to the ill-treatment that the animals are subjected to during an event.

An example is a feast locally called "Farra do Boi" (bullock racing), one of the most controversial cultural manifestations in Brazil. It is an old Iberian custom, transported to the state of Santa Catarina by immigrants who came to settle in the region. The cultural manifestation occurs in the period before Easter (Lent) and involves the torture of the animal beginning a few days before the feast, when the ox is isolated and left without food. Once the spree begins, the cattle are released in the so-called *mangueirões* (crops surrounded by woodlands) or even in the middle of a crowd. Men, women, children, and the elderly—armed with sticks, stones, whips, and sharp objects—constantly pursue the animal and try, at all costs, to injure or overcome it through fatigue (Felizola, 2011).

Cattle and horses are involved in different popular festivals, including the famous rodeos, very common in the United States, but also in Mexico, Chile, Australia, Canada, and Brazil. In such festivities, cowboys show their skills with horses and oxen, participating in trials such as lassoing, separating and marking cattle, calf roping, double roping, bulldogging, and The Little Rodeo. One of the most characteristic competitions of the rodeo is to ride untamed horses or wild oxen. Bull rides are considered to be the most difficult, but also the activity that excites the public the most. During a bull ride the cowboy must remain on top of the bull for at least 8 s, holding a rope with only with one hand (supporting hand), leaving the other free (style hand) and not touching anything. In Brazil, rodeos are associated with various activities, ranging from auctions and animal exhibitions to large music shows.

Horses and other equines have also been involved in fights promoted by humans. Combat between horses occurs naturally in the wild, but death or serious injury in naturally occurring animal fights is almost always avoided by ritualized behaviors or the retreat of one of the combatants (Eibl-Eibesfeldt, 1961). However, organized horse fights are common events in several countries. It is a blood sport between two stallions which is organized for the purposes of betting and/or entertainment (Treu, 2014). The event is part of the traditional observance of the Chinese New Year among the Miao people and has occurred for more than 500 years (Treu, 2014). It is also extensively practiced in the Philippines where, although illegal according to national law, during 2008 around 1000 horses were bred annually for horse fights (Vice Staff, 2008). Organized horse fighting has also been recorded in Thailand, in South Korea's Jeju Province, on Muna Island in Indonesia, and among medieval Norse settlers in Iceland (Tozer, 1908).

Other examples of animals used in fights for entertainment include domestic and wild birds, such as cockfighting. Cockfighting is an ancient spectator sport in which cocks, or more accurately gamecocks, fight in a ring called a cockpit. In some cases, the birds are equipped with either metal spurs (called gaffs) or knives, tied to the leg in the area where the bird's natural spur has been partially removed. Some fights force the animals to strive to the death (or almost so) to satisfy the audience.

FIGURE 20.12 Confrontation between two cocks as part of tabuh rah ritual taken (cockfighting) in Bali, Indonesia, 1971. *Reproduced from Tropenmuseum, part of the National Museum of World Cultures, via Wikimedia Commons.* http://creativecommons.org/licenses/by-sa/3.0.

Cockfighting is widespread in many countries around the world (Fig. 20.12), and has recently become an illegal practice in many of them; however, in many cases they occur clandestinely. Wild birds are also used in fights. In Brazil, for example, the wild bird species used most frequently in fights are the wild canary (*Sicalis flaveola*) and the red-cowled cardinal (*Paroaria dominicana*). *Sicalis flaveola* is known for its aggressiveness and so is used more frequently in such fights, which are locally called *rinhas* (Alves et al., 2010). Wild canaries with good fighting abilities command higher prices (Gama and Sassi, 2008). For a fight, the birds are placed in a large cage and the owners and the audience place bets on the bird they think will win the fight. When one bird gives up and tries to escape from its competitor, the fight is interrupted and their owners separate the birds (Alves et al., 2010); however, the birds often die as a result of their wounds (Gama and Sassi, 2008).

Dog fighting is another example of the use of fighting animals for entertainment. In these contests, two dogs fight each other, sometimes to death. Dogs such as the American pit bull terrier, bull terrier, and the Staffordshire bull terrier are bred especially for fighting because they are large, heavy-bodied dogs with a strongly developed head and a powerful build. Fights usually involve cash betting among the spectators, and the battle only ends when the owner of a dog surrenders or, in some cases, in what is called "till death do us part"; that is, the dispute ends only with the death of one of the dogs (Felizola, 2011). This practice has been banned in many localities throughout the world, but continues to occur clandestinely in some countries.

In addition to the aforementioned animal fights, camel, buffalo, pig, partridge, ram, and cricket have also been put to fight in some locations around the globe (Köhne and Ewigleben, 2000). All such activities of fighting between animals for sport or entertainment is considered a serious form of animal abuse by many people (Köhne and Ewigleben, 2000). As a result, the issue is the subject of intense legal and ethical debate, with animal advocacy groups arguing that human entertainment is not sufficient justification for the ill treatment, and even death, that the animals involved experience.

Animal Racing

Racing is one of the most popular forms of animal-related sport. Some racing events directly involve humans as riders while in others animals race alone. Among the animals most commonly involved in competitive racing are dogs, pigeons, and horses. One of the most well-known racing competitions is greyhound racing (Fig. 20.13), in which the dogs race around a track. Two types of races are generally used for these races, which involve different types of tracks: track racing, which uses an oval track; and coursing (Genders, 1981). Track racing uses some motorized stimulus that advances ahead of the dogs on a rail until they cross the finish line. In coursing, greyhounds also chase some sort of lure, but in this case it is usually a live hare, which can even be killed by the dog(s) even before the finish line. Many dogs are also involved in the very popular competition of sled dog racing in the Arctic regions of the United States, Ohio, Africa, Greenland, and some European countries (Lovell and Dublenko, 1999; Prebensen, 2012). Sled dog racing is a planned and timed competition among teams of sled dogs that run through a predetermined circuit,

FIGURE 20.13 Greyhound during a dog race. *From AngMoKio (Own work), via Wikimedia Commons.* http://creativecommons.org/licenses/by-sa/3.0; http://www.gnu.org/copyleft/fdl.html

each team comprising a driver, a sled, and a pack of dogs to pull the sled. The team to cross the finish line in the least amount of time is the winner of the competition.

Horse racing is considered one of the earliest entertainment competitions of human history. It originated in central Asia among nomadic tribesmen who had used wild horses since their domestication. For millennia, horse racing was the sport of monarchs and the nobility (AAWS, 2008). In the ancient Roman Empire horse races, both with mounted riders and pulling chariots, were considered great entertainment and offered to the populace during religious festivals or after battlefield victories. Contemporary horse racing dates back to the 12th century when English knights returned from the Crusades with Arabian horses of great racing lineages (Britannica, 2017). Currently, horse racing is widespread worldwide as a unique sport, and has developed during the modern era from a popular upper-class leisure attraction into a huge public-entertainment industry involving huge monetary investments (Fig. 20.14). Other animals are also used in races at traditional folk festivals in many countries. In Brazil, for example, donkey, mule, buffalo, goat, and even chicken races are part of the attractions of many popular festivals.

FIGURE 20.14 Horse racing in Deauville France. *Reproduced from Hippodrome de Deauville – Clairefontaine, via Wikimedia Commons.*

There are also some nonracing competitive events that use animals. Polo, for example, involves competitors who strike a ball with mallets while mounted on a horse. Another example is equestrianism, a type of competition that has existed for at least 2000 years from when the Greeks introduced training to prepare horses for war. The sport was first presented at the Olympic Games in 1900, was included as a permanent event at the Stockholm Games in 1912, and currently includes three disciplines of judgment (Haan and Dumbell, 2016).

In practical terms, track racing is considered the most common form of animal racing; however, there are other variations of this sport, such as pigeon racing. According to Johnes (2007), the practice of pigeon racing was very popular among male industrial workers in the 19th century in several European countries. The competition involves the release of fully trained racing pigeons at a certain distance from their birthplaces, and the time taken to travel from the point of release to arrival is compared among competitors, with the winning bird being the one with the shortest travel time.

Entertainment Shows With Trained Animals

There are a wide variety of animals that have been trained for entertainment purposes. Some animals are trained primarily as a hobby or for companionship with no economic objective. Others are trained for entertainment shows, the use of animals which dates back to the earliest human civilizations and reached its greatest popularity during the Roman Empire (Fig. 20.15) (Roth and Merz, 1997).

Trained animals are shown mainly in circus shows, which commonly include horses, monkeys, lions, elephants, tigers, bears, and snakes. Until the last century such shows exhibited animals on a large scale to attract public attention, but the practice has experienced a gradual decrease in popularity. Nowadays, more

FIGURE 20.15 Marble sculpture showing an African acrobat on a crocodile. Roman, 1st century BC or 1st century AD. Work of art exhibited at the British Museum, London. *Photo credits: Rômulo R.N. Alves.*

refined forms of animal show business have evolved, such as thematic aquatic parks with shows involving animals, especially marine mammals such as whales, dolphins, seals, and walruses. There were also showmen who specialize in wild animals as entertainment. Other forms of commercially oriented display of animals, which always entails their taming to some extent, involve crocodiles, snakes, and bird parks (Roth and Merz, 1997).

Animals in Documentaries, Cartoons, Films, and Television Series

Humans have always expressed strong interest in animals, which have become part of the cultural imaginary of civilizations worldwide. Unsurprisingly, animals have frequently become key elements in different types of media, including a constant presence on television and in the cinema. Historically, animals have appeared, with greater or lesser significance, in all genres of motion pictures; from wildlife films to Hollywood blockbusters, scientific films to animation, as well as surrealist, avant-garde, and experimental films, all of which use a multitude of different formats and technologies. Diversity of animal imagery is one measure of the extent to which movies reflect a cultural desire for such imagery, and demonstrates the extraordinarily rich potential of the image of animals (Ritvo, 2004).

In cinema, Robert J. Flaherty's classic 1922 film *Nanook of the North* is considered the earliest feature-length documentary produced involving animals. Years later (1948–60), the Walt Disney Company introduced the serial theatrical release of wild-documentaries with its production of the *True-Life Adventures* series, a collection of 14 full-length and short-topic wilderness movies (Cruz, 2012). The first full-length wilderness-documentary films using color underwater cinematography were the Italian film *Sesto Continente* (The Sixth Continent) and the French film *Le Monde du Silence* (The Silent World). The films stunned the world with incredible footage of fauna and flora never seen before, and spurred the interest of many people in scuba diving sport (Burt, 2002). Many other highly successful nature-documentary films followed in subsequent years, such as those made by Nicolas Vanier (*The Last Trapper*, 2004), Luc Jacquet (*March of the Penguins*, 2005), and Alastair Fothergill (*African Cats*, 2011), among others (Cruz, 2012).

Throughout the history of cinema, a number of animals have had significant roles in movies of varied genres, and have been seen by large audiences around the world. The superstar dogs of the 1920s, Strongheart and Rin Tin Tin, followed decades later by Lassie, Benji, and the Doberman Gang, achieved phenomenal success throughout the world. Animated films are also among the best known productions to exploit animal figures (e.g., *The Lion King*, *Ice Age*, *The Jungle Book*, *Bambi*, *Lady and the Tramp*, *Tarzan*, *Kung Fu Panda* and *Zootopia*). Although

FIGURE 20.16 Original poster from the 1933 film *King Kong*. *From Keye Luke (*www.widescreenmuseum.com*), via Wikimedia Commons.*

primarily intended for children, such movies have been gaining increasing popularity among adults as well.

The great media appeal of charismatic megafauna (large vertebrate species with widespread popular appeal) or targets of conflicts with humans have placed some animals as major protagonists, such as *King Kong* (Fig. 20.16), *Free Willy*, *Orca*, *Jaws*, *Anaconda*, *Planet of the Apes*, and *Piranha*, among others. These films generally portray interactions between wild animals and humans, with some focusing on harmonious relationships while others introducing animals and humans as adversaries. Some domesticated animals with which humans enjoy close relationships, such as dogs and horses, have been protagonists of several movies (e.g., *Hachi: A Dog's Tale*, *Marley & Me*, *Lassie*, *War horse*, *Seabiscuit*, and *Hidalgo*), which are, in many cases, based on real facts, thus reflecting the importance of these animals in different human activities. Although less frequent than vertebrates, invertebrates have also played parts in several films, particularly arthropods (e.g., *The Swarm*, *Arachnophobia*, *Empire of Ants*). A review by Castanheira et al. (2015) pointed out that insects and other arthropods have appeared in screenplay since the creation of movies, but especially by the beginning of the 20th century. These authors recorded at least 177 movies and TV series inspired by arthropods.

Movies about superheroes in many cases are connected to animals, a trend derived from successful comics that have ultimately inspired productions of cinematic success. In these cases, the hero is often represented by a character who possesses special powers, as is the case of the Spiderman who acquires special "spider-like" capabilities after being bitten by a radioactive spider, or an animal-like character superhero (anthropomorphized), as exemplified by the movie *The Teenage Mutant Ninja Turtles*, in which the ninja heroes are four fictional teenaged anthropomorphic turtles. Other examples of superheroes films that are strongly associated with animals are *Ant-man*, *Batman*, and *Catwoman*. Among television series, the successful American action–adventure series *Manimal* portrays the story of an agent who possesses the ability to transform himself into any animal he chooses, and uses this ability to assist police in solving crimes.

The cast of animals used in films even includes extinct species of dinosaurs, such as in *Jurassic Park* and the *Ice Age* movies. Mythological or fictional animals are also part of the wildlife cast in films. An illustrative example is the Harry Potter series *Fantastic Beasts and Where to Find Them*. The script tells the story of Newt Scamander (Eddie Redmayne), a magizoologist who carries a suitcase full of magical creatures to New York, but some of the animals escape and Newt's problems begin. Mythological animals are also abundant in films about Greek mythology, such as *Hercules* and *Clash of the Titans*.

It should be noted that in many media productions animals act directly in the films, and originate from different places. Actor animals have even been captured directly from the jungle. Ironically, in the movie *Free Willy*, the famous whale "Keiko" was captured from the wild specifically for use in the entertainment business. To have specific exotic animals as part of a show, many production companies sometimes work with zoos or legal "animal handlers" or "animal trainers." For instance, several films and TV shows featuring exotic and domestic animal species recruited animal actors via the California-based private company Birds and Animals Unlimited (https://nhes.org/animals-in-film-and-televison/).

On the other hand, fauna have long been a source of cultural conflict in a number of areas, which accounts in part for the strict codes of legislation that govern the appearance of animals on film. Violence toward animals in film dates as far back as Thomas Edison's 1903 silent short, *Electrocuting an Elephant* (Fig. 20.17). The film documented a real-life event of a fatal elephant electrocution in Luna Park, Coney Island. Edison Studios initially planned to hang the animal (seriously), before the American Society for the Prevention of Cruelty to Animals interfered. Around 1500 persons witnessed the event at the time (Artiquez, 2013). In 1966, a film directed by Andrey Tarkovsky called *Andrei Rublev* stirred up a controversy through the killing of a horse. The dramatic execution depicts a horse falling down a flight of stairs, frantic and unable to regain its balance from the fall. Soldiers then descend upon it and spear it to death. Another example of controversy regarding animal abuse and cinema was the *Monte Hellman* released in 1974 that showed the illicit world of cockfighting; the film remains banned in the United Kingdom, for infringing on the BBFC's 1937 Cinematograph Film (Animals) Act guidelines. In 1979, Francis Ford Coppola's seminal war epic, *Apocalypse Now*, portrays a real scene of a buffalo ritually slaughtered by indigenous Montagnard people. The film *Cannibal Holocaust* (1980) roused many controversies at that time, for showing scenes of real violence inflicted on fellow humans and animals. Although there are several counts of animal cruelty throughout the entire movie, one notorious scene stands out—the disemboweling and cooking of a yellow-spotted river turtle. Similarly, the 2014 movie *Still the Water* surprised many people when it showed a closeup shot of a goat's throat being sliced (Chapman, 2017).

FIGURE 20.17 A press photograph of the *Electrocuting an Elephant* (1903), showing the smoke rising from her feet before she falls dead to the ground. Reproduced from https://commons.wikimedia.org/wiki/File%3ATopsy_elephant_death_electrocution_at_luna_park_1903.png.

Miscellaneous

In some countries, people exploit animals in street performances, or simply display them as attractions for tourists, charging small amounts of money for photos or to feed them (Fig. 20.18). Another common attraction is carriage rides, with the coachman explaining the main stories and events of the city along the way. Such tours are very common in cities like Cartagena, Vienna, and Rome, for example. Elephants and camels are used to carry tourists in Africa, and mules and camels are used in tours in northeastern Brazil.

FIGURE 20.18 Owl exhibited to tourists on the streets of Edinburgh, Scotland (above) and people feeding domestic animals on the streets of Prague, Czech Republic. *Photo credits: Rômulo R.N. Alves.*

In several cities there are "Cat cafes" and "Dog cafes"; establishments that attract pet breeders and tourists from all over, and provide free access to basically any pet-friendly animal and fellowship encounters. Reconstructions of extinct animals, both fossil and taxidermy, stand out among the major attractions of natural history museums (Figs. 20.19 and 20.20). Dinosaurs, for example, are prominent attractions in major museums, often comprising a montage of skeletons, fossils, and reproductions of prehistoric fauna, including even their typical vocal sounds. It is no wonder that people worldwide seek out natural parks to spend part of their spare time. In these places animals constitute one of the main attractions, and there is often an opportunity for people to interact directly with animals or spend some of their time observing and photographing them.

Animal Welfare and Entertainment

As we have seen, a wide variety of animals are used by humans for entertainment, including observation (in the natural environment, in captivity and in competitions), direct interaction (diving, petting, feeding, etc.) or slaughtering (sport hunting and fishing). All of these activities have varied implications for the species involved and are therefore subject of intense debate regarding animal welfare. For example, Moorhouse et al. (2015) studied the direct impacts of 24 wildlife tourist attractions visited by nearly six million people per year and found that approximately 500,000 or more animals are being held in attractions that negatively influence their welfare. Just for comparison, Moorhouse concluded that only six types of wildlife attractions, involving only 1500–13,000 animals, were judged likely to have positive effects on animal welfare and conservation. That is, in general, the activities involving animals in entertainment can cause negative impacts that damage the well-being of the exploited species.

Among the activities that have been most criticized are the many violently exploitative animal sports, including horse fighting (Smith, 2014); dog fighting (Yilmaz, 2016); bullfighting (Ogorzaly, 2006), and pigeon shoots (Fielder, 2013), among others. These activities result in the death or mutilation of the animals involved, and so some of them have been, or are being, banned in several countries of the world. Similarly, keeping animals in captivity for entertainment purposes has also been the subject of intense debate in the field of animal welfare. An illustrative example is the case of the Luján Zoo,

FIGURE 20.19 Reconstructions of extinct animals in the Museum of Natural History, Vienna, Austria. *Photo credits: Rômulo R.N. Alves.*

FIGURE 20.20 Large mammals on display in the Natural History Museum, London. *Photo credits: Rômulo R.N. Alves.*

one of the main tourist attractions near Buenos Aires (Argentina). This institution is involved in polemics regarding the treatment of the animals it keeps, and is famous for allowing people to maintain direct contact with large animals such as elephants, lions, and tigers (cubs and adults). It is speculated that animals are drugged so that they can be caressed, fed, and photographed with.

The maintenance of vertebrates for shows has also been criticized because, in addition to the removal of animals from their natural environment, they are subject to mistreatment during training. Legal discussions regarding this issue are frequent around the globe, and have resulted in the prohibition of such practices in some places. Bottlenose dolphins and orcas are used to perform in very few acceptable aquariums in France (http://www.bbc.com/news/world-europe-39834098), while new legislation prohibits maintaining whales, other dolphins species, and porpoises in captivity throughout the entire country, as well as disallowing direct public contact with these animals, including allowing people to swim with them (http://focusingonwildlife.com/news/poll-should-more-countries-ban-breeding-captive-dolphins-and-orcas/).

Complaints of mistreatment and inadequate management of animals even target some wildlife programs where there are suspicions that the encounters between hosts and animals are fake. In addition, the broadcasts used during programs subject animals to stress, as does being handled by presenters who lack the necessary knowledge to deal with wild animals in their environment while exposing them in a sensationalist way. It should be emphasized that even seemingly passive activities, such as safaris and dives, can have direct impacts on the animals observed, and indirect impacts on other faunal groups that interact with these animals ecologically. Diving among coral reefs, for example, can damage a variety of invertebrate species that make up these ecosystems, even if the focus of the dive is to observe larger vertebrates such as sea turtles and fish.

CONCLUSION

Animals provide people with a variety of entertainment opportunities, providing evidence that the relationship between humans and animals is much more than simply material. Wild, tame, and domesticated animals have played different roles in human recreational activities, from being the subject of simple observation in captivity or in their natural environment, to sport activities and, in some cases, practices that involve extreme violence. These activities, in general, attract millions of people worldwide, and increasingly have been incorporated into the international tourism industry, which fosters millions of dollars.

Several animal species, mainly vertebrates, have been incorporated into our lives for recreational purposes. People interact with their own pets at home during their daily activities, which brings pleasure to both. Humans also obtain entertainment from wildlife by observing them in their habitats, on television, and in film; and they visit zoos, circuses, aquariums, and thematic parks to observe animals and attend shows with trained animals. Humans have also created sports in which animals compete with each other or with humans. Humans, whether in the most isolated regions or the most urbanized centers of the world, find animals to be a great source of entertainment.

The participation of animals in activities of human entertainment is a historical and global phenomenon. Many of these activities involve violence against animals causing mistreatment or death, and as a result have been heavily criticized by animal advocacy groups and thus are the subject of legal discussion in different countries, sometimes resulting in their decline or prohibition. A good example is circuses, an activity that arose in ancient Rome and which included animals among the main attractions, but which now has declined due to restrictions regarding the use of animals. This situation extends beyond circuses to other practices that cause mistreatment or violence against animals, and are prohibited in many countries.

Clearly, human activities of entertainment that involve animals raise a number of complex questions that need to be addressed by studies that combine both the natural and social sciences. Thus, ethnozoological studies toward understanding the social, economic, cultural, and ecological aspects involved in human–animal relationships in the context of entertainment has the potential of making a great contribution.

References

AAWS - Australian Animal Welfare Strategy, 2008. AAWS Education and Training Stocktake Work, Sport, Recreation on Display. Final Report. Australian Animal Welfare Strategy.

Adkins, L., Adkins, R.A., 1994. Handbook to Life in Ancient Rome. Oxford University Press, New York.

Albuquerque, N.S., Souto, A.S., 2013. Motorboat noise can potentially mask the whistle sound of estuarine dolphins (*Sotalia guianensis*). Ethnobiology and Conservation 2, 1–15.

Allaby, M., 2010. Animals: from Mythology to Zoology. Facts On File, Inc., New York.

Allsen, T.T., 2006. The Royal Hunt in Eurasian History. Cambridge University Press.

Alves, L.C.P.S., Zappes, C.A., Oliveira, R.G., Andriolo, A., Azevedo, A.F., 2013. Perception of local inhabitants regarding the socioeconomic impact of tourism focused on provisioning wild dolphins in Novo Airão, Central Amazon, Brazil. Anais da Academia Brasileira de Ciências 85, 1577–1591.

Alves, R.R.N., 2012. Relationships between fauna and people and the role of ethnozoology in animal conservation. Ethnobiology and Conservation 1, 1–69.

Alves, R.R.N., Mendonça, L.E.T., Confessor, M.V.A., Vieira, W.L.S., Lopez, L.C.S., 2009. Hunting strategies used in the semi-arid region of Northeastern Brazil. Journal of Ethnobiology and Ethnomedicine 5, 1–50.

Alves, R.R.N., Nogueira, E., Araujo, H., Brooks, S., 2010. Bird-keeping in the Caatinga, NE Brazil. Human Ecology 38, 147–156.

Alves, R.R.N., Souto, W.M.S., 2015. Ethnozoology: a brief introduction. Ethnobiology and Conservation 4, 1–13.

Arlinghaus, R., Cooke, S.J., Lyman, J., Policansky, D., Schwab, A., Suski, C., Sutton, S.G., Thorstad, E.B., 2007. Understanding the complexity of catch-and-release in recreational fishing: an integrative synthesis of global knowledge from historical, ethical, social, and biological perspectives. Reviews in Fisheries Science 15, 75–167.

Artiquez, B., 2013. Animal Holocaust in Film: Researching the Difference in Animal Welfare in Film from 1903 to 2013 with Regard to the Work of the American Humane Association, Established in 1943. Diblin Business School, School of Arts.

Barboza, R.R.D., Lopes, S.F., Souto, W.M.S., Fernandes-Ferreira, H., Alves, R.R.N., 2016. The role of game mammals as bushmeat in the Caatinga, Northeast Brazil. Ecology and Society 21, 1–11.

Bomgardner, D.L., 2013. The Story of the Roman Amphitheatre. Routledge.

Bowman, J.C., 1977. Animals for Man. Edward Arnold, London.

Britannica, E., 2017. Horse Racing. https://www.britannica.com/sports/horse-racing.

Brower, M., 2005. Trophy shots: early North American photographs of nonhuman animals and the display of masculine prowess. Society and Animals 13, 13–32.

Bryant, C.D., 1993. On the trail of the centaur: toward an amplified research agenda for the study of the human-animal interface. In: Hicks, E.K. (Ed.), Science and the Human Animal Relation. S.L.S.W.O., Amsterdam, pp. 13–38.

Burt, J., 2002. Animals in Film. Reaktion Books, London, UK.

Castanheira, P.S., Prado, A.W., Silva, E.R., Braga, R.B., 2015. Analyzing the 7th art–arthropods in movies and series. Vignettes of Research 3, 1–15.

Cater, C., 2007. Perceptions of and interactions with marine environments: diving attractions from great whites to pygmy seahorses. In: New Frontiers in Marine Tourism: Diving Experiences, Sustainability, Management. Elsevier, pp. 49–64. Oxford (forthcoming).

Chapman, A., 2017. A Brief History of Cinema's Most Controversial Scenes of Animal Abuse. http://lwlies.com/articles/controversial-animal-abuse-scenes-in-cinema/.

Coley, J., 2010. Roman Games: Playing with Animals, Heilbrunn Timeline of Art History. The Metropolitan Museum of Art, 2000, New York.

Cruz Jr., R., 2012. The Animated Roots of Wildlife Films: Animals, People, Animation and the Origin of Walt Disney's True-life Adventures. Montana State University, p. 64.

Eibl-Eibesfeldt, I., 1961. The fighting behavior of animals. Scientific American 205, 112–123.

Felizola, M.B., 2011. The culture of entertainment using animals and the posicion of the courts. Revista Brasileira de Direito Animal 9, 243–264.

Fernandes-Ferreira, H., Mendonça, S.V., Albano, C., Ferreira, F.S., Alves, R.R.N., 2012. Hunting, use and conservation of birds in Northeast Brazil. Biodiversity and Conservation 221–244.

Fielder, E., 2013. Fair or Foul? Pigeon Shoots Ruffle Feathers in Pennsylvania.

Franklin, A., 1999. Animals & Modern Cultures – a Sociology of Human-animal Relations in Modernity. Sage, London.

Freire, K.M.F., Tubino, R.A., Monteiro-Neto, C., Andrade-Tubino, M.F., Belruss, C.G., Tomás, A.R.G., Tutui, S.L.S., Castro, P.M.G., Maruyama, L.S., Catella, A.C., 2016. Brazilian recreational fisheries: current status, challenges and future direction. Fisheries Management and Ecology 23, 276–290.

Gama, T.F., Sassi, R., 2008. Aspectos do comércio Ilegal de Pássaros Silvestres na Cidade de João Pessoa, Paraíba, Brasil. Gaia Scientia 2, 1–20.

Garrod, B., Gössling, S., 2008. New Frontiers in Marine Tourism: Diving Experiences, Sustainability, Management. Routledge.

Genders, R., 1981. The Encyclopaedia of Greyhound Racing. Pelham Books.

Green, R., Jones, D.N., 2010. Practices, Needs and Attitudes of Bird-Watching Tourists in Australia. CRC for Sustainable Tourism Gold Coast.

Haan, D., Dumbell, L.C., 2016. Equestrian sport at the olympic games from 1900 to 1948. The International Journal of the History of Sport 33, 648–665.

Higham, J.E.S., Lück, M., 2007. Marine Wildlife and Tourism Management: Insights from the Natural and Social Sciences. CABI.

Hoppe, K.A., 2005. Simulated safaris: reading African landscapes in the US. In: Benesch, K., Schmidt, K. (Eds.), Space in America: Theory, History, Culture. Rodopi, Amsterdam, New York, pp. 179–194.

Hoyt, E., 2001. Whalewatching 2000: Worldwide Tourism Numbers, Expenditures and Expanding Socioeconomic Benefits. International Fund for Animal Welfare, Yarmouth, USA.

Hoyt, E., Hvenegaard, G.T., 2002. A review of whale-watching and whaling with applications for the Caribbean. Coastal Management 30, 381–399.

Huxley, J.S., 1916. Bird-watching and biological science. Some observations on the study of courtship in birds. The Auk 33, 256–270.

Hvenegaard, G.T., Butler, J.R., Krystofiak, D.K., 1989. Economic values of bird watching at point Pelee National Park, Canada. Wildlife Society Bulletin (1973–2006) 17, 526–531.

IFAW, International Fund for Animal Welfare, 2016. Killing for Trophies: an Analysis of Global Trophy Hunting Trade. International Fund for Animal Welfare, pp. 1–62.

Ihde, T.F., Wilberg, M.J., Loewensteiner, D.A., Secor, D.H., Miller, T.J., 2011. The increasing importance of marine recreational fishing in the US: challenges for management. Fisheries Research 108, 268–276.

Johnes, M., 2007. Pigeon racing and working-class culture in Britain, c. 1870–1950. Cultural and Social History 4, 361–383.

Kalof, L., 2011. Ancient animals. In: Kalof, L. (Ed.), A Cultural History of Animals in Antiquity. Berg, New York, pp. 1–16.

Kidd, E., 2012. "Beast-Hunts" in Roman Amphitheaters: the Impact of the Venationes on Animal Populations in the Ancient Roman World. The Eagle Feather.

Kiser, L.J., 2011. Animals in medieval sports, entertainment, and menageries. In: Resl, B. (Ed.), A Cultural History of Animals in the Medieval Age. Berg, New York, NY, pp. 103–126.

Köhne, E., Ewigleben, C., 2000. Gladiators and Caesars: the Power of Spectacle in Ancient Rome. University of California Press.

Kyle, D.G., 1998. Spectacles of Death in Ancient Rome. Routledge, London, New York.

Lindsey, P.A., Roulet, P.A., Romanach, S.S., 2007. Economic and conservation significance of the trophy hunting industry in sub-Saharan Africa. Biological Conservation 134, 455–469.

Lovell, N.C., Dublenko, A.A., 1999. Further aspects of fur trade life depicted in the skeleton. International Journal of Osteoarchaeology 9, 248–256.

Lucius, R., 2015. The Oppression of Non-human Animals as a Crisis of Social and Ecological Justice.

Moorhouse, T.P., Dahlsjö, C.A.L., Baker, S.E., D'Cruze, N.C., Macdonald, D.W., 2015. The customer isn't always right—conservation and animal welfare implications of the increasing demand for wildlife tourism. PLos One 10, e0138939.

Ogorzaly, M., 2006. When Bulls Cry: the Case Against Bullfighting. Authorhouse, Bloomington.

Orams, M.B., 1999. Marine Tourism: Development, Impacts and Management. Routledge, London.

Orams, M.B., 2002. Feeding wildlife as a tourism attraction: a review of issues and impacts. Tourism Management 23, 281–293.

Perrine, D., 1989. Reef fish feeding; amusement or nuisance? Sea Frontiers 35, 272–279.

Pitcher, T.J., Hollingworth, C.E., 2002. Fishing for fun: where's the catch? In: Pitcher, T.J., Hollingworth, C.E. (Eds.), Recreational Fisheries: Ecological, Economic and Social Evaluation. Blackwell Science, Oxford, pp. 1–16.

Prebensen, N.K., 2012. Value Creation in Experience-based Networks: a Case Study of Sport-events in Europe. INTECH. Open Access Publisher.

Ritvo, H., 2004. Animal planet. Environmental History 9, 204–220.

Roth, H.H., Merz, G., 1997. Wildlife Resources: a Global Account of Economic Use. Springer Science & Business Media.

Sá Alves, L.C.P., Andriolo, A., Orams, M.B., Freitas Azevedo, A., 2012. The growth of 'botos feeding tourism', a new tourism industry based on the boto (Amazon river dolphin) Inia geoffrensis in the Amazonas State, Brazil. Sitientibus Série Ciências Biológicas 11, 8–15.

Schwab, A., 2004. Dear Jim–reflections on the Beauty of Angling. Merlin Unwin Books, Ludlow, UK.

Shackley, M., 1996. Wildlife Tourism. International Thomson Business Press, London.

Shelton, J., 2007. Beastly spectacles in the ancient Mediterranean world. In: Kalof, L.A. (Ed.), Cultural History of Animals in Antiquity. Berg, Oxford, New York.

Smith, E.K., 2014. Stallions Fight to the Death in Illegal Horse-fighting Bout as Philippine Villagers Refuse to Abandon Tradition that was Banned in 1998.

Steven, R., Morrison, C., Arthur, J.M., Castley, J.G., 2015a. Avitourism and Australian important bird and biodiversity areas. PLos One 10, e0144445.

Steven, R., Morrison, C., Castley, J.G., 2015b. Birdwatching and avitourism: a global review of research into its participant markets, distribution and impacts, highlighting future research priorities to inform sustainable avitourism management. Journal of Sustainable Tourism 23, 1257–1276.

Ternes, M.L.F., Gerhardinger, L.C., Schiavetti, A., 2016. Seahorses in focus: local ecological knowledge of seahorse-watching operators in a tropical estuary. Journal of Ethnobiology and Ethnomedicine 12, 1–12.

Tilt, W.C., 1987. From Whaling to Whalewatching, 52nd North American Wildlife and Natural Resources Conference. Washington, DC.

Tozer, B., 1908. The Horse in History. Methuen.

Treu, Z., 2014. Villages in Southern China Ring in Chinese New Year with Horse Fighting.

U.S. Census Bureau, 2011. National Survey of Fishing, Hunting, and Wildlife-associated Recreation. U.S. Department of the Interior/U.S. Fish & Wildlife Service.

Urf, A.J., 2004. Birds: Beyond Watching. Universities Press.

Usman, J.O., 2009. The Game is Afoot: Constitutionalizing the Right to Hunt and Fish in the Tennessee Constitution.

Vice Staff, 2008. Horse Fights.

Yilmaz, O., 2016. Dog fighting in some European countries. International Journal of Livestock Research 6, 20–26.

Young, M.A.L., Foale, S., Bellwood, D.R., 2016. Why do fishers fish? A cross-cultural examination of the motivations for fishing. Marine Policy 66, 114–123.

CHAPTER

21

Animals as Ethnozooindicators of Weather and Climate

Rômulo Romeu Nóbrega Alves, Raynner Rilke Duarte Barboza
Universidade Estadual da Paraíba, Campina Grande, Brazil

INTRODUCTION

Many elements of numerous ethnic worldviews are rooted in a holistic foundation that connects land, air, and water; earth and sky; plants and animals; and people and spirit (Deloria and Wildcat, 2001; Wildcat, 2009). This perspective recognizes earth as a social–biophysical system in which all things are linked (Levin, 1999); so it is not surprising that many human cultures exert important feedback to the global ecosystem (ACIA, 2005; IPCC, 2007). Human communities that rely directly upon natural resources for subsistence usually possess a detailed knowledge of their local biota and its implicit environmental cycles (Nishida et al., 2006b,c). Knowledge resulting from interactions with a local environment is acquired through experience and observation. Such knowledge has been referred to by several different names in the literature, but for this chapter we have adopted the term local ecological knowledge (LEK).

Since time immemorial, human beings have been adapting to environmental conditions in their struggle for survival (Ayal1 et al., 2015; Ziervogel and Opere, 2010). Through observations of biotic indicators that forecast future weather conditions, many cultures throughout the world have developed refined local knowledge of when, where, and how animals, plants, algae, and fungi could be best used for their multiple needs (Alves et al., 2016; Alves et al., 2015; Barboza et al., 2016). Behavioral, physiological, and reproductive characteristics of many animal species are used by people as natural signs for predicting weather. The extent of LEK regarding animals as climate indicators reflects millennia of experiences, and has become culturally conserved and passed on from generation to generation (Blukis Onat, 2002; Huntington et al., 2005). This knowledge has been vital for human beings because, in combination with other stressors, climate change may affect societal relationships regarding water and food, such as access, availability, harvest, storage, processing, and the traditional use of these resources (Colombi, 2009; Jones et al., 2008).

Predicting climate and weather using animals has been part of numerous human cultures since time immemorial, and these relationships have persisted through generations

Ethnozoology
http://dx.doi.org/10.1016/B978-0-12-809913-1.00021-1

until today. The number of examples of animal behavior useful to human communities for predicting the weather is vast (Krupnik and Jolly, 2002; Voggesser et al., 2013). Indeed, numerous practical activities (including farming and hunting–gathering) are, to some extent, determined and circumscribed by meteorological events. For example, plant, animal, and human body conditions are used in Zimbabwe for forecasting the weather (Joshua et al., 2012; Shoko, 2012). In Nigeria and Kenya, native tribes make direct observations and interpretations of animal behaviors to gain insight into future weather conditions (Shukurat et al., 2012; Speranza et al., 2010; Ziervogel and Opere, 2010). Comparable practices are evident among ritual specialist elders in Burkina Faso and Swaziland, who also conjecture future climate forecasts based on physical indications from plants and animals (Roncoli et al., 2001). In South Africa, inhabitants rely on birds, toads, and white ants to forecast the summer season and the onset of rains (Merchant et al., 1987; Olbrich and King, 2003), while in northeastern Tanzania they look at behavioral patterns of birds and mammals (Prendergast et al., 1999). According to Selma and Fuentes (2015), in the municipality of Retirolândia, Bahia, Brazil, insects, birds, amphibians, and arachnids are considered indicators that can be used to predict changes in the weather through their behavior.

These human/animal relationships for predicting weather shape the core elements of fauna conservation and embody an emerging field of discourse about human LEK worldwide (Alves et al., 2016; Ayal et al., 2015). Information on this subject is often diluted in works of various academic fields including anthropology, geography, history, and ethnobiology, while research that deals specifically with this subject is rather scarce. In this chapter we have sought to assemble all the existing information on, and provide a brief overview of, the main forms of interactions between people and animals in the context of forecasting the weather and climate. We propose the term "climate ethnozooindicators" to designate animals used by humans as indicators of future weather and climate.

Natural Weather Forecasting: A Brief History

Observations of their immediate natural environment provided early humans the first signs for forecasting weather. Personal observations of the phenology of certain plants and/or the behaviors of certain animals have been among the most commonly used predictors of climatic events since prehistoric times.

Ancient written documents reveal that humankind has long had an awareness and understanding of ecological interactions between biotic and abiotic factors. For example, in his *Natural History Encyclopedia*, Gaius Plinius Secundus (AD 23–August 25, AD 79), a natural philosopher, provided detailed material relating astronomical phenomena to weather, and he did not discount the effects of celestial bodies and climate on terrestrial life. He recorded that the influence of the moon causes the shells of oysters, cockles, and all shellfish to grow bulkier, and then again smaller. Moreover, he held that the phases of the moon have an effect on the tissues of the shrew mouse and the smallest animal, the ant, since they are sensitive to the influence of the planet's weather and of the new moon (Pliny Natural History, BOOK II. VI. 32–36). Ethnobiological studies reinforce these earlier findings (Alves et al., 2005; Nishida et al., 2006a,b,c). Nishida et al. (2006c), for example, recorded that mollusk gatherers of northeastern Brazil recognize that mollusk meat production increases during spring tide and decreases during neap tide, which they confirmed experimentally by finding the condition index to increase during spring tide and to decrease during neap tide. Other examples of weather indicators and their use were recorded by Claudius Aelianus, a sophist of the first third of the 3rd century, in his extraordinary stories and anecdotes about animals in his work "On the Characteristics of Animals" (Taub, 2003). Some of these narratives report the abilities of different animals, including birds, to predict weather. Aelianus quotes some of his sources, highlighting Aristotle and giving him credit regarding the observation that "cranes flying in to land from the sea indicate to the intelligent man that a violent storm is threatening" and also "the Libyans and goats also give clear signs of impending rain" (Taub, 2003).

Moreover, the usage of myths and legends to elucidate this set of phenomena and the reliance on signs and omens to predict the weather are an important part of the fabric of many cultures including several that are ancient and which link astronomical events to animal behavior. For example, during medieval times in Europe many societies believed that thunderstorms were evil spirits, and so people often laid down on beds made of feathers and kept away from wet dogs and horses (Dennis and Wolff, 2013). In old German mythologies, it was commonly believed that winter was caused by an ancient spirit called Mother Frost, who would shake geese feathers from her bed that fell to earth as snow (Taub, 2003). Many of the Zulu warriors of ancient southeast Africa thought of rainbow as a serpent that drank from lakes on the ground and, according to the legend, the serpent would inhabit whatever lake it was drinking from and would devour anyone who happened to be bathing therein (Dennis and Wolff, 2013).

The earliest Indians of Canada believed that wind was formed by the flapping wings of an eagle, and that the harder its wings flapped, the stronger and more abundant the wind was. If there was too much wind the tribe's eldest climbed to the top of a mountain to tie the wings of the wind eagle; however, when the weather became too hot the elder had to climb again to cut the eagle's bond (Dennis and Wolff, 2013). The ancestors of modern Cherokee Indians of North America believed that spiders spun their web in the shape of the circularly rayed sun because the grandmother spider god stole a piece of the sun from a greedy band of people and hoarded its light on the other side of the world, thus producing the numerous changes in the weather throughout the year (Taub, 2003). Another lore from early North America comes from the Cree Indians, who believed that in the beginning of the world animals could talk and had the power of day and night. Daytime animals used all their power to keep the sun in the sky, while nighttime animals forced darkness to fall; each group arguing and bringing different seasons to the earth (Taub, 2003).

In Rome, ancient folk tales describe a faith in Sirius, the dog star, which caused very hot temperatures in July and August because it was the brightest summer star; from this belief came the expression "dogs day of summer" (Taub, 2003). In Baltic and Slavic mythology, the "fiery rooster," which is a sort of Slavic phoenix, was considered the thunder god for Pērkons, Perun. Also, black roosters were sacrificed to both the Slavic Perun and the Baltic Pērkons people, especially during droughts (Downden, 2000). Placing a rooster on top of a weather vane was supposed to protect a house from thunder (Taub, 2003). Interestingly, there is a belief in Serbia that the devil runs away from Christian cross, but he is even more fearful of roosters during the thunder. So it is possible that a rooster was placed on houses' roofs to also ward off evil spirits (Dennis and Wolff, 2013). About 100 years later, Pope Leo IV placed a rooster weather vane on the old St. Peter's Basilica, and Pope Nicholas the first did in fact order that all churches display the rooster on their dome or steeple (Downden, 2000). Currently, there are roosters on top of many churches in Europe.

The Use of Local Ecological Knowledge in Forecasting

All living things interact with the global ecosystem—the combination of land, atmosphere, and oceans—which includes our environment

(National Research Council, 2010). Thus, as part of the global ecosystem, humans are susceptible to its natural variation including that of climate, which is one of the most important variables of an ecosystem. Therefore, it is not surprising that for millennia humans have tried to forecast the weather, and in doing so they have looked for indicators or signs, both among biotic and abiotic resources. Flora, fauna, astrological constellations, lunar cycles, and winds are among the various indicators that have been used by humans to forecast seasonal climate (Alves and Nishida, 2002; Araujo et al., 2005; Bezerra et al., 2012, 2013; Gilles et al., 2013; Speranza et al., 2010). The observation and interpretation of these indicators certainly represents the oldest form of predicting the climate and weather, and which is still used to this day by local communities that continue to depend on "own knowledge of ecological systems for perceiving the environment and dealing with natural adversities." Currently, modern meteorological forecasting is performed by collecting instrumental data and analyzing parameters such as sea surface temperatures, wind direction and speed, temperature, humidity, and atmospheric pressure. Therefore, as pointed out by Ziervogel and Opere (2010), there are two main types of seasonal climate forecasts: meteorological climate forecasts and LEK-based climate forecasts.

Such forms of forecasting coexist in many regions of the world, demonstrating that they are not mutually exclusive and may even complement one another. This perspective is defended by some authors (Enock, 2013; Shumba, 1999), who support the integration of these two forms of forecasting, emphasizing that LEK can be significantly valuable and boost forecasting accuracy and reliability if it is systematically researched, documented, and subsequently integrated with conventional forecasting systems. As Tekwa and Belel (2009) highlight, the idea of integrating the know-how of contemporary science with insight from LEK for more rigorous weather forecasting is welcomed, since weather information is imperative to pastoral and agricultural decisions concerning planting crops, fertilizing, stocking, and rain-fed farm management (Ayal et al., 2015; Doherty et al., 2009; Field, 2005; Oba, 1997; Tekwa and Belel, 2009).

According to Stigter (2010) and Zuma-Netshiukhwi et al. (2013), local experiences with weather and climate include meticulous observations at different scales in time and space, and are useful as a complement to instrumental climatic data. Therefore, as a consequence of the decreasing popularity of traditional forecasting, some endemic species that were once used for weather forecasting are lost (Kipkorir et al., 2010; Roncoli et al., 2002). Furthermore, the development of modern education and monotheistic religions has also contributed to the decline in the view of TEK as being reasonable (Ayal et al., 2015; Joshua et al., 2012). In many places around the world, the unreliable and precarious subsistence of traditional weather and climate forecasting skills is further destabilized by poverty, famine, and continuous drought, as well as limited understanding of transmission mechanisms and insufficient research (Ayal et al., 2015; Chang'a et al., 2010; Chengula and Nyambo, 2016; Makwara, 2013; Nakashima et al., 2012; Shoko, 2012; Speranza et al., 2010). Thus far, contemporary science has not come up with a categorical position for, or contrary to, the information provided by traditional LEK for weather forecasting, even though some consider that modern science may gain valuable insights from LEK (Ayal et al., 2015; Mundy and Compton, 1991).

Local methods for forecasting weather and climatic events have been studied mainly within the context of ethnometeorology and ethnoclimatology (Cabrera et al., 2001; Lammel et al., 2008; Orlove et al., 2002; Sánchez-Cortés and Chavero, 2011). Understanding forecasting with LEK is interesting in its own right and is a way to preserve cultural traditions, but it can also help present-day providers of scientific weather information better communicate their findings to particular sectors of the public and thus lead to complementary exchanges of ideas (Peppler, 2008). Orlove et al. (2002), for example, uncovered a scientific basis (absence or presence of El Niño-produced sub-visual cirrus clouds) of the successful forecasts of coming rains by potato farmers in the Andes of Peru and Bolivia. Knowledge of nature may also have contemporary relevance in coping with, and adapting to, environmental extremes such as drought or climate change (e.g., Suzuki and Knudtson, 1992).

Animals are one of the most popular climate indicators. To guide their climate predictions, people observe behavioral, physiological, and reproductive aspects of these organisms. As a consequence, numerous animal species are used as climatic indicators throughout the world, which is the subject of discussion in the next section.

Ethnozooindicators: Animals as Climate Indicators

For centuries human communities have relied on natural indicators, such as plants and animals, for weather forecasting and climate prediction (Chisadza et al., 2015). Farmers and agro-pastoralists are familiar with the relationships between weather, crop suitability, crop selection, planting schedule, and raising livestock in a particular season (Mercer et al., 2007; Sillitoe, 2007). When and what to do is determined by integrated weather/climatic indicators and interpreting them within the context of the environment.

In this scenario, animals play a very important role as climatic indicators, and many species are used in this manner by people in the most varied regions of the world. These traditional weather-related faunal indicators differ across the cultural experience of different communities and are used to guide local choices regarding farming or hunting-gathering activities (Chisadza et al., 2013; Garay-Barayazarra and Puri, 2011; Hart, 2007; Zuma-Netshiukhwi et al., 2013).

Our review has revealed that at least 201 animal species of 48 orders and 10 classes are used in traditional climatic and weather forecasting worldwide. The taxonomic group with the largest number of species is birds, followed by insects and mammals (Table 21.1).

INVERTEBRATES

Our results identified at least 42 invertebrates that are used as climate ethnoindicators (Table 21.1). As a group, arthropods possess a great diversity of behavioral activities that are observed by humans for predicting weather and climatic events. According to Kihupi et al. (2003) and Dunn (2000), insects, such as houseflies, fleas, cockroaches, and tarantulas, among numerous others, are signals for the coming of summer in Japan. In some southern municipalities in Japan, peasants stated that cockroaches disappear into the ground before and during the winter (Kihupi et al., 2003). In some provinces of Burkina Faso, Africa, women also gain insight by observing insect behavior at water sources and in rubbish piles. For instance, the emergence of larval black insects, usually Orthoptera, from concave dirt nests predicts a good farming season and a resultant full granary (Roncoli et al., 2002). Variation in insect population dynamics has been proven to occur throughout the seasons of the year (Changa et al., 2010; Lowman, 1982; Zuma-Netshiukhwi et al., 2013). For example, in southwestern Free State, Africa, the appearance of ants and the mushrooming of anthills during the planting season indicate that daily temperatures are warm enough for ants to come out of hibernation and roam around in and on the soil. This also indicates that it is warm enough to plant crops that are sensitive to low temperatures (Zuma-Netshiukhwi et al., 2013).

Pareek and Trivedi (2011) studied how the local beliefs and traditional knowledge of the pastoral communities of Rajasthan (India) were used to predict weather change and natural

TABLE 21.1 Animals Used as Weather and Climate Indicators in Folk Knowledge According to Surveyed Literature

Animals	Region	Traditional Zoo Indicator Prediction	Decision Taken by People
MOLLUSCA			
Gastropoda			
Helix aspersa (O. F. Müller, 1774)—shelled snails/ garden snail	Tanzania	Occurrence during times of year when they are not expected is a sign that there will be much rain during the short rainy season	Prospect of a good season for crop
INSECTS			
Hymenoptera			
Ants	Southwestern Africa	Occurrence of red ants and rapidly increasing size of anthills, which are moist, are used to predict good rains	Preparation for (late) sowing season
	Bahia (Brazil)	Construction and position of anthill; larval transportation; and row movement behavior are a sign of good winter	Abundance of water to raise animals
	Santa Cruz do Sul—RS, Brazil	Whenever their work increases that's a sign of bad weather ahead	Preparation for good winter
	Native American Tribes	Preparing the ground and anthill building	Preparation of coming rains or hard winters
	Rajasthan, India	Occurrence of ants indicate imminent rainfall onset	Prospect for a very good season
	Tanzania	Their presence on *Albizia* trees with water dripping from them is an indication of a good season	Preparation for much water
	Zimbabwe	Ants searching food and sealing off the holes of anthill	Preparation to rainy season about to start
	India	Ants carrying their eggs and larvae to safe places predicts the occurrence of rain	Good season for plantation
	Philippines	Exit of ants from their caves usually carrying stored food	Preparation to onset of rainy season
	Brazil	When winged adults leave the anthills, it is a sign of imminent rainfall	Good rainy season
	Mexico	Moving in line across the width of a sidewalk or road predicts rain onset	Attention to the weather
Wasps	Philippines	Hide their honeycomb under the leaves	Good rainy season or drought
Polistes spp.—Marimbondo	Ceará, Brazil	When they get into houses, it is a sign of impending rain	Good winter season
Pogonomyrmex spp.	Southwestern highland of Tanzania	Indicators of forthcoming rainfall	Preparation for abundant water

TABLE 21.1 Animals Used as Weather and Climate Indicators in Folk Knowledge According to Surveyed Literature—cont'd

Animals	Region	Traditional Zoo Indicator Prediction	Decision Taken by People
Dorylus spp.	Uganda	Occurrence and uphill movement of African army ants is an indicator for the onset of the dry season	Preparations of field fertilization and livestock management
Dorylus wilverthi (Emery, 1899)	Southwestern highland of Tanzania	When ants and flying ants are seen during rainy season that's a prediction of abundant rainfall during the year	Herds will have permanent food for a year
Camponotus herculeanus (Linnaeus, 1758)—black ant	India	Row movement behavior of thousands of black ants in a stream, indicates rain	Attention to the short rainy season
Eciton burchellii(Westwood, 1842)—army ants	Tanzania	Appearance in great numbers almost everywhere including houses beavering agitatedly is a prediction of onset of rains and mostly heavy during short rainy season	Preparation for rains in 1 or 2 weeks
Apis mellifera (Linnaeus, 1758)—honey bee	Uganda	Occurrence and movement is an indicators for the onset of the dry season	Storage of resources
	Tanzania	Appearing in larger groups predicts that rains are forthcoming	Preparations for much rain during season
	Ceará, Brazil	When they get into houses it is a sign of impending rain	Good water season
Orthoptera			
Not identified	Burkina Faso	Larvae coming from concave nests symbolize a good harvest season	Preparation for a full granary
Crickets	Philippines	Appearance predicts droughts	Indicative of water shortage
	Nigeria	when crickets dig holes and make loud chirping courtship calls	Indicative that rains are well established
Melanoplus differentialis (Thomas, C., 1865)—green grasshopper	Tanzania	Occurrence of green grasshoppers in great numbers in the fields predicts the onset of rainfall especially during the short rainy season	Preparations for crop plantation
Hesperotettix speciosus (Scudder, 1872)—grasshopper	Southwestern highland of Tanzania	Occurrence of many grasshoppers in a particular area indicates drought and shortage	Food and water storage
Zonocerus variegatus (Linnaeus, 1758)—variegated grasshopper	Nigeria	its occurrence is a typical sign of weather changing	Attention to much rain or drought
Ruspolia baileyi (Otte, D., 1997)—bush cricket	Uganda	Occurrence and movement is an indicator for the onset of the dry season	Storage of resources
Brachytrupes membranaceus (Drury, 1770)—tobacco cricket	Zimbabwe	Abundance predicts rainfall and good rainy season	Preparation to water and food storage

Continued

TABLE 21.1 Animals Used as Weather and Climate Indicators in Folk Knowledge According to Surveyed Literature —cont'd

Animals	Region	Traditional Zoo Indicator Prediction	Decision Taken by People
Blattodea			
Blattaria spp.	Mount Kilimanjaro in Moshi rural district of Tanzania	Moving into houses avoiding wet conditions and lurking for shelter and food is a prediction of forthcoming rainy season	Crop preparations
Coleoptera			
Holotrichia spp.	Philippines	Becoming earthbound	Indicative of an upcoming rain
Ephemeroptera			
Mayflies	Burkina Faso	Occurrence of large numbers of flies during maize planting is a prediction of good season	Attention to production of staple grains
Diptera			
Musca domestica (Linnaeus, 1758)—fly	India	Swarms of house flies predict imminent rain	Attention to the short rainy season
	Ceará, Brazil	When they get into houses it is a sign of impending rain	Good water season
Odonata			
Dragonflies	India	When flying low, may indicate rain in the same day	Decisions on threshing floor, fodder; keeping the livestock under protection
Isoptera			
Termites	Rajasthan, India	Occurrence of termites indicate imminent rainfall onset	Prospect for a very good season
	Zimbabwe	Collecting and storing grass	Indicative that the summer season is just around
	India	Flying in the evening hours is a sign that there will be rain	Much water
	Philippines	Abundance of winged termites fluttering around light sources especially at night	Indicative of an upcoming rain
	Nigeria	Termites' nuptial flights	Indicative that rains are fully established
Ancistrotermes sp.—termites	Southwestern highland of Tanzania	Many termites is an indication of near rainfall onset	Crop preparations

TABLE 21.1 Animals Used as Weather and Climate Indicators in Folk Knowledge According to Surveyed Literature—cont'd

Animals	Region	Traditional Zoo Indicator Prediction	Decision Taken by People
Hemiptera			
Spittle bugs	Mount Kilimanjaro in Moshi rural district of Tanzania	Presence on trees is indicative of impending rains	Preparing for the planting season
Huechys sanguinea (De Geer, 1773)—cicada	Philippines	Beginning their incessant high-pitched droning call	Indicative of an upcoming rain
Hymenoptera			
Reticulitermes spp.	Tanzania	Appearing after strong sunshine predicts the star of rainy season	Indication of good year for arable rain-fed farming
Lepidoptera			
Butterflies	Uganda	Occurrence and movement is an indicator for the onset of the dry season	Storage of resources
	Rajasthan, India	Occurrence of many butterflies indicates early rainfall onset	Prospect for a very good season
Armyworms	Southwestern Africa	Occurrence is an indication of drought to come	Preparation for very dry season
Red caterpillars	Uganda	Occurrence and movement are indicators for the onset of the dry season	Storage of resources
Papilio spp.	Tanzania	Migration of butterflies is a premonition of rains	Preparations for much rain during season
Amsacta albistriga (Walker, 1864)—red hairy caterpillar	India	Quick movement is considered as an indicator of rain	Attention to coming of short rainy season
Acentrocneme hesperiaris (Walker, 1856)—maguey butterfly	Mexico	Great abundance predicts good rainy season	Good season for maize crop
Spodoptera exigua (Hübner, 1808)—locusts	Tanzania	Presence of many is a prediction of long rains	Prospect for a very good season
Charaxes pollux (Cramer, 1775)—black bordered charaxes	Southwestern Highland of Tanzania	Abundance of black butterflies in a particular area indicates great rainfall season	Attention to field fertilization and herds managing
Spodoptera exempta (Walker, 1856)—armyworms	Southwestern highland of Tanzania	Armyworms all over trees during October testifies abundant rainfall in the upcoming season	Water storage
	Nigeria	Appearance after planting, a period of dry spell is imminent	Water storage

Continued

TABLE 21.1 Animals Used as Weather and Climate Indicators in Folk Knowledge According to Surveyed Literature—cont'd

Animals	Region	Traditional Zoo Indicator Prediction	Decision Taken by People
Thysanoptera			
Various species	Tanzania	Insects near wetland areas indicate good rainy season	Preparation for abundant water
DIPLOPODA			
Millipedes	Uganda	Appearance and movement	Indicative for the onset of the rainy season
	Zimbabwe	Presence of millipedes indicate the start of the rain season	Preparation for much water
MALACOSTRACA			
Crabs	Cultures which live near sea	Crabs block the mouth of the hill and scratch (or not) the sand flat	Indicative of windy weather with high probability of rainfall
	Mount Kilimanjaro in Moshi rural district of Tanzania	Crabs migration behavior from rivers and streams looking for upper grounds predicts occurrence of floods and heavy storm	Preparation for water storage
	Philippines	Female native crabs migrating from rivers to brackish water predict onset of rainfall	Preparation to onset of rainy season
	India	If crab makes a bigger hole in its channel	Good moment for weeding and harvesting, plowing, and arranging seeds for sowing
ARACHNIDA			
Spiders	Native American Tribes	Spider webs in the air and in trees	Imminent hard and cold winter
	Zimbabwe	Spiders appearing around houses	Rains about to start
Araneae			
Acanthoscurria geniculata (C. L. Koch, 1841)—Brazilian whiteknee tarântula	Bahia (Brazil)	Occurrence in houses indicates great rainfall	Attention to field fertilization and herds managing
Scorpiones			
Scorpion	Tanzania	Occurrence of black scorpions during September and October is a prediction of much rain	Good year to feed animals

TABLE 21.1 Animals Used as Weather and Climate Indicators in Folk Knowledge According to Surveyed Literature—cont'd

Animals	Region	Traditional Zoo Indicator Prediction	Decision Taken by People
ACTINOPTERYGII			
Characiformes			
Hoplias malabaricus (Bloch, 1794)—Haimara; Traira	Ceará, Brazil	If they are ovate during the last months of the year, it is a prediction of good rainy season and that they will spawn in the new waters	Good rainy season
Prochilodus lineatus (Valenciennes, 1837)	Ceará, Brazil	If they are ovate during the last months of the year, it is a prediction of good rainy season and that they will spawn in the new waters	Good rainy season
AMPHIBIA			
Anura			
Frogs	Southwestern highland of Tanzania	When frogs vocalize that is a prediction of good rainfall in the coming rainy season	Abundance of water to raise animals
	Santa Cruz do Sul—RS, Brazil	When frogs and toads croak a lot, it is a sign of rain ahead	Good winter
	Uganda	When frogs in swampy areas start croaking at night, it is indicative of the onset of the rainy season	Preparations of field fertilization and livestock management
	Rajasthan, India	When frogs start to make a lot of noise	Prospect for a very good season
	Zimbabwe	When frogs start to make a "hiss," it's a typical sign of precipitation	Good season to raise animals
	India	Well or tank full of frogs making noise at night times clearly indicates heavy rain	Advice for plantation
	Kingdom of Swaziland, Africa	Continuous "cry" calls in a certain pattern is considered a sign of imminent rainfall	Preparing for the planting season
	Tanzania	Appearance of frogs making much of noise predicts coming rains and when delay occurs, the silence evidenciates rainy season is yet to start	Attentions to rainy season
	Philippines	Croaking calls of endemic frogs in swampy areas and hide their egg mass	Preparation to onset of rainy season
Xenopus laevis (Daudin, 1802)—African clawed frog	Tanzania	Making a lot of loud vocalization is a prediction of rainfall onset and good season	Preparations of field fertilization and livestock management
Lithobates pipiens (Schreber, 1782)—northern leopard frog	Bahia (Brazil)	When frogs "sing" it is a prediction of coming rain	Attention to field fertilization and herds managing

Continued

TABLE 21.1 Animals Used as Weather and Climate Indicators in Folk Knowledge According to Surveyed Literature—cont'd

Animals	Region	Traditional Zoo Indicator Prediction	Decision Taken by People
Hoplobatrachus tigerinus (Daudin, 1802)—Indian bullfrog	India	Croaking underneath stones, leaping over small frogs predicts rain onset	Attention to start of short rainy season
REPTILIA			
Squamata			
Snakes	Southwestern Africa	Some snakes species moving down the mountain is a prediction of great rains	Indicative to engage into different agricultural activities such as land preparation, planting, weeding, spraying, etc.
	Zimbabwe	Occurrence of mating predicts that winter is approaching	Preparation for hard winter
Chameleon sp.	Pakistan	Singing out loud is a sign of a good rainy season	Preparation to much rain
	Nigeria	Appearance on crops after rainfall commencement is a prediction of a period of dry spell	Preparation to water shortage
	Rio Grande do Norte, Brazil	When it buries underground, It is a sign of "early winter"	Preparations of field fertilization and livestock management
Salvator merianae (Duméril and Bibron, 1839)—black-and-white Tegu; Tejú	Rio Grande do Norte, Brazil	When they do not come out from within the burrow, signs of "early winter"	Preparations of field fertilization and livestock management
Heloderma suspectum (Cope, 1869)—gila monster	Native American Tribes	Roaming around	Indicative of a long and very cold winter
Testudines			
Tortoises	Southwestern Africa	Usual appearance of tortoises around is a sign of good rains	Indicative to engage into different agricultural activities such as land preparation, planting weeding, spraying, etc.
AVES			
Birds	Santa Cruz do Sul—RS, Brazil	Birds when they fly low and agitated perceive climate change	Indicative of rain ahead
	Rajasthan, India	When some birds "cry," it is a prediction of the rainy season onset	Prospect for a very good season
	Burkina Faso	Nests of the small quail-like bird hanging high or low of tree is a prediction of heavy or low rains during year	Preparation for much rain or storage

TABLE 21.1 Animals Used as Weather and Climate Indicators in Folk Knowledge According to Surveyed Literature—cont'd

Animals	Region	Traditional Zoo Indicator Prediction	Decision Taken by People
	Zimbabwe	When birds such as *hwidzi kwidzi* (black or blue birds) sing melodious songs, it continuously heralds the commencement of the rainy season; when migratory birds begin to surface in a particular environment, it predicts imminent rain	Much water during the season
	Tanzania	Occurrence of walking to the valley or wetlands indicates short rainy seasons	Preparation for rains
	Mount Kilimanjaro in Moshi rural district of Tanzania	Birds calling in the morning and evening near water lakes indicates onset of rains	Much water ahead
Passeriformes			
Crow/raven	Pakistan and India	They help predict rainfall through their behavior, movement, calls, and flight, between 12 a.m. and 6 p.m.	Preparation for good winter
	Zimbabwe	When the weather conditions are overcast and a crow calls, that is a prediction of clear day in the following morning	Expecting good weather condition to harvest
Swallows	Southwestern Africa	Flock of swallows preceeding dark clouds	Good sign of rainfall
	Tanzania	Flocks of swallows roaming from south to north in a particular area are an indication of onset of short rainy season	Preparation to start of rain in the next 2 or 3 days
	Zimbabwe	Flying at low altitude is a sign of imminent rain	Good for water storage
Ploceus spp.	Kingdom of Swaziland, Africa	The height of the nests on trees is a prediction of floods	Attention to imminent floods
Turdus spp.—Sabiá	Brazil	The bird song predicts drought or winter approaching	Preparation for much water or shortage
Corvus macrorhynchos philippinus (Bonaparte, 1853)—large-billed crow	Philippines	Making mournful calls or seem to be crying in tree branches	Preparation to onset of rainy season
Passer melanurus (Müller, 1776)—Sparrows	Southwestern Africa	First appearance of sparrows	Rainy season is very close
Artamus leucoryn (Linnaeus, 1771)—white-bellied wood swallow	Philippines	Birds flying low to capture insects predict upcoming rain	Preparation to onset of rainy season
Psalidoprocne pristoptera (Rüppell, 1836)—blue saw-wing	Kingdom of Swaziland, Africa	Abundance of flocks in the sky during the months of September and October is a sign of close rainfall	Preparing for the planting season

Continued

TABLE 21.1 Animals Used as Weather and Climate Indicators in Folk Knowledge According to Surveyed Literature—cont'd

Animals	Region	Traditional Zoo Indicator Prediction	Decision Taken by People
Oriolus auratus (Vieillot, 1817)—golden oriole	Tanzania	Singing out loud is a sign of good rainfall season	Preparation for rains
Furnarius rufus (Gmelin, 1788)—*Rufous Hornero*	Santa Cruz do Sul—RS, Brazil	When it sings, regardless of weather conditions, it is a sign of good weather	Indication of good year for arable rain-fed farming
Zonotrichia capensis (Müller, 1776)—rufous-collared sparrow; Salta-caminho	Santa Cruz do Sul—RS, Brazil	Usually sings during daytime. When it happens to sing at night it is a sign of rain	Good year for water storage
Motacilla aguimp (Dumont, 1821)—African pied wagtail	Uganda	Occurrence and movement is indicative of the onset of the dry season	Storage of resources
Ploceus philippinus (Linnaeus, 1766)—Baya weaver	India	Depending how the bird builds its nest on a well, it is believed that it predicts the start of a good or bad monsoon season	Attention to the start of monsoon
Laniarius aethiopicus (Gmelin, 1788)—tropical boubou	Mount Kilimanjaro in Moshi rural district of Tanzania	When it appears hovering over water bodies, it is a prediction of imminent rainfall	Preparation for field fertilization
Nectarinia famosa (Linnaeus, 1766)—malachite sunbird	Southwestern highland of Tanzania	The bird's song after a prolonged drought predicts an imminent onset of a good rainy season	Indicative of abundance of food and that crops will grow
Aethopyga saturata (Hodgson, 1836)—black-throated sunbird	India	Unusual chirping	Onset of rainy season
Alophoixus flaveolus (Gould, 1836)—white-throated bulbul	India	Unusual chirping and bathing with sand	Upcoming rain
Chloropsis hardwickii (Jardine and Selby, 1830)—orange-bellied leafbird	India	Unusual chirping and flying low to chase insects	Onset of rainy season
Copsychus malabaricus (Scopoli, 1788)—white-rumped shama	India	Unusual chirping	Onset of rainy season
Culicicapa ceylonensis (Swainson, 1820)—grey-headed canary-flycatcher	India	Unusual chirping and flying in the low catchment area	Onset of rainy season
Cyornis concretus (Müller, 1835)—white-tailed flycatcher	India	Unusual chirping and flying in the low catchment area	Adverse weather condition (typhoon or flood)

TABLE 21.1 Animals Used as Weather and Climate Indicators in Folk Knowledge According to Surveyed Literature—cont'd

Animals	Region	Traditional Zoo Indicator Prediction	Decision Taken by People
Gracula religiosa Linnaeus, 1758—hill myna	India	Unusual chirping and aggressive behavior	Adverse weather condition (typhoon or flood)
Lonchura striata (Linnaus, 1766)—white-rumped munia	India	Unusual chirping with shrill sound	Adverse weather condition (typhoon or flood)
Melanochlora sultanea (Hodgson, 1837)—sultan tit	India	Unusual chirping	Upcoming rain
Myiomela leucura (Hodgson, 1845)—white-tailed robin	India	Unusual chirping with very low tune	Upcoming rain
Myophonus caeruleus (Scopoli, 1786)—blue whistling-thrush	India	Unusual chirping with very low tune	Upcoming rain
Orthotomus atrogularis (Temminck, 1836)—dark-necked tailorbird	India	Unusual chirping and very fast movement	Upcoming rain
Orthotomus sutorius (Pennant, 1769)—common tailorbird	India	Unusual chirping and very fast movement	Upcoming rain
Pycnonotus jocosus (Linnaeus, 1758)—red-whiskered bulbul	India	Unusual chirping	Upcoming rain
Toxostoma curvirostre (Swainson, 1827)—curve-billed thrasher; Cuitlacoche	Mexico	Singing in the morning is considered an indication of frost	Attention to the weather
Onychognathus morio (Linnaeus, 1766)—red winged starling	Southwestern highland of Tanzania	The bird's song after a prolonged drought predicts an imminent onset of a good rainy season	Indicative of abundance of food and arable rain-fed farming
Troglodytes aedon (Vieillot, 1809)—house wren	Mexico	Singing on the trees in the morning is a prediction of air streams	Attention to the weather
Hirundo smithii (Leach, 1818)—wire-tailed swallow	Tanzania	Appearance of flocks predicts the onset of rains and leads to the prediction of forthcoming heavy rainfall	Preparations of good rainy season
Hirundo rustica (Linnaeus, 1758)—barn swallow; swallow	Mexico	If it flies at low altitudes near crop season, it means that it will rain soon	Quantity of rainfall will be favorable for crops
Furnarius leucopus (Swainson, 1837) - Pale-legged Hornero, João de Barro	Brazil	Building its nest with the entrance facing west predicts much rain for the season	Good year for water storage and crops

Continued

TABLE 21.1 Animals Used as Weather and Climate Indicators in Folk Knowledge According to Surveyed Literature—cont'd

Animals	Region	Traditional Zoo Indicator Prediction	Decision Taken by People
Cyanocompsa brissonii (Lichtenstein, 1823)—ultramarine grosbeak; Azulão	Brazil	Vocalizations like singing, crying, whistling, shouting, chirping are indicative signs of rain	Attention to crops
Paroaria dominicana (Linnaeus, 1758)—red-cowled cardinal; Galo de Campina	Brazil	Vocalizations like singing, crying, whistling, shouting, chirping are indicative signs of rain	Attention to crops
Turdus rufiventris (Vieillot, 1818)—rufous-bellied thrush; Sabiá	Brazil	Vocalizations like singing, crying, whistling, shouting, chirping are indicative signs of rain	Attention to crops
Fluvicola nengeta (Linnaeus, 1766)—masked water-tyrant; Lavandeira	Brazil	Vocalizations like singing, crying, whistling, shouting, chirping are indicative signsof rain	Attention to crops
Icterus jamacaii (Gmelin, 1788)—Campo Troupial; Concriz	Brazil	Behavior of breeding and egg laying is indicative of imminent rainfall	Good winter season
Icterus cayanensis (Linnaeus, 1766)—Epaulet oriole; Xexeu-de-bananeira	Brazil	When it makes a safe hole in dams and barriers, it indicates rainfall	Good water season
Volatinia jacarina (Linnaeus, 1766)—blue-black grassquit; Nego-Tziu	Brazil	Behavior of flight, height, and jump is a prediction of rainfall	Good winter season
Cyanocorax cyanopogon (Wied, 1821)—White-naped Jay, Cancão	Brazil	The bird's song indicates drought or winter is approaching	Preparation for much water or shortage
Luscinia megarhynchos (Brehm, 1831)—common nightingale; Rouxinol	Rio Grande do Norte—Brazil	Make nest on the roof of houses in November. Signs of "early winter"	Good season to fertilize crops
Mimus saturninus (Lichtenstein, 1823) - Chalk-browed mockingbird; Papa-sebo	Rio Grande do Norte—Brazil	Mating in the end of year predicts good winter season	Attention to field fertilization and herds managing
Cuculiformes			
Cuckoos	Uganda	Occurrence and vocalization is indicative of the onset of the rainy season	Preparations of field fertilization and livestock management
Cuculus solitarius (Stephens, 1815)—red-chested cuckoo	Kingdom of Swaziland, Africa	The "cry" call of the bird signals the start of the wet season in August–November	Preparations for the planting season
Cuculus clamosus (Latham, 1801)—black cuckoo	India	Melodious sounds are often taken as indicative of rain	Attention to start of short rainy season

TABLE 21.1 Animals Used as Weather and Climate Indicators in Folk Knowledge According to Surveyed Literature—cont'd

Animals	Region	Traditional Zoo Indicator Prediction	Decision Taken by People
Clamator jacobinus (Boddaert, 1783)—Jacobin cuckoo	Asia, Africa, Zimbabwe, Tanzania, and India	As the rainy season approaches, this bird starts to sings. By hearing it is singing people understand that the rainfall has come	Indicative of floods and monsoons
Cacomantis merulinus (Scopoli, 1786)—Plaintive cuckoo	Philippines	Making mournful calls or seem to be crying in tree branches	Preparation to onset of rainy season
Chalcites lucidus (Gmelin, 1788)—Pīpīwharauroa, shining bronze-cuckoo	New Zealand	Appearance at a specific area predicts the beginning of warm weather	Good time to fertilize crops
Urodynamis taitensis (Sparrman, 1787)—Koekoeä, long-tailed cuckoo	New Zealand	Appearance at a specific area predicts the onset of better weather	Good time to fertilize crops
Centropus superciliosus (Hemprich and Ehrenberg, 1833)—white-browed coucal	Southwestern highland of Tanzania	The bird's song after a prolonged drought predicts an imminent onset of a good rainy season	Indicative of abundance of food and that crops will grow
Centropus superciliosus burchellii (Swainson, 1838)—white-browed coucal	Kingdom of Swaziland, Africa	When the bird chirps from October to April, it is a sign of an approaching thunderstorm	Preparing for the planting season
Centropus bengalensis philippinensis (Mees, 1971)—Lesser coucal	Philippines	Unusual chirping of birds	Preparation to onset of rainy season
Guira guira (Gmelin, 1788)—Guira cuckoo; Anum-branco	Brazil	Vocalizations like singing, crying, whistling, shouting, chirping are indicativesigns of rain	Attention to crops
Crotophaga ani (Linnaeus, 1758)—Smooth-billed Ani, Anum Preto	Brazil	Singing in the afternoon is a sign of rain, if in the morning, it is a prediction of drought	Preparation for much water or shortage
Columbiformes			
Pigeon/Dove	Many countries	If pigeons sings from morning to evening, it is a prediction that rain will fall after 8–15 days	Preparation of field fertilization and arable rain-fed farming
Columbina spp.—Rolinhas	Brazil	Beginning of nest building predicts rain or droughts	Preparation for much water or shortage
Turtur afer (Linnaeus, 1766)—blue-spotted wood dove	Kingdom of Swaziland, Africa	The appearance predicts impending long drought and famine	Preparations for food and water storage

Continued

TABLE 21.1 Animals Used as Weather and Climate Indicators in Folk Knowledge According to Surveyed Literature—cont'd

Animals	Region	Traditional Zoo Indicator Prediction	Decision Taken by People
Ducula badia (Raffles, 1822)—mountain imperial pigeon	India	Unusual movement to take shelter in shadow of leaves	Adverse weather condition (typhoon or flood)
Spilopelia chinensis (Scopoli, 1786)—spotted dove	India	Unusual chirping, move in pairs, and take shelter in shadow of leaves	Upcoming rain
Treron curvirostra (Gmelin, 1789)—Thick-billed green pigeon	India	Unusual chirping and unusual movement to take shelter in shadow of leaves	Upcoming rain
Leptotila rufaxilla (Richard & Bernard, 1792)—grey-fronted dove; Juruti	Brazil	Vocalizations like singing, crying, whistling, shouting, chirping are indicative of rain	Attention to crops
Patagioenas picazuro (Temminck, 1813)—Picazuro pigeon; Asa branca	Brazil	Behavior of breeding and egg laying is indicative of imminent rainfall	Good winter season
Columbina minuta (Linnaeus, 1766)—plain-breasted ground-dove, Rolinha Cambute	Brazil	Behavior of breeding and egg laying is indicative of imminent rainfall	Good winter season
Zenaida auriculata (Des Murs, 1847)—eared dove; Ribaçã	Brazil	Occurrence in the region predicts rainfall	Preparation for good winter
Anseriformes			
Wild ducks	Uganda	Occurrence and vocalization is indicative of the onset of rainy season	Preparations of field fertilization and livestock management
Wild geese	Native American Tribes	Flying back south at high or low altitude	Indicative of a hard winter (high altitude); mild winter or fall (low altitude)
Anas platyrhynchos domesticus (Linnaeus, 1758)—domestic duck	Tanzania	Stretching their wings and playing in dust is a sign of the onset of rains	Preparation for field fertilization
	Kingdom of Swaziland, Africa	Restlessness and noisy behavior indicates an imminent heavy storm	Preparation for much water
Marmaronetta angustirostris (Ménétriés, 1832)—marbled teal	Asia, Africa, Zimbabwe, Tanzania, and India	Agitated behavior during morning and afternoon	Indicative of monsoons ahead
Dendrocygna viduata (Linnaeus, 1766)—white-faced whistling-duck; Marreca	Brazil	Occurrence in the region predicts rainfall	Preparation for good winter

TABLE 21.1 Animals Used as Weather and Climate Indicators in Folk Knowledge According to Surveyed Literature—cont'd

Animals	Region	Traditional Zoo Indicator Prediction	Decision Taken by People
Pelecaniformes			
Bubulcus ibis (Linnaeus, 1758)—cattle egret	Uganda	Occurrence and movement of this migratory bird is indicative of the onset of dry season	Storage of resources
	Tanzania	Occurrence during October and November is a prediction of imminent rainfall onset; appearance brings bad weather because it is linked to livestock diseases	Good rainfall season for resources storage; Preparation to rain shortage
Botaurus poiciloptilus (Wagler, 1827)—Matuku-hürepo, Australasian bittern	New Zealand	The continuous crying of the bird as it moves around at night predicts that a season of flood will follow	Attention to much water during season
Accipitriformes			
Eagles	Native American Tribes	Hovering around	Prediction of better weather conditions
Aquila verreauxii (Lesson, 1830)—black eagle	Uganda	Occurrence and vocalization is indicative of the onset of the dry season	Storage of resources
Terathopius ecaudatus (Daudin, 1800)—Bateleur	Uganda	Occurrence and vocalization is indicative of the onset of the dry season	Storage of resources
Psittaciformes			
Agapornis fischeri (Reichenow, 1887)—Fischer's lovebird	Tanzania	The "crying" call is a prediction of short rainy season	Preparation for rains
Loriculus vernalis (Sparrman, 1787)—vernal hanging parrot	India	Unusual chirping and flying in fleet	Upcoming rain
Nestor meridionalis (Gmelin, 1788)—Käkä, New Zealand Kaka	New Zealand	Twisting and squawking above the forest	Attention to a storm that is forthcoming
Amazona aestiva (Linnaeus, 1758)—Turquoise-fronted Amazon; Papagaio	Brazil	Building of nest predicts rains onset	Good winter season
Psittacula krameri manillensis (Bechstein, 1800)—rose-ringed parakeet	India	Migration of parakeet in N–S direction indicates a possible occurrence of rain	Attention to start of short rainy season
Gruiformes			
Cuckoos	Uganda	Occurrence and vocalization is indicative of the onset of rainy season	Preparations of field fertilization and livestock management
Balearica regulorum (Bennett, 1834)—grey crowned crane	Santa Cruz do Sul—RS, Brazil	Usually vocalizes at dusk, if it does at another time it is because it perceives climate change	Prediction of wind and winter

Continued

TABLE 21.1 Animals Used as Weather and Climate Indicators in Folk Knowledge According to Surveyed Literature—cont'd

Animals	Region	Traditional Zoo Indicator Prediction	Decision Taken by People
Aramides saracura (Spix, 1825)—slaty-breasted wood-rail	Philippines	Unusual chirping of birds	Preparation to onset of rainy season
Hypotaenidia torquata (Linnaeus, 1766)—barred rail	Philippines	Unusual chirping of birds	Preparation to onset of rainy season
Porphyrio porphyrio melanotus (Temminck, 1820)—Pükeko, Australasian swamphen	New Zealand	Looking for higher ground	Attention to imminent storm and flooding
Lewinia striata (Linnaeus, 1766)—Slaty-breasted rail	Philippines	Unusual chirping of birds	Preparation to onset of rainy season
Anthropoides virgo (Linnaeus, 1758)—Demoiselle crane	Pakistan	Agitated behavior such as movement and calls predict rainy season	During rainfall if this bird is seen hovering over in a triangle shape it is a sign of more rain
Aramides cajaneus (Müller, 1776)—grey-necked wood-rail, Três Cocos	Brazil	Singing at dusk is a sign of imminent rain	Good winter season
Aramus guarauna (Linnaeus, 1766)—Limpkin, Carão	Brazil	The bird's song predicts drought or winter approaching	Preparation for much water or shortage
Bucerotiformes			
Buceros bicornis (Linnaeus, 1758)—great hornbill	Tanzania	Flocks hovering in the sky are an indicator of short rainy season	Preparation for rains
Ciconiiformes			
Ciconia nigra (Linnaeus, 1758)—black stork	India	Parabolic flight behavior is a prediction of rain	Attention to rain season
Coraciiformes			
Halcyon smyrnensis gularis (Kuhl, 1820)—white-throated kingfisher	Philippines	Unusual chirping of birds	Preparation to onset of rainy season
Merops hirundineus (Lichtenstein, 1793)—swallow-tailed bee-eater	Tanzania	Occurrence of swallow flock all over sky during November is a sign of imminent heavy rain in one particular area	Good season to fertilize crops
Merops philippinus (Linnaeus, 1766)—blue-tailed bee-eater	Philippines	Migration to mountains	Preparation to onset of rainy season

TABLE 21.1 Animals Used as Weather and Climate Indicators in Folk Knowledge According to Surveyed Literature—cont'd

Animals	Region	Traditional Zoo Indicator Prediction	Decision Taken by People
Caprimulgiformes			
Aerodramus brevirostris (Horsfield, 1840)—Himalayan swiftlet	Philippines	Birds flying low to capture insects predicts an upcoming rain	Preparation to onset of rainy season
Nyctibius griseus (Gmelin, 1789)—common potoo; Mãe-da-lua	Brazil	Vocalizations like singing, crying, whistling, shouting, chirping are indicative signs of rain	Attention to crops
Charadriiformes			
Vanellus miles (Boddaert, 1783)—masked lapwing	Australia	Prediction based on where it lays its eggs on the field indicates good or poor rains	Good rainy season or drought
Limosa lapponica (Linnaeus, 1758)—Küaka, bar-tailed godwit	New Zealand	The bird's arrival on a specific area predicts the start of a warm season	Good weather
Vanellus chilensis (Molina, 1782)—southern lapwing; Teteu	Brazil	Vocalizations like singing, crying, whistling, shouting, chirping are indicativesigns of rain	Attention to crops
Galliformes			
Guinea fowls	Zimbabwe	Laying eggs predicts onset of summer season	Good food storage
	Nigeria	Laying eggs at onset of rains and lay daily when rains are fully established	Good rainy season
	Rio Grande do Norte—Brazil	Laying eggs in November	Attention to field fertilization and herds managing
Gallus gallus domesticus (Linnaeus, 1758)—rooster	Bahia (Brazil)	When rooster "sings" very much it is a prediction of coming rain	Indicative of good rainy season
	Santa Cruz do Sul—RS, Brazil	When hens clean their feathers, it is a sign of rain	Abundance of water to raise animals
	Zimbabwe	Feeding during rain is a sign that there will be more rainfall	Preparation for more precipitation
	Tanzania	Stretching their wings repeatedly is a prediction of short rains	Preparations of field fertilization and livestock management
	India	When poultry sit in a place for a long time inserting feathers in the soil, it is a prediction of forthcoming rain	Water storage
	Mexico	When the rooster calls after hours (afternoon or evening), it is an indicator that it will stop raining for a while during the rainy season	Attention to the weather

Continued

TABLE 21.1 Animals Used as Weather and Climate Indicators in Folk Knowledge According to Surveyed Literature—cont'd

Animals	Region	Traditional Zoo Indicator Prediction	Decision Taken by People
Pavo cristatus (Linnaeus, 1758)—peacock	Many countries	Dancing behavior	Rainy season is very close
	Kingdom of Swaziland, Africa	Restlessness and noisy behavior indicates an imminent heavy storm	Preparation for much water
	India	Making sound early in the morning and late in the evening predicts cool breeze or imminent rainfall	Attention to short rainy season
Arborophila atrogularis (Blyth, 1850)—white-cheeked partridge	India	Unusual chirping and flying up in the sky	Adverse weather condition (typhoon or flood)
Athene cunicularia (Molina, 1782)—burrowing owl	Ceará, Brazil	The high and persistent call of the owl in the early days of the year predicts rainy weather	Good season to crop
Piciformes			
Megalaima asiatica (Latham, 1790)—blue-throated barbet	India	Unusual chirping	Upcoming rain
Megalaima australis (Horsfield, 1821)—blue-eared barbet	India	Unusual chirping and flying low	Upcoming rain
Picus canus (Gmelin, 1788)—grey-headed woodpecker	India	Unusual activity with rotation around the tree	Upcoming rain
Nystalus maculatus (Gmelin, 1788)—Caatinga puffbird, Fura Barreira	Brazil	Building its nest in higher places is a prediction of forthcoming rainfall	Good rain season
Falconiformes			
Falco novaeseelandiae (Gmelin, 1788)—Kärearea, New Zealand falcon	New Zealand	Screaming on a fine day; on a rainy day	It will rain the next day; will be fine the next day
Herpetotheres cachinnans (Linnaeus, 1758)—laughing falcon, Acauã	Brazil	Singing at dusk it is a sign of near rain	Good winter season
Struthioniformes			
Nothura maculosa (Temminck, 1815)—spotted Nothura; Codorniz	Brazil	Vocalizations like singing, crying, whistling, shouting, chirping are indicative signsof rain	Attention to crops
Cariamiformes			
Cariama cristata (Linnaeus, 1766)—red-legged seriema, Siriema	Brazil	Singing at midday it is a sign of near rain	Good winter season

TABLE 21.1 Animals Used as Weather and Climate Indicators in Folk Knowledge According to Surveyed Literature—cont'd

Animals	Region	Traditional Zoo Indicator Prediction	Decision Taken by People
Strigiformes			
Owl	Pakistan	Depending on the direction of its flight it is a sign of winter or summer	Attention to field fertilization and herds managing
	Tanzania	Prediction of planting season	Preparation for field fertilization
Tyto alba pratincola (Bonaparte, 1838)—Western barn owl	India	Squeaking indicates rain onset	Attention to start of short rainy season
MAMMALIA			
Carnivora			
Foxes	India	howling in the morning and evening indicating impending rain	Short rainy season
Canis lupus familiaris—Dogs	Mount Kilimanjaro in Moshi Rural District of Tanzania	Barking and mating during night predicts near precipitation	Preparation for rainy day
	Philippines	Unusual behavior of dogs can also predict an upcoming storm	Predictions of rain ahead
	India	If dog jumps irregularly on the road at midday, it is a sign of imminent rain	Expecting good weather condition to harvest
	Mexico	Cheerful and impetuous behavior predicts good weather (heavy rain or hail frost) and a calm or sad behavior predicts cessation of rains and possible occurrence of frost or drought	Attention to weather
Felis catus domestica—Cats	England	If a cat washes behind its ears, it will rain	
	Indonesia	Pouring water on a cat will bring rain	
	Rio Grande do Norte—Brazil	Mating in December is a prediction of good winter	Attention to field fertilization and herds managing
Canis latrans (Say, 1823)—coyote	Native American Tribes	When coyotes howl most mournfully	Prediction of cold winds
	Mexico	When it howls melodically, it predicts next rains; or if obnoxious or clipped, it will not rain soon or even it will be dry during maize growth	Attention to the weather
Panthera pardus (Linnaeus, 1758)—leopard	Tanzania	Appearance in the village is a sign of a good rainfall season	Preparation for rains
Crocuta crocuta (Erxleben, 1777)—spotted hyena	Tanzania	Depending on the calls, it can predicts the rainfall season	Preparation to receive rains for crops fertilization

Continued

TABLE 21.1 Animals Used as Weather and Climate Indicators in Folk Knowledge According to Surveyed Literature—cont'd

Animals	Region	Traditional Zoo Indicator Prediction	Decision Taken by People
Artiodactyla			
Camel	Zimbabwe	Appearance of injuries on camel's legs is a prediction of rainfall ahead	Water storage
Bos taurus (Linnaeus, 1758)—cattle	Southwestern Africa	Well-fed calves jumping around during their foraging time near mountains and unwilling to graze the following morning indicates good rains are on the way	Preparation of field fertilization and good rains
	Native American Tribes	When herd head for high ridges even during a storm in winter	Indicative that storm will be over by the following morning
	Pakistan	When the cow lies down, it is a sign of abundant rainfall	Indicative of forthcoming rains
	India	When the cow licks each other, it is a prediction of drought	Preparation for food and water storage
	Uganda	Cattle are restless and start jumping	Indicative of the onset of the rainy season
	Tanzania	Ear flapping is a prediction of onset and prospects of a good season	Preparation for rainy day
Ovis aries (Linnaeus, 1758)/*Capra hircus* (Linnaeus, 1758)—sheep, goat	Southwestern Africa	Much libido in goats and sheep with increased mating is a sign for good rains	Indicative to engage into different agricultural activities such as land preparation, planting weeding, spraying, etc.
	Tanzania	Ear flapping is a prediction of onset and prospects of a good season	Preparation for rains
	Zimbabwe	Constant mating predicts onset of rain season	Good year for water storage
	Mexico	Cheerful and impetuous behavior predicts good weather (heavy rain or hail frost) and a calm or sad behavior predicts cessation of rains and possible occurrence of frost or drought	Attention to weather for crop management
Sus scrofa domesticus (Erxleben, 1777)	Southwestern Africa	Grunting is an indicative of low humidity and increase in temperature	Preparation for agricultural activities
	Native American Tribes	When a pig is butchered, usually in the fall, they look for its spleen	If found to be thick, then it's a prediction of a hard winter, and lots of snow
	Kingdom of Swaziland, Africa	Restlessness and noisy behavior indicates an imminent heavy storm	Preparation for much water

TABLE 21.1 Animals Used as Weather and Climate Indicators in Folk Knowledge According to Surveyed Literature—cont'd

Animals	Region	Traditional Zoo Indicator Prediction	Decision Taken by People
Perissodactyla			
Equus asinus (Linnaeus, 1758)—donkey	Santa Cruz do Sul—RS, Brazil	Rolling on the ground	Predictions of rain ahead
Equus (Equus) caballus (Linnaeus, 1758)—horse	Native American Tribes	When horses play with each other and stamp around	Indicative of a coming winter storm
	Santa Cruz do Sul—RS, Brazil	Approach each other	Predictions of rain ahead
	Mexico	When horse trembles in the evening, it is a sign that a frost is imminent on the fields	Attention to the weather
Cetartiodactyla			
Antilope cervicapra (Linnaeus, 1758)—antelopes, blackbuck	Tanzania	Appearance in the village is a sign of a good rainfall season	Preparation for rains
Lagomorpha			
Rabbit	Native American Tribes	Getting fat or not during the autumn	Fat in fall is a prediction of mild winter, if not it means cold winter
	Native American Tribes	If rabbits eat brushwood next to snow or at its height in a standing position	Indicative of a lot of snow during the winter
Hyracoidea			
Procavia capensis (Pallas, 1766)—rock rabbit	Zimbabwe	When it gives out a typical squeak, it is a prediction of imminent rainfall in a particular place	Much water during the season
Rodentia			
Sciurus spp.—fox squirrels	Native American Tribes	Not storing nuts and corn in previous autumn	Indicative that the snowfall would not last long and that the squirrels will be more energetic in search for food
Cynomys spp.—Prairie dogs	Native American Tribes	Standing at their den entrance	Prediction of a long and hard winter
Ondatra zibethicus (Linnaeus, 1766)—muskrat	Native American Tribes	Construction of an unusually high and large house	Indicative of a severe winter
Castor canadensis (Kuhl, 1820)—American beaver	Native American Tribes	Construction of an unusually high and large house	Indicative of a severe winter
Marmota (Marmota) monax (Linnaeus, 1758)—groundhog	North America	Its behavior can predict if winter will be heavy or not	Attention to crops

Continued

TABLE 21.1 Animals Used as Weather and Climate Indicators in Folk Knowledge According to Surveyed Literature—cont'd

Animals	Region	Traditional Zoo Indicator Prediction	Decision Taken by People
Kerodon rupestris (F. Cuvier, 1825)—rock cavy; Mocó	Rio Grande do Norte—Brazil	Grow the tail; If it is skinny; mating during November/December predicts good winter	Attention to field fertilization and herds managing
Galea spixii (Wagler, 1831)—Spix's yellow-toothed cavy; Preá	Rio Grande do Norte—Brazil	Grow the tail; If it is skinny; mating during November/December predicts good winter	Attention to field fertilization and herds managing
Chiroptera			
Bats	Zimbabwe	Occurrence of *muremwaremwa* is an indication of erratic rainfall or good season, if it's seen flying or on the ground	Attention for much water or low water for the fields
Primates			
Howler monkeys	Santa Cruz do Sul—RS, Brazil	The monkeys are sensitive to atmospheric pressure, making it howl drastically and insistently	Prediction of heavy rain
Lemurs	Mount Kilimanjaro in Moshi rural district of Tanzania	Occurrence of lemurs around farmhouses signs rainfall onset	Crop preparations
Colobus sp.—colobus monkey	Tanzania	Presence of many groups of colobus monkeys in the fields is a prediction of long rainy season	Much water during the season
Chlorocebus pygerythrus (F. Cuvier, 1821)—baboon, vervet monkey	Tanzania	Appearance in the village is a sign of a good rainfall season	Preparation for rains
Cingulata			
Armadillo	Brazil	Occurrence of armadillo pregnant in the month of December is a good sign of rain	Preparations for crop plantation
Tolypeutes tricinctus (Linnaeus, 1758)—Brazilian three-banded Armadillo	Ceará, Brazil	During November, the occurrence of three or four armadillo pups housed in the den is a good sign of rain	Good winter season
Pholidota			
Pangolin/scaly anteaters sp.	Tanzania	Occurrence is a prediction of rainfall	Much water during the season

disasters, and found them to be key elements of their environmental management and conservation efforts. For example, invertebrates such butterflies, ants, and termites can indicate the immediate onset of precipitation.

The presence of cicadas and termites is considered a traditional weather indicator for the locals of Mberengwa in Zimbabwe (Shoko, 2012). Similarly, elders in Masvingo and Manicaland (also Zimbabwe) rely on insect behavior (anthill and termite mound construction, leafcutter ants moving in line) to estimate the start and intensity of the rainy season (Muguti and Maposa, 2012; Risiro et al., 2012). Likewise, while exploring indigenous knowledge in the southwestern highlands of Tanzania, Changa et al. (2010) found that a considerable number of insects, such as grasshoppers, butterflies, armyworms, and termites, are used by local populations to predict the beginning of the rainy season. Furthermore, Mahoo et al. (2015) observed that locals of the Lushoto district in Tanzania forecast rains by observing the presence and migration of butterflies from south to north, especially black butterflies which are a sign of a good season. The appearance of large swarms of red ants from September to November and the occurrence of large swarms of butterflies is indicative of the imminent onset of rainfall and that the upcoming rainfall season will be good (Changa et al., 2010).

In various provinces of India, Anandaraja et al. (2008) recorded several local climate and weather forecasts linked to the folk practices that farmers and pastoralists use in their agricultural systems, which included the behaviors of some invertebrates, such as ants, dragon flies, and termites. The appearance and behavior of some insects, for instance, butterflies, red caterpillars, western honey bees, and bush crickets, are considered traditional indicators of the approach of the dry season in some districts of Uganda (Okonya and Kroschel, 2013). In northern Kenya, traditional communities of pastoralists make use of indigenous weather forecasting methods, such as the increased presence of crickets, butterflies, and other insects, in predicting the start of monsoons (Kagunyu et al., 2016).

Most traditional people in Africa also evaluate time based upon the behavior of several insects. For instance, the presence of black beetles is commonly related to the beginning of planting crops; according to tribal elders, these insects only occur during the harvest time (Mapara, 2009). Another example is the continuous singing of *mandere* (*Eulepida* sp.), which signals the imminence of precipitation (Pareek and Trivedi, 2011). Similarly, the singing of *nyenze* (cicadas) predicts the start of rains in 2–3 weeks. Other insects that inform some traditional African people about when to plant crops include *makugwe*, *Brachytrupes membranaceus* (Drury, 1770), and *mopani* worms (Emeagwali and Shizha, 2016). Muguti and Maposa (2012) found parallel patterns in their study of indigenous weather forecasting among the Shona people in Zimbabwe. Richards (1980) observed that farming groups in southern Nigeria used nature-based knowledge of the multicolored grasshopper *Zonocerus variegatus* (Linnaeus, 1758) to predict the severity and geographical extent of monsoon outbreaks. According to a traditional weather forecaster from the Philippines, a flood is believed to be coming when wasps build their honeycombs high in the tops of trees, and strong winds are anticipated when they build them near the ground (Galacgac and Balisacan, 2009).

The behavior of insects in relation to climatic parameters is not unknown to science. In a study conducted in Brazil, Pellegrino et al. (2013) observed variation in mating behaviors of three taxonomically unrelated insects, the curcurbit beetle, *Diabrotica speciosa* (Coleoptera), the true armyworm moth, *Pseudaletia unipuncta* (Lepidoptera), and the potato aphid, *Macrosiphum euphorbiae* (Hemiptera), when exposed to natural or experimentally manipulated alterations in atmospheric pressure. They demonstrated that insects, in fact, have the ability to detect changes in time by sensing reductions in atmospheric pressure, thus seeking shelter to protect themselves in anticipation of unfavorable climatic conditions such as thunderstorms and windstorms (Acharya, 2011). Similarly, it has been observed that most crabs react to certain environmental signals that accompany weather changes, and not to the weather itself. Sometimes the crabs, which live in sand hills, block the entrance to their home in anticipation of rainy weather. Other times they exhibit the behavior of scratching the sand, which is a signal of incoming strong winds (Alsaiari, 2012; Dudde and Apte, 2016; Hines et al., 2010). According to Chengula and Nyambo (2016), traditional weather forecasters of the southeastern slopes of Mount Kilimanjaro perceived that the migration of crabs from watercourses and rivers to higher land indicates probable flooding or heavy rainfall. Galacgac and Balisacan (2009) mentioned that Philippine people used crustaceans, such as shrimps and crabs, as indicators of upcoming typhoons, by observing them exhibit the behaviors of migrating from rivers to brackish water or crawling out of the water to riverbanks.

VERTEBRATES

Vertebrates stand out as the group of animals most widely used for evidence of climate change by traditional and local populations across the globe. Our review revealed that at least 158 species of vertebrates are used to predict weather conditions. Of these, birds were the most referenced, followed by mammals. Amphibians and reptiles have also been used as climate ethnozooindicators, in many cases inspiring beliefs associated with climatic factors.

In many folk beliefs, amphibians, including frogs and toads, are frequently related to rainstorms and blizzards. In India, for example, frogs were assumed to echo the thunder, and even the name frog in ancient Sanskrit means "cloud" (Sherman, 2008). In Mesoamerica, most amphibians were thought to be rain deities and were included in several rituals to bring precipitation. The *Aymara* natives of Bolivia and Peru placed small statues of toads on hilltops to call for the rainy season. If rains failed to arrive, such statues were usually broken because of their perceived failure (Pinch, 1995). In Australia, ancient aborigines also considered frogs to call upon rain (Robbins, 1996). Other local traditions, including those found in the Appalachian Mountains of the United States, hold that if someone kills a frog or toad, a downpour will ensue. It was even thought that frogs fell with the rain (Sherman, 2008).

Unsurprisingly, amphibians are commonly used as weather predictors in many regions. In Zimbabwe, for example, traditional communities use frogs to predict the intensity of rains (Mapara, 2009). Whenever they begin to produce an incessantly "*hiss*" sing, natives realize that the rainy season is "around the corner." Likewise, the congregation of huge, brownish bullfrogs, locally known as *machesi*, at a particular lake is interpreted as a prediction for much rain at that locality. Similarly, Emeagwali and Shizha (2016) explain how elders from African tribes identify frog sounds and celebrate the arrival of rainfall; they are able to associate rainfall patterns with the calls of certain species of frogs. For example, the croaking of bullfrogs with a high tone indicates that rains will arrive the following day. Their observations of the way other living organism behave are used to predict the quality of the impending season (Emeagwali and Shizha, 2016). Among the elders of the Mfereji village in the Monduli district of Tanzania, amphibian behaviors, in most cases frogs (*Africana* spp.) and toads (*Bufo* spp.), are clear indicators of seasonal variation. The absence of these animals is also used to indicate the coming of the dry season. According to the elders, when frogs stop croaking during the rainy season, even if it is still raining, it is an indication that the rains would soon cease (UNEP, 2008).

Reptiles play an important role in several cultures, including in mythology associated with climate and weather. In Australian aboriginal culture, Wollunqua is sanctified as the rainbow snake, a massive snake united with the rainbow and the entirety of creation itself. Another creature is *Eingana*, an aboriginal snake goddess who created the land–water living beings, and brought all seasons to the earth (Sherman, 2008). Many hunters become alert and cautious when snakes are seen more frequently while they hunt for prey, as it is an indication of the onset of the growing season. Abundant movements of snakes and tortoises are interpreted as a seasonal prediction of ensuing good rainfall (Changa et al., 2010; Gardner, 2011).

According to Marais (2008) and Zuma-Netshiukhwi et al. (2013), snakes normally come out of hibernation and descend mountains looking for prey and breeding partners early in the summer. In Tanzania, native tribes rely on snakes to predict the length of the fall, and if they hibernate later, the fall will be longer than normal (UNEP, 2008). On the other hand, in many areas of Kenya, the presence of snakes and other reptiles around homesteads in search of water and food indicate the prevalence and continuity of drought (UNEP, 2008). According to a peasant community in Tlaxcala, Mexico, water snakes indicate the arrival of good rains (Rivero-Romero et al., 2016). Lizards also are considered weather predictors in some places. For example, Ahmed et al. (2016) recorded that people of Tharparkar, Pakistan, believe that if chameleons climb trees and shake their heads, it is indicative of good rainfall. Furthermore, when another type of chameleon, which the Tharparkar people claim lives underground, senses imminent rainfall, as it emerges (Ahmed et al., 2016).

Perhaps, no other animal group is more famous for weather forecasting than birds. Birds are the group most cited by numerous traditional communities worldwide, and one of the most reliable with regard to the influence of climate change on behavior. Examples involving birds as climate ethnozooindicators are abundant. Zuma-Netshiukhwi et al. (2013) pointed out that South African farmers use the appearance of cape sparrows, *Passer melanurus* (Müller, 1776), as an indication of the onset of good rain in the next day or two, which stimulates farmers to prepare for field activities (De Swardt et al., 2004). Eastwood (1967) stressed that flocks of sparrows were commonly observed in southwestern Free State during the growing season, feeding on seeds and insects. Therefore, their presence flying around in the sky, along with scattered clouds, indicates that rain is coming in the afternoon (Koistinen, 2000). A group of sparrows over grayish clouds predicts heavy rainfall for the next hour (Kopij, 2002). Migration of certain bird species is associated with the change in seasons in response to temperature and rainfall (De Swardt et al., 2004; Koistinen, 2000; Kopij, 2002; Zuma-Netshiukhwi et al., 2013).

The behavior of animals, including the appearance and movements of birds and insects, is frequently used by Hehe and Nyakyusa elders in Tanzania to predict the weather and climate of their communities. In both Kilolo and Rungwe districts of Tanzania, the appearance of large swarms of Yangiyangi birds is indicative of a good rainfall season and the imminent onset of rainfall (Chang'a et al., 2010; Gissila et al., 2004). Peppler (2010) reported that Native American tribes were known for their long-term climate predictions, as a 1950–52 letter by Senator Robert S. Kerr explains. According to Peppler (2010), native Indians seemed to base their forecasts on natural phenomena, such as the flight altitude of wild geese or even eagles hovering over lakes.

Welty (1982) stressed that perhaps the most famous advantage provided by the observation of the migration of birds is the prediction that better weather is assured. It is well known that birds display different behaviors as weather fronts approach. Furthermore, birds are able to sense infrasonic frequencies (at least down to 0.05 Hz), which can travel hundreds of kilometers, thus permitting birds to orient to remote locations (e.g., distant mountains, canyons, or even shorelines). According to Anderson and Eriksson (2007), the appearance of birds and insects can assist in detecting meso-scale meteorological phenomena such as gust fronts. In fact, birds do have some ability to forecast weather over the short term. A very complex inner ear structure known as the Vitali or paratympanic organ is sensitive to variations in air pressure, which provides birds a warning of advancing "bad" weather, thus inducing them to indulge in feeding frenzies or short migrations to safe regions (Miller et al., 2016; Stach et al., 2016). Additionally, some bird species become very noisy or make mournful sounds when rainfall is forthcoming. According to Welty (1982), one of the functions of bird vocalization is to advertise an individual's sensitive state or mood, discharge anxious energy, and provide emotional release. For example, one of the behaviors of *Cacomantis merulinus* (Scopoli, 1786) involves expressing a call late in the day during cloudy and rainy days (Galacgac and Balisacan, 2009).

Numerous bird species are said to sing or call in advance of a windy day or a rainstorm. For centuries, many cultures worldwide believed that peacocks display a dance behavior prior to the arrival of rain, essentially foretelling when a place might experience rainfall (Ahmed et al., 2016). In Swaziland, Africa, where sporadic floods and drought are usual catastrophes, traditional communities would take particular caution in predicting such events. For instance, the agitation and noisy behavior of peacocks and ducks was interpreted to indicate an imminent heavy storm (UNEP, 2008). In many countries of Asia and Africa, the Jacobin cuckoo, *Clamator jacobinus* (Boddaert, 1783), is known by traditional communities to predict the arrival of rain (Ahmed et al., 2016). A 16th-century English antiquarian noted that the green woodpecker (*Picus viridis*, Linnaeus, 1758) was used by the druids for divination (Warren-Chadd and Taylor, 2016) due to its laughing vocal sound that supposedly heralds rain (the bird is laughing at the sun). In Orkney, among archipelagos located in the North Sea and near the north of Scotland, and in Shetland, locals named *Gavia stellate* (Pontoppidan, 1763) the "rain goose" because its drawn-out wailing vocalization indicates imminent rain, while a different, more exuberant, kind of call is a prediction that the clouds will clear (Warren-Chadd and Taylor, 2016). In Delaware Indian folklore, the red-throated loon plays a key role because it helped survivors of a great flood by finding them land, and has since served to predict rain (Trail, 2007). The marbled teal, *Marmaronetta angustirostris* (Ménétriés, 1832), is thought to indicate early monsoons by exhibiting a distinct agitated behavior during different periods of the day (Ahmed et al., 2016). In many traditional African cultures, the booming vocalization of the southern ground hornbill, *Bucorvus leadbeateri* (Vigors, 1825), is an assured predictor of precipitation, and due to its large size and loud calls, it has become a target of hunters (Warren-Chadd and Taylor, 2016). The same is true for the shouts of the yellow-tailed black-cockatoo, *Zanda funereal* (Shaw, 1794), in Australia (Warren-Chadd and Taylor, 2016). According to Shoko and Shoko (2013), the Australian aborigines consult the masked lapwing, *Vanellus miles* (Boddaert, 1783), by looking to see whether it lays its eggs or not on the upper part of a field. By doing so, the natives know if it will be a good rainy season or heavy drought. Pigeons have also served as a source of prediction for many cultures, but especially in Asia where people recognize that when a pigeon sings from morning to evening, it will rain in around 15 days (Ahmed et al., 2016). Interestingly, the slaty-breasted wood-rail, *Aramides saracura* (Spix, 1825) is considered a good predictor of heavy rain because of its vocalization behavior (Ruoso, 2012). Indians and Pakistanis consider crows and owls to be an omen or a sign of luck, depending on their behavior, bringing a good rainy season or bad luck (Ahmed et al., 2016).

According to some Scottish mythology, when an owl frequently whoops loudly, it is deemed, by most, as a sign of pleasant weather (Saxby and Clouston, 1892). The cry of the owl, like that of the raven, is also significant to the weather-lore of several districts of England. According to a local statement, when the owl screeches during foul weather it is a sign of fair weather ahead; others state that the owls cry is sometimes taken as an indication of approaching hail or rain, accompanied by lightning. Because of this link between owls and bad weather, it was a custom, until the twentieth century to nail the body of an owl to a barn door to ward off lightning (Sherman, 2008). The Chinese associate owls with the prediction of thunder and lightning, and owl effigies are placed in each corner of a home to protect it from lightning strikes (Sherman, 2008; Weinstein, 1988).

In an old chapbook entitled *The Shepherd's Prognostication for the Weather*, printed in 1573, it is said that "if ravens be seen to stand gaping toward the sun, it is a manifest sign of extreme heat to follow" (Godfridus, 1983). In some districts of ancient Scotland, ravens build their nests in sea-cliffs, and it was believed that if they make short flights inland, it is an indication of stormy weather, but if they make a strong flight inland for a considerable distance, it is a token of fair weather (Gregor, 1881).

Freeland (2006) observed that agricultural workers of 19th century United Kingdom, relied on typical rhymes and sayings about animals to forecast the weather. For instance, "*An early cuckoo heralds a fine summer*" predicts a good summer if one hears a cuckoo before St George's Day on 23 April, and "*Seagull, seagull, stay on the sand; it's never fair weather if you're over land,*" means that in calm weather seagulls obtain their food from the sea and shoreline; therefore, seagulls are believed to predict good weather when hovering over coastal areas (Freeland, 2006). The early coming of other migratory birds in late April, May, and June, such as swallows and martins, was taken as a promising omen, which assured dry, warm conditions throughout their nesting seasons (Giles, 1990). Simpson (1973) stressed that nonmigratory or solitary species, such as crows, wood pigeons, and thrushes, were rarely observed by peasants of northeastern United Kingdom during the nesting season, and according to local folklores, the higher these birds built their nests, the greater the probability that the summer would occur. Another example according to the local peasants would be if swallows and martins were seen in groups of 10 or more flying high and repeatedly making circular flights, they were said to be "wheeling" or "drawing water from the well," a typical sign that rain was on the way.

In Brazil, many bird species are recognized as indicators of coming rainfall in the dry northeast region (Araujo et al., 2005; Bezerra et al., 2013; Marques, 1999). One example is *Turdus rufiventris* (rufous-bellied thrush), whose vocalization is believed to have the power of predicting weather and climatic events. The call of this species, as well as others, have been cited by Araujo et al. (2005) and Marques (1999) as being used by local populations to predict rainfall.

Similar to birds, wild mammals constitute a key group of predictors of weather conditions for many societies. The behavior of livestock animals, for example, is considered to forecast weather and the onset of seasons. Observation of the behavioral and/or morphological characteristics of mammals, whether domestic or wild, yields basic information used in the construction of weather forecasts in various locations. For example, some mammal species have the ability to change color to blend into their environments. Rabbits, for instance, change the color of their fur from white to brown prior to the beginning of snowmelt (Sandoval Salinasa et al., 2016). Indeed, some European cultures have used animal oracles to predict weather for centuries, such as watching hibernating animals, including bears, badgers, and hedgehogs, for signs of spring (Ring, 2008). Medieval Germans often relied on the shadow of a badger to make predictions about the coming spring. If the badger saw his shadow on a particular day, it meant that more winter would be forthcoming; if he did not see his shadow, there would be an early onset of spring. The Europeans brought this type of weather forecasting to North America, and replaced the badger with the groundhog in a tradition celebrated every February 2nd in the state of Pennsylvania, United States (Ring, 2008). Still today, North Americans observe the way muskrat, *Ondatra zibethicus* (Linnaeus, 1766), and beaver, *Castor canadensis* (Kuhl, 1820), build their huts and how rabbits feed during autumn (Klemm and McPherson, 2017; Peppler, 2010). According to Huntington et al. (2005), native settlements of the Arctic observed that some species of deer and reindeer would seek cover among low-branched trees, such as hemlocks and pines, in advance of a rain or snow storm. Some American tales predict that when squirrels gather a large supply of nuts, a cold winter can be expected (Mallery, 2000).

Local communities of the Lushoto district of Tanzania use the occurrence of wild animals in villages to predict weather and climate. For example, baboons entering a village during the dry season indicates a good forthcoming rainy season (Mahoo et al., 2015). Even the sight of lemurs foraging around farmhouses in villages near Mount Kilimanjaro, in the Moshi rural district of Tanzania, is considered a good predictor of the onset rainfall (Chengula and Nyambo, 2016). Remarkably, in Santa Cruz do Sul, Brazil, the vocal behavior of howler monkeys is considered a predictor of good heavy rain (Ruoso, 2012). Even though they are used by Tanzanian natives for multiple purposes, pangolins are recognized as rainfall indicators by locals of the Mahenge and Ismani wards of the Morogoro and Iringa regions (Kijazi et al., 2013). Some traditional communities of the Caatinga biome, a semiarid region of Brazil, believe that if a pregnant armadillo was seen during the month of December, there would be good rains ahead (Abrantes et al., 2011). In some regions of England and the United States, it is believed that a bat circling a house at least three times is a warning of impending death. However, a bat flying playfully is a good omen, forecasting fair weather; the latter contains a bit of truth, since better atmospheric conditions mean easier flying for bats (Sherman, 2008). According to Muguti and Maposa (2012), when Zimbabweans see bats, locally known as muremwaremwa, flying in a certain area, it indicates that rainfall will be irregular. On the other hand, if they occasionally land on the ground, particularly as a swarm, it signifies good rainfall throughout the season (Muguti and Maposa, 2012). Carnivores, such as the leopard, *Panthera pardus* (Linnaeus, 1758), and the spotted hyena, *Crocuta crocuta* (Erxleben, 1777), are considered climate ethnozooindicators by some traditional cultures in Tanzania. Mahoo et al. (2015) observed that locals of the Lushoto district considered the appearance of leopards as a prediction of a good rainy season. In the village of Maluga, in central Tanzania, subsistence farmers commonly distinguish and use certain calls produced by hyenas to foretell the rainy season (Elia et al., 2014).

It is worth mentioning that the behavioral displays of some mammal species in response to changes in weather are scientifically demonstrated to be accurate climate forecasters around the globe. For example, Bartlam-Brooks et al. (2013) observed that zebras in Botswana heed subtle weather and vegetation clues when choosing when and how to move to greener pastures. The animals seemed to anticipate food and water availability at their annual migration destination and fine-tune their movements accordingly, for example, by delaying departure or reversing direction when rainfall was unseasonably late (Bartlam-Brooks et al., 2013).

In many traditional cultures worldwide, domestic animals are widely used for climate prediction. For example, cattle are symbols of fertility and worship, as well as predictors of climate (Egeru, 2012; Nyong et al., 2007; UNEP, 2008). In Pakistan, the way a cow sits is used as a weather predictor, such that having three legs under itself and one leg out, or even if it sits on the sand, is taken as an indication that rain is forthcoming. In India, the smell of cow urine can predict rain, while if cows lick each other it is an indication of drought (Ahmed et al., 2016). Berkeley and Linklater (2010) pointed out that in some provinces of Africa, farmers rely on some sort of behavior by cattle herds for predicting the weather. For instance, an ox hesitant to go into a veld for grazing indicates the onset of rains within a few hours, as does the suddenly lying down in a field by cows. Goats and sheep can predict the weather by exhibiting a similar behavior. For example, in some Pakistani communities, goats entering properties at night time and roaming toward the eastern side is taken as a prediction of rainfall in the next week or two (Ahmed et al., 2016). Locals of the Lushoto district of Tanzania interpret the frequent flapping of ears of sheep, goats, and cattle, at any time during the dry season, as indicative that the onset of rain is near, and that there are good prospects for the coming season (Mahoo et al., 2015). Shoko and Shoko (2013) observed that in the Mberengwa district of Zimbabwe, locals believe that injuries, such as swellings, on the lower extensions of a camel's leg are indications that rainfall is ahead, and water must be stored as quickly as possible. Mapara (2009) stresses that by simple observation of the behavior of fowl, including chickens, people are able to predict whether rains are going to stop or not. If fowl continue to feed when it is raining, it is expected that it will rain for the next few days. For traditional cultures in regions of southwestern Africa and Tanzania, goats and domestic dogs circling around themselves give signs of onset of rains and also their mating and doing much noise during night time is a typical sign of coming rain (Chengula and Nyambo, 2016).

In contrast to most local communities, who observe environmental indicators for the prediction of weather and climate, some ethnic groups use divination and other spiritual prophesies. In the province of Namcntenga in Burkina Faso (Africa), *Tengsoba* is the eldest descendant of an ancient clan and is considered by the locals to be a shaman authority. Furthermore, other traditional experts similarly receive forecast insights from deities via dreams. Traditionally, animal sacrifices are made to appease earth spirits and influence the occurrence of rains. Predictions are drawn from the behavior of slaughtered animals, including the length of time it takes for a slaughtered animal to fall, which direction it falls, and where the blood spills (Roncoli et al., 2001). Rain fall after a ritual is considered to be well-accepted offering by the deities. Among the Hausa people in some pastoral areas of northern Nigeria, offerings and sacrifices, by means of slaughtering bulls are performed to ward off events of famine, drought, plague, and unproductive fields (Musa and Omokore, 2011; Musa, 2006). Peppler (2010) stated that during autumn, some American Indian tribes slaughtered pigs to look at their spleen, and if it was thick and rugose, the natives would believe a harsh winter was ahead that year. Similarly, the pastoralists of northern Kenya developed their sense of predicting the weather by observing the intestines of slaughtered animals (wild and domestic), and if their color was dark, it was taken as a sign of coming drought or war (Kagunyu et al., 2016). Also in Kenya, and northern Tanzania, the elders of Maasai regularly predict weather using the behavior of certain livestock animals. Goat intestines would be examined by an expert Maasai and if watery cysts were found on them during the month of August it was considered a sign of an imminent rainy season, but if the stomach was found partially or completely empty, it indicated drought, famine, hostility or war (UNEP, 2008). Risiro et al. (2012) pointed out that rain-making ceremonies are performed in sacred forests of some districts of Manicaland, Zimbabwe. These not only involved the brewing of beer, roasting meat, singing, and dancing, but the prohibition of killing sacred animals, such as cuckoo birds and frogs under circumstances of drought, so as not to displease ancestral spirits. In the Teso subregion of eastern Uganda, the killing of an Abyssinian hornbill, *Bucorvus abyssinicus* (Boddaert, 1783), is considered a fault and brings severe droughts to villagers by deities;

the offender would then be "buried" in the mud of a wetland in a ritual ceremony to bring back the rainy season (Egeru, 2012).

FINAL REMARKS

Since the beginning of human history, variation in climate has always received special attention, especially considering its influence on the life of plants and animals on which humankind depends. Human activities linked to these resources, such as hunting, fishing, agriculture, and the raising of livestock, depend directly on climate, and so attempts to predict weather to coexist with natural adversities has always been part of human history.

Therefore, animals have provided a fundamental service to humans since they are sensitive to variations in climate, which influences their behavioral, physiological, and reproductive characteristics. Human populations worldwide have come to appreciate the particularities of animals through observations from daily interactions. The result has been an accumulation of knowledge that helps humans develop climatic predictions, which influence various aspects of human life.

In this chapter we proposed the term "climate ethnozooindicators" in reference to animals that are used by humans to predict climatic events. There is a considerable richness of species that are used as ethnozooindicators, according to the peculiarities of the fauna of a given region or country. Species of invertebrates (mainly arthropods), and vertebrates, especially amphibians and birds, are taxa that are, as we have seen, widely used as ethnozooindicators. The predictions these animals provide are important to a great variety of human activities, and provide insight into the understanding of wildlife and ecology by local peoples. From an ethnozoological perspective, the use of animals as ethnozooindicators represents one of the oldest and most widespread forms of interaction between humans and nonhuman animals. However, the need for additional studies that seek to document this knowledge is urgent, especially considering that in many cases this knowledge has been lost, either due to a lack of interest on behalf of younger generations, or due to the extinction of important ethnozooindicator species.

References

Abrantes, P.M., Sousa, R.F., Lucena, C.M., Lucena, R.F.P., Pereira, D.D., 2011. Aviso de chuva e de seca na memória do povo: O caso do Cariri Paraibano. BioFar 5, 18–24.

Acharya, S., 2011. Presage Biology: lessons from nature in weather forecasting. Indian Journal of Traditional Knowledge 10, 114–124.

ACIA - Arctic Climate Impact Assessment, 2005. Arctic Climate Impact Assessment. Cambridge University Press.

Ahmed, C.I., Hu, W.B., Kumar, S., 2016. Indigenous knowledge about prediction in climate change. International Journal of Humanities and Social Science 5, 45–62.

Alsaiari, N.O., 2012. An expert system for weather prediction based on animal behavior. In: The International Conference on Informatics and Applications. King Abdulaziz University, Jeddah, Saudi Arabia.

Alves, R.R.N., Nishida, A.K., 2002. A ecdise do caranguejo-uçá, *Ucides cordatus* L. (Decapoda, Brachyura) na visão dos caranguejeiros. Interciencia 27, 110–117.

Alves, R.R.N., Nishida, A., Hernandez, M., 2005. Environmental perception of gatherers of the crab 'caranguejo-uca' (*Ucides cordatus*, Decapoda, Brachyura) affecting their collection attitudes. Journal of Ethnobiology and Ethnomedicine 1, 1–8.

Alves, R.R.N., Melo, M.F., Ferreira, F.S., Trovão, D.M.B.M., Dias, T.L.P., Oliveira, J.V., Lucena, R.F.P., Barboza, R.R.D., 2015. Healing with animals in a semiarid northeastern area of Brazil. Environment. Development and Sustainability 18, 1733–1747.

Alves, R.R.N., Feijó, A., Barboza, R.R.D., Souto, W.M.S., Fernandes-Ferreira, H., Cordeiro-Estrela, P., Langguth, A., 2016. Game mammals of the Caatinga biome. Ethnobiology and Conservation 5, 1–51.

Anandaraja, N., Rathakrishnan, T., Ramasubramanian, M., Saravanan, P., Suganthi, N.S., 2008. Indigenous weather and forecast practices of Coimbatore district farmers of Tamil Nadu. Indian Journal of Traditional Knowledge 7, 630–633.

Anderson, D., Eriksson, L.O., 2007. Effects of temporal aggregation in integrated strategic/tactical and strategic forest planning. Forest Policy and Economics 9, 965–981.

Araujo, H.F.P., Lucena, R.F.P., Mourão, J.S., 2005. Prenúncio de chuvas pelas aves na percepção de moradores de comunidades rurais no município de Soledade-PB, Brasil. Interciencia 30, 764–769.

Ayal, D.Y., Desta, S., Gebru, G., Kinyangi, J., Recha, J., Radeny, M., 2015. Opportunities and challenges of indigenous biotic weather forecasting among the Borena herders of southern Ethiopia. SpringerPlus 4, 606–617.

Barboza, R.R.D., Lopes, S.F., Souto, W.M.S., Fernandes-Ferreira, H., Alves, R.R.N., 2016. The role of game mammals as bushmeat in the Caatinga, northeast Brazil. Ecology and Society 21, 1–11.

Bartlam-Brooks, H.L.A., Beck, P.S.A., Bohrer, G., Harris, S., 2013. In search of greener pastures: using satellite images to predict the effects of environmental change on zebra migration. Journal of Geophysical Research 118, 1427–1437.

Berkeley, E.V., Linklater, W.L., 2010. Annual and seasonal rainfall may influence progeny sex ratio in the black rhinoceros. South African Journal of Wildlife Research 40, 53–57.

Bezerra, D.M.M., Araújo, H.F.P., Alves, A.G.C., Alves, R.R.N., 2013. Birds and people in semiarid Northeastern Brazil: symbolic and medicinal relationships. Journal of Ethnobiology and Ethnomedicine 9, 1–11.

Bezerra, D.M.M., Nascimento, D.M., Ferreira, E.N., Rocha, P.D., Mourão, J.S., 2012. Influence of tides and winds on fishing techniques and strategies in the Mamanguape River Estuary, Paraíba State, NE Brazil. Anais da Academia Brasileira de Ciências 84, 775–788.

Blukis Onat, A.R., 2002. Resource cultivation on the northwest coast of North America. Journal of Northwest Anthropology 36, 125–144.

Cabrera, A., Incháustegui, C., García, A., Toledo, V.M., 2001. Etnoecología Mazateca: una aproximación al complejo cosmos-corpus-praxis. Etnoecológica 6, 61–83.

Chang'a, L.B., Yanda, P.Z., Ngana, J., 2010. Indigenous knowledge in seasonal rainfall prediction in Tanzania: a case of the south-western Highland of Tanzania. Journal of Geography and Regional Planning 3, 66–72.

Chengula, F., Nyambo, B., 2016. The significance of indigenous weather forecast knowledge and practices under weather variability and climate change: a case study of smallholder farmers on the slopes of Mount Kilimanjaro. International Journal of Agricultural Extension 2, 31–43.

Chisadza, B., Tumbare, M.J., Nyabeze, W.R., Nhapi, I., 2015. Linkages between local knowledge drought forecasting indicators and scientific drought forecasting parameters in the Limpopo River Basin in Southern Africa. International Journal of Disaster Risk Reduction 12, 226–233.

Chisadza, B., Tumbare, M.J., Nhapi, I., Nyabeze, W.R., 2013. Useful traditional knowledge indicators for drought forecasting in the Mzingwane Catchment area of Zimbabwe. Disaster Prevention and Management 22, 312–325.

Colombi, B.J., 2009. Salmon nation: climate change and tribal sovereignty. In: Crate, S.A., Nuttall, M. (Eds.), Anthropology and Climate Change: From Encounters to Actions. Left Coast Press, Inc, New York, NY, pp. 186–196.

De Swardt, D.H., Grobler, G.P.J., Oschadleus, H.D., 2004. Bird ringing in the Free State National Botanical Gardens, Bloemfontein, with notes on recaptures. Afring News 33, 65–70.

Deloria, V., Wildcat, D., 2001. Power and Place: Indian Education in America. Fulcrum Resources, Golden, CO. 168 pp.

Dennis, J., Wolff, G., 2013. It's Raining Frogs and Fishes: Four Seasons of Natural Phenomena and Oddities of the Sky (The Wonders of Nature). DCA, Inc., USA. 276 pp.

Doherty, R., Sitch, S., Smith, B., Lewis, S., Thornton, P., 2009. Implications of future weather and atmospheric CO_2 content for regional biogeochemistry, biogeography and ecosystem services across East Africa. Global Change Biology 16, 617–640.

Downden, K., 2000. European Paganism: The Realities of Cult from Antiquity to the Middle Ages, first ed. Routledge, United Kingdom.

Dudde, N.B., Apte, S.S., 2016. Arbitrary decision tree for weather prediction. International Journal of Scientific Research 5, 87–89.

Dunn, R.R., 2000. Poetic entomology: insects in Japanese haiku. American Entomologist 46, 70–72.

Eastwood, E., 1967. Radar Ornit. Methuen and Co., London, UK.

Egeru, A., 2012. Role of indigenous knowledge in climate change adaptation: a case study of the Teso sub-region, Eastern Uganda. Indian Journal of Traditional Knowledge 11, 217–224.

Elia, E.F., Mutula, S., Stilwell, C., 2014. Indigenous Knowledge use in seasonal weather forecasting in Tanzania: the case of semi-arid central Tanzania. South African Journal of Libraries and Information Science 80, 18–27.

Emeagwali, G., Shizha, E., 2016. African Indigenous Knowledge and the Sciences: Journeys into the Past and Present, second ed. Sense Publishers, Netherlands.

Enock, M., 2013. Indigenous Knowledge Systems and Modern Weather Forecasting: Exploring the Linkages. Journal of Agriculture and Sustainability 2, 98–141.

Field, C., 2005. Where There Is No Development Agency. A Manual for Pastoralists and Their Promoters. Aylesford: NR International.

Freeland, P., 2006. An early cuckoo heralds a fine summer. School Science Review 88, 99–111.

Galacgac, E.S., Balisacan, C.M., 2009. Traditional weather forecasting for sustainable agroforestry practices in Ilocos Norte Province, Philippines. Forest Ecology and Management 257, 2044–2053.

Garay-Barayazarra, G., Puri, R., 2011. Smelling the monsoon: senses and traditional weather forecasting knowledge among the Kenyah Badeng Farmers of Sarawak, Malaysia. Indian Journal of Traditional Knowledge 10, 21–30.

Gardner, D., 2011. Future Babble: Why Expert Predictions Fail and Why We Believe Them Anyway. Virgin Books, London, UK.

Giles, B., 1990. Story of Weather. BBC Publications, London.

Gilles, J.L., Thomas, J.L., Valdivia, C., Yucra, E.S., 2013. Laggards or leaders: conservers of traditional agricultural knowledge in Bolivia. Rural Sociology 78 , 51–74.

Gissila, T., Black, E., Grimes, D.I.F., Slingo, J.M., 2004. Seasonal forecastingof the Ethiopian summer rains. International Journal of Climatology 24, 1345–1358.

Godfridus, M., 1983. The Shepherds Prognostication for the Weather. The Knowledge [of] Things Unknown: Shewing the Effects of the Planets and Other Astronomical Constellations: With the Strange Events that Befal Men, Women and Children Born under Them. W.T. and are sold J. Ho[se], London.

Gregor, W., 1881. Notes on the Folk-lore of the North-east of Scotland. London: Folk-lore Society by E. Stock, London.

Hart, T.G.B., 2007. Local knowledge and agricultural application: lessons from an Ugandian parish. South African Journal of Agricultural Extension 36, 229 268.

Hines, A.H., Johnson, E.G., Darnell, M.Z., Rittschof, D., Miller, T.J., Bauer, L.J., Rodgers, P., Aguilar, R., 2010. Predicting effects of climate change on blue crabs in Chesapeake Bay. In: Kruse, G.H., Eckert, G.L., Foy, R.J., Lipcius, R.N., Sainte-Marie, B., Stram, D.L., Woodby, D. (Eds.), Biology and Management of Exploited Crab Populations under Climate Change. Alaska Sea Grant, University of Alaska, Fairbanks, pp. 109–127.

Huntington, H.P., Fox, S., Berks, F., Krupnik, I., 2005. The changing Arctic: indigenous perspectives. In: ACIA (Ed.), Arctic Climate Impact Assessment. Cambridge University Press, Cambridge, pp. 61–98.

IPCC, 2007. Fourth Assessment Report on Climate Change. Geneva, Switzerland, 104 pp.

Jones, P.D., Lister, D.H., Li, Q., 2008. Urbanization effects in large-scale temperature records, with an emphasis on China. Journal of Geophysical Research 113, 1–12.

Joshua, R., Dominic, M., Doreen, T., Elias, R., 2012. Weather forecasting and indigenous knowledge systems in Chimanimani District of Manicaland, Zimbabwe. Journal of Emerging Trends in Educational Research 3, 561–566.

Kagunyu, A., Wandibba, S., Wanjohi, J.G., 2016. The use of indigenous climate forecasting methods by the pastoralists of Northern Kenya. Research for Policy and Practice 6, 1–7.

Kihupi, N., Kingamkono, R., Dihenga, H., Kingamkono, M., Rwamugira, W., 2003. Integrating indigenous knowledge and climate forecasts in Tanzania. In: Coping with Climate Variability: The Use of Seasonal Climate Forecasts in Southern Africa. Ashgate Publishing, Hampshire, UK and Burlington, USA.

Kijazi, A.L., Chang'a, L.B., Liwenga, E.T., Kanemba, A., Nindi, S.J., 2013. The use of indigenous knowledge in weather and climate prediction in Mahenge and Ismani wards, Tanzania. Journal of Geography and Regional Planning 6, 274–280.

Klemm, T., McPherson, R.A., 2017. The development of seasonal climate forecasting for agricultural producers. Agricultural and Forest Meteorologyis 232, 384–399.

Kipkorir, E., Mugalavai, E., Songok, C., 2010. Integrating indigenous and scientific knowledge systems on seasonal rainfall characteristics prediction and utilization. Kenya Journal of Science Technology and Innovation 2, 19–29.

Koistinen, J., 2000. Bird migration patterns on weather radar. Physics and Chemistry of the Earth 25, 1185–1193.

Kopij, G., 2002. Birds of Ooseinde and Bloemspruit sewage dams, Bloemfontein. Mirafra 19, 2–7.

Krupnik, I., Jolly, D., 2002. The earth is Faster Now: Indigenous Observations of Arctic Environmental Change. Arctic Research Consortium of the United States, Fairbanks press.

Lammel, A., Goloubinoff, M., Katz, E., 2008. Aires y lluvias. Antropología del clima en México. CIESAS/CEMCA/ IRD, México.

Levin, S.A., 1999. Fragile Dominion: Complexity and the Commons. Perseus Books, Reading, MA.

Lowman, M.D., 1982. Seasonal variation in insect abundance among three Australian rain forests, with particular reference to phytophagous types. Australian Journal of Ecology 7, 353–361.

Mahoo, H., Mbungu, W., Yonah, I., Recha, J., Radeny, M., Kimeli, P., Kinyangi, J., 2015. Integrating Indigenous Knowledge with Scientific Seasonal Forecasts for Climate Risk Management in Lushoto District in Tanzania. CCAFS Working Paper no. 103. CGIAR Research Program on Climate Change. Agriculture and Food Security (CCAFS), Copenhagen, Denmark.

Makwara, E., 2013. Indigenous knowledge systems and modern weather forecasting: exploring the linkages. International Journal of Agricultural Sustainability 2, 98–141.

Mallery, R.L., 2000. Nuts about Squirrels: A Guide to Coexisting With-and Even Appreciating-your Bushy-tailed Friends. Grand Central Publishing, New York.

Mapara, J., 2009. Indigenous knowledge systems in Zimbabwe: juxtaposing postcolonial theory. Journal of Pan African Studies 3, 139–155.

Marais, J., 2008. What's that Snake? a Starter's Guide to Snakes of Southern Africa. Struik Publishers, Cape Town, South Africa.

Marques, J.G.W., 1999. Da gargalhada ao pranto. Inserção Etnoecológica da Vocalização de Aves em Ecossistemas Rurais do Brasil (Ph.D thesis). Universidade Estadual de Feira de Santana, Feira de Santana. 212 pp.

Merchant, M.E., Flanders, R.V., Williams, R.E., 1987. Seasonal abundance and parasitism of house fly (Diptera: Muscidae) pupae in enclosed, shallow-pit poultry houses in Indiana. Environmental Entomology 16, 716–721.

Mercer, J., Dominey-Howes, D., Kelman, I., Lloyd, K., 2007. The potential for combining indigenous and western knowledge in reducing vulnerability to environmental hazards in small island developing states. Environmental Hazards 7, 245–256.

Miller, T.A., Brooks, R.P., Lanzone, M.J., Brandes, D., Cooper, J., Tremblay, J.A., Wilhelm, J., Duerr, A., Katzner, T.E., 2016. Limitations and mechanisms influencing the migratory performance of soaring birds. International Journal of Avian Science 158, 116–134.

Muguti, T., Maposa, S.R., 2012. Indigenous weather forecasting: a phenomenological study engaging the Shona of Zimbabwe. Journal of Pan African Studies 4, 102–112.

Musa, M.W., 2006. Indigenous Resource Management Systems (IRMS) Among Rural Communities in North-West Zone of Nigeria and Their Relevance for Participatory Poverty Reduction (Ph.D. thesis). Ahmadu Bello University, Zaria, Nigeria.

Musa, M.W., Omokore, D.F., 2011. Reducing vulnerability and increasing resiliency to climate change: learning from rural communities. International Journal of Agricultural Extension 15, 1–9.

Mundy, P., Compton, L., 1991. Indigenous communication and indigenous knowledge. Development Communication Report 74, 1–3.

Nakashima, D., Galloway, M., Thulstrup, H., Ramos, C., Rubis, J., 2012. Weathering Uncertainty: Traditional Knowledge for Climate Change Assessment and Adaptation. UNESCO and UNU, Paris and Darwin.

National Research Council, 2010. Understanding Climate's Influence on Human Evolution. The National Academies Pres, Washington, DC.

Nishida, A.K., Nordi, N., Alves, R.R.N., 2006a. Mollusc gathering in northeast Brazil: an ethnoecological approach. Human Ecology 34, 133–145.

Nishida, A.K., Nordi, N., Alves, R.R.N., 2006b. The lunar-tide cycle viewed by crustacean and mollusc gatherers in the State of Paraíba, Northeast Brazil and their influence in collection attitudes. Journal of Ethnobiology and Ethnomedicine 2, 1–12.

Nishida, A.K., Nordi, N., Alves, R.R.N., 2006c. Molluscs production associated to lunar-tide cycle: a case study in Paraíba State under ethnoecology viewpoint. Journal of Ethnobiology and Ethnomedicine 2, 1–6.

Nyong, A., Adesina, F., Osman Elasha, B., 2007. The value of indigenous knowledge in climate change mitigation and adaptation strategies in the African Sahel. Mitigation and Adaptation Strategies for Global Change 12, 787–797.

Oba, G., 1997. Pastoralists' Traditional Drought Coping Strategies in Northern Kenya. A Report for the Government of the Netherlands and the Government of Kenya. Euroconsult BV, Arnheim and Acacia Consultants Ltd, Nairobi.

Okonya, J.S., Kroschel, J., 2013. Indigenous knowledge of seasonal weather forecasting: a case study in six regions of Uganda. Indian Journal of Agricultural Sciences 4, 641–648.

Olbrich, D.L., King, B.H., 2003. Host and habitat use by parasitoids (Hymenoptera: Pteromalidae) of house fly and stable fly (Diptera: Muscidae) pupae. Great Lakes Entomologist 36, 179–190.

Orlove, B., Chiang, J., Cane, M., 2002. Ethnoclimatology in the Andes A cross-disciplinary study uncovers a scientific basis for the scheme Andean potato farmers traditionally use to predict the coming rains. American Scientist 90, 428–435.

Pareek, A., Trivedi, P.C., 2011. Cultural values and indigenous knowledge of climate change and disaster prediction. Indian Journal of Traditional Knowledge 10, 183–189.

Pellegrino, A.C., Penaflor, M.F.G.V., Nardi, C., Bezner-Kerr, W., Guglielmo, C.G., Bento, J.M.S., McNeil, J.N., 2013. Weather forecasting by insects: modified sexual behaviour in response to atmospheric pressure changes. PLoS One 8, 1–5.

Peppler, R.A., 2008. Knowing weather and climate: how do peoples with extended interaction histories with the natural environment recognize or forecast meteorological events? In: Third Symposium on Policy and Socio-economic Research.

Peppler, R.A., 2010. Old Indian ways of predicting the weather: senator Robert S. Kerr and the winter predictions of 1950–1951 and 1951–1952. Weather, Climate and Society 2, 200–209.

Pinch, G., 1995. Magic in Ancient Egypt, second ed. University of Texas Press, Austin, Texas.

Pliny, 1938. Natural History Volume 1: Books 1-2 (H. Rackham, Trans. (Loeb Classical Library 330)). Harvard University Press, Cambridge, MA. 2014, 236 pp.

Prendergast, H.D.V., Davis, S.D., Way, M., 1999. Dryland plants and their uses. In: Posey, D.A. (Ed.), Spiritual Values of Biodiversity, first ed. Intermediate Technology Publications, London, UK.

Richards, P., 1980. Community environmental knowledge in African rural development. In: Brokensha, D., Warren, D.M., Werner, O. (Eds.), Indigenous Knowledge Systems and Development. University Press of America, Washington, DC, pp. 54–63.

Ring, K., 2008. Predicting Weather by the Moon. Hazard Press, Titirangi, Auckland, New Zealand.

Risiro, J., Mashoko, D., Tshuma, D.T., Rurinda, E., 2012. Weather forecasting and indigenous knowledge systems in Chimanimani District of Manicaland, Zimbabwe. Journal of Emerging Trends in Educational Research and Policy Studies 3, 561–566.

Rivero-Romero, A.D., Moreno-Calles, A.I., Casas, A., Castillo, A., Camou-Guerrero, A., 2016. Journal of Ethnobiology and Ethnomedicine 12, 1–11.

Robbins, M.E., 1996. The truculent toad in the middle ages. In: Flores, N. (Ed.), Animals in the Middle Ages, New York and London, pp. 27–47.

Roncoli, C., Ingram, K., Kirshen, P., Jost, C., 2001. Reading the rains: local knowledge and rainfall forecasting in Burkina Faso. Society & Natural Resources: An International Journal 15, 409–427.

Roncoli, C., Ingram, K., Kirshen, P., Jost, C., 2002. Burkina Faso A: integrating indigenous and scientifc rainfall forecasting. Society & Natural Resources: An International Journal 15, 409–427.

Ruoso, D., 2012. The climatic perception of urban population of Santa Cruz do Sul/RS. Raega 25, 64–91.

Sánchez-Cortés, M.S., Chavero, E.L., 2011. Indigenous perception of changes in climate variability and its relationship with agriculture in a Zoque community of Chiapas, Mexico. Climatic Change 107, 363–389.

Sandoval Salinasa, M.L., Barquez, R.M., Colomboa, E.M., Sandovala, J.D., 2016. Intra-specific pelage color variation in a South American small rodent species. Brazilian Journal of Biology 77, 1–11.

Saxby, J.M.E., Clouston, W.A., 1892. Birds of Omen in Shetland. Viking Society for Northern Research, New York.

Sherman, J., 2008. Storytelling: An Encyclopedia of Mythology and Folklore, first ed. Myron E. Sharpe, Armonk, New York.

Shoko, K., 2012. Indigenous weather forecasting systems: a case study of the biotic weather forecasting indicators for wards 12 and 13 in Mberengwa district Zimbabwe. Journal of Sustainable Development in Africa 14, 1520–5509.

Shoko, K., Shoko, N., 2013. Indigenous weather forecasting systems: a case study of the abiotic weather forecasting indicators for wards 12 and 13 in Mberengwa district Zimbabwe. Asian Social Science 9, 285–297.

Shukurat, A., Kolapo, O., Nnadozie, O., 2012. Traditional capacity for weather forecast, variability and coping strategies in the front line states of Nigeria. Agricultural Science 3, 625–630.

Shumba, O., 1999. Coping with Drought: Status of Integrating Contemporary and Indigenous Climate/Drought Forecasting in Communal Areas of Zimbabwe Consultancy Report. UNDP/UNSO. p. 72.

Sillitoe, P., 2007. Local science vs. global science: an overview. In: Sillitoe, P. (Ed.), Local Science Vs. Global Science: Approaches to Indigenous Knowledge in International Development. Berghahn, New York, pp. 1–22.

Simpson, J., 1973. The Folklore of Sussex. Batsford, London.

Speranza, C., Kiteme, B., Ambenje, P., Wiesmann, U., Makali, S., 2010. Indigenous knowledge related to weather variability and change: insights from droughts in semi-arid areas of former Makueni District, Kenya. Climatic Change 100, 295–315.

Stach, R., Kullberg, C., Jakobsson, S., Strom, K., Fransson, T., 2016. Migration routes and timing in a bird wintering in south Asia, the common rosefinch *Carpodacus erythrinus*. Journal of Ornithology 157, 671–679.

Stigter, K., 2010. Applied Agrometeorology. Springer. Heidelberg/Berlin, Germany and New York, NY, USA.

Suzuki, D.T., Knudtson, P., 1992. Wisdom of the Elders Honoring Sacred Native Visions of Nature. Bantam Books.

Taub, L., 2003. In: French, R. (Ed.), Ancient Meteorology. Routledge, New York, NY, USA. 286 pp.

Tekwa, I., Belel, M., 2009. Impacts of traditional soil conservation practices in sustainable food production. Journal of Agriculture and Social Sciences 5, 128–130.

Trail, P.W., 2007. African hornbills: keystone species threatened by habitat loss, hunting and international trade. Ostrich 78, 609–613.

UNEP, 2008. Indigenous People in Disaster Management in Africa. United Nation Environment Program, Nairobi, Kenya.

Voggesser, G., Lynn, K., Daigle, J., Lake, F.K., Ranco, D., 2013. Cultural impacts to tribes from climate change influences on forests. Climatic Change 120, 615–626.

Warren-Chadd, R., Taylor, M., 2016. Birds: Myth, Lore and Legend. Bloomsbury Natural History, London.

Weinstein, K., 1988. The Owl: In Art, Myth and Legend. Random House, New York.

Welty, J.C., 1982. The Life of Birds, third ed. Saunders College Publishing, Philadelphia.

Wildcat, D., 2009. Red alert! Saving the Planet with Indigenous Knowledge. Fulcrum Publishing, Golden.

Ziervogel, G., Opere, A., 2010. Integrating Meteorological and Indigenous Knowledge-based Seasonal Weather Forecasts in the Agricultural Sector. International Development Research Centre, Ottawa, Canada. Weather Change Adaptation in Africa learning paper series.

Zuma-Netshiukhwi, G., Stigter, S., Walker, S., 2013. Use of traditional weather/climate knowledge by farmers in the south-Western free state of South Africa: Agrometeorological learning by scientists. Atmosphere 4, 383–410.

CHAPTER

22

Understanding Human–Wildlife Conflicts and Their Implications

Denise Freitas Torres[1], *Eduardo Silva Oliveira*[1], *Rômulo Romeu Nóbrega Alves*[2]

[1]Universidade Federal Rural de Pernambuco, Recife, Brazil; [2]Universidade Estadual da Paraíba, Campina Grande, Brazil

INTRODUCTION

Human survival and reproduction depend on obtaining food and water and often some form of protection against parasites and predators (Kormondy and Brown, 2002). In this context, hunting, one of the oldest human activities, has been fundamental since it allowed the first hominids to obtain animal protein at the same time as it served to provide protection against potential predators. Alves and Souto (2010) point out that animals have always been hunted for their utilitarian value and for humans' need to defend themselves against large predators.

The first evidence of the occurrence of predation on human beings was obtained from the skull of a hominid child who had been killed by a leopard one million years ago (Kruuk, 2002). According to this author, all the evidence suggests that due to the large number of species of predators existed during the Pliocene and Pleistocene, human ancestors were more threatened because they were same size as other predated animals, and their hunter–gatherer way of life probably left them more exposed to predation by large carnivores. During this period, the first hominids represented potential prey at the same time as they competed for resources with other animals. Hunter–gatherer prehistoric humans fed on seeds and small animals, but with the advent of tool usage they began to communicate with each other and started hunting in groups, thereby becoming more efficient predators (Conover, 2002). This behavior not only allowed early humans to slaughter large animals for consumption, but also provided protection against predators (McCade and McCade, 1984). It is in this scenario that the first conflicts between our ancestors and wild animals certainly emerged, and subsequently, it was perpetuated to this day with varying causes.

Conflicts between humans and wild animals are characterized by the interactions that occur when an action by one has a negative effect on the other (Conover, 2002), or when wild animals directly cause damage to human possessions (Dickman, 2010). Thus, for a conflict to occur there needs to be damage to an object, an animal that caused the damage, and a person affected by this interaction (Conover, 2002). Alves et al. (2010b) point out that conflicts occur when the needs and behavior of

Ethnozoology
http://dx.doi.org/10.1016/B978-0-12-809913-1.00022-3

wild animals clash with the interests of humans, or when human activities generate negative impacts on wildlife. It is clear that this theme is of broad interest to humans and, not surprisingly, academic publications addressing the complex conflicting interactions between humans and wild animals have intensified (Redpath et al., 2015).

Conflicts between wild animals and humans occur throughout the world (Greentree et al., 2000; Mishra, 1997; Naughton-Treves, 1997; Polisar et al., 2003; Redpath et al., 2013), involve a wide variety of animal taxa and have a diversity of causes. These complex conflicting interactions can be viewed in a variety of ways depending on the location, the cause, and the animals involved, and so this chapter focuses on the series of factors that influence these problems.

THE PRINCIPAL TAXA INVOLVED IN CONFLICTS

In terrestrial environments, most of the animals involved in conflicting interactions are vertebrates. The taxa that stand out include, among mammals, the order Carnivora (mainly bears, felids, and canids) and large herbivores such as elephants and hippopotamus (Figs. 22.1 and 22.2); among reptiles, crocodilians and snakes; and among birds, birds of prey and granivorous birds. In aquatic environments the most prominent taxa involved in conflicts are sharks (Fischer et al., 2009; Gadig, 1998), marine turtles (Nogueira and Alves, 2016; Pinedo and Polacheck, 2004), otters (Rosas-Ribeiro et al., 2012), manatees (Franzini et al., 2013), and dolphins (Loch et al., 2009; Plagányi and

FIGURE 22.1 Examples of animals involved in human–carnivore conflicts (tigers, lions, jaguars, and wolves). *Photo credits: (A) Itamar Barbosa Lima, (B) Franciany Braga-Pereira, (C) André Pessoa, and (D) Rômulo Romeu Nóbrega Alves.*

FIGURE 22.2 (A) Elephants, (B) buffaloes, and (C and D) hippopotamus are among large herbivores associated with human/wildlife conflict. *Photo credits: (A, C, and D) Franciany Braga-Pereira, (B) Anthony B. Cunningham.*

Butterworth, 2002; Toledo et al., 2010). Conflicts with animals also occur in the aerial environment, and include bats (Manville, 2016) and, mainly, birds (Manville, 2016; Thorpe, 2016).

In general, conflicts in terrestrial ecosystems are caused by the predation of domestic animals (Holmern et al., 2007; Mishra, 1997; Oli et al., 1994; Woodroffe et al., 2005), crop degradation (Basili and Temple, 1999; Kebede et al., 2016; Kendall, 2011), human injury, or death (Amarasinghe et al., 2015; Dunham et al., 2010; Fukuda et al., 2014, 2015) and the potential transmission of zoonoses from wild animals to humans and/or domestic animals (Dunham et al., 2010). It should be emphasized that diseases are also transmitted from domestic to wild animals such as in Zimbabwe where domestic dogs have become agents for disease transmission to large wild carnivores (Butler et al., 2003). On one hand, conflicts can generate economic losses and severe damage for the people involved, but on the other they also end up generating negative impacts to wildlife through the hunting and slaughter of the target species of conflicts, which can drastically reduce populations. This impact becomes even more problematic when endangered animals, such as elephants, rhinos, lions, tigers, and snow leopards are involved (Bulte and Rondeau, 2005).

As mentioned previously, conflicts involving people and wild carnivorous mammals began very early in human history and certainly intensified with the advent of domestication of animals and plants, which, as Grupta (2004) points out, made it possible for human hunters to become selective hunters and finally pastoralists. Although domestication provides a range of products for human subsistence, it also has a significant cost because of the need for the maintenance and protection of domestic herds and agricultural crops. Thus, as the loss of investment caused by predation on crops or domestic animals by wild animals intensifies, so will the occurrence of conflicts, which results in more retaliatory actions by humans. Pitman et al. (2002) point out that many of the management practices used to reduce predation on domestic livestock have been used since cattle were domesticated 10,000 years ago.

Likewise, conflicts involving humans and herbivores began early in the domestication of plant species. The development of agriculture allowed people to become sedentary and establish towns and cities (Grupta, 2004). With the expansion of agricultural activities and cultivated pastures, there was an increase in habitat loss and fragmentation, enhancing the competition between domestic and wild animals for foraging areas (Gordon, 2009), which resulted in herbivores becoming targets of conflicts with farmers and livestock ranchers.

Several reptiles are also significant targets of human conflict. These animals, while having a utilitarian value, are often regarded as harmful, are associated with evil, and are considered true pests in many parts of the world, and so they are often hunted and killed (Alves et al., 2009). The possible damage caused by the death of domestic animals has been an additional reason for the persecution of snakes, for example, which adds to the negative stereotype commonly associated with this group of animals (Mendonça et al., 2011). Venomous snakes can cause the death of goats, sheep, and cattle, leading to significant losses for farmers; a factor that encourages their slaughter whenever they are encountered (Alves et al., 2010a).

Currently, wildlife live in substantially reduced numbers within the boundaries of protected areas, such as national parks, and are generally actively excluded by humans from agricultural areas (Gordon, 2009). However, in many cases agricultural areas are established around protected areas, so the growth of activities such as agriculture, forestry, and energy development tend to intensify conflicts (Young et al., 2010). These authors also stress that increasing pressure on natural systems can increase the significance and magnitude of conflicts, thereby negatively affecting both biodiversity and people.

Human shark attacks stand out as one of the most severe conflicts in aquatic environments. According to data compiled by the International Shark Attack File (ISAF), an institution linked with the Florida Museum of Natural History, which was last updated in February 2016, there have been about 5800 shark attacks on humans throughout the world since the mid-16th century. There have been attacks in 88 countries, most notably in the United States (1301 attacks), Australia (593), Republic of South Africa (249), and Brazil (102) (Fig. 22.3).

Conflicts between humans and aquatic animals, such as turtles and cetaceans, in search of common food sources have increased (Plagányi and Butterworth, 2002), especially with the intensification of fishing activities for meeting an increasing demand. Such situations are common in the Amazon region, for example, where porpoises are considered direct competitors for certain species of fish, leading to conflicts with fishermen that has caused high mortality of porpoises by accidental ensnarement in fishing nets (Da Silva and Best, 1996; Loch et al., 2009).

FACTORS THAT LEAD TO CONFLICT

Conflicts with animals can be direct, as when they directly attack people (e.g., tigers, lions, elephants, hippopotamus, pumas (cougars), crocodiles, snakes, sharks, etc.), or indirect, as when animals attack domestic animals, cause damage to crops, damage fishing gear and vessels,

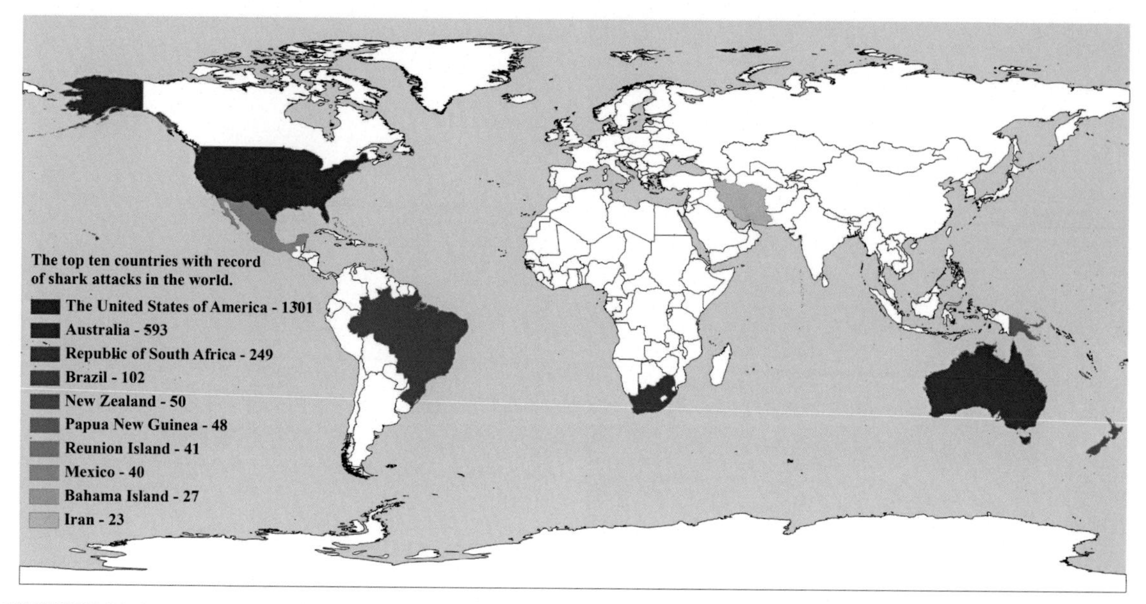

FIGURE 22.3 The 10 countries with the greatest number of recorded human shark attacks in the world according to data from the International Shark Attack File (ISAF).

and cause accidents in airspace and on roads, among other reasons. These conflicts mainly emerge as a result of factors such as a decreased abundance of prey (Bhattarai and Fischer, 2014; Hoogesteijn, 2001; Polisar et al., 2003), loss of wild animal habitat (Fahrig, 2003; Fischer and Lindenmayer, 2007; Hoogesteijn, 2001), competition for shared resources (Graham et al., 2005; Thirgood et al., 2000), management of livestock (Datiko and Bekele, 2013; Polisar et al., 2003), financial losses (Datiko and Bekele, 2013; Oli et al., 1994), and human intolerance of wildlife (Datiko and Bekele, 2013; Foloma, 2005; Oli et al., 1994).

Attacks on Human Beings

Attacks of wild animals on humans may be a result of predatory, territorial, or defensive instincts (Conover, 2002). Conflicts between coyotes, wolves, and foxes and humans, for example, are common in many parts of the world (Baker and Timm, 1998; Bisi et al., 2007; Bjerke et al., 1998; Farrar, 2007) and are motivated by interspecific competition for livestock and other domestic animals, as well as direct attacks on humans (Fritts et al., 2003; Kellert et al., 1996; Thirgood et al., 2005). In California (The United States) for example, over a period of 10 years (1988–97), 53 individual people were attacked by coyotes resulting in 21 of them being injured (Baker and Timm, 1998).

Bears, members of the so-called charismatic fauna, are also responsible for attacks on humans. Many of these attacks are of a defensive nature and are carried out by black bears (*Ursus americanus*), brown bears (*Ursus arctos*), polar bears (*Ursus maritimus*) (Conover, 2002), sloth bears (*Melursus ursinus*) (Rajpurohit and Krausman, 2000), and Asian black bears (*Ursus thibetanus*) (Liu et al., 2011), among other species (Fig. 22.4). According to Conover (2002), bears usually attack after losing their fear of humans. In the Indian state of Madhya Pradesh, between 1989 and 1994, 735 people were injured or killed by bear attacks (*M. ursinus*) (Rajpurohit

FIGURE 22.4 Bear species involved in conflicts with humans. (A) The polar bear (*Ursus maritimus*), (B) The spectacled bear (*Tremarctos ornatus*), (C) The brown bear (*Ursus arctos*), and (D) The sloth bear (*Melursus ursinus*). *Photo credits: (A–C) Rômulo Romeu Nóbrega Alves, (D) L. Shyamal (Own work), via Wikimedia Commons.*

and Krausman, 2000), and it is estimated that in North America, around 30 people are attacked every year by bears, especially black bears (Conover, 2002). A study of puma (cougar) attacks in the United States and Canada found that between 1890 and 1990 there were 58 victims, of which 64% were children and of those 78% of the attacks were on children who were not accompanied by adults or were accompanied by other children (Beier, 1991). Examples such as the latter reinforce the idea that attacks on adult humans should be less frequent than on children because the body size of adults is too large to be attacked by small- and medium-sized animals (Conover, 2002).

Wolves are another example of animals targeted in conflicts, especially when their natural populations become abundant (Bisi et al., 2007). These authors pointed out that in Finland where there is a considerable abundance of wolves, there have been heated discussions between people living in rural areas and nongovernmental organizations about the application of population control measures to wolves and how to solve the problems they cause.

Similar to wolves, species of fox are targets of control through hunting because of their conflicting relationships with humans, especially when there are attacks on domestic animals on farms (Alves et al., 2009, 2012; Mendonça et al., 2011;

Souza and Alves, 2014). In areas of the Atlantic Forest in the state of Paraíba in northeastern Brazil, *Cerdocyon thous* (fox) and *Leopardus tigrinus* (tiger cat) are commonly hunted to prevent attacks on domestic animals (Souza and Alves, 2014).

In coastal areas, wild animals have to share the environment with people who frequent these areas for recreational activities, including practising sports such as surfing, diving, and fishing, thus creating a situation conducive to conflict. Sharks exemplify this well. According to Burgess (2015), attacks by sharks on humans can be classified as either unprovoked attacks, which are attacks that occur in the natural environment and without provocation by the human victim; and provoked attacks, which usually occur when humans make physical contact with sharks outside of their natural habitat, such as in aquariums and research ponds.

Venomous animals are also at the center of conflicting interactions with people. Normally, when an animal attacks a human it would be acting defensively (Conover, 2002), which is mainly valid for small and venomous animals, such as snakes, spiders, and scorpions. Although in many of these cases the attacks occur accidentally, the rates of this type of conflict is alarming. Throughout the world many humans are stung by venomous animals annually, causing countless losses of human lives. In Mexico, about 1000 deaths from scorpion stings occur each year (Cheng, 2016); in Brazil, in the year 2015, the number of recorded stings by these animals reached 74.5 per 1000 people, with 119 deaths (Ministério da Saúde, 2016); in Turkey, the annual incidence is about 36 scorpion stings per 100,000 inhabitants and the estimated annual mortality is 0.01 death per 100,000 (Ozkan et al., 2008); in some states of central and western Mexico, the incidence reaches 1350 scorpion stings per 100,000 inhabitants (Chowell et al., 2006).

Of the approximately 1500 species of scorpions in the world, 50 are potentially dangerous to humans (Cheng, 2016), mainly the genera *Tityus*, *Leiurus*, *Androctonus*, *Centruroides*, and *Buthus* (Chippaux and Goyffon, 2008). Known species of scorpions are distributed in tropical and subtropical zones of the planet, with areas with the greatest incidence of scorpion stings being: northern Saharan Africa, Near- and Middle East, Mexico, Brazil, southern and eastern Africa, the Amazon Basin (Guyanas, Venezuela, and northern Brazil), and southern India (Chippaux and Goyffon, 2008).

In a review paper, Kasturiratne et al. (2008) estimated that at least 421,000 envenomations and 20,000 deaths occur annually around the world as a result of snakebites. According to these authors, and based on the estimate that the total number of snakebites is two to three times the number of envenomations, they estimated that 1,200,000–5,500,000 snakebites of humans occur throughout the world annually (Table 22.1). It is estimated that most snakebites occur in southern and southeastern Asia, sub-Saharan Africa, and Central and South America, with India having the highest estimated number of snakebites and deaths (Kasturiratne et al., 2008).

TABLE 22.1 Animal Attacks on Humans in the World

Animals	Number of Attacks per Year
Snakes[c]	1,200,000–5,500,000
Scorpions[b]	1,200,000
Crocodiles[a]	2,500
Bees[a]	1,250
Hippopotamus[a]	400
Elephants[a]	250
Sharks[a]	90

[a] *Szpilman, M., 1998. Seres marinhos perigosos: Guia prático de identificação, prevenção e tratamento. Instituto Ecológico Aqualang, Rio de Janeiro.*
[b] *Chippaux, J.P., Goyffon, M., 2008. Epidemiology of scorpionism: a global appraisal. Acta Tropica 107, 71–79.*
[c] *Kasturiratne, A., Wickremasinghe, A.R., Silva, N., Gunawardena, N.K., Pathmeswaran, A., Premaratna, R., Savioli, L., Lalloo, D.G., Silva, H.J., 2008. The global burden of snakebite: a literature analysis and modelling based on regional estimates of envenoming and deaths. PLoS Medicine 5, e218.*

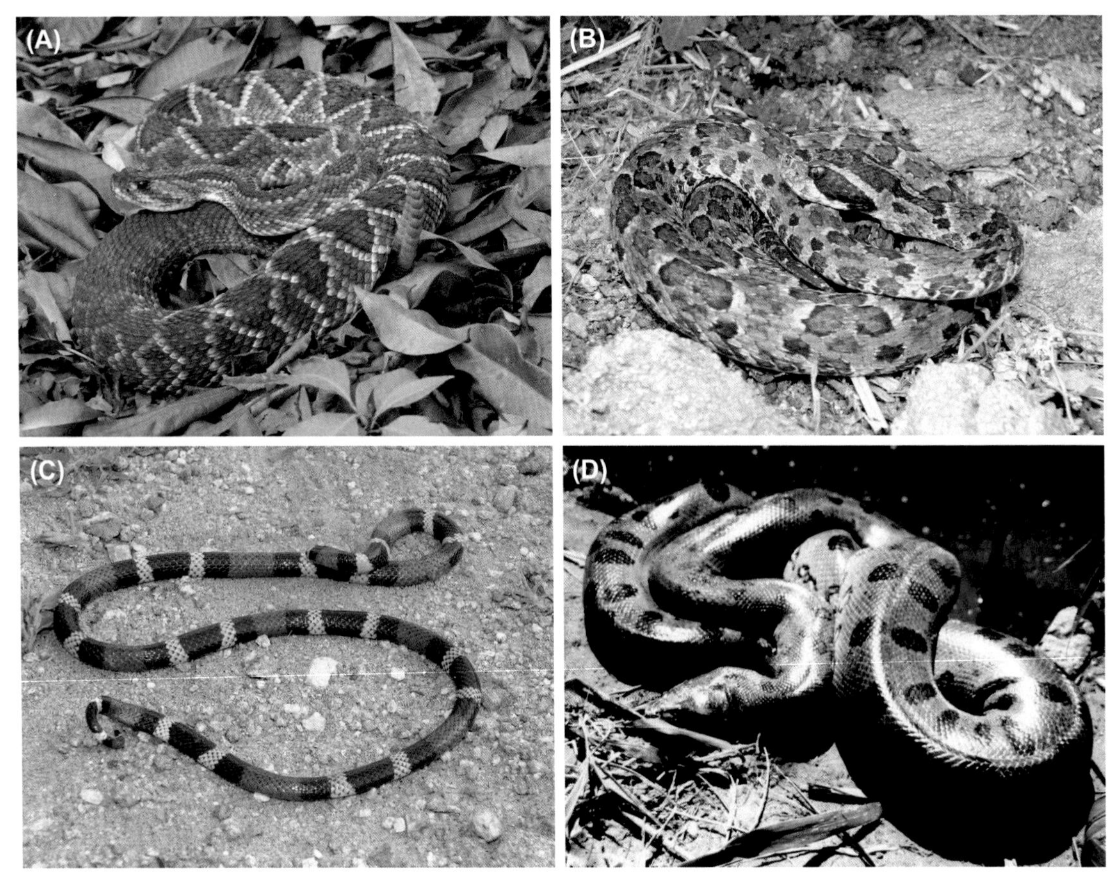

FIGURE 22.5 Examples of snake species commonly killed by humans due to being considered dangerous for people and domestic animals. (A) *Crotalus durissus* Linnaeus, 1758, (B) *Bothropoides erythromelas* (Amaral, 1923), (C) *Micrurus ibiboboca* (Merrem, 1820), and (D) *Eunectes murinus* (Linnaeus, 1758). *Photo credits: (A) Gentil Pereira Filho, (B and C) Washington Vieira, and (D) Yuri Lima.*

Among venomous snakes, species of the families *Elapidae* and *Viperidae* stand out as having the greatest impact on the health of humans on a global scale (Fig. 22.5). These venomous snake species are distributed throughout most of the planet, are common in densely populated areas, and can cause numerous snakebite accidents, resulting in high levels of morbidity, disability, and mortality among the victims (WHO, 2010). Table 22.2 lists the most medically important species of venomous snakes in the world, according to data from the WHO (WHO, 2010).

Although not venomous, snakes of the families *Boidae* and *Pythonidae* are potential predators of domestic animals (Fig. 22.5) (Bhupathy et al., 2014; Goursi et al., 2012; Strüssmann, 1997). In South America, anacondas can have high rates of predation on domestic animals, such as *Bos taurus* (cattle), *Gallus gallus* (chicken), *Cairina moschata* (duck), and *Felis catus* (domestic cat), and are consequently killed in retaliation and for the perceived risk they pose to humans (Miranda et al., 2016). However, direct attacks on humans are rare, suggesting that the killing of the largest anacondas is a preventive action because of

TABLE 22.2 Main Venomous Snakes of Highest Medical Importance in the World

AFRICA AND THE MIDDLE EAST	
Family *Atractaspididae*	*Atractaspis andersonii*
Family *Elapidae*	*Dendroaspis angusticeps, Dendroaspis jamesoni, Dendroaspis polylepis, Dendroaspis viridis, Naja anchietae, Naja annulifera, Naja arabica, Naja ashei, Naja haje, Naja katiensis, Naja oxiana, Naja melanoleuca, Naja mossambica, Naja naja, Naja nigricollis, Naja nigricincta, Naja nivea, Naja senegalensis, Pseudocerastes persicus*
Family *Viperidae*	*Bitis arietans, Bitis gabonica, Bitis nasicornis, Bitis rhinoceros, Cerastes cerastes, Cerastes gasperettii, Daboia mauritanica, Daboia palaestinae, Echis borkini, Echis carinatus, Echis coloratusi, Echis jogeri, Echis leucogaster, Echis ocellatus, Echis omanensis, Echis pyramidum, Macrovipera lebetina, Montivipera xanthina, Pseudocerastes persicus*
ASIA AND AUSTRALASIA	
Family *Elapidae*	*Acanthophis laevis, Bungarus caeruleus, Bungarus candidus, Bungarus magnimaculatus, Bungarus multicinctus, Bungarus niger, Bungarus sindanus, Bungarus walli, Deinagkistrodon acutus, Daboia siamensis, Naja atra, Naja kaouthia, Naja mandalayensis, Naja naja, Naja oxiana, Naja philippinensis, Naja samarensis, Naja siamensis, Naja sputatrix, Naja sumatrana, Notechis scutatus, Oxyuranus scutellatus, Pseudonaja affinis, Pseudechis australis, Pseudonaja mengdeni, Pseudonaja nuchalis, Pseudonaja textilis*
Family *Viperidae*	*Cryptelytrops albolabris, Cryptelytrops erythrurus, Daboia russelii, Daboia siamensis, Deinagkistrodon acutus, Echis carinatus, Gloydius blomhoffii, Gloydius brevicaudus, Gloydius halys, Hypnale hypnale, Macrovipera lebetina, Protobothrops flavoviridis, Protobothrops mucrosquamatus, Viridovipera stejnegeri*
EUROPE	
Family Viperidae	*Vipera ammodytes, Vipera aspis, Vipera berus*
THE AMERICAS	
Family Viperidae	*Agkistrodon bilineatus, Agkistrodon contortrix, Agkistrodon piscivorus, Agkistrodon taylori, Bothrops alternatus, Bothrops asper, Bothrops atrox, Bothrops bilineatus, Bothrops brazili, Bothrops caribbaeus* (St Lucia), *Bothrops diporus, Bothrops jararaca, Bothrops jararacussu, Bothrops leucurus, Bothrops lanceolatus* (Martinique), *Bothrops mattogrossensis, Bothrops moojeni, Bothrops pictus, Bothrops venezuelensis, Crotalus adamanteus, Crotalus atrox, Crotalus durissus, Crotalus horridus, Crotalus oreganus, Crotalus simus, Crotalus scutulatus, Crotalus totonacus, Crotalus viridis, Lachesis muta*

people's fear of these reptiles (see Murphy and Henderson, 1997).

Reduction of Natural Prey and Habitat Loss

Predation is a natural and fundamental behavior that maintains biological diversity and ecological processes (Pitman et al., 2002). The ecological interactions of competition and predation are fundamental to the regulation of prey availability in terrestrial ecosystems. Throughout the world, predators play an important ecological role in maintaining community structure and ecosystem equilibrium (Marchini et al., 2011). These authors also mention that carnivores will substitute their natural prey for domestic species when there is a decrease in the prey's abundance due to hunting, resulting in an increase in contact between predators and domestic animals. In Nepal, for example, rates of predation by tigers were high in areas with low abundances of natural prey (Bhattarai and Fischer, 2014). A decline in prey can occur as a

FIGURE 22.6 *Puma yagouaroundi* (É. Geoffroy 1803) (left), wild cat frequently considered a livestock predator in Brazilian semiarid region. On the right, specimen killed by retaliation due to attack on poultry. *Photo credits: (Left) John Philip Medcraft, (right) Raynner Rilke Duarte Barboza.*

result of predatory hunting, disorderly deforestation, or some form of epidemic transmitted by domestic animals being in contact with natural preys (Hoogesteijn et al., 1993).

In terrestrial environments, habitat loss and fragmentation due to agricultural expansion, timber harvesting, ranching, and urban sprawl have strongly impacted terrestrial ecosystems (Fahrig, 2003; Fischer and Lindenmayer, 2007), leading to extreme fragmentation of forest cover, isolation of forest patches, increased edge effect, and reduction of native fragments (Fischer and Lindenmayer, 2007). This environmental degradation causes wild animals to hunt outside of forest areas, and approach human inhabited environments, where the chance for conflicts is enhanced.

Habitat reduction results in severe competition for natural resources between wild animals and local communities (Ertiban, 2016). Competition for area is intensified in times of scarcity and when these areas represent relevant sources of water. In Africa, for example, in areas where water availability becomes progressively limited, areas with water sources become places of conflict and competition between hippopotamus and domestic animals (Ertiban, 2016).

In Brazil, livestock farming has increased deforestation in the Cerrado, replacing natural areas with extensive pastures, which are connected with forest fragments, and as a consequence pumas (*Puma concolor*) and jaguars (*Panthera onca*) have to coexist with cattle, which results in predation, justifying, to the farmers, retaliatory action (Palmeira et al., 2008). Most attacks on domestic animals, such as attacks to chicken coops, are carried out by *Cerdocyon thous* (forest fox), *Leopardus pardalis* (ocelot), *Puma yagouaroundi* (jaguarundi) (Fig. 22.6), *Lycalopex gymnocercus* (pampas fox), and *Procyon cancrivorus* (crab-eating racoon) (Alves et al., 2016; Pitman et al., 2002).

Marine Environment

The loss of habitat and overlapping use of aquatic environments has also intensified conflicts involving marine animals and humans. In 2015 alone, ISAF investigated 164 incidents, of which 98 were confirmed to be cases of shark

FIGURE 22.7 *Galeocerdo cuvier* (Tiger shark—left) and *Carcharhinus leucas* (Bull shark—right) are among the shark species related to human attacks. *Photo credits: Albert Kok, via Wikimedia Commons.*

attacks on humans. In the state of Pernambuco in northeastern Brazil, Gadig (1998) reports that a work project undertaken along the coast resulted in closing the mouths of rivers, which provide nutrients to the coastal environment, thus disrupting the food chain and promoting shark attacks by *Carcharhinus leucas* (bull shark) and *Galeocerdo cuvier* (tiger shark) on people, mainly surfers (Fig. 22.7). According to ISAF, surfers and other participants in sports with boards were involved in 49% of shark attacks in 2015. Surfers have been the group most affected by shark attacks these days, most likely due to the large amount of time they spend in the surf while kicking their feet and paddling with their hands, movements that catch the attention of sharks.

Accidental catch, also called subcapture or bycatch, directly affects animal species such as sea turtles, porpoises, and dolphins (Fig. 22.8). Accidental catches of these animals can be avoided by not placing fishing nets near the surf zone where dolphin species frequent, or by not using fishing gear in regions near rocky bottoms that are suitable for the occurrence of turtles (Domit et al., 2011). In Brazil, accidental catches have also been reported in the Amazon region, mainly for piscivorous species such as dolphins (Loch et al., 2009; Iriarte and Marmontel, 2013).

FIGURE 22.8 The olive ridley sea turtle (*Lepidochelys olivacea* (Eschscholtz, 1829)) entangled in a fishing net on the coast of Angola, West Africa. *Photo credits: Franciany Braga-Pereira/Projecto Kitabanga.*

Conflicts Associated With Transportation Systems

Other types of conflicts between fauna and humans are linked to transportation systems. Though the highways are important for the development of social, economic, and cultural relations of a nation (Bager and Fontoura, 2012), the implementation of linear structures (roads and highways) in the landscape results in roadkills. Roadkills generally occur for two main reasons. First, the habitat of a species is divided, impacting movements for watering, foraging, migration, and reproduction. Second, there is often a supply of food on highways and along their margins due to trash and carcasses of road-kill animals, which serve as an attraction to other species. Therefore, these linear structures begin to generate conflicts between the fauna and humans, as well as cause the direct death of animals and potentially, depending on the size of the animal, death to people.

The search for strategies to promote safety for the users of roads and highways, as well as to ensure the maintenance of biodiversity through the reduction of roadkill, using approaches such as road ecology, has been a challenge discussed by researchers (Beckmann et al., 2010). In 2007, in the United States, between one and two million accidents involving motor vehicles and large mammals were recorded, causing a loss of around US$ 8.3billion (Huijser et al., 2007); in Germany in 1988, 145,636 collisions with damages totaling US$ 210million were recorded (Hartwig, 1991); in France, around 50 people lose their lives each year in accidents with ungulates on the roads, while almost 2500 people are injured (Bruinderink and Hazebroek, 1996). In Brazil, according to estimates by the Centro Brasileiro de Estudos em Ecologia de Estradas (Brazilian Center for Studies in Road Ecology), more than 15 animals die on Brazilian roads every second (Fig. 22.9), and up to 475million wild animals are run over each year with approximately 430million being small animals, 40million being medium, and 5million being large, the latter including species such as the puma (*Puma concolor*), maned wolf (*Chrysocyon brachyurus*), jaguar (*Panthera onca*), tapir (*Tapirus terrestris*), and capybara (*Hydrochoerus hydrochaeris*) (CBEE, 2016).

Another cause of conflict with fauna is the collision between birds and aircraft both in airspace and on land during landing and takeoff. The collision between birds and aircraft is a constant and imminent danger to aviation throughout the world. These collisions may not only cause the death of birds, but they can also seriously damage aircraft and, in some cases, cause accidents that are fatal to humans (Thorpe, 2012). In the United States, 2059 collisions of civil aircraft with birds were reported in 1991(Conover et al., 1995), yet this is an underestimate because only 25% of bird collisions were reported (Linnell et al., 1996, 1999a,b). Therefore, by extrapolation, there would be about 8000 collisions per year (Conover, 2002). It is estimated that collisions of birds and other wild animals with airplanes causes more than US$ 900million in civil and military aviation damage annually in the United States (BSC USA, 2016). According to the Bird Strike Committee USA (BSC USA, 2016), these collisions can put the life of the aircraft crew and its passengers at risk during flight, a fact corroborated by Thorpe (2016), who reports that worldwide more than 450 people have been killed and more than 500 civilian and military aircraft destroyed as a result of these types of collisions.

FIGURE 22.9 Carnivorous mammals' road-killed in northeast region of Brazil: *Leopardus tigrinus* (above) and *Procyon cancrivorus* (below). *Photo credits: John Philip Medcraft.*

FIGURE 22.10 Examples of bird species involved in aircraft collisions in Brazil. *Photo credits: John Philip Medcraft.*

In the United Kingdom, the Central Science Laboratory estimated that worldwide the cost of collisions between birds and aircraft to airlines is about US$ 1.28 billion annually, including the direct repair cost and the revenue opportunities lost while the damaged aircraft was out of commission (Allan and Orosz, 2001). In Brazil, direct costs to the main air transport companies exceeds US$ 6 million annually, although there has never been an accident related to collisions between airplanes and birds in the country that resulted in fatality (Mendonça, 2009).

Among more than 7500 collisions between birds and aircraft reported by European airlines over a span of 5 years, gulls were involved in almost 40%, lapwings (14%), species of swifts, swallows, and martins (11%), as well as several birds of prey (10%) (Thorpe, 2016). In a study carried out in the Parnaíba International Airport, Brazil, Cardoso et al. (2014) reported that the species that are likely to be at highest risk of collisions with airplanes are *Vanellus chilensis*, *Athene cunicularia*, *Caracara plancus*, *Coragyps atratus* (Fig. 22.10), *Cathartes burrovianus*, *Rosthramus sociabilis*, *Sturnella superciliaris*, *Cathartes aura*, *Bubulcus ibis*, *Nothura maculosa*, *Egretta thula*, *Columbina passerina*, *Phalacrocorax brasilianus*, *Columba livia*, *Phaetusa simplex*, *Charadrius collaris*, *Hirundo rustica*, *Pitangus sulphuratus*, and *Tyrannus savana*.

Some of the main species involved in aircraft collisions throughout the world are *Larus* sp. (gulls), *Cygnus columbianus* (whistling swan), *Grus* sp.(cranes), *Vanellus vanellus* (lapwings), *Gavia immer* (loon), *Pelecanus erythrorhynchos* (white pelican), *Branta canadensis* (canada goose), *Gyps* sp. (vultures), *Milvus migrans* (black kites), *Falco sparverius* (sparrowhawk), *Columba* sp. (pigeons), and *Sturnus vulgaris* (european starlings) (Thorpe, 2012).

Competition for Resources and Poaching

In many parts of the world, habitat loss and fragmentation and the expansion of human populations have forced wild carnivores into close contact with domestic ungulates, accentuating competition between humans and large carnivores (Conover, 2002). Humans have been hunters for thousands of years, and the incidence of species extinction from hunting activities has been low; however, hunting is now considered the second greatest threat to wildlife, behind only loss of habitat (Vié et al., 2009). Hunting causes population declines of the natural prey of wild predators, which then take to attacking domestic animals as an alternative food source, especially in border areas between conservation units and rural properties (Hoogesteijn, 2001). It is also necessary to consider the increase in availability of domestic animals close to forest remnants, and thus a greater likelihood of a predator attack (Marchini et al., 2011). Another important factor is that domestic animals do not exhibit all the antipredator behavior of their ancestors, and thus represent relatively easier prey than wild prey of a similar size (Linnell et al., 1999a). It should be noted that attempts to eliminate wild predators can often leave them injured, making them more aggressive, and perhaps unable to hunt normally, thus leading them to search for easier prey such as domestic animals (Pitman et al., 2002).

Another type of conflict linked to competition for resources is that between fishermen and turtles and dolphins. Conflicts with these animals differ considerably from conflicts with sharks (when humans generally die or are mutilated), as it is the wild animals that are killed, and the fishermen suffer material losses (fishing equipment) (Fig. 22.9). The accidental capture of turtles and dolphins can either occur when animals feed directly on the bait used by fishermen, or when the animals are incidentally caught by fishing equipment (Loch et al., 2009; Nogueira and Alves, 2016; Pinedo and Polacheck, 2004). In this context, accidental captures in coastal or continental areas represent a serious conservation problem for the animals involved.

Management of Domestic Animals

The management of domestic animals is closely related to the likelihood of attacks by wild animals. In neotropical savannas, for example, most cattle farms under extensive conditions practice rudimentary management of herds where they are exposed to drought, floods, disease epidemics, parasitism, and malnutrition (Hoogesteijn, 2001). In addition, the proximity of farms to protected areas results in high levels of conflict between wildlife and humans (Datiko and Bekele, 2013).

Linnell et al. (1996) argue that grazing techniques are the primary factor responsible for individual problems with specialized predators, because when animals are constantly pastured, grazed in open areas, or kept confined at night, predation requires the development of specialized behavior by a predator since to succeed in predation these animals need to avoid detection by attendants and their dogs, and deal with physical barriers and open habitats. On the other hand, these authors argue that when animals are allowed to graze freely and unattended in the natural habitat of carnivores, the rate of carnivorous encounters increases, and individual problems are less likely to occur because the majority of predators have the opportunity to prey on animals without the need to develop specialized behaviors.

Intolerance

In some countries of the African continent, herbivores, such as elephants, hippopotamus, and buffalo, among others, are the most problematic animals for agricultural crops (Datiko and Bekele, 2013; Foloma, 2005), while lions, leopards, hyenas, and crocodiles can be threats to humans and livestock (Foloma, 2005). Conflicts between Asian black bears and humans have been documented in some Asian countries, involving damage to agriculture and livestock, and injuries and even death to humans (Chauhan, 2003; Liu et al., 2011; Sangay and Vernes, 2008; Stubblefield and Shrestha, 2007; Yamada and Fujioka, 2010). In Sichuan Province in southeastern China, bears attack crops such as *Zea mays* (corn), *Triticum aestivum* (wheat), *Glycine max* (beans), *Solanum tuberosum* (potato) and kill domestic animals such as

Capra hircus (goat), *Ovis aries* (sheep), *Bos taurus* (cattle), *Bos grunniens* (yak), *Sus scrofa domestica* (pig), and *Equus caballus* (horse) (Liu et al., 2011). According to these authors, bears are killed by firearms (42.0%), traps (24.3%), and poisoning (23.1%) in retaliation for losses or to prevent future losses.

Despite the diversity of the types of conflicts, some of them are more significant in terms of losses than others. From a financial point of view, carnivore predation on domestic animals is one of the most significant types of conflicts (Linnell et al., 1999a). In many cases people are expected to be less tolerant of carnivores than herbivores, based on fear and/or the higher cost of losses of livestock than crops (Treves and Naughton-Treves, 2005). However, on a worldwide scale, more people are injured by large herbivores than by large predators (Conover, 2002). In countries such as Mozambique, attacks of wild animals on humans are considered rare compared to attacks on crops and domestic animals (Dunham et al., 2010).

The degree of human tolerance varies greatly, depending on the species and the groups of humans involved in conflicts. In Nepal, for example, in the buffer zone of the Bardia National Park, the local population generally possesses a positive attitude toward the conservation of *Panthera tigris* and is willing to tolerate the loss of some domestic animals, but not human losses (Bhattarai and Fischer, 2014). According to the authors, such behavior shows an opportunity for the implementation of adequate conservation measures and mitigation strategies, such as education, monetary compensation, and monitoring of tigers. Another illustrative example was recorded in the state of Mato Grosso, Brazil, by Trinca and Ferrari (2006), where local hunters reported that they favored the extermination of all species of predators that have the potential to predate domestic animals or attack people even though the animals live in natural environments away from humans.

IMPLICATIONS FOR CONSERVATION OF SPECIES

Human-caused changes to natural environments can lead to the persecution or exclusion of wild animals from areas where humans live and grow plants or raise animals, thereby restricting them to areas that are inappropriate for agricultural production and unsuitable for the maintenance of wildlife, which may even lead to extinction (Gordon, 2009).

Among the species that in some way maintain a conflicting relationship with humans, some deserve mentioning because they are listed among the categories of worrisome levels of threat of the red list of the International Union for the Conservation of Nature (IUCN 2016-2), such as the following: *Loxodonta africana* (Vulnerable), *Hippopotamus amphibius* (Vulnerable), *Pa. tigris* (Endangered), *Panthera leo* (Vulnerable), *Uncia uncia* (Endangered), *Panthera pardus* (Vulnerable), *Leopardus tigrinus* (Vulnerable), *Lycaon pictus* (Endangered), *Acinonyx jubatus* (Vulnerable), *Crocodylus acutus* (Vulnerable), among others.

Some species that appear in worrisome categories of the IUCN red list, such as *Puma concolor*, *Panthera onca*, *Leopardus wiedii*, *Alcelaphus buselaphus*, *Papio anubis*, *Chlorocebus pygerythrus*, *Hystrix cristata*, *Sylvicapra grimmia*, *Tragelaphus scriptus*, among others, have different regional or local conservation status. For example, in some countries, such as Brazil, local red lists classify some of these species as threatened with extinction, such as *Puma concolor*, *Panthera onca*, and *L. wiedii*, which are all listed as "Vulnerable." Thus, even though they are not considered threatened with extinction globally, some species are regionally or locally threatened and thus require appropriate conservation policies in regional and local contexts.

Other species that maintain a conflicting relationship with humans, and that are included on the IUCN red list (Version 2016-2), are sharks of the species *Carcharhinus leucas* (Near Threatened), *Galeocerdo cuvier* (Near Threatened), *Carcharhinus limbatus* (Near Threatened), *Carcharodon carcharias* (Vulnerable), *Sphyrna zygaena* (Vulnerable); the marine turtle species *Caretta caretta* (Vulnerable), *Chelonia mydas* (Endangered), *Dermochelys coriacea* (Vulnerable), *Eretmochelys imbricata* (Critically Endangered), *Lepidochelys kempii* (Critically Endangered), *Lepidochelys olivácea* (Vulnerable). and *Natator depressus* (Data Deficient). The conflict between wildlife and humans is a matter of widespread conservation and of increasing interest to conservationists (Kaltenborn et al., 2006).

There are, of course, direct impacts on the target animals of conflicts, which are pursued and slaughtered, but there are also a number of indirect ecological consequences from the loss or decline of target species, especially since some of them play key roles in ecosystems such as large carnivores that are top predators and play an important ecological role in keeping the ecosystems in which they live stable and balanced (Marchini et al., 2011).

The effects of predators on their ecosystem, however, vary considerably among species and ecosystems (Ferretti et al., 2010). Large predators regulate the populations of all the species that comprise their food spectrum. In this way, predators prevent excessively large populations of prey, maintain their vigor by eliminating old and diseased individuals, and intervene in the spread of diseases that affect these species (Hoogesteijn, 2001). In the absence of predators, natural prey such as herbivorous mammals, rodents, birds, reptiles, and insects tend to multiply exponentially, causing considerable damage to agriculture and significant financial losses to producers (Pitman et al., 2002).

Paradoxically, many animals involved with conflicts are charismatic fauna, which attract the attention of, and arouse the admiration by, people, such as lions, jaguars, tigers, elephants, hippopotamus, gorillas, among others. This particularity can be explored for solutions to conflicts that favor the capture of resources for the purpose of conservation of the species. In addition, large mammals can be used as flagship species (Simberloff, 1998), and they are usually used in conservation campaigns because they arouse public interest and sympathy.

PROPOSALS FOR CONFLICT MITIGATION

Proposals for conflict mitigation should take into account the fact that the level of harm is directly related to the existing conflict and that the severity of the conflict generates a proportional response (Dickman, 2010). Forms of conflict mitigation include nonlethal and lethal methods. Lethal methods should be only used in cases when the problem with a particular animal persists in spite of attempts with other methods, or when the animal poses a clear and imminent danger to human life (Sechele and Nzehengwa, 2002). However, lethal methods are often used indiscriminately, even in the absence of conflict. Since lethal methods have significant negative impacts on species conservation, the search for an incentive to use nonlethal methods should be encouraged.

Several conflict mitigation strategies have been proposed with the aim of reducing the financial losses generated by conflicts, while guaranteeing the conservation of the animal species involved. Among the main forms of mitigation used by livestock and agriculture farmers are the use of electric fences; lighting systems (Vidolin et al., 2004); human surveillance (Ohrens et al., 2016); disturbance mechanisms such as shouts, torches, and shots (Okello and D'Amour, 2008); confinement of animals (Palmeira and Barrella, 2007; Vidolin et al., 2004); reintroduction of natural prey (Hoogesteijn and Hoogesteijn, 2011); and translocation of wild animals (Bradley et al., 2005; Conforti and Azevedo, 2003; Massei et al., 2010).

Electric fences have been shown to be a successful method for preventing predation of domestic animals by felines. They are mainly used in small areas, such as nocturnal enclosures, because of the probability of the occurrence of faults and malfunctions is greater in larger electrified fences, and so the incidence of predation may persist in larger electrified areas (Hoogesteijn and Hoogesteijn, 2011). Even if fences are not electrified, the investment for deployment and maintenance can be high. In areas where animals such as elephants or wild boars are plentiful, fence maintenance costs can be higher yet, and without sufficient investment to maintain them, the value of the method is quickly lost (Stander, 1990).

Employing different methods of frightening animals can be adequate for some target species of conflicts. Conflicts with hippopotamus, for example, can be avoided by the use of fire, colored objects, sounds, or other disturbances to which these animals are sensitive (Ertiban, 2016). Another way to prevent attacks on animals is by replacing them with animals that exhibit defensive behaviors or by breeding animals with others that have more hostile behaviors. For example, the implementation of Indian buffalo (*Bubalus bubalis*) instead of cattle in savanna environments proved to be a good alternative and reduced predation without the need for other control measures for the felines involved (Hoogesteijn and Hoogesteijn, 2011). Another method that is considered friendly in mitigating elephant attacks on crops is the planting of peppers around the crops since elephants tend to avoid areas with them present (Okello and D'Amour, 2008). However, some mitigation methods are not peaceful and may even lead to the death of wild animals, such as the use of firearms and the construction of moats around farms (Okello and D'Amour, 2008); for example, moats have been used to trap and kill hippopotamus (Ertiban, 2016).

While on the one hand there is a considerable array of methods that can be used to reduce the incidence of conflicts, on the other hand many of these methods have proven to be ineffective or inappropriate. These negative results are often derived from the use of, for example, fences that, while appropriate to contain livestock within pastures, are ineffective in preventing the entrance of wildlife. According to Ertiban (2016), the use of low fences and ditches are mechanisms that are capable of preventing hippopotamus from invading cultivated areas. Conversely, in work done in Sierra Leone, fences used to keep hippopotamus out of local farms was ineffective because these animals were able to break most of the fences (Conway et al., 2015). The correct identification of the predator of domestic animals is an important step in determining suitable control methods, as these will depend on the characteristics of the species in question (Pitman et al., 2002). Another problem is that even if appropriate techniques and methods are used, the maintenance of agricultural areas and pastures relatively close to forests or protected areas tends to intensify the occurrence of conflicts. Damage to crops and crop loss is often associated with proximity to protected areas (Karanth et al., 2012).

The translocation of wild animals is another widely used mitigation measure. Work done in Namibia has shown that the translocation of lions that occasionally attacked wildlife of the Etosha National Park was successful as only 1 of the 12 translocated animals attacked an animal in the park again, while the other individuals remained away for least 1 year after translocation and without further attacks (Stander, 1990). According to the author, the success of the translocation method is directly related to the distance from the area where the conflict occurred to the new area to which the animal is being translocated.

Another form of conflict mitigation is compensation. Given that on many occasions the solution found to solve problems of predation is the death of the problem animal, compensation schemes seek to encourage tolerance of wild animals by reducing the impact of economic losses. Thus, the application of measures to mitigate conflicts between humans and wildlife includes direct payment to individuals or their families to compensate for the threat of wildlife to their crops, livestock, property, or personal safety (Nyhus et al., 2003). According to these authors, a total or partial payment is made in cash or other assistance, or other methods of damage prevention are provided or, in some cases, people are compensated only to tolerate these animals on their land.

The value of compensation is directly related to the species involved (Karanth et al., 2012). However, in cases where compensatory measures are already in place, the farmers themselves may cease to employ preventive actions since the compensation removes the impetus to protect crops and livestock (Bulte and Rondeau, 2005). Another negative point concerns cases in which producers induce conflicts unlawfully receive compensation. Bulte and Rondeau (2005) argue that the provision of compensation can lead to excessive damage as people may put their animals at risk of predation to benefit from compensation. In addition, compensation for damage may make the owners feel that they do not need to adjust their property to prevent further predator attacks; the compensation could also benefit owners who do not have adequate management of domestic animals, and so the predator responsible for the attacks would remain with the same pattern of predation on domestic animals without reaching a solution to the problem other than the unfortunate possibility that the predator would have to be killed (Pitman et al., 2002).

In the case of sharks, knowledge about the biological characteristics of these large predators along beaches and throughout the oceans, including identification of the most abundant species, as well as ontogenetic and physiological conditions, is fundamental for an adequate understanding of the causes and conditions under which shark attacks occur (Fischer et al., 2009). According to these authors, without an exact understanding of the reasons for attacks, the adoption of measures to reduce their incidence and, consequently, the reduction of socioeconomic impacts associated with them, becomes much more difficult.

Some recommendations to reduce the chances of shark attacks in areas with previous records of attacks on humans were presented by Gadig (1998) and Szpilman (1998). Among these are the following: avoid entering the sea during dawn, dusk, and at night (most shark species that attack humans are more active and feed at these times); avoid swimming in estuaries or bays with sewage discharges (garbage or sewage emitters are great attractions for sharks); do not swim in the presence of aquatic mammals such as dolphins, sea lions, or porpoises as they are very attractive food for sharks; avoid swimming beyond the area of the surf break because, theoretically, the risk of attack is higher; and avoid swimming or diving in murky or dark water, because the sharks have a favorable sensory advantage in these environments.

As for sea turtles, several actions can be taken to mitigate the effects of bycatch in fishing activities, according to Santos et al. (2011), including: the establishment of priority areas for the conservation of sea turtles; propose areas and periods of restriction to fishing according to region; develop and implement marine turtle capture reduction devices; encourage and qualify certification programs for fisheries with low impacts on sea turtle populations; encourage the use of fishing gear alternatives with less impact on sea turtle populations (e.g., floating fences and fishing pens), considering the ecosystem approach (marine mammals, marine birds, and elasmobranchs); carry out awareness-raising and environmental education campaigns with fishing communities;

train fishermen for the correct management of accidentally caught turtles and for the use of mitigation measures in artisanal and industrial fishing.

With regard to venomous animals, the vast majority of their victims are humans and domestic animals, and mainly in rural areas. Snakebites are a threat to rural communities and to public health in general, and it is crucial to know the species found in a given area so that there can be appropriate management of ophidian cases with humans and livestock (WHO, 2010). It is interesting to note that almost a quarter of the world's snakebites occur in Central and South America, with the number of snakebite deaths being relatively low compared to other regions of the world (Kasturiratne et al., 2008). According to these authors, this lower mortality may be due to better management of the system responsible for ophidian accidents, including the local development of anti-ophidian serum in Latin American countries.

According to the World Health Organization (WHO) (WHO, 2010), preventive campaigns at the community level, with interdisciplinary programs aimed at generating a database with epidemiological records of ophidian accidents, preclinical evaluation of the effectiveness of antivenoms, programs for the acquisition and effective distribution of antivenoms, among other actions, may mitigate the damage cause by accidents with snakes.

As for roadkills, one of the ways to mitigate this is to reestablish the connection of fragmented landscapes by the construction of highways which will allow the movement of fauna between different habitat patches, favoring dispersion and genetic flow within and among populations, thus facilitating ecological processes and metapopulation dynamics (Corlatti et al., 2009). One of the most commonly used ways to link a fragmented landscape is through the construction of wildlife passages (see Bond and Jones, 2008; Corlatti et al., 2009; Grilo et al., 2008). These should be planned by considering the wildlife species of conservation interest such as: threatened or endangered species and abundant species that present a risk of collision with motor vehicles with possible loss of human life (Beckmann et al., 2010).

Among the ways of mitigating wildlife roadkill, aerial or subterranean passages with guide fences for the associated fauna (Bond and Jones, 2008; Grilo et al., 2008) constructed according to the target group, signal signs that warn about areas of crossing fauna (Ascensão and Mira, 2007), velocity reducers (Bager, 2003; Mastro et al., 2008), reflectors that illuminate the road and provide a view of the fauna on the highway with vehicle lights (D'Angelo et al., 2006), among other solutions, can contribute to the minimization of conflicts between fauna and humans on roads and highways. Faunal viaducts are structures that enable, and encourage, the passage of large animals such as Highway 93 in the state of Montana in the United States, where deer (*Odocoileus virginianus, Odocoileus hemionus*), moose (*Alces alces*), and bear (*Ursus arctus, Ursus americanos*) use these types of structures to cross the highway safely (Beckmann et al., 2010). Although designed for large animals, according to the authors, small- and medium-sized animals can also make use of faunal viaducts, as long as there is adequate ground and vegetation cover.

Regarding ways to reduce accidents caused by the collision of wild animals with aircraft, Thorpe (2016) provides some considerations, including producing airplanes, engines, and windshields that support the impact of birds; manage area near airports to be unattractive to birds and not sources of food, water, breeding grounds, and shelters. International rules suggest that the airport security area is a 13 km radius from the airport where dumps, crops, and other land uses that attract birds are eliminated (International Birdstrike Committee Organization, 2006). Cities should also invest in landfills and basic sanitation to help reduce conflicts between humans and wildlife in areas around airports.

FINAL CONSIDERATIONS

Conflicting interactions between wild animals and humans have occurred for centuries and have negative implications for both. These interactions demonstrate the need for measures to minimize these problems and should be based on multidisciplinary studies that understand the context of conflicts, and consider all the actors involved, from local populations directly affected to managers and conservations.

Measures that reduce contact between potentially dangerous wild animals and people (and their domestic animals) are important in reducing conflict and conserving wild species. There is also a need to implement education programs to enlighten human populations about the importance of wildlife to the balance of ecosystems and the need for animal conservation.

In this context, ethnozoology proves to be an important line of research for understanding conflicts involving humans and wild animals. Ethnozoological studies can also mediate the dialogue between social actors who suffer directly from the damages caused by conflicts and conservationists who aim to maintain the species involved in the conflicts, especially species that are threatened or endangered with extinction. In addition, studies of this nature could also assess which mitigation measures best apply to a particular local reality, help to disseminate measures that prove to be more effective in solving local problems, and provide fundamental support for the implementation of public policies for the conservation of species and the compensation for damages.

References

Allan, J.R., Orosz, A.P., 2001. The costs of bird strikes to commercial aviation. In: Bird Strike 2001. Proceedings of the Bird Strike Committee-USA/Canada Meeting. Transport Canada. Ottawa, Ontario, Canada, pp. 218–226.

Alves, R.R.N., Feijó, A., Barboza, R.R.D., Souto, W.M.S., Fernandes-Ferreira, H., Cordeiro-Estrela, P., Langgutj, A., 2016. Game mammals of the caatinga biome. Ethnobiology and Conservation 5, 1–51.

Alves, R.R.N., Gonçalves, M.B.R., Vieira, W.L.S., 2012. Caça, uso e conservação de vertebrados no semiárido Brasileiro. Tropical Conservation Science 5, 394–416.

Alves, R.R.N., Mendonça, L.E.T., Confessor, M.V.A., Vieira, W.L.S., Lopes, L.C.S., 2009. Hunting strategies used in the semi-arid region of northeastern Brazil. Journal of Ethnobiology and Ethnomedicine 5, 1–50.

Alves, R.R.N., Nogueira, E., Araújo, H., Brooks, S., 2010a. Bird-keeping in the caatinga, NE, Brazil. Human Ecology 38, 147–156.

Alves, R.R.N., Pereira-Filho, G.A., Vieira, K.S., Santana, G.G., Vieira, W.L.S., Almeida, W.O., 2010b. Répteis e as populações humanas no Brasil: uma abordagem etnoherpetológica. In: Alves, R.R.N., Souto, W.M.S., Mourão, J.S. (Eds.), A etnozoologia no Brasil: Importância, status atual e perspectivas. NUPEEA, Recife, pp. 123–147.

Alves, R.R.N., Souto, W.M.S., 2010. Etnozoologia: conceitos, considerações históricas. In: Alves, R.R.N., Souto, W.M.S., Mourão, J.S. (Eds.), A etnozoologia no Brasil: Importância, Status atual e Perspectivas. NUPEEA, Recife, pp. 21–40.

Amarasinghe, A.A.T., Madawala, M.B., Karunarathna, D.M.S.S., Manolis, S.C., Silva, A., Sommerlad, R., 2015. Human-crocodile conflict and conservation implications of saltwater crocodiles *Crocodylus porosus* (Reptilia: Crocodylia: Crocodylidae) in Sri Lanka. Journal of Threatened Taxa 7, 7111–7130.

Ascensão, F., Mira, A., 2007. Factors affecting culvert use by vertebrates along two stretches of road in southern Portugal. Ecological Research 22, 57–66.

Bager, A., 2003. Repensando as medidas mitigadoras impostas aos empreendimentos viários associados às unidades de conservação. In: Bager, A. (Ed.), Áreas Protegidas—Conservação no âmbito do Cone Sul. Edição do editor, Pelotas, pp. 159–172.

Bager, A., Fontoura, V., 2012. Ecologia de estradas no Brasil—contexto histórico e perspectivas futuras. In: Bager, A. (Ed.), Ecologia de Estradas—Tendências e pesquisas. Editora UFLA, Lavras, pp. 13–33.

Baker, R.O., Timm, R.M., 1998. Management of conflicts between urban coyotes and humans in Southern California. In: Baker, R.O., Crabb, A.C. (Eds.), Proceedings of the 18th Vertebrate Pest Conference. University of Califronia, Davis, pp. 299–312.

Basili, G.D., Temple, S.A., 1999. Dickcissels and crop damage in Venezuela: defining the problem with ecological models. Ecological Applications 9, 732–739.

Beckmann, J.P., Clevenger, A.P., Huijser, M.P., Hilty, J.A., 2010. Safe Passages: Highways, Wildlife and Habitat Connectivity. Island Press, Washington.

Beier, P., 1991. Cougar attacks on humans in the United States and Canada. Wildlife Society Bulletin 19, 403–412.

Bhattarai, B.R., Fischer, K., 2014. Human-tiger *Panthera tigris* conflict and its perception in Bardia national park, Nepal. Oryx 48, 522–528.

Bhupathy, S., Ramesh, C., Bahuguna, A., 2014. Feeding habits of Indian rock pythons in Keoladeo National Park, Bharatpur, India. Herpetological Journal 24, 59–64.

Bisi, J., Kurki, S., Svensberg, M., Liukkonen, T., 2007. Human dimensions of wolf (*Canis lupus*) conflicts in Finland. European Journal of Wildlife Research 53, 304–314.

Bjerke, T., Reitan, O., Kellert, R., 1998. Attitudes towards wolves in southeastern Norway. Society and Natural Resource 11, 169–178.

Bond, A.R., Jones, N.J., 2008. Temporal trends in use of fauna friendly underpasses and overpasses. Wildlife Research 35, 103–112.

Bradley, E.H., Pletscher, D.H., Bangs, E.E., Kunkel, K.E., Smith, D.W., Mack, C.M., Meier, T.J., Fontaine, J.A., Niemeyer, C.C., Jimenez, M.D., 2005. Evaluating wolf translocation as a non-lethal method to reduce livestock conflicts in the northwestern United States. Conservation Biology 19, 1498–1508.

Bruinderink, G.W.T.A.G., Hazebroek, E., 1996. Ungulate traffic collisions in Europe. Conservation Biology 10, 1059–1067.

BSC–Bird Strike Committee USA, 2016. Bird Strike Committee USA Statistics on Birdstrikes. Available at: http://www.birdstrike.org/.

Bulte, E.H., Rondeau, D., 2005. Why compensating wildlife damages may be bad for conservation. Journal of Wildlife Management 69, 14–19.

Burgess, G.H., 2015. ISAF 2015 Worldwide Shark Attack Summary. International Shark Attack File. Available at: http://www.flmnh.ufl.edu/fish/isaf/worldwide-summary.

Butler, J.R.A., Toit, J.T., Bingham, J., 2003. Free-ranging domestic dogs (*Canis familiaris*) as predator and prey in rural Zimbabwe: threats of competition and disease to large wild carnivores. Biological Conservation 115, 369–378.

Cardoso, C.O., Gomes, D.N., Santos, A.G.S., Tavares, A.A., Guzzi, A., 2014. Risco de colisão de aves com aeronaves no aeroporto internacional de Parnaíba, Piauí, Brasil. Ornitologia Neotropical 25, 179–193.

CBEE, 2016. Centro Brasileiro de Estudos em Ecologia de Estradas. Available at: http://cbee.ufla.br/portal/atropelometro/.

Chauhan, N.P.S., 2003. Human casualties and livestock depredation by black and brown bears in the Indian Himalaya, 1989–1998. Ursus 14, 84–87.

Cheng, D., 2016. Scorpion Envenomation: Background, Pathophysiology, Epidemiology. Available at: http://emedicine.medscape.com/article/168230-overview#showall.

Chippaux, J.P., Goyffon, M., 2008. Epidemiology of scorpionism: a global appraisal. Acta Tropica 107, 71–79.

Chowell, G., Díaz-Duenas, P., Bustos-Saldana, R., Mireles, A.A., Fet, V., 2006. Epidemiological and clinical characteristics of scorpionism in Colima, Mexico (2000–2001). Toxicon 47, 738–753.

Conforti, V.A., Azevedo, F.C.C., 2003. Local perceptions of jaguars (*Panthera onca*) and pumas (*Puma concolor*) in the Iguaçu National Park area, south Brazil. Biological Conservation 111, 215–221.

Conover, M.R., 2002. Resolving Human-Wildlife Conflicts. The Science of Wildlife Damage Management. Lewis Publishers, Florida.

Conover, M.R., Pitt, W.C., Kessler, K.K., DuBow, T.J., Sanborn, W.A., 1995. Review of human injuries, illnesses, and economic losses caused by wildlife in the United States. Wildlife Society Bulletin 23, 407–414.

Conway, A.L., Hernandez, S.M., Carroll, J.P., Green, G.T., Larson, L.R., 2015. Local awareness of and attitudes toward the pygmy hippopotamus (*Choeropsis liberiensis*) in the Moa river island complex, Sierra Leone. Oryx 49, 550–558.

Corlatti, L., Hacklaender, K., Frey-roos, F., 2009. Ability of wildlife overpasses to provide connectivity and prevent genetic isolation. Conservation Biology 23, 548–556.

D'Angelo, G.J., D'Angelo, J.G., Gallagher, G.R., Osborn, D.A., Miller, K.V., Warren, R.J., 2006. Evaluation of wildlife warning reflectors for altering white tailed deer behavior along roadways. Wildlife Society Bulletin 34, 1175–1183.

Da Silva, V.M.F., Best, R.C., 1996. Freshwater dolphin/fisheries interaction in the Central Amazon (Brazil). Amazoniana 14, 165–175.

Datiko, D., Bekele, A., 2013. Conservation challenge: human-herbivore conflict in Chebera Churchura National Park, Ethiopia. Pakistan Journal of Biological Sciences 16, 1758–1764.

Dickman, A., 2010. Complexities of conflict: the importance of considering social factors for effectively resolving human–wildlife conflict. Animal Conservation 13, 458–466.

Domit, C., Robert, M.C., Araújo, V., 2011. Integração entre a pesca artesanal Paranaense e a conservação de golfinhos, botos e tartarugas Marinhas. Available at: http://www.terrabrasilis.org.br/ecotecadigital/images/abook/pdf/2sem2015/novembro/Nov.15.54.pdf.

Dunham, K.M., Ghiurghi, A., Cumbi, R., Urbano, F., 2010. Human–wildlife conflict in Mozambique: a national perspective, with emphasis on wildlife attacks on humans. Oryx 44, 185–193.

Ertiban, S.M., 2016. Population status and human conflict of common hippopotamus (*Hippopotamus amphibius*, Linnaeus, 1758) in Boye wetland, Jimma, Ethiopia. American Journal of Scientific and Industrial Research 7, 32–40.

Fahrig, L., 2003. Effects of habitat fragmentation on biodiversity. Annual Review of Ecology, Evolution, and Systematics 487–515.

Farrar, R.O., 2007. Assessing the impact of urban coyote on people ad pets in Austin, Travis County, Texas. In: Nolte, D.L., Arjo, W.M., Stalman, D.H. (Eds.), Proceedings of the 12th Wildlife Damage Management Conference Available at: http://digitalcommons.unl.edu/cgi/viewcontent.cgi?article=1061&context=icwdm_wdmconfproc.

Ferretti, F., Worm, B., Britten, G.L., Heithaus, M.R., Lotze, H.K., 2010. Patterns and ecosystem consequences of shark declines in the ocean. Ecology Letters 13, 1055–1071.

Fischer, A.F., Hazin, F.H.V., Carvalho, F., Viana, D.L., Rêgo, M.G., Wor, C., 2009. Biological aspects of sharks caught off the coast of Pernambuco, Northeast Brazil. Brazilian Journal of Biology 69, 1173–1181.

Fischer, J., Lindenmayer, D.B., 2007. Landscape modification and habitat fragmentation: a synthesis. Global Ecology and Biogeography 16, 265–280.

Foloma, M., 2005. Impacto do conflito Homem e animais selvagens na segurança alimentar na Província de Cabo Delgado, Moçambique. Food and Agriculture Organization of the United Nations, Rome.

Franzini, A.M., Castelblanco-Martínez, D.N., Rosas, F.C.W., da Silva, V.M.F., 2013. What do local people know about Amazonian manatees? Traditional ecological knowledge of *Trichechus inunguis* in the oil province of Urucu, AM, Brazil. Brazilian Journal of Nature Conservation 11, 75–80.

Fritts, S., Stephenson, R., Hayes, R., Boitani, L., 2003. Wolves and humans. In: Mech, D., Boitani, L. (Eds.), Wolves: Behavior, Ecology, and Conservation. University of Chicago Press, Chicago.

Fukuda, Y., Manolis, C., Appel, K., 2014. Management of human-crocodile conflict in the northern territory, Australia: review of crocodile attacks and removal of problem crocodiles. The Journal of Wildlife Management 78, 1239–1249.

Fukuda, Y., Manolis, C., Saalfeld, K., Zuur, A., 2015. Dead or alive? Factors affecting the survival of victims during attacks by saltwater crocodiles (*Crocodylus porosus*) in Australia. PLoS One 10, e0126778.

Gadig, O., 1998. Tubarões. Ática, São Paulo.

Gordon, I.J., 2009. What is the future for wild, large herbivores in human-modified agricultural landscapes? Wildlife Biology 15, 1–9.

Goursi, U., Awan, M., Minhas, R., Ali, U., Kabir, M., Dar, N., 2012. Status and conservation of indian rock python (*Python molurus molurus*) in Deva Vatala National Park, Azad Jammu and Kashmir, Pakistan. Pakistan Journal of Zoology 44, 1507–1514.

Graham, K., Beckerman, A.P., Thirgood, S., 2005. Human-predator-prey conflicts: ecological correlates, prey losses and patterns of management. Biological Conservation 122, 159–171.

Greentree, C., Saunders, G., McLeod, L., Hone, J., 2000. Lamb predation and fox control in south-eastern. Australia Journal of Applied Ecology 37, 935–943.

Grilo, C., Bissonette, J.A., Santos-Reis, M., 2008. Response of carnivores to existing highway culverts and underpasses: implications for road planning and mitigation. Biodiversity and Conservation 17, 1685–1699.

Grupta, A.K., 2004. Origin of agriculture and domestication of plants and animals linked to early Holocene climate amelioration. Current Science 87, 1.

Hartwig, D., 1991. Erfassung der Verkehrsunfälle mit Wild im Jahre 1989 in Nordrhein-Westfalen im Bereich der Polizeibehörden. Zeitschrift für Jagdwissenschaft 37, 55–62.

Holmern, T., Nyahongo, J., Røskaft, E., 2007. Livestock loss caused by predators outside the Serengeti National Park, Tanzania. Biological Conservation 135, 518–526.

Hoogesteijn, R., 2001. Manual on the Problems of Depredation Caused by Jaguars and Pumas on Cattle Ranches. Wildlife Conservation Society, Tikal.

Hoogesteijn, R., Hoogesteijn, A., 2011. Estratégias anti-predação para fazendas de pecuária na América Latina: um guia. Gráfica Editora Microart Ltda, Campo Grande.

Hoogesteijn, R., Hoogesteijn, A., Mondolfi, E., 1993. Jaguar predation vs. conservation: cattle mortality by felines on three ranches in the Venezuelan Llanos. In: Dunstone, N., Gorman, M.L. (Eds.), Mammals as Predators. Proceedings of the Zoological Society of London, vol. 65 , pp. 391–407.

Huijser, M.P., McGowen, P.T., Mfuller, J., Hardy, A., Kociolek, A., Clevenger, A.P., Smith, D., Ament, R., 2007. Wildlife Vehicle Collision Reduction Study. Report to US Congress. US Department of Transportation, Federal Highway Adminstrations, Washington, DC.

International Bird Strike Committee (IBSC), 2006. Recommended Practices No. 1. Standards for Aerodrome Bird/wildlife Control. International Civil Aviation Organisation. Available at: http://www.intbirdstrike.org/Standards_for_Aerodrome_bird_wildlife%20control.pdf.

Iriarte, V., Marmontel, M., 2013. River dolphin (*Inia geoffrensis, Sotalia fluviatilis*) mortality events attributed to artisanal fisheries in the Western Brazilian Amazon. Aquatic Mammals 39, 116–124.

Kaltenborn, B.P., Bjerke, T., Nyahongo, J., 2006. Living with problem animals—self-reported fear of potentially dangerous species in the serengeti region, Tanzania. Human Dimensions of Wildlife 11, 397–409.

Karanth, K.K., Gopalaswamy, A.M., DeFries, R., Ballal, N., 2012. Assessing patterns of human-wildlife conflicts and compensation around a central Indian protected area. PLoS One 7, e50433.

Kasturiratne, A., Wickremasinghe, A.R., Silva, N., Gunawardena, N.K., Pathmeswaran, A., Premaratna, R., Savioli, L., Lalloo, D.G., Silva, H.J., 2008. The global burden of snakebite: a literature analysis and modelling based on regional estimates of envenoming and deaths. PLoS Medicine 5, e218.

Kebede, Y., Tekalign, W., Menale, H., 2016. Conservation challenge: human-herbivore conflict in sodo community managed conservation forest, Wolaita Sodo Zuriya district, southern Ethiopia. Journal of Culture, Society and Development 18, 7–16.

Kellert, S., Black, M., Rush, C., Bath, A., 1996. Human culture and large carnivore conservation in North America. Conservation Biology 10, 977–990.

Kendall, C.J., 2011. The spatial and agricultural basis of crop raiding by the vulnerable common hippopotamus *Hippopotamus amphibius* around Ruaha National Park, Tanzania. Oryx 45, 28–34.

Kormondy, E.J., Brown, D.E., 2002. Ecologia Humana. Editora Atheneu, São Paulo.

Kruuk, H., 2002. Hunter and Hunted: Relationships between Carnivores and People. Cambridge University Press, Cambridge.

Linnell, J.D.C., Odden, J., Smith, M.E., Aanes, R., Swenson, J.E., 1999a. Large carnivores that kill livestock: do "problem indivíduals" really exist? Wildlife Society Bulletin 27, 698–705.

Linnell, M.A., Conover, M.R., Ohashi, T.J., 1996. Analysis of bird strikes at a tropical airport. The Journal of Wildlife Management 60, 935–945.

Linnell, M.A., Conover, M.R., Ohashi, T.J., 1999b. Biases in bird strike statistics based on pilot reports. The Journal of Wildlife Management 27, 997–1003.

Liu, F., McShea, W.J., Garshelis, D.L., Zhu, X., Wang, D., Shao, L., 2011. Human-wildlife conflicts influence attitudes but not necessarily behaviors: factors driving the poaching of bears in China. Biological Conservation 144, 538–547.

Loch, C., Marmontel, M., Simoes-Lopes, P.C., 2009. Conflicts with fisheries and intentional killing of freshwater dolphins (Cetacea: Odontoceti) in the western Brazilian Amazon. Biodiversity and Conservation 18, 3979–3988.

Manville, A.M., 2016. Impacts to birds and bats due to collisions and electrocutions from some tall structures in the United States—wires, towers, turbines, and solar arrays: state of the art in addressing the problems. In: Angelici, F.M. (Ed.), Problematic Wildlife—a Cross-disciplinary Approach. Springer, New York, pp. 415–442.

Marchini, S., Cavalcante, S.M.C., Paula, R.C., 2011. Predadores silvestres e animais domésticos: guia prático de convivência. ICMBio, Brasília.

Massei, G., Quy, R.J., Gurney, J., Cowan, D.P., 2010. Can translocations be used to mitigate human–wildlife conflicts? Wildlife Research 37, 428–439.

Mastro, L., Conover, M.R., Frey, S.N., 2008. Deer–vehicle collision prevention techniques. Human–Wildlife Conflicts 2, 80–92.

McCade, R.E., McCade, T.R., 1984. Of slings and arrows: a historical retrospection. In: Halls, L.K. (Ed.), White-Tailed Deer: Ecology and Management. Stackpole Books, Harrisburg, pp. 19–72.

Mendonça, F.A.C., 2009. Gerenciamento do perigo aviário em aeroportos. Conexão Sipaer 1, 153–174.

Mendonça, L.E.T., Souto, C.M., Andrelino, L.L., Souto, W.M.S., Vieira, W.L.S., Alves, R.R.N., 2011. Conflitos entre pessoas e animais silvestres no semiárido paraibano e suas implicações para conservação. Sitientibus Série Ciências Biológicas 11, 185–199.

Ministério da Saúde, 2016. Incidência de acidentes por escorpiões. Brasil, Grandes Regiões e Unidades Federadas. 2000 a 2015. Available at: http://portalsaude.saude.gov.br/index.php/o-ministerio/principal/leia-mais-o-ministerio/1019-secretaria-svs/vigilancia-de-a-a-z/animais-peconhentos-escorpioes/12-animais-peconhentos-escorpioes/13692-situacao-epidemiologica-dados.

Miranda, E.B.P., Ribeiro Jr., R.P., Strüssmann, C., 2016. The ecology of human-anaconda conflict: a study using internet videos. Tropical Conservation Science 9, 43–77.

Mishra, C., 1997. Livestock depredation by large carnivores in the Indian trans-Himalaya: conflict perceptions and conservation prospects. Environmental Conservation 24, 338–343.

Murphy, J.C., Henderson, R.W., 1997. Tales of Giant Snakes: A Historical Natural History of Anacondas and Pythons. Krieger Publishing Company, Florida.

Naughton-Treves, L., 1997. Farming the forest edge: vulnerable places and people around Kibale National Park, Uganda. The Geographical Review 87, 27–46.

Nogueira, M., Alves, R.R.N., 2016. Assessing sea turtle bycatch in Northeast Brazil through an ethnozoological approach. Ocean & Coastal Management 133, 37–42.

Nyhus, P., Fischer, H., Madden, F., Osofsky, S.A., 2003. Taking the bite out of wildlife damage. Conservation in Practice 4, 37–40.

Ohrens, O., Treves, A., Bonacic, C., 2016. Relationship between rural depopulation and puma-human conflict in the high Andes of Chile. Environmental Conservation 43, 24–33.

Okello, M.M., D'Amour, D.E., 2008. Agricultural expansion within Kimana electric fences and implications for natural resource conservation around Amboseli National Park, Kenya. Journal of Arid Environments 72, 2179–2192.

Oli, M.K., Taylor, I.R., Rogers, G., 1994. Snow leopard *Panthera uncia* predation of livestock: an assessment of local perceptions in the Annapurna conservation area, Nepal. Biological Conservation 68, 63–68.

Ozkan, O., Uzun, R., Adiguzel, S., Cesaretli, Y., Ertek, M., 2008. Evaluation of scorpion sting incidence in Turkey. Journal of Venomous Animals and Toxins Including Tropical Diseases 14, 128–140.

Palmeira, F.B.L., Barrella, W., 2007. Conflitos causados pela predação de rebanhos domésticos por grandes felinos em comunidades quilombolas na Mata Atlântica. Biota Neotropica 7, 119–128.

Palmeira, F.B.L., Crawshaw Jr., P.G., Haddad, C.M., Ferraz, K.M., Verdade, L.M., 2008. Cattle depredation by puma (*Puma concolor*) and jaguar (*Panthera onca*) in south-western Brazil. Biological Conservation 141, 118–125.

Pinedo, M.C., Polacheck, T., 2004. Sea turtle bycatch in pelagic longline sets off southern Brazil. Biological Conservation 119, 335–339.

Pitman, M.R.P.L., Oliveira, T.G., Paula, R.C., Indrusiak, C., 2002. Manual de identificação, prevenção e controle de predação por carnívoros. Edições IBAMA, Brasília.

Plagányi, É.E., Butterworth, D.S., 2002. Competition with fisheries. In: Perrin, W.F., Würsig, B., Thewissen, H. (Eds.), Encyclopedia of Marine Mammals. Academic Press, San Diego, pp. 268–273.

Polisar, J., Maxit, I., Scognamillo, D., Farrell, L., Sunquist, M.E., Eisenberg, J.F., 2003. Jaguars, pumas, their prey base, and cattle ranching: ecological interpretations of a management problem. Biological Conservation 109, 297–310.

Rajpurohit, K.S., Krausman, P.R., 2000. Human–sloth-bear conflicts in Madhya Pradesh, India. Wildlife Society Bulletin 28, 393–399.

Redpath, S.M., Bhatia, S., Young, J., 2015. Tilting at wildlife: reconsidering human–wildlife conflict. Oryx 49, 222–225.

Redpath, S.M., Young, J., Evely, A., Adams, W.M., Sutherland, W.J., Whitehouse, A., Amar, A., Lambert, R.A., Linnell, J.D.C., Watt, A., Gutierrez, R.J., 2013. Understanding and managing conservation conflicts. Trends in Ecology & Evolution 28, 100–109.

Rosas-Ribeiro, P.F., Fernando Rosas, F.C.W., Zuanon, J., 2012. Conflict between fishermen and giant otters *Pteronura brasiliensis* in western Brazilian Amazon. Biotropica 44, 437–444.

Sangay, T., Vernes, K., 2008. Human–wildlife conflict in the kingdom of Bhutan: patterns of livestock predation by large mammalian carnivores. Biological Conservation 141, 1272–1282.

Santos, A.J.B., Gallo, B., Giffoni, B., Baptistotte, C., Lima, E., Sales, G., Lopes, G.G., Becker, H., Castilho, J.C., Thomé, J.C.A., Marcovaldi, M.C.A., Mendilaharsu, M.L.M.L., Barata, P.C.R., Sforza, R., 2011. Plano de ação nacional para conservação das tartarugas marinhas, primeira. ICMBio, Brasilia.

Sechele, M.L., Nzehengwa, D.M., 2002. Human predator conflicts and control measures in North-west district, Botswana. In: Loveridge, A.J., Lynam, T., Macdonald, D.W. (Eds.), Lion Conservation Research, Workshop 2: Modelling Conflict, vol. 2. WildCRU, Oxford.

Simberloff, D., 1998. Flagships, umbrellas, and keystones: is single-species management passé in the landscape era? Biological Conservation 83, 247–257.

Souza, J.B., Alves, R.R.N., 2014. Hunting and wildlife use in an Atlantic Forest remnant of northeastern Brazil. Tropical Conservation Science 7, 145–160.

Stander, P.E., 1990. A suggested management strategy for stock-raiding lions in Namibia. South African Journal of Wildlife Research 20, 37–43.

Strüssmann, C., 1997. Hábitos alimentares da sucuri-amarela, *Eunectes notaeus* Cope, 1862, no Pantanal matogrossense. Biociencias 5, 35–52.

Stubblefield, C.H., Shrestha, M., 2007. Status of Asiatic black bears in protected areas of Nepal and the effects of political turmoil. Ursus 18, 101–108.

Szpilman, M., 1998. Seres marinhos perigosos: Guia prático de identificação, prevenção e tratamento. Instituto Ecológico Aqualang, Rio de Janeiro.

Thirgood, S., Woodroffe, R., Rabinowitz, A., 2005. The impact of human–wildlife conflict on human lives and livelihoods. In: Woodroffe, R., Thirgood, S., Rabinowitz, A. (Eds.), People and Wildlife. Conflict or Coexistence?. Cambridge University Press, Cambridge, pp. 13–26.

Thirgood, S.J., Redpath, S.M., Newton, I., Hudson, P., 2000. Raptors and red grouse: conservation conflicts and management solutions. Conservation Biology 14, 95–104.

Thorpe, J., 2012. 100 Years of Fatalities and Destroyed Civil Aircraft Due to Bird Strikes. World Birdstrike Association, Amsterdam. Available at: http://worldbirdstrike.com/Stavanger/100%20years%20of%20fatalities%20and%20destroyed%20civil%20aircfaft%20due%20to%20bird%20strikes%20Paper.pdf.

Thorpe, J., 2016. Conflict of wings-birds versus aircraft. In: Angelici, F.M. (Ed.), Problematic Wildlife-Across-Disciplinary Approach. Springer, New York, pp. 443–463.

Toledo, G.A.C., Campos, B.A.T.P., Feitosa, I.C.S., Souto, A.S., Alves, R.R.N., 2010. Interações entre pescadores artesanais e o boto-cinza (Sotalia guianensis–Van Bénéden, 1864) na região de Baía Formosa, Rio Grande do Norte – Brasil. In: Alves, R.R.N., Souto, W.M.S., Mourão, J.S. (Eds.), A Etnozoologia no Brasil: importância, status atual e perspectivas, primeira ed. NUPEEA, Recife, pp. 277–296.

Treves, A., Naughton-Treves, L., 2005. Evaluating lethal control in the management of human-wildlife conflict. In: Woodroof, R., Thirgood, S., Rabinowitz, A. (Eds.), People and Wildlife, Conflict or Coexistence?. Cambridge University, Cambridge, pp. 86–106.

Trinca, C.T., Ferrari, S.F., 2006. Caça em assentamento rural na Amazônia matogrossense. In: Jacobi, P., Ferreira, L.C. (Eds.), Diálogos em ambiente e sociedade no Brasil. Annablume, Indaiatuba, pp. 155–167.

Vidolin, G.P., Moura-Britto, M., Braga, F.G., Cabeças-Filho, A., 2004. Avaliação da Predação a animais domésticos por felinos de grande porte no Estado do Paraná: implicações e estratégias conservacionistas. Cadernos da Biodiversidade 4, 50–58.

Vié, J.C., Hilton-Taylor, C., Stuart, S.N., 2009. Wildlife in a Changing World – an Analysis of the 2008 IUCN Red List of Threatened Species, first ed. Lynx Edicions, Barcelona.

WHO, 2010. Guidelines for the Production, Control and Regulation of Snake Antivenom Immunoglobulins. Available at: http://www.who.int/bloodproducts/snake_antivenoms/snakeantivenomguideline.pdf.

Woodroffe, R., Lindsey, P., Romanãch, S., Stein, A., Ranah, S.M.K., 2005. Livestock predation by endangered African wild dogs (*Lycaon pictus*) in northern Kenya. Biological Conservation 124, 225–234.

Yamada, A., Fujioka, M., 2010. Features of planted cypress trees vulnerable to damage by Japanese black bears. Ursus 21, 72–80.

Young, J.C., Marzano, M., White, R.M., Mccracken, D.I., Redpath, S.M., Carss, D.N., Quine, C.P., Watt, A.D., 2010. The emergence of biodiversity conflicts from biodiversity impacts: characteristics and management strategies. Biodiversity and Conservation 19, 3973–3990.

CHAPTER

23

Biological Predispositions and Individual Differences in Human Attitudes Toward Animals

Pavol Prokop[1,2], *Christoph Randler*[3]

[1]Trnava University, Trnava, Slovakia; [2]Slovak Academy of Sciences, Bratislava, Slovakia; [3]Didaktik der Biologie, Tübingen, Germany

Animals play inseparable roles in human life. Everywhere around us animals are presented in clips, books, movies, or on symbols of various cultures since ancient times (Walsh, 2009). Animals are also significant sources of human foods for at least the past two million years (Ungar and Teaford, 2002). Cultures without any doubt play an important role in perception of animals, although our biological predispositions seems to be more than significant (Herzog and Burghardt, 1988). The focus here will be specifically on the biological influences of human attitudes to animals, as well as on the role of education in forming children's attitudes and their perception of living creatures.

HOW DOES EVOLUTION SHAPE HUMAN ATTITUDES TO ANIMALS?

Although the genus *Homo* exists from roughly about 2.8 million years (Villmoare et al., 2015), the history of interactions with animals that could have influenced the evolution of the human brain is much older. Approximately 150 million years ago, snakes hunted for small placental mammals, from which the order Primates originated around 85 million years later (Zhang et al., 2008). It is reasonable to assume that individuals who were able to detect and escape snakes could have a survival advantage because they were able to reproduce and transfer their genes to the following generation. It is believed that the origin of certain emotions, such as the emotion of fear, which plays an important role in human attitudes to animals, lies somewhere in these early interactions with snakes (Isbell, 2006, 2009; Öhman and Mineka, 2001, 2003). Indeed, the amygdala, the ancient part of the brain located deep and medially within the temporal lobes of the brain, could have evolved under pressure from snakes on small mammals. A number of studies have shown that the amygdala is involved in fear conditioning in a variety of mammals, including rats, mice, rabbits, and monkeys (see LeDoux, 2012 for a review) providing evidence for the origin

Ethnozoology
http://dx.doi.org/10.1016/B978-0-12-809913-1.00023-5

of fear in our evolutionary past. About 60 million years ago, snakes introduced venom, as a new powerful weapon in the coevolutionary arms race with their mammalian prey. Although actual estimates suggest that human mortality caused by snake bite is relatively low (roughly about 20,000–125,000 deaths worldwide, see Chippaux, 1998; Kasturiratne et al., 2008), it is argued that these estimates are based on incomplete data (Warrell, 2010), partly because reporting is not mandatory in many regions in the world. In contrast to modern humans living in cities, however, our ancestors were in daily contact with the natural environment, making the likelihood of snake bite quite high. Moreover, a lack of medical help necessarily led to higher mortality rates than at present.

COEVOLUTION WITH SNAKES ENHANCES VISUAL ATTENTION TO DANGEROUS STIMULI

In light of the deadly threat represented by snakes on their prey, it is reasonable to assume that visual attention providing self-protection should have evolved in mammals (Öhman et al., 2001a). Neurobiological evidence supports this idea because amygdala tunes visual brain areas for rapid perception of fear-related stimuli (Phelps et al., 2006). In current research, Van Le et al. (2013) implanted electrodes in the individual neurons of the brains of three macaque monkeys (*Macaca fuscata*) and then made recordings from the neurons where the monkeys were exposed to four sets of pictures: snakes, angry monkey faces, monkey hands, and other geometrical shapes, such as circles. Researchers have hypothesized that the neurons placed in a specific part of the brain, unique in primates, will respond particularly quickly to images of snakes. Of the 91 neurons tested with all stimuli, 37 were more sensitive to snake images, 26 to angry face images, 17 to hand images, and 11 to shapes images. In line with their hypothesis, the neurons that responded specifically to snakes were more numerous and more sensitive to the relevant stimuli than neurons in the other three groups. Interestingly, the abundance of neurons sensitive to angry faces was in the second place suggesting that recognition of emotions by other members of the social group is advantageous in terms of enhanced survival and reproduction. Humans, similar to nonhuman primates, are also sensitive to detection of visual cues in the faces of conspecifics. In all probability, accurate recognition of the presence of anger, fear, disgust, and other emotions is necessary since significant parts of violent deaths can be attributed to conflicts between conspecifics (Daly and Wilson, 1988)—similarly as with chimpanzees (Wrangham and Peterson, 1996). Threatening faces are actually detected by humans more quickly than neutral or friendly faces (for a review see LoBue and Rakison, 2013; Öhman, 2009; Öhman et al., 2001a,b).

Snakes promote rapid visual detection by humans and this innate ability was in all probability shaped by the predator (the snake)–prey coevolution. In a series of experiments, Öhman et al. (2001a) compared the detection of a fear-relevant target (a snake or a spider) among eight fear-irrelevant distractors (flowers or mushrooms) with that of a fear-irrelevant target (a flower or a mushroom) among fear-relevant distractors (snakes or spiders). Snakes and spiders were detected more quickly than fear-irrelevant targets. These experiments were successfully replicated several times (reviewed by LoBue and Rakison, 2013), including children (LoBue and DeLoache, 2008) who also detect snakes sooner than, for example, flowers, albeit children's detection time is longer than that of adults. In addition, infants at 5 months of age look longer at a schematic image of a spider than a partly or completely scrambled image of a spider but do not do so for schematic and scrambled images of a flower (Rakison and Derringer, 2008).

Previous neurobiological evidence clearly suggests that nonhuman primates should be afraid of snakes. Indeed, wild monkeys are still attacked by snakes and do manifest strong fearful responses to them (Cook and Mineka, 1991). These responses seem to be learned, rather than innate, because laboratory reared monkeys lacking any experience with snakes show no fear (Mineka et al., 1980). Now classic experiments performed by Cook and Mineka (1990, 1991) determined that monkeys were able to quickly acquire fear of a predator by observing other monkeys expressing fear in interaction with the predator. When laboratory reared monkeys observed a wild-reared monkey displaying fear of a live and toy snake, they were rapidly conditioned to fear snakes (Cook and Mineka, 1990). Additional experiments with toy snakes and crocodiles as predators, and flowers and rabbits as harmless objects, provided further evidence for a predisposition to quickly learn fearful responses in dangerous situations: fear responses were conditioned only to predators, but not to harmless stimuli (Cook and Mineka, 1991). Research on humans also revealed exciting findings: 7 to 18-month-old infants looked longer at clips of snakes paired with a frightened human voice than at clips of snakes paired with a happy human voice (DeLoache and LoBue, 2009). In light of the fact that attention enhances learning (Shirey and Reynolds, 1988), children would learn from stories with frightened voices, which are typical for stories with predators, more than from other stories. Clearly, learned avoidance of snakes could have an adaptive advantage, because children could more easily avoid potential predators. These results are in agreement with earlier suggestions by Martin Seligman (1971) who predicted that common fears reflect evolutionarily prepared learning to fear events and situations that have provided survival threats in our evolutionary past. These results also highlight evolutionary significance in the development of fear of prototypical predators such as snakes.

AESTHETIC PREFERENCES FOR ANIMALS

Everyday life is accompanied by spontaneous ratings of common things as "beautiful" or "ugly." We rate almost everything around us: buildings, flowers, landscapes, works of art, other humans, as well as nonhuman animals. Visual perception of beautiful objects activates the prefrontal cortex (Cela-Conde et al., 2004), providing neurobiological evidence that ugly objects are perceived differently from beautiful objects. From an evolutionary perspective, a preference for beauty is advantageous, because beauty is associated with fitness-enhancing traits (for a review, see Rusch and Voland, 2013; Voland and Grammer, 2003). Colorfulness can be associated, for example, with parasite avoidance in birds (Hamilton and Zuk, 1982). A peacock's train can be a costly signal of a male quality (Zahavi and Zahavi, 1997) and female barn swallows (*Hirundo rustica*) have an esthetic preference for male tail feathers that are symmetrical (Møller, 1992). Similarly, people have an esthetic preference for symmetrical faces (Thornhill and Gangestad, 1999) and more physically attractive people have a higher reproductive success (Jokela, 2009; Pawlowski et al., 2000; Prokop and Fedor, 2011). Perceived beauty influences feeding preferences—what is beautiful should taste good (Prokop and Fančovičová, 2012a), suggesting that these preferences also work in nonsexual contexts. Esthetic preferences of animals may also have a possible fitness-relevant value, by favoring harmless animals over dangerous ones.

WHY ARE SOME ANIMALS UGLY?

Research suggests that phylogenetically and behaviorally distant or dissimilar animals are often perceived negatively (Batt, 2009; Bjerke and Østdahl, 2004; Kellert, 1984, 1993). Negative perception is associated with ugliness. Ugly animals receive lower preference scores compared with cute animals (Driscoll, 1995; Prokop et al., 2010a,b). Specifically, rats, snakes, cockroaches, mosquitoes, spiders, slugs, earthworms, bats, frogs, crocodiles, or hyenas received lower preference scores than, for example, birds, squirrels, butterflies, horses, hedgehogs, zebras, or gazelles (Almeida et al., 2014; Batt, 2009; Bennett-Levy and Marteau, 1984; Bjerke and Østdahl, 2004; de Pinho et al., 2014; Driscoll, 1995; Kaltenborn et al., 2006b; Knight, 2008; Schlegel and Rupf, 2010). These preferences can be seen among kindergarten children suggesting that the predisposition to have a preference for animals sharing certain similarities (phylogenetic or behavioral) with humans emerges early in childhood (Borgi and Cirulli, 2015). As concerns invertebrates, parasites clearly show the lowest scores compared with nonparasitic species (Prokop et al., 2010a,b) with spiders always scoring the lowest while butterflies receive the top rankings (Batt, 2009; Schlegel and Rupf, 2010; Schlegel et al., 2015).

The ultimate explanation of the role of perceived beauty in the ranking of animals can be partly explained by the potential threat that ugly animals possess to humans. Rats, cockroaches, and mosquitoes transmit certain serious diseases to humans such as plague, leptospirosis, malaria, or salmonella (Baumholtz et al., 1997; Morelli et al., 2010; Murray et al., 2012). It can therefore be argued that people are more prone to consider these harmful animals as disgusting and/or ugly thereby reducing the risk of contamination (Prokop et al., 2010a,b). It may be that we are able to associate these animals with danger more easily than with other animals (prepared learning, Seligman, 1971). To support this idea, it was repeatedly found that people who are more vulnerable to diseases consider vectors/reservoirs of diseases as more dangerous than those who are healthier and/or less vulnerable to diseases (Prokop et al., 2010a,b,c). Negative attitudes to spiders are still disputable. Although spider phobia is one of the most common phobias in the world (about 3.5% of US people, Fredrikson et al., 1997); mortality rates caused by spiders are extremely small (99 deaths over 20 years in the United States, Forrester and Stanley, 2004). Despite this fact, spiders, similar to snakes, meet with the strong attention of both children and adults (LoBue, 2010; Öhman et al., 2001a). The possibility that spiders caused danger to humans in our evolutionary past cannot be excluded, of course, which may be the reason why South African students living in places where early humans originated and where the diversity of spiders is high, manifest a stronger fear of spiders than Slovak students living in Europe where the diversity of spiders is much lower (Prokop et al., 2010d). Furthermore, it appears that humans do not misidentify spiders with other arthropods, but the fear of spiders is very specific (Gerdes et al., 2009). There are also cultural explanations suggesting, of course, that spiders are associated with disgust, illness, and infection (Davey, 1994a) by Europeans from the Middle Ages onward. Van Strien et al. (2014) recently found that pictures of snakes received earlier visual attention than pictures of spiders suggesting that neural circuitry for defense behavior was initially designed to deal with snakes, as dangerous predators of mammals (Isbell, 2006). This evidence suggests that the origin of spider fear is still unexplored and no firm conclusions can be made as yet.

Terrestrial predators such as hyenas receive low attractiveness/high fear scores, while lions (de Pinho et al., 2014; Kaltenborn et al., 2006b) or leopards (Batt, 2009) are much more valued by people. These predators obviously do not cause phobias (in contrast to snakes), despite being dangerous for humans (Treves and Naughton-Treves, 1999). This apparent inconsistency can be explained by an evolutionary perspective. While carnivore predators occurred only about 40–60 million years ago (Heinrich et al., 2008; Polly et al., 2006), snakes are phylogenetically much older. It may be that the coevolution between humans and our predecessors with carnivore predators is too short for developing specific phobias. It does not mean, however, that carnivores do not enhance attention. By recording eye movements, Yorzinski et al. (2014), for example, determined that participants visually detected dangerous animals (snakes and lions) faster than nondangerous animals (lizards and impalas). Similarly as it was found in detecting snakes by LoBue and DeLoache (2008), the participants were slower to locate nondangerous animals because they spent more time looking at dangerous distractors, a process known as delayed disengagement, and looked at a larger number of dangerous distractors. In summary, dangerous animals capture and maintain attention in humans, suggesting that coevolution with predators shaped the evolution of neural architecture and visual abilities in modern humans.

Evolutionary forces underlying the (dis)liking of certain animals are further strengthened by cultural beliefs. Traditional beliefs are passed on from generation to generation and designate specific attitudes and behaviors toward certain groups of animals (Herzog and Burghardt, 1988). Cultural beliefs about animals often include fear, respect, and abhorrence (Costa Rego et al., 2015; Davey et al., 1998; Fredrikson et al., 1997; Herzog and Burghardt, 1988). Mofu people in Africa, for example, treat ants in genus *Dorylus* with respect and fear, by bending over and touching their chest when they encounter them, because they are considered the prince of the insect fauna (Santos and Antonini, 2008). Ugly animals such as reptiles, frogs, bats, or spiders are often emblazoned with myths (Ceríaco, 2012; Costa Rego et al., 2015; Prokop et al., 2009a,b; Tarrant et al., 2016), which can itself be extremely dangerous for the survival of the particular species. First, unpopular animals receive less funding for conservation than more popular animals (Tarrant et al., 2016 and references therein). Second, the killing of ugly animals such as frogs or reptiles by people has been documented by various researchers (e.g., Alves et al., 2012; Ballouard et al., 2013; Ceríaco, 2012; Pagani et al., 2007; Prokop and Fančovičová, 2012b; Whitaker and Shine, 2000). It is not actually apparent how education is associated with beliefs concerning myths about animals since the results are mixed. In certain cases less educated people hold more beliefs about myths (Tarrant et al., 2016), while others found opposite patterns (Ceríaco, 2012) or that there was no difference in beliefs between biology majors and nonmajors (Prokop et al., 2009a). According to certain authors, less educated people engage in direct persecution of animals more than people with a higher education level (Ceríaco, 2012; but see Prokop and Fančovičová, 2012b) suggesting that education is associated with positive behavior toward animals. It seems that belief in myths also has an adaptive context, because more disease-vulnerable children manifest stronger beliefs for untrue myths than less disease-vulnerable children (Prokop and Kubiatko, 2014), which can be explained by the stronger avoidance of potentially dangerous animals by those people who are more vulnerable to contamination or injury from animals.

WHY CERTAIN HARMLESS ANIMALS ARE CONSIDERED UGLY?

A number of researchers have observed that harmless snakes (Arrindell et al., 2003; Bennett-Levy and Marteau, 1984) or worm-like invertebrates (Batt, 2009; Prokop and Fančovičová, 2010a; Prokop et al., 2010a,b; Schlegel et al., 2015) receive low attractiveness/high fear scores despite not being dangerous to humans. In a similar vein, Rozin et al. (1986) demonstrated that if a heat sterilized dead cockroach was briefly dipped into a glass containing a previously desirable beverage, that beverage became psychologically contaminated and hence unacceptable. This suggests that some cues of contamination significantly influence people's behavior. Superficial perception of potentially harmful stimuli is favored by natural selection in order to minimize the likelihood of failing to register the presence of actual danger (smoke detector principle, Nesse, 2005). Although the avoidance of a worm-like invertebrates could be perceived as "erroneous" (a false positive error), it is still less risky than the erroneous detection of a truly harmful snake (a false negative error). The smoke detector principle is helpful in understanding negative attitudes to certain harmless animals, particularly if they superficially resemble certain harmful animals.

WHY CERTAIN ANIMALS ARE CONSIDERED CUTE?

Humans have an innate propensity to affiliate with other organisms (biophilia hypothesis, see Wilson, 1984). Although the biophilia hypothesis is not restricted to animals (also to plants and habitats), the focus of this chapter will be on the animal domain.

From early life children are introduced to animals through cartoons, books, and everyday experiences. Their preferences for animals seem to be innate, rather than taught (Jacobs, 2009). Simion et al. (2008) found, for example, that 2-day-old babies had a preference for looking at biological motion as opposed to nonbiological motion. One film clip shown to the babies depicted a dozen spotlights representing the joints and contours of a walking hen. Another clip depicting a dozen spotlights that moved with the same characteristics (in terms of angles and speed) was generated randomly. As expected, most babies had a preference for the clip representing the walking hen. Similar results were obtained with 11- to 40-month-old children. By allowing children (and their parents) to freely interact with various toys and certain live animals, Lobue et al. (2013) found that children, as well as their parents interacted with the animals more often than with the toys. Children also talked about the animals more than the toys and asked more questions about them. These examples clearly support the idea that humans have an innate tendency to interact with live animals more than with attractive toys. Interestingly, animals are helpful to children with autistic-spectrum disorders (Katcher, 2002; O'Haire et al., 2013).

Phylogenetic and behavioral similarity is not the only mechanism that can explain human affinity to certain animals (Herzog and Burghardt, 1988; Serpell, 2004). Some anthropometric features, particularly big eyes, a round face and a small nose and mouth, are morphological traits associated with "cuteness." Lorenz (1943) was the first to recognize that these traits (baby schema) enhance parental care over infants. Innate responses to baby faces are adaptive because caregiving enhances infant survival. Interestingly, positive responses to infantile facial configurations emerge early during development (evidence come from 3- to 6-year-old children, see Borgi and Cirulli, 2013; Borgi et al., 2014).

In line with the original suggestions of Lorenz (1943), positive responses to infantile conspecifics are not restricted to conspecifics. Several empirical works repeatedly demonstrated that humans show a cute response to nonhuman animals with these responses being measured by showing real animals, representations of animals in cartoons, as well as with stuffed or toy animals (for a review, see Borgi and Cirulli, 2016). By comparing various bird species, Lišková et al. (2015) found that shorter necks and bigger eyes were associated with a higher attractiveness ranking, once again suggesting the prominent role of baby schema in perceived attractiveness. Furthermore, esthetic perception of animals correlates with an individual's willingness to protect them (Gunnthorsdottir, 2001; Knight, 2008; Martín-López et al., 2007; Prokop and Fančovičová, 2013a). Martín-López et al. (2008) in their metaanalysis of 60 recent papers on the economic valuation of biodiversity showed that a willingness to pay for biodiversity conservation increases in favor of conservation species with anthropomorphic and anthropocentric (e.g., larger eyes and weight) characteristics instead of more relevant, scientific factors.

Another interesting example is the story of pets (Hinde and Barden, 1985). These authors measured toy teddy bears between 1900 and 1980. From the beginning, the snout became shorter, this being measured by the distance between the tip of the snout and the back of the head. This "evolutionary" change was nearly complete by the 1930s. A larger forehead also developed over time. Although teddy bears do not reproduce themselves, this kind of evolution of traits can be seen as a form of selection because the cuter teddy bears sell better.

Positive responses to infantile cues on nonhuman animals could be by-products of parental care originally focused on their own infants and can be particularly strong in humans, since parental care is prolonged as compared with other nonhuman primates (Kramer, 2011). Apart from mechanisms underlying positive perception of infantile cues in humans, baby schema is helpful in understanding why certain animals receive high likability ratings and why some animals are more frequently kept as pets compared with others. Prokop et al. (2008) found, for example, with a large sample of Slovak children that about 30%–40% of children of various age groups reported having a dog at home, whereas less than 1% of children reported having an invertebrate as a pet. Clearly, the baby schema cannot be applied to the perception of the majority of invertebrates.

Mammalian facial traits fit with positive rankings of cuteness. Mammals are generally highly ranked by people (Prokop and Fančovičová, 2013a) and are frequently used as flagship species in nature protection (Clucas et al., 2008). Being a mammal can also be an advantage because mammals are phylogenetically similar to humans and humans tend to be more prosocial toward others who are more similar to them (e.g., Allen et al., 2002a,b; Stephan and Finlay, 1999). A similar expression of emotions can be observed in nonhuman mammals (Darwin, 1872), which in all probability allow humans to better recognize them, thus promoting human–animal bonds. Russell (2003) found, for example, that humans are able to recognize the emotional expressions of horses. Correct recognition of animal emotions could be advantageous for hunters who have to actively plan their strategies for capturing animal prey. Moreover, the tendency to anthropomorphize animals (that is, assigning human traits and states to animals) can be at least partly explained by the ability to recognize the same emotions in animals (Jacobs, 2009). Anthropomorphism could have later enabled the domestication of companion and agricultural animals (Serpell, 2003). In summary, successful encoding of animal emotions along with anthropomorphism provides a meaningful basis for why humans have a preference for animals similar to them.

THE ROLE OF COLORS IN ATTITUDES TOWARD ANIMALS

Colors play important roles in human life (Elliot and Maier, 2014); it is therefore reasonable to suggest that they can influence human perception and attitudes toward animals. Evidence suggests that red and black colors are associated with aggression, dominance (Hill and Barton, 2005; Little and Hill, 2007), and physical attractiveness (Elliot and Niesta, 2008; Roberts et al., 2010) in humans. Certain animals have aposematic, warning coloration by means of which they advertise defensive mechanisms to predators, while others are inconspicuous and cryptic (Ruxton et al., 2004). Although experimental data are scarce, humans seem to perceive warning coloration as more highly conspicuous, than (confusing) natural predators (Bohlin et al., 2012). Aposematically colored species (Marešová et al., 2009b; Prokop and Fančovičová, 2013a), brightly colored butterflies (Barua et al., 2012), and penguins with a warm color (Stokes, 2007) are perceived as more beautiful than other species. These preferences seem to be cross-culturally universal (Frynta et al., 2011; Marešová et al., 2009a). Frynta and his colleagues (2010; Lišková and Frynta, 2013; Lišková et al., 2015) have determined that, apart from body shape and body mass, blue and yellow colors in particular make birds more attractive than other colors. The green color, in contrast, was perceived negatively, which can be explained by its cryptic function at least in certain birds. Interestingly, the color red only played an important role in esthetic preferences of snakes (Marešová et al., 2009b), not of birds (Lišková et al., 2015). It is probable that colorful objects capture human attention thereby increasing its esthetic value. Preferences for certain colors can be derived from preferences for ripe fruits (Prokop and Fančovičová, 2012a); consider, for example, low preferences for green fruits (a parallel with green birds, see above) can be explained by their low edibility (Prokop and Fančovičová, 2014). These observations have implications for the selection of flagship species (Prokop and Fančovičová, 2013a), individual willingness to protect animals (Knight, 2008; Prokop and Fančovičová, 2013a), and their being kept in zoo populations (Frynta et al., 2010; Marešová and Frynta, 2008).

RARITY AND ATTITUDES TOWARD ANIMALS

Humans manifest a tendency to favor rare animals over common ones (Herzog and Burghardt, 1988). Being rare and endemic are often viewed as the most important species attributes (Morse-Jones et al., 2012; Takahashi et al., 2012; Veríssimo et al., 2009). Endangerment and rarity have been significantly associated with an individual's reported willingness to pay for species preservation (DeKay and McClelland, 1996; Samples et al., 1986) with one study determining that endemic species were ranked as the highest conservation priority, followed by species with declining numbers and species of economic importance (Meuser et al., 2009). Certain additional factors such as, for example, age (Frew et al., 2017) and education can moderate preferences for endangered species (Kellert, 1984/1985). Species rarity may also be a factor utilized by zoos, which tend to keep and breed rare species preferentially in their collections (Frynta et al., 2010).

The origin of individual preferences for endangered species has received, however, limited attention. Herzog and Burghardt (1988) have speculated about certain evolutionary components underlying these preferences although evidence demonstrating that rare species played a role in the rituals of traditional societies still does not explain their ultimate origin. One can speculate that preferences for rare species can be a product of frequency-dependent sexual selection. This process suggests that the fitness of a phenotype increases as it becomes rarer (Allen and Clarke, 1984). Janif et al. (2014), for example, found that clean shaven male faces were the least attractive for females when clean shaven faces were the most common compared with males with full beards. It may be that preferences for rare phenotypes influence our preferences for gems, gold, pearls, or exceptional clothes, as well as for rare animals. Although this idea is currently purely speculative, further research may provide some empirical or experimental tests concerning its validity.

GENDER DIFFERENCES IN PREFERENCES FOR ANIMALS

Preferences for animals are gender specific (Kellert and Berry, 1987). The differences, however, might only be small and may be generally overstated (see for discussion: Herzog, 2007). In most large-scale studies, females have higher scores in pro-animal attitudes than males (e.g., Herzog, 2007; Herzog et al., 1991; Kellert and Berry, 1987; Pagani et al., 2007; Stanisstreet et al., 1993). Females manifest stronger preferences for cute, popular animals such as common pets while males obviously score higher in preferences for less attractive, unpopular animals (Bjerke et al., 2001; de Pinho et al., 2014; Lindemann-Matthies, 2005; Prokop and Tunnicliffe, 2010). Preferences for cute animals can stem from higher parental investment of human females compared with males, thereby enhancing preferences for baby-like faces. In contrast, females' low preferences for unpopular animals could be influenced by their potential danger. Females show a higher fear of predators, for example, compared with males (Alves et al., 2014; de Pinho et al., 2014; Kaltenborn et al., 2006a; Prokop and Fančovičová, 2013b). These fears possibly result from their lower exposure to predators (de Pinho et al., 2014) which can have a long-lasting evolutionary history considering that females were more sedentary and males as hunters were more frequently exposed to various predators (Prokop and Fančovičová, 2010b; Røskaft et al., 2003).

Certain researchers have suggested that a higher fear of large carnivore predators in females can be explained by females' lower physical condition compared with the physical condition of males (Prokop and Fančovičová, 2010b, 2013b; Røskaft et al., 2003). Indeed, females and children who are physically weaker than males are less likely to survive predatory attacks of large carnivores (Treves and Naughton-Treves, 1999). The disease avoidance hypothesis suggests that vulnerability to infectious diseases makes people more sensitive to potential danger. Indeed, females are more disease-vulnerable than males and more disease-vulnerable people perceive animals as more dangerous than less disease-vulnerable people (Prokop and Fančovičová, 2013b). These two hypotheses can be viewed as compatible, because disease vulnerability is associated with a poorer perceived physical condition (Prokop and Fančovičová, 2013b). The parental investment hypothesis proposes that females take care of children and have to be more sensitive to danger in order to protect their offspring (Røskaft et al., 2003). This hypothesis has not received empirical support as yet (Prokop and Fančovičová, 2010b).

Another domain explaining certain gender differences in attitudes toward animals is empathy. Females score higher on measures of empathy (Alterman et al., 2003) and are able to encode and decode facial emotions (McLure, 2000) better than males. These abilities were possibly favored by natural selection more in females, considering their higher vulnerability to attacks by predators (see above) or by other people (particularly by males). Apart from the evolutionary origins of gender differences in socioemotional behavior, it is reasonable to predict that females will be more sensitive to animals' needs. Females actually scored higher in attitudes on the Animal Attitudes Scale (AAS) (Herzog, 2011; Mathews and Herzog, 1997) and empathy scores positively correlated with the AAS scores, particularly in females (Taylor and Signal, 2005). Dogs, similar to humans, are able to discriminate between the emotional expressions of human faces (Müller et al., 2015) and thus their affinity to socially support owners is predictable (Wedl et al., 2010).

DEVELOPMENTAL ASPECTS AND CORRELATES WITH ANIMAL ATTITUDES

During adolescence, young people experience strong changes in their attitudes and behavior, with this also being related to animal attitudes. Vegetarianism usually develops, for example, during adolescence. Starting with primary school children, Kellert (1983) reported increasing respect toward animals with increasing age. In contrast, positive attitudes toward animals decreases, once again, during adolescence (Binngießer et al., 2013; Bjerke et al., 1998;

Randler et al., 2017; Stanisstreet et al., 1993). Similarly, younger children were more opposed to animals in zoos/circuses, while they manifested a more negative attitude toward hunting (Pagani et al., 2007).

Concerning personality, there is only scarce evidence for a relationship between animal attitudes and personality. Mathews and Herzog (1997) based their study on the 16 Personality Factor Inventory and their AAS. Two personality factors, sensitivity and imaginativeness, were significantly correlated with attitudes toward animals. Their sample was based, however, on 99 college students, which is fairly small for correlational studies. Studies based on the Big Five personality measures are slightly different. The Big Five ranks among the most frequently used instruments in order to assess personality of late. It is based on five factors: extraversion, agreeableness, conscientiousness, neuroticism (or emotional instability), and openness to new experiences. Furnham et al. (2003) found that agreeableness was the most consistent predictor of animal welfare attitudes, followed by extraversion and openness. Hanna et al. (2009) assessed the animal attitudes of 311 dairy stock people and found that agreeableness and conscientiousness most strongly correlated with positive attitudes toward working with dairy cows. Randler et al. (2017) reported significant correlations between agreeableness and conscientiousness on the one hand and pro-animal attitudes on the other hand. Although it is difficult to draw general conclusions from these studies, it seems a general result that people who are more agreeable to other people also seem to have a higher pro-animal attitude.

COMPANION ANIMALS

Certain studies have demonstrated that adolescents with companion animals have higher pro-animal attitudes (Binngießer et al., 2013; Prokop and Tunnicliffe, 2010). In adults, however, Signal and Taylor (2006) failed to find the significant effect of companion animals on animal respect in a community sample while Taylor and Signal (2005) found a significant difference for current people with companion animals but not when they had companion animals in childhood. This unequivocal pattern requires further research. It also seems unclear whether people have basically a higher pro-animal attitude and in consequence, opt for a companion animal or if the companion animal itself provides higher attitude scores. This could be investigated in a long-term prospective study and would seem to be rewarding.

INFLUENCES OF KEEPING ANIMALS AS PETS ON ATTITUDES TOWARD ANIMALS

Pet ownership correlates with certain positive physiological measures, such as lower blood pressure, serum triglycerides, and cholesterol levels (Walsh, 2009), which ameliorate the cardiovascular effects of stress (Allen et al., 2002a,b). Furthermore, close relationships with pets positively influence oxytocin release which is one of the body's "feel good" chemicals and also plays a role in social bonding. In particular, oxytocin is boosted in both the dog and the human when a dog owner stares into the eyes of the dog (Nagasawa et al., 2015). Oxytocin physically facilitates childbirth and nursing for women, and this is probably the reason why it was found to be associated with increased caregiving, displays of affection, and empathy even toward members of other species (Handlin et al., 2012; Odendaal and Meintjes, 2003). This suggests that humans coevolved in their relationships with their pets and that these relationships are beneficial for both of them.

It is suggested that physical interactions with pets (Fig. 23.1) and the resulting improvement of attitudes to animals are generalized

FIGURE 23.1 Children playing with the rabbit.

to concerns for a broader species of animals (Amiot and Bastian, 2015). Based on this view, pets may act as ambassadors for other animals (Serpell, 1995, 2000; Serpell and Paul, 1994). Certain studies have provided support for the generalizing effect. In a large study of Slovak school children aged between 10 and 15 years, Prokop and Tunnicliffe (2010) compared children's knowledge and attitudes toward popular animals (i.e., ladybird, beetle, rabbit, and squirrel) with unpopular animals that are considered pests, predators, or those posing a disease threat to humans (i.e., potato beetle, wolf, and mouse, respectively). The results showed that children with pets at home were associated with more positive attitudes to, and a better knowledge of, both popular and unpopular animals. Bowd (1984) demonstrated that children who had pets had less negative attitudes toward nonpets such as lions, pigs, chickens, and snakes compared with non–pet owners. Keeping a pet was found to be associated with more empathic feeling for animals used in the fur and leather industry and for zoo animals (Pagani et al., 2007). Children with pets at home read more animal-related stories and reported enjoying animal-related films, visits to zoos and wildlife parks, and television programs to a greater extent than children who did not have pets (Kidd and Kidd, 1990). Conversely, school children who more frequently watched natural history films and walked in the outdoors were less fearful of wolves (Prokop et al., 2011) suggesting that attitudes to animals can be influenced by various activities supporting a connection with the outdoors among children.

INFLUENCES OF MEAT CONSUMPTION ON ATTITUDES TOWARD ANIMALS

Meat consumption evolved around 2.6 million years ago (Domínguez-Rodrigo et al., 2005) and humans are seen as omnivores. Vegetarianism is consequently a new development (ancient India, Greece, and Italy, see Spencer, 1993). Certain studies have reported that meat consumption as measured on the Likert scale is related to animal attitudes in adolescents (Binngießer et al., 2013; Randler et al., 2017) and adults (Schröder and McEachern, 2004). People that score high on pro-animal attitudes report lower meat consumption. Based on the dichotomous classification, vegetarians had a more positive attitude toward animal welfare than nonvegetarians (Furnham et al., 2003; Herzog and Golden, 2009). These studies do not reveal cause and effect; however, it could be possible that high pro-animal attitudes lead to vegetarianism or vice versa, or even that they develop in a parallel fashion. This could be investigated in a prospective study.

HUMAN EMOTIONS AND ANIMAL CONSERVATION

Research has highlighted the role of certain emotions (traditionally labeled as "negative") in solving environmental issues. Disgust is mainly related to avoidance of certain animals, ill humans, feces, vomit, sexual substances, and other harmful events (Rozin et al., 2000). Tybur et al. (2009, 2013) have demonstrated that there are at least three distinct types of disgust: *pathogen*, *sexual disgust*, and *moral* disgust. Pathogen disgust refers to disgust elicitors caused by the sources of various pathogens (e.g., stepping in dog excrement). Moral disgust refers to disgust that pertains to social transgressions (e.g., deceiving a friend). These social transgressions broadly include nonnormative, often antisocial activities, such as cheating, stealing, etc. Sexual disgust refers to disgust, which motivates sexual avoidance of an unsuitable mating partner or other reproductively costly behavior (e.g., performing anal sex or being in a situation with a high probability of having sex with a stranger). Research on animals is centered particularly on the pathogen domain of disgust.

Disgust is an adaptive (rather than "negative") emotion because it reduces the probability of transmission of infectious diseases (Curtis and Biran, 2001; Curtis et al., 2004; Navarrete et al., 2007; Oaten et al., 2009; Prokop et al., 2010a,b,c; Stevenson et al., 2009; Tybur et al., 2009). Interestingly, disgust is easily conditioned and it often takes only one single event to condition it. It is also, however, rigid and resistant to conventional changes (see Curtis and Biran, 2001 for discussion).

Certain animals are considered ugly and ugliness is associated with the emotion of disgust and fear. Indeed, the physical appearance of animals and attitudes toward them have received increasing attention from researchers (e.g., de Pinho et al., 2014; Gunnthorsdottir, 2001; Jimenez and Lindemann-Matthies, 2015a,b; Knight, 2008; Prokop et al., 2016). Jacobs et al. (2014), for example, found that disgust for wolves was consistently and significantly associated with the acceptability of lethal control of wolves both among Dutch and Canadian University students. Prokop and Fančovičová (2012b) and Prokop et al. (2016) found that the emotion of disgust was negatively related to tolerance for frogs. In another study, higher fear of bears correlated with the willingness of participants to exterminate bears by shooting (Prokop and Fančovičová, 2010b). All this evidence strongly highlight the role of human emotions in conservation efforts.

ATTITUDE CHANGE: THE ROLE OF EDUCATION

Higher educated people have more positive attitudes toward animals (Kellert, 1993); thus, education is viewed as one of the primary tools in influencing attitude change (Kellert, 1996). The emotion of disgust, for example, influences learning and retention among school pupils and students (Holstermann et al., 2012; Randler et al., 2012a; Štefaniková and Prokop, 2013). It is believed that physical contact with animals may inhibit disgust/fear thereby positively influencing attitudes toward animals (Ballouard et al., 2012; Johansson et al., 2016; Morgan and Gramann, 1989; Tomažič, 2008, 2011a,b; Randler et al., 2005, 2012b). During an intervention using live animals (wood lice, snails, and mice), for example, disgust for these animals was reduced compared to a control group, suggesting that educational programs are a means of reducing disgust (Randler et al., 2012b). 11- to 13-year-old children who interacted with unpopular live animals (i.e., wood louse, snail, mouse) became less disgusted and fearful of these animals after interacting with them compared to those who did not have such contact (Randler et al., 2012b). In a study with snails, as examples of slimy, disgusting animals (Davey et al., 1998), Prokop and Fančovičová (2017) demonstrated that hands-on experience with snails not only reduces specific disgust of snails but also a tendency to inhibit disgust from the other, unfamiliar but disgusting animals (Fig. 23.2). Though the drop of disgust scores was not clearly significant ($P = .09$), it provides some preliminary evidence about the possible positive influence of hands-on activities with disgusting animals on attitudes toward other unpopular animals.

In a farm animal education research, adolescents learned about chicken biology, welfare, and food labeling. Both knowledge of and positive behaviors toward poultry species increased immediately after the intervention, although then tended to diminish 3 months following the event (Jamieson et al., 2012).

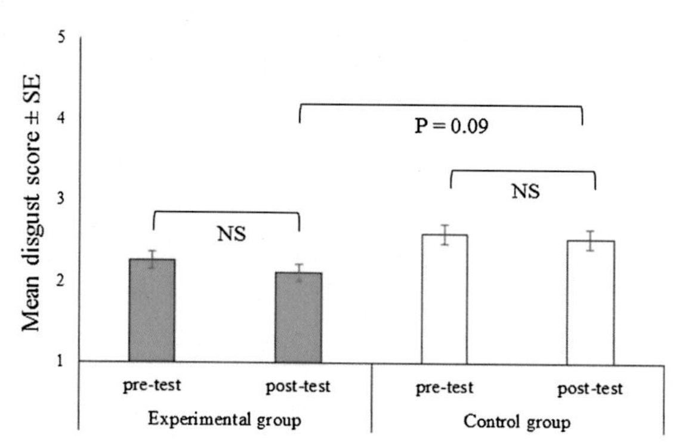

FIGURE 23.2 Disgust for animals dropped down after hands-on activities in the experimental group more than in the control group. *NS*, not statistically significant.

Although a great deal of work is needed before definite conclusions can be reached, these research reports reveal promising results suggesting that physical contact with animals has the potential to improve human–animal relationships in a more general sense. Schools, as well as conservation organizations, have the potential to apply these findings to intervention programs in order to improve children's attitudes toward animals in particular and environmental attitudes in general (Binngießer and Randler, 2015; Prokop and Kubiatko, 2014).

CONCLUSION

Increasing environmental problems have forced people to take active steps in saving biodiversity. Research on human–animal relationships may therefore provide certain helpful recommendations as to how perception of certain animals by people can be improved. Science helps us understand the nature of our own sympathies and antipathies toward animals around us. Animal appearance, particularly colors, genetic closeness to humans, its abundance in nature, potential threats from animals, and a number of other variables make human–animal relationships extremely complicated, but important to investigate. Research findings cannot be, however, used as a justification of our negative attitudes toward unpopular animals. Instead, an understanding of those evolutionary processes that have favored our fear and disgust of snakes or slimy animals should help us better control our behavior and future decisions. An improved understanding of the role of biological predispositions in individual attitudes toward animals may help us protect animals and finally save the biodiversity of the earth.

Acknowledgments

David Livingstone improved the English of the text. PP was partly funded by grant VEGA no. 1/0104/16.

References

Allen, J.A., Clarke, B.C., 1984. Frequency-dependent selection: homage to E.B. Poulton. Biological Journal of the Linnean Society 23, 15–18.

Allen, K.M., Blascovich, J., Mendes, W.B., 2002a. Cardiovascular reactivity in the presence of pets, friends, and spouses: the truth about cats and dogs. Psychosomatic Medicine 64, 727–739.

Allen, M.W., Hunstone, M., Waerstad, J., Foy, E., Hobbins, T., Wikner, B., Wirrel, J., 2002b. Human-to-animal similarity and participant mood influence punishment recommendations for animal abusers. Society & Animals 10, 267–284.

Almeida, A., Vasconcelos, C., Strecht-Ribeiro, O., 2014. Attitudes toward animals: a study of Portuguese children. Anthrozoös 27, 173–190.

Alterman, A.I., McDermott, P.A., Cacciola, J.S., Rutherford, M.J., 2003. Latent structure of the Davis interpersonal reactivity index in methadone maintenance patients. Journal of Psychopathology and Behavioral Assessment 25, 257–265.

Alves, R.R.N., Pereira Filho, G.A., Silva Vieira, K., Souto, W.M.S., Mendonças, L.E.T., Montenegro, P.F.G.P., Almeida, W.O., Vieira, W.L.S., 2012. A zoological catalogue of hunted reptiles in the semiarid region of Brazil. Journal of Ethnobiology and Ethnomedicine 8, 27.

Alves, R.R.N., Silva, V.N., Trovão, D.M., Oliveira, J.V., Mourão, J.S., Dias, T.L., Alves, A.G.C., Lucena, R.F.P., Barboza, R.R.D., Montenegro, P.F.G.P., Vieira, W.L., Souto, W.M.S., 2014. Students' attitudes toward and knowledge about snakes in the semiarid region of Northeastern Brazil. Journal of Ethnobiology and Ethnomedicine 10, 1.

Amiot, C.E., Bastian, B., 2015. Toward a psychology of human–animal relations. Psychological Bulletin 141, 6–47.

Arrindell, W.A., Eisemann, M., Richter, J., Oei, T.P., Caballo, V.E., van der Ende, J., Sanavio, E., Bgés, N., Feldman, L., Torres, B., Sica, C., Iwawaki, S., Edelmann, R.J., Crozier, W.R., Furnham, A., Hudson, B.L., 2003. Phobic anxiety in 11 nations: part I: dimensional constancy of the five-factor model. Behaviour Research and Therapy 41, 461–479.

Ballouard, J.M., Provost, G., Barré, D., Bonnet, X., 2012. Influence of a field trip on the attitude of schoolchildren toward unpopular organisms: an experience with snakes. Journal of Herpetology 46, 423–428.

Ballouard, J.M., Ajtic, R., Balint, H., Brito, J., Crnobrnja-Isailovic, J., Elmouden, E.L., Erdoğan, M., Feriche, M., Pleuguezuelos, J.M., Prokop, P., Sanchez, A., Santos, X., Slimani, T., Sterijovski, B., Tomovic, L., Uşak, M., Zuffi, M., Bonnet, X., 2013. Schoolchildren and one of the most unpopular animals: are they ready to protect snakes? Anthrozoös 26, 93–109.

Barua, M., Gurdak, D.J., Ahmed, R.A., Tamuly, J., 2012. Selecting flagships for invertebrate conservation. Biodiversity and Conservation 21, 1457–1476.

Batt, S., 2009. Human attitudes towards animals in relation to species similarity to humans: a multivariate approach. Bioscience Horizons 2, 180–190.

Baumholtz, M.A., Parish, S.C., Witkowski, J.A., Nutting, W.B., 1997. The medical importance of cockroaches. International Journal of Dermatology 36, 90–96.

Bennett-Levy, J., Marteau, T., 1984. Fear of animals: what is prepared? British Journal of Psychology 75, 37–42.

Binngießer, J., Randler, C., 2015. Association of the environmental attitudes "preservation" and "utilization" with pro-animal attitudes. International Journal of Environmental and Science Education 10, 477–492.

Binngießer, J., Wilhelm, C., Randler, C., 2013. Attitudes toward animals among German children and adolescents. Anthrozoös 26, 325–339.

Bjerke, T., Østdahl, T., 2004. Animal-related attitudes and activities in an urban population. Anthrozoös 17, 109–129.

Bjerke, T., Ødegårdstuen, T.S., Kaltenborn, B.P., 1998. Attitudes toward animals among Norwegian adolescents. Anthrozoös 11, 79–86.

Bjerke, T., Kaltenborn, B.P., Odegardstuen, T.S., 2001. Animal-related activities and appreciation of animals among children and adolescents. Anthrozoös 14, 86–94.

Bohlin, T., Gamberale-Stille, G., Merilaita, S., Exnerová, A., Štys, P., Tullberg, B., 2012. The detectability of the colour pattern in the aposematic firebug, *Pyrrhocoris apterus*: an image-based experiment with human 'predators'. Biological Journal of the Linnean Society 105, 806–816.

Borgi, M., Cirulli, F., 2013. Children's preferences for infantile features in dogs and cats. Human – Animal Interaction Bulletin 1, 1–15.

Borgi, M., Cirulli, F., 2015. Attitudes toward animals among kindergarten children: species preferences. Anthrozoös 28, 45–59.

Borgi, M., Cirulli, F., 2016. Pet face: mechanisms underlying human-animal relationships. Frontiers in Psychology 7, 298.

Borgi, M., Cogliati-Dezza, I., Brelsford, V., Meints, K., Cirulli, F., 2014. Baby schema in human and animal faces induces cuteness perception and gaze allocation in children. Frontiers in Psychology 5, 411.

Bowd, A.D., 1984. Fears and understanding of animals in middle childhood. Journal of Genetic Psychology: Research and Theory on Human Development 145, 143–144.

Cela-Conde, C.J., Marty, G., Maestu, F., Ortiz, T., Munar, E., Fernández, A., Roca, M., Rosselló, J., Quesney, F., 2004. Activation of the prefrontal cortex in the human visual aesthetic perception. Proceedings of the National Academy of Sciences of the United States of America 101, 6321–6325.

Ceríaco, L.M.P., 2012. Human attitudes towards herpetofauna: the influence of folklore and negative values on the conservation of amphibians and reptiles in Portugal. Journal of Ethnobiology and Ethnomedicine 8, 8.

Chippaux, J.P., 1998. Snake-bites: appraisal of the global situation. Bulletin of the World Health Organization 76, 515.

Clucas, B., McHugh, K., Caro, T., 2008. Flagship species on covers of US conservation and nature magazines. Biodiversity and Conservation 17, 1517–1528.

Cook, M., Mineka, S., 1990. Selective associations in the observational conditioning of fear in rhesus monkeys. Journal of Experimental Psychology: Animal Behavior Processes 16, 372–389.

Cook, M., Mineka, S., 1991. Selective associations in the origins of phobic fears and their implications for behavior therapy. In: Martin, P. (Ed.), Handbook of Behavior Therapy and Psychological Science: An Integrative Approach. Pergamon Press, Oxford, pp. 413–434.

Costa Rego, K.M., Zeppelini, C.G., Serramo Lopez, L.C., Alves, R.R.N., 2015. Assessing human-bat interactions around a protected area in northeastern Brazil. Journal of Ethnobiology and Ethnomedicine 11, 80.

Curtis, V., Biran, A., 2001. Dirt, disgust, and disease: is hygiene in our genes? Perspectives in Biology and Medicine 44, 17–31.

Curtis, V., Aunger, R., Rabie, T., 2004. Evidence that disgust evolved to protect from risk of disease. Proceedings of the Royal Society, Series B 271, S131–S133.

Daly, M., Wilson, M., 1988. Homicide. Aldine, Hawthorne, NY.

Darwin, C., 1872. The Expression of the Emotions in Man and Animals. John Murray, London.

Davey, G.C.L., 1994a. The "disgusting" spider: the role of disease and illness in the perpetuation of fear of spiders. Society & Animals 2, 17–25.

Davey, G.C.L., McDonald, A.S., Hirisave, U., Prabhu, G.G., Iwawaki, S., Jim, C.I., Merckelbach, H., de Jong, P.J., Lejny, P.W.L., Reimann, L., 1998. A cross-cultural study of animal fears. Behaviour Research and Therapy 36, 735–750.

de Pinho, J.R., Grilo, C., Boone, R.B., Galvin, K.A., Snodgrass, J.G., 2014. Influence of aesthetic appreciation of wildlife species on attitudes towards their conservation in Kenyan agropastoralist communities. PLoS One 9, e88842.

DeKay, M.L., McClelland, G.H., 1996. Probability and utility components of endangered species preservation programs. Journal of Experimental Psychology: Applied 2, 60–83.

DeLoache, J.S., LoBue, V., 2009. The narrow fellow in the grass: human infants associate snakes and fear. Developmental Science 12, 201–207.

Domínguez-Rodrigo, M., Pickering, T.R., Semaw, S., Rogers, M.J., 2005. Cutmarked bones from Pliocene archaeological sites at Gona, Afar, Ethiopia: implications for the function of the world's oldest stone tools. Journal of Human Evolution 48, 109–121.

Driscoll, J.W., 1995. Attitudes toward animals: species ratings. Society & Animals 3, 139–150.

Elliot, A.J., Maier, M.A., 2014. Color psychology: effects of perceiving color on psychological functioning in humans. Annual Review of Psychology 65, 95–120.

Elliot, A.J., Niesta, D., 2008. Romantic red: red enhances men's attraction to women. Journal of Personality and Social Psychology 95, 1150–1164.

Forrester, M.B., Stanley, S.K., 2004. Epidemiology of spider bites in Texas, 1998–2002. Public Health 118, 506–507.

Fredrikson, M., Annas, P., Wik, G., 1997. Parental history, aversive exposure and the development of snake and spider phobia in women. Behaviour Research and Therapy 35, 23–28.

Frew, K., Peterson, M.N., Steveson, K., 2017. Are we working to save the species our children want to protect? Evaluating species attribute preferences among children. Oryx (in press).

Frynta, D., Lišková, S., Bültmann, S., Burda, H., 2010. Being attractive brings advantages: the case of parrot species in captivity. PLoS One 5, e12568.

Frynta, D., Marešová, J., Reháková-Petru, M., Šklíba, J., Šumbera, R., Krása, A., 2011. Cross-cultural agreement in perception of animal beauty: boid snakes viewed by people from five continents. Human Ecology 39, 829–834.

Furnham, A., McManus, C., Scott, D., 2003. Personality, empathy and attitudes to animal welfare. Anthrozoös 16, 135–146.

Gerdes, A.B.M., Uhl, G., Alpers, G.W., 2009. Spiders are special: fear and disgust evoked by pictures of arthropods. Evolution and Human Behavior 30, 66–72.

Gunnthorsdottir, A., 2001. Physical attractiveness of an animal species as a decision factor for its preservation. Anthrozoös 14, 204–215.

Hamilton, W.D., Zuk, M., 1982. Heritable true fitness and bright birds: a role for parasites? Science 218, 384–387.

Handlin, L., Nilsson, A., Ejdebäck, M., Hydbring-Sandberg, E., Uvnäs-Moberg, K., 2012. Associations between the psychological characteristics of the human-dog relationship and oxytocin and cortisol levels. Anthrozoös 25, 215–228.

Hanna, D., Sneddon, I.A., Beattie, V.E., 2009. The relationship between the stockperson's personality and attitudes and the productivity of dairy cows. Animal 3, 737–743.

Heinrich, R.E., Strait, S.G., Houde, P., 2008. Earliest Eocene *Miacidae* (Mammalia: Carnivora) from northwestern Wyoming. Journal of Paleontology 82, 154–162.

Herzog, H.A., 2007. Gender differences in human–animal interactions: a review. Anthrozoös 20, 7–21.

Herzog, H.A., 2011. The impact of pets on human health and psychological well-being fact, fiction, or hypothesis? Current Directions in Psychological Science 20, 236–239.

Herzog, H.A., Burghardt, G.M., 1988. Attitudes toward animals: origins and diversity. Anthrozoös 1, 214–222.

Herzog, H.A., Golden, L.L., 2009. Moral emotions and social activism: the case of animal rights. Journal of Social Issues 65, 485–498.

Herzog, H.A., Betchart, N.S., Pittman, R.B., 1991. Gender, sex role orientation, and attitudes toward animals. Anthrozoös 4, 184–191.

Hill, R.A., Barton, R.A., 2005. Psychology: red enhances human performance in contests. Nature 435, 293.

Hinde, R.A., Barden, L.A., 1985. The evolution of the teddy bear. Animal Behaviour 33, 1371–1373.

Holstermann, N., Ainley, M., Grube, D., Roick, T., Bögeholz, S., 2012. The specific relationship between disgust and interest: relevance during biology class dissections and gender differences. Learning and Instruction 22, 185–192.

Isbell, L.A., 2006. Snakes as agents of evolutionary change in primate brains. Journal of Human Evolution 51, 1–35.

Isbell, L.A., 2009. The Fruit, the Tree, and the Serpent. Harvard University Press, Cambridge.

Jacobs, M.H., 2009. Why do we like or dislike animals? Human Dimensions of Wildlife 14, 1–11.

Jacobs, M.H., Vaske, J.J., Dubois, S., Fehres, P., 2014. More than fear: role of emotions in acceptability of lethal control of wolves. European Journal of Wildlife Research 60, 589–598.

Jamieson, J., Reiss, M.J., Allen, D., Asher, L., Wathes, C.M., Abeyesinghe, S.M., 2012. Measuring the success of a farm animal welfare education event. Animal Welfare–the UFAW Journal 21, 65–75.

Janif, J.Z., Brooks, R.C., Dixson, B.J., 2014. Negative frequency dependent preferences and variation in male facial hair. Biology Letters 10, 20130958.

Jimenez, J.N., Lindemann-Matthies, P., 2015a. Public knowledge of, and attitudes to, frogs in Colombia. Anthrozoös 28, 319–332.

Jimenez, J.N., Lindemann-Matthies, P., 2015b. Public knowledge and perception of toads and frogs in three areas of subtropical Southeast China. Society & Animals 23, 166–192.

Johansson, M., Støen, O.G., Flykt, A., 2016. Exposure as an intervention to address human fear of bears. Human Dimensions of Wildlife 21, 311–327.

Jokela, M., 2009. Physical attractiveness and reproductive success in humans: evidence from the late 20th century United States. Evolution and Human Behavior 30, 342–350.

Kaltenborn, B.P., Bjerke, T., Nyahongo, J., 2006a. Living with problem animals-self-reported fear of potentially dangerous species in the Serengeti region, Tanzania. Human Dimensions of Wildlife 11, 397–409.

Kaltenborn, B.P., Bjerke, T., Nyahongo, J.W., Williams, D.R., 2006b. Animal preferences and acceptability of wildlife management actions around Serengeti National Park, Tanzania. Biodiversity and Conservation 15, 4633–4649.

Kasturiratne, A., Wickremasinghe, A.R., de Silva, N., Gunawardena, N.K., Pathmeswaran, A., Premaratna, R., Savioli, L., Lalloo, D.G., de Silva, H.J., 2008. The global burden of snakebite: a literature analysis and modelling based on regional estimates of envenoming and deaths. PLoS Medicine 5, e218.

Katcher, A., 2002. Animals in therapeutic education: guides into the liminal state. In: Kahn, P.H., Kellert, S.R. (Eds.), Children and Nature: Psychological, Sociocultural, and Evolutionary Investigations. MIT Press, Massachusetts, pp. 179–198.

Kellert, S.R., 1983. Affective, cognitive, and evaluative perceptions of animals. In: Altman, I., Wohwill, J.F. (Eds.), Behavior and the Natural Environment. Plenum Press, New York, pp. 241–267.

Kellert, S.R., 1984. American attitudes toward and knowledge of animals: an update. In: Fox, M.W., Mickley, L.D. (Eds.), Advances in Animal Welfare Science 1984/85. The Humane Society of the United States, Washington, DC, pp. 177–213.

Kellert, S.R., 1993. Values and perceptions of invertebrates. Conservation Biology 7, 845–855.

Kellert, S.R., 1996. The Value of Life: Biological Diversity and Human Society. Island Press, Washington, DC.

Kellert, S.R., Berry, J.K., 1987. Attitudes, knowledge, and behaviors toward wildlife as affected by gender. Wildlife Society Bulletin 15, 363–371.

Kidd, A.H., Kidd, R.M., 1990. Social and environmental influences on children's attitudes toward pets. Psychological Reports 67, 807–818.

Knight, A.J., 2008. "Bats, snakes and spiders, Oh my!" How aesthetic and negativistic attitudes, and other concepts predict support for species protection. Journal of Environmental Psychology 28, 94–103.

Kramer, K.L., 2011. The evolution of human parental care and recruitment of juvenile help. Trends in Ecology and Evolution 26, 533–540.

LeDoux, J.E., 2012. Evolution of human emotion: a view through fear. Progress in Brain Research 195, 431–442.

Lindemann-Matthies, P., 2005. "Loveable" mammals and "lifeless" plants: how children's interest in common local organisms can be enhanced through observation of nature. International Journal of Science Education 27, 655–677.

Lišková, S., Frynta, D., 2013. What determines bird beauty in human eyes? Anthrozoös 26, 27–41.

Lišková, S., Landová, E., Frynta, D., 2015. Human preferences for colorful birds: vivid colors or pattern? Evolutionary Psychology 13, 339–359.

Little, A.C., Hill, R.A., 2007. Attribution to red suggests special role in dominance signalling. Journal of Evolutionary Psychology 5, 161–168.

LoBue, V., 2010. And along came a spider: superior detection of spiders in children and adults. Journal of Experimental Child Psychology 107, 59–66.

LoBue, V., DeLoache, J.S., 2008. Detecting the snake in the grass. Attention to fear-relevant stimuli by adults and young children. Psychological Science 19, 284–289.

LoBue, V., Rakison, D.H., 2013. What we fear most: a developmental advantage for threat-relevant stimuli. Developmental Review 33, 285–303.

Lobue, V., Bloom Pickard, M., Sherman, K., Axford, C., DeLoache, J.S., 2013. Young children's interest in live animals. British Journal of Developmental Psychology 31, 57–69.

Lorenz, K., 1943. Die angeborenen formen möglicher erfahrung. Zeitschrift für Tierpsychologie 5, 235–409.

Marešová, J., Frynta, D., 2008. Noah's ark is full of common species attractive to humans: the case of bold snakes in zoos. Ecological Economics 64, 554–558.

Marešová, J., Krása, A., Frynta, D., 2009a. We all appreciate the same animals: cross-cultural comparison of human aesthetic preferences for snake species in Papua New Guinea and Europe. Ethology 115, 297–300.

Marešová, J., Landová, E., Frynta, D., 2009b. What makes some species of milk snakes more attractive to humans than others? Theory Biosciences 128, 227–235.

Martín-López, B., Montes, C., Benayas, J., 2007. The non-economic motives behind the willingness to pay for biodiversity conservation. Biological Conservation 139, 67–82.

Martín-López, B., Montes, C., Benayas, C., 2008. Economic valuation of biodiversity conservation: the meaning of numbers. Conservation Biology 22, 624–635.

Mathews, S., Herzog, H., 1997. Personality and attitudes towards the treatment of animals. Society & Animals 5, 57–63.

McLure, E.B., 2000. A meta-analytic review of sex differences in facial expression processing and their development in infants, children, and adolescents. Psychological Bulletin 126, 424–453.

Meuser, E., Harshaw, H.W., Moers, A.Ø., 2009. Public preference for endemism over other conservation-related species attributes. Conservation Biology 23, 1041–1046.

Mineka, S., Keir, R., Price, V., 1980. Fear of snakes in wild- and laboratory-reared rhesus monkeys (*Macaca mulatta*). Animal Learning & Behavior 8, 653–663.

Møller, A.P., 1992. Female swallow preference for symmetrical male sexual ornaments. Nature 357, 238–240.

Morelli, G., Song, Y., Mazzoni, C.J., Eppinger, M., Roumagnac, P., Wagner, D.M., Feldkamp, M., Kusecek, B., Vogler, A.J., Li, Y., Cui, Y.J., Thomson, N.R., Jombart, T., Leblois, R., Lichtner, P., Rahalison, L., Petersen, J.M., Balloux, F., Keim, P., Wirth, T., Ravel, J., Yang, R.F., Carniel, E., Achtman, M., 2010. *Yersinia pestis* genome sequencing identifies patterns of global phylogenetic diversity. Nature Genetics 42, 1140–1143.

Morgan, J.M., Gramann, J.H., 1989. Predicting effectiveness of wildlife education programs: a study of students' attitudes and knowledge toward snakes. Wildlife Society Bulletin 17, 501–509.

Morse-Jones, S., Bateman, I.J., Kontoleon, A., Ferrini, S., Burgess, N.D., Turner, R.K., 2012. Stated preferences for tropical wildlife conservation amongst distant beneficiaries: charisma, endemism, scope and substitution effects. Ecological Economics 78, 9–18.

Müller, C.A., Schmitt, K., Barber, A.L., Huber, L., 2015. Dogs can discriminate emotional expressions of human faces. Current Biology 25, 601–605.

Murray, C.J.L., Rosenfeld, L.C., Lim, S.S., Andrews, G.C., Foreman, K.J., Haring, D., Fullman, N., Naghavi, M., Lozano, R., Lopez, A.D., 2012. Global malaria mortality between 1980 and 2010: a systematic analysis. Lancet 379, 413–431.

Nagasawa, M., Mitsui, S., En, S., Ohtani, N., Ohta, M., Sakuma, Y., Onaka, T., Mgi, K., Kikusui, T., 2015. Oxytocin-gaze positive loop and the coevolution of human-dog bonds. Science 348, 333–336.

Navarrete, C.D., Fessler, D.M., Eng, S.J., 2007. Elevated ethnocentrism in the first trimester of pregnancy. Evolution and Human Behavior 28, 60–65.

Nesse, R.M., 2005. Natural selection and the regulation of defenses: a signal detection analysis of the smoke detector principle. Evolution and Human Behavior 26, 88–105.

Oaten, M., Stevenson, R.J., Case, T.I., 2009. Disgust as a disease-avoidance mechanism. Psychological Bulletin 135, 303–321.

Odendaal, J.S.J., Meintjes, R.A., 2003. Neurophysiological correlates of affiliative behaviour between humans and dogs. Veterinary Journal 165, 296–301.

Öhman, A., 2009. Of snakes and faces: an evolutionary perspective on the psychology of fear. Scandinavian Journal of Psychology 50, 543–552.

Öhman, A., Mineka, S., 2001. Fear, phobias and preparedness: toward an evolved module of fear and fear learning. Psychological Review 108, 483–522.

Öhman, A., Mineka, S., 2003. The malicious serpent snakes as a prototypical stimulus for an evolved module of fear. Current Directions in Psychological Science 12, 5–9.

Öhman, A., Flykt, A., Esteves, F., 2001a. Emotion drives attention: detecting the snake in the grass. Journal of Experimental Psychology: General 130, 466–478.

Öhman, A., Lundqvist, D., Esteves, F., 2001b. The face in the crowd revisited: a threat advantage with schematic stimuli. Journal of Personality and Social Psychology 80, 381–396.

O'Haire, M.E., McKenzie, S.J., Beck, A.M., Slaughter, V., 2013. Social behaviors increase in children with autismin the presence of animals compared to toys. PLoS One 8, e57010.

Pagani, C., Robustelli, F., Ascione, F.R., 2007. Italian youths' attitudes toward, and concern for, animals. Anthrozoös 20, 275–293.

Pawlowski, B., Dunbar, R.I.M., Lipowicz, A., 2000. Evolutionary fitness: tall men have more reproductive success. Nature 403, 156.

Phelps, E.A., Ling, S., Carrasco, M., 2006. Emotion facilitates perception and potentiates the perceptual benefits of attention. Psychological Science 17, 292–299.

Polly, P.D., Wesley-Hunt, G.D., Heinrich, R.E., Davis, G., Houde, P., 2006. Earliest known carnivoran auditory bulla and support for a recent origin of crown-group Carnivora (Eutheria, Mammalia). Palaeontology 49, 1019–1027.

Prokop, P., Fančovičová, J., 2010a. The association between disgust, danger and fear of macroparasites and human behaviour. Acta Ethologica 13, 57–62.

Prokop, P., Fančovičová, J., 2010b. Perceived body condition is associated with fear of a large carnivore predator in humans. Annales Zoologici Fennici 47, 417–425.

Prokop, P., Fančovičová, J., 2012a. Beautiful fruits taste good: the aesthetic influences of fruit preferences in humans. Anthropologischer Anzeiger 69, 71–83.

Prokop, P., Fančovičová, J., 2012b. Tolerance of amphibians in Slovakian people: a comparison of pond owners and non-owners. Anthrozoös 25, 277–288.

Prokop, P., Fančovičová, J., 2013a. Does colour matter? The influence of animal warning colouration in human emotions and willingness to protect them. Animal Conservation 16, 458–466.

Prokop, P., Fančovičová, J., 2013b. Self-protection versus disease avoidance: the perceived physical condition is associated with fear of predators in humans. Journal of Individual Differences 34, 15–23.

Prokop, P., Fančovičová, J., 2014. Seeing coloured fruits: utilization of the theory of adaptive memory in teaching botany. Journal of Biological Education 48, 127–132.

Prokop, P., Fančovičová, J., 2017. The effect of hands-on activities on children's knowledge and disgust for animals. Journal of Biological Education (in press).

Prokop, P., Fedor, P., 2011. Physical attractiveness influences reproductive success of modern men. Journal of Ethology 29, 453–458.

Prokop, P., Kubiatko, M., 2014. Perceived vulnerability to disease predicts environmental attitudes. Eurasia Journal of Mathematics, Science and Technology Education 10, 3–11.

Prokop, P., Tunnicliffe, S.D., 2010. Effects of keeping pets at home on children's attitudes toward popular and unpopular animals. Anthrozoös 23, 21–35.

Prokop, P., Prokop, M., Tunnicliffe, S.D., 2008. Effects of keeping animals as pets on children's concepts of vertebrates and invertebrates. International Journal of Science Education 30, 431–449.

Prokop, P., Fančovičová, J., Kubiatko, M., 2009a. Vampires are still alive: Slovakian students' attitudes toward bats. Anthrozoös 22, 19–30.

Prokop, P., Özel, M., Uşak, M., 2009b. Cross-cultural comparison of student attitudes toward snakes. Society & Animals 17, 224–240.

Prokop, P., Uşak, M., Fančovičová, J., 2010a. Health and the avoidance of macroparasites: a preliminary cross-cultural study. Journal of Ethology 28, 345–351.

Prokop, P., Uşak, M., Fančovičová, J., 2010b. Risk of parasite transmission influences perceived vulnerability to disease and perceived danger of disease-relevant animals. Behavioural Processes 85, 52–57.

Prokop, P., Fančovičová, J., Fedor, P., 2010c. Health is associated with anti-parasite behavior and fear of disease-relevant animals in humans. Ecological Psychology 22, 222–237.

Prokop, P., Tolarovičová, A., Camerik, A., Peterková, V., 2010d. High school students' attitudes towards spiders: a cross-cultural comparison. International Journal of Science Education 32, 1665–1688.

Prokop, P., Uşak, M., Erdoğan, M., 2011. Good predators in bad stories: cross-cultural comparison of children's attitudes toward wolves. Journal of Baltic Science Education 10, 229–242.

Prokop, P., Medina-Jerez, W., Coleman, J., Fančovičová, J., Özel, M., Fedor, P., 2016. Tolerance of frogs among high school students: influences of disgust and culture. Eurasia Journal of Mathematics, Science and Technology Education 12, 1499–1505.

Rakison, D.H., Derringer, J.L., 2008. Do infants possess an evolved spider-detection mechanism? Cognition 107, 381–393.

Randler, C., Ilg, A., Kern, J., 2005. Cognitive and emotional evaluation of an amphibian conservation program for elementary school students. Journal of Environmental Education 37, 43–52.

Randler, C., Wust-Ackermann, P.P., Vollmer, C., Hummel, E., 2012a. The relationship between disgust, state-anxiety and motivation during a dissection task. Learning and Individual Differences 22, 419–424.

Randler, C., Hummel, E., Prokop, P., 2012b. Practical work at school reduces disgust and fear of unpopular animals. Society & Animals 20, 61–74.

Randler, C., Binngießer, J., Vollmer, C., 2017. The composite respect for animals scale: full and brief version. Society & Animals (in press).

Roberts, S.C., Owen, R.C., Havlíček, J., 2010. Distinguishing between perceiver and wearer effects in clothing color-associated attributions. Evolutionary Psychology 8, 350–364.

Røskaft, E., Bjerke, T., Kaltenborn, B.P., Linnell, J.D.C., Andersen, R., 2003. Patterns of self-reported fear towards large carnivores among the Norwegian public. Evolution and Human Behavior 24, 184–198.

Rozin, P., Millman, L., Nemeroff, C., 1986. Operation of the laws of sympathetic magic in disgust and other domains. Journal of Personality and Social Psychology 50, 703–712.

Rozin, P., Haidt, J., McCauley, C., 2000. Disgust. In: Lewis, M., Haviland, J.M. (Eds.), Handbook of Emotions, second ed. Guilford Press, New York, pp. 637–653.

Rusch, H., Voland, E., 2013. Evolutionary aesthetics: an introduction to key concepts and current issues. Aisthesis. Pratiche, linguaggi e saperi dell'estetico 6, 113–133.

Russell, L., 2003. Decoding equine emotions. Society & Animals 13, 265–266.

Ruxton, G.D., Sherratt, T.N., Speed, M.P., 2004. Avoiding Attack. The Evolution of Crypsis, Warning Signals and Mimicry. Oxford University Press, New York.

Samples, K.C., Dixon, J.A., Gowen, M.M., 1986. Information disclosure and endangered species valuation. Land Economics 62, 306–312.

Santos, G.M., Antonini, Y., 2008. The traditional knowledge on stingless bees (Apidae: Meliponina) used by the Enawene-Nawe tribe in western Brazil. Journal of Ethnobiology and Ethnomedicine 4, 19–25.

Schlegel, J., Rupf, R., 2010. Attitudes towards potential animal flagship species in nature conservation: a survey among students of different educational institutions. Journal for Nature Conservation 18, 278–290.

Schlegel, J., Breuer, G., Rupf, R., 2015. Local insects as flagship species to promote nature conservation? A survey among primary school children on their attitudes toward invertebrates. Anthrozoös 28, 229–245.

Schröder, M.J., McEachern, M.G., 2004. Consumer value conflicts surrounding ethical food purchase decisions: a focus on animal welfare. International Journal of Consumer Studies 28, 168–177.

Seligman, M., 1971. Phobias and preparedness. Behaviour Therapy 2, 307–320.

Serpell, J., 1995. The Domestic Dog: Its Evolution, Behaviour, and Interactions with People. Cambridge University Press, Cambridge.

Serpell, J.A., 2000. Creatures of the unconscious: companion animals as mediators. In: Podberscek, A.L., Paul, E.S., Serpell, J.A. (Eds.), Companion Animals and Us:

Exploring the Relationships between People and Pets. Cambridge University Press, Cambridge, pp. 108–121.
Serpell, J.A., 2003. Anthropomorphism and anthropomorphic selection – beyond the 'cute response'. Society & Animals 11, 83–100.
Serpell, J., 2004. Factors influencing human attitudes to animals and their welfare. Animal Welfare 13, 145–151.
Serpell, J.A., Paul, E.S., 1994. Pets and the development of positive attitudes to animals. In: Manning, A., Serpell, J.A. (Eds.), Animals and Human Society: Changing Perspectives. Routledge, London, pp. 127–144.
Shirey, L.L., Reynolds, R.E., 1988. Effect of interest on attention and learning. Journal of Educational Psychology 80, 159–166.
Signal, T.D., Taylor, N., 2006. Attitudes to animals: demographics within a community sample. Society & Animals 14, 147–157.
Simion, F., Regolin, L., Bulf, H., 2008. A predisposition for biological motion in the newborn baby. Proceedings of the National Academy of Sciences of the United States of America 105, 809–813.
Spencer, C., 1993. The Heretic's Feast. A History of Vegetarianism. Fourth Estate, London.
Stanisstreet, M., Spofforth, N., Williams, T., 1993. Attitudes of children to the uses of animals. International Journal of Science Education 15, 411–425.
Štefaniková, S., Prokop, P., 2013. Introducing the concept of adaptive memory to science education: does survival threat influence our knowledge about animals? Journal of Environmental Protection and Ecology 14, 1403–1414.
Stephan, W.G., Finlay, K., 1999. The role of empathy in improving intergroup relation. Journal of Social Issues 55, 729–743.
Stevenson, R.J., Case, T.I., Oaten, M.J., 2009. Frequency and recency of infection and their relationship with disgust and contamination sensitivity. Evolution and Human Behavior 30, 363–368.
Stokes, D.L., 2007. Things we like: human preferences among similar organisms and implications for conservation. Human Ecology 35, 361–369.
Van Strien, J.W., Franken, I.H.A., Huijding, J., 2014. Testing the snake-detection hypothesis: larger early posterior negativity in humans to pictures of snakes than to pictures of other reptiles, spiders and slugs. Frontiers in Human Neuroscience 8, 691.
Takahashi, Y., Veríssimo, D., MacMillan, D.C., Godbole, A., 2012. Stakeholder perceptions of potentil flagship species for the sacred groves of the North Western Ghats, India. Human Dimensions of Wildlife 17, 257–259.
Tarrant, J., Kruger, D., Du Preez, L.H., 2016. Do public attitudes affect conservation effort? Using a questionnaire-based survey to assess perceptions, beliefs and superstitions associated with frogs in South Africa. African Zoology 51, 13–20.
Taylor, N., Signal, T.D., 2005. Empathy and attitudes to animals. Anthrozoös 18, 18–27.
Thornhill, R., Gangestad, S.W., 1999. Facial attractiveness. Trends in Cognitive Sciences 3, 452–460.
Tomažič, I., 2008. The influence of direct experience on students' attitudes to, and knowledge about amphibians. Acta Biologica Slovenica 51, 39–49.
Tomažič, I., 2011a. Reported experiences enhance favourable attitudes toward toads. Eurasia Journal of Mathematics, Science and Technology Education 7, 253–262.
Tomažič, I., 2011b. Seventh graders' direct experience with, and feelings toward, amphibians and some other nonhuman animals. Society & Animals 19, 225–247.
Treves, A., Naughton-Treves, L., 1999. Risk and opportunity for humans coexisting with large carnivores. Journal of Human Evolution 36, 275–282.
Tybur, J.M., Lieberman, D., Griskevicius, V., 2009. Microbes, mating, and morality: individual differences in three functional domains of disgust. Journal of Personality and Social Psychology 97, 103–122.
Tybur, J.M., Lieberman, D., Kurzban, R., DeScioli, P., 2013. Disgust: evolved function and structure. Psychological Review 120, 65–84.
Ungar, P.S., Teaford, M.F., 2002. Human Diet: Its Origin and Evolution, first ed. Praeger, Westport, CT.
Van Le, Q., Isbell, L.A., Matsumoto, J., Nguyen, M., Hori, E., Maior, R.S., Tomaz, C., Tran, A.H., Ono, T., Nishijo, H., 2013. Pulvinar neurons reveal neurobiological evidence of past selection for rapid detection of snakes. Proceedings of the National Academy of Sciences of the United States of America 110, 19000–19005.
Veríssimo, D., Fraser, I., Groombridge, J., Bristol, R., Macmillan, D.C., 2009. Birds as tourism flagship species: a case study of tropical islands. Animal Conservation 12, 549–558.
Villmoare, B., Kimbel, H., Seyoum, C., Campisano, C., DiMaggio, E., Rowan, J., Braun, D., Arrowsmith, J., Reed, K., 2015. Early *Homo* at 2.8 Ma from Ledi-Geraru, Afar, Ethiopia. Science 1352–1355.
Voland, E., Grammer, K., 2003. Evolutionary Aesthetics. Springer-Verlag, Berlin.
Walsh, F., 2009. Human-animal bonds I: the relational significance of companion animals. Family Process 48, 462–480.
Warrell, D.A., 2010. Snake bite. The Lancet 375, 77–88.
Wedl, M., Schöberl, I., Bauer, B., Day, J., Kotrschal, K., 2010. Relational factors affecting dog social attraction to human partners. Interaction Studies 11, 482–503.
Whitaker, P.B., Shine, R., 2000. Sources of mortality of large elapid snakes in an agricultural landscape. Journal of Herpetology 34, 121–128.
Wilson, E.O., 1984. Biophilia: The Human Bond with Other Species. Harvard University Press, Harvard, MA.

Wrangham, R.W., Peterson, D., 1996. Demonic Males: Apes and the Origins of Human Violence. Houghton Mifflin Harcourt, Boston, MA.

Yorzinski, J.L., Penkunas, M.J., Platt, M.L., Coss, R.G., 2014. Dangerous animals capture and maintain attention in humans. Evolutionary Psychology 12, 534–548.

Zahavi, A., Zahavi, A., 1997. The Handicap Principle: A Missing Piece of Darwin's Puzzle. Oxford University Press, Oxford.

Zhang, R., Wang, Y.Q., Su, B., 2008. Molecular evolution of a primate-specific microRNA family. Molecular Biology and Evolution 25, 1493–1502.

Further Reading

Davey, G.C.L., 1994b. Self-reported fears to common indigenous animals in an adult UK population: the role of disgust sensitivity. British Journal of Psychology 85, 541–554.

Öhman, A., Juth, P., Lundqvist, D., 2010. Finding the face in a crowd: relationships between distractor redundancy, target emotion, and target gender. Cognition and Emotion 24, 1216–1228.

Tomkins, S.S., 1962. Affect/Imagery. Consciousness VI. Springer, New York.

CHAPTER

24

The Role of Ethnozoology in Animal Studies

Rômulo Romeu Nóbrega Alves, Sérgio de Faria Lopes
Universidade Estadual da Paraíba, Campina Grande, Brazil

INTRODUCTION

People have always been interested in other animals since the beginning of the history shared between them. Driven by necessity, the earliest human beings must have been observant of animals, generating a zoological knowledge that was essential to increase their chances of survival. When an animal is hunted or fished, knowledge about it tends to be even more accurate, since capture success by hunters and fishermen depends on knowing the biology and ecology of the target species. Also, when humans are potential prey, it is necessary to understand the biology of large predators. Zoological knowledge is not limited to people who directly exploit animals, extending to all who interact with them in some way. Magner (2002), for example, points out that healers and shamans develop knowledge about aspects of animal biology, because they have served the totems of tribes or clans, and that the appearance and behavior of animals influence their practices of divination and prophecy. The prescription of animal-derived medicines for them may also be related to aspects of the morphology and behavior of animals used in folk medicine (Alves and Alves, 2011).

Empirical knowledge about animals is obtained on a day-to-day basis, on the basis of experiences that are accumulated through extensive exposure over a period of time, or transmitted from person to person. In the literature, several terms have been used to refer to this type of knowledge, including traditional ecological knowledge (TEK) (Berkes, 1999; Berkes et al., 2000; Huntington, 2000), indigenous knowledge (Gadgil et al., 1993; Sillitoe et al., 1998), folk knowledge (Berkes, 1999), traditional zoological knowledge (Longo, 2004), and local ecological knowledge (LEK) (Berkes, 1993). Ruddle (1994) considered that the term "local" is less problematic, and thus a more practical description or identifier of the relevant people and their knowledge. Brook and McLachlan (2008) also considered LEK as a more inclusive term. We will use it for the purposes of this chapter.

Since the emergence of zoology as a science, the knowledge of local people had already caught the attention of pioneering naturalist philosophers, making an essential contribution to the construction of zoological knowledge of

Ethnozoology
http://dx.doi.org/10.1016/B978-0-12-809913-1.00024-7

the time. In many cases, localization, collection, preparation, and preservation of various new animal species were carried out from the knowledge of native peoples (Pierotti and Wildcat, 2000), which were systematized by naturalists, purified within a predominant scientific view, and incorporated into the universal scientific reserve (Moreira, 2002). Although this contribution is recognized by the naturalists themselves in their writings, it is generally disregarded by science historians (Moreira, 2002).

LEK exists parallel to academic knowledge, but both are derived from the same source—the systematic observation of nature—although these observations are interpreted within unique cultural contexts (Alves and Souto, 2015). Both knowledge systems produce detailed empirical information about natural phenomena and the relationships between ecosystem components (Alves and Nishida, 2002; Kimmerer, 2002; Nishida et al., 2006; Dantas et al., 2016). Thus, it is evident that LEK can contribute to academic research on animals in areas such as zoology, ecology, evaluation of environmental impacts, resource management, and sustainable development (Alves and Nishida, 2002; Alves and Souto, 2015; Anadón et al., 2010; Johannes, 1993; Kotschwar Logan et al., 2015; Nishida et al., 2006; Sillitoe et al., 1998; Silvano et al., 2006; Ziembicki et al., 2013; Zuercher et al., 2003).

Some researchers have been reluctant to accept the use of LEK in academic research because they consider information derived from LEK to be "anecdotal," "imprecise," "unsubstantiated," or "inaccurate" (Gilchrist et al., 2005; Hobson, 1992; Johannes, 1993; Rist et al., 2010). Now, however, it has been increasingly recognized that the application of LEK has many obvious advantages in understanding and responding to ecological problems (Bart, 2006), and its incorporation into contemporary ecological and conservation research has intensified (Brook and McLachlan, 2008; Capistrano and Lopes, 2012; Mackinson and Nottestad, 1998; Nascimento et al., 2016, 2017; Silvano et al., 2006). For example, community-based monitoring and participatory research have been shown to be effective in providing reliable scientific data and to be cost-effective if well designed (Carvalho et al., 2009; Holck, 2008), particularly for small populations or low encounter rates (Gaidet et al., 2003; Humber et al., 2016, 2011). Thus, depending on the objective, LEK is an important tool in ecological research (Alves and Souto, 2015), and its possible biases and uncertainties can be minimized through appropriate data collection and analytical procedures for robust interpretation (Turvey et al., 2015). Importantly, it must be highlighted that potential biases influence the validity of extracting and translating LEK into scientific formats (Berkes, 2009), but all forms of information carry uncertainty (Aylesworth et al., 2017).

In this chapter, we discuss how LEK can contribute to academic research involving wildlife and sources of scientific insights and complement zoological/ecological and taxonomic studies.

USE OF TRADITIONAL KNOWLEDGE IN ECOLOGICAL/ ZOOLOGICAL STUDIES

Integrating different sources of knowledge to investigate a particular phenomenon allows its greater understanding. Accordingly, the use of LEK represents an important methodological tool, serving as a source of information in research involving biological resources, where it is considered an important complement to scientific ecological knowledge (SEK), especially in studies on the management and conservation of biodiversity. LEK has been recognized as important avenue for discussion and for fostering dialogue between scientists and the communities in which they work (Turner, 2000). Consequently, it is not surprising that there is an increasing number of scholars actively employing LEK and working collaboratively with communities, which helps accumulate a critical mass of evidence and support for the value of local knowledge (Brook and McLachlan, 2008; Kimmerer, 2002; Maffi et al., 1999; Moller et al., 2004; Tidemann and Gosler, 2010).

LEK exists in all cultures and arises from the material or spiritual relationships between humans and biodiversity (independent of the ethnic group involved) (Alves and Souto, 2015). People who directly use animals (such as hunters, fishermen, harvesters/collectors) tend to possess a considerable knowledge associated with them, considering that their success at harvesting or capturing animals is intimately linked to the quality and reliability of their ecological observations (Alves et al., 2005; Begossi and Silvano, 2008; Capistrano and Lopes, 2012; López-Arévalo et al., 2011; Marques, 1995; Nordi et al., 2009; Rosa et al., 2005). As a result, these people retain a wide range of biological information that can complement scientific knowledge in ecology, zoology, and biological conservation studies, and could be especially useful to studies of population biology, ethology, resource evaluation and management, patterns of climate and resource variations, interactions between species, relationships between abiotic factors and fauna, ethnotaxonomy, and the sustainable use and adaptive administration of natural resources (Alves and Nishida, 2002; Berkes, 1999; Brinkman et al., 2009; Huntington, 2000; Huntington et al., 2004).

Surveys conducted with fishermen, for example, have provided relevant biological data on the distribution and relative abundance of species in a shorter time frame and at a lower cost (Bundy and Davis, 2013; Silvano et al., 2006; Turvey et al., 2014; White et al., 2005), especially at the local scale (Zukowski et al., 2011). Also, LEK serves as a tool for determining threats and conservation status of wild species (Turvey et al., 2014). Rojchanaprasart et al. (2014), for example, studied local residents' knowledge of dugongs in Thailand and concluded that LEK was significantly correlated to SEK. They pointed out that LEK should be considered in planning conservation programs for dugong populations which show continuous decrease.

LEK has also been useful in obtaining quantitative estimates of wild animal populations. A good example of this situation was undertaken by Anadón et al. (2010) who worked with shepherds in southeastern Spain. These authors found that local knowledge provided high-quality and low-cost information about the distribution and abundance of the tortoise *Testudo graeca* and estimates of its abundance in an area much larger than that covered by linear transects used in their standard methods. The abundance estimates of both methods were closely related, and cost analyses revealed that the information obtained through local knowledge was 100 times less costly than setting up and monitoring linear transects. In another study focusing on mammals, López-Arévalo et al. (2011) used LEK as a tool for testing predictive models of the spatial distribution of medium-sized mammalian richness, identifying local patterns of species richness and evaluating the role of local protected areas in the conservation of these animals. The authors concluded that the differences between estimated richness in the different scenarios and observed richness indicated local inhabitants' ability to identify rare and extinct species, as well as abundant pest species. These examples reinforce the role of LEK as complementary in quantitative abundance monitoring programs of a large variety of taxa, particularly when their population densities are low and traditional field sampling methods are expensive or difficult to implement (Bundy and Davis, 2013).

HOW CAN LOCAL ECOLOGICAL KNOWLEDGE HELP IN FAUNISTIC SURVEYS AND TAXONOMIC STUDIES?

Knowing the fauna of a certain locality has always been one of the topics most investigated by zoologists and other researchers interested in zoological studies (Greene, 1994; Silveira et al., 2010). The early works of naturalists already involve the compilation of lists of native animals, with local and scientific names (Sillitoe, 2006). With the consolidation of zoology as a science, knowing the richness of animal species has been one of the main focuses of zoological research. As pointed out by Greene (1994), studies on diversity inventories and natural history increase the knowledge of a regional fauna, its interaction with other organisms, and the environment in general. These studies are also an essential tool for decision-making aimed at animal conservation (Gilchrist et al., 2005; Silveira et al., 2010).

LEK has always been and continues to be important in the performance of faunal inventories and determination of geographical distributions of animals (Beaudreau and Levin, 2014; Kotschwar Logan et al., 2015). Accordingly, local people can contribute, among other things, information on the ecology of animals, their location and geographic distribution, habitat use, indication of collection points, and species capture strategies (Irvine et al., 2009). Silveira et al. (2010) emphasized that the results of any faunal sampling are the sum of the techniques used, the ability of the investigator to detect the organisms in the sample space, and the temporal component. In this scenario, local residents possess a series of information about the composition and biology of the fauna present in a given area, mainly related to the species that they exploit in some way and with which they interact more frequently. Thus, LEK can be an important tool as a form of methodological complementarity, and it is necessary that the information obtained be compared and calibrated with data from other methods.

Of course, the animal group to be studied has a direct influence on the ways of obtaining the inventory data, and in general, the knowledge of people who exploit the fauna can complement and assist in the zoological research (Caruso et al., 2017). In ethnoichthyological studies, for example, fish can be purchased directly from local fishermen (Pinnegar and Engelhard, 2008; Pinto et al., 2013). Similarly, according to Alves et al. (2009), the knowledge of hunters on the species and capture techniques used can maximize the collection of terrestrial vertebrate specimens, and can serve as a complement in faunistic inventories. In some cases, an animal that has not been collected using zoological techniques can be captured through techniques adopted by local hunters, who even know the most suitable place for a trap, for example, contributing to increased capture success (Alves et al., 2009; Bezerra et al., 2012). However, the quality of information provided by local people varies according to the animal group. Caruso et al. (2017), for example, analyzed the abundance of three species of large mammals through interviews and camera trapping surveys, and found that the reliability of results collected from surveys varies between species. The interviews indicated an overestimation of the presence of the puma, a elusive but charismatic species that often comes into conflict with people. Msoffe et al. (2007) noted that local people have better knowledge of species that tend to cause more problems than of more rare and shy species.

Vieira et al. (2014) emphasize that interviews with hunters will often supply interesting zoological information, although the accuracy of their observations will depend on obtaining supportive elements from them that will allow more precise identifications of the species and serve as testimonial material. For some animal groups, especially vertebrates, parts derived from animals can in many cases enable the identification of the species, serving as testimonial material. Very often, hunters, fishermen, and users of wildlife products stock products derived from animals, such as skull and skins, which can be used for medicinal, ornamental, and magical-religious purposes (Alves et al., 2012; Barboza et al., 2016; Dias et al., 2011; Fernandes-Ferreira et al., 2012, 2013; Ferreira et al., 2012;

Ferreira et al., 2013; Léo Neto et al., 2009). Such products can be purchased from those interviewed in ethnozoological research. Skeletal remains, shells, and skins that allow the identification of many species of mammals, reptiles, birds, and fish are commonly accepted as testimonial material in zoological collections. In the case of invertebrates, shells of mollusks and exoskeleton of corals are examples of materials that serve as proof (Alves and Souto, 2010; Rosa et al., 2011; Vieira et al., 2014).

Hunters or residents interacting with the fauna of a certain area may also contribute to the search for indirect evidence on the occurrence of animal species. They can locate and identify, for example, traces such as footprints, feces, nests, burrows, hair, regurgitation pellets, etc., which can be confirmed as indicative of the presence of a certain species. An example of how the recognition of indirect evidence of occurrence of fauna can complement/assist in zoological inventories is the work of Zuercher et al. (2003). These authors used molecular analyses of cytochrome-b found in the feces of animals collected by Amerindians and local inhabitants in the Mbaracayu Forest Natural Reserve in Paraguay to identify carnivorous mammals in the area, demonstrating the complementarity of the methods used, which is particularly important in research involving animal groups such as large mammals; such mammals are often elusive and sparsely distributed, which requires large investments of time and money to adequately sample a given population (Sunquist and Sunquist, 2002; Zuercher et al., 2003) and which generally requires a combination of more complete sampling. Although it was not evaluated in this study, the possibility of obtaining information from the carnivorous diet inventories is evident, since from the feces it is possible to identify the animals that make up its diet. In this context, hunters who capture various animals and generally gut them before ingesting them, possess accurate information on the diet of game animals and can provide material to confirm the identification of the prey, being potential partners in studies on the diet of vertebrates.

It should be emphasized that the components of the animal diversity of a biome or locality, in a given space and time, will never be completely sampled (Silveira et al., 2010). Thus, to access the diversity of a given area, it is recommended that more than one sampling method be used and they should be used in a complementary way (Gilchrist and Mallory, 2007). Obviously, in the same way, local knowledge should not be used as the only method in research of this type. Can and Togan (2009) emphasized that information derived from local people and authorities should complement (not replace) information of field surveys. Of course, like other zoological methods, faunal inventories should not be done using only local knowledge but should be used in consortium with other complementary methods (Ainsworth et al., 2008). For example, hunters may indicate locations best suited for the installation of traps, enhancing sampling (Can and Togan, 2009).

Studies involving LEK can provide information on species distribution. One example is the work by Cosham (2015) on the patterns and factors that shape local-scale distributions of the invasive green crab (*Carcinus maenas* L.) in Nova Scotia, which demonstrated that LEK had a greater predilection toward environmental features or landmarks in comparison to the scientific literature, which infers on the patterns of distributions on regional scales. The key advantage of local knowledge is that it covers longer time spans than most scientific studies (which often focus on a detailed snapshot of the situation instead) (Cosham, 2015).

Another fundamental ecological indicator is the basic presence or absence of species in an ecosystem (Coll and Lotze, 2016), and LEK has great contribution potential in this field. Hunters, for example, have extensive knowledge about the occurrence sites of species, indicating even those that occurred before and that no longer occur in a particular area, but also indicating the local declines in abundance of individuals. In the semiarid Brazilian northeast, for example, local hunters recognize the disappearance of medium- and large-sized mammals in various locations; this is only recorded through personal accounts and old pictures. This reinforces the view that factors such as capture and excessive slaughter, coupled with habitat loss, have caused the decrease of large populations and apparent local extinction of many animals in several areas of the region (Alves et al., 2016; Barboza et al., 2016).

In summary, LEK has recognized potential as a source of zoological information for different taxa, from charismatic and large-bodied species (Mallory et al., 2003; Ziembicki et al., 2013) to smaller-bodied, elusive, and noncharismatic animals (Turvey et al., 2014, 2015). It should be emphasized, however, that knowledge about animal species can vary according to taxa and cultural, ecological, and socioeconomic factors, so that the use of LEK as a complementary method in biological inventories and ecological studies should consider this aspect (Bentley, 1994). Large mammals, for example, are commonly described with greater ease and detail by local/traditional communities. On the other hand, with regard to small mammals or insects, the ease of observation or the utility of environmental services provided for them favor popular taxonomy and observation of behavioral aspects and ecology (Starr et al., 2011). For example, Bentley and Rodríguez (2001), in a study on rural Honduran folk entomology, found that local people showed more in-depth knowledge of insects that are culturally important and easily observed.

DISCOVERING NEW SPECIES THROUGH ETHNOZOOLOGICAL STUDIES

Ethnozoological studies have also been important for the discovery of new species, as well as records of geographic distribution and documentation of the local extinction of some species (Jenkins et al., 2005; Ziembicki et al., 2013). Sillitoe (2006) pointed out that the discovery of the hylid frog *Litoria bulmeri* was associated with the ethnoherpetological work of the anthropologist Ralph Bulmer (who was also honored in the naming of the new species). Even when the purpose of ethnozoological research is not directly associated with zoological inventories, new species may be discovered. A good example is the work of Sanamxay et al. (2013). These authors described a new species of the flying squirrel genus *Biswamoyopterus* from Lao PDR. Description was based on a single specimen collected from a local market. The authors observed several species of *Pteromyini* including *Petaurista elegans* Müller, *Hylopetes phayrei* Blyth, and *Petaurista philippensis* Elliot for sale as bushmeat. On subsequent evaluation in the university, this individual was also found to have cranial and dental characters that clearly differentiated it not just from *P. philippensis* but from the genus *Petaurista* itself (Jenkins et al., 2005). Also in Lao PDR, specimens of an unknown species of hystricognathous rodent were discovered in local markets, being sold for food (Jenkins et al., 2005). These findings are evidence that important taxonomic information can be obtained in ethnozoological studies, which, for example, investigate the trade of medicinal animals in public markets or street fairs (Alves and Rosa, 2010; Alves et al., 2013, 2012; Ferreira et al., 2012; Nascimento et al., 2015; Roldán-Clarà et al., 2014; Rosa et al., 2011). Local market surveys can provide an informative overview of the exploitation of animals occurring in the surrounding area and occasionally provide interesting species records (Jenkins et al., 2005) and represent small-scale reproductions of useful fauna in a particular region (Albuquerque et al., 2007; Alves et al., 2013).

LEK AS SOURCE OF SCIENTIFIC INSIGHTS

Since time immemorial, humans have observed the natural world around them, something that was essential for their survival and became the focus of the first scientific developments (Ushakov, 2007). Animals are among the main organisms observed, which is reflected in the knowledge about them and the environment where they live, fundamental in the human activities that involve the exploitation of these resources. This knowledge can provide not only a series of new insights for biological research, but also opportunities for the cross-validation of scientific hypotheses (Alves and Nishida, 2002; Kimmerer, 2002; Nishida et al., 2006). Literature offers several examples that support such assertion. Alves and Nishida (2002) carried out a study based on the knowledge of the harvesters of crabs of the species *Ucides cordatus* and formulated a hypothesis concerning the influence of tides on ecdysis (molting) of this species. Their results indicated that the process of ecdysis in this species in its natural environment lasts from 28 to 29 days—a period very different from that previously obtained in laboratory studies by Nascimento (1993) (who estimated that the molting process lasted 15–20 days). The data obtained in the natural environment for ecdysis are certainly more accurate than those obtained in the laboratory, where it is very difficult to simulate the environmental factors characteristic of the mangrove/estuary environment that influence crab ecdysis. The crab harvesters also furnished a good deal of information about the behavior of *U. cordatus* during ecdysis, as well as during other important stages of their life cycle, evidencing how local zoological knowledge can not only subsidize the formulation of scientific hypotheses, but also complement academic knowledge (Alves and Nishida, 2002; Nascimento et al., 2016, 2017).

Other examples of the importance of LEK have been seen in studies with fishermen, confirming that ethnoecology can provide relevant biological data more rapidly and at lower cost than traditional field research techniques (Lavides et al., 2009; Lopes et al., 2010). A study by Silvano et al. (2006), for example, indicated that Brazilian fishermen had intimate knowledge of the diets and habitats of various local species of fish (some of which were barely known to conventional science). Marques (1991) also recorded an accurate knowledge among fishermen on the diet of an ariid catfish and elaborated a hypothesis (which did not initially seem very plausible) about an important item in the diet of these fish. The author was able, however, to confirm this hypothesis during his work and to add that information about the trophic ecology of that species.

The contribution of traditional zoological knowledge has also been an important source of information in research on abundance and possible changes in populations of animal species (Coll and Lotze, 2016). Such research has been carried out mainly in the fishing area. For example, Eddy et al. (2010) used archival information, fishermen's ecological knowledge, and underwater surveys to reconstruct a 400-year timeline of lobsters (*Jasus frontalis*) in Chile, suggesting modern lobster biomass is only 15% of historical abundance from 1550 to 1750. Similar approaches have been used to document depletions in many marine populations over past centuries or millennia (Coll and Lotze, 2016). In a study of cetaceans regarding their inventory and distribution in China, Turvey et al. (2013) used interview information from fishing communities across the middle-lower Yangtze drainage to investigate spatial and temporal patterns of porpoise abundance, mortality, and population change. The authors demonstrated that data on porpoise numbers and decline collected through interviews with Yangtze fishers show spatial patterns congruent with data from boat-based surveys. Cetacean survey techniques are typically boat based and can be labor- and cost-intensive (Taylor et al., 2007; Turvey et al., 2014), placing restrictions on survey regularity and limiting the ability to detect population trends. In contrast, community interviews are a relatively inexpensive approach for collecting data across wide geographical areas (Moore et al., 2010). Another interesting study was conducted by Beaudreau and Levin (2014) in Puget

Sound, Washington. These authors used LEK to create a historical record of abundance for 22 marine species and quantified the variation in perceptions of abundance trends among fishers, divers, and researchers. As a result, they found a pattern of decline in the abundance of the species investigated over time and a high degree of agreement between respondents. Lastly, the authors concluded that the use of quantitative approaches in studies involving LEK is of extreme importance to evaluate variance in ecological observations of resource users, which can be broadly applied to other species and systems.

RELEVANCE OF LOCAL ECOLOGICAL KNOWLEDGE IN STUDIES ON THE CONSERVATION AND MANAGEMENT OF FAUNA

The main conservation problems affecting several animal species are associated with human activities, whether through direct exploitation (hunting or fishing) or indirect exploitation, such as deforestation and urbanization. Nevertheless, there are also examples of human populations that exploit wildlife resources in a sustainable way. In this scenario, it is evident that conservation and management plans for fauna should consider the interactions of humans with animals and the zoological knowledge of humans.

Several studies have pointed out the contribution of LEK in the evaluation and planning directed to the conservation and management of fauna. The application of LEK and customary ecological management practices to conservation issues has reemerged now and, complemented with scientific information, expands the knowledge base on the status of marine resources (Ainsworth et al., 2008; Drew, 2005). LEK can contribute, for example, to historical information on the population decline of marine and freshwater species (Bender et al., 2013; Saenz-Arroyo et al., 2005), indicating the potential causes of their depletion.

It is important to note that in some cases, fishers failed to provide accurate information on species decline (Lima et al., 2010). Differences in the interpretation of fishery decline may be related to species importance and value (e.g., commercial species), fishing gear and exploited habitat (water depth, coral/rocky reef, mangroves, estuaries), and social position of respondents (Bender et al., 2013; Bush and Hirsch, 2005). In addition, the age of respondents may be correlated with a more accurate perception of species decline over time (Pauly, 1995). This trend has been recorded in several studies. An example is the study by Bender et al. (2013), conducted in Porto Seguro, Brazil, where it was found that knowledge acquired by elder fishermen (>50 years old) can provide valuable insights into the conservation status of reef fish and the adjustment of environmental baselines for proper management of a marine park. Pan et al. (2016) pointed out that the age of interviewees has also been indicated as a variable more correlated with knowledge of the distribution and ecology of animal species. Capistrano and Lopes (2012), in studying the mangrove crab *U. cordatus*, an important resource for the livelihood of many poor families living on the Brazilian coast, found that crab gatherers were able to provide accurate information on crab life history traits, but overestimated their catch. This suggests that information provided by gatherers may have some limitations, although it could provide inexpensive and reliable information, depending on the topic investigated. For this, the use of LEK to rank environmental factors shaping species distributions may be possible but should be undertaken with caution.

In a study of the distribution of carnivore species in central–southeastern Madagascar, LEK provided important conservation insights and revealed novel distribution patterns, besides being congruent and complementary with other data collected in the same region (Kotschwar Logan et al., 2015). According to the authors, a total of 182 interviewed in 17 different communities indicated distinct distribution patterns for two native and two exotic carnivore species, suggesting a range of tolerance to the human-dominated landscape. In this sense, the negative experience of 20% of the local population with wild carnivores (predation in poultry breeding) negatively influenced the perception of carnivore conservation in that region (Kotschwar Logan et al., 2015) and emphasized the importance of local participation in projects and management plans of native animals (Msoffe et al., 2007; Shen et al., 2012).

It is important to point out that comanagement requires contributions from both resource users and government management agencies. Just as a community should not simply accept that a species is declining without knowing how scientists arrived at that interpretation, so too should researchers be able to evaluate community perspectives on an issue (Gilchrist and Mallory, 2007). Ethnozoological studies may also help in promoting dialogue and cooperation between those with LEK and scientists, aimed at promoting the sustainable exploitation of faunistic resources. On-site communities with extensive LEK may be useful for collaborative conservation and management. Examples of such practices can be found at the Mamirauá Sustainable Development Reserve, Brazil (Castello, 2004; Gillingham, 2001), in Canada (Pomeroy and Berkes, 1997), in Zimbabwe (Child, 1996), in the Philippines (Pomeroy and Carlos, 1997), and in Pacific Islands (Michael and Lambeth, 2000). Exclusion of local communities from consultation/decision processes, on the other hand, may lead to the construction of public policies devoid of historical information and without much social resonance (Alves and Rosa, 2007).

FINAL REMARKS

Animals have always aroused the interest of humans and have certainly been observation targets since time immemorial. From the daily observations and interactions with fauna emerged an informed knowledge that was fundamental for human survival and has been an important source of information in zoological research. Knowledge about fauna is more pronounced among people who interact more frequently with animals, such as hunters, fishermen, and breeders and includes various aspects of animal biology, and therefore, it can support scientific studies related to wildlife. In this scenario, ethnozoology represents an important mediator in the dialogue between LEK and SEK within a collaborative framework.

References

Ainsworth, C.H., Pitcher, T.J., Rotinsulu, C., 2008. Evidence of fishery depletions and shifting cognitive baselines in Eastern Indonesia. Biological Conservation 141, 848–859.

Albuquerque, U.P., Monteiro, J.M., Ramos, M.A., Amorim, E.L.C., 2007. Medicinal and magic plants from a public market in northeastern Brazil. Journal of Ethnopharmacology 110, 76–91.

Alves, R.R.N., Nishida, A., Hernandez, M., 2005. Environmental perception of gatherers of the crab 'caranguejo-uca' (*Ucides cordatus*, Decapoda, Brachyura) affecting their collection attitudes. Journal of Ethnobiology and Ethnomedicine 1, 1–10.

Alves, R.R.N., Nishida, A., 2002. A ecdise do caranguejo-uçá, *Ucides cordatus* L.\(Decapoda, Brachyura) na visão dos caranguejeiros. Interciencia 27, 110–117.

Alves, R.R.N., Alves, H.N., 2011. The faunal drugstore: animal-based remedies used in traditional medicines in Latin America. Journal of Ethnobiology and Ethnomedicine 7 (9), 1–43.

Alves, R.R.N., Feijó, A., Barboza, R.R.D., Souto, W.M.S., Fernandes-Ferreira, H., Cordeiro-Estrela, P., Langguth, A., 2016. Game mammals of the Caatinga biome. Ethnobiology and Conservation 5, 1–51.

Alves, R.R.N., Mendonça, L.E.T., Confessor, M.V.A., Vieira, W.L.S., Lopez, L.C.S., 2009. Hunting strategies used in the semi-arid region of northeastern Brazil. Journal of Ethnobiology and Ethnomedicine 5, 1–50.

Alves, R.R.N., Rosa, I.L., 2007. Biodiversity, traditional medicine and public health: where do they meet? Journal of Ethnobiology and Ethnomedicine 3 (14), 1–9.

Alves, R.R.N., Rosa, I.L., 2010. Trade of animals used in Brazilian traditional medicine: trends and implications for conservation. Human Ecology 38, 691–704.

Alves, R.R.N., Rosa, I.L., Albuquerque, U.P., Cunningham, A.B., 2013. Medicine from the Wild: An Overview of the Use and Trade of Animal Products in Traditional Medicines, Animals in Traditional Folk Medicine. Springer, pp. 25–42.

Alves, R.R.N., Rosa, I.L., Léo Neto, N.A., Voeks, R., 2012. Animals for the gods: magical and religious faunal use and trade in Brazil. Human Ecology 40, 751–780.

Alves, R.R.N., Souto, W.M.S., 2010. Desafios e dificuldades associadas as pesquisas etnozoológicas no Brasil. In: Alves, R.R.N., Souto, W.M.S., Mourão, J.S. (Eds.), A Etnozoologia no Brasil: Importância, Status atual e Perspectivas. NUPEEA, Recife.

Alves, R.R.N., Souto, W.M.S., 2015. Ethnozoology: a brief introduction. Ethnobiology and Conservation 4, 1–13.

Anadón, J.D., Giménez, A., Ballestar, R., 2010. Linking local ecological knowledge and habitat modelling to predict absolute species abundance on large scales. Biodiversity and Conservation 19, 1443–1454.

Aylesworth, L., Phoonsawat, R., Suvanachai, P., Vincent, A.C.J., 2017. Generating spatial data for marine conservation and management. Biodiversity and Conservation 26, 383–399.

Barboza, R.R.D., Lopes, S.F., Souto, W.M.S., Fernandes-Ferreira, H., Alves, R.R.N., 2016. The role of game mammals as bushmeat in the Caatinga, northeast Brazil. Ecology and Society 21, 1–11.

Bart, D., 2006. Integrating local ecological knowledge and manipulative experiments to find the causes of environmental change. Frontiers in Ecology and the Environment 4, 541–546.

Beaudreau, A.H., Levin, P.S., 2014. Advancing the use of local ecological knowledge for assessing data-poor species in coastal ecosystems. Ecological Applications 24, 244–256.

Begossi, A., Silvano, R.A.M., 2008. Ecology and ethnoecology of dusky grouper [garoupa, *Epinephelus marginatus* (Lowe, 1834)] along the coast of Brazil. Journal of Ethnobiology and Ethnomedicine 4.

Bender, M.G., Floeter, S.R., Hanazaki, N., 2013. Do traditional fishers recognise reef fish species declines? Shifting environmental baselines in Eastern Brazil. Fisheries Management and Ecology 20, 58–67.

Bentley, J.W., 1994. Stimulating peasant farmer experiments in non-chemical pest control in Central America. In: Scoones, I., Thompson, J. (Eds.), Beyond Farmer First: Rural People's Knowledge, Agricultural Research and Extension Practice. Intermediate Technology Publications, London, pp. 147–150.

Bentley, J.W., Rodríguez, G., 2001. Honduran folk entomology. Current Anthropology 42, 285–300.

Berkes, F., 1993. Traditional ecological knowledge in perspective. In: Inglis, J.T. (Ed.), Traditional Ecological Knowledge: Concepts and Cases, first ed. International Program on Traditional Ecological Knowledge/ International Development Research Centre, Ottawa, Canadá, pp. 1–10.

Berkes, F., 1999. Sacred Ecology: Traditional Ecological Knowledge and Resource Management, first ed. Taylor & Francis, Philadelphia, USA.

Berkes, F., 2009. Evolution of co-management: role of knowledge generation, bridging organizations and social learning. Journal of Environmental Management 90, 1692–1702.

Berkes, F., Colding, J., Folke, C., 2000. Rediscovery of traditional ecological knowledge as adaptative management. Ecological Applications 10, 1251–1262.

Bezerra, D.M.M., Araujo, H.F.P., Alves, R.R.N., 2012. Captura de aves silvestres no semiárido brasileiro: técnicas cinegéticas e implicações para conservação. Tropical Conservation Science 5, 50–66.

Brinkman, T.J., Chapin, T., Kofinas, G., Person, D.K., 2009. Linking hunter knowledge with forest change to understand changing deer harvest opportunities in intensively logged landscapes. Ecology and Society 14.

Brook, R.K., McLachlan, S.M., 2008. Trends and prospects for local knowledge in ecological and conservation research and monitoring. Biodiversity and Conservation 17, 3501–3512.

Bundy, A., Davis, A., 2013. Knowing in context: an exploration of the interface of marine harvesters' local ecological knowledge with ecosystem approaches to management. Marine Policy 38, 277–286.

Bush, S.R., Hirsch, P., 2005. Framing fishery decline. Aquatic Resources, Culture and Development 1, 79–90.

Can, Ö.E., Togan, İ., 2009. Camera trapping of large mammals in Yenice Forest, Turkey: local information versus camera traps. Oryx 43, 427–430.

Capistrano, J.F., Lopes, P.F.M.L., 2012. Crab gatherers perceive concrete changes in the life history traits of *Ucides cordatus* (Linnaeus, 1763), but overestimate their past and current catches. Ethnobiology and Conservation 1, 1–21.

Caruso, N., Vidal, E.L., Guerisoli, M., Lucherini, M., 2017. Carnivore occurrence: do interview-based surveys produce unreliable results? Oryx 51, 240–245.

Carvalho, A.R., Williams, S., January, M., Sowman, M., 2009. Reliability of community-based data monitoring in the Olifants river estuary (South Africa). Fisheries Research 96, 119–128.

Castello, L., 2004. A method to count pirarucu *Arapaima gigas*: fishers, assessment, and management. North American Journal of Fisheries Management 24, 379–389.

Child, B., 1996. The practice and principles of community-based wildlife management in Zimbabwe: the CAMPFIRE programme. Biodiversity and Conservation 5, 369–398.

Coll, M., Lotze, H.K., 2016. Ecological indicators and food-web models as tools to study historical changes in marine ecosystems. In: Máñez, S.K., Poulsen, B. (Eds.), Perspectives on Oceans Past. Springer, Dordrecht, pp. 103–132.

Cosham, J.A., 2015. Using Local and Scientific Perspectives to Understand Factors Affecting the Distribution of Invasive Green Crab (*Carcinus maenas* L.) (Master's thesis). Nova Scotia, Halifax.

Dantas, C.A., Silva, R.R.V., Silveira, P.C.B., Silva, A.C.B.L., Alves, A.G.C., 2016. "They call me a woodsman": cognitive and social aspects on the relationships between woodsmen and forest researchers. Ethnobiology and Conservation 5, 1–10.

Dias, T.L.P., Leo Neto, N.A., Alves, R.R.N., 2011. Molluscs in the marine curio and souvenir trade in NE Brazil: species composition and implications for their conservation and management. Biodiversity and Conservation 20, 2393–2405.

Drew, J.A., 2005. Use of traditional ecological knowledge in marine conservation. Conservation Biology 19, 1286–1293.

Eddy, T.D., Gardner, J.P.A., Pérez-Matus, A., 2010. Applying fishers' ecological knowledge to construct past and future lobster stocks in the Juan Fernández Archipelago, Chile. PLoS One 5, e13670.

Fernandes-Ferreira, H., Mendonça, S.V., Albano, C., Ferreira, F.S., Alves, R.R.N., 2012. Hunting, use and conservation of birds in Northeast Brazil. Biodiversity and Conservation 21 (1), 221–244.

Fernandes-Ferreira, H., Mendonca, S.V., Cruz, R.L., Borges-Nojosa, D.M., Alves, R.R.N., 2013. Hunting of herpetofauna in montane, coastal, and dryland areas of Northeastern Brazil. Herpetological Conservation and Biology 8, 652–666.

Ferreira, F.S., Albuquerque, U.P., Coutinho, H.D.M., Almeida, W.O., Alves, R.R.N., 2012. The trade in medicinal animals in northeastern Brazil. Evidence-Based Complementary and Alternative Medicine 2012, 1–20.

Ferreira, F.S., Fernandes-Ferreira, H., Leo Neto, N., Brito, S.V., Alves, R.R.N., 2013. The trade of medicinal animals in Brazil: current status and perspectives. Biodiversity and Conservation 22, 839–870.

Gadgil, M., Berkes, F., Folke, C., 1993. Indigenous knowledge for biodiversity conservation. Ambio 22, 151–156.

Gaidet, N., Fritz, H., Nyahuma, C., 2003. A participatory counting method to monitor populations of large mammals in non-protected areas: a case study of bicycle counts in the Zambezi Valley, Zimbabwe. Biodiversity and Conservation 12, 1571–1585.

Gilchrist, G., Mallory, M., Merkel, F., 2005. Can local ecological knowledge contribute to wildlife management? Case studies of migratory birds. Ecology and Society 10, 20.

Gilchrist, G., Mallory, M.L., 2007. Comparing expert-based science with local ecological knowledge: what are we afraid of. Ecology and Society 12, r1.

Gillingham, S., 2001. Social organization and participatory resource management in Brazilian ribeirinho communities: a case study of the Mamiraua Sustainable Development Reserve, Amazonas. Society & Natural Resources 14, 803–814.

Greene, H.W., 1994. Systematics and natural history, foundations for understanding and conserving biodiversity. American Zoologist 34, 48–56.

Hobson, G., 1992. Traditional knowledge is science. Northern Perspectives 20, 2–3.

Holck, M.H., 2008. Participatory forest monitoring: an assessment of the accuracy of simple cost–effective methods. Biodiversity and Conservation 17, 2023–2036.

Humber, F., Godley, B.J., Nicolas, T., Raynaud, O., Pichon, F., Broderick, A., 2016. Placing Madagascar's marine turtle populations in a regional context using community-based monitoring. Oryx 1–12. http://dx.doi.org/10.1017/S0030605315001398.

Humber, F., Godley, B.J., Ramahery, V., Broderick, A.C., 2011. Using community members to assess artisanal fisheries: the marine turtle fishery in Madagascar. Animal Conservation 14, 175–185.

Huntington, H.P., 2000. Using traditional ecological knowledge in science: methods and applications. Ecological Applications 10, 1270–1274.

Huntington, H.P., Suydam, R.S., Rosenberg, D.H., 2004. Traditional knowledge and satellite tracking as complementary approaches to ecological understanding. Environmental Conservation 31, 177–180.

Irvine, R.J., Fiorini, S., Yearley, S., McLeod, J.E., Turner, A., Armstrong, H., White, P.C.L., Van Der Wal, R., 2009. Can managers inform models? Integrating local knowledge into models of red deer habitat use. Journal of Applied Ecology 46, 344–352.

Jenkins, P.D., Kilpatrick, C.W., Robinson, M.F., Timmins, R.J., 2005. Morphological and molecular investigations of a new family, genus and species of rodent (Mammalia: Rodentia: Hystricognatha) from Lao PDR. Systematics and Biodiversity 2, 419–454.

Johannes, R.E., 1993. Integrating traditional ecological knowledge and management with environmental impact assessment. In: Inglis, J.T. (Ed.), Traditional Ecological Knowledge: Concepts and Cases. International Program on Traditional Ecological Knowledge and International Development Research Centre, Ottwa, pp. 33–39.

Kimmerer, R.W., 2002. Weaving traditional ecological knowledge into biological education: a call to action. BioScience 52, 432–438.

Kotschwar Logan, M., Gerber, B.D., Karpanty, S.M., Justin, S., Rabenahy, F.N., 2015. Assessing carnivore distribution from local knowledge across a human-dominated landscape in central-southeastern Madagascar. Animal Conservation 18, 82–91.

Lavides, M.N., Polunin, N.V.C., Stead, S.M., Tabaranza, D.G., Comeros, M.T., Dongallo, J.R., 2009. Finfish disappearances around Bohol, Philippines inferred from traditional ecological knowledge. Environmental Conservation 36, 235–244.

Léo Neto, N.A., Brooks, S.E., Alves, R.R.N., 2009. From Eshu to Obatala: animals used in sacrificial rituals at Candomble "terreiros" in Brazil. Journal of Ethnobiology and Ethnomedicine 5, 1–23.

Lima, F.P., Latini, A.O., Júnior, M.P., 2010. How are the lakes? Environmental perception by fishermen and alien fish dispersal in Brazilian tropical lakes. Interciencia 35, 84–90.

Longo, O., 2004. Tackling Aristotelian ethnozoology. In: Sanga, G., Ortalli, G. (Eds.), Nature Knowledge. Bergham, New York, Oxford, pp. 57–67.

Lopes, P.F.M., Silvano, R., Begossi, A., 2010. Da Biologia a Etnobiologia – Taxonomia e etnotaxomia, ecologia e etnoecologia. In: Alves, R.R.N., Souto, W.M.S., Mourão, J.S. (Eds.), A Etnozoologia no Brasil: Importância, Status atual e Perspectivas, first ed. NUPEEA, Recife, PE, Brazil, pp. 67–94.

López-Arévalo, H.F., Gallina, S., Landgrave, R., Martínez-Meyer, E., Muñoz-Villers, L.E., 2011. Local knowledge and species distribution models' contribution towards mammalian conservation. Biological Conservation 144, 1451–1463.

Mackinson, S., Nottestad, L., 1998. Combining local and scientific knowledge. Reviews in Fish Biology and Fisheries 8, 481–490.

Maffi, L., Skutnabb-Kangas, T., Andrianarivo, J., 1999. Language and diversity. In: Posey, D.A. (Ed.), Cultural and Spiritual Values of Biodiversity. Intermediate Technology Publications Ltd./UNEP, London.

Magner, L.N., 2002. A History of the Life Sciences, Revised and Expanded, third ed. CRC Press, New York, Basel.

Mallory, M.L., Gilchrist, H.G., Fontaine, A.J., Akearok, J.A., 2003. Local ecological knowledge of ivory gull declines in Arctic Canada. Arctic 56, 293–298.

Marques, J.G.W., 1991. Aspectos ecológicos na etnoictiologia dos pescadores do complexo estuarino-lagunar Mundaú-Manguaba, Alagoas. Universidade Estadual de Campinas 274.

Marques, J.G.W., 1995. Pescando pescadores: etnoecologia abrangente no baixo São Francisco alagoano. NUPAUB-USP, São Paulo, BR.

Michael, K., Lambeth, L., 2000. Fisheries management by communities: a manual on promoting the management of subsistence fisheries by Pacific Island communities. Journal of Ethnopharmacology. Secretariat of the Pacific Community.

Moller, H., Berkes, F., Lyver, P.O.B., Kislalioglu, M., 2004. Combining science and traditional ecological knowledge: monitoring populations for co-management. Ecology and Society 9.

Moore, J.E., Cox, T.M., Lewison, R.L., Read, A.J., Bjorkland, R., McDonald, S.L., Crowder, L.B., Aruna, E., Ayissi, I., Espeut, P., 2010. An interview-based approach to assess marine mammal and sea turtle captures in artisanal fisheries. Biological Conservation 143, 795–805.

Moreira, I.C., 2002. O escravo do naturalista – O papel do conhecimento nativo nas viagens científicas do século 19. Ciência Hoje 31, 40–48.

Msoffe, F., Mturi, F.A., Galanti, V., Tosi, W., Wauters, L.A., Tosi, G., 2007. Comparing data of different survey methods for sustainable wildlife management in hunting areas: the case of Tarangire–Manyara ecosystem, northern Tanzania. European Journal of Wildlife Research 53, 112–124.

Nascimento, C.A.R., Czaban, R.E., Alves, R.R.N., 2015. Trends in illegal trade of wild birds in Amazonas state, Brazil. Tropical Conservation Science 8, 1098–1113.

Nascimento, D.M., Alves, A.G.C., Alves, R.R.N., Barboza, R.R.D., Diele, K., Mourão, J.S., 2016. An examination of the techniques used to capture mangrove crabs, *Ucides cordatus*, in the Mamanguape River estuary, northeastern Brazil, with implications for management. Ocean & Coastal Management 130, 50–57.

Nascimento, D.M., Alves, R.R.N., Barboza, R.R.D., Schmidt, A.J., Diele, K., Mourão, J.S., 2017. Commercial relationships between intermediaries and harvesters of the mangrove crab *Ucides cordatus* (Linnaeus, 1763) in the Mamanguape river estuary, Brazil, and their socio-ecological implications. Ecological Economics 131, 44–51.

Nascimento, S.A., 1993. Biologia do caranguejo-uçá *Ucides cordatus*. Administração Estadual do Meio Ambiente (ADEMA), Sergipe, p. 45.

Nishida, A.K., Nordi, N., Alves, R.R.N., 2006. Molluscs production associated to lunar-tide cycle: a case study in Paraíba state under ethnoecology viewpoint. Journal of Ethnobiology and Ethnomedicine 2, 1–6.

Nordi, N., Nishida, A.K., Alves, R.R.N., 2009. Effectiveness of two gathering techniques for *Ucides cordatus* in northeast Brazil: implications for the sustainability of mangrove ecosystems. Human Ecology 37, 121–127.

Pan, Y., Wei, G., Cunningham, A.A., Li, S., Chen, S., Milner-Gulland, E.J., Turvey, S.T., 2016. Using local ecological knowledge to assess the status of the critically endangered Chinese giant salamander *Andrias davidianus* in Guizhou province, China. Oryx 50, 257–264.

Pauly, D., 1995. Anecdotes and the shifting baseline syndrome of fisheries. Trends in Ecology and Evolution 10, 430.

Pierotti, R., Wildcat, D., 2000. Traditional ecological knowledge: the third alternative (commentary). Ecological Applications 10, 1333–1340.

Pinnegar, J.K., Engelhard, G.H., 2008. The 'shifting baseline' phenomenon: a global perspective. Reviews in Fish Biology and Fisheries 18, 1–16.

Pinto, M.F., Mourão, J.S., Alves, R.R.N., 2013. Ethnotaxonomical considerations and usage of ichthyofauna in a fishing community in Ceará State, Northeast Brazil. Journal of Ethnobiology and Ethnomedicine 9, 1–11.

Pomeroy, R.S., Berkes, F., 1997. Two to tango: the role of government in fisheries co-management. Marine Policy 21, 465–480.

Pomeroy, R.S., Carlos, M.B., 1997. Community-based coastal resource management in the Philippines: a review and evaluation of programs and projects, 1984–1994. Marine Policy 21, 445–464.

Rist, L., Shaanker, R.U., Milner-Gulland, E.J., Ghazoul, J., 2010. The use of traditional ecological knowledge in forest management: an example from India. Ecology and Society 15, 3.

Rojchanaprasart, N., Tongnunui, P., Tinnungwattana, W., 2014. Comparison between traditional ecological knowledge of coastal villagers in Thailand and scientific ecological knowledge regarding Dugong. Kasetsart Journal: Social Sciences 35, 36.

Roldán-Clarà, B., Lopez-Medellín, X., Espejel, I., Arellano, E., 2014. Literature review of the use of birds as pets in Latin-America, with a detailed perspective on Mexico. Ethnobiology and Conservation 3, 1–18.

Rosa, I.L., Oliveira, T.P.R., Osório, F.M., Moraes, L.E., Castro, A.L.C., Barros, G.M.L., Alves, R.R.N., 2011. Fisheries and trade of seahorses in Brazil: historical perspective, current trends, and future directions. Biodiversity and Conservation 20, 1951–1971.

Rosa, I.M.L., Alves, R.R.N., Bonifácio, K.M., Mourão, J.S., Osório, F.M., Oliveira, T.P.R., Nottingham, M.C., 2005. Fishers' knowledge and seahorse conservation in Brazil. Journal of Ethnobiology and Ethnomedicine 1, 1–15.

Ruddle, K., 1994. Local knowledge in the folk management of fisheries and coastal marine environments. In: Dyer, C.L., McGoodwin, J.R. (Eds.), Folk Management in the World's Fisheries. University Press of Colorado, Niwot, Colorado.

Saenz-Arroyo, A., Roberts, C.M., Torre, J., Cariño-Olvera, M., Enríquez-Andrade, R.R., 2005. Rapidly shifting environmental baselines among fishers of the Gulf of California. Proceedings of the Royal Society of London B: Biological Sciences 272, 1957–1962.

Sanamxay, D., Douangboubpha, B., Bumrungsri, S., Xayavong, S., Xayaphet, V., Satasook, C., Bates, P.J.J., 2013. Rediscovery of *Biswamoyopterus* (Mammalia: Rodentia: Sciuridae: Pteromyini) in Asia, with the description of a new species from Lao PDR. Zootaxa 3686, 471–481.

Shen, X., Li, S., Chen, N., Li, S., McShea, W.J., Lu, Z., 2012. Does science replace traditions? Correlates between traditional Tibetan culture and local bird diversity in Southwest China. Biological Conservation 145, 160–170.

Sillitoe, P., 2006. Ethnobiology and applied anthropology: rapprochement of the academic with the practical. Journal of the Royal Anthropological Institute 12, S119–S142.

Sillitoe, P., Bentley, J.W., Brokensha, D., Cleveland, D.A., Ellen, R., Ferradas, C., Forsyth, T., Posey, D.A., Stirrat, R.L., Stone, M.P., 1998. The development of indigenous knowledge: a new applied anthropology. Current Anthropology 39, 223–252.

Silvano, R.A.M., MacCord, P.F.L., Lima, R.V., Begossi, A., 2006. When does this fish spawn? Fishermen's local knowledge of migration and reproduction of Brazilian coastal fishes. Environmental Biology of Fishes 76, 371–386.

Silveira, L.F., Beisiegel, B.M., Curcio, F.F., Valdujo, P.H., Dixo, M., Verdade, V.K., Mattox, G.M.T., Cunningham, P.T.M., 2010. Para que servem os inventários de fauna? Estudos Avançados 24, 173–207.

Starr, C., Nekaris, K.A.I., Streicher, U., Leung, L.K.P., 2011. Field surveys of the vulnerable pygmy slow loris *Nycticebus pygmaeus* using local knowledge in Mondulkiri province, Cambodia. Oryx 45, 135–142.

Sunquist, M., Sunquist, F., 2002. Wild Cats of the World. The University of Chicago Press, Chicago, USA.

Taylor, B.L., Martinez, M., Gerrodette, T., Barlow, J., Hrovat, Y.N., 2007. Lessons from monitoring trends in abundance of marine mammals. Marine Mammal Science 23, 157–175.

Tidemann, S., Gosler, A., 2010. Ethno-ornithology: Birds, Indigenous People, Culture and Society, first ed. Earthscan/James & James.

Turner, R.K., 2000. Integrating natural and socio-economic science in coastal management. Journal of Marine Systems 25, 447–460.

Turvey, S.T., Fernández-Secades, C., Nuñez-Miño, J.M., Hart, T., Martinez, P., Brocca, J.L., Young, R.P., 2014. Is local ecological knowledge a useful conservation tool for small mammals in a Caribbean multicultural landscape? Biological Conservation 169, 189–197.

Turvey, S.T., Risley, C.L., Moore, J.E., Barrett, L.A., Yujiang, H., Xiujiang, Z., Kaiya, Z., Ding, W., 2013. Can local ecological knowledge be used to assess status and extinction drivers in a threatened freshwater cetacean? Biological Conservation 157, 352–360.

Turvey, S.T., Trung, C.T., Quyet, V.D., Nhu, H.V., Thoai, D.V., Tuan, V.C.A., Hoa, D.T., Kacha, K., Sysomphone, T., Wallate, S., 2015. Interview-based sighting histories can inform regional conservation prioritization for highly threatened cryptic species. Journal of Applied Ecology 52, 422–433.

Ushakov, I., 2007. Histories of Scientific Insights. Lulu.com.

Vieira, K.S., Vieira, W.L.S., Alves, R.R.N., 2014. An introduction to zoological taxonomy and the collection and preparation of zoological specimens. In: Albuquerque, U.P., Cunha, L.V.F.C., Lucena, R.F.P., Alves, R.R.N. (Eds.), Methods and Techniques in Ethnobiology and Ethnoecology. Springer New York,Heidelberg, Dordrecht, London, pp. 175–196.

White, P.C.L., Jennings, N.V., Renwick, A.R., Barker, N.H.L., 2005. Questionnaires in ecology: a review of past use and recommendations for best practice. Journal of Applied Ecology 42, 421–430.

Ziembicki, M.R., Woinarski, J.C.Z., Mackey, B., 2013. Evaluating the status of species using Indigenous knowledge: novel evidence for major native mammal declines in northern Australia. Biological Conservation 157, 78–92.

Zuercher, G.L., Gipson, P.S., Stewart, G.C., 2003. Identification of carnivore feces by local peoples and molecular analyses. Wildlife Society Bulletin 31, 961–970.

Zukowski, S., Curtis, A., Watts, R.J., 2011. Using fisher local ecological knowledge to improve management: the Murray crayfish in Australia. Fisheries Research 110, 120–127.

CHAPTER

25

Ethnozoology and Animal Conservation*

Rômulo Romeu Nóbrega Alves[1], Josivan Soares Silva[2], Leonardo da Silva Chaves[2], Ulysses Paulino Albuquerque[3]

[1]Universidade Estadual da Paraíba, Campina Grande, Brazil; [2]Universidade Federal Rural de Pernambuco, Recife, Brazil; [3]Universidade Federal de Pernambuco, Recife, Brazil

INTRODUCTION

Modern humans are but one of the many species of animals that exist on the planet, yet they interact with numerous others in myriad ways, forming networks of relationships that influence the natural environments they share. Historically, humans have interacted with their environment for a variety of reasons related directly to their requirements for subsistence, such as for food or medicine, or in the search for new natural sites (Alves and Souto, 2015). Like any other animal, of course, humans have always affected the environment in which they live, but for much of their existence the impact had always been local or regional (Artaxo, 2014).

As a result of both genetic and cultural evolution, humans have become the dominant species on earth today (Ehrlich and Ehrlich, 2008), a condition that has magnified their influence on biodiversity. These authors emphasize that "*Homo sapiens* has become so powerful that it can significantly undermine the ability of the Earth's environment to support much of life—including its own." As the human population has grown, and the power of its technology has expanded, the scope and nature of its impact has changed drastically. Until recently, the term "human-dominated ecosystems" elicited images of agricultural fields, pastures, or urban landscapes; now it applies, with greater or lesser force, to the entire earth. Many ecosystems are directly dominated by humanity, and no ecosystem on the earth's surface is free from its pervasive influence (Vitousek et al., 1997). The impact of humans on the planet has been so marked that in the last two centuries researchers have assimilated a new term, the Anthropocene (the Human Age), to refer to the geological epoch that they dominate, dating from when human activities started to have a significant global impact on the earth's geology and ecosystems (Monastersky, 2015).

Humanities' actions of destroying, degrading, and polluting the earth's natural habitats have placed them in a deep environmental

*This chapter is a revised and updated version of the section "Ethnozoology and conservation" from the paper: "Alves, R.R.N., 2012. Relationships between fauna and people and the role of ethnozoology in animal conservation. Ethnobiology and Conservation 1, 1–69. "

Ethnozoology
http://dx.doi.org/10.1016/B978-0-12-809913-1.00025-9

crisis—indeed, virtually all habitats on earth have felt the influence of humans as the dominant species. As a result, the vast majority of populations and species of plants and animals—which are key working parts of human life-support systems—are in decline, and many are already extinct (Sodhi and Ehrlich, 2010). Furthermore, population growth and technological development have enabled humans to increase their efficiency at obtaining natural resources (Alves et al., 2009; Bezerra et al., 2012; Diamond, 2002).

In general, humans commonly interact with natural resources in a variety of ways, which vary according to the culture and the environment in which they occur. From a biological perspective, it is not surprising that human beings have become agents of ecological change and environmental disturbance in the environments in which they live, considering that all species transform their surroundings. What is striking, however, is the success humans have had adapting to other species, which reflects their incredible ability to influence virtually all biodiversity on the planet. Therefore, understanding the context in which human beings and biodiversity interact is critical to the establishment of successful conservation strategies. Regarding fauna in particular, ethnozoologists have proven successful at providing fundamental contributions to conservation and thus human survival given the high degree of human dependence on these resources. In this chapter we address these direct interactions between humans and their co-occurring fauna, and the subsequent impacts they have. We conclude by addressing the potential role of ethnozoology as a mediator between human exploitation of fauna and its conservation.

THE ROLE OF ETHNOBIOLOGY IN BIODIVERSITY CONSERVATION

The global biodiversity crisis, caused mainly by anthropogenic activities, makes it abundantly clear that *ethnobiology* can contribute considerably to *conservation biology*, especially considering that these disciplines share many similarities. As pointed out by Soule (1985), conservation biology was originally conceptualized as a "crisis" discipline, with the goal of providing the principles and tools for preserving biodiversity. Newing (2010) noted that, like ethnobiology, conservation biology has a long history as a concept, but it emerged as an academic discipline in the 1980s, largely in response to the increasingly urgent need to address the perceived biodiversity crisis (Noss, 1999).

The mission to preserve biodiversity is clearly value driven and implies urgency, yet the techniques of conservation biology are scientific, with research questions and methods being derived from a broad range of pure and applied fields (Saunders, 2003). Indeed, many early conservation biologists were field biologists whose study sites faced an immediate threat of destruction from the impact of human activities; yet, their professional training in ecology left them ill-equipped to deal with these threats. Conservation biology was then, from the beginning, an overtly mission-oriented discipline (Meine et al., 2006; Noss, 1999). It aimed to develop new, applied, and interdisciplinary perspectives and to produce a generation of professionals who were better equipped to address the "human dimensions" of biodiversity conservation in a changing global context (Buscher and Wolmer, 2007). Most of the disciplines contributing to conservation biology have been related to the natural sciences, but conservation biologists had long acknowledged that biological knowledge alone is not sufficient for solving conservation problems (Mascia et al., 2003). In fact, Lidicker et al. (1998) emphasized that "conservation needs conservation biologists for sure, but it also needs conservation sociologists, conservation political scientists, conservation chemists, conservation economists, conservation psychologists, and conservation humanitarians." We add to this list conservation ethnobiologists, as they fit perfectly within this context.

HUMAN INFLUENCE ON ANIMAL BIODIVERSITY

Throughout their history, humans have acquired from their environment the natural resources necessary for their survival, thus influencing their surroundings to a greater or lesser extent, depending on the particular activity being undertaken. The influence of anthropic activities on other animals varies according to its intensity over time and depends on the particular human cultures involved, and has not always been what was expected. The exploitation of fauna, including by hunting and the use of primitive fishing techniques, which are still perpetuated by some traditional communities, produces minimal pressure on the populations of the exploited animals, and, in most cases, is sustainable.

In the last few decades, however, the human population has grown, and the ways in which biodiversity is exploited have changed to meet increasing demands. According to the United Nations (2015) report "World Population Prospects: The 2015 Revision," the human population is set to reach 8.5 billion by 2030, 9.7 billion in 2050, and 11.2 billion in 2100, meaning a greater demand for food, water, and energy. As a consequence, pressures on wild species and natural ecosystems will become increasingly severe, resulting in more hunting, fishing, and other forms of exploitation of faunal resources. Unfortunately, overexploitation by humans, combined with loss of suitable habitat, has resulted in many species of animals becoming endangered in the wild. Practically all threats that affect wild animal species included in the IUCN Red List (see http://www.iucnredlist.org/) are direct or indirectly associated with anthropogenic activities. Among the human activities that have had the greatest impacts on a variety of animal taxa are direct exploitation through hunting and fishing; illegal trade; loss and fragmentation of habitats; conflicts with people; invasions by exotic species; introduction of exotic diseases; contamination of the soil, water, and atmosphere by pollutants; and climate change.

An example of a direct impact from human exploitation of animal biodiversity is defaunation. Defaunation is the reduction in abundance of individuals of animal populations on a local or regional scale, or even their complete extinction (Van Vliet et al., 2016). The consequences of a reduction in biodiversity can be equated to other large-scale global deleterious changes (Hooper et al., 2012), since negative impacts can occur both directly and through a cascade of effects that cause problems that extend beyond a simple reduction in species diversity. The various levels that make up an ecosystem are interdependent, and changes in any one of the levels can cause disturbances throughout the system (Terborgh et al., 2010), but the consequences are more visible when the diversity of consumed organisms is affected (Cardinale et al., 2012).

Defaunation can have direct consequences for the health of a large part of the human population due to the reduction of ecological goods and services (Myers et al., 2013). The overexploitation of wild animals to meet the world demand for traditional medicine, for example, threatens the natural populations of many species of important medical interest (Alves and Rosa, 2013; Still, 2003). The hunting of tigers and rhinos to meet the intense demand of traditional Asian medicine is perhaps the most dramatic example of how the unsustainable exploitation of these resources can threaten wildlife conservation (Ellis, 2005).

The hunting of wild animals, whether for food, medicine (Figs. 25.1 and 25.2), or other purposes, also causes significant changes in ecosystems due to cascade effects. In tropical forests, birds and mammals in particular play key roles in the reproductive cycle of angiosperms through seed dispersal (Fleming and Kress, 2011), since the form of dispersal most commonly found in tropical forests, and even in semiarid regions, is zoochory

FIGURE 25.1 Some examples of lizards hunted for food and medicine in Latin America. *Salvator merianae* (Duméril and Bibron, 1839) (left) and *Iguana iguana* (Linnaeus, 1758) (right). *Photo credits: John Philip Medcraft.*

FIGURE 25.2 Examples of birds and primates used as pets. (A) Orangutan living in captivity in Selimbau village, Sentarum lake, West Kalimantan, Indonesia; animals kept as pets in Brazil: *Callithrix jacchus* (Linnaeus, 1758)—White tufted ear marmoset (B), *Paroaria dominicana* (Linnaeus, 1758)—red-cowled Cardinal (C), and (D) *Eupsittula cactorum* (Kuhl, 1820)—(cactus parakeet). *Photo credits: (A) Ramadian Bachtiar for Center for International Forestry Research (CIFOR), (B–D) John Philip Medcraft.*

(Bullock, 1995; Griz and Machado, 2001). Thus, variation in the population size or behavior of dispersers will have a direct effect on plant populations (Bullock, 1995). For example, in Nigeria, Effiom et al. (2014) showed that areas subjected to different intensities of hunting exhibited not only variation in the abundance of the hunted animals, but also clear alterations in the recruitment of seedlings dispersed by game species. Likewise, similar results were found when comparing protected areas with areas with intense hunting of primates in Peru, since the hunted areas experienced a reduction of approximately 55% in species richness of plants whose seeds are dispersed by primates (Nuñez-Iturri and Howe, 2007).

The cascade effects of defaunation may also be jeopardizing another natural process of great ecological and economic importance; the global reduction of pollinators has been associated with human activities related to changes in soil use, habitat destruction and fragmentation, pesticide use and invasive species (Potts et al., 2010). This reduction in pollinators can have profound economic consequences and drastic ecological impacts, since about 75% of agricultural products cultivated by humankind and 80% of wild plants are dependent on insect pollination (Klein et al., 2007).

Furthermore, the effects of defaunation are likely to extend well beyond its ecological consequences, because drastic changes in ecological functions can trigger rapid evolutionary changes, especially in short-lived organisms (Estes et al., 2013). Some studies have demonstrated that in systems subjected to human exploitation, target organisms can exhibit phenotypic changes about 300% faster than in a natural undisturbed system (Darimont et al., 2009). Galetti et al. (2013) provided evidence of this phenomenon in a study on different populations of the palm tree *Euterpe edulis* Mart. in areas of the Atlantic forest with different levels of hunting for birds that disperse this palm's seeds (toucans and cotingas). The results showed that the local extinction of large birds may have caused new interactions to take place that favored individuals with significantly smaller seeds (Galetti et al., 2013).

By-products of captured animals are frequently used in different regions as adornments, in crafts and in religious rites, and mammals, birds, reptiles, and fish, as well as some invertebrates, are currently used for both making handicrafts and as pets (Alves, 2012). Birds, for example, are an especially important group of such captured animals, being exploited for their by-products (Atuo et al., 2015) and as pets, primarily because of their exuberant coloring, with about 295 native species having been recorded exploited in this manner in Brazil (Alves et al., 2016b). These different forms of animal appropriation demonstrate the importance of the role that some animals have played throughout human evolution and how human actions may have drastically affected the ecology and evolution of these species that are the object of their attention.

This scenario presents the challenge of finding ways to exploit animal resources in a manner that minimizes negative impacts on species and thus achieve sustainability. The challenge becomes more complicated, however, as the human population grows, and new technologies facilitate even greater exploitation of aquatic and terrestrial animals to meet the growing demand for food. Considering that, in some cases, human beings have lived with and exploited fauna in ways that have produced minimal pressure on species, while in other cases their exploitation has caused serious impacts, it is clear that animal conservation strategies must try to reconcile human and conservation needs. It is therefore necessary to understand the multidimensional context (biological, social, economical, cultural, political, and institutional) within which interactions between humans and animals have evolved, and continue to evolve. In this context, ethnozoological research is of indispensable usefulness to conservation efforts, because it can aid the evaluation of human impacts on native animal species and support the development of sustainable management plans.

THE ROLE OF ETHNOZOOLOGY IN ANIMAL CONSERVATION

Embedded within the scenario described above, the human species is a key element of the environment, and possesses an unequaled ability to interfere in the most diverse of ecological functions, which often results in modifications of an entire ecosystem (Boivin et al., 2016). Therefore, it can be assumed that for animal conservation, understanding human complexity is equally as important as understanding ecosystems; even conservation biologists recognize that biological knowledge alone is not sufficient to solve conservation problems (Mascia et al., 2003). However, many conservation studies have understood that the actions of humans are as important in predicting environmental change (Effiom et al., 2014; Galetti et al., 2013), studies regarding sociocultural dynamics, although fundamental, are still relatively rare (see Dickman, 2010).

Since humans are the source of conservation problems, as well as the hope for their solutions (Saunders, 2003), we cannot talk about biodiversity conservation without incorporating human dimensions. In this way, ethnozoological surveys can generate support for conservation efforts. In the case of fauna, a number of examples can be cited that illustrate the application of ethnozoological research to assisting animal conservation strategies. One such example is the application of ethnozoology to solving problems regarding wildlife–human conflict, which is a widespread conservation issue and of increasing concern to conservationists (Table 25.1 and Chapter 22). Ethnozoological studies that investigate human perception of species involved in conflicts can provide information crucial for making conservation efforts more effective. As pointed out by Dickman (2010), social factors can play an extremely important role in wildlife–human conflicts, yet they are rarely considered.

Animals play important roles in the folklore of almost all cultures, which can substantially influence attitudes toward particular species. For instance, mythology regarding vampirism has been shown to be related to negative attitudes toward bats (Prokop et al., 2009), while beliefs that the aye-aye, *Daubentonia madagascariensis*, is a harbinger of doom has often lead to it being killed on sight, with some people believing that an entire village should be burned down and abandoned if an aye-aye is seen nearby (Glaw et al., 2008). These perceptions of certain species as innately evil or harmful mean that even if wildlife damage is entirely mitigated, residual fear and antagonism can still lead to continued persecution (Dickman, 2010). Another interesting example is found in the research carried out by Ceríaco et al. (2011) on the subject of folklore and traditional ecological knowledge of geckos in southern Portugal, where it is recorded that local populations possess misconceptions about geckos being poisonous or carriers of dermatological diseases. The persistence of these ideas has led to continued fear and aversion of geckos by the population, resulting in their direct persecution and one of the major conservation problems facing these animals. Education can help lessen hostility, but such deep-seated preconceptions tend to be hard to overcome and must be taken into account in studies of wildlife–human conflicts (Dickman, 2010). Ceríaco (2010) points out that it is essential, from both a scientific and conservationist perspective, to understand the knowledge and perceptions that people have of animals, since only then may hitherto unrecognized pertinent information and conservation problems be detected and resolved.

It is important to emphasize that, in some circumstances, folk beliefs, religious doctrines, and species-specific taboos can be important in the conservation of declining or threatened species, and therefore such practices should be stimulated (Alves et al., 2010). Primates exemplify this situation (Alves et al., 2010; Shepherd et al., 2004).

TABLE 25.1 Examples of Animal Species Involved in Conflicts With Human Beings in Different Parts of the World

Animal	Conflict	Country	Source
Jackal (*Canis mesomelas*)	Attack domestic animals (livestock)	South Africa	Humphries et al. (2016)
Monkey—Chacma Baboons (*Papio ursinus*)	Crop destruction	South Africa	Kansky et al. (2016)
Hippopotamus, Elephant	Crop destruction	South Africa	Massé (2016)
Lion	Attack domestic animals (livestock)	South Africa	Massé (2016)
Crane (*Anthropoides paradiseus*)	Crop destruction	South Africa	van Velden et al. (2016)
Dingo (*Canis lupus* ssp. *dingo*)	Attack humans and domestic animals (livestock), disease transmission	Australia	Allen et al. (2016) and Johnson and Wallach (2016)
Carnaby's Black-Cockatoo (*Calyptorhynchus latirostris*)	Crop destruction	Australia	Johnston et al. (2016)
Cheetah (*Acinonyx jubatus*)	Attack domestic animals (livestock)	Botswana	Boast et al. (2016)
Giant otter (*Pteronura brasiliensis*)	Destruction of fishing equipments	Brazil	Marchand (2016) and Rosas-Ribeiro et al. (2012)
Jaguar (*Panthera onca*)	Attack domestic animals (livestock)	Brazil	Marchand (2016)
Anaconda	Attack domestic animals (livestock)	Brazil	Miranda et al. (2016)
Puma (*Puma concolor*)	Attack domestic animals (livestock)	Brazil	Palmeira et al. (2008) and Schulz et al. (2014)
Jaguar (*P. onca*)	Attack domestic animals (livestock)	Brazil	Palmeira et al. (2008) and Zimmermann et al. (2005)
Bear (*Ursus arctos horribilis*)	Attack humans	Canada	Artelle et al. (2016)
Snow leopard (*Panthera uncia*)	Attack domestic animals (livestock)	China	Acharya et al. (2016)
Eurasian lynx, grey wolf	Attack domestic animals (livestock)	China	Alexander et al. (2015)
Brown bear	Attack humans	China	Alexander et al. (2015)
Elephant (*Elephas maximus*)	Crop destruction	China	Chen et al. (2016)
Bats (*Desmodus rotundus*)	Attack domestic animals (livestock)	Costa Rica	Reid (2016)
Bear andean (*Tremarctos ornatus*)	Attack domestic animals (livestock)	Ecuador	Jampel (2016)
Bear (*Ursus americanus*)	Attack humans	United States	Boonman-Berson et al. (2016)

Continued

TABLE 25.1 Examples of Animal Species Involved in Conflicts With Human Beings in Different Parts of the World—cont'd

Animal	Conflict	Country	Source
Coyote (*Canis latrans*)	Attack domestic animals (livestock)	United States	Elliot et al. (2016)
Canada geese (*Branta canadensis*)	Aircraft collisions	United States	Dolbeer (2011) and Dolbeer et al. (2014)
Seal	Destruction of fishing equipments	Finland	Varjopuro (2011)
Bears—Pyrenean brown bears (*Ursus arctos*)	Attack domestic animals (livestock)	France	Piédallu et al. (2016)
Elephant, gaur, wild pig, monkey	Crop destruction	India	Senthilkumar et al. (2016)
Fox (*Vulpes vulpes*)	Attack domestic animals (livestock)	England	Baker and Macdonald (2000)
Badgers (*Meles meles*)	Property damage	England	Ward et al. (2016)
Bornean Orangutan (*Pongo pygmaeus*)	Crop destruction	Malaysia	Abram et al. (2015)
Cheetah (*Acinonyx jubatus*) Leopard (*Panthera pardus*) Jackals (*Canis mesomelas*)	Attack domestic animals (livestock)	Malaysia	Potgieter et al. (2016)
Tiger	Attack humans and domestic animals (livestock)	Nepal (China)	Acharya et al. (2016)
Elephant, Leopard, Bear, Rhinoceros	Attack humans	Nepal (China)	Acharya et al. (2016)
Tiger	Attack humans and domestic animals (livestock)	Nepal (China)	Dhungana et al. (2016)
Wild boar (*Sus scrofa*)	Crop destruction	Nepal (China)	Pandey et al. (2016)
Otter (*Lutra lutra*), Cormorants (*Phalacrocorax carbo*), Grey heron (*Ardea cinerea*), beavers (*Castor fiber*)	Aquaculture losses	Poland	Kloskowski (2011)
Cane rat (*Thryonomys swinderianus*)	Crop destruction	Sierra Leone	Larson et al. (2016)
Red river hog (*Potamochoerus porcus*)	Crop destruction	Sierra Leone	Larson et al. (2016)
Rodents	Damage to crops and property	Tanzania	Bencin et al. (2016)
Hyena	Attack domestic animals (livestock e pets), property damage and disease transmission	Tanzania	Bencin et al. (2016)
Birds of Prey	Attack domestic animals (livestock)	Tanzania	Bencin et al. (2016)

TABLE 25.1 Examples of Animal Species Involved in Conflicts With Human Beings in Different Parts of the World—cont'd

Animal	Conflict	Country	Source
Elephant	Crop destruction, property damage and disease transmission	Tanzania	Bencin et al. (2016)
Jackal, Lion	Attack domestic animals (livestock)	Tanzania	Bencin et al. (2016)
Lion, Leopard, Hyena, Cheetah	Attack domestic animals (livestock)	Tanzania	Koziarski et al. (2016) and Lyamuya et al. (2016)
Wild dog	Attack domestic animals (livestock e pets), property damage and attack humans	Tanzania	Koziarski et al. (2016) and Lyamuya et al. (2016)
Wolf (*Canis lupus*)	Attack domestic animals (livestock)	Turkey	Capitani et al. (2016)
Gorilla (*Gorilla beringei beringei*)	Crop destruction	Uganda	Seiler and Robbins (2016)

Some chimpanzee (*Pan troglodytes*) populations are not hunted because of their physical similarity to humans or because of folklore regarding an ancestral relationship with humans (Alves et al., 2016b; Kortlandt, 1986; Putra et al., 2008; Silva et al., 2005). In the village of Bossou (Republic of Guinea), the Manon people consider chimpanzees to be sacred, embodying the reincarnation of their ancestors, and believe that their ancestors' souls rest on the sacred hill of Gaban. As the chimpanzee is a totem of the most influential family of Bossou, it is strictly forbidden to hunt or eat the chimpanzee (Yamakoshi, 2005).

Human values and wildlife folklore strongly influence the effectiveness of conservation efforts (Alves et al., 2012; Bezerra et al., 2013; Ceriaco, 2012; Fernandes-Ferreira et al., 2012). Together with values and norms, world views underlie resource management systems and form the basis for decision making and action (Claus et al., 2010). Failure to recognize the importance of the human dimension in animal conservation may be implicated in negative consequences for management plans. The implementation of a successful management strategy fundamentally requires involvement of the main stakeholders, who must be made aware of the need to conserve natural resources as a guarantee for its sustainable exploitation.

An ethnozoological approach can help guide programs that aim for sustainable exploitation and conservation of species, such as for hunting and fishing for food. Since meat from wild animals is often the only source of animal protein available to some people (Fig. 25.3), it seems that economics is the rather obvious motivation for this choice (Brashares et al., 2011; Jenkins et al., 2011). However, ethnobiological studies have shown that many other factors motivate people to feed on wild animals even in areas where other food options are available. For example, in describing hunting consumption in urban areas of the Republic of Cameroon, King (1994) points out that preference is the predominant factor for the consumption of game meat. According to this author, game meat is consumed in the region in spite of the great availability of meat from domestic animals, even at costs generally lower than that of hunting. Other studies have pointed out how hunting consumption seems to be linked to habits or traditions (Schenck et al., 2006) (Mfunda and Røskaft, 2010). In the

FIGURE 25.3 Some vertebrate species hunted for bushmeat in tropical regions. (A) Western Lowland gorilla (*Gorilla gorilla gorilla* (Savage, 1847), Cameroon, Africa; (B) White-lipped peccaries (*Tayassu pecari*) (Link, 1795), Brazil; (C) Yellow armadillo (*Euphractus sexcinctus* (Linnaeus, 1758)), Brazil; (D) Rock Cavy (*Kerodon rupestris* (Wied, 1820)), Brazil; (E) White-browed Guan (*Penelope jacucaca* (Spix, 1825)), Brazil; (F) Razor-billed curassow (*Pauxi tuberosa* (Spix, 1825)), Brazil. *Photo credits: (A) Edmond Dounias for Center for International Forestry Research (CIFOR), (B and F) Flávio Bezerra Barros, (C–E) John Philip Medcraft.*

northeast region of Brazil, for example, hunting seems to be more associated with traditions than to meet a demand for animal protein (Alves et al., 2016a, 2009; Barboza et al., 2016). Thus, understanding the different realities held by people can help in the design of effective strategies customized to different socioecological contexts.

It should be mentioned that it is not only subsistence hunting that relates to the scenario outlined above. Game hunting, common in North American countries, has resulted in a significant number of wild animals being slaughtered annually. This activity, when regulated, has been identified as an important management tool, particularly when restricted areas of use are established to mitigate harm to animal populations (Van Vliet et al., 2016). These strategies, however, need to be analyzed from conservationist and ethnozoological perspectives, thus reiterating the need for multidisciplinary research involving all agents (humans and other animals) involved and their interrelationships.

Local communities have extensive knowledge of their environment, which can be used for promoting collaborative conservation and management efforts. Examples of such practices can be found at the Mamirauá Sustainable Development Reserve, Brazil (Castello, 2004; Gillingham, 2001), in Zimbabwe (Child, 1996), in the Philippines (Pomeroy and Carlos, 1997), and on the Pacific Islands (Michael and Lambeth, 2000). On the other hand, the exclusion of local communities from consultation and the decision-making process may lead to the construction of public policies devoid of historical information, and without much social resonance (Alves and Rosa, 2007). Ommer and Perry (2011) emphasize that the management of the world's fish populations, and the activities of fishermen, remain deeply problematic, not in the least due to separating fish from fishermen, by not recognizing their interdependence and treating these interdependent problems separately.

In many countries, but especially those in tropical regions that have great faunal diversity, illegal commerce in wild animals removes many species from their natural environments. This is certainly one of the gravest threats to many populations of native species, and ethnozoological studies constitute an invaluable tool for understanding the socioeconomic and cultural contexts in which the commercialization of wild fauna is embedded—an essential aspect of the development of conservation proposals. Lopes et al. (2010) noted that ethnozoological studies have made many contributions to conservation efforts, including initiating dialogues between local communities involved in, or affected by, conservation initiatives; suggesting better resource-use strategies (and management alternatives); monitoring the abundance of resources being used by human populations and the practical results of conservation management strategies; and a better understanding and interpretation of both general and complex ecological phenomena and environmental impacts and alterations.

FINAL CONSIDERATIONS

Humans are but only one of the thousands of species that make up biodiversity, but their actions affect the great majority of other species. Therefore, understanding the context of exploitation and interaction between humans and natural resources is imperative for the development and implementation of conservation measures. Considering human needs and their dependence on biological resources, ethnobiology (including ethnozoology) stands out as a scientific discipline that focuses on the relationship between the two main elements of the current environmental crisis: humans and natural resources (Albuquerque and Alves, 2016).

Thus, ethnozoology becomes essential to efforts of faunal conservation because in addition to addressing relevant biological factors, its research also considers cultural, economic, social, and traditional roles of animals in human societies (Alves, 2012). The human species is not only an agent of destruction (as often understood through classical approaches to ecology), but also a species that interacts with the environment and other species, not only to reproduce and give continuity to its way of life, but especially for its continued biological existence. Ignoring the needs of humanity, especially those with a heavy reliance on these resources, is to undermine conservation initiatives.

The complexity of the interactions between humans and nature also makes it difficult to assess the effect human actions have on faunal resources. It is necessary to take into account social, economic, cultural, and environmental factors for a better understanding of all the dynamics of use and management (Gubbi and Linkie, 2012; Morsello et al., 2015), and to increase considerably the efficiency of conservation efforts.

Given the emergency to find solutions to deal with the current crisis of biodiversity loss, it is necessary to adopt strategies that address the problem in all its complexity. In this sense, ethnozoology presents itself as an interdisciplinary tool, approaching the issue in a more comprehensive way than traditional biological disciplines. Furthermore, the connection between ethnozoology and the social and human sciences, means that it can act as a bridge between managers, scientists, and local communities in the development of conservation plans.

References

Abram, N.K., Meijaard, E., Wells, J.A., Ancrenaz, M., Pellier, A.-S., Runting, R.K., Gaveau, D., Wich, S., Nardiyono, Tjiu, A., Nurcahyo, A., Mengersen, K., 2015. Mapping perceptions of species' threats and population trends to inform conservation efforts: the Bornean orangutan case study. Diversity and Distributions 21, 487–499.

Acharya, K.P., Paudel, P.K., Neupane, P.R., Köhl, M., 2016. Human-wildlife conflicts in Nepal: patterns of human fatalities and injuries caused by large mammals. PLoS One 11.

Albuquerque, U.P., Alves, A.G.C., 2016. What Is Ethnobiology? In: Albuquerque, U.P., da Alves, R.R. (Eds.), Introduction to Ethnobiology. Springer International Publishing, New York, pp. 3–7.

Alexander, J., Chen, P., Damerell, P., Youkui, W., Hughes, J., Shi, K., Riordan, P., 2015. Human wildlife conflict involving large carnivores in Qilianshan, China and the minimal paw-print of snow leopards. Biological Conservation 187, 1–9.

Allen, B.L., Carmelito, E., Amos, M., Goullet, M.S., Allen, L.R., Speed, J., Gentle, M., Leung, L.K.-P., 2016. Diet of dingoes and other wild dogs in peri-urban areas of north-eastern Australia. Scientific Reports 6.

Alves, R.R.N., 2012. Relationships between fauna and people and the role of ethnozoology in animal conservation. Ethnobiology and Conservation 1, 1–69.

Alves, R.R.N., Rosa, I.L., 2007. Biodiversity, traditional medicine and public health: where do they meet? Journal of Ethnobiology and Ethnomedicine 3, 1–9.

Alves, R.R.N., Rosa, I.L., 2013. Animals in Traditional Folk Medicine: Implications for Conservation. Springer-Verlag, Berlin, Heidelberg.

Alves, R.R.N., Souto, W.M.S., 2015. Ethnozoology: a brief introduction. Ethnobiology and Conservation 4, 1–13.

Alves, R.R.N., Mendonça, L.E.T., Confessor, M.V.A., Vieira, W.L.S., Lopez, L.C.S., 2009. Hunting strategies used in the semi-arid region of northeastern Brazil. Journal of Ethnobiology and Ethnomedicine 5, 1–50.

Alves, R.R.N., Souto, W.M.S., Barboza, R.R.D., 2010. Primates in traditional folk medicine: a world overview. Mammal Review 40, 155–180.

Alves, R.R.N., Rosa, I.L., Léo Neto, N.A., Voeks, R., 2012. Animals for the Gods: magical and religious faunal use and trade in Brazil. Human Ecology 40, 751–780.

Alves, R.R.N., Feijó, A., Barboza, R.R.D., Souto, W.M.S., Fernandes-Ferreira, H., Cordeiro-Estrela, P., Langguth, A., 2016a. Game mammals of the Caatinga biome. Ethnobiology and Conservation 5, 1–51.

Alves, R.R.N., Souto, W.M.S., Barboza, R.R.D., 2016b. The role of nonhuman primates in religious and folk medicine beliefs. In: Waller, M.T. (Ed.), Ethnoprimatology: Primate Conservation in the 21st Century. Springer, Switzerland, pp. 117–135.

Atuo, F.A., O'Connell, T.J., Abanyam, P.U., 2015. An assessment of socio-economic drivers of avian body parts trade in West African rainforests. Biological Conservation. 191, 614–622. http://dx.doi.org/10.1016/j.biocon.2015.08.013.

Artaxo, P., 2014. Uma nova era geológica em nosso planeta: o Antropoceno? Revista USP 13–24.

Artelle, K.A., Anderson, S.C., Reynolds, J.D., Cooper, A.B., Paquet, P.C., Darimont, C.T., 2016. Ecology of conflict: marine food supply affects human-wildlife interactions on land. Scientific Reports 6.

Baker, S.E., Macdonald, D.W., 2000. Foxes and foxhunting on farms in Wiltshire: a case study. Journal of Rural Studies 16, 185–201.

Barboza, R.R.D., Lopes, S.F., Souto, W.M.S., Fernandes-Ferreira, H., Alves, R.R.N., 2016. The role of game mammals as bushmeat in the Caatinga, northeast Brazil. Ecology and Society 21, 1–11.

Bencin, H., Kioko, J., Kiffner, C., 2016. Local people's perceptions of wildlife species in two distinct landscapes of Northern Tanzania. Journal for Nature Conservation 34, 82–92. http://dx.doi.org/10.1016/j.jnc.2016.09.004.

Bezerra, D.M.M., Araujo, H.F.P., Alves, R.R.N., 2012. Captura de aves silvestres no semiárido brasileiro: técnicas cinegéticas e implicações para conservação. Tropical Conservation Science 5, 50–66.

Bezerra, D.M.M., Araújo, H.F.P., Alves, Â.G.C., Alves, R.R.N., 2013, 1–11. Birds and people in semiarid northeastern Brazil: symbolic and medicinal relationships. Journal of Ethnobiology and Ethnomedicine 9.

Boast, L.K., Good, K., Klein, R., 2016. Translocation of problem predators: is it an effective way to mitigate conflict between farmers and cheetahs *Acinonyx jubatus* in Botswana? Oryx 50, 537–544.

Boivin, N.L., Zeder, M.A., Fuller, D.Q., Crowther, A., Larson, G., Erlandson, J.M., Denham, T., Petraglia, M.D., 2016. Ecological consequences of human niche construction: Examining long-term anthropogenic shaping of global species distributions. Proceedings of the National Academy of Sciences 113, 6388–6396.

Boonman-Berson, S., Turnhout, E., Carolan, M., 2016. Common sensing: human-black bear cohabitation practices in Colorado. Geoforum 74, 192–201.

Brashares, J.S., Golden, C.D., Weinbaum, K.Z., Barrett, C.B., Okello, G.V., 2011. Economic and geographic drivers of wildlife consumption in rural Africa. Proceedings of the National Academy of Sciences 108, 13931–13936.

Bullock, S.H., 1995. Plant reproduction in neotropical dry forests. In: Bullock, S.H., Mooney, H.A., Medina, E. (Eds.), Seasonally Dry Tropical Forests. Cambridge University Press, New York, pp. 277–303.

Buscher, B., Wolmer, W., 2007. Introduction: the politics of engagement between biodiversity conservation and the social sciences. Conservation and Society 5, 1–21.

Capitani, C., Chynoweth, M., Kusak, J., Çoban, E., Sekercioğlu, C.H., 2016. Wolf diet in an agricultural landscape of north-eastern Turkey. Mammalia 80, 329–334.

Cardinale, B.J., Duffy, J.E., Gonzalez, A., Hooper, D.U., Perrings, C., Venail, P., Narwani, A., Mace, G.M., Tilman, D., Wardle, D.A., Kinzig, A.P., Daily, G.C., Loreau, M., Grace, J.B., Larigauderie, A., Srivastava, D.S., Naeem, S., 2012. Biodiversity loss and its impact on humanity. Nature 489, 326 326.

Castello, L., 2004. A method to count pirarucu *Arapaima gigas*: fishers, assessment, and management. North American Journal of Fisheries Management 24, 379–389.

Ceríaco, L.M.P., 2010. Human Attitudes Towards Herpetofauna: How Preferences, Fear and Beliefs Can Influence the Conservation of Reptiles and Amphibians. Universidade de Évora, Évora, p. 164.

Ceriaco, L.M.P., 2012. Human attitudes towards herpetofauna: the influence of folklore and negative values on the conservation of amphibians and reptiles in Portugal. Journal of Ethnobiology and Ethnomedicine 8, 1–12.

Ceríaco, L.M.P., Marques, M.P., Madeira, N.C., Vila-Viçosa, C.M., Mendes, P., 2011. Folklore and traditional ecological knowledge of geckos in Southern Portugal: implications for conservation and science. Journal of Ethnobiology and Ethnomedicine 7, 1–9.

Chen, Y., Marino, J., Chen, Y., Tao, Q., Sullivan, C.D., Shi, K., Macdonald, D.W., 2016. Predicting hotspots of human-elephant conflict to inform mitigation strategies in Xishuangbanna, Southwest China. PLoS One 11.

Child, B., 1996. The practice and principles of community-based wildlife management in Zimbabwe: the CAMPFIRE programme. Biodiversity and Conservation 5, 369–398.

Claus, C.A., Chan, K.M.A., Satterfield, T., 2010. The roles of people in conservation. In: Sodhi, N.S., Ehrlich, P.R. (Eds.), Conservation Biology for All. Oxford University Press, New York, pp. 262–283.

Darimont, C.T., Carlson, S.M., Kinnison, M.T., Paquet, P.C., Reimchen, T.E., Wilmers, C.C., 2009. Human predators outpace other agents of trait change in the wild. Proceedings of the National Academy of Sciences 106, 952–954.

Dhungana, R., Savini, T., Karki, J.B., Bumrungsri, S., 2016. Mitigating human-tiger conflict: An assessment of compensation payments and tiger removals in Chitwan national park, Nepal. Tropical Conservation Science 9, 776–787.

Diamond, J., 2002. Evolution, consequences and future of plant and animal domestication. Nature 418, 700–707.

Dickman, A.J., 2010. Complexities of conflict: the importance of considering social factors for effectively resolving human–wildlife conflict. Animal Conservation 13, 458–466.

Dolbeer, R.A., 2011. Increasing trend of damaging bird strikes with aircraft outside the airport boundary: implications for mitigation measures. Human-Wildlife Interactions 5, 235–248.

Dolbeer, R.A., Seubert, J.L., Begier, M.J., 2014. Population trends of resident and migratory *Canada geese* in relation to strikes with civil aircraft. Human-Wildlife Interactions 8, 88–99.

Effiom, E.O., Birkhofer, K., Smith, H.G., Olsson, O., 2014. Changes of community composition at multiple trophic levels due to hunting in Nigerian tropical forests. Ecography (Cop.). 37, 367–377.

Ehrlich, P.R., Ehrlich, A.H., 2008. The Dominant Animal: Human Evolution and the Environment. Island Pr.

Elliot, E.E., Vallance, S., Molles, L.E., 2016. Coexisting with coyotes (*Canis latrans*) in an urban environment. Urban Ecosystems 19, 1335–1350.

Ellis, R., 2005. Tiger Bone & Rhino Horn: The Destruction of Wildlife for Traditional Chinese Medicine. Island Press, Washington DC.

Estes, J.A., Brashares, J.S., Power, M.E., 2013. Predicting and Detecting Reciprocity between Indirect Ecological Interactions and Evolution. The American Naturalist 181, S76–S99.

Fernandes-Ferreira, H., Cruz, R.L., Borges-Nojosa, D.M., Alves, R.R.N., 2012. Folklore concerning snakes in the Ceará State, northeastern Brazil. Sitientibus Série Ciências Biológicas 11, 153–163.

Fleming, T.H., Kress, W.J., 2011. A brief history of fruits and frugivores. Acta Oecologica 37, 521–530.

Galetti, M., Guevara, R., Côrtes, M.C., Fadini, R., Von Matter, S., Leite, A.B., Labecca, F., Ribeiro, T., Carvalho, C.S., Collevatti, R.G., Pires, M.M., Guimarães, P.R., Brancalion, P.H., Ribeiro, M.C., Jordano, P., 2013. Functional extinction of birds drives rapid evolutionary changes in seed size. Science 340, 1086–1090.

Gillingham, S., 2001. Social organization and participatory resource management in Brazilian ribeirinho communities: A case study of the Mamiraua Sustainable Development Reserve, Amazonas. Society & Natural Resources 14, 803–814.

Glaw, F., Vences, M., Randrianiania, R.D., 2008. Killed aye-aye (*Daubentonia madagascariensis*) exposed on the gallows in northeastern Madagascar. Lemur News 13, 6–7.

Griz, L.M.S., Machado, I.C.S., 2001. Fruiting phenology and seed dispersal syndromes in Caatinga, a tropical dry forest in the northeast of Brazil. Journal of Tropical Ecology 17, 303–321.

Gubbi, S., Linkie, M., 2012. Wildlife hunting patterns, techniques, and profile of hunters in and around Periyar Tiger Reserve. Journal of the Bombay Natural History Society 109, 165–172.

Hooper, D.U., Adair, E.C., Cardinale, B.J., Byrnes, J.E.K., Hungate, B.A., Matulich, K.L., Gonzalez, A., Duffy, J.E., Gamfeldt, L., O'Connor, M.I., 2012. A global synthesis reveals biodiversity loss as a major driver of ecosystem change. Nature 486, 105–108.

Humphries, B.D., Ramesh, T., Downs, C.T., 2016. Diet of black-backed jackals (*Canis mesomelas*) on farmlands in the KwaZulu-Natal Midlands, South Africa. Mammalia 80, 405–412.

Jampel, C., 2016. Cattle-based livelihoods, changes in the taskscape, and human-bear conflict in the Ecuadorian Andes. Geoforum 69, 84–93.

Jenkins, R.K.B., Keane, A., Rakotoarivelo, A.R., Rakotomboavonjy, V., Randrianandrianina, F.H., Razafimanahaka, H.J., Ralaiarimalala, S.R., Jones, J.P.G., 2011. Analysis of patterns of bushmeat consumption reveals extensive exploitation of protected species in eastern Madagascar. PLoS One 6.

Johnson, C.N., Wallach, A.D., 2016. The virtuous circle: predator-friendly farming and ecological restoration in Australia. Restoration Ecology 24, 821–826. http://dx.doi.org/10.1111/rec.12396.

Johnston, T.R., Stock, W.D., Mawson, P.R., 2016. Foraging by Carnaby's black-cockatoo in Banksia woodland on the Swan Coastal Plain, Western Australia. Emu 116, 284–293.

Kansky, R., Kidd, M., Knight, A.T., 2016. A wildlife tolerance model and case study for understanding human wildlife conflicts. Biological Conservation 201, 137–145..

King, S., 1994. Utilisation of Wildlife in Bakossiland, West Cameroon, with particular reference to primates. Traffic Bulletin. 14, 63–73.

Klein, A.-M., Vaissière, B.E., Cane, J.H., Steffan-Dewenter, I., Cunningham, S.A., Kremen, C., Tscharntke, T., 2007. Importance of pollinators in changing landscapes for world crops. Proceedings of the Royal Society of London B: Biological Sciences 274, 303–313.

Kloskowski, J., 2011. Human-wildlife conflicts at pond fisheries in eastern Poland: perceptions and management of wildlife damage. European Journal of Wildlife Research 57, 295–304. http://dx.doi.org/10.1007/s10344-010-0426-5.

Kortlandt, A., 1986. The use of stone tools by wild-living chimpanzees and earliest hominids. Journal of Human Evolution 15, 77–132.

Koziarski, A., Kissui, B., Kiffner, C., 2016. Patterns and correlates of perceived conflict between humans and large carnivores in Northern Tanzania. Biological Conservation 199, 41–50.

Larson, L.R., Conway, A.L., Hernandez, S.M., Carroll, J.P., 2016. Human-wildlife conflict, conservation attitudes, and a potential role for citizen science in Sierra Leone, Africa. Conservation and Society 14, 205–217.

Lidicker Jr., W.Z., Meffe, G.K., Noss, R.F., Jacobson, S.K., 1998. Revisiting the human dimension in conservation biology. Conservation Biology 12, 1168–1172.

Lopes, P.F.M., Silvano, R., Begossi, A., 2010. Da Biologia a Etnobiologia – Taxonomia e etnotaxomia, ecologia e etnoecologia. In: Alves, R.R.N., Souto, W.M.S., Mourão, J.S. (Eds.), A Etnozoologia no Brasil: Importância, Status atual e Perspectivas, first ed. NUPEEA, Recife, PE, Brazil, pp. 67–94.

Lyamuya, R.D., Masenga, E.H., Fyumagwa, R.D., Mwita, M.N., Røskaft, E., 2016. Pastoralist herding efficiency in dealing with carnivore-livestock conflicts in the eastern Serengeti, Tanzania. International Journal of Biodiversity Science, Ecosystems Services and Management 12, 202–211.

Marchand, G., 2016. Analyzing the spatial dimension of the human/wildlife conflicts in the sustainable development reserve of Uatumã (Amazonas, Brasil). CyberGeo.

Mascia, M.B., Brosius, J.P., Dobson, T.A., Forbes, B.C., Horowitz, L., McKean, M.A., Turner, N.J., 2003. Conservation and the social sciences. Conservation Biology 17, 649–650.

Massé, F., 2016. The political ecology of human-wildlife conflict: producing wilderness, insecurity, and displacement in the Limpopo National Park. Conservation and Society 14, 100–111.

Meine, C., Soule, M., Noss, R.F., 2006. "A mission driven discipline": the growth of conservation biology. Conservation Biology 20, 631–651.

Mfunda, I.M., Røskaft, E., 2010. Bushmeat hunting in Serengeti, Tanzania: An important economic activity to local people. International Journal of Biodiversity and Conservation 2, 263–272.

Michael, K., Lambeth, L., 2000. Fisheries management by communities: a manual on promoting the management of subsistence fisheries by Pacific island communities. Journal of Ethnopharmacology. Secretariat of the Pacific Community.

Miranda, E.B.P., Ribeiro Jr., R.P., Strüssmann, C., 2016. The ecology of human-anaconda conflict: A study using internet videos. Tropical Conservation Science 9, 43–77.

Monastersky, R., 2015. Anthropocene: the human age. Nature 519, 144–147.

Morsello, C., Yagüe, B., Beltreschi, L., van Vliet, N., Adams, C., Schor, T., Quiceno-Mesa, M.P., Cruz, D., 2015. Cultural attitudes are stronger predictors of bushmeat consumption and preference than economic factors among urban amazonians from Brazil and Colombia. Ecology and Society 20 (4).

Myers, S.S., Gaffikin, L., Golden, C.D., Ostfeld, R.S., Redford, K.H., Ricketts, T.H., Turner, W.R., Osofsky, S.A., 2013. Human health impacts of ecosystem alteration. Proceedings of the National Academy of Sciences 110, 18753–18760.

Newing, H., 2010. Bridging the gap: interdisciplinarity, biocultural diversity and conservation. In: Pretty, J., Pilgrim, S. (Eds.), Nature and Culture: Revitalising the Connection. Earthscan, London, UK, pp. 23–40.

Noss, R., 1999. Is there a special conservation biology? Ecography 22, 113–122.

Nuñez-Iturri, G., Howe, H.F., 2007. Bushmeat and the fate of trees with seeds dispersed by large primates in a lowland rain forest in Western Amazonia. Biotropica 39, 348–354.

Ommer, R.E., Perry, R.I., 2011. Introduction. In: Ommer, R.E., Perry, R.I., Cochrane, K., Cury, P. (Eds.), World Fisheries: A Social-Ecological Analysis. Blackwell Publishing Ltd, pp. 3–8.

Palmeira, F.B.L., Crawshaw Jr., P.G., Haddad, C.M., Ferraz, K.M.P.M.B., Verdade, L.M., 2008. Cattle depredation by puma (*Puma concolor*) and jaguar (*Panthera onca*) in central-western Brazil. Biological Conservation 141, 118–125.

Pandey, P., Shaner, P.-J.L., Sharma, H.P., 2016. The wild boar as a driver of human-wildlife conflict in the protected park lands of Nepal. European Journal of Wildlife Research 62, 103–108.

Piédallu, B., Quenette, P.-Y., Mounet, C., Lescureux, N., Borelli-Massines, M., Dubarry, E., Camarra, J.-J., Gimenez, O., 2016. Spatial variation in public attitudes towards brown bears in the French Pyrenees. Biologial Conservation 197, 90–97.

Pomeroy, R.S., Carlos, M.B., 1997. Community-based coastal resource management in the Philippines: a review and evaluation of programs and projects, 1984–1994. Marine Policy 21, 445–464.

Potgieter, G.C., Kerley, G.I.H., Marker, L.L., 2016. More bark than bite? the role of livestock guarding dogs in predator control on Namibian farmlands. Oryx 50, 514–522.

Potts, S.G., Biesmeijer, J.C., Kremen, C., Neumann, P., Schweiger, O., Kunin, W.E., 2010. Global pollinator declines: Trends, impacts and drivers. Trends in Ecology & Evolution 25, 345–353.

Prokop, P., Fancovicova, J., Kubiatko, M., 2009. Vampires are still alive: Slovakian students' attitudes toward bats. Anthrozoos: A Multidisciplinary Journal of the Interactions of People & Animals 22, 19–30.

Putra, Y., Masy'ud, B., Ulfah, M., 2008. Diversity of medicinal animals in Betung Kerihun National Park, West Kalimantan, Indonesia. Media Konservasi 13, 8–15.

Reid, J.L., 2016. Knowledge and experience predict indiscriminate bat-killing intentions among Costa Rican men. Biotropica 48, 394–404.

Rosas-Ribeiro, P.F., Rosas, F.C.W., Zuanon, J., 2012. Conflict between fishermen and giant otters *Pteronura brasiliensis* in Western Brazilian Amazon. Biotropica 44, 437–444.

Saunders, C.D., 2003. The emerging field of conservation psychology. Human Ecology Review 10, 137–149.

Schenck, M., Nsame Effa, E., Starkey, M., Wilkie, D., Abernethy, K., Telfer, P., Godoy, R., Treves, A., 2006. Why people eat bushmeat: results from two-choice, taste tests in Gabon, Central Africa. Human Ecology 34, 433–445.

Schulz, F., Printes, R.C., Oliveira, L.R., 2014. Depredation of domestic herds by pumas based on farmer's information in Southern Brazil. Journal of Ethnobiology and Ethnomedicine 10.

Seiler, N., Robbins, M.M., 2016. Factors influencing ranging on community land and crop raiding by mountain gorillas. Animal Conservation 19, 176–188.

Senthilkumar, K., Mathialagan, P., Manivannan, C., Jayathangaraj, M.G., Gomathinayagam, S., 2016. A study on the tolerance level of farmers toward human-wildlife conflict in the forest buffer zones of Tamil Nadu. Veterinary World 9, 747–752.

Shepherd, C.R., Sukumaran, J., Wich, S.A., 2004. Open Season: An Analysis of the Pet Trade in Medan, Sumatra 1997–2001. TRAFFIC Southeast Asia, Kuala Lumpur, Malaysia.

Silva, M.N.F., Shepard Jr., G.H., Yu, D.W., 2005. Conservation implications of primate hunting practices among the Matsigenka of Manu National Park. Neotropical Primates 13, 31–36.

Sodhi, N.S., Ehrlich, P.R., 2010. Conservation Biology for All. Oxford University Press Oxford, UK.

Soule, M.E., 1985. What is conservation biology? BioScience 35, 727–734.

Still, J., 2003. Use of animal products in traditional Chinese medicine: environmental impact and health hazards. Complementary Therapies in Medicine 11, 118–122.

Terborgh, J., Holt, R.D., Estes, J.A., 2010. Trophic cascades: what they are, how they work, and why they matter. In: Terborgh, J., Estes, J.A. (Eds.), Trophic Cascades: Predators, Prey, and the Changing Dynamics of Nature, first ed. Island Press, New York, pp. 1–18.

United Nations, 2015. World Population Prospects: The 2015 Revision, Data Booklet. United Nations, Department of Economic and Social Affairs, Population Division.

Varjopuro, R., 2011. Co-existence of seals and fisheries? Adaptation of a coastal fishery for recovery of the Baltic grey seal. Marine Policy 35, 450–456.

van Velden, J.L., Smith, T., Ryan, P.G., 2016. Cranes and crops: investigating farmer tolerances toward crop damage by threatened blue cranes (*Anthropoides paradiseus*) in the Western Cape, South Africa. Environmental Management 58, 972–983.

Van Vliet, N., Cornelis, D., Beck, H., Lindsey, P., Nasi, R., LeBel, S., Moreno, J., Fragoso, J., 2016. Current Trends in Wildlife Research. http://dx.doi.org/10.1007/978-3-319-27912-1.

Vitousek, P.M., Mooney, H.A., Lubchenco, J., Melillo, J.M., 1997. Human domination of Earth's ecosystems. Science 277, 494–499.

Ward, A.I., Finney, J.K., Beatham, S.E., Delahay, R.J., Robertson, P.A., Cowan, D.P., 2016. Exclusions for resolving urban badger damage problems: outcomes and consequences. PeerJ.

Yamakoshi, G., 2005. What Is Happening on the Border between Humans and Chimpanzees? Wildlife Conservation in West African Rural Landscapes, 7th Kyoto University International Symposium. Kyoto University, Kyoto, Japan.

Zimmermann, A., Walpole, M.J., Leader-Williams, N., 2005. Cattle ranchers' attitudes to conflicts with jaguar *Panthera onca* in the Pantanal of Brazil. Oryx 39, 406–412.

CHAPTER

26

The Use of Traditional Ecological Knowledge in the Context of Participatory Wildlife Management: Examples From Indigenous Communities in Puerto Nariño, Amazonas-Colombia

Nathalie van Vliet[1], *Laurane L'haridon*[2], *Juanita Gomez*[2], *Liliana Vanegas*[2], *François Sandrin*[2], *Robert Nasi*[1]

[1]Center for International Forestry Research (CIFOR), Bogor, Indonesia; [2]Fundación SI, Bogotá, Colombia

INTRODUCTION

In the Amazon Basin, given that more than 25% of the territory is under legal or de facto management by indigenous peoples (Peres and Terborgh, 1995), the ways in which Amazonia's indigenous peoples manage resources on their extensive territories have implications for the biome as a whole. Over the history, the cultural values of many indigenous societies have been eroded first through the long history of Christianity and missionary activities, later though in-migration from *colonos*, trade in primary products (rubber, pelts, etc.), and displacement, then through modern communication means, expansion of market economies, and urbanization. These changes have rapidly transformed traditional indigenous cultures, local economies, and dietary patterns (Popkin, 2006; Popkin and Gordon-Larsen, 2004) and produced new forms of urban and rural indigenous knowledge, innovations, and practices, which have become a part and parcel of their culture (Oviedo and Maffi, 2000).

Despite these rapid transformations that often pull indigenous livelihoods away from the dependency on natural resources, even in rural communities of the Amazon (Nardoto et al., 2011), wildlife remains a key element in diet and income diversification, and plays

Ethnozoology
http://dx.doi.org/10.1016/B978-0-12-809913-1.00026-0

important social and cultural roles (Koster, 2011; Sirén, 2012; van Vliet et al., 2015a,b; Sarti et al., 2015; Parry et al., 2014). The links between contemporary wildlife use and local culture have inspired research in Africa (Ngoufo et al., 2014; Horsthemke, 2015; Bobo et al., 2015) and Madagascar (Golden and Comaroff, 2015). In the Amazon, some evidence shows that wildlife consumption is also still intimately linked to cultural values, either because culture shapes taste preferences for wildlife species (Pezzuti et al., 2010; Luzar et al., 2012; Morsello et al., 2015) or because wildlife use practices consolidate cultural identities (Sirén, 2012; Santos-Fita et al., 2015; van Vliet et al., 2015a,b) or because cultural norms and practices still guide wildlife use (Colding and Folke, 2001).

In a context where wildlife conservation is increasingly approached through socioecological lenses, understanding how traditional cultural beliefs still influence the use of biodiversity in a changing world is essential to build ecologically sustainable management and conservation strategies (Seixas and Begossi, 2001). Article 8 of the Convention of Biological Diversity actually recognizes that signatory nations should "respect, preserve and maintain knowledge, innovations and practices of indigenous and local communities embodying traditional lifestyles relevant for the conservation and sustainable use of biological diversity and promote their wider application with the approval and involvement of the holders of such knowledge." Careful investigation and official recognition of these practices, along with community awareness and stakeholder involvement, may indeed heighten the success of conservation programs.

In this chapter we illustrate the importance of integrating traditional ecological knowledge in the management of wildlife use in indigenous communities from Puerto Nariño, Amazonas, Colombia. We describe taboos and beliefs affecting the use of terrestrial wildlife and discuss how these may guide the definition of wildlife management agreements. We also illustrate how a traditional practice to imitate animals during hunting trips may be adapted to the context of participatory monitoring of wildlife. Finally, we discuss the challenges and opportunities offered by a more integrated approach where formal and traditional local knowledge are integrated for more successful sustainable use initiatives.

STUDY SITE

Description of the Study Location

The study was carried out in the municipality of Puerto Nariño—Amazonas (03° 24′-85′S; 70° 28′-70′W), the second largest municipality in the Colombian Amazon region after the regional capital, Leticia, located 87 km downstream along the Amazon River (Fig. 26.1). Three types of forests are found in the Puerto Nariño: *terra firme* forest (not subject to flooding regimes); varzea forest (subject to periodic flooding by white waters), and swamp forests (which are seasonally flooded by black waters) (Moreno Arocha, 2014).

The climate of the region is warm and humid, with rainfall causing four distinct periods: high waters (February to April), decreasing waters (May to July), low waters (August to October), and rising waters (November to January) (Fig. 26.2). The average annual temperature is 26°C (although it can reach 38°C) and the relative humidity is around 87% (Rangel and Luengas, 1997). The study was carried out as part of a project that started in September 2014 to support local communities in their wildlife management plan. The specific activities described in this chapter took place from January to December 2015 in six communities (San Francisco, Ticoya, Casco Urbano, Patrullero, 20 de Julio, and Puerto Esperanza) from the municipality of Puerto Nariño, located in the Amazonas region of Colombia. The communities from Puerto Nariño are indigenous communities from the Ticuna, Yagua, and Cocama ethnical groups, sharing

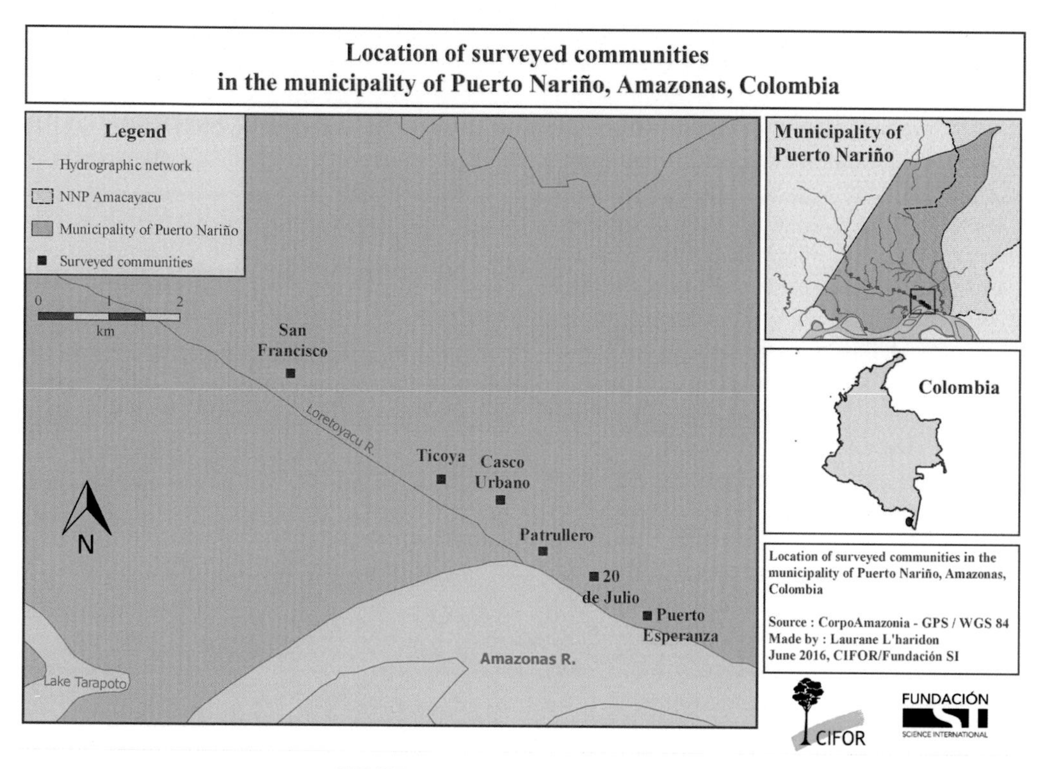

FIGURE 26.1 Map of the study área.

a territory of 1876 km^2. According to the census conducted by the National Administrative Department of Statistics (DANE for its initials in Spanish), the population of the municipality of Puerto Nariño in 2010 was 2015 inhabitants, with 323 inhabitants in the town of Puerto Nariño and the rest in neighboring communities.

History of Cultural Transformation Among the Ticuna

The first contact of Ticunas with Western society dates back to the early-seventeenth-century expeditions. During these expeditions, the indigenous people were forced to learn new technologies and beliefs in exchange for new iron tools for agricultural practices. These exchanges generated a dependency between missionaries and the indigenous population (Riaño Umbarila, 2003).

FIGURE 26.2 Sunset in Puerto Nariño, Amazonas. *Photo: Laurane L'haridon.*

The colonial regime that existed between the seventeenth and nineteenth centuries, generated for the Ticunas a series of changes concerning their territorial organization and a reconfiguration of social relations, mainly shaped by the evangelical missions and slavers companies (Vieco and Oyuela-Caycedo, 1999; Ullán de la Rosa, 2000). Later, with the departure of the missions from the Amazon, rubber companies raided in the region, imposing semislavery systems for the exploitation of latex. Ticunas were forced to leave their "*malokas*" (traditional housing), disperse into the jungle, and entirely transform their ways of production, becoming for the first time dependent on a cash economy (Ullán de la Rosa, 2000). Finally, during the 1930s, a sovereignty dispute between Colombia and Peru resulted in the Colombian government recognizing the Amazonian indigenous people as part of the nation-state. It is during this period that the Ticunas—together with other tribes of the Amazon region—went through a phase of cultural and social integration promoted by the Colombian state (López, 2000). After the conflict with Peru, the Colombian State and the Catholic Church joined forces in the "savage reduction" process, evangelizing and integrating indigenous people to the Colombian state standards (Valencia Llano, 1987).

In the 1980s, drug trafficking was a flowering economy in the region and indigenous people began to manage significant amounts of money, with which they bought goods that were previously out of their budgets (outboard motors, televisions, refrigerators, etc.). This period of rapid economic growth was slowed down in the late 1980s, with drug eradication policies in the Amazon. The region remains an important route for drugs, but the trade has been significantly reduced, at least on the Colombian border. This period of illegal economic growth generated significant socioeconomical changes, based on a mentality of short-term gratification. Since the 1990s, a double and contradicting social process took place: on the one hand, the Ticuna community experienced undeniable changes as a result of globalization and Westernization; on the other hand, the promotion of indigenous territories allowed for increased initiatives for the conservation of Ticuna's culture and language (Ullán, 2000).

Nowadays, almost the total area of Puerto Nariño is composed by indigenous reserves, covering 93% of the total area. Ticunas are part of the TICOYA reserve, which was established in 1990 by the Resolution 021, covering an area of 140,623 Ha (Ruiz, 2008).

Contemporary Livelihoods in Indigenous Communities From Puerto Nariño

The main livelihood activities of these communities include shifting cultivation (31%), fishing (24%), timber extraction (7%), hunting (4%), the collection of nontimber forest products (4%), livestock (4%), as well as some salaried jobs (18%) (Trujillo, 2008; data from three communities) (Figs. 26.3 and 26.4).

Tourism has become an important livelihood activity providing jobs for guides, cooks, cleaning

FIGURE 26.3 Woman carrying a paca in an indigenous community. *Photo: Laurane L'haridon.*

FIGURE 26.4 Seasonal floodings drive fishing activities in Puerto Nariño. *Photo: Laurane L'haridon.*

FIGURE 26.6 Grilled paca: a traditional meal among indigenous communities of Puerto Nariño.

FIGURE 26.5 Hunter showing a traditional hunting tool.

services in hostels, and handicraft. Studies indicate the cultural importance of wildlife and its contribution to food security especially in situations where there is a tendency for nutritional transitions (van Vliet et al., 2015a,b). Fish and bushmeat are important sources of animal protein for Ticuna indigenous families (Fig. 26.5). Hunting, fishing, and crops contribute to preserve their self-sufficiency of food and to rely less on products coming from other regions (Muñoz, 2012).

Because of the importance of bushmeat in the region, in 2015, a group of hunters and leaders decided to create the Airumaküchi Association. The aim of this association is to improve the quality of life and food security of indigenous communities, especially hunters and their families and strengthen the culture of indigenous people through activities based on traditional knowledge related with hunting of wild animals (Fig. 26.6).

TABOOS AND BELIEFS AFFECTING THE USE OF WILDLIFE

Studying Taboos and Beliefs in Indigenous Communities From Puerto Nariño

In order to understand the taboos and beliefs affecting the use of wildlife in Puerto Nariño, we carried out open discussions with two key traditional experts and 82 semistructured interviews (out of a total adult population of 110 people) with 16 hunters and 19 women members of their families, 26 nonhunter men and 18 women members of their families, and 3 women involved in selling wildmeat in the small town of Puerto Nariño. The questionnaire meant to obtain information on beliefs and taboos about hunting, terrestrial wildlife consumption, and the use of wildlife parts for zoo-therapeutic purposes. The questions aimed at describing the type of belief [temporal (concerning different seasons), segment (concerning a given fraction of the population), practice (regulating a given practice), species (regulating the use of a given species), life history (regulating the use at a given stage of the life history of an animal), habitat (regulating the use on certain habitats)], the species, and part of the animal involved, whether it regulated hunting/handling or consumption, the target for the taboo, the level of compliance with the belief, and the consequences of not complying.

Taboos Regulating the Use of Wildlife in Puerto Nariño

In total we identified 67 taboos including 49 taboos related to species, each one concerning a specific animal; and 18 taboos related to hunting practices, consumption, and/or use of bushmeat in general. We identified 24 animal species related to the different taboos (one species can be subject to more than one taboo). According to IUCN 2016, most of the species (16 species) are classified as "Least concern," four species as "Vulnerable," four species as "Data deficient," and one species as "Near threatened" (Table 26.1).

Tapirus terrestris (tapir), *Panthera onca* (jaguar), and tortoises (*Podocnomis* sp. *Chelonoidis denticulata*, *Chelus fimbriatus*) are the species for which most beliefs on hunting, consumption, and handling exist (Fig. 26.7).

Most beliefs regulate consumption (69%), hunting (13%), and handling of bushmeat (11%). 25% of the taboos mentioned applied to hunters, 23% to pregnant women, 10% to children. If taboos are not respected, the consequences can be illness (in 59% of the cases), bad luck in hunting trips (12% of the cases), less animals in the forest (11%), getting lost in the forest (10%). Illnesses mainly referred to skin irritations and spots (63% of the taboos) and diarrhea or stomach pain (19%). The level of compliance was similar between men (49% of them complied) and women (53% of them complied) not between class ages. Beliefs about wildlife also included several therapeutic uses of wildlife, mainly to strengthen bones and muscles, cure or prevent respiratory problems, improve growth and general strength, or cure and prevent dysfunctions in the stomach and digestive system. The parts most used are the grease, bones, hooves, gall, and penis.

Implications of Taboos for Wildlife Management

We found potential evidence of taboo erosion from a lack of adherence to particular taboos both among men and women and across generations. Drivers of taboo nonadherence may include migration (in-migration from *colonos*, in/out migration in dynamic rural-urban systems), decreased dependency on wildlife for local livelihoods, decreased nature-relatedness of the younger generation. However, without a proper understanding of the dynamic nature of taboos and the patterns of their transmission, there is no proper evidence of erosion. Other studies have shown that over the broad course of human history, particular elements of traditional ecological knowledge eventually disappear while new elements of knowledge are generated as cultures evolve and adapt to new conditions (Boyd and Richerson, 2005). Indeed, in our study area, several taboos continue to exist and influence behaviors around the use of wildlife. A historical perspective would allow understanding the dynamic nature of belief systems leading to the parallel loss of some beliefs and the creation of new ones.

Current taboos may not be used as such to protect endangered species, but the persistence

TABLE 26.1 Species Related With Ticuna Taboos and Their Conservation Status According to IUCN

Species	Scientific Name	IUCN Conservation Status
MAMMALS		
Tapir	*Tapirus terrestris* (Linnaeus, 1758)	Vulnerable
Red duiker	*Mazama americana* (Erxleben, 1777)	Data deficient
Red howler monkey	*Alouatta seniculus* (Linnaeus, 1776)	Least concern
Colombian night monkey	*Aotus vociferans* (Spix, 1823)	Least concern
Jaguar	*Panthera onca* (Linnaeus, 1758)	Near threatened
Puma	*Puma concolor* (Linnaeus, 1771)	Least concern
Ocelot	*Leopardus pardalis* (Linnaeus, 1758)	Least concern
Brown-throated sloth	*Bradypus variegatus* (Schinz, 1825)	Least concern
Capybara	*Hydrochoerus hydrochaeris* (Linnaeus, 1776)	Least concern
Spotted paca	*Cuniculus paca* (Linnaeus, 1776)	Least concern
Giant anteater	*Myrmecophaga tridactyla* (Linnaeus, 1758)	Vulnerable
White-lipped peccary	*Tayassu pecari* (Link, 1795)	Vulnerable
Tayra	*Eira barbara* (Linnaeus, 1758)	Least concern
Bush dog	*Speothos venaticus* (Lund, 1842)	Least concern
Black agouti	*Dasyprocta fuliginosa* (Wagler, 1832)	Least concern
Dolphin	*Inia geoffrensis* (Blainville, 1817) *Sotalia fluviatilis* (Gervais et Deville, 1853)	Data deficient
REPTILES		
Yellow-footed tortoise	*Chelonoidis denticulata* (Linnaeus, 1766)	Vulnerable
Tortuga mata-mata	*Chelus fimbriatus* (Schneider, 1783)	Least concern
Black caimán	*Melanosuchus niger* (Spix, 1825)	Least concern
Boa	*Eunectes murinus* (Linnaeus, 1758)	Data deficient
BIRDS		
Bush pigeon	*Crypturellus cinereus* (Gmelin, 1789)	Least concern
Hawk	*Buteogallus schistaceus* (Sundevall, 1851)	Least concern
Eagle	*Pandion haliaetus* (Linnaeus, 1758)	Least concern
Razor-billed Curassow	*Mitu tuberosum* (Spix, 1825)	Least concern
FISH		
Arapaima, Pirarucu	*Arapaima gigas* (Cuvier, 1829)	Data deficient

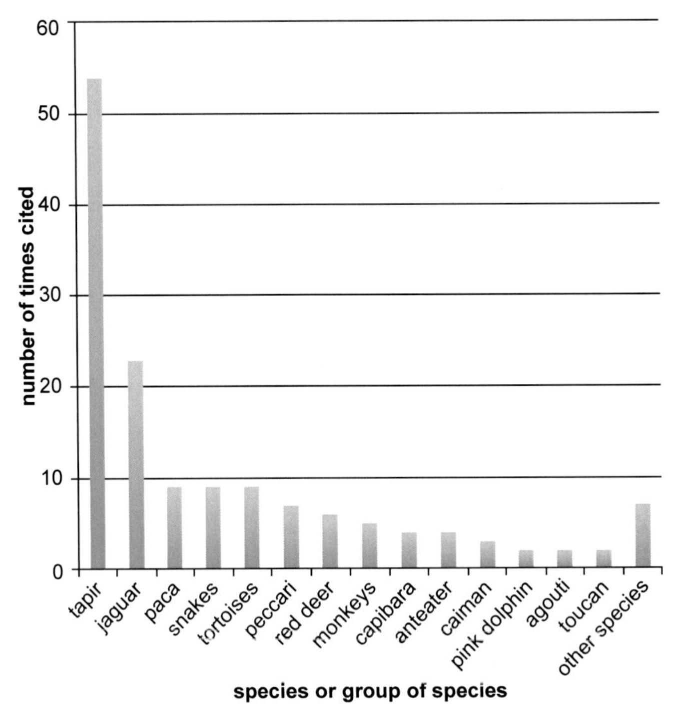

FIGURE 26.7 Number of times a species or group of species is cited in a taboo.

of a local belief system may offer some opportunities to enhance sustainable use. The jaguar, tapir, and tortoise species are among the most cited in taboos and are also listed as vulnerable or near threatened in IUCN categories. In more than 22% of the cases analyzed in this study, taboos protect the abundance of wildlife in the forest, and lacking respect to those taboos may imply bad luck in hunting or reduced animals in the forest. It is examples such as these that may facilitate mechanisms for finding synergies in informal institutions and formal conservation strategies. The taboos and beliefs about wildlife described here have been included in the diagnosis document that will be used as a basis to prepare the hunting management plan of Airumaküchi. Besides, a leaflet describing those taboos will be disseminated to community members as a way to raise awareness about those taboos, particularly to the younger generation.

MONITORING WILDLIFE POPULATIONS USING A TRADITIONAL PRACTICE TO IMITATE THE ANIMALS

The Traditional Practice of Calling Animals

In tropical forest regions, one technique commonly used by the hunters to attract their prey is to call the animals. This technique consists in imitating a call usually produced by the animals to communicate among their pairs (Alves et al., 2009). For example, in Central Africa, hunters imitate a distressed blue duiker in order to attract different species of duikers (*Philantomba monticola, Cephalophus callipygus,* and *Cephalophus dorsalis*) and shoot. Scientific surveys have also used the call method, particularly for birds (Brandes, 2008) or large carnivores (Ogutu and Dublin, 1998; Van Vliet et al., 2009)

FIGURE 26.8 Hunters of the Airumaküchi Association in a meeting.

and explored the possibility to standardize the duiker call to improve daytime transect counts. However, to our knowledge there are no other studies describing the traditional practice of calling animals and adapting them to the context of local monitoring. In this section we describe the diversity of call methods used by indigenous hunters in Puerto Nariño, Amazonas, Colombia, and test the possibility of using the call of agouti (*Dasyprocta fuliginosa*) in the context of a standardized monitoring protocol.

Studying "Animal Calls" in Indigenous Communities From Puerto Nariño

In order to understand the local practice of animal imitations by the hunters in Puerto Nariño, we carried out semistructured interviews with all the hunters that acknowledged practising the calls (N=23 out of 34) (Fig. 26.8). The semistructured questionnaire included questions about the hunter (age, community, ethnical group, hunting techniques used), the way knowledge on animal calls was transmitted, the species called, and the circumstances in which the calls were used (species, potential instruments used to make the call, type of habitat, other characteristics of the place, moment of the day, period of the year, number of repetitions required, reactions of the animal, distance of efficiency of the call). All calls were recorded in order to keep a library of all animal calls that hunters imitate.

The application of the call in the field was only carried out for the most commonly called species: agouti (*Dasyprocta fuliginosa*). Agouti is also among the most hunted species in the studied communities. The call of agoutis was tested to monitor abundance based on a standardized protocol. The protocol covered uniformly the section of the hunting ground that is most frequently used by the hunters of Puerto Nariño and Puerto Esperanza for hunting activities (at less than 6 km from the village). We geolocated 12 calling sites at about 1.4 km from one another and carried out two calls (at a minimum distance of 50 m away from each other) in each calling site (Fig. 26.9).

The protocol was repeated five times (each week from end of September to end of November 2015) for a total of 120 calls. The calls were carried out during the day between 6 a.m. and 5 p.m. by a team composed of one hunter (who imitated the call) and one researcher who observed the responses. For each call, the researcher located the calling site using the GPS and the hunter decided on the most appropriate place to carry out the call within the vicinity (e.g., a fruiting tree). The call consists of a whistle that imitates an agouti in distress. The whistle was repeated up to 10 times in each calling location. For each calling session we described the weather conditions, the presence of human disturbances close by, and the responses obtained (number of animals observed, species, type of observation (sound of movement/whistle responses/direct observation), and number of whistles before a reaction was observed).

Local Knowledge on the Call of Forest Animals

Species Called

From the 34 hunters present in the six communities surveyed, 23 practised the calls in their

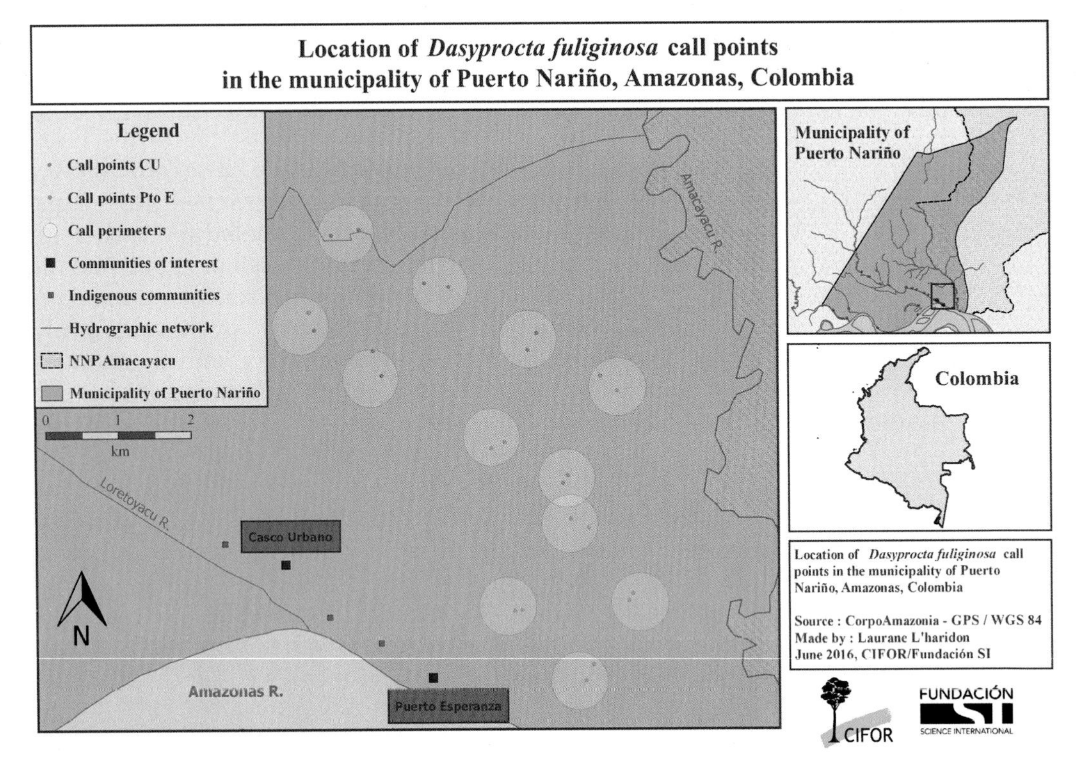

FIGURE 26.9 Map with the location of the "call" points.

hunting trips. The hunters who practise the calls are 48 years old on average and are mostly Ticuna or Yagua. They hunt mainly with guns. Their knowledge on animal calls was transmitted to them by their father or grandfather, and they also transmit their knowledge to their sons. However, many of them state that they have shared less than one quarter of their knowledge to the next generation. Indeed, according to the hunters, the younger generation shows an increased lack of interest in traditional activities and the traditional knowledge associated with them. This could come from the diversification of the economic activities in the region, generating other preferred opportunities for the youth. Each hunter is able to imitate eight animals on average, but the knowledge on animal calls may vary significantly from one hunter to the other. In total, 28 animal species are imitated by the hunters from the studied communities (18 mammals; 9 birds; 1 reptile) (Fig. 26.10).

The most popular call is that of agouti (*Dasyprocta fuliginosa*), and it is the only one that all hunters are able to imitate. The other two well-known calls are that of tapir (*Tapirus terrestris*) and that of caiman (*Caiman crocodilus*). The call of the "panguana" or tinamou (Crypturellus sp., *Tinamus* sp.) and that of the curassow (*Mitu* sp.) are known by half of the hunters. The other 23 calls are only known by a few hunters. Some calls may be practised with a tool, a leaf, a whistle, a gourd, etc. The whistle may be used for different animals and can be made out of turtle bones, sticks, iron, among others. It produces a cleaner and louder sound than just using the mouth. Out of 23 hunters, 8 use the whistle, particularly to imitate the agouti and the tapir (Fig. 26.11).

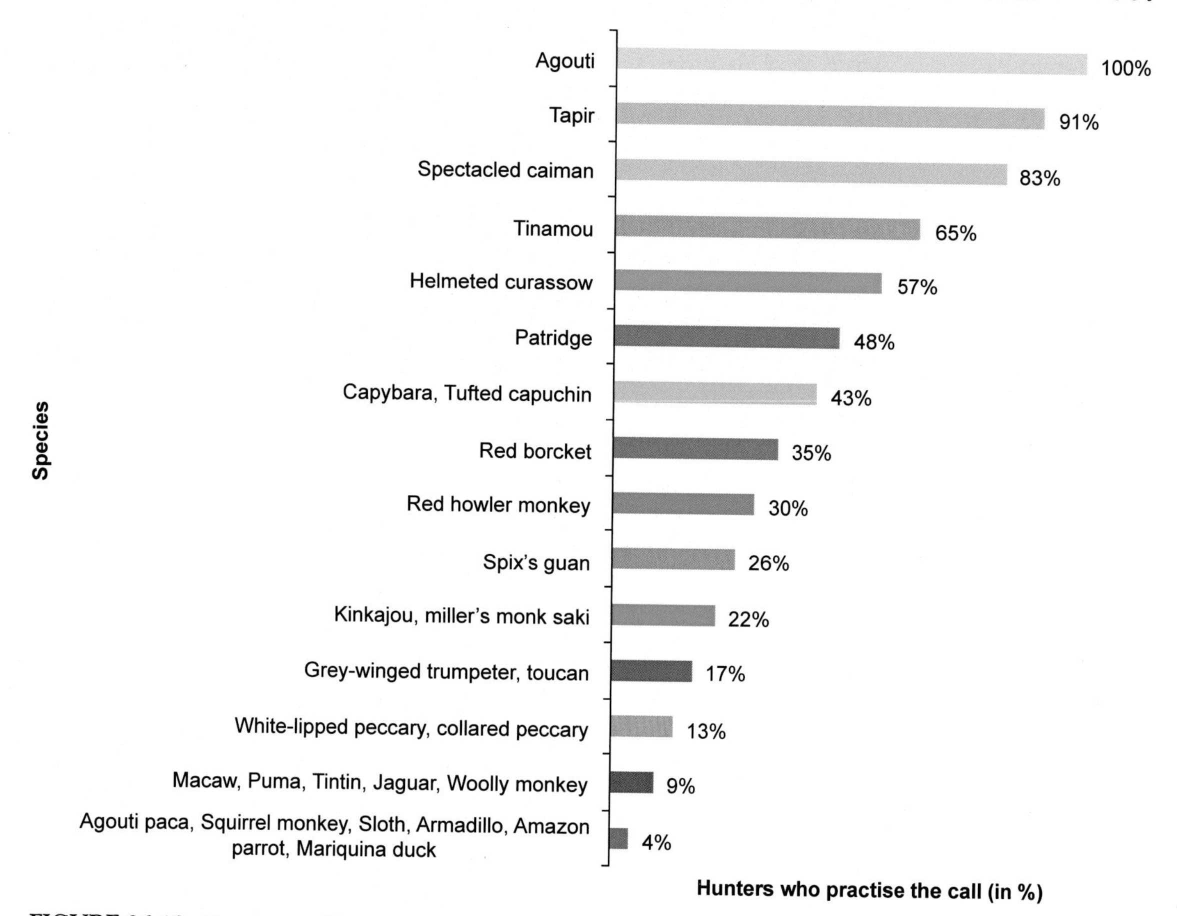

FIGURE 26.10 Percentage of hunters interviewed who know how to reproduce the sound of each of the species listed.

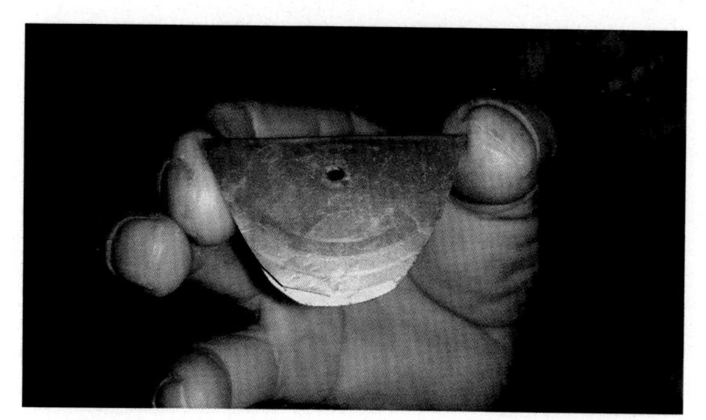

FIGURE 26.11 Whistle used for calling the paca.

Although all calls target specific species, some may also attract other species. For example, the calls for capuchin (*Cebus apella*), woolly monkey (*Lagothrix lagotricha*), and howler monkey (*Alouatta seniculus*) may also attract predators such as cats (*Puma concolor, Leopardus wiedii, Leopardus pardalis, Panthera onca*) and hawks (*Accipiter* sp.). Some species may also be attracted by curiosity, for example, small monkeys (*Saimiri sciureus, Saguinus nigricollis*) may respond to the call for agouti and other terrestrial mammal species.

Different Types of Animal Calls

According to the hunters, the calls may imitate different situations: animals in distress (e.g., agouti), offspring calling its mother (e.g., tapirs), mating calls (e.g., caimans), communication calls between members of the same group (e.g., birds), etc. Moreover, a same species can use different calls (e.g., tapirs use also mating calls). According to the type of calls, the hunters suggest that the reaction of the animal differs. The simple communication call that is practised between members of the same group generates a cautious approach and a call back by the animal trying to get closer. On the contrary, the three other calls that we have mentioned above generate an aggressive attitude and rapid approach toward the call point (e.g., agouti, tapir), especially the distress call. The waiting time before an answer depends on the type of call. An animal with an aggressive attitude will arrive swiftly, while an animal responding to a simple communication call will approach slowly and might only be recognized by the hunter because of the noise made or because it answers back, but does not necessarily get visible to the hunter (e.g., tinamou and curassow). The call may be heard by animals located as far as 50 m from the calling point, although there might be a few exceptions: for example, the caiman call may be heard as far as 200 m from the calling point.

Most Suitable Time and Place for Calling Animals

In the following section, we will focus on the five most commonly called species: agouti, tapir, caiman, tinamou, and curassow.

The calls are used all along the year, no matter the season. Most of the time it must be repeated at least 10 times in the same site before it is effective. The agouti, tinamou, and curassow are diurnal species, and as such, the call is practised during the day, most often at dawn or dusk. The tapir may be called both at night and during the day. As for the caiman, the call is most often practised at night for hunting purposes because its red eyes can be easily located at night using a torch. However, some hunters mentioned that it was potentially dangerous to practise the call at night due to the potential presence of the black boa (Boidae sp.), that could attack the hunter; especially in case of imitating animals characterized as easy prey, like the agouti.

FIGURE 26.12 Hunter waiting for response in "call" point.

FIGURE 26.13 White-nosed monkey on an "aguaje" tree.

FIGURE 26.14 Fruiting spot used as a calling point.

According to the hunter, the call may be practised in any place where animal signs such as footprints, eaten fruits, sounds were observed (Figure 26.12, 26.13 and 26.14)

However, there are specific places that are thought to be strategic for success: fruiting trees, salt licks, and watersheds. The "aguajales" [places where moriche palms (*Mauritia flexuosa*) grow] were mentioned by some hunters as favorite areas to call agoutis and tapirs, due to the great availability of food present in these hydromorphic soil areas. Agoutis and tinamous may be called close to the village due to their abundance in disturbed areas, while tapir, caimans, and curassows are mainly called in more remote areas. The caimans are called from the canoe in lakes and small streams.

Adaptation of the Call of Agouti to the Context of Participatory Monitoring

From a total of 120 calls, 18 yielded the presence of agouti in 15 different call points. Among those, 6 yielded a visual response and the rest were only heard. Among the agoutis seen, two arrived quickly very close to the call point, and the other four approached cautiously. Other species were also attracted by the call for agoutis: *Crypturellus* sp., *Tinamus* sp., *Penelope jacquacu*, *Ramphastos tucanus*, *Pithecia monachus*, *Accipiter* sp., *Pecari tajacu*. Calls were more effective between 7 and 9 h and from 15 to 16 h. According to our protocol, the probability to observe agoutis using the call was higher near fruiting trees (during our study period (dry season) the only fruiting trees available were the "aguajales"), small streams, below fruiting trees, and near fallen trees. The agoutis were most often observed after the sixth call, equivalent to 5–8 minutes after the initial call. In most cases, the agoutis were first heard and then seen, for some of them.

DISCUSSION AND CONCLUSIONS

Previous research on traditional ecological knowledge has mainly focused on (1) documenting the erosion of local knowledge (e.g., Ferguson and Messier, 1997; Pieroni et al., 2004), (2) understanding the feedback effects of decrease of cultural diversity on decrease of biological diversity (Maffi, 2005; Harmon and Loh, 2010), and (3) assessing the processes and drivers of change that lead to the loss of traditional ecological knowledge (Godoy et al., 2005). In this study, our focus is to illustrate how traditional ecological knowledge in modern indigenous communities can contribute to guide decision-making regarding natural resource management and biodiversity conservation practices. Our assumption is that initiatives to support wildlife management in indigenous territories may prove more effective if they are inspired or based on traditional ecological knowledge. This knowledge can take various forms. Our chapter provides two original ways in which traditional ecological knowledge can be taken into account in formal management strategies. These results come to complement an important body of literature that describes how local practices (including monitoring, temporal or total protection of species or habitats, multiple-species management, resource rotation, succession management), and the social mechanisms behind them such as cross-scale institutions, taboos and regulations, rituals or ceremonies, and social and religious sanctions contribute to improve natural resources management (Agrawal, 2001; Colding et al., 2003; Mishra et al., 2003; Grant and Berkes, 2007; Rai, 2007).

Our results indicate that current taboos and beliefs about wildlife in Puerto Nariño may provide a positive ground to build conservation strategies. As shown elsewhere, indigenous belief systems and practices play a key role in shaping the exploitation of natural resources, including wild game (Koster et al., 2010). When applied to hunting, indigenous beliefs can affect both hunting pressure in specific areas of the landscape (Read et al., 2010) and the targeting or avoidance of specific classes and species of wildlife (Pezzuti et al., 2010; Luzar et al., 2012). Dietary taboos have been widely documented among various indigenous peoples in Amazonia and elsewhere and often appear as cultural responses to specific health maladies (Meyer- Rochow, 2009; Pezzuti et al., 2010). In many cases, taboos do not apply across the society but to specific individuals or demographic subsets such as women and children (Begossi et al., 2004; Pezzuti et al., 2010). Game meat taboos are often believed to be essential to preserving good health, especially those taboos applying to children, menstruating women, and parents of newborn infants (Begossi et al., 2004; Meyer-Rochow, 2009; Koster et al., 2010). In Puerto Nariño, the belief and taboo system about wildlife may contribute to build a holistic approach where food security, human health, and conservation can be taken into consideration in wildlife management planning.

Our results also show that traditional techniques may be used for the monitoring of wildlife. Here we illustrate how the traditional "call" method, used for hunting, may be adapted to the context of wildlife monitoring. The monitoring of prey population trends is a critical first component of the sustainable management of customary harvests of wildlife (Moller et al., 2004). Indeed, in a dynamic context, with constant changes, the adaptive management approach must include a monitoring system to evaluate the impacts of the actions put in place and constantly adapt in the search for improved outcomes. For example, reliable monitoring could signal potential overharvesting and then lead to an adjustment of hunting pressure to safeguard sustainability. Several approaches and tools have been used over the last 15 years to develop monitoring of local resources in a participatory context (Borrini-Feyerabend, 1996; Nguinguiri, 1999; Wollenberg et al., 2005). This study contributes to the existing literature on the process of designing participatory monitoring systems, trying to combine scientific information with traditional knowledge (Fraser et al., 2006; Mendoza and Prabhu 2000, 2003) to further involve locals in the sustainable management of resources, by giving due importance to their knowledge and facilitating communication between them and other stakeholders, especially administrations and state agencies. As shown by Moller et al., 2004, methodologies such as those presented in this study in which traditional monitoring methods are calibrated against population abundance make it possible to mesh traditional ecological knowledge with scientific inferences of prey population dynamics.

Acknowledgments

We are grateful to all informants, hunters, and stakeholders of the bushmeat trade who actively participated in this study and facilitated our visits to the field. This work was possible thanks to the financial support from USAID and UKAID grants through the Bushmeat Research Initiative from CIFOR. This research is supported by CGIAR Fund Donors. For a list of Fund donors, please see: www.cgiar.org/about-us/our-funders.

References

Agrawal, A., 2001. Common property institutions and sustainable governance of resources. World Development 29 (10), 1649–1672.

Alves, R.R.N., Mendonça, L.E., Confessor, M.V., Vieira, W.L., Lopez, L.C., 2009. Hunting strategies used in the semi-arid region of northeastern Brazil. Journal of Ethnobiology and Ethnomedicine 5 (12), 1–16.

Begossi, A., Hanazaki, N., Ramos, R.M., 2004. Food chain and the reasons for fish food taboos among Amazonian and Atlantic Forest fishers (Brazil). Ecological Applications 14 (5), 1334–1343.

Bobo, K.S., Aghomo, F.F.M., Ntumwel, B.C., 2015. Wildlife use and the role of taboos in the conservation of wildlife around the Nkwende Hills Forest Reserve; South-west Cameroon. Journal of Ethnobiology and Ethnomedicine 11 (1), 1.

Borrini-Feyerabend, G., 1996. Co-management-a new approach to conserving Uganda's forests. Plant Talk 22–25.

Boyd, R., Richerson, P.J., 2005. The Origin and Evolution of Cultures. Oxford University Press, Oxford.

Brandes, T.S., 2008. Automated sound recording and analysis techniques for bird surveys and conservation. Bird Conservation International 18 (S1), S163–S173.

Colding, J., Folke, C., 2001. Social taboos:"invisible" systems of local resource management and biological conservation. Ecological Applications 11 (2), 584–600.

Colding, J., Folke, C., Elmqvist, T., 2003. Social institutions in ecosystem management and biodiversity conservation. Tropical Ecology 44 (1), 25–41.

Ferguson, M.A., Messier, F., 1997. Collection and analysis of traditional ecological knowledge about a population of Arctic tundra caribou. Arctic 17–28.

Fraser, D.J., Coon, T., Prince, M.R., Dion, R., Bernatchez, L., 2006. Integrating traditional and evolutionary knowledge in biodiversity conservation: a population level case study. Ecology and Society 11 (2), 4.

Godoy, R., Reyes-García, V., Byron, E., Leonard, W.R., Vadez, V., 2005. The effect of market economies on the well-being of indigenous peoples and on their use of renewable natural resources. Annual Review of Anthropology 34, 121–138.

Golden, C.D., Comaroff, J., 2015. The human health and conservation relevance of food taboos in northeastern Madagascar. Ecology and Society 20 (2), 42.

Grant, S., Berkes, F., 2007. Fisher knowledge as expert system: a case from the longline fishery of Grenada, the Eastern Caribbean. Fisheries Research 84 (2), 162–170.

Harmon, D., Loh, J., 2010. The index of linguistic diversity: a new quantitative measure of trends in the status of the world's languages. Language, Documentation and Conservation 4, 97–151.

Horsthemke, K., 2015. Animals and the law in east, west and southern Africa. In: Animals and African Ethics. Palgrave Macmillan, UK, pp. 101–117.

Koster, J., 2011. Interhousehold meat sharing among Mayangna and Miskito horticulturalists in Nicaragua. Human Nature 22 (4), 394–415.

Koster, J.M., Hodgen, J.J., Venegas, M.D., Copeland, T.J., 2010. Is meat flavor a factor in hunters' prey choice decisions? Human Nature 21 (3), 219–242.

López, C.L., 2000. Ticunas Brasileros, Colombianos Y Peruanos. Etnicidad y nacionalidad en la región de fronteras del Alto Amazonas/Solimões. Brasília: Ceppac.

Luzar, J.B., Silvius, K.M., Fragoso, J.M., 2012. Church affiliation and meat taboos in indigenous communities of guyanese Amazonia. Human Ecology 40 (6), 833–845.

Maffi, L., 2005. Linguistic, cultural, and biological diversity. Annual Review of Anthropology 34, 599–617.

Mendoza, G.A., Prabhu, R., 2000. Development of a methodology for selecting criteria and indicators of sustainable forest management: a case study on participatory assessment. Environmental Management 26 (6), 659–673.

Mendoza, G.A., Prabhu, R., 2003. Qualitative multi-criteria approaches to assessing indicators of sustainable forest resource management. Forest Ecology and Management 174 (1), 329–343.

Meyer-Rochow, V.B., 2009. Food taboos: their origins and purposes. Journal of Ethnobiology and Ethnomedicine 5 (1), 1.

Mishra, C., Allen, P., McCarthy, T.O.M., Madhusudan, M.D., Bayarjargal, A., Prins, H.H., 2003. The role of incentive programs in conserving the snow leopard. Conservation Biology 17 (6), 1512–1520.

Moller, H., Berkes, F., O'brian Lyver, P., Kislalioglu, M., 2004. Combining science and traditional ecological knowledge: monitoring populations for co-management. Ecology and Society 9 (3), 15.

Moreno Arocha, 2014. Descripción Geográfica. In: Trujillo, F., Duque, S. (Eds.), Los humedales de Tarapoto: aportes al conocimiento sobre su biodiversidad y uso. Serie humedales de la Amazonia y Orinoquia, Fundación Omacha, Corpoamazonia, Universidad Nacional Sede Leticia. 400 p.

Morsello, C., Yagüe, B., Beltreschi, L., Van Vliet, N., Adams, C., Schor, T., Cruz, D., 2015. Cultural attitudes are stronger predictors of bushmeat consumption and preference than economic factors among urban Amazonians from Brazil and Colombia. Ecology and Society 20 (4).

Muñoz, L.E.A., 2012. Los sistemas de producción de la etnia Ticuna del resguardo de Puerto Nariño, sur del Trapecio Amazónico: una aproximación socioeconómica. Cuadernos de Desarrollo Rural 46.

Nardoto, G.B., Murrieta, R.S.S., Prates, L.E.G., Adams, C., Garavello, M.E.P., Schor, T., Duarte-Neto, P.J., Schor, T., De Moraes, A., Rinaldi, F.D., Gragnani, J.G., Moura, E.A., Duarte-Neto, P.J., Martinelli, L.A., 2011. Frozen chicken for wild fish: nutritional transition in the Brazilian Amazon region determined by carbon and nitrogen stable isotope ratios in fingernails. American Journal of Human Biology 23 (5), 642–650.

Ngoufo, R., Yongyeh, N.K., Obioha, E.E., Bobo, K.S., Jimoh, S.O., Waltert, M., 2014. Social norms and cultural services-community belief system and use of wildlife products in the Northern periphery of the Korup National Park, South-West Cameroon. Change and Adaptation in Socio-ecological Systems 1 (1), 26–34.

Nguinguiri, J.C., 1999. Les approches participatives dans la gestion des écosystemes forestiers d'Afrique centrale: Revue des initiatives existantes (No. CIFOR Occasional Paper no. 23, p. 24p). CIFOR, Bogor, Indonesia.

Ogutu, J.O., Dublin, H.T., 1998. The response of lions and spotted hyaenas to sound playbacks as a technique for estimating population size. African Journal of Ecology 36 (1), 83–95.

Oviedo, G., Maffi, K., 2000. Toward a Bio-cultural Approach to Conserving the Diversity of Life in the World's Eco-regions. Indigenous and traditional peoples of the world and eco-region conservation.

Parry, L., Barlow, J., Pereira, H., 2014. Wildlife harvest and consumption in Amazonia's urbanized wilderness. Conservation Letters 7 (6), 565–574.

Peres, C.A., Terborgh, J.W., 1995. Amazonian nature reserves: an analysis of the defensibility status of existing conservation units and design criteria for the future. Conservation Biology 9 (1), 34–46.

Pezzuti, J.C., Lima, J.P., da Silva, D.F., Begossi, A., 2010. Uses and taboos of turtles and tortoises along Rio Negro, Amazon Basin. Journal of Ethnobiology 30 (1), 153–168.

Pieroni, A., Howard, P., Volpato, G., Santoro, R.F., 2004. Natural remedies and nutraceuticals used in ethnoveterinary practices in inland southern Italy. Veterinary Research Communications 28 (1), 55–80.

Popkin, B.M., 2006. Global nutrition dynamics: the world is shifting rapidly toward a diet linked with noncommunicable diseases. The American Journal of Clinical Nutrition 84 (2), 289–298.

Popkin, B.M., Gordon-Larsen, P., 2004. The nutrition transition: worldwide obesity dynamics and their determinants. International Journal of Obesity 28, S2–S9.

Rai, S.C., 2007. Traditional ecological knowledge and community-based natural resource management in northeast India. Journal of Mountain Science 4 (3), 248–258.

Rangel, E., Luengas, B., 1997. Clima y aguas del eje Apaporis–Tabatinga, pp. 49–68 IGAC–SINCHI–Universidad Nacional de Colombia. Zonificación ambiental para el plan modelo colombo–brasilero (eje Apaporis–Tabatinga). Santafé de Bogotá.

Read, J.M., Fragoso, J.M., Silvius, K.M., Luzar, J., Overman, H., Cummings, A., Gery, S.T., de Oliveira, L.F., 2010. Space, place, and hunting patterns among indigenous peoples of the Guyanese Rupununi region. Journal of Latin American Geography 9 (3), 213–243.

Riaño Umbarila, E., 2003. Organizando su espacio, construyendo su territorio: transformaciones de los asentamientos ticuna en la ribera del Amazonas colombiano (No. 980.4616 R481). Universidad Nacional de Colombia, Bogotá (Colombia).

Ruiz, A., 2008. Coordinación General del Plan de Desarrollo: diseño, elaboración y formulación año 2008. Alcaldía Municipal de Puerto Nariño.

Santos-Fita, D., Naranjo, E.J., Estrada, E.I., Mariaca, R., Bello, E., 2015. Symbolism and ritual practices related to hunting in Maya communities from central Quintana Roo, Mexico. Journal of Ethnobiology and Ethnomedicine 11 (1), 1.

Sarti, F.M., Adams, C., Morsello, C., Van Vliet, N., Schor, T., Yagüe, B., Tellez, L., Quiceno, M.P., Cruz, D., 2015. Beyond protein intake: bushmeat as source of micronutrients in the Amazon. Ecology and Society 20 (4).

Seixas, C.S., Begossi, A., 2001. Ethnozoology of fishing communities from Ilha Grande (Atlantic forest coast, Brazil). Journal of Ethnobiology 21 (1), 107–135.

Sirén, A., 2012. Festival hunting by the kichwa people in the Ecuadorian amazon. Journal of Ethnobiology 32 (1), 30–50.

Trujillo, C., 2008. Selva y mercado: exploración cuantitativa de los ingresos en hogares indígenas. MSc, 161. Universidad Nacional de Colombia, Leticia, Amazonas, Colombia.

Ullán de la Rosa, F., 2000. Plurimorfología del fenómeno mesiánico-milenarista: la secuencia histórica de los movimientos ticuna. Estudios del Hombre 11, 13–40.

Valencia Llano, A., 1987. La política de colombianización de los salvajes. El caso huitoto. Revista Palabra 3–4 Popayán.

Van Vliet, N., Kaniowska, E., Bourgarel, M., Fargeot, C., Nasi, R., 2009. Answering the call! Adapting a traditional hunting practice to monitor duiker populations. African Journal of Ecology 47 (3), 393–399.

van Vliet, N., Quiceno-Mesa, M.P., Cruz-Antia, D., Tellez, L., Martins, C., Haiden, E., Oliveira, M., Adams, C., Morsello, C., Valencia, L., Bonilla, T., Yague, B., Bonilla, T., Nasi, R., 2015a. From fish and bushmeat to chicken nuggets: the nutrition transition in a continuum from rural to urban settings in the Tri frontier Amazon region. Ethnobiology and Conservation 4.

Van Vliet, N., Quiceno, M.P., Cruz, D., Neves de Aquino, L.J., Yagüe, B., Schor, T., Hernández, S., Nasi, R., 2015b. Bushmeat networks link the forest to urban areas in the trifrontier region between Brazil, Colombia, and Peru. Ecology and Society 20 (3).

Vieco, J.J., Oyuela-Caycedo, A., 1999. La pesca entre los ticuna: historia, técnicas y ecosistemas. Boletín de Antropología de la Universidad de Antioquia 13, 30.

Wollenberg, E., López, C., Anderson, J., 2005. Though All Things Differ: Pluralism as a Basis for Cooperation in Forests. CIFOR, Bogor.

CHAPTER

27

Ethnozoology: An Overview and Current Perspectives

Rômulo Romeu Nóbrega Alves[1], Josivan Soares Silva[2], Leonardo da Silva Chaves[2], Ulysses Paulino Albuquerque[3]

[1]Universidade Estadual da Paraíba, Campina Grande, Brazil; [2]Universidade Federal Rural de Pernambuco, Recife, Brazil; [3]Universidade Federal de Pernambuco, Recife, Brazil

INTRODUCTION

Relationships between humans and other animals are as ancient as the appearance of our own species. This affirmation is completely logical—as we have evolved together with the world's other organisms—although our species stands out because of its incredible capacity to alter the natural environment through intentional interactions with the physical and biological landscape. The first known records of these interrelationships are widely distributed as pictographs and petroglyphs (Bahn and Vertut, 1997) that reveal the interest of early humans in many of the animals that cohabited diverse ecosystems. Analyses of large numbers of historical documents directly or indirectly recording relationships between humans and other animals (Bahn and Vertut, 1997) have shown that those interactions evolved and became extremely complex over time. Gradually, our vision of animals appears to have transitioned from a more simplistic view of predators or prey to a plane of mystical significance and, slightly later, humans took on animals to become pets and companions (Mithen, 1999)—often being considered true family members by their human keepers—which represents perhaps the most intimate relationship between humans and other animals.

Recording and understanding these relationships became the motivation for the first "ethnozoologists." Although that term was only coined in 1889 (Mason, 1899), botanists, zoologists, naturalists, missionaries, and adventurers, often in the service of great museums, had traveled to "savage regions" around the world in the early 1800s (Clément, 1998) to describe the biodiversity known to local native populations (Hunn, 2007). More than two centuries later, researchers began to transcend, in incipient manner, merely descriptive approaches of purely utilitarian character, to become the first true ethnobiologists (see Hunn, 2007).

Ethnozoology
http://dx.doi.org/10.1016/B978-0-12-809913-1.00027-2

The current growth in the volume of information gathered and the greater facility for accessing data have promoted ramifications in the lines of ethnozoological research and, consequently, attracted greater numbers of researchers. The increased numbers of published papers focusing on ethnozoology have resulted in the increased visibility of that specialty (see Albuquerque et al., 2013), now allowing detailed investigations of its advance as an area of scientific investigation (Lyra-Neves et al., 2015).

The growth and consolidation of any academic area is associated with the knowledge it generates and its subsequent diffusion. As such, scientific publications represent a fundamental metric for analyzing the growth and perspectives of a given area of investigation and, in that sense, the numbers of publications can be considered the most relevant indicators of the productivities of researchers in their respective areas of specialization. Therefore, in order to investigate the development of ethnozoology throughout the world and describe its current tendencies, we investigated the history of ethnozoological publications on a global scale and examined their evolution, tendencies, and perspectives.

COLLECTING DATA CONCERNING ETHNOZOOLOGICAL PUBLICATIONS

To analyze the global scenario of ethnozoological research, we conducted searches for scientific publications related to human/faunal interactions using the Scopus (http://www.scopus.com) and Web of Science (http://www.webofknowledge.com) databases, accessing all of the publications indexed up until the year 2016 that contained the words: ethnozoology, ethnoentomology, ethnoichthyology, historic ethnozoology, cinegetic activities, hunting, poaching, ethnocarcinology, ethnoornithology, ethnospongiology, ethnotaxonomy, ethnozoology, ethnoherpetology, ethnomalacology, ethnoelephantology, or ethnoprimatology in their titles, abstracts, or lists of key words. Additionally, we researched the terms "Local zoological knowledge," "Traditional zoological knowledge," "Local ecological knowledge," and "Traditional ecological knowledge," combined with the terms "fauna" and "animal."

We included only studies that directly investigated the relationships between human groups and faunal resources, and studies of the historical relationships between humans and the fauna (considered an area of historical ethnozoology). As such, in examining the titles and abstracts of scientific articles, we discarded: (1) purely pharmacological publications focusing exclusively on investigations of the effectiveness of animal-derived products; (2) ecological publications focusing on human actions only as factors impacting animal communities; (3) bibliographic reviews. The following information was extracted from the data banks accessed: (1) the country and continent where the study was undertaken (considering that a given article could include fieldwork undertaken in more than one country); (2) the year of publication (as indicated within the text); (3) the thematic area of ethnozoology (the works being principally grouped according to the classifications used by the authors themselves, but also taking into consideration their titles and key words, as well as a detailed reading of the text to identify the central themes. Studies focusing on hunting mammals, for example, where classified within the theme "cinegetic activities" and not "ethnozoology"); (4) the taxonomic group investigated (whenever possible, we recorded the animal groups considered in the research paper at the phylum or class level. We also noted cases in which a given publication investigated interactions between more than one taxonomic group).

The data obtained from this literature survey were then analyzed using the chi-square test to identify significant differences between the numbers of studies published on different continents, as well as the numbers of studies focusing on different taxa—which allowed us to categorize the articles according to the different subareas of ethnozoology. We also use the G-test to identify significant differences between the numbers of published works in the different "subareas" of ethnozoology. Bioestat 5.0 software was used for these analyses (Ayres et al., 2007).

GLOBAL SCIENTIFIC PRODUCTION IN ETHNOZOOLOGY

We identified 540 articles compatible with our search and selection criteria. In terms of the numbers of published articles in relation to the world's continents, we observed that the greatest number of texts were from Africa ($X^2 = 294.641$; $P < .001$). A total of 181 publications focused on the African continent, followed by South America (140), Asia (92), North America (71), Europe (50), Oceania (17), and Central America (11), those texts being distributed among different ethnozoological thematic areas (Fig. 27.1). In relation to the number of publications in the different countries on each continent, we observed that the greatest number of texts from Latin America focused on Brazil (n = 107), representing 76% of the scientific production of ethnozoology for that region, also putting that country in first place globally, considering our search criteria. The United States (n = 31) was responsible for 44% of the publications from North America; India (n = 21) for 23% of the publications in Asia; Tanzania (n = 20) for 11% of the African publications; Spain (n = 11) for 22% of the European texts; Australia (n = 6) for 35% of the publications from Oceania; while Haiti and Honduras (n = 3) produced 27% of the studies in Central America.

The large number of articles published in Africa reflects its enormous faunal biodiversity that is extensively exploited by local human populations living there—drawing the attention of researchers from all over the world. Additionally, Africa demonstrated a strong population growth between the years 1980 and 2000, increasing from 469 million to approximately 798

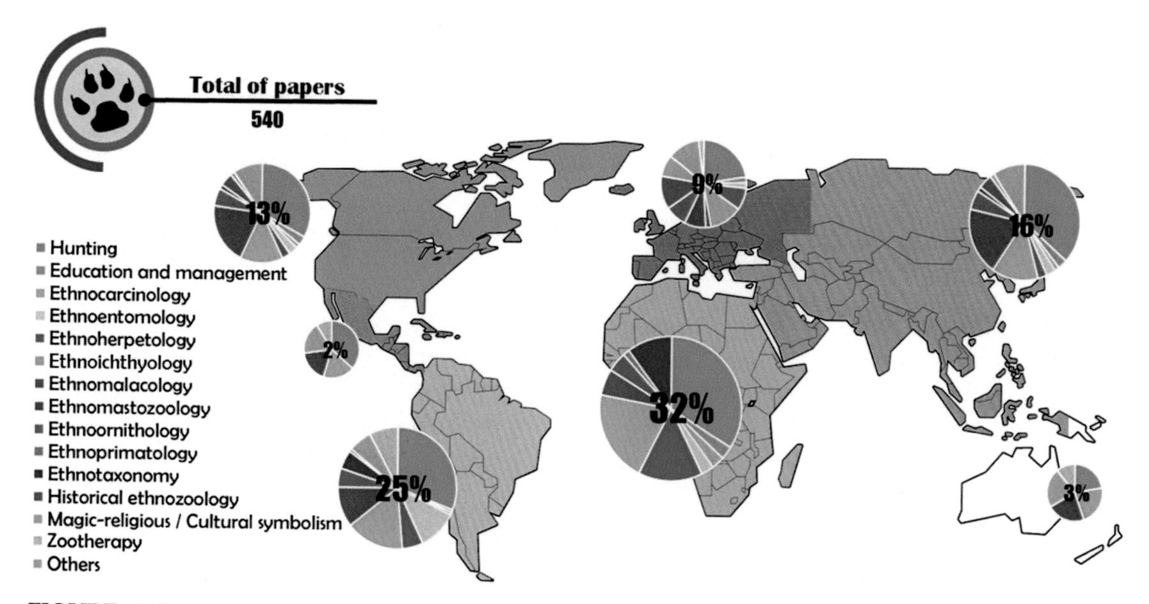

FIGURE 27.1 Number of publications per continent distributed among the different thematic areas of ethnozoology.

million inhabitants, which represented fully 13% of the world's population in 2000. Associated with this, the unequal distribution of wealth on that continent (40% of the population receives only 11% of the income) represents a large and continuous socioeconomic challenge (Kinzig and McShane, 2015). The high African population growth, associated with high poverty indices, represents an important factor intensifying their dependence on regional natural resources for survival—with the local fauna representing an important alternative nutritional resource for families, as well as additional sources of income (Van Vliet et al., 2016). The African continent harbors large numbers of medium- and large-sized vertebrates that are utilized by diverse traditional communities (Martin et al., 2012). Those animals are cited in the literature as the most sought-after targets of hunting activities, as they have large biomasses and present large energetic rewards (Van Vliet et al., 2016). Large animals, however, demonstrate lower reproduction rates than smaller animals (Cowlishaw et al., 2005), with consequent higher risks of extinction—which has generated great interest among researchers who investigated the dynamics of local communities and evaluated the natural stocks of those large animals to generate proactive plans to minimize anthropogenic impacts. The large contribution of publications from Tanzania in relation to other African countries may be related to the large areas used for hunting in that country (Van Vliet et al., 2016). A number of important national parks are found in Tanzania containing large numbers of vertebrates (such as the Serengeti National Park), which were set aside for both tourism and legal hunting by local populations (Martin et al., 2012)—factors that favor ethnozoological research, as well as investigations, in many other areas of the natural sciences focusing on the conservation of nature.

The large numbers of research reports in ethnobiology that have originated from South America in the last 15 years reflect the increasing activities of researchers there, principally among Brazilians. This interest, much like that seen in Africa, can be associated not only with the mega-diversity of that continent but also with the complex social situations encountered there. In terms of Brazil, the country that produced the largest numbers of studies for the continent (as well as for the world), important theoretical/practical advances have been introduced now that reflect the success of incentives for training professionals and facilitating contact between researchers—thus, strengthening their interactions and promoting greater visibility of that discipline (Alves and Souto, 2011; Albuquerque et al., 2013).

Alves and Souto (2011) noted that quantitative advances in ethnozoological studies in Brazil have been associated with additional new researchers working with ethnozoology in many research and teaching institutes who, together with the pioneer researchers, have contributed to many of the observed advances in that field. Likewise, a number of zoologists and ecologists who did not rigidly consider ethnozoology as their principal line of research have come to work in that field and orient undergraduate and graduate students in ethnobiology. The first graduate program in Latin America at the PhD level (the graduate program in ethnobiology and the conservation of nature) was created in 2012, together with the first Brazilian periodical: *Ethnobiology and Conservation* (Albuquerque et al., 2013). It is also important to note that scientific production as a whole has increased expressively in Brazil in the last two decades (Van Noorden, 2014), representing more than two-thirds of the total scientific output of all of South America (although the per capita outputs from that country are very similar to those of Argentina, Uruguay, and Chile). One factor that contributed to this high output was increased investments in research; Brazil was the only country on the continent at that time that allocated more than 1% of its GNP to research and development (Van Noorden, 2014). Unfortunately, however, government investments in science and technology have tapered off significantly in the last 2 years, leaving Brazilian science in a fog of uncertainty (Angelo, 2016, 2017) that has stifled scientific production in all areas. Other South American countries have historically invested very little in research (Van Noorden, 2014), which is certainly one of the factors associated with their small scientific production in ethnozoology (and other areas of research). This situation reinforces the call by Albuquerque et al. (2013) for greater investments in scientific research that would favor ethnobiological studies throughout South America.

Asia was identified in our results as the region producing the third largest number of publications in ethnozoology. Asian countries tend to consume many zoological taxa, or their subproducts, which can readily be seen for sale in open markets (Van Vliet et al., 2016). Hunting in that region has become a serious problem, and has led to the extinction of many large animals. Additionally, the extensive fragmentation of natural habitats in Asia, associated with the strong growth of human populations and infrastructure, has facilitated access to remnant forest areas and the economic exploitation of their remaining wildlife (Corlett, 2007)—factors that argue for the intensification of ethnozoological studies. Traditional Chinese medicine counts on a wide repertoire of animal species to treat an even wider variety of illnesses and diseases (Alves and Rosa, 2013; Alves, 2012). In India, the country with the largest numbers of publications concerning ethnozoology on the Asian continent, the relationships between humans and the native fauna transcend purely utilitarian aspects. Cultural forces notoriously permeate the relationships between humans and animals in that country and are extremely important in determining interactions with the local fauna. In addition to their importance for regional biodiversity, the animals there have taken on religious significance that tightens even more the links between them and human communities.

In relation to the publication of research undertaken in North America and Europe, we identified important contributions to ethnozoological investigations from the United States and Spain, respectively. Those continents have important historical roles in the development of ethnobiology (Hunn, 2011), and many projects in both Africa and Latin America have been undertaken by researchers from North America and Europe. The low numbers of publications originating from Oceania and Central America reflect the developing ethnozoological programs in those regions.

A temporal examination of the publications from each continent showed an important increase in the numbers of ethnozoological studies after the year 2000 (Fig. 27.2), with more

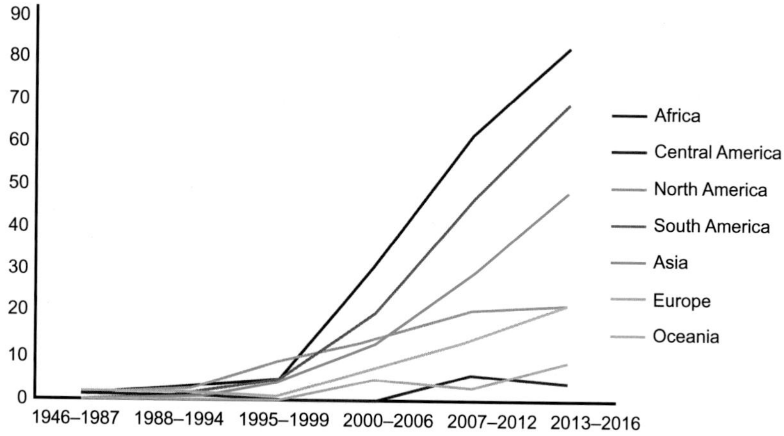

FIGURE 27.2 Number of ethnozoological publications per continent between the years 1946 and 2016.

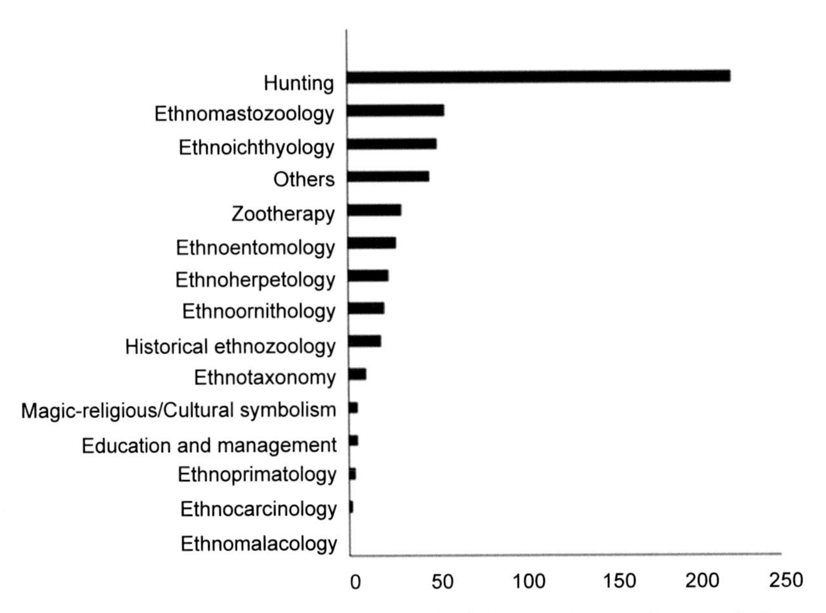

FIGURE 27.3 Number of published articles in each of the general thematic areas of ethnozoology.

publications in just the first decade of that century than in all previous years together. Alves and Souto (2011) noted the same dramatic increase in the numbers of ethnozoological publications in Brazil during that same period. Those authors reported a considerable increase in publications during the 1990s, with an even greater production at the beginning of the twenty-first century that reflected the increasing circulation of scientific information.

Our data indicated that studies involving cinegetic activities contributed most of the published articles (223) (G-test: 738.7262, $P = <.0001$), followed by ethnomastozoology (57) and ethnoichthyology (53) (Fig. 27.3).

The multidisciplinary nature of ethnozoology makes it difficult to precisely evaluate its scientific production, a fact that could introduce bias into bibliographic searches. Nonetheless, our data, which is based only on publications indexed in the two most important scientific databases, shows intense ethnozoological investigation in the field of hunting activities (Fig. 27.4). This scientific interest is related to the fact that hunting represents an important means of subsistence in many localities throughout the world, as well as one of the activities that most impacts terrestrial vertebrates globally (Alves et al., 2016; Barboza et al., 2016; Fernandes-Ferreira et al., 2012; Mesquita and Barreto, 2015; Nasi et al., 2011; Redford, 1992; Ripple et al., 2016; van Vliet et al., 2015). That same spike in cinegetic studies is probably also related to the significant number of ethnozoological investigations focusing on vertebrates ($X^2 = 726.678$, $P < .0001$) (especially mammals)—fully 88% of all the publications analyzed—followed by studies of invertebrates [8%] and studies that examined both taxonomic groups [4%]. Hunting activities have historically been practised by human populations throughout the world, although population growth, increased facility of access to natural sites, and the increased availability of more advanced technologies for capturing and killing animals have intensified pressure on target species (Robinson, 1999). In light of its importance to both humans and the natural environment, hunting has consistently attracted the attention of researchers from

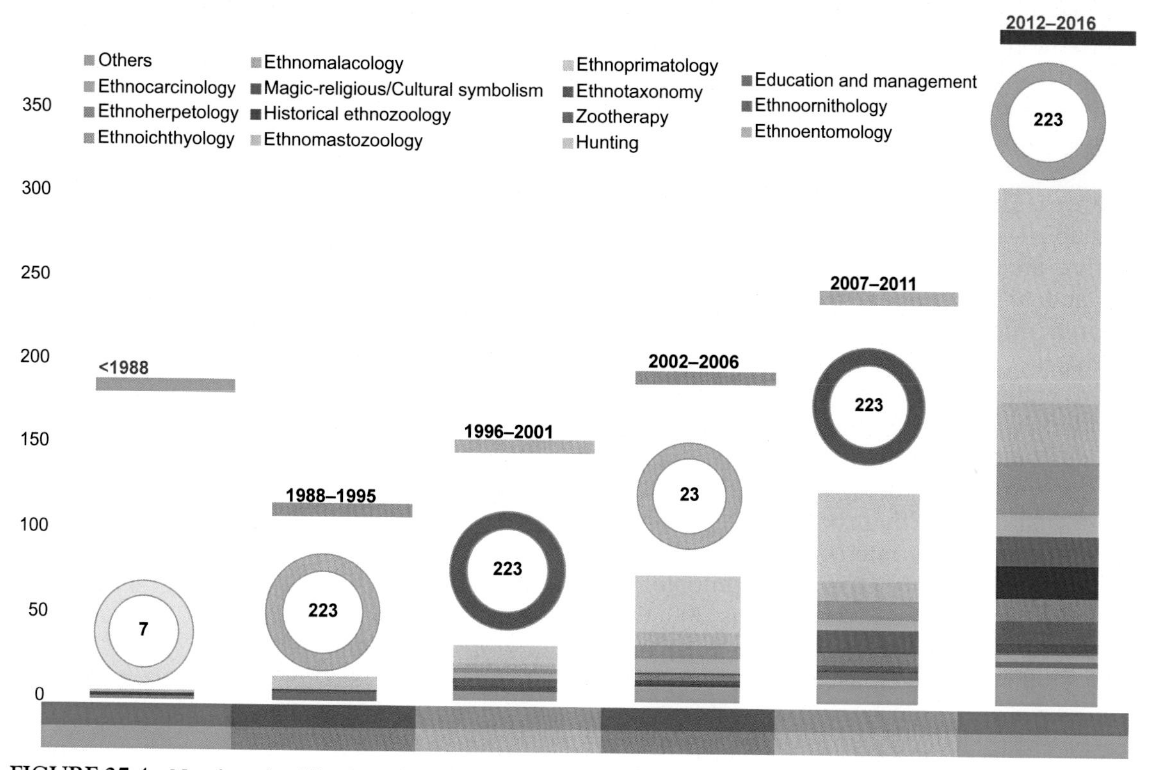

FIGURE 27.4 Number of publications throughout the world between 1964 and 2016 in the different areas of specialization of ethnozoology.

many academic areas besides ethnozoology. The urgency to better understand the relationships between humans and the animals they hunt is driven by what has been called the "hunting crisis," which has become more prominent now (Bennett et al., 2007).

FINAL CONSIDERATIONS

The large number of publications compiled in this review reflects the theoretical/methodological advances that ethnozoology has incorporated over the years. These advances have been possible due to the greater ease and rapidity of disseminating scientific discoveries, as well as increases in the numbers of academic events and encounters that facilitate the exchange of ideas between researchers—favoring the incorporation of new approaches to the area. The low numbers of investigations on some continents, and in some areas of ethnozoology, demonstrate that this discipline still has challenges to overcome, and that strategies will need to be developed that can aid research programs and the efficient and effective dissemination of their results.

The construction and transformation of ethnozoology has occurred in regions of great cultural and faunal diversity, reflecting the importance of fauna resources to humans—but not just from a utilitarian point of view—as ethnozoology also recognizes the historical and cultural importance of wildlife. The dynamic character of ethnozoological investigations has revealed the outlines of the wide spectra of interactions between human cultures with the fauna in very diverse natural environments, as well as in rural and urban areas.

Alves and Souto (2011) noted that ethnozoology now faces numerous challenges that will serve to strengthen and consolidate its advances. Those authors pointed out the need to establish efficient dialogues between investigative areas that interface with ethnozoology, and to increase the scientific rigor of its research efforts and improve its qualitative approaches. Oliveira et al. (2009) pointed out that dynamism is fundamental to any area of knowledge, and that periodic reformulations of academic areas help assure their maturation and growth. Ethnozoology will be no different in that sense. Ethnozoological research is as diverse as human interactions with regional faunas, making that subarea of ethnobiology rich in possibilities for investigations. At a time when the importance of wildlife is increasingly evident—as well as the severe impacts of humans on global biodiversity (Alves and Albuquerque, 2012; Alves and Souto, 2015)—ethnozoological research represents a field of intellectual activity that can contribute to our understanding of the world from both ecological and social perspectives.

References

Alves, R.R.N., Albuquerque, U.P., 2012. Ethnobiology and conservation: why do we need a new journal? Ethnobiology and Conservation 1, 1–3. http://dx.doi.org/10.15451/ec2012-8-1.1-1-03.

Albuquerque, U.P., Silva, J.S., Campos, J.L.A., Sousa, R.S., Silva, T.C., Alves, R.R.N., 2013, 1–9. The current status of ethnobiological research in Latin America: gaps and perspectives. Journal of Ethnobiology and Ethnomedicine 9. http://dx.doi.org/10.1186/1746-4269-9-72.

Alves, R.R.N., 2012. Relationships between fauna and people and the role of ethnozoology in animal conservation. Ethnobiology and Conservation 1–69. http://dx.doi.org/10.15451/ec2012-8-1.2-1-69.

Alves, R.R.N., Souto, W.M.S., 2011. Ethnozoology in Brazil: current status and perspectives. Journal of Ethnobiology and Ethnomedicine 7 (22), 1–18. http://dx.doi.org/10.1186/1746-4269-7-22.

Alves, R.R.N., Souto, W.M.S., 2015. Ethnozoology: a brief introduction. Ethnobiology and Conservation 4, 1–13. http://dx.doi.org/10.15451/ec2015-1-4.1-1-13.

Alves, R.R.N., Feijó, A., Barboza, R.R.D., Souto, W.M.S., Fernandes-Ferreira, H., Cordeiro-Estrela, P., Langguth, A., 2016. Game mammals of the Caatinga biome. Ethnobiology and Conservation 5, 1–51. http://dx.doi.org/10.15451/ec2016-7-5.5-1-51.

Alves, R.R.N., Rosa, I.L., 2013. Animals in Traditional Folk Medicine: Implications for Conservation. Springer-Verlag, Berlin Heidelberg.

Ayres, M., Ayres Júnior, M., Ayres, D.L., Santos, A.A., 2007. BIOESTAT – Aplicações estatísticas nas áreas das ciências bio-médicas. Ong Mamiraua, Belém, PA.

Angelo, C., 2016. Brazil's scientists battle to escape 20-year funding freeze. Nature 539, 480.

Angelo, C., 2017. Brazilian scientists reeling as federal funds slashed by nearly half. Nature. http://dx.doi.org/10.1038/nature.2017.21766.

Bahn, P.G., Vertut, J., 1997. Journey Through the Ice Age, first ed. University of California Press, California.

Barboza, R.R.D., Lopes, S.F., Souto, W.M.S., Fernandes-Ferreira, H., Alves, R.R.N., 2016. The role of game mammals as bushmeat in the Caatinga, northeast Brazil. Ecology and Society 21, 1–11.

Bennett, E.L., Blencowe, E., Brandon, K., Brown, D., Burn, R.W., Cowlishaw, G., Davies, G., Dublin, H., Fa, J.E., Milner-Gulland, E.J., Robinson, J.G., Rowcliffe, J.M., Underwood, F.M., Wilkie, D.S., 2007. Hunting for consensus: reconciling bushmeat harvest, conservation, and development policy in West and Central Africa. Conservation Biology 21, 884–887. http://dx.doi.org/10.1111/j.1523-1739.2006.00595.x.

Clément, D., 1998. The historical foundations of ethnobiology (1860–1899). Journal of Ethnobiology 18, 161–187.

Corlett, R.T., 2007. The impact of hunting on the mammalian fauna of tropical asian forests. Biotropica 39, 292–303.

Cowlishaw, G., Mendelson, S., Rowcliffe, J.M., 2005. Evidence for post-depletion sustainability in a mature bushmeat market. Journal of Applied Ecology 42, 460–468.

Fernandes-Ferreira, H., Mendonça, S.V., Albano, C., Ferreira, F.S., Alves, R.R.N., 2012. Hunting, use and conservation of birds in Northeast Brazil. Biodiversity and Conservation 221–244. http://dx.doi.org/10.1007/s10531-011-0179-9.

Hunn, E., 2007. Ethnobiology in four phases. Journal of Ethnobiology 27, 1–10. http://dx.doi.org/10.2993/0278-0771(2007)27[1:EIFP]2.0.CO;2.

Hunn, E.S., 2011. Ethnozoology. In: Ethnobiology. Department of Anthropology, University of Washington, United States, pp. 83–96. http://dx.doi.org/10.1002/9781118015872.ch6.

Kinzig, A.P., McShane, T.O., 2015. Conservation in Africa: exploring the impact of social, economic and political drivers on conservation outcomes. Environmental Research Letters 10, 090201. http://dx.doi.org/10.1088/1748-9326/10/9/090201.

Lyra-Neves, R.M., Santos, E.M., Medeiros, P.M., Alves, R.R.N., Albuquerque, U.P., 2015. Ethnozoology in Brazil: analysis of the methodological risks in published studies. Brazilian Journal of Biology 75, S184–S191. http://dx.doi.org/10.1590/1519-6984.09314.

Martin, A., Caro, T., Mulder, M.B., 2012. Bushmeat consumption in western Tanzania: a comparative analysis from the same ecosystem. Tropical Conservation Science 5, 352–364.

Mason, O.T., 1899. Aboriginal American Zoötechny, vol. 1, pp. 45–81.

Mesquita, G.P., Barreto, G.P., 2015. Evaluation of mammals hunting in indigenous and rural localities in Eastern Brazilian Amazon. Ethnobiology and Conservation 4, 1–14. http://dx.doi.org/10.15451/ec2015-1-4.2-1-14.

Mithen, S., 1999. The hunter-gatherer prehistory of human-animal interactions. Anthrozoos 12, 195–204. http://dx.doi.org/10.2752/089279399787000.

Nasi, R., Taber, A., van Vliet, N., 2011. Empty forests, empty stomachs? Bushmeat and livelihoods in the Congo and Amazon Basins. International Forestry Review 13, 355–368.

Oliveira, F.C., Albuquerque, U.P., Fonseca-Kruel, V.S., Hanazaki, N., 2009. Avanços nas pesquisas etnobotânicas no Brasil. Acta Botanica Brasilica 23, 590–605.

Redford, K.H., 1992. The empty forest. BioScience 42, 412–422.

Ripple, W.J., Abernethy, K., Betts, M.G., Chapron, G., Dirzo, R., Galetti, M., Levi, T., Lindsey, P.A., Macdonald, D.W., Machovina, B., Newsome, T.M., Peres, C.A., Wallach, A.D., Wolf, C., Young, H., 2016. Bushmeat hunting and extinction risk to the world's mammals. Royal Society Open Science 3, 1–16.

Robinson, J.G., 1999. Conservation: wildlife harvest in logged tropical forests. Science 284, 595–596. http://dx.doi.org/10.1126/science.284.5414.595.

Van Noorden, R., 2014. The impact gap: South America by the numbers. Nature 510, 202–203.

Van Vliet, N., Cornelis, D., Beck, H., Lindsey, P., Nasi, R., LeBel, S., Moreno, J., Fragoso, J., 2016. Meat from the wild: extractive uses of wildlife and alternatives for sustainability. In: Mateo, R., Arroyo, B., Garcia, J.T. (Eds.), Current Trends in Wildlife Research, Spinger, Switzerland, pp. 225–265. http://dx.doi.org/10.1007/978-3-319-27912-1.

van Vliet, N., Quiceno-Mesa, M.P., Cruz-Antia, D., Tellez, L., Martins, C., Haiden, E., Oliveira, M.R., Adams, C., Morsello, C., Valencia, L., 2015. From fish and bushmeat to chicken nuggets: the nutrition transition in a continuum from rural to urban settings in the Tri frontier Amazon region. Ethnobiology and Conservation 4, 1–12. http://dx.doi.org/10.15451/ec2015-7-4.6-1-12.

Index

'*Note*: Page numbers followed by "f" indicate figures, "t" indicate tables.'

F

I

J

K

L

Q

R

T

Y

Z